INTEGRALS (An arbitrary constant may be added to each integral.)

1. $\displaystyle\int x^n\,dx = \frac{1}{n+1}x^{n+1} \quad (n \neq -1)$

2. $\displaystyle\int \frac{1}{x}\,dx = \log|x|$

3. $\displaystyle\int e^x\,dx = e^x$

4. $\displaystyle\int a^x\,dx = \frac{a^x}{\log a}$

5. $\displaystyle\int \sin x\,dx = -\cos x$

6. $\displaystyle\int \cos x\,dx = \sin x$

7. $\displaystyle\int \tan x\,dx = -\log|\cos x|$

8. $\displaystyle\int \cot x\,dx = \log|\sin x|$

9. $\displaystyle\int \sec x\,dx = \log|\sec x + \tan x| = \log\left|\tan\left(\tfrac{1}{2}x + \tfrac{1}{4}\pi\right)\right|$

10. $\displaystyle\int \csc x\,dx = \log|\csc x - \cot x| = \log\left|\tan\tfrac{1}{2}x\right|$

11. $\displaystyle\int \arcsin\frac{x}{a}\,dx = x\arcsin\frac{x}{a} + \sqrt{a^2 - x^2} \quad (a > 0)$

12. $\displaystyle\int \arccos\frac{x}{a}\,dx = x\arccos\frac{x}{a} - \sqrt{a^2 - x^2} \quad (a > 0)$

13. $\displaystyle\int \arctan\frac{x}{a}\,dx = x\arctan\frac{x}{a} - \frac{a}{2}\log(a^2 + x^2) \quad (a > 0)$

14. $\displaystyle\int \sin^2 mx\,dx = \frac{1}{2m}(mx - \sin mx\cos mx)$

15. $\displaystyle\int \cos^2 mx\,dx = \frac{1}{2m}(mx + \sin mx\cos mx)$

16. $\displaystyle\int \sec^2 x\,dx = \tan x$

17. $\displaystyle\int \csc^2 x\,dx = -\cot x$

18. $\displaystyle\int \sin^n x\,dx = -\frac{\sin^{n-1}x\cos x}{n} + \frac{n-1}{n}\int \sin^{n-2}x\,dx$

19. $\displaystyle\int \cos^n x\,dx = \frac{\cos^{n-1}x\sin x}{n} + \frac{n-1}{n}\int \cos^{n-2}x\,dx$

20. $\displaystyle\int \tan^n x\,dx = \frac{\tan^{n-1}x}{n-1} - \int \tan^{n-2}x\,dx \quad (n \neq 1)$

21. $\displaystyle\int \cot^n x\,dx = -\frac{\cot^{n-1}x}{n-1} - \int \cot^{n-2}x\,dx \quad (n \neq 1)$

22. $\displaystyle\int \sec^n x\,dx = \frac{\tan x\sec^{n-2}x}{n-1} + \frac{n-2}{n-1}\int \sec^{n-2}x\,dx \quad (n \neq 1)$

23. $\displaystyle\int \csc^n x\,dx = -\frac{\cot x\csc^{n-2}x}{n-1} + \frac{n-2}{n-1}\int \csc^{n-2}x\,dx \quad (n \neq 1)$

(Continued on next page)

24. $\int \sinh x \, dx = \cosh x$

25. $\int \cosh x \, dx = \sinh x$

26. $\int \tanh x \, dx = \log |\cosh x|$

27. $\int \coth x \, dx = \log |\sinh x|$

28. $\int \operatorname{sech} x \, dx = \arctan(\sinh x)$

29. $\int \operatorname{csch} x \, dx = \log \left| \tanh \frac{x}{2} \right| = -\frac{1}{2} \log \frac{\cosh x + 1}{\cosh x - 1}$

30. $\int \sinh^2 x \, dx = \frac{1}{4} \sinh 2x - \frac{1}{2} x$

31. $\int \cosh^2 x \, dx = \frac{1}{4} \sinh 2x + \frac{1}{2} x$

32. $\int \operatorname{sech}^2 x \, dx = \tanh x$

33. $\int \sinh^{-1} \frac{x}{a} \, dx = x \sinh^{-1} \frac{x}{a} - \sqrt{x^2 + a^2} \quad (a > 0)$

34. $\int \cosh^{-1} \frac{x}{a} \, dx = \begin{cases} x \cosh^{-1} \frac{x}{a} - \sqrt{x^2 - a^2} & \left[\cosh^{-1}\left(\frac{x}{a}\right) > 0, a > 0\right] \\ x \cosh^{-1} \frac{x}{a} + \sqrt{x^2 - a^2} & \left[\cosh^{-1}\left(\frac{x}{a}\right) < 0, a > 0\right] \end{cases}$

35. $\int \tanh^{-1} \frac{x}{a} \, dx = x \tanh^{-1} \frac{x}{a} + \frac{a}{2} \log |a^2 - x^2|$

36. $\int \frac{1}{\sqrt{a^2 + x^2}} \, dx = \log(x + \sqrt{a^2 + x^2}) = \sinh^{-1} \frac{x}{a} \quad (a > 0)$

37. $\int \frac{1}{a^2 + x^2} \, dx = \frac{1}{a} \arctan \frac{x}{a} \quad (a > 0)$

38. $\int \sqrt{a^2 - x^2} \, dx = \frac{x}{2} \sqrt{a^2 - x^2} + \frac{a^2}{2} \arcsin \frac{x}{a} \quad (a > 0)$

39. $\int (a^2 - x^2)^{3/2} \, dx = \frac{x}{8}(5a^2 - 2x^2)\sqrt{a^2 - x^2} + \frac{3a^4}{8} \arcsin \frac{x}{a} \quad (a > 0)$

40. $\int \frac{1}{\sqrt{a^2 - x^2}} \, dx = \arcsin \frac{x}{a} \quad (a > 0)$

41. $\int \frac{1}{a^2 - x^2} \, dx = \frac{1}{2a} \log \left| \frac{a + x}{a - x} \right|$

42. $\int \frac{1}{(a^2 - x^2)^{3/2}} \, dx = \frac{x}{a^2 \sqrt{a^2 - x^2}}$

43. $\int \sqrt{x^2 \pm a^2} \, dx = \frac{x}{2} \sqrt{x^2 \pm a^2} \pm \frac{a^2}{2} \log |x + \sqrt{x^2 \pm a^2}|$

44. $\int \frac{1}{\sqrt{x^2 - a^2}} \, dx = \log |x + \sqrt{x^2 - a^2}| = \cosh^{-1} \frac{x}{a} \quad (a > 0)$

45. $\int \frac{1}{x(a + bx)} \, dx = \frac{1}{a} \log \left| \frac{x}{a + bx} \right|$

46. $\int x \sqrt{a + bx} \, dx = \frac{2(3bx - 2a)(a + bx)^{3/2}}{15b^2}$

47. $\int \frac{\sqrt{a + bx}}{x} \, dx = 2\sqrt{a + bx} + a \int \frac{1}{x\sqrt{a + bx}} \, dx$

(Continued at the back of the book)

Vector Calculus

Fifth Edition

Jerrold E. Marsden
California Institute of Technology, Pasadena

Anthony J. Tromba
University of California, Santa Cruz

W. H. Freeman and Company
New York

Executive Editor: Craig Bleyer
Acquisitions Editor: Terri Ward
Marketing Manager: Jeffrey Rucker
Project Editor: Vivien Weiss
Cover and Text Designer: Diana Blume
Production Manager: Julia DeRosa
Editorial Assistant: Kristy Cates
Media and Supplements Editor: Brian Donnellan
Illustration Coordinator: Shawn Churchman
Illustrations: The GTS Companies/York, PA Campus
Compositor: The GTS Companies/York, PA Campus
Manufacturer: RR Donnelly & Sons Company
Cover Photo: Isaac Newton (1642–1727). Painting by I. Vanderbank (1725). London, National Portrait Gallery.

Politics is for the moment.
An equation is for eternity.

A. EINSTEIN

Some calculus tricks are quite easy.
Some are enormously difficult. The fools
who write the textbooks of
advanced mathematics seldom take the trouble
to show you how easy the easy calculations are.

SILVANUS P. THOMPSON, *CALCULUS MADE EASY,* MACMILLAN (1910)

Library of Congress Cataloging-in-Publication Data

Marsden, Jerrold E.
Vector calculus/Jerrold E. Marsden, Anthony J. Tromba.—5th ed.
p. cm.
Includes bibliographical references and index.
ISBN 0-7167-4992-0 (pbk.)
1. Calculus. 2. Vector analysis. I. Tromba, Anthony. II. Title.

QA303.M338 2003
515′.63—dc21 2003049184

Printed in the United States of America

Contents

To Barbara and Inga for all their love and support

Preface

This text is intended for a one-semester course in the calculus of functions of several variables and vector analysis, which is normally taught at the sophomore level. In addition to making changes and improvements throughout the text, in this new edition we have added considerable material that presents the historical development of the subject and have also attempted to convey a sense of excitement, relevance, and importance of the subject matter.

Prerequisites

Sometimes courses in vector calculus are preceded by a first course in linear algebra, but this is not an essential prerequisite. We require only the bare rudiments of matrix algebra, and the necessary concepts are developed in the text. If this course is preceded by a course in linear algebra, the instructor will have no difficulty enhancing the material. However, we do assume a knowledge of the fundamentals of one-variable calculus—the process of differentiation and integration and their geometric and physical meaning as well as a knowledge of the standard functions, such as the trigonometric and exponential functions.

The Role of Theory

The text includes much of the basic theory as well as many concrete examples and problems. Some of the technical proofs for theorems in Chapters 2 and 5 are given in optional sections that are readily available on the book's Web site at www.whfreeman.com/MarsdenVC5e (see the description on the next page). Section 2.2, on limits and continuity, is designed to be treated lightly and is deliberately brief. More sophisticated theoretical topics, such as compactness and delicate proofs in integration theory, have been omitted, because they usually belong to a more advanced course in real analysis.

Concrete and Student-Oriented

Computational skills and intuitive understanding are important at this level, and we have tried to meet this need by making the book concrete and student-oriented. For example, although we formulate the definition of the derivative correctly, it is done by using matrices of partial derivatives rather than abstract linear transformations. We also include a number of physical illustrations such as fluid mechanics, gravitation, and electromagnetic theory, and from economics as well, although knowledge of these subjects is not assumed.

Order of Topics

A special feature of the text is the early introduction of vector fields, divergence, and curl in Chapter 4, before integration. Vector analysis often suffers in a course of this type, and the present arrangement is designed to offset this tendency. To go even further, one might consider teaching Chapter 3 (Taylor's theorems, maxima and minima, Lagrange multipliers) after Chapter 8 (the integral theorems of vector analysis).

This fifth edition was completely reset, but retains and improves on the balance between theory, applications, optional material, and historical notes that was present in earlier editions.

Supplements

One of the main changes in this edition is in the supplement. They are as follows:

1. **Web Site.** The book's Web site contains the following materials:

    - **Internet Supplement,** a PDF file containing additional material suitable for projects as well as technical proofs and sample examinations with complete solutions.
    - **PowerPoint and KeyNote Slides** for instructors to use in presentations of the text's figures, as well as section-by-section summaries.
    - **LaTeX and PDF Files of Sample Exams** (on instructor's protected site)
    - **Updates**

    It is available to everyone and can be found at www.whfreeman.com/MarsdenVC5e.

2. **Student Study Guide with Solutions.** This student guide, written by Karen Pao and Fred Soon, contains helpful hints and summaries for the material in each section, contains the solutions to selected problems, and contains sample exams to help students in exam preparation. Problems whose solutions appear in the Student Study Guide have a colored number in the text, for easy reference. The guide has been revised and reset for the Fifth Edition of *Vector Calculus*. ISBN 0-7167-0528-1

3. **Instructor's Manual with Solutions.** This supplement contains material available only to instructors. This includes summaries of material and additional worked-out examples that are helpful in the preparation of lectures. It also contains additional solutions to problems and sample exams (some of them with complete solutions). ISBN 0-7167-0646-6

Final Exam Questions

There are practice exams available in the Student Study Guide, the Internet supplement, as well as in the Instructor's Manual. We also include some final exam questions (some of them challenging) for the reader's convenience on the book's Web site.

Of course, the level and choice of topics and the lengths of final exams will vary from instructor to instructor. Working these problems requires a knowledge of most of the main material of the book, and solving 10 of these problems should take the reader about 3 hours to complete. Some solutions are also given on the book's Web site.

We are excited about this new edition of *Vector Calculus*, especially the inclusion of the new historical material as well as the new discussions of interesting applications of vector analysis, both mathematical and physical. We hope that the reader will be equally pleased.

Jerry Marsden and Tony Tromba,
Caltech and UC Santa Cruz, Summer 2003.

Acknowledgements

Many colleagues and students in the mathematical community have made valuable contributions and suggestions since this book was begun. An early draft of the book was written in collaboration with Ralph Abraham. We thank him for allowing us to draw upon his work. It is impossible to list all those who assisted with this book, but we wish especially to thank Michael Hoffman and Joanne Seitz for their help on earlier editions. We also received valuable comments from Mary Anderson, John Ball, Patrick Brosnan, Andrea Brose, David Drasin, Gerald Edgar, Michael Fischer, Frank Gerrish, Mohammad Gohmi, Jenny Harrison, Jan Hogendijk, Jan-Jaap Oosterwijk, and Anne van Weerden (Uterecht), David Knudson, Richard Kock, Andrew Lenard, William McCain, Gordon McLean, David Merriell, Jeanette Nelson, Dan Norman, Keith Phillips, Anne Perleman, Oren Walter Rosen, Kenneth Ross, Ray Sachs, Diane Sauvageot, Joel Smoller, Francis Su, Melvyn Tews, Ralph and Bob Tromba, Steve Wan, Alan Weinstein, John Wilker, and Peter Zvengrowski. The students and faculty of Austin Community College deserve a special note of thanks, as do our students at both Caltech and UC Santa Cruz.

We owe a very special thanks to Stefan Hildebrandt for his historical advice.

We are grateful to the following instructors who provided detailed reviews of the manuscript. Dr. Michael Barbosu, SUNY Brockport; Brian Bradie, Christopher Newport University; Mike Daven, Mount Saint Mary; Elias Deeba, University of Houston–Downtown; John Feroe, Vassar; David Gurari, Case Western Reserve; Alan Horowitz, Penn State; Rhonda Hughes, Bryn Mawr; Frank Jones, Rice University; Leslie Kay, Virginia Tech; Richard Laugesen, University of Michigan; Namyong Lee, Minnesota State University; Tanya Leiese, Rose Hullman Institute; John Lott, University of Michigan; Gerald Paquin, Université du Québec à Montréal; Joan Rand Moschovakis, Occidental College; A. Shadi Tahvildar-Zadeh, Princeton University; Dr. Stuart Smith, California State University at Hayward; Howard Swann, San Jose State University; Denise Szecsei, Stetson University; Edward Taylor, Wesleyan; and Chaogui Zhang, Case Western Reserve. For the fifth edition, we want to thank all the reviewers, but especially Andrea Brose, UCLA, for her detailed and valuable comments. Most important of all are the readers and users of this book whose loyalty for over a quarter of a century has made the fifth edition possible.

A final word of thanks goes to those who helped in the preparation of the manuscript and the production of the book. For the earlier editions, we thank Connie Calica, Nora Lee, Marnie McElhiney, Ruth Suzuki, Ikuko Workman, and Esther Zack for their excellent typing of various versions and revisions of the manuscript; Herb Holden of Gonzaga University and Jerry Kazdan of the University of Pennsylvania for suggesting and preparing early versions of the computer-generated figures; Jerry Lyons and Holly Hodder for their roles as our previous mathematics editors; Christine Hastings for editorial supervision; and Trumbull Rogers for his expert copyediting. For this fifth edition, we thank Matt Haigh and Wendy McKay for their help with TeX and Mathematica preparation of the material and also Terri Ward, the Mathematics Acquisitions Editor at W. H. Freeman for her excellent stewardship of the project, and Vivien Weiss for her excellent handling of production matters.

We will be maintaining an up-to-date web-based list of corrections and suggestions for the fifth edition and will be happy to receive any additional suggestions and corrections from our readers. Please send your request to either Jerrold Marsden (marsden@cds.caltech.edu) or Anthony Tromba (tromba@cats.ucsc.edu).

Historical Introduction:
A Brief Account

This, therefore, is Mathematics; *she reminds you of the invisible form of the soul; she gives life to her own discoveries; she awakens the mind and purifies the intellect; she brings light to our intrinsic ideas; she abolishes oblivion and ignorance which is ours by birth.*

Proclus, c. 450

Cum Deus Calculat Fit Mundus.
(As God calculates, so the world is created).

Leibniz, c. 1700

The word *mathematics* derives from the Greek word *mathema*, meaning knowledge, cognition, understanding, or perception, suggesting that the study of what we now call mathematics began by asking questions about the world. In fact, the historical evidence suggests that mathematics began about 2700 years ago as an attempt to comprehend nature. Unfortunately, in most mathematical expositions, historical motivations and contexts are often sacrificed. In this new edition, the authors continue to address this problem by increasing the discussion of historical and contextual material where appropriate. Therefore, before we dive into the mathematics of *Vector Calculus*, we briefly discuss the development of mathematics prior to and including the discovery of calculus.

Egyptian, Babylonian, and Greek Mathematics

It is generally acknowledged that mathematics developed in the seventh and sixth centuries B.C., somewhat after the Greeks had developed a uniform alphabet. This is not to say, however, that mathematical knowledge did not exist before the Greeks. In fact, the Egyptians and Babylonians knew many empirical facts centuries before the rise of the Greek civilization. For example, they could solve quadratic equations, compute the areas of certain geometric figures, such as squares, rectangles, and triangles, and they possessed a reasonably good formula for the area of a circle, using the value of 3.16 for π. They also knew how to compute certain volumes like the size of cubes, rectangles, rectangular solids, cones, cylinders, and (not surprisingly) pyramids. The ancients were also acquainted with the pythagorean theorem (at least empirically).

The Greeks, who settled throughout the Mediterranean, must have played an important role in preserving and spreading the mathematical knowledge of the Egyptians and the Babylonians. However, the Greeks were aware that there were different formulas for the same area or volumes. For example, the Babylonians had one formula for the volume of a frustum of a pyramid with a square base, and the Egyptians had another (see Figure 1).

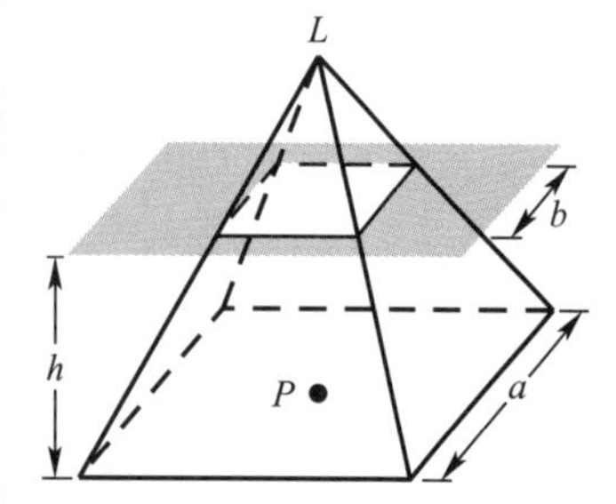

Figure 1 Volume of a frustum of a pyramid with a square base: $V = \frac{1}{3}h(a^2 + ab + b^2)$.

It is not surprising that the Egyptians (with the experience in pyramid construction) had the correct formula. Now, given two formulas, it was clear that only one could be correct. But how could one decide such an answer? Certainly it is not a question for debate, as would be the question of the quality of works of art. It is likely that the necessity to determine the answers to such questions is what led to the development of mathematical proof and to the method of deductive reasoning.

The person usually credited for the invention of rigorous mathematical proof was a merchant named Thales of Miletus (624–548 B.C.). It is Thales who is said to be the creator of Greek geometry, and it was this geometry (earth measure) as an abstract mathematical theory (rather than a collection of empirical facts) supported by rigorous deductive proofs that was one of the turning points of scientific thinking. It led to the creation of the first mathematical model for physical phenomena.

For example, one of the most beautiful geometric theories developed during antiquity was that of conic sections. See Figure 2.

Conics include the straight line, circle, ellipse, parabola, and hyperbola. Their discovery is attributed to Menaechmus, a member of the school of the great Greek philosopher Plato. Plato, a student of Socrates, founded his school *The Academy* (see Figure 3) in a sacred area of the ancient city of Athens, called Hekadameia (after the hero Hekademos). All later academies obtained their name from this institution, which existed without interruption for about 1000 years until it was dissolved by the Roman Emperor Justinian in A.D. 529.

Plato suggested the following problem to his students:

Explain the motion of the heavenly bodies by some geometrical theory.

Why was this a question of interest and puzzlement for the Greeks? Observed from the Earth, these motions appear to be quite complicated. The motions of the sun and the moon can be roughly described as circular with constant speed, but the deviations from the circular orbit were troublesome to the Greeks and they felt challenged to find an explanation for these irregularities. The observed orbits of the

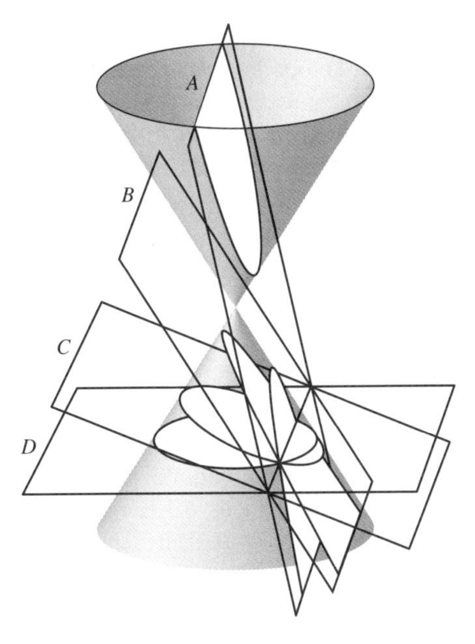

Figure 2 The conic sections: (A) hyperbola, (B) parabola, (C) ellipse, (D) circle.

planets are even more complicated, because as they go through a revolution, they appear to reverse direction several times.

The Greeks sought to understand this apparently wild motion by means of their geometry. Eudoxus, Hipparchus, and then Apollonius of Perga (262–190 B.C.)

Figure 3 *Plato's Academy* (mosaic found in Pompeii, Villa of T. Siminius Stephanus, 86 × 85 cm, Naples, Archaeological Museum). With certainty the seven men have been identified as Plato (third from the left) and six other philosophers, who are talking about the universe, the celestial spheres, and the stars. The mosaic shows Plato's Academy, with the city of Athens in the background. It is probably a copy (from the first century B.C.) of a Hellenistic painting.

suggested that the celestial orbits could be explained by combinations of circular motion (that is, through the construction of curves called epicycles traced out by circles moving on other circles). This idea was to become the most important astronomical theory of the next two thousand years. This theory, known by us through the writings of the Greek astronomer Ptolemy of Alexandria, ultimately becomes known as the "Ptolemaic theory." See Figures 4 and 5.

Figure 4 Woodcut from Georg von Peurbach's *Theoricae novae planetarum,* edited by Oronce Fine as a teaching text for the University of Paris (1515). It was the canonical description of the heavens until the end of the sixteenth century, and even Copernicus was to a large extent under the influence of this work. Peurbach described the solid sphere representations of Ptolemaic planetary models, which he probably based on Ibn al-Haytham's work "On the configuration of the world" (translated into Latin in the thirteenth century). The same frontispiece was used for the Sacrosbosco edition of the first four books of Euclid's *Elements* (in excerpts), which appeared under the title *Textus de Sphaera* in Paris (1521).

Figure 5 Ptolemy observing the stars with a quadrant, together with an allegoric Astronomia. (From Gregorius Reish, *Margarita Philosophica nova*, Strasbourg, 1512, an early compendium of philosophy and science.) In those days, Ptolemy was often depicted as a king, because he was erroneously thought to be descended from the Ptolemaic dynasty that ruled Egypt after Alexander.

Most of Greek geometry was codified by Euclid in his *Elements* (of Mathematics). Actually the *Elements* consist of thirteen books, in which Euclid collected most of the mathematical knowledge of his age (circa 300 B.C.), transforming it into a lucid, logically developed masterpiece. In addition to the *Elements*, some of Euclid's other writings were also handed down to us, including his *Optics* and the *Catoptrica* (theory of mirrors).

The success of Greek mathematics had a profound effect on views of nature. The Platonists, or followers of Plato, distinguished between the world of ideas and the world of physical objects. Plato was the first to propose that ultimate truth or understanding could not come from the material world, which is constantly subject to change, but only from mathematical models or constructs. Thus, infallible knowledge could be attained only through mathematics. Plato not only wished to use mathematics in the study of nature, but he actually went so far as to attempt to substitute mathematics *for* nature. For Plato, reality lies only within the realm of ideas, especially mathematical ideas.

Not everyone in antiquity agreed with this point of view. Aristotle, a student of Plato, criticized Plato's reduction of science to the study of mathematics. Aristotle thought that the study of the material world was one's primary source of reality. Despite Aristotle's critique, the view that mathematical laws governed the universe took a firm hold on classical thought. *The search for the mathematical laws of nature was underway.*

After the death of Archimedes in 212 B.C., Greek civilization went into a period of slow decline. The final blow to Greek civilization came in 640 A.D. with the Moslem conquest of Egypt. The remaining Greek texts housed in the great library in Alexandria were burned. Those scholars who survived migrated to Constantinople (now part of Turkey), which had become the capitol of the Eastern Roman Empire. It was in this great city that what survived of Greek civilization was preserved for its rediscovery by European civilization some five hundred years later.

Indian and Arabian Mathematics

Mathematical activity did not, however, cease with the decline of Greek civilization. In the middle of the sixth century, somewhere in the Indus Valley in India, our modern system of numeration evolved. The Indians developed a number system based on ten, with ten rather abstract symbols from zero to nine looking "roughly" as they do today. They developed rules for addition, multiplication, and division (as we have today), a system infinitely superior to the Roman abacus, which was used (by a special class of servants called *arithmeticians*) throughout Europe until the fifteenth century. See Figure 6.

After the fall of Egypt, came the rise of Arab civilization centered in Baghdad. Scholars from Constantinople and India were invited to study and to share their knowledge. It was through these contacts that the Arabs came to acquire the learning of the ancients as well as the newly discovered Indian system of numeration. See Figure 7.

It was the Arabs who gave us the name *Algebra*, which comes from the book by the astronomer Mohammed ibn Musa al-Khuwarizmi titled "Al-Jabr w'al muqabola,"

Figure 6 Arithmetician performing a calculation on a counter-abacus.

which means "restoring" or "balancing" (equations). Al-Khuwarizmi is also responsible for a second profoundly influential book entitled "Kitab al jami' wa'l tafriq bi hisab al hind" (Indian Technique of Addition and Subtraction), which described and clarified the Indian decimal place value system.

Al-Khuwarizmi also gave us another name for a fundamental branch of science, the word *algorithm.* Latinized, his name became Algorism, then Algorismus, and finally Algorithm. The term initially represented the Indian system of numeration, but ultimately came to be used in its modern computational sense.

The decline of Arab civilization coincided with the rise of European civilization. The dawn of the modern age began when Richard the Lionhearted reached the walls of Jerusalem. From approximately 1192 through around 1270, the Christian knights brought the learning of the "infidels" back to Europe. Around 1200–1205, Leonardo of Pisa (also known as Fibonacci), who had traveled extensively in Africa and Asia Minor, wrote his interpretation (in Latin) of Arabic and Greek mathematics. His

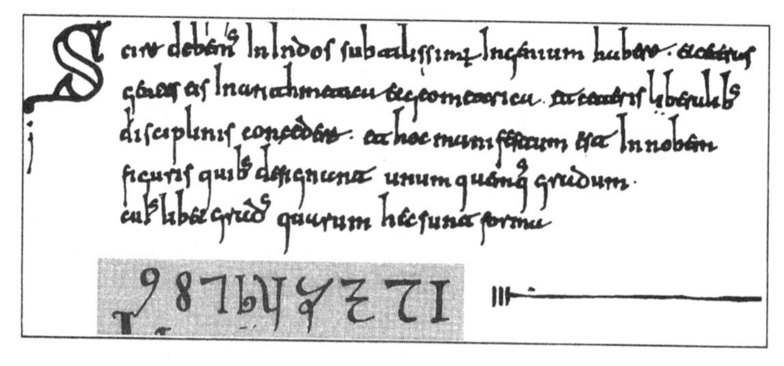

Figure 7 Detail from the *Codex Vigilanus* (976 A.D. northern Spain). The first known occurrence of the nine Indo-Arabic numerals in Western Europe. (Escurial Library, Madrid.)

historic texts brought the work of al-Khuwarizmi and Euclid to the attention of a large audience in Europe.

European Mathematics

Around 1450 Johann Gutenberg invented the printing press with movable type. This, combined with the advent of linen and cotton paper obtained from the Chinese, dramatically increased the rate of the dissemination of knowledge. The steep rise in trade and manufacturing fueled the growth of wealth and dramatic change in European societies from feudal to city-states. In Italy, the mother of the Renaissance, we see the rise of extraordinarily wealthy states such as Venice under the Doges and Florence under the Medicis.

The needs of the rising merchant class accelerated the adoption of the Indian system of numeration. The teachings of the Catholic Church, which rested on absolute authority and dogma, began to be challenged by the ideas of Plato. From Plato, scholars learned that *the world was rational and could be understood*, and that the means of understanding nature was through mathematics. But this sharply contradicted the teachings of the church, which taught that God designed the universe. The only possible resolution of this apparent contradiction was that "God designed the universe mathematically" or that "God is a mathematician."

It is perhaps surprising how much this point of view inspired the work of many sixteenth- to eighteenth-century mathematicians and scientists. For if this were indeed the case, then by understanding the mathematical laws of the universe, one could come closer to an understanding of the Creator himself. Believe it or not, this point of view survives to this day. The following is a quote from Paul Dirac, a Nobel Prize–winning physicist and a creator of modern quantum mechanics.

> *It seems to be one of the fundamental features of nature that fundamental physical laws are described in terms of a mathematical theory of great beauty and power, needing quite a high standard of mathematics for one to understand it. You may wonder: Why is nature constructed along these lines? One can only answer that our present knowledge seems to show that nature is so constructed. We simply have to accept it. One could perhaps describe the situation by saying that God is a mathematician of a very high order, and He used very advanced mathematics in constructing the universe. Our feeble attempts at mathematics enable us to understand a bit of the universe, and as we proceed to develop higher and higher mathematics we can hope to understand the universe better.*

Mathematics began to see further advances and applications. In the sixteenth and seventeenth centuries, al-Khuwarizmi's algebra was significantly advanced by Cardano, Vieta, and Descartes. The Babylonians had solved the quadratic equation, but now two thousand years later, del Ferro and Tartaglia solved the cubic equation, which in turn led to the discovery of imaginary numbers. These imaginary numbers were later to play a fundamental role, as we shall see, in the development of vector calculus. In the early seventeenth century, Descartes, perhaps motivated by the grid technique used by Italian fresco painters to locate points on a wall or canvas, created, in a moment of great mathematical inspiration, coordinate (or analytic) geometry.

This new mathematical model enables one to reduce Euclid's geometry to algebra and provides a precise and quantitative method to describe and calculate with space curves and surfaces.

Early on, Archimedes' great work in statics and equilibrium (centers of gravity, the principle of the lever—which we study in this book) was absorbed and improved upon, leading to dramatic engineering achievements. In a building spree that remains astonishing to this day, engineering advances made possible the rise of an incredible number of cathedrals throughout Europe, including the stunning Duomo in Florence, Notre Dame in Paris, and the Great Cathedral in Cologne, to mention a few. See Figure 8.

Figure 8 Duomo.

Filippo Brunelleschi (1377–1446) studied the works of Euclid and Hipparchus and was the first artist to employ mathematics extensively. The mathematical principles of perspective were eventually completed by Piero della Francesca (1410–1492). Mathematicians and engineers were recruited by warring princes to fuel the development of advanced weapons and ballistic science. The most famous among these was none other than Leonardo da Vinci, who in the last years of his life was employed by the Duke of Milan. It was in these final years that he painted the "Mona Lisa," now housed in the Louvre in Paris. See Figure 9.

However, as in Greek times, it was astronomy that was to give mathematics its greatest impetus. It is not surprising that the Greek astronomers placed the Earth and not the sun at the center of our universe, because on a daily basis we see the sun both rise and set. Still, it is interesting to ask if the Greeks, who were such marvelous thinkers, at least tested the heliocentric theory, which places the sun at the center of the

Figure 9 Leonardo, self portrait.

universe. In fact, they did. In the third century B.C., Aristarchus of Samus taught that the Earth and other planets move in circular orbits around a fixed sun. His hypotheses were, for several reasons, rejected. First, the opposing astronomers reasoned that if the Earth were indeed moving, one should be able to sense it. Second, how would objects, circulating with us, be able to stay on a moving Earth? Third, why are the clouds not lagging behind the moving Earth?

Such arguments were to be used again in the sixteenth century against the Polish astronomer Nicolas Copernicus (see Figure 10), who in 1543 introduced the heliocentric theory (the planets move in orbit around the sun). His book *Revolutionibus Orbium Coelestium* (*On the Revolution of the Heavenly Orbits*) was to initiate the "Copernican revolution" in science and to give the world a new word, *revolutionary*.

In 1619, the German astronomer Johannes Kepler (see Figure 11), using the astronomical calculations of the Danish astronomer Tycho Brahe, showed that the planetary

Figure 10 Nicolaus Copernicus (1473–1543).

orbits were in fact elliptical, the same ellipses that the Greeks had studied as abstract forms some 2000 years earlier (see Figure 12).

But Kepler's law of elliptical orbits was only one of three laws he discovered governing planetary motion. Kepler's second law states that if a planet moves from a point A to another point B in a certain amount of time T, and also moves from A' to B' in the same time, and if S is a focus of the orbital ellipse, then the sections SAB are $SA'B'$ have equal areas (see Figure 13). Kepler's third law was that the square of time T a planetary body requires to complete an orbit is proportional to a^3, where a is the great axis of the elliptical orbit. In equation form, $T^2 = Ka^3$, where K is some constant (we shall derive this law for circular orbits in Chapter 4).

Figure 11 Johannes Kepler (1571–1630).

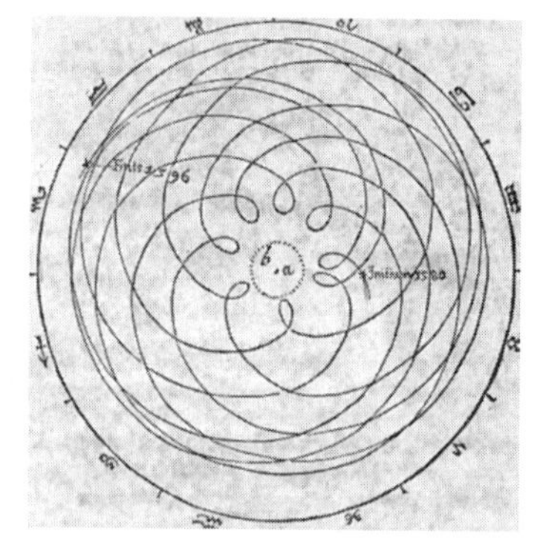

Figure 12 The motion of Mars. From Kepler's *Astronomia Nova* (1609).

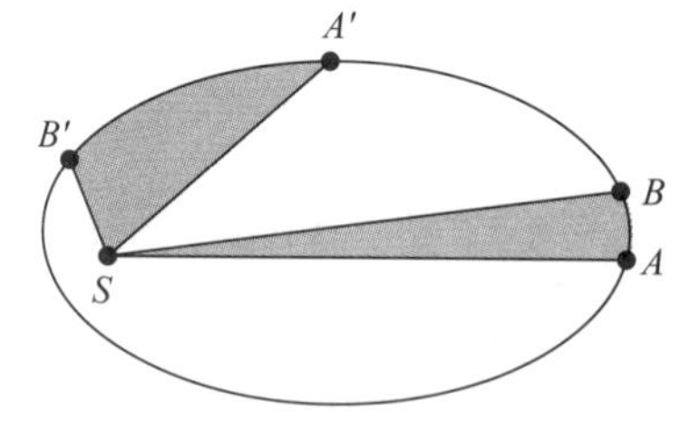

Figure 13 Kepler's second law.

Profound as these observations were, an explanation of why these laws held was lacking. However, by the middle of the seventeenth century, it was fully understood that a change of velocity requires the action of forces, but how these forces influenced motion was not at all clear. In 1674 Robert Hooke, in an attempt to explain Kepler's laws, assumed the existence of an attractive force the sun must exert on the planets, a force that decreased with planetary distance. Hooke's theory, however, was only qualitative.

Newton

What was also seriously lacking was a quantitative, precise definition of both velocity and acceleration. This was ultimately solved by the invention of calculus by both Isaac Newton and Gottfried Wilhelm Leibniz (see Figure 14). Hooke was never able to achieve an understanding of the profound ideas behind the infinitesimal calculus. However, during the period of 1679–1680 Hooke discussed his ideas with Newton, including his conjecture that the force the sun exerts on the planets was actually inversely proportional to the square of the planetary distance.

Figure 14 Gottfried Wilhelm Leibniz (1646–1716).

After Sir Christopher Wren, amateur astronomer, architect of the city of London and London's magnificent St. Paul's Cathedral, issued a public challenge to "theoretically determine" the orbits of the planets, Isaac Newton took a serious interest in the problem. Perhaps acting on rumors, the great British astronomer Edmund Halley (1656–1743) in August 1684 visited Newton in Cambridge and asked him directly what the orbit of a planet would be under an inverse square force. Newton answered that it had to be an ellipse. As the stunned Halley asked him how he knew this, Newton's famous reply was "*Why I have calculated it.*" Halley ultimately urged Newton to publish his results as a book, and these appeared in 1686 in Newton's now legendary *Principia*. See Figure 15.

This book, often and justly referred to as the foundation of modern science, had an immediate dramatic impact. Alexander Pope wrote:

Nature and nature's laws lay hid at night,
God said, "Let Newton be" and all was light.

On the front cover of this text, we see Newton holding open a copy of his *Principia*.

Although Newton did not use calculus in the *Principia*, convincing arguments have been put forward that Newton originally used his calculus to derive the

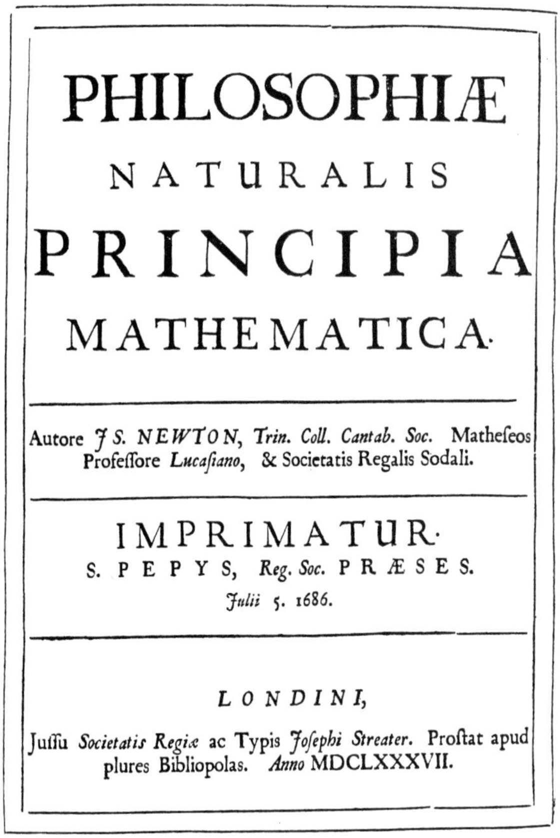

PHILOSOPHIÆ
NATURALIS
PRINCIPIA
MATHEMATICA.

Autore *J S. NEWTON*, *Trin. Coll. Cantab. Soc.* Matheſeos Profeſſore *Lucaſiano*, & Societatis Regalis Sodali.

IMPRIMATUR.
S. PEPYS, *Reg. Soc.* PRÆSES.
Julii 5. 1686.

LONDINI,

Juſſu *Societatis Regiæ* ac Typis *Joſephi Streater.* Proſtat apud plures Bibliopolas. *Anno* MDCLXXXVII.

Figure 15 The frontispiece of the two-lines print of the *Principia*, carrying the imprint "Prostat apud plures Bibliopolas," which is sometimes called the "first issue" of the first edition. The "export copy" (with the three lines "Prostant Venales apud Sam Smith... aliosq; nonnullos Bibliopolas") is called the second issue of the first edition. This distinction between the first and second issues seems to be quite unfounded. It has been suggested that Halley made an agreement with Smith concerning foreign sales; in fact, most of Smith's fifty copies were apparently sold on the continent.

trajectories of the planetary orbits from the inverse square law.* The *Principia* provided profound evidence that the universe, as the early Greeks had understood, was indeed designed mathematically. Incidentally, it was Newton who first conceptualized force as a *vector*, although he provided no formal definition of what a vector was. Such a formal definition had to wait for William Rowan Hamilton, a century and a half after the *Principia*. It was for this achievement and his creation of calculus itself that we chose Newton for our cover.

The invention of the calculus and the subsequent development of vector calculus was the true beginning of modern science and technology, which has changed our world so dramatically. From the mathematics of Newton's mechanics to the profound intellectual constructs of Maxwell's electrodynamics, Einstein's relativity, and Heisenberg's and Schrödinger's quantum mechanics, we have seen the discoveries of radio, television, wireless communications, flight, computers, space travel, and countless engineering marvels.

Underlying all these developments was mathematics, an exciting adventure of the mind and a celebration of the human spirit. It is in this context that we begin our account of Vector calculus.

---

*We shall study the problem of planetary orbits in Section 4.1 and further in the Internet supplement.

Prerequisites and Notation

We assume that students have studied the calculus of functions of a real variable, including analytic geometry in the plane. Some students may have had some exposure to matrices as well, although what we shall need is given in Sections 1.3 and 1.5.

We also assume that students are familiar with functions of elementary calculus, such as $\sin x$, $\cos x$, e^x, and $\log x$ (we write $\log x$ or $\ln x$ for the natural logarithm, which is sometimes denoted $\log_e x$). Students are expected to know, or to review as the course proceeds, the basic rules of differentiation and integration for functions of one variable, such as the chain rule, the quotient rule, integration by parts, and so forth.

We now summarize the notations to be used later. Students can read through these quickly now, then refer to them later if the need arises.

The collection of all real numbers is denoted $\mathbb{R}$. Thus $\mathbb{R}$ includes the ***integers***, ..., $-3, -2, -1, 0, 1, 2, 3, \ldots$; the ***rational numbers***, p/q, where p and q are integers ($q \neq 0$); and the ***irrational numbers***, such as $\sqrt{2}$, π, and e. Members of $\mathbb{R}$ may be visualized as points on the real-number line, as shown in Figure P.1.

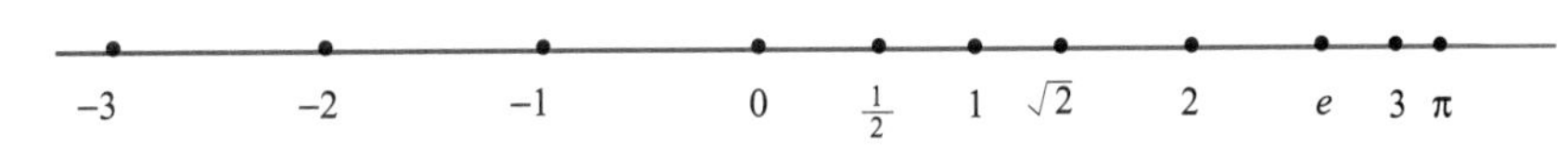

Figure P.1 The geometric representation of points on the real-number line.

When we write $a \in \mathbb{R}$ we mean that a is a member of the set $\mathbb{R}$, in other words, that a is a real number. Given two real numbers a and b with $a < b$ (that is, with a less than b), we can form the ***closed interval*** $[a, b]$, consisting of all x such that $a \leq x \leq b$, and the ***open interval*** (a, b), consisting of all x such that $a < x < b$. Similarly, we can form half-open intervals $(a, b]$ and $[a, b)$ (Figure P.2).

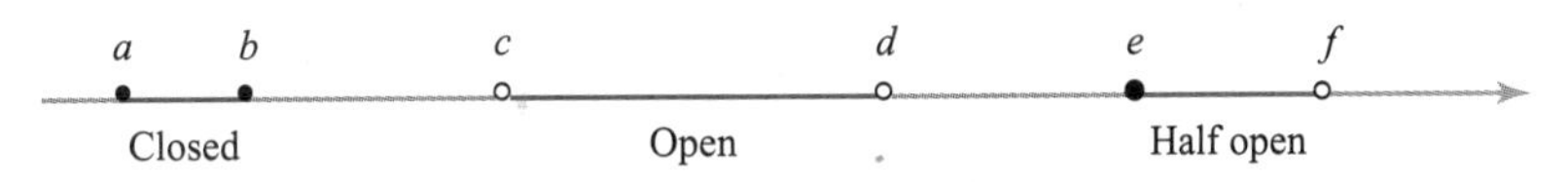

Figure P.2 The geometric representation of the intervals $[a, b]$, (c, d), and $[e, f)$.

The ***absolute value*** of a number $a \in \mathbb{R}$ is written $|a|$ and is defined as

$$|a| = \begin{cases} a & \text{if } a \geq 0 \\ -a & \text{if } a < 0. \end{cases}$$

For example, $|3| = 3$, $|-3| = 3$, $|0| = 0$, and $|-6| = 6$. The inequality $|a + b| \leq |a| + |b|$ always holds. The ***distance from*** a ***to*** b is given by $|a - b|$. Thus, the distance from 6 to 10 is 4 and from -6 to 3 is 9.

If we write $A \subset \mathbb{R}$, we mean A is a ***subset*** of $\mathbb{R}$. For example, A could equal the set of integers $\{\ldots, -3, -2, -1, 0, 1, 2, 3, \ldots\}$. Another example of a subset of $\mathbb{R}$ is the set $\mathbb{Q}$ of rational numbers. Generally, for two collections of objects (that is, sets) A and B, $A \subset B$ means A is a subset of B; that is, every member of A is also a member of B.

The symbol $A \cup B$ means the ***union*** of A and B, the collection whose members are members of either A or B (or both). Thus,

$$\{\ldots, -3, -2, -1, 0\} \cup \{-1, 0, 1, 2, \ldots\} = \{\ldots, -3, -2, -1, 0, 1, 2, \ldots\}.$$

Similarly, $A \cap B$ means the ***intersection*** of A and B; that is, this set consists of those members of A and B that are in both A and B. Thus, the intersection of the two sets above is $\{-1, 0\}$.

We shall write $A \backslash B$ for those members of A that are not in B. Thus,

$$\{\ldots, -3, -2, -1, 0\} \backslash \{-1, 0, 1, 2, \ldots\} = \{\ldots, -3, -2\}.$$

We can also specify sets as in the following examples:

$$\{a \in \mathbb{R} \mid a \text{ is an integer}\} = \{\ldots, -3, -2, -1, 0, 1, 2, \ldots\}$$
$$\{a \in \mathbb{R} \mid a \text{ is an even integer}\} = \{\ldots, -2, 0, 2, 4, \ldots\}$$
$$\{x \in \mathbb{R} \mid a \leq x \leq b\} = [a, b].$$

A ***function*** $f\colon A \to B$ is a rule that assigns to each $a \in A$ one specific member $f(a)$ of B. We call A the ***domain*** of f and B the ***target*** of f. The set $\{f(x) \mid x \in A\}$ consisting of all the values of $f(x)$ is called the ***range*** of f. Denoted by $f(A)$, the range is a subset of the target B. It may be all of B, in which case f is said to be ***onto*** B. The fact that the function f sends a to $f(a)$ is denoted by $a \mapsto f(a)$. For example, the function $f(x) = x^3/(1 - x)$ that assigns the number $x^3/(1 - x)$ to each $x \neq 1$ in $\mathbb{R}$ can also be defined by the rule $x \mapsto x^3/(1 - x)$. Functions are also called ***mappings***, ***maps***, or ***transformations***. The notation $f\colon A \subset \mathbb{R} \to \mathbb{R}$ means that A is a subset of $\mathbb{R}$ and that f assigns a value $f(x)$ in $\mathbb{R}$ to each $x \in A$. The ***graph*** of f consists of all the points $(x, f(x))$ in the plane (Figure P.3).

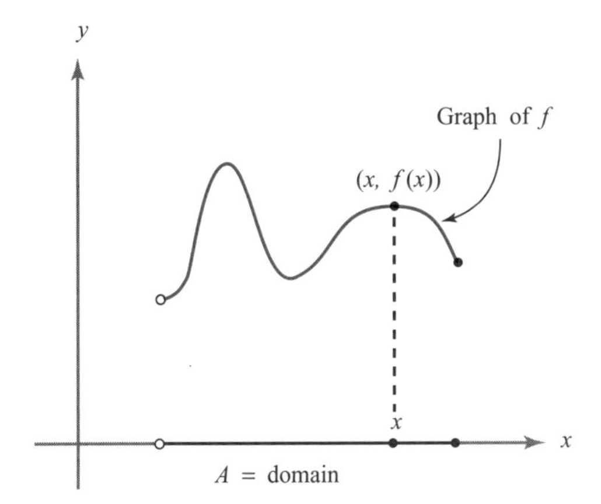

Figure P.3 The graph of a function with the half-open interval A as domain.

The notation $\sum_{i=1}^{n} a_i$ means $a_1 + \cdots + a_n$, where $a_1, \ldots, a_n$ are given numbers. The sum of the first n integers is

$$1 + 2 + \cdots + n = \sum_{i=1}^{n} i = \frac{n(n+1)}{2}.$$

The ***derivative*** of a function $f(x)$ is denoted $f'(x)$, or

$$\frac{df}{dx},$$

and the ***definite integral*** is written

$$\int_a^b f(x)\,dx.$$

If we set $y = f(x)$, the derivative is also denoted by

$$\frac{dy}{dx}.$$

Readers are assumed to be familiar with the chain rule, integration by parts, and other basic facts from the calculus of functions of one variable. In particular, they should know how to differentiate and integrate exponential, logarithmic, and trigonometric functions. Short tables of derivatives and integrals, which are adequate for the needs of this text, are printed at the front and back of the book.

The following notations are used synonymously: $e^x = \exp x$, $\ln x = \log x$, and $\sin^{-1} x = \arcsin x$.

The end of a proof is denoted by the symbol ■, while the end of an example or remark is denoted by the symbol ▲.

1

The Geometry of Euclidean Space

Quaternions came from Hamilton . . . and have been an unmixed evil to those who have touched them in any way. Vector is a useless survival . . . and has never been of the slightest use to any creature.

Lord Kelvin

In this chapter we consider the basic operations on vectors in two- and three-dimensional space: vector addition, scalar multiplication, and the dot and cross products. In Section 1.5 we generalize some of these notions to n-space and review properties of matrices that will be needed in Chapters 2 and 3.

1.1 Vectors in Two- and Three-Dimensional Space

Points P in the plane are represented by ordered pairs of real numbers (a_1, a_2); the numbers a_1 and a_2 are called the ***Cartesian coordinates of*** P. We draw two perpendicular lines, label them as the x and y axes, and then drop perpendiculars from P to these axes, as in Figure 1.1.1. After designating the intersection of the x and y axes as the origin and choosing units on these axes, we produce two signed distances a_1 and a_2 as shown in the figure; a_1 is called the x ***component*** of P, and a_2 is called the y ***component***.

Points in space may be similarly represented as ordered triples of real numbers. To construct such a representation, we choose three mutually perpendicular lines that meet at a point in space. These lines are called x ***axis***, y ***axis***, and z ***axis***, and the point at which they meet is called the ***origin*** (this is our reference point). We choose a scale on these axes, as shown in Figure 1.1.2.

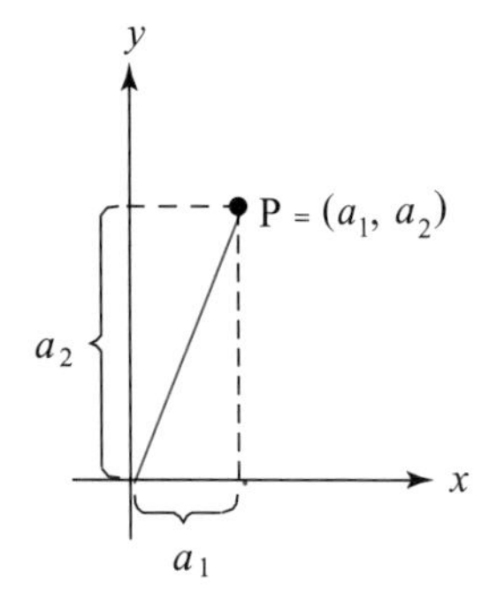

Figure 1.1.1 Cartesian coordinates in the plane.

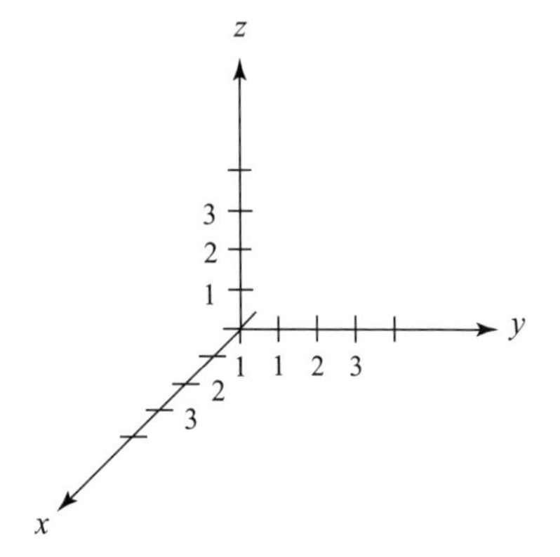

Figure 1.1.2 Cartesian coordinates in space.

The triple (0, 0, 0) corresponds to the origin of the coordinate system, and the arrows on the axes indicate the positive directions. For example, the triple (2, 4, 4) represents a point 2 units from the origin in the positive direction along the x axis, 4 units in the positive direction along the y axis, and 4 units in the positive direction along the z axis (Figure 1.1.3).

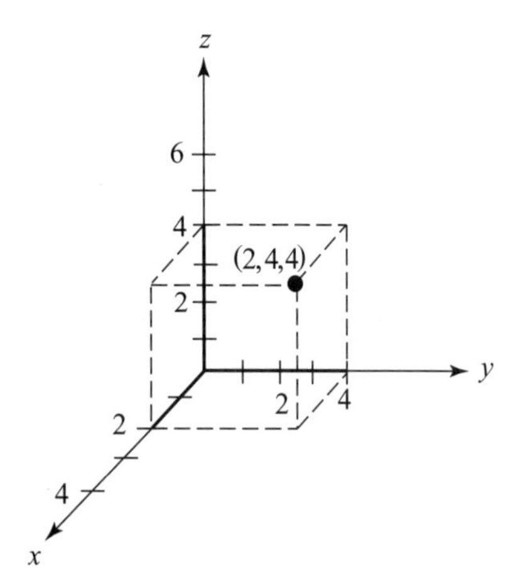

Figure 1.1.3 Geometric representation of the point (2, 4, 4) in Cartesian coordinates.

Because we can associate points in space with ordered triples in this way, we often use the expression "the point (a_1, a_2, a_3)" instead of the longer phrase "the point P that corresponds to the triple (a_1, a_2, a_3)." We say that a_1 is the ***x coordinate*** (or first coordinate), a_2 is the ***y coordinate*** (or second coordinate), and a_3 is the ***z coordinate*** (or third coordinate) of P. It is also common to denote points in space with the letters x, y, and z in place of a_1, a_2, and a_3. Thus, the triple (x, y, z) represents a point whose first coordinate is x, second coordinate is y, and third coordinate is z.

We employ the following notation for the line, the plane, and three-dimensional space:

(i) The real number line is denoted $\mathbb{R}^1$ or simply $\mathbb{R}$.

(ii) The set of all ordered pairs (x, y) of real numbers is denoted $\mathbb{R}^2$.

(iii) The set of all ordered triples (x, y, z) of real numbers is denoted $\mathbb{R}^3$.

When speaking of $\mathbb{R}^1$, $\mathbb{R}^2$, and $\mathbb{R}^3$ simultaneously, we write $\mathbb{R}^n$, where $n = 1, 2$, or 3; or $\mathbb{R}^m$, where $m = 1, 2, 3$. Starting in Section 1.5 we will also study $\mathbb{R}^n$ for $n = 4, 5, 6, \ldots$, but the cases $n = 1, 2, 3$ are closest to our geometric intuition and will be stressed throughout the book.

Vector Addition and Scalar Multiplication

The operation of addition can be extended from $\mathbb{R}$ to $\mathbb{R}^2$ and $\mathbb{R}^3$. For $\mathbb{R}^3$, this is done as follows. Given the two triples (a_1, a_2, a_3) and (b_1, b_2, b_3), we define their ***sum*** to be

$$(a_1, a_2, a_3) + (b_1, b_2, b_3) = (a_1 + b_1, a_2 + b_2, a_3 + b_3).$$

EXAMPLE 1

$$(1, 1, 1) + (2, -3, 4) = (3, -2, 5),$$
$$(x, y, z) + (0, 0, 0) = (x, y, z),$$
$$(1, 7, 3) + (a, b, c) = (1 + a, 7 + b, 3 + c). \quad \blacktriangle$$

The element $(0, 0, 0)$ is called the ***zero element*** (or just ***zero***) of $\mathbb{R}^3$. The element $(-a_1, -a_2, -a_3)$ is the ***additive inverse*** (or ***negative***) of (a_1, a_2, a_3), and we will write $(a_1, a_2, a_3) - (b_1, b_2, b_3)$ for $(a_1, a_2, a_3) + (-b_1, -b_2, -b_3)$.

The additive inverse, when added to the vector itself, of course produces zero:

$$(a_1, a_2, a_3) + (-a_1, -a_2, -a_3) = (0, 0, 0).$$

There are several important product operations that we will define on $\mathbb{R}^3$. One of these, called the *inner product*, assigns a real number to each pair of elements of $\mathbb{R}^3$. We shall discuss it in detail in Section 1.2. Another product operation for $\mathbb{R}^3$ is called *scalar multiplication* (the word "scalar" is a synonym for "real number"). This product combines scalars (real numbers) and elements of $\mathbb{R}^3$ (ordered triples) to yield elements of $\mathbb{R}^3$ as follows: Given a scalar α and a triple (a_1, a_2, a_3), we define the ***scalar multiple*** by

$$\alpha(a_1, a_2, a_3) = (\alpha a_1, \alpha a_2, \alpha a_3).$$

EXAMPLE 2

$$\begin{aligned}2(4, e, 1) &= (2 \cdot 4, 2 \cdot e, 2 \cdot 1) = (8, 2e, 2),\\ 6(1, 1, 1) &= (6, 6, 6),\\ 1(u, v, w) &= (u, v, w),\\ 0(p, q, r) &= (0, 0, 0). \quad \blacktriangle\end{aligned}$$

Addition and scalar multiplication of triples satisfy the following properties:

(i) $(\alpha\beta)(a_1, a_2, a_3) = \alpha[\beta(a_1, a_2, a_3)]$ (associativity)

(ii) $(\alpha + \beta)(a_1, a_2, a_3) = \alpha(a_1, a_2, a_3) + \beta(a_1, a_2, a_3)$ (distributivity)

(iii) $\alpha[(a_1, a_2, a_3) + (b_1, b_2, b_3)] = \alpha(a_1, a_2, a_3) + \alpha(b_1, b_2, b_3)$ (distributivity)

(iv) $\alpha(0, 0, 0) = (0, 0, 0)$ (property of zero)

(v) $0(a_1, a_2, a_3) = (0, 0, 0)$ (property of zero)

(vi) $1(a_1, a_2, a_3) = (a_1, a_2, a_3)$ (property of the unit element)

The identities are proven directly from the definitions of addition and scalar multiplication. For instance,

$$\begin{aligned}(\alpha + \beta)(a_1, a_2, a_3) &= ((\alpha + \beta)a_1, (\alpha + \beta)a_2, (\alpha + \beta)a_3)\\ &= (\alpha a_1 + \beta a_1, \alpha a_2 + \beta a_2, \alpha a_3 + \beta a_3)\\ &= \alpha(a_1, a_2, a_3) + \beta(a_1, a_2, a_3).\end{aligned}$$

For $\mathbb{R}^2$, addition and scalar multiplication are defined just as in $\mathbb{R}^3$, with the third component of each vector dropped off. All the properties (i) to (vi) still hold.

EXAMPLE 3 Interpret the chemical equation $2NH_2 + H_2 = 2NH_3$ as a relation in the algebra of ordered pairs.

SOLUTION We think of the molecule N_xH_y (x atoms of nitrogen, y atoms of hydrogen) as represented by the ordered pair (x, y). Then the chemical equation given is equivalent to $2(1, 2) + (0, 2) = 2(1, 3)$. Indeed, both sides are equal to $(2, 6)$. ▲

Geometry of Vector Operations

Let us turn to the geometry of these operations in $\mathbb{R}^2$ and $\mathbb{R}^3$. For the moment, we define a ***vector*** to be a directed line segment beginning at the origin, that is, a line segment with specified magnitude and direction, and initial point at the origin. Figure 1.1.4 shows several vectors, drawn as arrows beginning at the origin. In print, vectors are usually denoted by boldface letters such as **a**. By hand, we usually write them as $\vec{a}$ or simply as a, possibly with a line or wavy line under it.

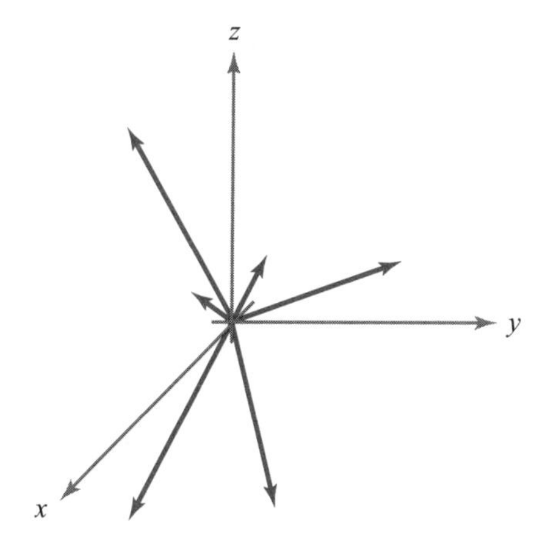

Figure 1.1.4 Geometrically, vectors are thought of as arrows emanating from the origin.

Using this definition of a vector, we associate with each vector **a** the point (a_1, a_2, a_3) where **a** terminates, and conversely, we can associate a vector **a** with each point (a_1, a_2, a_3) in space. Thus, we shall identify **a** with (a_1, a_2, a_3) and write $\mathbf{a} = (a_1, a_2, a_3)$. For this reason, the elements of $\mathbb{R}^3$ not only are ordered triples of real numbers, but are also regarded as vectors. The triple (0, 0, 0) is denoted **0**. We call a_1, a_2, and a_3 the ***components*** of **a**, or when we think of **a** as a point, its ***coordinates***.

Two vectors $\mathbf{a} = (a_1, a_2, a_3)$ and $\mathbf{b} = (b_1, b_2, b_3)$ are equal if and only if $a_1 = b_1$, $a_2 = b_2$, and $a_3 = b_3$. Geometrically this means that **a** and **b** have the same direction and the same length (or "magnitude").

Geometrically, we define vector addition as follows. In the plane containing the vectors $\mathbf{a} = (a_1, a_2, a_3)$ and $\mathbf{b} = (b_1, b_2, b_3)$ (see Figure 1.1.5), form the parallelogram having **a** as one side and **b** as its adjacent side. The sum $\mathbf{a} + \mathbf{b}$ is the directed line segment along the diagonal of the parallelogram.

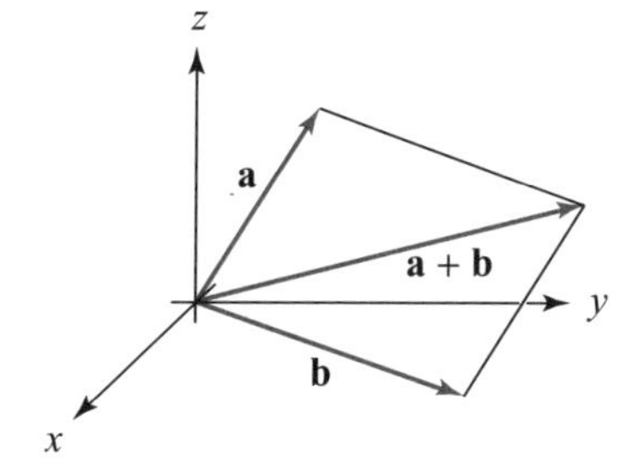

Figure 1.1.5 The geometry of vector addition.

This geometric view of vector addition is useful in many physical situations, as we shall see in the next section. For an easily visualized example, consider a bird or an airplane flying through the air with velocity $\mathbf{v}_1$, but in the presence of a wind with velocity $\mathbf{v}_2$. The resultant velocity, $\mathbf{v}_1 + \mathbf{v}_2$, is what one sees; see Figure 1.1.6.

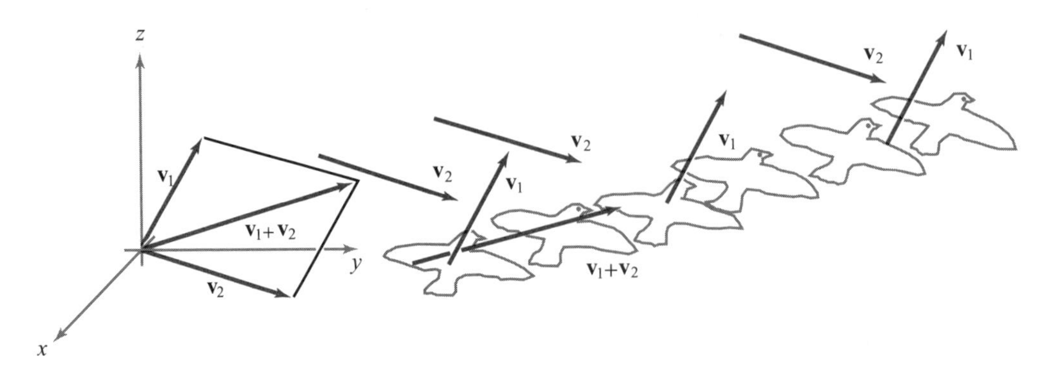

Figure 1.1.6 A physical interpretation of vector addition.

To show that our geometric definition of addition is consistent with our algebraic definition, we demonstrate that $\mathbf{a}+\mathbf{b}=(a_1+b_1, a_2+b_2, a_3+b_3)$. We shall prove this result in the plane and leave the proof in three-dimensional space to the reader. Thus, we wish to show that if $\mathbf{a}=(a_1, a_2)$ and $\mathbf{b}=(b_1, b_2)$, then $\mathbf{a}+\mathbf{b}=(a_1+b_1, a_2+b_2)$.

In Figure 1.1.7 let $\mathbf{a}=(a_1, a_2)$ be the vector ending at the point A, and let $\mathbf{b}=(b_1, b_2)$ be the vector ending at point B. By definition, the vector $\mathbf{a}+\mathbf{b}$ ends at the vertex C of parallelogram OBCA. To verify that $\mathbf{a}+\mathbf{b}=(a_1+b_1, a_2+b_2)$, it suffices to show that the coordinates of C are (a_1+b_1, a_2+b_2). The sides of the triangles OAD and BCG are parallel, and the sides OA and BC have equal lengths, which we write as OA = BC. These triangles are congruent, so BG = OD; since BGFE is a rectangle, EF = BG. Furthermore, OD = a_1 and OE = b_1. Hence, EF = BG = OD = a_1. Since OF = EF + OE, it follows that OF = a_1+b_1. This shows that the x coordinate of $\mathbf{a}+\mathbf{b}$ is a_1+b_1. The proof that the y coordinate is a_2+b_2 is analogous. This argument assumes A and B to be in the first quadrant, but similar arguments hold for the other quadrants.

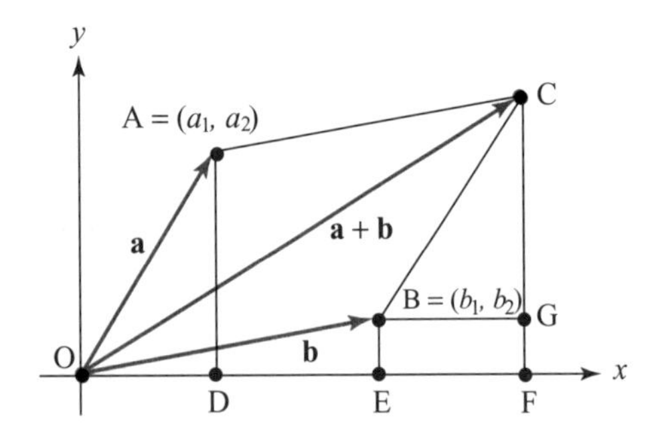

Figure 1.1.7 The construction used to prove that $(a_1, a_2)+(b_1, b_2)=(a_1+b_1, a_2+b_2)$.

Figure 1.1.8(a) illustrates another way of looking at vector addition: in terms of triangles rather than parallelograms. That is, we translate (without rotation) the directed line segment representing the vector $\mathbf{b}$ so that it begins at the end of the vector $\mathbf{a}$. The endpoint of the resulting directed segment is the endpoint of the vector

a + **b**. We note that when **a** and **b** are collinear, the triangle collapses to a line segment, as in Figure 1.1.8(b).

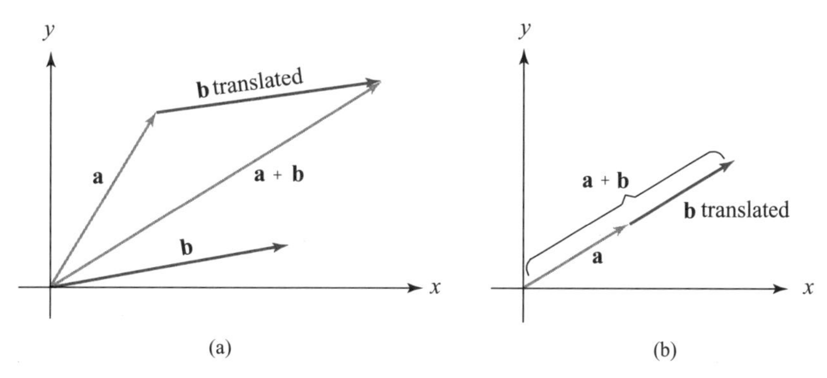

Figure 1.1.8 (a) Vector addition may be visualized in terms of triangles as well as parallelograms. (b) The triangle collapses to a line segment when **a** and **b** are collinear.

In Figure 1.1.8 we have placed **a** and **b** *head to tail*. That is, the tail of **b** is placed at the head of **a**, and the vector **a** + **b** goes from the tail of **a** to the head of **b**. If we do it in the other order, **b** + **a**, we get the same vector by going around the parallelogram the other way. Consistent with this figure, it is useful to let vectors "glide" or "slide," keeping the same magnitude and direction. We want, in fact, to regard two vectors as the *same* if they have the same magnitude and direction. When we insist on vectors beginning at the origin, we will say that we have ***bound vectors***. If we allow vectors to begin at other points, we will speak of ***free vectors*** or just ***vectors***.

Vectors Vectors (also called *free vectors*) are directed line segments in [the plane or] space represented by directed line segments with a beginning (tail) and an end (head). Directed line segments obtained from each other by parallel translation (but not rotation) represent the same vector.

The components (a_1, a_2, a_3) of **a** are the (signed) lengths of the projections of **a** along the three coordinate axes; equivalently, they are defined by placing the tail of **a** at the origin and letting the head be the point (a_1, a_2, a_3). We write $\mathbf{a} = (a_1, a_2, a_3)$.

Two vectors are added by placing them head to tail and drawing the vectors from the tail of the first to the head of the second, as in Figure 1.1.8.

Scalar multiplication of vectors also has a geometric interpretation. If α is a scalar and **a** a vector, we define $\alpha\mathbf{a}$ to be the vector that is $|\alpha|$ times as long as **a**, with the same direction as **a** if $\alpha > 0$, but with the opposite direction if $\alpha < 0$. Figure 1.1.9 illustrates several examples.

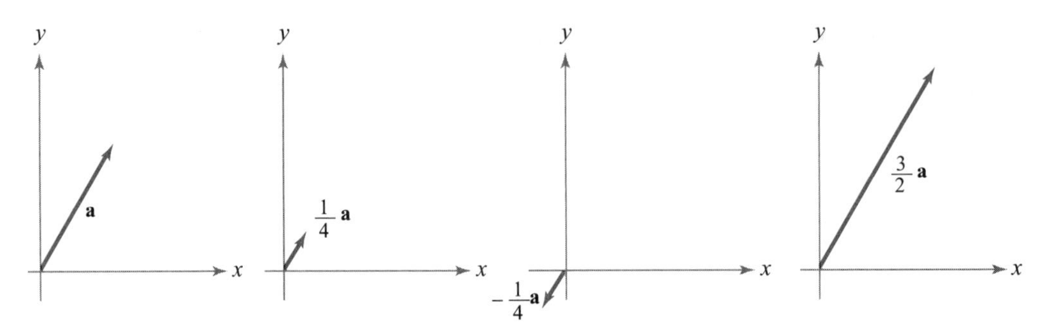

Figure 1.1.9 Some scalar multiples of a vector **a**.

Using an argument based on similar triangles, one finds that if $\mathbf{a} = (a_1, a_2, a_3)$, and α is a scalar, then

$$\alpha\mathbf{a} = (\alpha a_1, \alpha a_2, \alpha a_3).$$

That is, the geometric definition coincides with the algebraic one.

Given two vectors **a** and **b**, how do we represent the vector $\mathbf{b} - \mathbf{a}$ geometrically, that is, what is the geometry of vector subtraction? Because $\mathbf{a} + (\mathbf{b} - \mathbf{a}) = \mathbf{b}$, we see that $\mathbf{b} - \mathbf{a}$ is the vector that one adds to **a** to get **b**. In view of this, we may conclude that $\mathbf{b} - \mathbf{a}$ is the vector parallel to, and with the same magnitude as, the directed line segment beginning at the endpoint of **a** and terminating at the endpoint of **b** when **a** and **b** begin at the same point (see Figure 1.1.10).

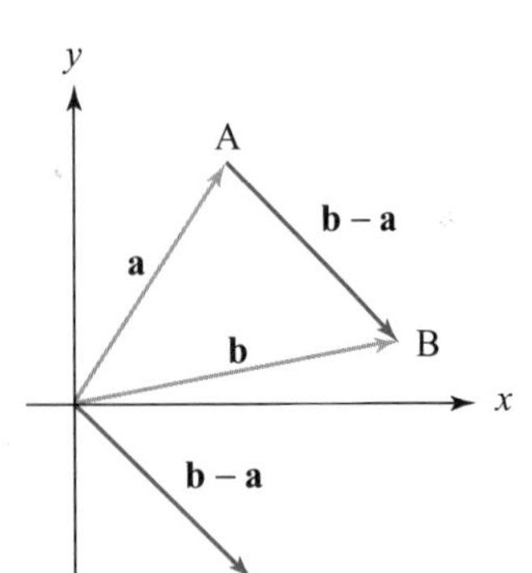

Figure 1.1.10 The geometry of vector subtraction.

EXAMPLE 4 Let **u** and **v** be the vectors shown in Figure 1.1.11. Draw the two vectors $\mathbf{u} + \mathbf{v}$ and $-2\mathbf{u}$. What are their components?

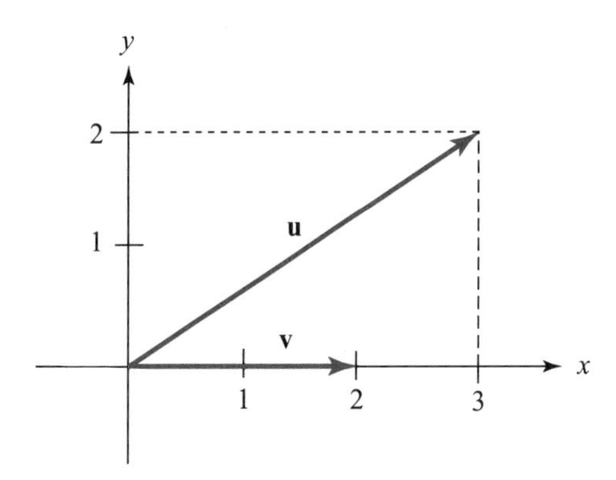

Figure 1.1.11 Find $\mathbf{u} + \mathbf{v}$ and $-2\mathbf{u}$.

SOLUTION Place the tail of $\mathbf{v}$ at the tip of $\mathbf{u}$ to obtain the vector shown in Figure 1.1.12.

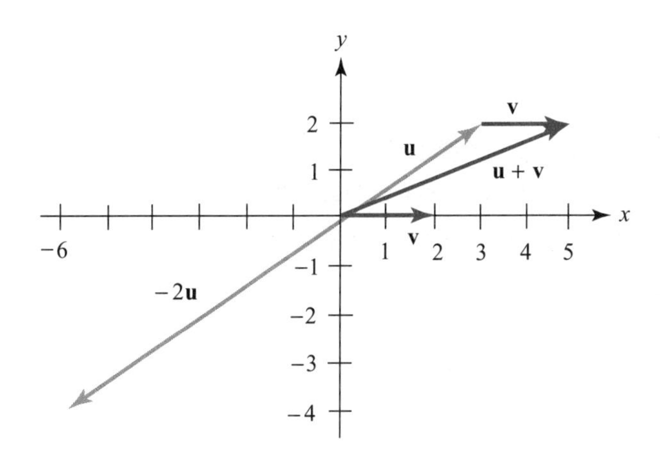

Figure 1.1.12 Computing $\mathbf{u} + \mathbf{v}$ and $-2\mathbf{u}$.

The vector $-2\mathbf{u}$, also shown, has length twice that of $\mathbf{u}$ and points in the opposite direction. From the figure, we see that the vector $\mathbf{u} + \mathbf{v}$ has components $(5, 2)$ and $-2\mathbf{u}$ has components $(-6, -4)$. ▲

EXAMPLE 5

(a) Sketch $-2\mathbf{v}$, where $\mathbf{v}$ has components $(-1, 1, 2)$.

(b) If $\mathbf{v}$ and $\mathbf{w}$ are any two vectors, show that $\mathbf{v} - \frac{1}{3}\mathbf{w}$ and $3\mathbf{v} - \mathbf{w}$ are parallel.

SOLUTION

(a) The vector $-2\mathbf{v}$ is twice as long as $\mathbf{v}$, but points in the opposite direction (see Figure 1.1.13).

(b) $\mathbf{v} - \frac{1}{3}\mathbf{w} = \frac{1}{3}(3\mathbf{v} - \mathbf{w})$; vectors that are multiples of one another are parallel. ▲

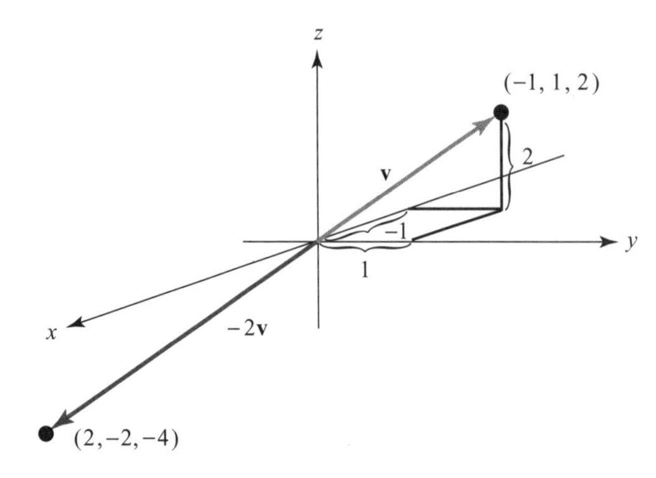

Figure 1.1.13 Multiplying $(-1, 1, 2)$ by -2.

The Standard Basis Vectors

To describe vectors in space, it is convenient to introduce three special vectors along the x, y, and z axes:

$\mathbf{i}$: the vector with components $(1, 0, 0)$

$\mathbf{j}$: the vector with components $(0, 1, 0)$

$\mathbf{k}$: the vector with components $(0, 0, 1)$.

These ***standard basis vectors*** are illustrated in Figure 1.1.14. In the plane one has the standard basis $\mathbf{i}$ and $\mathbf{j}$ with components $(1, 0)$ and $(0, 1)$.

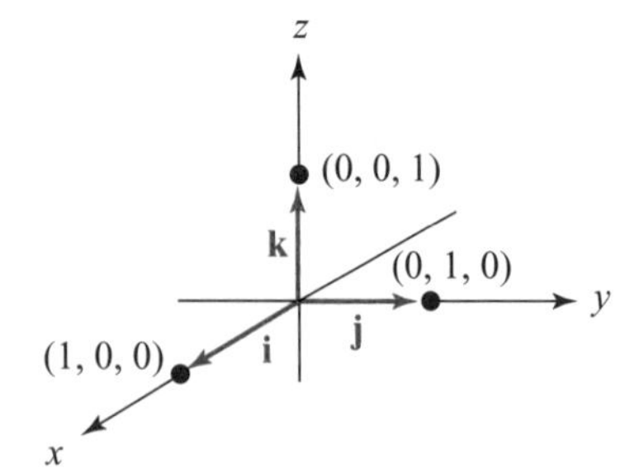

Figure 1.1.14 The standard basis vectors.

Let $\mathbf{a}$ be any vector, and let (a_1, a_2, a_3) be its components. Then

$$\mathbf{a} = a_1\mathbf{i} + a_2\mathbf{j} + a_3\mathbf{k},$$

because the right-hand side is given in components by

$$\begin{aligned} a_1(1, 0, 0) + a_2(0, 1, 0) + a_3(0, 0, 1) &= (a_1, 0, 0) + (0, a_2, 0) + (0, 0, a_3) \\ &= (a_1, a_2, a_3). \end{aligned}$$

Thus, we can express every vector as a sum of scalar multiples of **i**, **j**, and **k**.

The Standard Basis Vectors

1. The vectors **i**, **j**, and **k** are unit vectors along the three coordinate axes, as shown in Figure 1.1.14.
2. If **a** has components (a_1, a_2, a_3), then

$$\mathbf{a} = a_1\mathbf{i} + a_2\mathbf{j} + a_3\mathbf{k}.$$

EXAMPLE 6 Express the vector whose components are $(e, \pi, -\sqrt{3})$ in the standard basis.

SOLUTION Substituting $a_1 = e$, $a_2 = \pi$, and $a_3 = -\sqrt{3}$ into $\mathbf{a} = a_1\mathbf{i} + a_2\mathbf{j} + a_3\mathbf{k}$ gives

$$\mathbf{v} = e\mathbf{i} + \pi\mathbf{j} - \sqrt{3}\mathbf{k}. \quad \blacktriangle$$

EXAMPLE 7 The vector (2, 3, 2) equals $2\mathbf{i} + 3\mathbf{j} + 2\mathbf{k}$, and the vector $(0, -1, 4)$ is $-\mathbf{j} + 4\mathbf{k}$. Figure 1.1.15 shows $2\mathbf{i} + 3\mathbf{j} + 2\mathbf{k}$; the student should draw in the vector $-\mathbf{j} + 4\mathbf{k}$. ▲

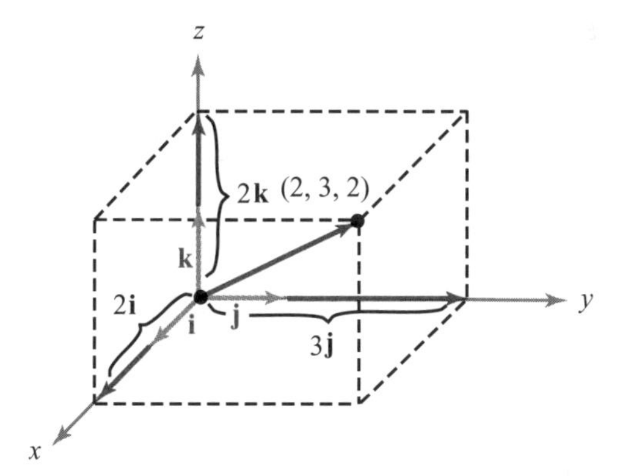

Figure 1.1.15 Representation of (2, 3, 2) in terms of the standard basis vectors **i**, **j**, and **k**.

Addition and scalar multiplication may be written in terms of the standard basis vectors as follows:

$$(a_1\mathbf{i} + a_2\mathbf{j} + a_3\mathbf{k}) + (b_1\mathbf{i} + b_2\mathbf{j} + b_3\mathbf{k}) = (a_1 + b_1)\mathbf{i} + (a_2 + b_2)\mathbf{j} + (a_3 + b_3)\mathbf{k}$$

and

$$\alpha(a_1\mathbf{i} + a_2\mathbf{j} + a_3\mathbf{k}) = (\alpha a_1)\mathbf{i} + (\alpha a_2)\mathbf{j} + \alpha(a_3)\mathbf{k}.$$

The Vector Joining Two Points

To apply vectors to geometric problems, it is useful to assign a vector to a *pair* of points in the plane or in space, as follows. Given two points P and P′, we can draw the vector **v** with tail P and head P′, as in Figure 1.1.16, where we write $\overrightarrow{PP'}$ for **v**.

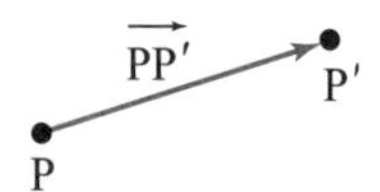

Figure 1.1.16 The vector from P to P′ is denoted $\overrightarrow{PP'}$.

If $P = (x, y, z)$ and $P' = (x', y', z')$, then the vectors from the origin to P and P′ are $\mathbf{a} = x\mathbf{i} + y\mathbf{j} + z\mathbf{k}$ and $\mathbf{a}' = x'\mathbf{i} + y'\mathbf{j} + z'\mathbf{k}$, respectively, so the vector $\overrightarrow{PP'}$ is the difference $\mathbf{a}' - \mathbf{a} = (x' - x)\mathbf{i} + (y' - y)\mathbf{j} + (z' - z)\mathbf{k}$. (See Figure 1.1.17.)

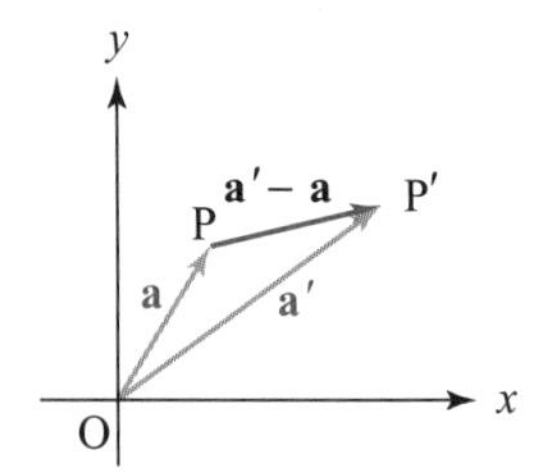

Figure 1.1.17 $\overrightarrow{PP'} = \overrightarrow{OP'} - \overrightarrow{OP}$.

> **The Vector Joining Two Points** If the point P has coordinates (x, y, z) and P′ has coordinates (x', y', z'), then the vector $\overrightarrow{PP'}$ from the tip of P to the tip of P′ has components $(x' - x, y' - y, z' - z)$.

EXAMPLE 8

(a) Find the components of the vector from (3, 5) to (4, 7).

(b) Add the vector **v** from $(-1, 0)$ to $(2, -3)$ and the vector **w** from (2, 0) to (1, 1).

(c) Multiply the vector **v** in (b) by 8. If the resulting vector is represented by the directed line segment from (5, 6) to Q, what is Q?

SOLUTION

(a) As in the preceding box, we subtract the ordered pairs: $(4, 7) - (3, 5) = (1, 2)$. Thus the required components are (1, 2).

(b) The vector **v** has components $(2, -3) - (-1, 0) = (3, -3)$, and **w** has components $(1, 1) - (2, 0) = (-1, 1)$. Therefore, the vector $\mathbf{v} + \mathbf{w}$ has components $(3, -3) + (-1, 1) = (2, -2)$.

(c) The vector $8\mathbf{v}$ has components $8(3, -3) = (24, -24)$. If this vector is represented by the directed line segment from $(5, 6)$ to Q, and Q has coordinates (x, y), then $(x, y) - (5, 6) = (24, -24)$, so $(x, y) = (5, 6) + (24, -24) = (29, -18)$. ▲

EXAMPLE 9 Let $\mathrm{P} = (-2, -1)$, $\mathrm{Q} = (-3, -3)$, and $\mathrm{R} = (-1, -4)$ in the xy plane.

(a) Draw these vectors: $\mathbf{v}$ joining P to Q; $\mathbf{w}$ joining Q to R; $\mathbf{u}$ joining R to P.

(b) What are the components of $\mathbf{v}$, $\mathbf{w}$, and $\mathbf{u}$?

(c) What is $\mathbf{v} + \mathbf{w} + \mathbf{u}$?

SOLUTION

(a) See Figure 1.1.18.

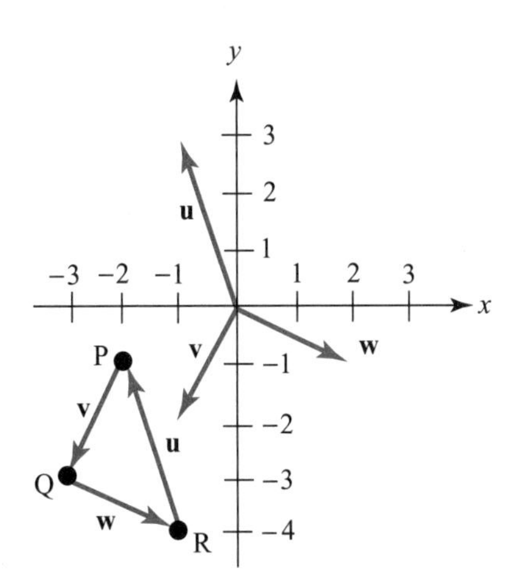

Figure 1.1.18 The vector $\mathbf{v}$ joins P to Q; $\mathbf{w}$ joins Q to R; and $\mathbf{u}$ joins R to P.

(b) Because $\mathbf{v} = \overrightarrow{\mathrm{PQ}}$, $\mathbf{w} = \overrightarrow{\mathrm{QR}}$, and $\mathbf{u} = \overrightarrow{\mathrm{RP}}$, we get

$$\mathbf{v} = (-3, -3) - (-2, -1) = (-1, -2),$$

$$\mathbf{w} = (-1, -4) - (-3, -3) = (2, -1),$$

$$\mathbf{u} = -(-1, -4) + (-2, -1) = (-1, 3).$$

(c) $\mathbf{v} + \mathbf{w} + \mathbf{u} = (-1, -2) + (2, -1) + (-1, 3) = (0, 0)$. ▲

Geometry Theorems by Vector Methods

Many of the theorems of plane geometry can be proved by vector methods. Here is one example.

EXAMPLE 10 Use vectors to prove that the diagonals of a parallelogram bisect each other.

SOLUTION Let OPRQ be the parallelogram, with two adjacent sides represented by the vectors $\mathbf{a} = \overrightarrow{OP}$ and $\mathbf{b} = \overrightarrow{OQ}$. Let M be the midpoint of the diagonal OR, N the midpoint of the other diagonal, PQ. (See Figure 1.1.19.)

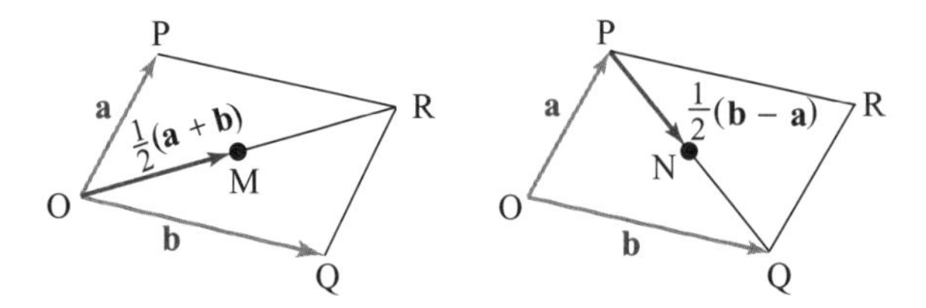

Figure 1.1.19 If the midpoints M and N coincide, then the diagonals OR and PQ bisect each other.

Observe that $\overrightarrow{OR} = \overrightarrow{OP} + \overrightarrow{OQ} = \mathbf{a} + \mathbf{b}$ by the parallelogram rule for vector addition, so $\overrightarrow{OM} = \frac{1}{2}\overrightarrow{OR} = \frac{1}{2}(\mathbf{a} + \mathbf{b})$. On the other hand,

$$\overrightarrow{PQ} = \overrightarrow{OQ} - \overrightarrow{OP} = \mathbf{b} - \mathbf{a}, \qquad \text{so} \qquad \overrightarrow{PN} = \tfrac{1}{2}\overrightarrow{PQ} = \tfrac{1}{2}(\mathbf{b} - \mathbf{a}),$$

and hence

$$\overrightarrow{ON} = \overrightarrow{OP} + \overrightarrow{PN} = \mathbf{a} + \tfrac{1}{2}(\mathbf{b} - \mathbf{a}) = \tfrac{1}{2}(\mathbf{a} + \mathbf{b}).$$

Because $\overrightarrow{OM}$ and $\overrightarrow{ON}$ are equal vectors, the points M and N coincide, so the diagonals bisect each other. ▲

Equations of Lines

Planes and lines are geometric objects that can be represented by equations. We shall defer until Section 1.3 a study of equations representing planes. However, using the geometric interpretation of vector addition and scalar multiplication, we will now find the *equation of a line l that passes through the endpoint of the vector* **a**, *with the direction of a vector* **v** (see Figure 1.1.20).

As t varies through all real values, the points of the form $t\mathbf{v}$ are all scalar multiples of the vector **v**, and therefore exhaust the points of the line *passing through the origin* in the direction of **v**. Because every point on l is the endpoint of the diagonal of a parallelogram with sides **a** and $t\mathbf{v}$ for some real value of t, we see that all the points on l are of the form $\mathbf{a} + t\mathbf{v}$. Thus, the line l may be expressed by the equation $\mathbf{l}(t) = \mathbf{a} + t\mathbf{v}$. We say that l is expressed ***parametrically***, with t the parameter. At $t = 0$, $\mathbf{l}(t) = \mathbf{a}$. As t increases, the point $\mathbf{l}(t)$ moves away from **a** in the direction of **v**.

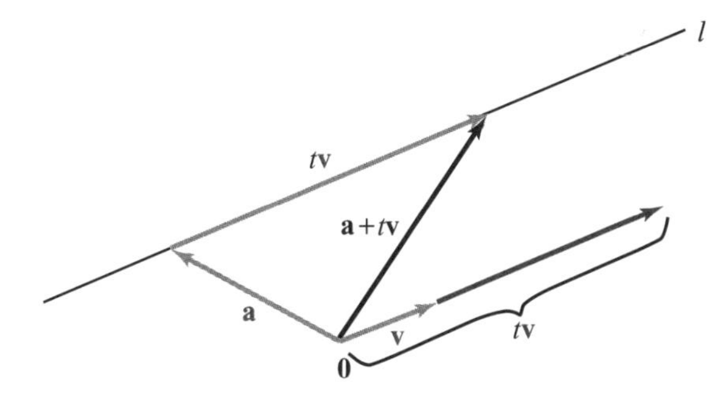

Figure 1.1.20 The line l, parametrically given by $\mathbf{l}(t) = \mathbf{a} + t\mathbf{v}$, lies in the direction $\mathbf{v}$ and passes through the tip of $\mathbf{a}$.

As t decreases from $t = 0$ through negative values, $\mathbf{l}(t)$ moves away from $\mathbf{a}$ in the direction of $-\mathbf{v}$.

Point-Direction Form of a Line The equation of the line l through the tip of $\mathbf{a}$ and pointing in the direction of the vector $\mathbf{v}$ is $\mathbf{l}(t) = \mathbf{a} + t\mathbf{v}$, where the parameter t takes on all real values. In coordinate form, the equations are

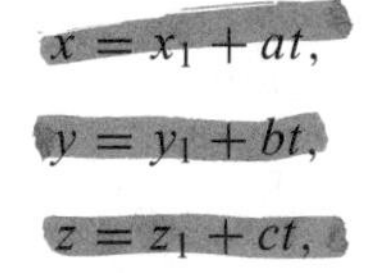

$$x = x_1 + at,$$
$$y = y_1 + bt,$$
$$z = z_1 + ct,$$

where $\mathbf{a} = (x_1, y_1, z_1)$ and $\mathbf{v} = (a, b, c)$. For lines in the xy plane, one simply drops the z component.

EXAMPLE 11 Determine the equation of the line l passing through (1, 0, 0) in the direction $\mathbf{j}$. See Figure 1.1.21.

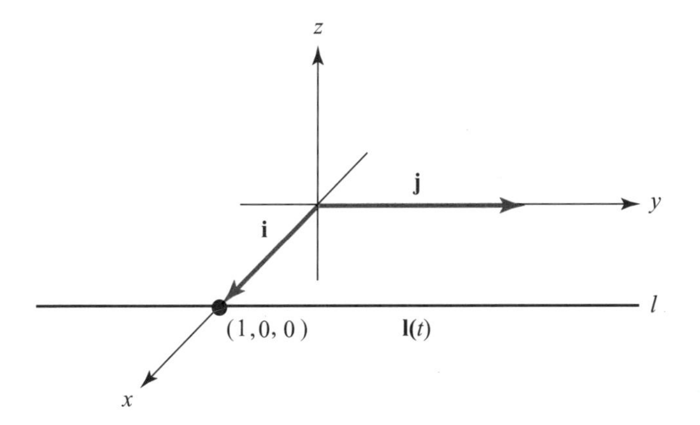

Figure 1.1.21 The line l passes through the tip of $\mathbf{i}$ in the direction $\mathbf{j}$.

SOLUTION The desired line can be expressed parametrically as $\mathbf{l}(t) = \mathbf{i} + t\mathbf{j}$. In terms of coordinates,

$$\mathbf{l}(t) = (1, 0, 0) + t(0, 1, 0) = (1, t, 0). \quad \blacktriangle$$

EXAMPLE 12

(a) Find the equations of the line in space through the point $(3, -1, 2)$ in the direction $2\mathbf{i} - 3\mathbf{j} + 4\mathbf{k}$.

(b) Find the equation of the line in the plane through the point $(1, -6)$ in the direction of $5\mathbf{i} - \pi\mathbf{j}$.

(c) In what direction does the line $x = -3t + 2$, $y = -2(t - 1)$, $z = 8t + 2$ point?

SOLUTION

(a) Here $\mathbf{a} = (3, -1, 2) = (x_1, y_1, z_1)$ and $\mathbf{v} = 2\mathbf{i} - 3\mathbf{j} + 4\mathbf{k}$, so $a = 2$, $b = -3$, and $c = 4$. From the box above, the equations are

$$x = 3 + 2t, \qquad y = -1 - 3t, \qquad z = 2 + 4t.$$

(b) Here $\mathbf{a} = (1, -6)$ and $\mathbf{v} = 5\mathbf{i} - \pi\mathbf{j}$, so the required line is

$$\mathbf{l}(t) = (1, -6) + (5t, -\pi t) = (1 + 5t, -6 - \pi t);$$

that is,

$$x = 1 + 5t, \qquad y = -6 - \pi t.$$

(c) Using the preceding box, we construct the direction $\mathbf{v} = a\mathbf{i} + b\mathbf{j} + c\mathbf{k}$ from the coefficients of t: $a = -3$, $b = -2$, $c = 8$. Thus, the line points in the direction of $\mathbf{v} = -3\mathbf{i} - 2\mathbf{j} + 8\mathbf{k}$. $\blacktriangle$

EXAMPLE 13 Do the two lines $(x, y, z) = (t, -6t + 1, 2t - 8)$ and $(x, y, z) = (3t + 1, 2t, 0)$ intersect?

SOLUTION If the lines intersect, there must be numbers t_1 and t_2 such that the corresponding points are equal:

$$(t_1, -6t_1 + 1, 2t_1 - 8) = (3t_2 + 1, 2t_2, 0);$$

that is, all three of the following equations hold:

$$\begin{aligned} t_1 &= 3t_2 + 1, \\ -6t_1 + 1 &= 2t_2, \\ 2t_1 - 8 &= 0. \end{aligned}$$

From the third equation, $t_1 = 4$. The first equation then becomes $4 = 3t_2 + 1$; that is, $t_2 = 1$. We must check whether these values satisfy the middle equation:

$$-6t_1 + 1 \stackrel{?}{=} 2t_2.$$

Since $t_1 = 4$ and $t_2 = 1$, this reads

$$-24 + 1 \stackrel{?}{=} 2,$$

which is false, so the lines do not intersect. ▲

Notice that there can be many equations of the same line. Some may be obtained by choosing instead of $\mathbf{a}$, a different point on the given line, and forming the parametric equation of the line beginning at that point and in the direction of $\mathbf{v}$. For example, the endpoint of $\mathbf{a} + \mathbf{v}$ is on the line $\mathbf{l}(t) = \mathbf{a} + t\mathbf{v}$, and thus, $\mathbf{l}_1(t) = (\mathbf{a} + \mathbf{v}) + t\mathbf{v}$ represents the same line. Still other equations may be obtained by observing that if $\alpha \neq 0$, the vector $\alpha\mathbf{v}$ has the same (or opposite) direction as $\mathbf{v}$. Thus, $\mathbf{l}_2(t) = \mathbf{a} + t\alpha\mathbf{v}$ is another equation of $\mathbf{l}(t) = \mathbf{a} + t\mathbf{v}$.

For example, both $\mathbf{l}(t) = (1, 0, 0) + (t, t, 0)$ and $\mathbf{l}_1(s) = (0, -1, 0) + (s, s, 0)$ represent the same line since both are in the direction $\mathbf{i} + \mathbf{j}$ and both pass through the point $(1, 0, 0)$; $\mathbf{l}$ passes through $(1, 0, 0)$ at $t = 0$ and $\mathbf{l}_1$ passes through $(1, 0, 0)$ at $s = 1$.

Therefore, the equation of a line is not uniquely determined. Nevertheless, it is customary to use the term "the" equation of a line. Keeping this in mind, let us derive *the equation of a line passing through the endpoints of two given vectors* $\mathbf{a}$ *and* $\mathbf{b}$. Because the vector $\mathbf{b} - \mathbf{a}$ is parallel to the directed line segment from $\mathbf{a}$ to $\mathbf{b}$, we calculate the parametric equation of the line passing through $\mathbf{a}$ in the direction of $\mathbf{b} - \mathbf{a}$ (Figure 1.1.22). Thus,

$$\mathbf{l}(t) = \mathbf{a} + t(\mathbf{b} - \mathbf{a}); \qquad \text{that is,} \qquad \mathbf{l}(t) = (1 - t)\mathbf{a} + t\mathbf{b}.$$

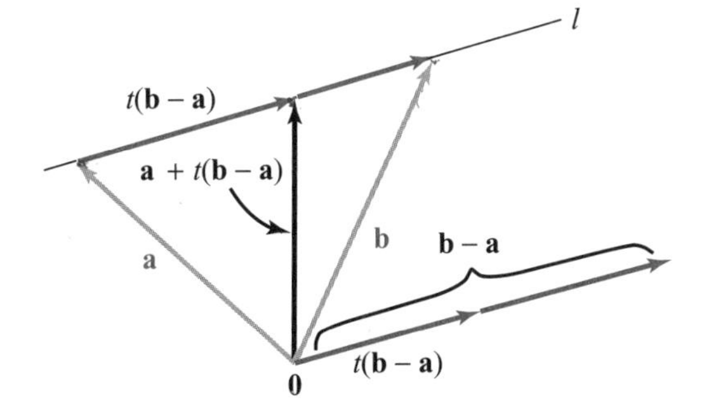

Figure 1.1.22 The line l, parametrically given by $\mathbf{l}(t) = \mathbf{a} + t(\mathbf{b} - \mathbf{a}) = (1 - t)\mathbf{a} + t\mathbf{b}$, passes through the tips of $\mathbf{a}$ and $\mathbf{b}$.

As t increases from 0 to 1, $t(\mathbf{b} - \mathbf{a})$ starts as the zero vector and increases in length (remaining in the direction of $\mathbf{b} - \mathbf{a}$) until at $t = 1$ it is the vector $\mathbf{b} - \mathbf{a}$. Thus, for

$\mathbf{l}(t) = \mathbf{a} + t(\mathbf{b} - \mathbf{a})$, as t increases from 0 to 1, the vector $\mathbf{l}(t)$ moves from the endpoint of $\mathbf{a}$ to the endpoint of $\mathbf{b}$ along the directed line segment from $\mathbf{a}$ to $\mathbf{b}$.

If $\mathrm{P} = (x_1, y_1, z_1)$ is the tip of $\mathbf{a}$ and $\mathrm{Q} = (x_2, y_2, z_2)$ is the tip of $\mathbf{b}$, then $\mathbf{v} = (x_2 - x_1)\mathbf{i} + (y_2 - y_1)\mathbf{j} + (z_2 - z_1)\mathbf{k}$, and so the equations of the line are

$$x = x_1 + (x_2 - x_1)t,$$
$$y = y_1 + (y_2 - y_1)t,$$
$$z = z_1 + (z_2 - z_1)t.$$

By eliminating t, these can be written as

$$\frac{x - x_1}{x_2 - x_1} = \frac{y - y_1}{y_2 - y_1} = \frac{z - z_1}{z_2 - z_1}.$$

Parametric Equation of a Line: Point–Point Form The parametric equations of the line l through the points $\mathrm{P} = (x_1, y_1, z_1)$ and $\mathrm{Q} = (x_2, y_2, z_2)$ are

$$x = x_1 + (x_2 - x_1)t,$$
$$y = y_1 + (y_2 - y_1)t,$$
$$z = z_1 + (z_2 - z_1)t,$$

where (x, y, z) is the general point of l, and the parameter t takes on all real values.

EXAMPLE 14 Find the equation of the line through $(2, 1, -3)$ and $(6, -1, -5)$.

SOLUTION Using the preceding box, we choose $(x_1, y_1, z_1) = (2, 1, -3)$ and $(x_2, y_2, z_2) = (6, -1, -5)$, so the equations are

$$x = 2 + (6 - 2)t = 2 + 4t,$$
$$y = 1 + (-1 - 1)t = 1 - 2t,$$
$$z = -3 + (-5 - (-3))t = -3 - 2t. \quad \blacktriangle$$

EXAMPLE 15 Find the equation of the line passing through $(-1, 1, 0)$ and $(0, 0, 1)$ (see Figure 1.1.23).

SOLUTION Letting $\mathbf{a} = -\mathbf{i} + \mathbf{j}$ and $\mathbf{b} = \mathbf{k}$ represent the given points, we have

$$\mathbf{l}(t) = (1 - t)(-\mathbf{i} + \mathbf{j}) + t\mathbf{k} = -(1 - t)\mathbf{i} + (1 - t)\mathbf{j} + t\mathbf{k}.$$

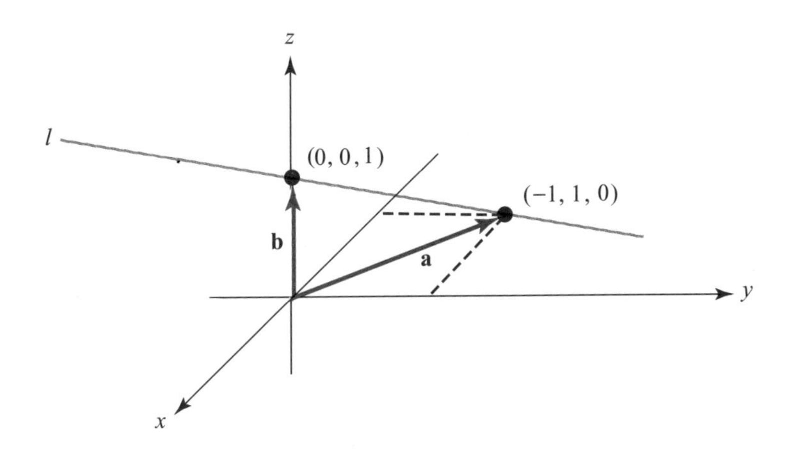

Figure 1.1.23 Finding the equation of the line through two points.

The equation of this line may thus be written as

$$\mathbf{l}(t) = (t-1)\mathbf{i} + (1-t)\mathbf{j} + t\mathbf{k},$$

or, equivalently, if $\mathbf{l}(t) = x\mathbf{i} + y\mathbf{j} + z\mathbf{k}$,

$$x = t-1, \qquad y = 1-t, \qquad z = t. \quad \blacktriangle$$

The description of a line *segment* requires that the domain of the parameter t be restricted, as in the following example.

EXAMPLE 16 Find the equation of the line segment between (1, 1, 1) and (2, 1, 2).

SOLUTION The *line* through (1, 1, 1) and (2, 1, 2) is described in parametric form by $(x, y, z) = (1+t, 1, 1+t)$, as t takes on all real values. When $t = 0$, the point (x, y, z) is (1, 1, 1), and when $t = 1$, the point (x, y, z) is (2, 1, 2). Thus, the point (x, y, z) lies between (1, 1, 1) and (2, 1, 2) when $0 \le t \le 1$, so the line *segment* is described by the equations

$$x = 1 + t,$$
$$y = 1,$$
$$z = 1 + t,$$

together with the inequalities $0 \le t \le 1$. ▲

We can also give parametric descriptions of geometric objects other than lines.

EXAMPLE 17 Describe the points that lie within the parallelogram whose adjacent sides are the vectors **a** and **b** based at the origin ("within" includes points on the edges of the parallelogram).

SOLUTION Consider Figure 1.1.24. If P is any point within the given parallelogram and we construct lines l_1 and l_2 through P parallel to the vectors **a** and **b**, respectively, we see that l_1 intersects the side of the parallelogram determined by the vector **b** at some point $t\mathbf{b}$, where $0 \le t \le 1$. Likewise, l_2 intersects the side determined by the vector **a** at some point $s\mathbf{a}$, where $0 \le s \le 1$.

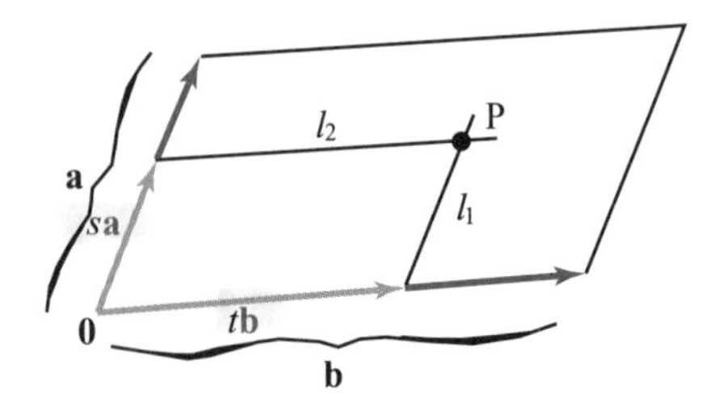

Figure 1.1.24 Describing points within the parallelogram formed by vectors **a** and **b**, with vertex **0**.

Note that P is the endpoint of the diagonal of a parallelogram having adjacent sides $s\mathbf{a}$ and $t\mathbf{b}$; hence, if **v** denotes the vector $\overrightarrow{\mathrm{OP}}$, we see that $\mathbf{v} = s\mathbf{a} + t\mathbf{b}$. We conclude that all the points in the given parallelogram are endpoints of vectors of the form $s\mathbf{a} + t\mathbf{b}$ for $0 \le s \le 1$ and $0 \le t \le 1$. Reversing our steps, we see that all vectors of this form end within the parallelogram. ▲

As two different lines through the origin determine a plane through the origin, so do two nonparallel vectors. If we apply the same reasoning as in Example 17, we see that the entire plane formed by two nonparallel vectors **v** and **w** consists of all points of the form $s\mathbf{v} + t\mathbf{w}$ where s and t can be any real numbers, as in Figure 1.1.25.

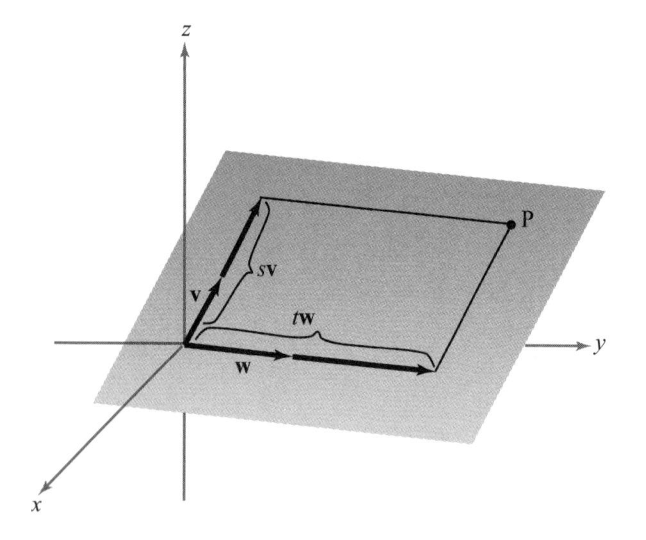

Figure 1.1.25 Describing points P in the plane formed from vectors **v** and **w**.

We have thus described the points in the plane by two parameters. For this reason, we say the plane is ***two-dimensional***. Similarly, a line is called ***one-dimensional*** whether it lies in the plane or in space or is the real number line itself.

The plane determined by $\mathbf{v}$ and $\mathbf{w}$ is called the plane ***spanned by*** $\mathbf{v}$ and $\mathbf{w}$. When $\mathbf{v}$ is a scalar multiple of $\mathbf{w}$ and $\mathbf{w} \neq \mathbf{0}$, then $\mathbf{v}$ and $\mathbf{w}$ are parallel and the plane degenerates to a straight line. When $\mathbf{v} = \mathbf{w} = \mathbf{0}$ (that is, both are zero vectors), we obtain a single point.

There are three particular planes that arise naturally in a coordinate system and that will be useful to us later. We call the plane spanned by vectors $\mathbf{i}$ and $\mathbf{j}$ the xy plane, the plane spanned by $\mathbf{j}$ and $\mathbf{k}$ the yz plane, and the plane spanned by $\mathbf{i}$ and $\mathbf{k}$ the xz plane. These planes are illustrated in Figure 1.1.26.

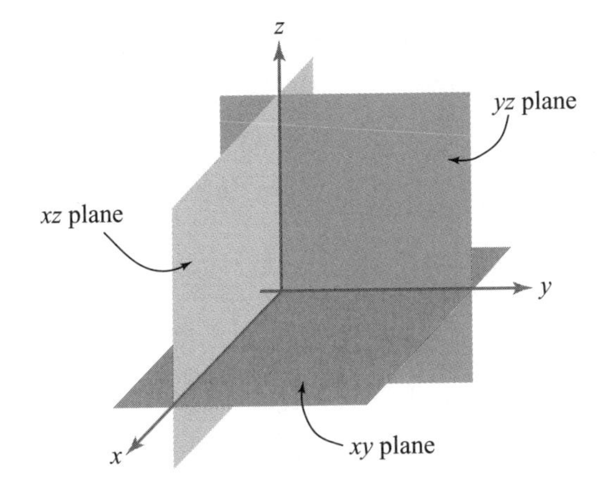

Figure 1.1.26 The three coordinate planes.

EXERCISES

(Exercises with colored numbers are solved in the Study Guide.)

Complete the computations in Exercises 1 to 4.

1. $(-21, 23) - (?, 6) = (-25, ?)$

2. $3(133, -0.33, 0) + (-399, 0.99, 0) = (?, ?, ?)$

3. $(8a, -2b, 13c) = (52, 12, 11) + \frac{1}{2}(?, ?, ?)$

4. $(2, 3, 5) - 4\mathbf{i} + 3\mathbf{j} = (?, ?, ?)$

In Exercises 5 to 8, sketch the given vectors $\mathbf{v}$ *and* $\mathbf{w}$. *On your sketch, draw in* $-\mathbf{v}$, $\mathbf{v} + \mathbf{w}$, *and* $\mathbf{v} - \mathbf{w}$.

5. $\mathbf{v} = (2, 1)$ and $\mathbf{w} = (1, 2)$

6. $\mathbf{v} = (0, 4)$ and $\mathbf{w} = (2, -1)$

7. $\mathbf{v} = (2, 3, -6)$ and $\mathbf{w} = (-1, 1, 1)$

8. $\mathbf{v} = (2, 1, 3)$ and $\mathbf{w} = (-2, 0, -1)$

9. What restrictions must be made on x, y, and z so that the triple (x, y, z) will represent a point on the y axis? On the z axis? In the xz plane? In the yz plane?

10. (a) Generalize the geometric construction in Figure 1.1.7 to show that if $\mathbf{v}_1 = (x, y, z)$ and $\mathbf{v}_2 = (x', y', z')$, then $\mathbf{v}_1 + \mathbf{v}_2 = (x + x', y + y', z + z')$.

(b) Using an argument based on similar triangles, prove that $\alpha\mathbf{v} = (\alpha x, \alpha y, \alpha z)$ when $\mathbf{v} = (x, y, z)$.

In Exercises 11 to 17, use set theoretic or vector notation or both to describe the points that lie in the given configurations.

11. The plane spanned by $\mathbf{v}_1 = (2, 7, 0)$ and $\mathbf{v}_2 = (0, 2, 7)$

12. The plane spanned by $\mathbf{v}_1 = (3, -1, 1)$ and $\mathbf{v}_2 = (0, 3, 4)$

13. The line passing through $(-1, -1, -1)$ in the direction of $\mathbf{j}$

14. The line passing through $(0, 2, 1)$ in the direction of $2\mathbf{i} - \mathbf{k}$

15. The line passing through $(-1, -1, -1)$ and $(1, -1, 2)$

16. The line passing through $(-5, 0, 4)$ and $(6, -3, 2)$

17. The parallelogram whose adjacent sides are the vectors $\mathbf{i} + 3\mathbf{k}$ and $-2\mathbf{j}$

18. Find the points of intersection of the line $x = 3 + 2t$, $y = 7 + 8t$, $z = -2 + t$, that is, $\mathbf{l}(t) = (3 + 2t, 7 + 8t, -2 + t)$, with the coordinate planes.

19. Show that there are no points (x, y, z) satisfying $2x - 3y + z - 2 = 0$ and lying on the line $\mathbf{v} = (2, -2, -1) + t(1, 1, 1)$.

20. Show that every point on the line $\mathbf{v} = (1, -1, 2) + t(2, 3, 1)$ satisfies the equation $5x - 3y - z - 6 = 0$.

21. Determine whether the lines $x = 3t + 2$, $y = t - 1$, $z = 6t + 1$, and $x = 3s - 1$, $y = s - 2$, $z = s$ intersect.

22. Do the lines $(x, y, z) = (t + 4, 4t + 5, t - 2)$ and $(x, y, z) = (2s + 3, s + 1, 2s - 3)$ intersect?

In Exercises 23 to 25, use vector methods to describe the given configurations.

23. The parallelepiped with edges the vectors $\mathbf{a}$, $\mathbf{b}$, and $\mathbf{c}$ emanating from the origin.

24. The points within the parallelogram with one corner at (x_0, y_0, z_0) whose sides extending from that corner are equal in magnitude and direction to vectors $\mathbf{a}$ and $\mathbf{b}$.

25. The plane determined by the three points (x_0, y_0, z_0), (x_1, y_1, z_1), and (x_2, y_2, z_2).

Prove the statements in Exercises 26 to 28.

26. The line segment joining the midpoints of two sides of a triangle is parallel to and has half the length of the third side.

27. If PQR is a triangle in space and $b > 0$ is a number, then there is a triangle with sides parallel to those of PQR and side lengths b times those of PQR.

28. The medians of a triangle intersect at a point, and this point divides each median in a ratio of $2:1$.

Problems 29 and 30 require some knowledge of chemical notation.

29. Write the chemical equation $CO + H_2O = H_2 + CO_2$ as an equation in ordered triples (x_1, x_2, x_3) where x_1, x_2, x_3 are the number of carbon, hydrogen, and oxygen atoms, respectively, in each molecule.

30. (a) Write the chemical equation $pC_3H_4O_3 + qO_2 = rCO_2 + sH_2O$ as an equation in ordered triples with unknown coefficients p, q, r, and s.

(b) Find the smallest positive integer solution for p, q, r, and s.

(c) Illustrate the solution by a vector diagram in space.

31. Find a line that lies entirely in the set defined by the equation $x^2 + y^2 - z^2 = 1$.

1.2 The Inner Product, Length, and Distance

In this section and the next we shall discuss two products of vectors: the inner product and the cross product. These are very useful in physical applications and have interesting geometric interpretations. The first product we shall consider is called the *inner product*. The name *dot product* is often used instead.

The Inner Product

Suppose we have two vectors **a** and **b** in $\mathbb{R}^3$ (Figure 1.2.1) and we wish to determine the angle between them, that is, the smaller angle subtended by **a** and **b** in the plane

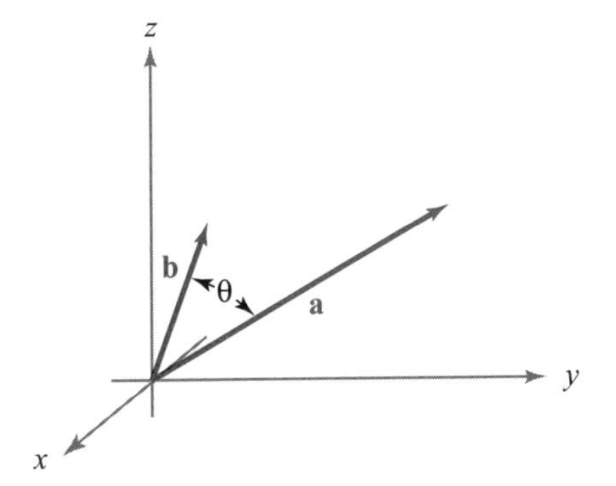

Figure 1.2.1 θ is the angle between the vectors **a** and **b**.

that they span. The inner product enables us to do this. Let us first develop the concept formally and then prove that this product does what we claim. Let $\mathbf{a} = a_1\mathbf{i} + a_2\mathbf{j} + a_3\mathbf{k}$ and $\mathbf{b} = b_1\mathbf{i} + b_2\mathbf{j} + b_3\mathbf{k}$. We define the ***inner product*** of $\mathbf{a}$ and $\mathbf{b}$, written $\mathbf{a} \cdot \mathbf{b}$, to be the real number

$$\mathbf{a} \cdot \mathbf{b} = a_1 b_1 + a_2 b_2 + a_3 b_3.$$

Note that the inner product of two vectors is a scalar quantity. Sometimes the inner product is denoted $\langle \mathbf{a}, \mathbf{b} \rangle$; *thus,* $\langle \mathbf{a}, \mathbf{b} \rangle$ *and* $\mathbf{a} \cdot \mathbf{b}$ *mean exactly the same thing.*

EXAMPLE 1

(a) If $\mathbf{a} = 3\mathbf{i} + \mathbf{j} - 2\mathbf{k}$ and $\mathbf{b} = \mathbf{i} - \mathbf{j} + \mathbf{k}$, calculate $\mathbf{a} \cdot \mathbf{b}$.

(b) Calculate $(2\mathbf{i} + \mathbf{j} - \mathbf{k}) \cdot (3\mathbf{k} - 2\mathbf{j})$.

SOLUTION

(a) $\mathbf{a} \cdot \mathbf{b} = 3 \cdot 1 + 1 \cdot (-1) + (-2) \cdot 1 = 3 - 1 - 2 = 0.$

(b)
$$\begin{aligned}(2\mathbf{i} + \mathbf{j} - \mathbf{k}) \cdot (3\mathbf{k} - 2\mathbf{j}) &= (2\mathbf{i} + \mathbf{j} - \mathbf{k}) \cdot (0\mathbf{i} - 2\mathbf{j} + 3\mathbf{k}) \\ &= 2 \cdot 0 - 1 \cdot 2 - 1 \cdot 3 = -5. \quad \blacktriangle\end{aligned}$$

Certain properties of the inner product follow from the definition. If $\mathbf{a}$, $\mathbf{b}$, and $\mathbf{c}$ are vectors in $\mathbb{R}^3$ and α and β are real numbers, then

(i) $\mathbf{a} \cdot \mathbf{a} \geq 0$;

$\mathbf{a} \cdot \mathbf{a} = 0$ if and only if $\mathbf{a} = \mathbf{0}$.

(ii) $\alpha\mathbf{a} \cdot \mathbf{b} = \alpha(\mathbf{a} \cdot \mathbf{b})$ and $\mathbf{a} \cdot \beta\mathbf{b} = \beta(\mathbf{a} \cdot \mathbf{b})$.

(iii) $\mathbf{a} \cdot (\mathbf{b} + \mathbf{c}) = \mathbf{a} \cdot \mathbf{b} + \mathbf{a} \cdot \mathbf{c}$ and $(\mathbf{a} + \mathbf{b}) \cdot \mathbf{c} = \mathbf{a} \cdot \mathbf{c} + \mathbf{b} \cdot \mathbf{c}$.

(iv) $\mathbf{a} \cdot \mathbf{b} = \mathbf{b} \cdot \mathbf{a}$.

To prove the first of these properties, observe that if $\mathbf{a} = a_1\mathbf{i} + a_2\mathbf{j} + a_3\mathbf{k}$, then $\mathbf{a} \cdot \mathbf{a} = a_1^2 + a_2^2 + a_3^2$. Because a_1, a_2, and a_3 are real numbers, we know $a_1^2 \geq 0$, $a_2^2 \geq 0, a_3^2 \geq 0$. Thus, $\mathbf{a} \cdot \mathbf{a} \geq 0$. Moreover, if $a_1^2 + a_2^2 + a_3^2 = 0$, then $a_1 = a_2 = a_3 = 0$; therefore, $\mathbf{a} = \mathbf{0}$ (zero vector). The proofs of the other properties of the inner product are also easily obtained.

It follows from the Pythagorean theorem that the ***length*** of the vector $\mathbf{a} = a_1\mathbf{i} + a_2\mathbf{j} + a_3\mathbf{k}$ is $\sqrt{a_1^2 + a_2^2 + a_3^2}$ (see Figure 1.2.2). The length of the vector $\mathbf{a}$ is denoted by $\|\mathbf{a}\|$. This quantity is often called the ***norm*** of $\mathbf{a}$. Because $\mathbf{a} \cdot \mathbf{a} = a_1^2 + a_2^2 + a_3^2$, it follows that

$$\|\mathbf{a}\| = (\mathbf{a} \cdot \mathbf{a})^{1/2}.$$

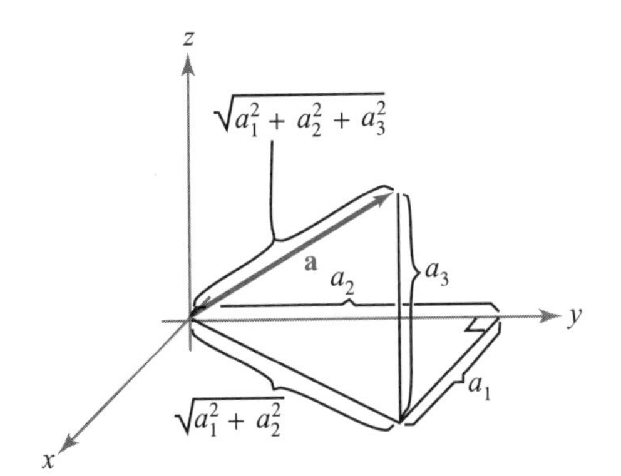

Figure 1.2.2 The length of the vector $\mathbf{a} = (a_1, a_2, a_3)$ is given by the Pythagorean formula: $\sqrt{a_1^2 + a_2^2 + a_3^2}$.

Unit Vectors

Vectors with norm 1 are called ***unit vectors***. For example, the vectors $\mathbf{i}, \mathbf{j}, \mathbf{k}$ are unit vectors. Observe that for any nonzero vector $\mathbf{a}$, $\mathbf{a}/\|\mathbf{a}\|$ is a unit vector; when we divide $\mathbf{a}$ by $\|\mathbf{a}\|$, we say that we have ***normalized*** $\mathbf{a}$.

EXAMPLE 2

(a) Normalize $\mathbf{v} = 2\mathbf{i} + 3\mathbf{j} - \frac{1}{2}\mathbf{k}$.

(b) Find unit vectors $\mathbf{a}$, $\mathbf{b}$, and $\mathbf{c}$ in the plane such that $\mathbf{b} + \mathbf{c} = \mathbf{a}$.

SOLUTION

(a) We have $\|\mathbf{v}\| = \sqrt{2^2 + 3^2 + (1/2)^2} = (1/2)\sqrt{53}$, so the normalization of $\mathbf{v}$ is

$$\mathbf{u} = \frac{1}{\|\mathbf{v}\|}\mathbf{v} = \frac{4}{\sqrt{53}}\mathbf{i} + \frac{6}{\sqrt{53}}\mathbf{j} - \frac{1}{\sqrt{53}}\mathbf{k}.$$

(b) Because all three vectors are to have length 1, a triangle with sides $\mathbf{a}$, $\mathbf{b}$, and $\mathbf{c}$ must be equilateral, as in Figure 1.2.3. Orienting the triangle as in the figure, we take $\mathbf{a} = \mathbf{i}$, then necessarily

$$\mathbf{b} = \frac{1}{2}\mathbf{i} + \frac{\sqrt{3}}{2}\mathbf{j}, \qquad \text{and} \qquad \mathbf{c} = \frac{1}{2}\mathbf{i} - \frac{\sqrt{3}}{2}\mathbf{j}.$$

Note that indeed $\|\mathbf{a}\| = \|\mathbf{b}\| = \|\mathbf{c}\| = 1$ and that $\mathbf{b} + \mathbf{c} = \mathbf{a}$. ▲

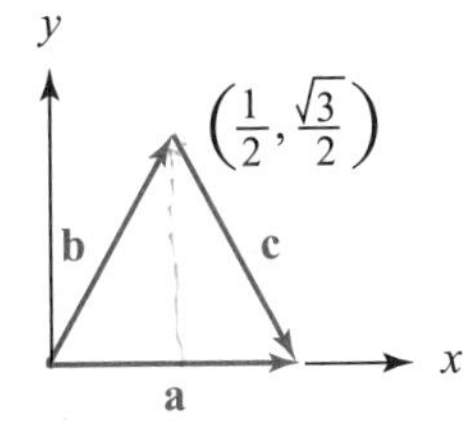

Figure 1.2.3 The vectors $\mathbf{a}$, $\mathbf{b}$, and $\mathbf{c}$ are represented by the sides of an equilateral triangle.

In the plane, define the vector $\mathbf{i}_\theta = (\cos\theta)\mathbf{i} + (\sin\theta)\mathbf{j}$, which is the unit vector making an angle θ with the x axis (see Figure 1.2.4).

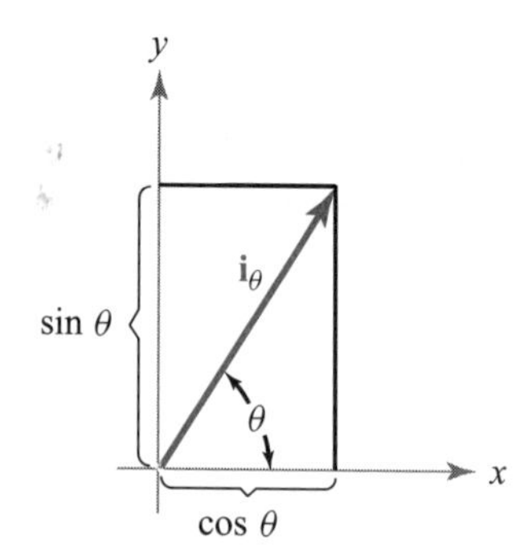

Figure 1.2.4 The coordinates of $\mathbf{i}_\theta$ are $\cos\theta$ and $\sin\theta$; it is a unit vector because $\cos^2\theta + \sin^2\theta = 1$.

Distance

If $\mathbf{a}$ and $\mathbf{b}$ are vectors, we have seen that the vector $\mathbf{b} - \mathbf{a}$ is parallel to and has the same magnitude as the directed line segment from the endpoint of $\mathbf{a}$ to the endpoint of $\mathbf{b}$. It follows that the distance from the endpoint of $\mathbf{a}$ to the endpoint of $\mathbf{b}$ is $\|\mathbf{b} - \mathbf{a}\|$ (see Figure 1.2.5).

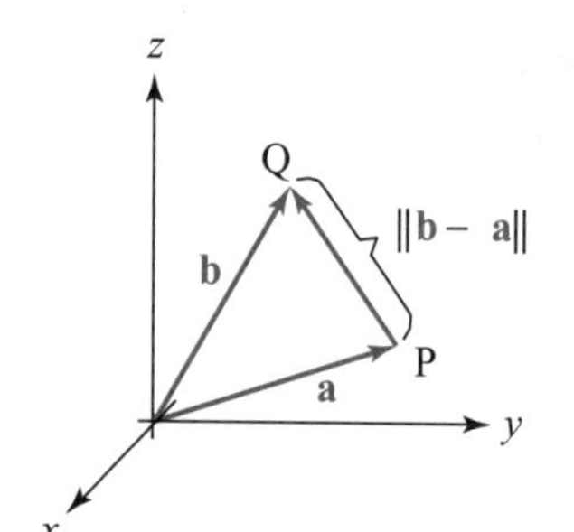

Figure 1.2.5 The distance between the tips of $\mathbf{a}$ and $\mathbf{b}$ is $\|\mathbf{b} - \mathbf{a}\|$.

Inner Product, Length, and Distance Letting $\mathbf{a} = a_1\mathbf{i} + a_2\mathbf{j} + a_3\mathbf{k}$ and $\mathbf{b} = b_1\mathbf{i} + b_2\mathbf{j} + b_3\mathbf{k}$, their ***inner product*** is

$$\mathbf{a} \cdot \mathbf{b} = a_1b_1 + a_2b_2 + a_3b_3,$$

while the ***length*** of $\mathbf{a}$ is

$$\|\mathbf{a}\| = \sqrt{\mathbf{a} \cdot \mathbf{a}} = \sqrt{a_1^2 + a_2^2 + a_3^2}.$$

To ***normalize*** a vector $\mathbf{a}$, form the vector

$$\frac{\mathbf{a}}{\|\mathbf{a}\|}.$$

The ***distance between*** the endpoints of $\mathbf{a}$ and $\mathbf{b}$ is $\|\mathbf{a} - \mathbf{b}\|$, and the ***distance between*** P and Q is $\|\overrightarrow{\mathrm{PQ}}\|$.

EXAMPLE 3 Find the distance from the endpoint of the vector $\mathbf{i}$, that is, the point $(1, 0, 0)$, to the endpoint of the vector $\mathbf{j}$, that is, the point $(0, 1, 0)$.

SOLUTION $\|\mathbf{j} - \mathbf{i}\| = \sqrt{(0-1)^2 + (1-0)^2 + (0-0)^2} = \sqrt{2}$. ▲

The Angle Between Two Vectors

Let us now show that the inner product does indeed measure the angle between two vectors.

THEOREM 1 Let $\mathbf{a}$ and $\mathbf{b}$ be two vectors in $\mathbb{R}^3$ and let θ, where $0 \leq \theta \leq \pi$, be the angle between them (Figure 1.2.6). Then

$$\mathbf{a} \cdot \mathbf{b} = \|\mathbf{a}\| \|\mathbf{b}\| \cos\theta.$$

It follows from the equation $\mathbf{a} \cdot \mathbf{b} = \|\mathbf{a}\| \|\mathbf{b}\| \cos\theta$ that if $\mathbf{a}$ and $\mathbf{b}$ are nonzero, we may express the angle between them as

$$\theta = \cos^{-1}\left(\frac{\mathbf{a} \cdot \mathbf{b}}{\|\mathbf{a}\| \|\mathbf{b}\|}\right).$$

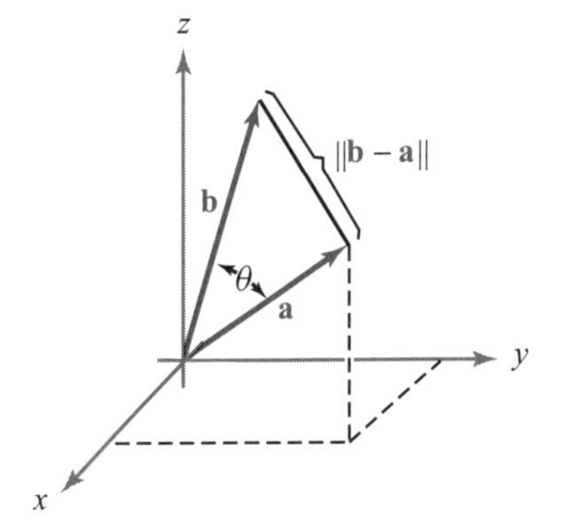

Figure 1.2.6 The vectors $\mathbf{a}$, $\mathbf{b}$, and the angle θ between them; the geometry for Theorem 1 and its proof.

PROOF If we apply the law of cosines from trigonometry to the triangle with one vertex at the origin and adjacent sides determined by the vectors $\mathbf{a}$ and $\mathbf{b}$ (as in the figure), it follows that

$$\|\mathbf{b} - \mathbf{a}\|^2 = \|\mathbf{a}\|^2 + \|\mathbf{b}\|^2 - 2\|\mathbf{a}\| \|\mathbf{b}\| \cos\theta.$$

Because $\|\mathbf{b}-\mathbf{a}\|^2 = (\mathbf{b}-\mathbf{a})\cdot(\mathbf{b}-\mathbf{a})$, $\|\mathbf{a}\|^2 = \mathbf{a}\cdot\mathbf{a}$, and $\|\mathbf{b}\|^2 = \mathbf{b}\cdot\mathbf{b}$, we can rewrite the above equation as

$$(\mathbf{b}-\mathbf{a})\cdot(\mathbf{b}-\mathbf{a}) = \mathbf{a}\cdot\mathbf{a} + \mathbf{b}\cdot\mathbf{b} - 2\|\mathbf{a}\|\,\|\mathbf{b}\|\cos\theta.$$

We can also expand $(\mathbf{b}-\mathbf{a})\cdot(\mathbf{b}-\mathbf{a})$ as follows:

$$\begin{aligned}(\mathbf{b}-\mathbf{a})\cdot(\mathbf{b}-\mathbf{a}) &= \mathbf{b}\cdot(\mathbf{b}-\mathbf{a}) - \mathbf{a}\cdot(\mathbf{b}-\mathbf{a})\\ &= \mathbf{b}\cdot\mathbf{b} - \mathbf{b}\cdot\mathbf{a} - \mathbf{a}\cdot\mathbf{b} + \mathbf{a}\cdot\mathbf{a}\\ &= \mathbf{a}\cdot\mathbf{a} + \mathbf{b}\cdot\mathbf{b} - 2\mathbf{a}\cdot\mathbf{b}.\end{aligned}$$

Thus,

$$\mathbf{a}\cdot\mathbf{a} + \mathbf{b}\cdot\mathbf{b} - 2\mathbf{a}\cdot\mathbf{b} = \mathbf{a}\cdot\mathbf{a} + \mathbf{b}\cdot\mathbf{b} - 2\|\mathbf{a}\|\,\|\mathbf{b}\|\cos\theta.$$

That is,

$$\mathbf{a}\cdot\mathbf{b} = \|\mathbf{a}\|\,\|\mathbf{b}\|\cos\theta. \quad ■$$

EXAMPLE 4 Find the angle between the vectors $\mathbf{i}+\mathbf{j}+\mathbf{k}$ and $\mathbf{i}+\mathbf{j}-\mathbf{k}$ (see Figure 1.2.7).

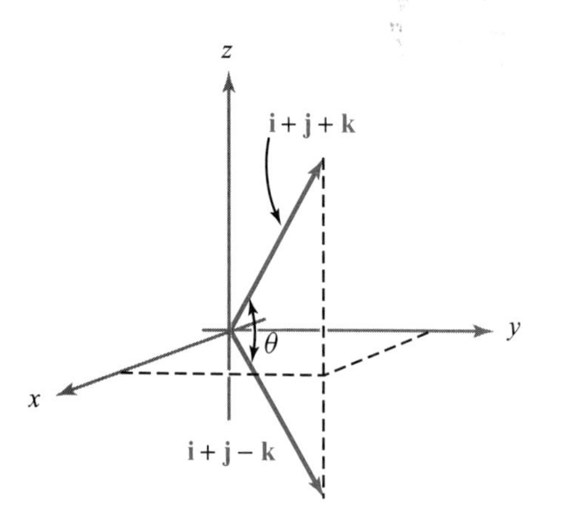

Figure 1.2.7 Finding the angle between $\mathbf{a} = \mathbf{i}+\mathbf{j}+\mathbf{k}$ and $\mathbf{b} = \mathbf{i}+\mathbf{j}-\mathbf{k}$.

SOLUTION Using Theorem 1, we have

$$(\mathbf{i}+\mathbf{j}+\mathbf{k})\cdot(\mathbf{i}+\mathbf{j}-\mathbf{k}) = \|\mathbf{i}+\mathbf{j}+\mathbf{k}\|\,\|\mathbf{i}+\mathbf{j}-\mathbf{k}\|\cos\theta,$$

and so

$$1+1-1 = (\sqrt{3})(\sqrt{3})\cos\theta.$$

Hence,

$$\cos\theta = \tfrac{1}{3}.$$

That is,

$$\theta = \cos^{-1}(\tfrac{1}{3}) \approx 1.23 \text{ radians } (71^\circ). \quad \blacktriangle$$

The Cauchy–Schwarz Inequality

Theorem 1 shows that the inner product of two vectors is the product of their lengths times the cosine of the angle between them. This relationship is often of value in problems of a geometric nature. An important consequence of Theorem 1 is:

COROLLARY: Cauchy–Schwarz Inequality For any two vectors $\mathbf{a}$ and $\mathbf{b}$, we have

$$|\mathbf{a}\cdot\mathbf{b}| \leq \|\mathbf{a}\|\|\mathbf{b}\|$$

with equality if and only if $\mathbf{a}$ is a scalar multiple of $\mathbf{b}$, or one of them is $\mathbf{0}$.

PROOF If $\mathbf{a}$ is not a scalar multiple of $\mathbf{b}$, then θ, the angle between them, is not zero or π, and so $|\cos\theta| < 1$, and thus the inequality holds; in fact, if $\mathbf{a}$ and $\mathbf{b}$ are both nonzero, *strict* inequality holds in this case. When $\mathbf{a}$ is a scalar multiple of $\mathbf{b}$, then $\theta = 0$ or π and $|\cos\theta| = 1$, so equality holds in this case. ■

EXAMPLE 5 Verify the Cauchy–Schwarz inequality for $\mathbf{a} = -\mathbf{i} + \mathbf{j} + \mathbf{k}$ and $\mathbf{b} = 3\mathbf{i} + \mathbf{k}$.

SOLUTION The dot product is $\mathbf{a}\cdot\mathbf{b} = -3 + 0 + 1 = -2$, so $|\mathbf{a}\cdot\mathbf{b}| = 2$. Also, $\|\mathbf{a}\| = \sqrt{1+1+1} = \sqrt{3}$ and $\|\mathbf{b}\| = \sqrt{9+1} = \sqrt{10}$, and it is true that $2 \leq \sqrt{3}\cdot\sqrt{10}$ because $\sqrt{3}\cdot\sqrt{10} > \sqrt{3}\cdot\sqrt{3} = 3 \geq 2$. ▲

If $\mathbf{a}$ and $\mathbf{b}$ are nonzero vectors in $\mathbb{R}^3$ and θ is the angle between them, we see that $\mathbf{a}\cdot\mathbf{b} = 0$ if and only if $\cos\theta = 0$. Thus, *the inner product of two nonzero vectors is zero if and only if the vectors are perpendicular*. Hence, the inner product provides us with a convenient method for determining whether two vectors are perpendicular. Often we say that perpendicular vectors are ***orthogonal***. The standard basis vectors $\mathbf{i}$, $\mathbf{j}$, and $\mathbf{k}$ are mutually orthogonal and of length 1; any such system is called ***orthonormal***. We shall adopt the convention that the zero vector is orthogonal to all vectors.

EXAMPLE 6 The vectors $\mathbf{i}_\theta = (\cos\theta)\mathbf{i} + (\sin\theta)\mathbf{j}$ and $\mathbf{j}_\theta = -(\sin\theta)\mathbf{i} + (\cos\theta)\mathbf{j}$ are orthogonal, because

$$\mathbf{i}_\theta \cdot \mathbf{j}_\theta = -\cos\theta\sin\theta + \sin\theta\cos\theta = 0$$

(see Figure 1.2.8). ▲

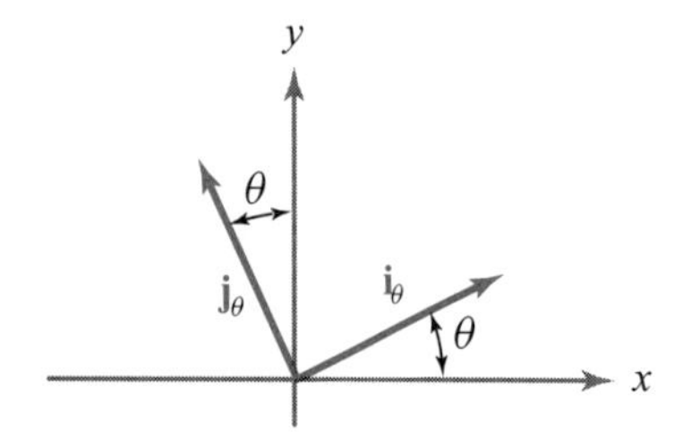

Figure 1.2.8 The vectors $\mathbf{i}_\theta$ and $\mathbf{j}_\theta$ are orthogonal and of unit length, that is, they are orthonormal.

EXAMPLE 7 Let $\mathbf{a}$ and $\mathbf{b}$ be two nonzero orthogonal vectors. If $\mathbf{c}$ is a vector in the plane spanned by $\mathbf{a}$ and $\mathbf{b}$, then there are scalars α and β such that $\mathbf{c} = \alpha\mathbf{a} + \beta\mathbf{b}$. Use the inner product to determine α and β (see Figure 1.2.9).

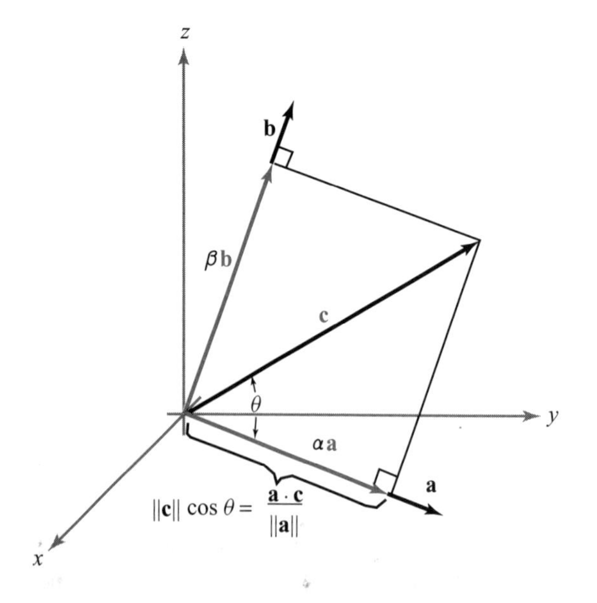

Figure 1.2.9 The geometry for finding α and β, where $\mathbf{c} = \alpha\mathbf{a} + \beta\mathbf{b}$.

SOLUTION Taking the inner product of $\mathbf{a}$ and $\mathbf{c}$, we have

$$\mathbf{a}\cdot\mathbf{c} = \mathbf{a}\cdot(\alpha\mathbf{a} + \beta\mathbf{b}) = \alpha\mathbf{a}\cdot\mathbf{a} + \beta\mathbf{a}\cdot\mathbf{b}.$$

Because $\mathbf{a}$ and $\mathbf{b}$ are orthogonal, $\mathbf{a} \cdot \mathbf{b} = 0$, and so

$$\alpha = \frac{\mathbf{a} \cdot \mathbf{c}}{\mathbf{a} \cdot \mathbf{a}} = \frac{\mathbf{a} \cdot \mathbf{c}}{\|\mathbf{a}\|^2}.$$

Similarly,

$$\beta = \frac{\mathbf{b} \cdot \mathbf{c}}{\mathbf{b} \cdot \mathbf{b}} = \frac{\mathbf{b} \cdot \mathbf{c}}{\|\mathbf{b}\|^2}. \quad \blacktriangle$$

Orthogonal Projection

In the preceding example, the vector $\alpha\mathbf{a}$ is called the ***projection of* c *along* a**, and $\beta\mathbf{b}$ is its ***projection along* b**. Let us formulate this idea more generally. If $\mathbf{v}$ is a vector, and l is the line through the origin in the direction of a vector $\mathbf{a}$, then the ***orthogonal projection of* v *on* a** is the vector $\mathbf{p}$ whose tip is obtained by dropping a perpendicular line to l from the tip of $\mathbf{v}$, as in Figure 1.2.10.

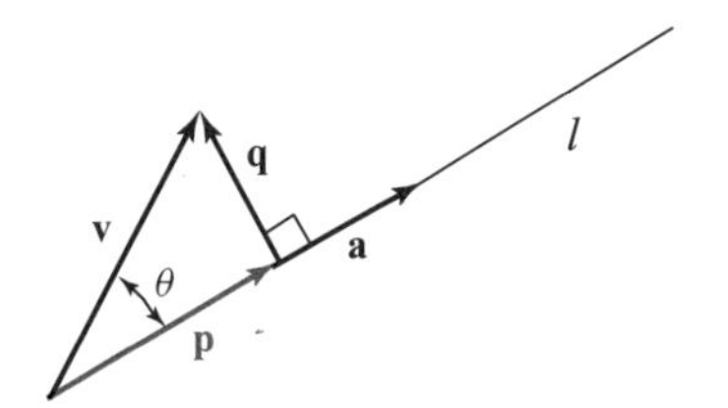

Figure 1.2.10 $\mathbf{p}$ is the orthogonal projection of $\mathbf{v}$ on $\mathbf{a}$.

Referring to the figure, we see that $\mathbf{p}$ is a multiple of $\mathbf{a}$ and that $\mathbf{v}$ is the sum of $\mathbf{p}$ and a vector $\mathbf{q}$ perpendicular to $\mathbf{a}$. Thus,

$$\mathbf{v} = c\mathbf{a} + \mathbf{q},$$

where $\mathbf{p} = c\mathbf{a}$ and $\mathbf{a} \cdot \mathbf{q} = 0$. Taking the dot product of $\mathbf{a}$ with both sides of $\mathbf{v} = c\mathbf{a} + \mathbf{q}$, we find $\mathbf{a} \cdot \mathbf{v} = c\mathbf{a} \cdot \mathbf{a}$, so $c = (\mathbf{a} \cdot \mathbf{v})/(\mathbf{a} \cdot \mathbf{a})$, and hence

$$\mathbf{p} = \frac{\mathbf{a} \cdot \mathbf{v}}{\|\mathbf{a}\|^2}\mathbf{a}.$$

The length of $\mathbf{p}$ is

$$\|\mathbf{p}\| = \frac{|\mathbf{a} \cdot \mathbf{v}|}{\|\mathbf{a}\|^2}\|\mathbf{a}\| = \frac{|\mathbf{a} \cdot \mathbf{v}|}{\|\mathbf{a}\|} = \|\mathbf{v}\| \cos\theta.$$

Orthogonal Projection The ***orthogonal projection*** of $\mathbf{v}$ on $\mathbf{a}$ is the vector

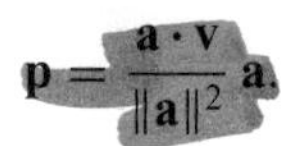

$$\mathbf{p} = \frac{\mathbf{a} \cdot \mathbf{v}}{\|\mathbf{a}\|^2}\mathbf{a}.$$

EXAMPLE 8 Find the orthogonal projection of $\mathbf{i}+\mathbf{j}$ on $\mathbf{i}-2\mathbf{j}$.

SOLUTION With $\mathbf{a}=\mathbf{i}-2\mathbf{j}$ and $\mathbf{v}=\mathbf{i}+\mathbf{j}$, the orthogonal projection of $\mathbf{v}$ on $\mathbf{a}$ is

$$\frac{\mathbf{a}\cdot\mathbf{v}}{\mathbf{a}\cdot\mathbf{a}}\mathbf{a}=\frac{1-2}{1+4}(\mathbf{i}-2\mathbf{j})=-\frac{1}{5}(\mathbf{i}-2\mathbf{j})$$

(see Figure 1.2.11). ▲

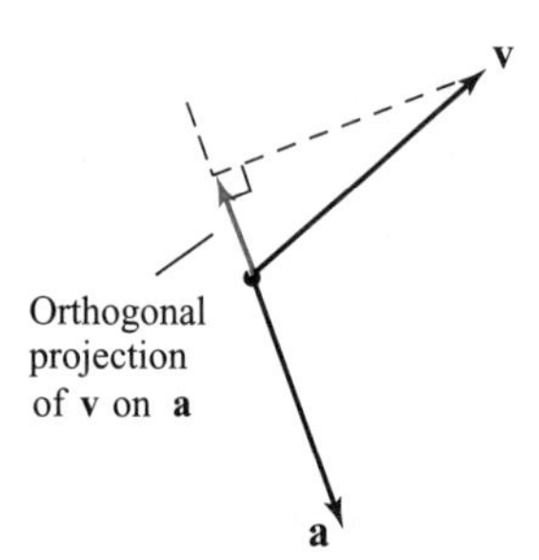

Figure 1.2.11 The orthogonal projection of $\mathbf{v}$ on $\mathbf{a}$ equals $-\frac{1}{5}\mathbf{a}$.

The Triangle Inequality

A useful consequence of the Cauchy–Schwarz inequality, which is called the ***triangle inequality***, relates the lengths of vectors $\mathbf{a}$ and $\mathbf{b}$ and of their sum $\mathbf{a}+\mathbf{b}$. Geometrically, the triangle inequality says that the length of any side of a triangle is no greater than the sum of the lengths of the other two sides (see Figure 1.2.12).

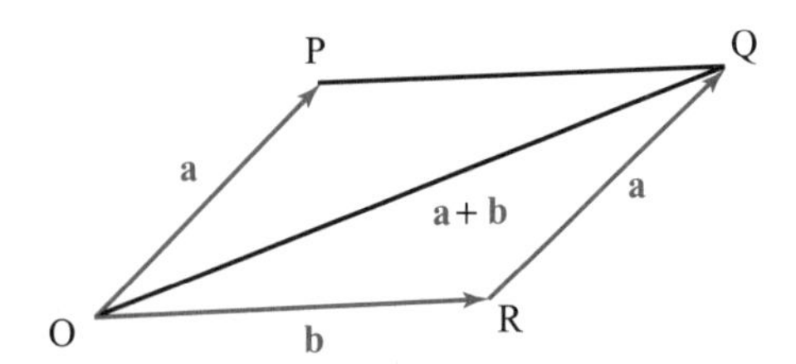

Figure 1.2.12 This geometry shows that $\|OQ\| \le \|OR\| + \|RQ\|$ or, in vector notation, that $\|\mathbf{a}+\mathbf{b}\| \le \|\mathbf{a}\| + \|\mathbf{b}\|$, which is the triangle inequality.

THEOREM 2: Triangle Inequality For vectors $\mathbf{a}$ and $\mathbf{b}$ in space,

$$\|\mathbf{a}+\mathbf{b}\| \le \|\mathbf{a}\| + \|\mathbf{b}\|.$$

PROOF While this result may be clear geometrically, it is useful to give a proof using the Cauchy–Schwarz inequality, as it will generalize to n-dimensional vectors. We consider the square of the left-hand side:

$$\|\mathbf{a}+\mathbf{b}\|^2 = (\mathbf{a}+\mathbf{b})\cdot(\mathbf{a}+\mathbf{b}) = \|\mathbf{a}\|^2 + 2\mathbf{a}\cdot\mathbf{b} + \|\mathbf{b}\|^2.$$

By the Cauchy–Schwarz inequality, we have

$$\|\mathbf{a}\|^2 + 2\mathbf{a}\cdot\mathbf{b} + \|\mathbf{b}\|^2 \le \|\mathbf{a}\|^2 + 2\|\mathbf{a}\|\,\|\mathbf{b}\| + \|\mathbf{b}\|^2 = (\|\mathbf{a}\| + \|\mathbf{b}\|)^2.$$

Thus,

$$\|\mathbf{a}+\mathbf{b}\|^2 \le (\|\mathbf{a}\| + \|\mathbf{b}\|)^2;$$

taking square roots proves the result. ■

EXAMPLE 9

(a) Verify the triangle inequality for $\mathbf{a} = \mathbf{i} + \mathbf{j}$ and $\mathbf{b} = 2\mathbf{i} + \mathbf{j} + \mathbf{k}$.

(b) Prove that $\|\mathbf{u} - \mathbf{v}\| \le \|\mathbf{u} - \mathbf{w}\| + \|\mathbf{w} - \mathbf{v}\|$ for any vectors $\mathbf{u}$, $\mathbf{v}$, and $\mathbf{w}$. Illustrate with a figure in which $\mathbf{u}$, $\mathbf{v}$, and $\mathbf{w}$ have the same base point.

SOLUTION

(a) We have $\mathbf{a} + \mathbf{b} = 3\mathbf{i} + 2\mathbf{j} + \mathbf{k}$, so $\|\mathbf{a} + \mathbf{b}\| = \sqrt{9+4+1} = \sqrt{14}$. On the other hand, $\|\mathbf{a}\| = \sqrt{2}$ and $\|\mathbf{b}\| = \sqrt{6}$, so the triangle inequality asserts that $\sqrt{14} \le \sqrt{2} + \sqrt{6}$. The numbers bear this out: $\sqrt{14} \approx 3.74$, while $\sqrt{2} + \sqrt{6} \approx 1.41 + 2.45 = 3.86$.

(b) We find that $\mathbf{u} - \mathbf{v} = (\mathbf{u} - \mathbf{w}) + (\mathbf{w} - \mathbf{v})$, so the result follows from the triangle inequality with $\mathbf{a}$ replaced by $\mathbf{u} - \mathbf{w}$ and $\mathbf{b}$ replaced by $\mathbf{w} - \mathbf{v}$. Geometrically, we are considering the shaded triangle in Figure 1.2.13. ▲

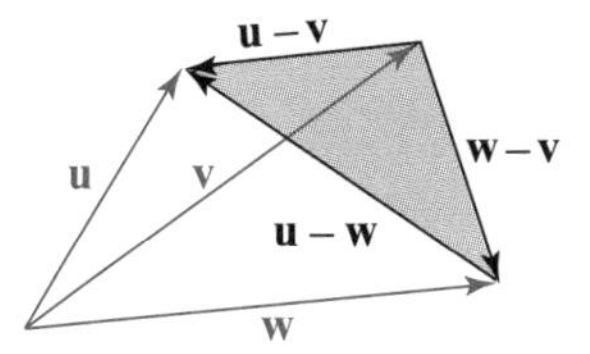

Figure 1.2.13 Illustrating the inequality $\|\mathbf{u} - \mathbf{v}\| \le \|\mathbf{u} - \mathbf{w}\| + \|\mathbf{w} - \mathbf{v}\|$.

Physical Applications of Vectors

A simple example of a physical quantity represented by a vector is a displacement. Suppose that, on a part of the earth's surface small enough to be considered flat, we introduce coordinates so that the x axis points east, the y axis points north, and the unit of length is the kilometer. If we are at a point P and wish to get to a point Q, the ***displacement vector*** $\mathbf{d} = \overrightarrow{PQ}$ joining P to Q tells us the direction and distance we have to travel. If x and y are the components of this vector, the displacement of P to Q is "x kilometers east, y kilometers north."

EXAMPLE 10 Suppose that two navigators who cannot see one another but can communicate by radio wish to determine the relative position of their ships. Explain how they can do this if they can each determine their displacement vector to the same lighthouse.

SOLUTION Let P_1 and P_2 be the positions of the ships, and let Q be the position of the lighthouse. The displacement of the lighthouse from the ith ship is the vector $\mathbf{d}_i$ joining P_i to Q. The displacement of the second ship from the first is the vector $\mathbf{d}$ joining P_1 to P_2. We have $\mathbf{d} + \mathbf{d}_2 = \mathbf{d}_1$ (Figure 1.2.14), and so $\mathbf{d} = \mathbf{d}_1 - \mathbf{d}_2$. That is, the displacement from one ship to the other is the difference between the displacements from the ships to the lighthouse. ▲

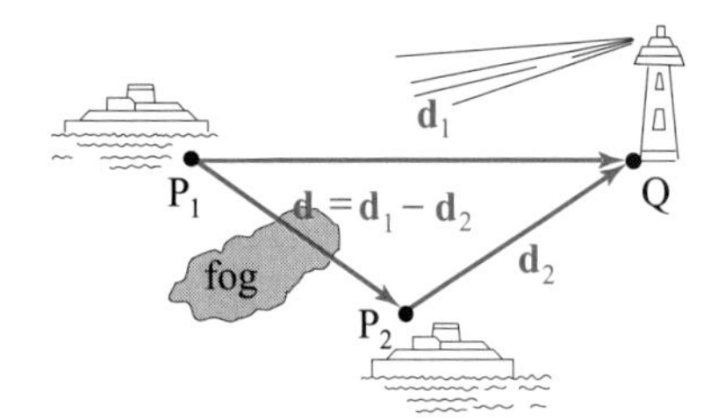

Figure 1.2.14 Vector methods can be used to locate objects.

We can also represent the velocity of a moving object as a vector. For the moment, we will consider only objects moving at uniform speed along straight lines. Suppose, for example, that a boat is steaming across a lake at 10 kilometers per hour (km/h) in the northeast direction. After 1 hour of travel, the displacement is $(10/\sqrt{2}, 10/\sqrt{2}) \approx (7.07, 7.07)$; see Figure 1.2.15.

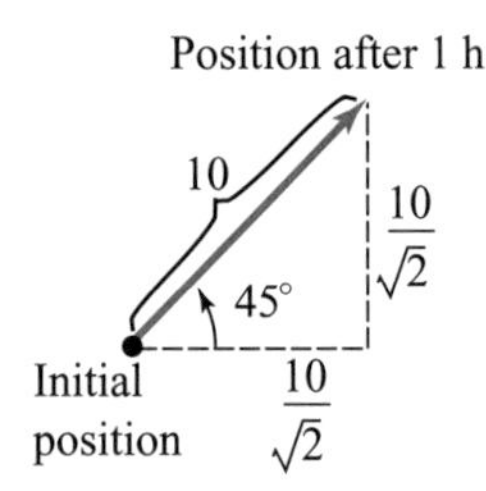

Figure 1.2.15 If an object moves northeast at 10 km/h, its velocity vector has components $(10/\sqrt{2}, 10/\sqrt{2}) = 10(1/\sqrt{2}, 1/\sqrt{2})$, where $(1/\sqrt{2}, 1/\sqrt{2})$ is the unit vector of the northeast direction.

The vector whose components are $(10/\sqrt{2}, 10/\sqrt{2})$ is called the *velocity vector* of the boat. In general, if an object is moving uniformly along a straight line, *its* ***velocity vector*** *is the displacement vector from the position at any moment to the position* 1 *unit of time later*. If a current appears on the lake, moving due eastward at 2 km/h, and the boat continues to point in the same direction with its engine running at the same rate, its displacement after 1 hour will have components given by $(10/\sqrt{2}+2, 10/\sqrt{2})$; see Figure 1.2.16. The new velocity vector, therefore, has components $(10/\sqrt{2}+2, 10/\sqrt{2})$. We note that this is the sum of the original velocity vector $(10/\sqrt{2}, 10/\sqrt{2})$ of the boat and the velocity vector (2, 0) of the current.

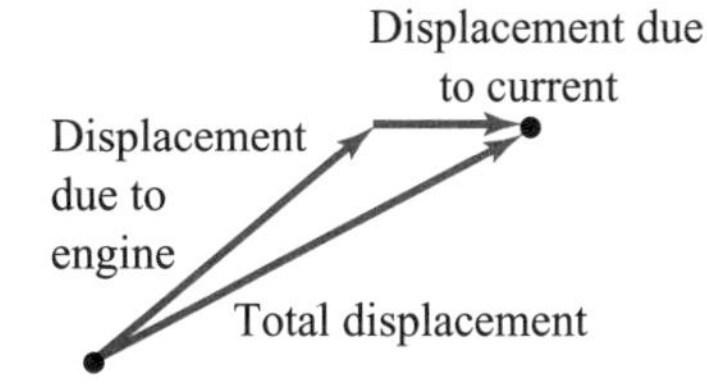

Figure 1.2.16 The total displacement is the sum of the displacements due to the engine and the current.

> **Displacement and Velocity** If an object has a (constant) velocity vector $\mathbf{v}$, then in t units of time the resulting displacement vector of the object is $\mathbf{d} = t\mathbf{v}$; thus, after time $t = 1$, the displacement vector equals the velocity vector. See Figure 1.2.17.

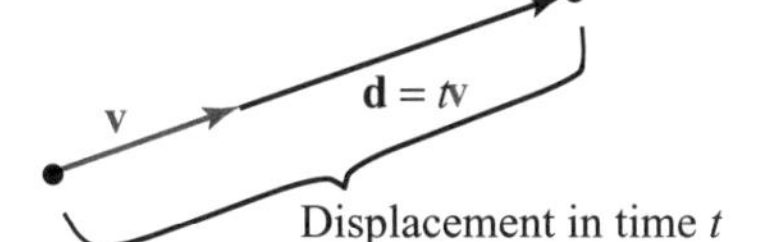

Figure 1.2.17 Displacement = time × velocity.

EXAMPLE 11 A bird is flying in a straight line with velocity vector $10\mathbf{i}+6\mathbf{j}+\mathbf{k}$ (in kilometers per hour). Suppose that (x, y) are its coordinates on the ground and z is its height above the ground.

(a) If the bird is at position (1, 2, 3) at a certain moment, what is its location 1 hour later? 1 minute later?

(b) How many seconds does it take the bird to climb 10 meters?

SOLUTION (a) The displacement vector from (1, 2, 3) after 1 hour is given by $10\mathbf{i}+6\mathbf{j}+\mathbf{k}$, so the new position is $(1, 2, 3)+(10, 6, 1)=(11, 8, 4)$. After 1 minute, the displacement vector from (1, 2, 3) is $\frac{1}{60}(10\mathbf{i}+6\mathbf{j}+\mathbf{k})=\frac{1}{6}\mathbf{i}+\frac{1}{10}\mathbf{j}+\frac{1}{60}\mathbf{k}$, and so the new position is $(1, 2, 3)+(\frac{1}{6}, \frac{1}{10}, \frac{1}{60})=(\frac{7}{6}, \frac{21}{10}, \frac{181}{60})$.

(b) After t seconds $(= t/3600$ hours), the displacement vector from (1, 2, 3) is $(t/3600)(10\mathbf{i}+6\mathbf{j}+\mathbf{k})=(t/360)\mathbf{i}+(t/600)\mathbf{j}+(t/3600)\mathbf{k}$. The increase in altitude is the z component, namely, $t/3600$. This will equal 10 m $(=\frac{1}{100}$ km) when $t/3600=\frac{1}{100}$, that is, when $t = 36$ s. ▲

EXAMPLE 12 Physical forces have magnitude and direction and may thus be represented by vectors. If several forces act at once on an object, the resultant force is represented by the sum of the individual force vectors. Suppose that forces $\mathbf{i}+\mathbf{k}$ and $\mathbf{j}+\mathbf{k}$ are acting on a body. What third force $\mathbf{F}$ must we impose to counteract the two—that is, to make the total force equal to zero?

SOLUTION The force $\mathbf{F}$ should be chosen so that $(\mathbf{i}+\mathbf{k})+(\mathbf{j}+\mathbf{k})+\mathbf{F}=\mathbf{0}$; that is, $\mathbf{F}=-(\mathbf{i}+\mathbf{k})-(\mathbf{j}+\mathbf{k})=-\mathbf{i}-\mathbf{j}-2\mathbf{k}$. (Recall that $\mathbf{0}$ is the *zero vector*, the vector whose components are all zero.) ▲

EXERCISES

1. Calculate $(3\mathbf{i}+2\mathbf{j}+\mathbf{k})\cdot(\mathbf{i}+2\mathbf{j}-\mathbf{k})$.

2. Calculate $\mathbf{a}\cdot\mathbf{b}$ where $\mathbf{a}=2\mathbf{i}+10\mathbf{j}-12\mathbf{k}$ and $\mathbf{b}=-3\mathbf{i}+4\mathbf{k}$.

3. Find the angle between $7\mathbf{j}+19\mathbf{k}$ and $-2\mathbf{i}-\mathbf{j}$ (to the nearest degree).

4. Compute $\mathbf{u}\cdot\mathbf{v}$, where $\mathbf{u}=\sqrt{3}\mathbf{i}-315\mathbf{j}+22\mathbf{k}$ and $\mathbf{v}=\mathbf{u}/\|\mathbf{u}\|$.

5. Is $\|8\mathbf{i}-12\mathbf{k}\|\cdot\|6\mathbf{j}+\mathbf{k}\|-|(8\mathbf{i}-12\mathbf{k})\cdot(6\mathbf{j}+\mathbf{k})|$ equal to zero? Explain.

In Exercises 6 to 11, compute $\|\mathbf{u}\|$, $\|\mathbf{v}\|$, *and* $\mathbf{u}\cdot\mathbf{v}$ *for the given vectors in* $\mathbb{R}^3$.

6. $\mathbf{u}=15\mathbf{i}-2\mathbf{j}+4\mathbf{k}$, $\mathbf{v}=\pi\mathbf{i}+3\mathbf{j}-\mathbf{k}$

7. $\mathbf{u}=2\mathbf{j}-\mathbf{i}$, $\mathbf{v}=-\mathbf{j}+\mathbf{i}$

8. $\mathbf{u}=5\mathbf{i}-\mathbf{j}+2\mathbf{k}$, $\mathbf{v}=\mathbf{i}+\mathbf{j}-\mathbf{k}$

9. $\mathbf{u}=-\mathbf{i}+3\mathbf{j}+\mathbf{k}$, $\mathbf{v}=-2\mathbf{i}-3\mathbf{j}-7\mathbf{k}$

10. $\mathbf{u}=-\mathbf{i}+3\mathbf{k}$, $\mathbf{v}=4\mathbf{j}$

11. $\mathbf{u}=-\mathbf{i}+2\mathbf{j}-3\mathbf{k}$, $\mathbf{v}=-\mathbf{i}-3\mathbf{j}+4\mathbf{k}$

12. Normalize the vectors in Exercises 6 to 8. (Only the solution corresponding to Exercise 7 is in the Student Guide.)

13. Find the angle between the vectors in Exercises 9 to 11. If necessary, express your answer in terms of $\cos^{-1}$.

14. Find the projection of $\mathbf{u}=-\mathbf{i}+\mathbf{j}+\mathbf{k}$ onto $\mathbf{v}=2\mathbf{i}+\mathbf{j}-3\mathbf{k}$.

15. Find the projection of $\mathbf{v}=2\mathbf{i}+\mathbf{j}-3\mathbf{k}$ onto $\mathbf{u}=-\mathbf{i}+\mathbf{j}+\mathbf{k}$.

16. What restrictions must be made on the scalar b so that the vector $2\mathbf{i}+b\mathbf{j}$ is orthogonal to (a) $-3\mathbf{i}+2\mathbf{j}+\mathbf{k}$ and (b) $\mathbf{k}$?

17. Find two nonparallel vectors both orthogonal to $(1, 1, 1)$.

18. Find the line through (3, 1, −2) that intersects and is perpendicular to the line $x = -1 + t, y = -2 + t, z = -1 + t$. [HINT: If (x_0, y_0, z_0) is the point of intersection, find its coordinates.]

19. A ship at position (1, 0) on a nautical chart (with north in the positive y direction) sights a rock at position (2, 4). What is the vector joining the ship to the rock? What angle θ does this vector make with due north? (This is called the ***bearing*** of the rock from the ship.)

20. Suppose that the ship in Exercise 19 is pointing due north and traveling at a speed of 4 knots relative to the water. There is a current flowing due east at 1 knot. The units on the chart are nautical miles; 1 knot = 1 nautical mile per hour.

(a) If there were no current, what vector $\mathbf{u}$ would represent the velocity of the ship relative to the sea bottom?

(b) If the ship were just drifting with the current, what vector $\mathbf{v}$ would represent its velocity relative to the sea bottom?

(c) What vector $\mathbf{w}$ represents the total velocity of the ship?

(d) Where would the ship be after 1 hour?

(e) Should the captain change course?

(f) What if the rock were an iceberg?

21. An airplane is located at position (3, 4, 5) at noon and traveling with velocity $400\mathbf{i} + 500\mathbf{j} - \mathbf{k}$ kilometers per hour. The pilot spots an airport at position (23, 29, 0).

(a) At what time will the plane pass directly over the airport? (Assume that the plane is flying over flat ground and that the vector $\mathbf{k}$ points straight up.)

(b) How high above the airport will the plane be when it passes?

22. The wind velocity $\mathbf{v}_1$ is 40 miles per hour (mi/h) from east to west while an airplane travels with air speed $\mathbf{v}_2$ of 100 mi/h due north. The speed of the airplane relative to the ground is the vector sum $\mathbf{v}_1 + \mathbf{v}_2$.

(a) Find $\mathbf{v}_1 + \mathbf{v}_2$.

(b) Draw a figure to scale.

23. A force of 50 lb is directed 50° above horizontal, pointing to the right. Determine its horizontal and vertical components. Display all results in a figure.

24. Two persons pull horizontally on ropes attached to a post, the angle between the ropes being 60°. Person A pulls with a force of 150 lb, while B pulls with a force of 110 lb.

(a) The resultant force is the vector sum of the two forces. Draw a figure to scale that graphically represents the three forces.

(b) Using trigonometry, determine formulas for the vector components of the two forces in a conveniently chosen coordinate system. Perform the algebraic addition, and find the angle the resultant force makes with A.

25. A 1-kilogram (1-kg) mass located at the origin is suspended by ropes attached to the two points (1, 1, 1) and (−1, −1, 1). If the force of gravity is pointing in the direction of the vector $-\mathbf{k}$, what is the vector describing the force along each rope? [HINT: Use the symmetry of the problem. A 1-kg mass weighs 9.8 newtons (N).]

26. Suppose that an object moving in direction $\mathbf{i}+\mathbf{j}$ is acted on by a force given by the vector $2\mathbf{i}+\mathbf{j}$. Express this force as a sum of a force in the direction of motion and a force perpendicular to the direction of motion.

27. A force of 6 N (newtons) makes an angle of $\pi/4$ radian with the y axis, pointing to the right. The force acts against the movement of an object along the straight line connecting (1, 2) to (5, 4).

(a) Find a formula for the force vector $\mathbf{F}$.

(b) Find the angle θ between the displacement direction $\mathbf{D}=(5-1)\mathbf{i}+(4-2)\mathbf{j}$ and the force direction $\mathbf{F}$.

(c) The ***work done*** is $\mathbf{F}\cdot\mathbf{D}$, or equivalently, $\|\mathbf{F}\|\|\mathbf{D}\|\cos\theta$. Compute the work from both formulas and compare.

1.3 Matrices, Determinants, and the Cross Product

In Section 1.2 we defined a product of vectors that was a scalar. In this section we shall define a product of vectors that is a vector; that is, we shall show how, given two vectors **a** and **b**, we can produce a third vector $\mathbf{a}\times\mathbf{b}$, called the *cross product* of **a** and **b**. This new vector will have the pleasing geometric property that it is perpendicular to the plane spanned (determined) by **a** and **b**. The definition of the cross product is based on the notions of the matrix and the determinant, and so these are developed first. Once this has been accomplished, we can study the geometric implications of the mathematical structure we have built.

2 × 2 Matrices

We define a 2×2 ***matrix*** to be an array

$$\begin{bmatrix} a_{11} & a_{12} \\ a_{21} & a_{22} \end{bmatrix}$$

where a_{11}, a_{12}, a_{21}, and a_{22} are four scalars. For example,

$$\begin{bmatrix} 2 & 1 \\ 0 & 4 \end{bmatrix}, \quad \begin{bmatrix} -1 & 0 \\ 1 & 1 \end{bmatrix}, \quad \text{and} \quad \begin{bmatrix} 13 & 7 \\ 6 & 11 \end{bmatrix}$$

are 2×2 matrices. The ***determinant***

$$\begin{vmatrix} a_{11} & a_{12} \\ a_{21} & a_{22} \end{vmatrix}$$

of such a matrix is the real number defined by the equation

$$\begin{vmatrix} a_{11} & a_{12} \\ a_{21} & a_{22} \end{vmatrix} = a_{11}a_{22} - a_{12}a_{21}. \tag{1}$$

EXAMPLE 1

$$\begin{vmatrix} 1 & 1 \\ 1 & 1 \end{vmatrix} = 1 - 1 = 0; \qquad \begin{vmatrix} 1 & 2 \\ 3 & 4 \end{vmatrix} = 4 - 6 = -2; \qquad \begin{vmatrix} 5 & 6 \\ 7 & 8 \end{vmatrix} = 40 - 42 = -2. \quad \blacktriangle$$

3×3 Matrices

A 3×3 ***matrix*** is an array

$$\begin{bmatrix} a_{11} & a_{12} & a_{13} \\ a_{21} & a_{22} & a_{23} \\ a_{31} & a_{32} & a_{33} \end{bmatrix}$$

where, again, each a_{ij} is a scalar; a_{ij} denotes the entry in the array that is in the ith row and the jth column. We define the ***determinant*** of a 3×3 matrix by the rule

$$\begin{vmatrix} a_{11} & a_{12} & a_{13} \\ a_{21} & a_{22} & a_{23} \\ a_{31} & a_{32} & a_{33} \end{vmatrix} = a_{11}\begin{vmatrix} a_{22} & a_{23} \\ a_{32} & a_{33} \end{vmatrix} - a_{12}\begin{vmatrix} a_{21} & a_{23} \\ a_{31} & a_{33} \end{vmatrix} + a_{13}\begin{vmatrix} a_{21} & a_{22} \\ a_{31} & a_{32} \end{vmatrix}. \tag{2}$$

Without some mnemonic device, formula (2) would be difficult to memorize. The rule to learn is that you move along the first row, multiplying a_{1j} by the determinant of the 2×2 matrix obtained by canceling out the first row and the jth column, and then you add these up, remembering to put a minus in front of the a_{12} term. For example, the determinant multiplied by the middle term of formula (2), namely,

$$\begin{vmatrix} a_{21} & a_{23} \\ a_{31} & a_{33} \end{vmatrix},$$

is obtained by crossing out the first row and the second column of the given 3×3 matrix:

$$\begin{bmatrix} \cancel{a_{11}} & \cancel{a_{12}} & \cancel{a_{13}} \\ a_{21} & \cancel{a_{22}} & a_{23} \\ a_{31} & \cancel{a_{32}} & a_{33} \end{bmatrix}.$$

EXAMPLE 2

$$\begin{vmatrix} 1 & 0 & 0 \\ 0 & 1 & 0 \\ 0 & 0 & 1 \end{vmatrix} = 1\begin{vmatrix} 1 & 0 \\ 0 & 1 \end{vmatrix} - 0\begin{vmatrix} 0 & 0 \\ 0 & 1 \end{vmatrix} + 0\begin{vmatrix} 0 & 1 \\ 0 & 0 \end{vmatrix} = 1.$$

$$\begin{vmatrix} 1 & 2 & 3 \\ 4 & 5 & 6 \\ 7 & 8 & 9 \end{vmatrix} = 1\begin{vmatrix} 5 & 6 \\ 8 & 9 \end{vmatrix} - 2\begin{vmatrix} 4 & 6 \\ 7 & 9 \end{vmatrix} + 3\begin{vmatrix} 4 & 5 \\ 7 & 8 \end{vmatrix} = -3 + 12 - 9 = 0. \quad \blacktriangle$$

Properties of Determinants

An important property of determinants is that interchanging two rows or two columns results in a change of sign. For 2×2 determinants, this is a consequence of the definition as follows: For rows, we have

$$\begin{vmatrix} a_{11} & a_{12} \\ a_{21} & a_{22} \end{vmatrix} = a_{11}a_{22} - a_{21}a_{12} = -(a_{21}a_{12} - a_{11}a_{22}) = -\begin{vmatrix} a_{21} & a_{22} \\ a_{11} & a_{12} \end{vmatrix}$$

and for columns,

$$\begin{vmatrix} a_{11} & a_{12} \\ a_{21} & a_{22} \end{vmatrix} = -(a_{12}a_{21} - a_{11}a_{22}) = -\begin{vmatrix} a_{12} & a_{11} \\ a_{22} & a_{21} \end{vmatrix}.$$

We leave it to the reader to verify this property for the 3×3 case.

A second fundamental property of determinants is that *we can factor scalars out of any row or column.* For 2×2 determinants, this means

$$\begin{vmatrix} \alpha a_{11} & a_{12} \\ \alpha a_{21} & a_{22} \end{vmatrix} = \begin{vmatrix} a_{11} & \alpha a_{12} \\ a_{21} & \alpha a_{22} \end{vmatrix} = \alpha \begin{vmatrix} a_{11} & a_{12} \\ a_{21} & a_{22} \end{vmatrix} = \begin{vmatrix} \alpha a_{11} & \alpha a_{12} \\ a_{21} & a_{22} \end{vmatrix} = \begin{vmatrix} a_{11} & a_{12} \\ \alpha a_{21} & \alpha a_{22} \end{vmatrix}.$$

Similarly, for 3×3 determinants we have

$$\begin{vmatrix} \alpha a_{11} & \alpha a_{12} & \alpha a_{13} \\ a_{21} & a_{22} & a_{23} \\ a_{31} & a_{32} & a_{33} \end{vmatrix} = \alpha \begin{vmatrix} a_{11} & a_{12} & a_{13} \\ a_{21} & a_{22} & a_{23} \\ a_{31} & a_{32} & a_{33} \end{vmatrix} = \begin{vmatrix} a_{11} & \alpha a_{12} & a_{13} \\ a_{21} & \alpha a_{22} & a_{23} \\ a_{31} & \alpha a_{32} & a_{33} \end{vmatrix},$$

and so on. These results follow from the definitions. In particular, if any row or column consists of zeros, then the value of the determinant is zero.

A third fundamental fact about determinants is the following: *If we change a row (or column) by adding another row (or, respectively, column) to it, the value of the determinant remains the same.* For the 2×2 case, this means that

$$\begin{vmatrix} a_1 & a_2 \\ b_1 & b_2 \end{vmatrix} = \begin{vmatrix} a_1 + b_1 & a_2 + b_2 \\ b_1 & b_2 \end{vmatrix} = \begin{vmatrix} a_1 & a_2 \\ b_1 + a_1 & b_2 + a_2 \end{vmatrix}$$

$$= \begin{vmatrix} a_1 + a_2 & a_2 \\ b_1 + b_2 & b_2 \end{vmatrix} = \begin{vmatrix} a_1 & a_1 + a_2 \\ b_1 & b_1 + b_2 \end{vmatrix}.$$

For the 3×3 case, this means

$$\begin{vmatrix} a_1 & a_2 & a_3 \\ b_1 & b_2 & b_3 \\ c_1 & c_2 & c_3 \end{vmatrix} = \begin{vmatrix} a_1+b_1 & a_2+b_2 & a_3+b_3 \\ b_1 & b_2 & b_3 \\ c_1 & c_2 & c_3 \end{vmatrix} = \begin{vmatrix} a_1+a_2 & a_2 & a_3 \\ b_1+b_2 & b_2 & b_3 \\ c_1+c_2 & c_2 & c_3 \end{vmatrix},$$

and so on. Again, this property can be proved using the definition of the determinant.

EXAMPLE 3 Suppose

$$\mathbf{a} = \alpha\mathbf{b} + \beta\mathbf{c}; \qquad \text{that is, } \mathbf{a} = (a_1, a_2, a_3) = \alpha(b_1, b_2, b_3) + \beta(c_1, c_2, c_3).$$

Show that

$$\begin{vmatrix} a_1 & a_2 & a_3 \\ b_1 & b_2 & b_3 \\ c_1 & c_2 & c_3 \end{vmatrix} = 0.$$

SOLUTION We shall prove the case $\alpha \neq 0$, $\beta \neq 0$. The case $\alpha = 0 = \beta$ is trivial, and the case where exactly one of α, β is zero is a simple modification of the case we prove. Using the fundamental properties of determinants, the determinant in question is

$$\begin{vmatrix} \alpha b_1 + \beta c_1 & \alpha b_2 + \beta c_2 & \alpha b_3 + \beta c_3 \\ b_1 & b_2 & b_3 \\ c_1 & c_2 & c_3 \end{vmatrix}$$

$$= -\frac{1}{\alpha}\begin{vmatrix} \alpha b_1 + \beta c_1 & \alpha b_2 + \beta c_2 & \alpha b_3 + \beta c_3 \\ -\alpha b_1 & -\alpha b_2 & -\alpha b_3 \\ c_1 & c_2 & c_3 \end{vmatrix}$$

(factoring $-1/\alpha$ out of the second row)

$$= \left(-\frac{1}{\alpha}\right)\left(-\frac{1}{\beta}\right)\begin{vmatrix} \alpha b_1 + \beta c_1 & \alpha b_2 + \beta c_2 & \alpha b_3 + \beta c_3 \\ -\alpha b_1 & -\alpha b_2 & -\alpha b_3 \\ -\beta c_1 & -\beta c_2 & -\beta c_3 \end{vmatrix}$$

(factoring $-1/\beta$ out of the third row)

$$= \frac{1}{\alpha\beta}\begin{vmatrix} \beta c_1 & \beta c_2 & \beta c_3 \\ -\alpha b_1 & -\alpha b_2 & -\alpha b_3 \\ -\beta c_1 & -\beta c_2 & -\beta c_3 \end{vmatrix} \qquad \text{(adding the second row to the first row)}$$

$$= \frac{1}{\alpha\beta}\begin{vmatrix} 0 & 0 & 0 \\ -\alpha b_1 & -\alpha b_2 & -\alpha b_3 \\ -\beta c_1 & -\beta c_2 & -\beta c_3 \end{vmatrix} \qquad \text{(adding the third row to the first row)}$$

$= 0.$ ▲

Closely related to these properties is the fact that *we can expand a* 3×3 *determinant along any row or column* using the signs in the following checkerboard pattern:

$$\begin{vmatrix} + & - & + \\ - & + & - \\ + & - & + \end{vmatrix}$$

For instance, the reader can check that we can expand "by minors" along the middle row:

$$\begin{vmatrix} a_{11} & a_{12} & a_{13} \\ a_{21} & a_{22} & a_{23} \\ a_{31} & a_{32} & a_{33} \end{vmatrix} = -a_{21}\begin{vmatrix} a_{12} & a_{13} \\ a_{32} & a_{33} \end{vmatrix} + a_{22}\begin{vmatrix} a_{11} & a_{13} \\ a_{31} & a_{33} \end{vmatrix} - a_{23}\begin{vmatrix} a_{11} & a_{12} \\ a_{31} & a_{32} \end{vmatrix}.$$

Let us redo the second determinant in Example 2 using this formula:

$$\begin{vmatrix} 1 & 2 & 3 \\ 4 & 5 & 6 \\ 7 & 8 & 9 \end{vmatrix} = -4\begin{vmatrix} 2 & 3 \\ 8 & 9 \end{vmatrix} + 5\begin{vmatrix} 1 & 3 \\ 7 & 9 \end{vmatrix} - 6\begin{vmatrix} 1 & 2 \\ 7 & 8 \end{vmatrix} = (-4)(-6) + (5)(12) + (-6)(6) = 0.$$

Historical Note

Determinants appear to have been invented and first used by Leibniz in 1693, in connection with solutions of linear equations. Maclaurin and Cramer developed their properties between 1729 and 1750; in particular, they showed that the solution of the system of equations

$$a_{11}x_1 + a_{12}x_2 + a_{13}x_3 = b_1$$
$$a_{21}x_1 + a_{22}x_2 + a_{23}x_3 = b_2$$
$$a_{31}x_1 + a_{32}x_2 + a_{33}x_3 = b_3$$

is

$$x_1 = \frac{1}{\Delta}\begin{vmatrix} b_1 & a_{12} & a_{13} \\ b_2 & a_{22} & a_{23} \\ b_3 & a_{32} & a_{33} \end{vmatrix}, \qquad x_2 = \frac{1}{\Delta}\begin{vmatrix} a_{11} & b_1 & a_{13} \\ a_{21} & b_2 & a_{23} \\ a_{31} & b_3 & a_{33} \end{vmatrix}, \qquad x_3 = \frac{1}{\Delta}\begin{vmatrix} a_{11} & a_{12} & b_1 \\ a_{21} & a_{22} & b_2 \\ a_{31} & a_{32} & b_3 \end{vmatrix}$$

where

$$\Delta = \begin{vmatrix} a_{11} & a_{12} & a_{13} \\ a_{21} & a_{22} & a_{23} \\ a_{31} & a_{32} & a_{33} \end{vmatrix},$$

a fact now known as ***Cramer's rule***. While this method is rather inefficient from a numerical point of view, it is of theoretical importance in matrix theory. Later, Vandermonde (1772) and Cauchy (1812), treating determinants as a separate topic worthy of special attention, developed the field more systematically, with contributions by Laplace, Jacobi, and others. Formulas for volumes of parallelepipeds in terms of determinants are due to Lagrange (1775). We shall study these later in this section. Although during the nineteenth century mathematicians studied matrices and determinants, the subjects were considered separate. For the full history up to 1900, see T. Muir, *The Theory of Determinants in the Historical Order of Development* (reprinted by Dover, New York, 1960).

Cross Products

Now that we have established the necessary properties of determinants and discussed their history, we are ready to proceed with the cross product of vectors.

DEFINITION: The Cross Product Suppose that $\mathbf{a} = a_1\mathbf{i} + a_2\mathbf{j} + a_3\mathbf{k}$ and $\mathbf{b} = b_1\mathbf{i} + b_2\mathbf{j} + b_3\mathbf{k}$ are vectors in $\mathbb{R}^3$. The ***cross product*** or ***vector product*** of **a** and **b**, denoted $\mathbf{a} \times \mathbf{b}$, is defined to be the vector

$$\mathbf{a} \times \mathbf{b} = \begin{vmatrix} a_2 & a_3 \\ b_2 & b_3 \end{vmatrix} \mathbf{i} - \begin{vmatrix} a_1 & a_3 \\ b_1 & b_3 \end{vmatrix} \mathbf{j} + \begin{vmatrix} a_1 & a_2 \\ b_1 & b_2 \end{vmatrix} \mathbf{k},$$

or, symbolically,

$$\mathbf{a} \times \mathbf{b} = \begin{vmatrix} \mathbf{i} & \mathbf{j} & \mathbf{k} \\ a_1 & a_2 & a_3 \\ b_1 & b_2 & b_3 \end{vmatrix}.$$

Even though we only defined determinants for arrays of *real* numbers, this formal expression involving *vectors* is a useful memory aid for the cross product.

EXAMPLE 4 Find $(3\mathbf{i} - \mathbf{j} + \mathbf{k}) \times (\mathbf{i} + 2\mathbf{j} - \mathbf{k})$.

SOLUTION

$$(3\mathbf{i} - \mathbf{j} + \mathbf{k}) \times (\mathbf{i} + 2\mathbf{j} - \mathbf{k}) = \begin{vmatrix} \mathbf{i} & \mathbf{j} & \mathbf{k} \\ 3 & -1 & 1 \\ 1 & 2 & -1 \end{vmatrix} = -\mathbf{i} + 4\mathbf{j} + 7\mathbf{k}. \quad \blacktriangle$$

Certain algebraic properties of the cross product follow from the definition. If **a**, **b**, and **c** are vectors and α, β, and γ are scalars, then

(i) $\mathbf{a} \times \mathbf{b} = -(\mathbf{b} \times \mathbf{a})$

(ii) $\mathbf{a} \times (\beta\mathbf{b} + \gamma\mathbf{c}) = \beta(\mathbf{a} \times \mathbf{b}) + \gamma(\mathbf{a} \times \mathbf{c})$ and $(\alpha\mathbf{a} + \beta\mathbf{b}) \times \mathbf{c} = \alpha(\mathbf{a} \times \mathbf{c}) + \beta(\mathbf{b} \times \mathbf{c})$.

Note that $\mathbf{a} \times \mathbf{a} = -(\mathbf{a} \times \mathbf{a})$, by property (i). Thus, $\mathbf{a} \times \mathbf{a} = \mathbf{0}$. In particular,

$$\mathbf{i} \times \mathbf{i} = \mathbf{0}, \qquad \mathbf{j} \times \mathbf{j} = \mathbf{0}, \qquad \mathbf{k} \times \mathbf{k} = \mathbf{0}.$$

Also,

$$\mathbf{i} \times \mathbf{j} = \mathbf{k}, \qquad \mathbf{j} \times \mathbf{k} = \mathbf{i}, \qquad \mathbf{k} \times \mathbf{i} = \mathbf{j},$$

which can be remembered by cyclicly permuting **i**, **j**, **k** like this:

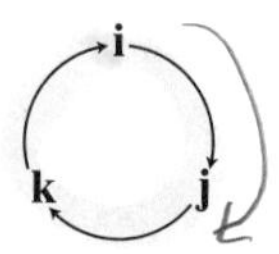

To give a geometric interpretation of the cross product, we first introduce the triple product. Given three vectors **a**, **b**, and **c**, the real number

$$(\mathbf{a} \times \mathbf{b}) \cdot \mathbf{c}$$

is called the ***triple product*** of **a**, **b**, and **c** (in that order). To obtain a formula for it, let $\mathbf{a} = a_1\mathbf{i} + a_2\mathbf{j} + a_3\mathbf{k}$, $\mathbf{b} = b_1\mathbf{i} + b_2\mathbf{j} + b_3\mathbf{k}$, and $\mathbf{c} = c_1\mathbf{i} + c_2\mathbf{j} + c_3\mathbf{k}$. Then

$$\begin{aligned}(\mathbf{a} \times \mathbf{b}) \cdot \mathbf{c} &= \left(\begin{vmatrix} a_2 & a_3 \\ b_2 & b_3 \end{vmatrix} \mathbf{i} - \begin{vmatrix} a_1 & a_3 \\ b_1 & b_3 \end{vmatrix} \mathbf{j} + \begin{vmatrix} a_1 & a_2 \\ b_1 & b_2 \end{vmatrix} \mathbf{k} \right) \cdot (c_1\mathbf{i} + c_2\mathbf{j} + c_3\mathbf{k}) \\ &= \begin{vmatrix} a_2 & a_3 \\ b_2 & b_3 \end{vmatrix} c_1 - \begin{vmatrix} a_1 & a_3 \\ b_1 & b_3 \end{vmatrix} c_2 + \begin{vmatrix} a_1 & a_2 \\ b_1 & b_2 \end{vmatrix} c_3.\end{aligned}$$

This is the expansion by minors of the third row of the determinant, so

$$(\mathbf{a} \times \mathbf{b}) \cdot \mathbf{c} = \begin{vmatrix} a_1 & a_2 & a_3 \\ b_1 & b_2 & b_3 \\ c_1 & c_2 & c_3 \end{vmatrix}.$$

If **c** is a vector in the plane spanned by the vectors **a** and **b**, then the third row in the determinant expression for $(\mathbf{a} \times \mathbf{b}) \cdot \mathbf{c}$ is a linear combination of the first and second rows, and therefore $(\mathbf{a} \times \mathbf{b}) \cdot \mathbf{c} = 0$. In other words, *the vector* $\mathbf{a} \times \mathbf{b}$ *is orthogonal to any vector in the plane spanned by* **a** *and* **b**, *in particular to both* **a** *and* **b**.

Next, we calculate the length of $\mathbf{a} \times \mathbf{b}$. Note that

$$\begin{aligned}\|\mathbf{a} \times \mathbf{b}\|^2 &= \begin{vmatrix} a_2 & a_3 \\ b_2 & b_3 \end{vmatrix}^2 + \begin{vmatrix} a_1 & a_3 \\ b_1 & b_3 \end{vmatrix}^2 + \begin{vmatrix} a_1 & a_2 \\ b_1 & b_2 \end{vmatrix}^2 \\ &= (a_2b_3 - a_3b_2)^2 + (a_1b_3 - b_1a_3)^2 + (a_1b_2 - b_1a_2)^2.\end{aligned}$$

If we expand the terms in the last expression, we can recollect them to give

$$(a_1^2 + a_2^2 + a_3^2)(b_1^2 + b_2^2 + b_3^2) - (a_1b_1 + a_2b_2 + a_3b_3)^2,$$

which equals

$$\|\mathbf{a}\|^2\|\mathbf{b}\|^2 - (\mathbf{a}\cdot\mathbf{b})^2 = \|\mathbf{a}\|^2\|\mathbf{b}\|^2 - \|\mathbf{a}\|^2\|\mathbf{b}\|^2\cos^2\theta = \|\mathbf{a}\|^2\|\mathbf{b}\|^2\sin^2\theta$$

where θ is the angle between $\mathbf{a}$ and $\mathbf{b}$, $0 \le \theta \le \pi$. Taking square roots and using $\sqrt{k^2} = |k|$, we find that $\|\mathbf{a}\times\mathbf{b}\| = \|\mathbf{a}\|\|\mathbf{b}\||\sin\theta|$.

Combining our results, we conclude that $\mathbf{a}\times\mathbf{b}$ is a *vector perpendicular to the plane* $\mathcal{P}$ *spanned by* $\mathbf{a}$ *and* $\mathbf{b}$ *with length* $\|\mathbf{a}\|\|\mathbf{b}\||\sin\theta|$. We see from Figure 1.3.1 that this length is also the area of the parallelogram (with base $\|\mathbf{a}\|$ and height $\|\mathbf{b}\sin\theta\|$) spanned by $\mathbf{a}$ and $\mathbf{b}$. There are still two possible vectors that satisfy these conditions, because there are two choices of direction that are perpendicular (or normal) to $\mathcal{P}$. This is clear from Figure 1.3.1, which shows the two choices $\mathbf{n}_1$ and $-\mathbf{n}_1$ perpendicular to $\mathcal{P}$, with $\|\mathbf{n}_1\| = \|-\mathbf{n}_1\| = \|\mathbf{a}\|\|\mathbf{b}\||\sin\theta|$.

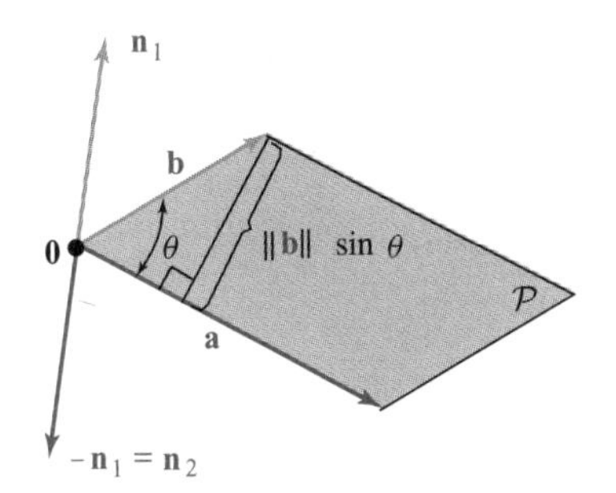

Figure 1.3.1 $\mathbf{n}_1$ and $\mathbf{n}_2$ are the two possible vectors orthogonal to both $\mathbf{a}$ and $\mathbf{b}$, and with norm $\|\mathbf{a}\|\ \|\mathbf{b}\|\ |\sin\theta|$.

Which vector represents $\mathbf{a}\times\mathbf{b}$, $\mathbf{n}_1$ or $-\mathbf{n}_1$? The answer is $\mathbf{n}_1$. Try a few cases such as $\mathbf{k} = \mathbf{i}\times\mathbf{j}$ to verify this. The following "right-hand rule" determines the direction of $\mathbf{a}\times\mathbf{b}$ in general. Take your *right hand* and place it so your fingers curl from $\mathbf{a}$ toward $\mathbf{b}$ through the *acute* angle θ, as in Figure 1.3.2. Then your thumb points in the direction of $\mathbf{a}\times\mathbf{b}$.

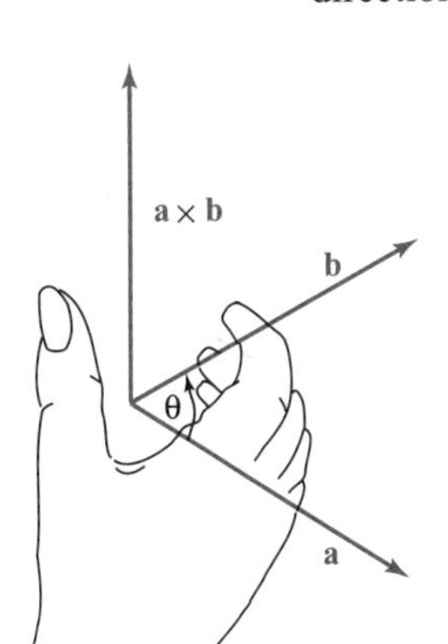

Figure 1.3.2 The right-hand rule for determining in which of the two possible directions $\mathbf{a}\times\mathbf{b}$ points.

The Cross Product

Geometric definition: $\mathbf{a} \times \mathbf{b}$ is the vector such that:

(1) $\|\mathbf{a} \times \mathbf{b}\| = \|\mathbf{a}\|\|\mathbf{b}\| \sin\theta$, the area of the parallelogram spanned by **a** and **b** (θ is the angle between **a** and **b**; $0 \le \theta \le \pi$); see Figure 1.3.3.

(2) $\mathbf{a} \times \mathbf{b}$ is perpendicular to **a** and **b**, and the triple $(\mathbf{a}, \mathbf{b}, \mathbf{a} \times \mathbf{b})$ obeys the right-hand rule.

Component formula:

$$(a_1\mathbf{i} + a_2\mathbf{j} + a_3\mathbf{k}) \times (b_1\mathbf{i} + b_2\mathbf{j} + b_3\mathbf{k}) = \begin{vmatrix} \mathbf{i} & \mathbf{j} & \mathbf{k} \\ a_1 & a_2 & a_3 \\ b_1 & b_2 & b_3 \end{vmatrix}$$

$$= (a_2b_3 - a_3b_2)\mathbf{i} - (a_1b_3 - a_3b_1)\mathbf{j} + (a_1b_2 - a_2b_1)\mathbf{k}.$$

Algebraic rules:

1. $\mathbf{a} \times \mathbf{b} = \mathbf{0}$ if and only if **a** and **b** are parallel or **a** or **b** is zero.
2. $\mathbf{a} \times \mathbf{b} = -\mathbf{b} \times \mathbf{a}$.
3. $\mathbf{a} \times (\mathbf{b} + \mathbf{c}) = \mathbf{a} \times \mathbf{b} + \mathbf{a} \times \mathbf{c}$.
4. $(\mathbf{a} + \mathbf{b}) \times \mathbf{c} = \mathbf{a} \times \mathbf{c} + \mathbf{b} \times \mathbf{c}$.
5. $(\alpha\mathbf{a}) \times \mathbf{b} = \alpha(\mathbf{a} \times \mathbf{b})$.

Multiplication table:

	$\times$	Second factor **i**	**j**	**k**
	i	**0**	**k**	$-$**j**
First	**j**	$-$**k**	**0**	**i**
factor	**k**	**j**	$-$**i**	**0**

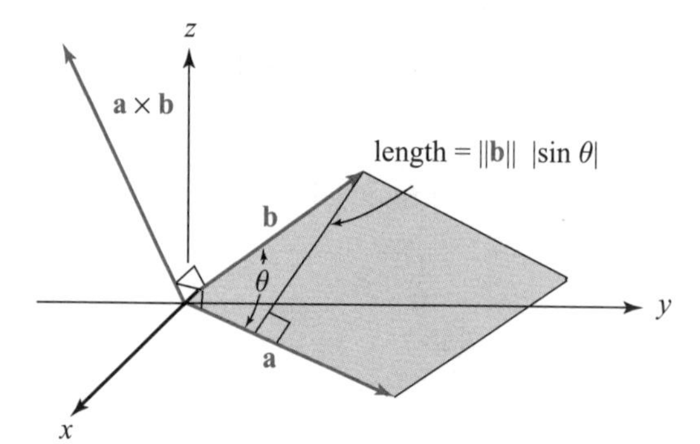

Figure 1.3.3 The length of $\mathbf{a} \times \mathbf{b}$ is the area of the parallelogram formed by **a** and **b**.

EXAMPLE 5 Find the area of the parallelogram spanned by the two vectors $\mathbf{a} = \mathbf{i} + 2\mathbf{j} + 3\mathbf{k}$ and $\mathbf{b} = -\mathbf{i} - \mathbf{k}$.

SOLUTION We calculate the cross product of $\mathbf{a}$ and $\mathbf{b}$ by applying the component or determinant formula, with $a_1 = 1$, $a_2 = 2$, $a_3 = 3$, $b_1 = -1$, $b_2 = 0$, $b_3 = -1$:

$$\begin{aligned}\mathbf{a} \times \mathbf{b} &= [(2)(-1) - (3)(0)]\mathbf{i} + [(3)(-1) - (1)(-1)]\mathbf{j} + [(1)(0) - (2)(-1)]\mathbf{k} \\ &= -2\mathbf{i} - 2\mathbf{j} + 2\mathbf{k}.\end{aligned}$$

Thus, the area is

$$\|\mathbf{a} \times \mathbf{b}\| = \sqrt{(-2)^2 + (-2)^2 + (2)^2} = 2\sqrt{3}. \quad \blacktriangle$$

EXAMPLE 6 Find a unit vector orthogonal to the vectors $\mathbf{i} + \mathbf{j}$ and $\mathbf{j} + \mathbf{k}$.

SOLUTION A vector perpendicular to both $\mathbf{i} + \mathbf{j}$ and $\mathbf{j} + \mathbf{k}$ is their cross product, namely, the vector

$$(\mathbf{i} + \mathbf{j}) \times (\mathbf{j} + \mathbf{k}) = \begin{vmatrix} \mathbf{i} & \mathbf{j} & \mathbf{k} \\ 1 & 1 & 0 \\ 0 & 1 & 1 \end{vmatrix} = \mathbf{i} - \mathbf{j} + \mathbf{k}.$$

Because $\|\mathbf{i} - \mathbf{j} + \mathbf{k}\| = \sqrt{3}$, the vector

$$\frac{1}{\sqrt{3}}(\mathbf{i} - \mathbf{j} + \mathbf{k})$$

is a unit vector perpendicular to $\mathbf{i} + \mathbf{j}$ and $\mathbf{j} + \mathbf{k}$. ▲

EXAMPLE 7 Derive an identity relating the dot and cross products from the formulas

$$\|\mathbf{u} \times \mathbf{v}\| = \|\mathbf{u}\|\,\|\mathbf{v}\| \sin\theta \qquad \text{and} \qquad \mathbf{u} \cdot \mathbf{v} = \|\mathbf{u}\|\,\|\mathbf{v}\| \cos\theta$$

by eliminating θ.

SOLUTION Seeing $\sin\theta$ and $\cos\theta$ multiplied by the same expression suggests squaring the two formulas and adding the results. We get

$$\|\mathbf{u} \times \mathbf{v}\|^2 + (\mathbf{u} \cdot \mathbf{v})^2 = \|\mathbf{u}\|^2 \|\mathbf{v}\|^2 (\sin^2\theta + \cos^2\theta) = \|\mathbf{u}\|^2 \|\mathbf{v}\|^2,$$

so

$$\|\mathbf{u} \times \mathbf{v}\|^2 = \|\mathbf{u}\|^2 \|\mathbf{v}\|^2 - (\mathbf{u} \cdot \mathbf{v})^2.$$

This identity is interesting because it establishes a link between the dot and cross products. ▲

Geometry of Determinants

Using the cross product, we may obtain a basic geometric interpretation of 2×2 and 3×3 determinants. Let $\mathbf{a} = a_1\mathbf{i} + a_2\mathbf{j}$ and $\mathbf{b} = b_1\mathbf{i} + b_2\mathbf{j}$ be two vectors in the plane. If θ is the angle between $\mathbf{a}$ and $\mathbf{b}$, we have seen that $\|\mathbf{a} \times \mathbf{b}\| = \|\mathbf{a}\|\|\mathbf{b}\||\sin\theta|$ is the area of the parallelogram with adjacent sides $\mathbf{a}$ and $\mathbf{b}$. The cross product as a determinant is

$$\mathbf{a} \times \mathbf{b} = \begin{vmatrix} \mathbf{i} & \mathbf{j} & \mathbf{k} \\ a_1 & a_2 & 0 \\ b_1 & b_2 & 0 \end{vmatrix} = \begin{vmatrix} a_1 & a_2 \\ b_1 & b_2 \end{vmatrix} \mathbf{k}.$$

Thus, the area $\|\mathbf{a} \times \mathbf{b}\|$ is the absolute value of the determinant

$$\begin{vmatrix} a_1 & a_2 \\ b_1 & b_2 \end{vmatrix} = a_1b_2 - a_2b_1.$$

Geometry of 2×2 Determinants The absolute value of the determinant $\begin{vmatrix} a_1 & a_2 \\ b_1 & b_2 \end{vmatrix}$ is the area of the parallelogram whose adjacent sides are the vectors $\mathbf{a} = a_1\mathbf{i} + a_2\mathbf{j}$ and $\mathbf{b} = b_1\mathbf{i} + b_2\mathbf{j}$. The sign of the determinant is $+$ when, rotating in the counterclockwise direction, the angle from $\mathbf{a}$ to $\mathbf{b}$ is less than π.

EXAMPLE 8 Find the area of the triangle with vertices at the points (1, 1), (0, 2), and (3, 2) (Figure 1.3.4).

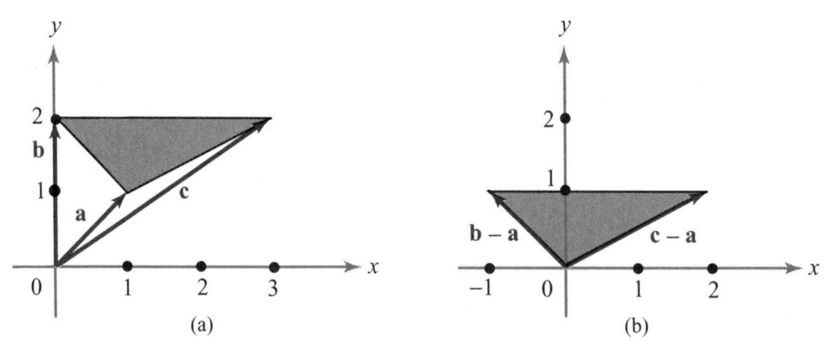

Figure 1.3.4 (a) Find the area A of the shaded triangle by expressing the sides as vector differences (b) to get $A = \|(\mathbf{b} - \mathbf{a}) \times (\mathbf{c} - \mathbf{a})\|/2$.

SOLUTION Let $\mathbf{a} = \mathbf{i} + \mathbf{j}$, $\mathbf{b} = 2\mathbf{j}$, and $\mathbf{c} = 3\mathbf{i} + 2\mathbf{j}$. It is clear that the triangle whose vertices are the endpoints of the vectors $\mathbf{a}$, $\mathbf{b}$, and $\mathbf{c}$ has the same area as the triangle with vertices at $\mathbf{0}$, $\mathbf{b} - \mathbf{a}$, and $\mathbf{c} - \mathbf{a}$ (Figure 1.3.4). Indeed, the latter is merely a translation of the former triangle. Because the area of this translated triangle is one-half the area of the parallelogram with adjacent sides $\mathbf{b} - \mathbf{a} = -\mathbf{i} + \mathbf{j}$, and $\mathbf{c} - \mathbf{a} = 2\mathbf{i} + \mathbf{j}$, we find that the area of the triangle with vertices $(1, 1)$, $(0, 2)$, and $(3, 2)$ is the absolute value of

$$\frac{1}{2}\begin{vmatrix} -1 & 1 \\ 2 & 1 \end{vmatrix} = -\frac{3}{2},$$

that is, 3/2. ▲

There is an interpretation of determinants of 3×3 matrices as volumes that is analogous to the interpretation of determinants of 2×2 matrices as areas.

Geometry of 3×3 Determinants The absolute value of the determinant

$$D = \begin{vmatrix} a_1 & a_2 & a_3 \\ b_1 & b_2 & b_3 \\ c_1 & c_2 & c_3 \end{vmatrix}$$

is the volume of the parallelepiped whose adjacent sides are the vectors

$$\mathbf{a} = a_1\mathbf{i} + a_2\mathbf{j} + a_3\mathbf{k}, \qquad \mathbf{b} = b_1\mathbf{i} + b_2\mathbf{j} + b_3\mathbf{k}, \qquad \text{and} \qquad \mathbf{c} = c_1\mathbf{i} + c_2\mathbf{j} + c_3\mathbf{k}.$$

To prove the statement in the box above, we refer to Figure 1.3.5 and note that the length of the cross product, namely, $\|\mathbf{a} \times \mathbf{b}\|$, is the area of the parallelogram with adjacent sides $\mathbf{a}$ and $\mathbf{b}$. Moreover, $(\mathbf{a} \times \mathbf{b}) \cdot \mathbf{c} = \|\mathbf{a} \times \mathbf{b}\| \|\mathbf{c}\| \cos\psi$, where ψ is the angle that $\mathbf{c}$ makes with the normal to the plane spanned by $\mathbf{a}$ and $\mathbf{b}$. Because the volume of the parallelepiped with adjacent sides $\mathbf{a}$, $\mathbf{b}$, and $\mathbf{c}$ is the product of the area of the base $\|\mathbf{a} \times \mathbf{b}\|$ and the altitude $\|\mathbf{c}\| |\cos\psi|$, it follows that the volume is $|(\mathbf{a} \times \mathbf{b}) \cdot \mathbf{c})|$. We saw earlier that $(\mathbf{a} \times \mathbf{b}) \cdot \mathbf{c} = D$, so the volume equals the absolute value of D.

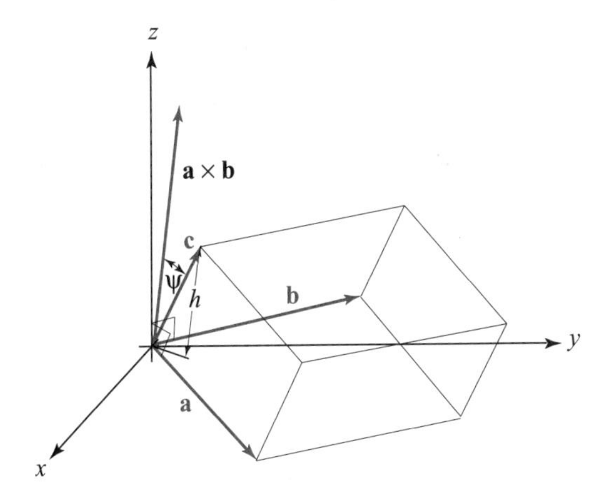

Figure 1.3.5 The volume of the parallelepiped spanned by **a**, **b**, **c** is the absolute value of the determinant of the 3×3 matrix having **a**, **b**, **c** as its rows.

EXAMPLE 9 Find the volume of the parallelepiped spanned by the three vectors $\mathbf{i} + 3\mathbf{k}$, $2\mathbf{i} + \mathbf{j} - 2\mathbf{k}$, and $5\mathbf{i} + 4\mathbf{k}$.

SOLUTION The volume is the absolute value of

$$\begin{vmatrix} 1 & 0 & 3 \\ 2 & 1 & -2 \\ 5 & 0 & 4 \end{vmatrix}.$$

If we expand this determinant by minors by going down the second column, the only nonzero term is

$$\begin{vmatrix} 1 & 3 \\ 5 & 4 \end{vmatrix}(1) = -11,$$

so the volume equals 11. ▲

Equations of Planes

Let $\mathcal{P}$ be a plane in space, $\mathrm{P}_0 = (x_0, y_0, z_0)$ a point on that plane, and suppose that $\mathbf{n} = A\mathbf{i} + B\mathbf{j} + C\mathbf{k}$ is a vector normal to that plane (see Figure 1.3.6). Let $\mathrm{P} = (x, y, z)$ be a point in $\mathbb{R}^3$. Then P lies on the plane $\mathcal{P}$ if and only if the vector $\overrightarrow{\mathrm{P}_0\mathrm{P}} = (x - x_0)\mathbf{i} + (y - y_0)\mathbf{j} + (z - z_0)\mathbf{k}$ is perpendicular to $\mathbf{n}$, that is, $\overrightarrow{\mathrm{P}_0\mathrm{P}} \cdot \mathbf{n} = 0$, or, equivalently,

$$(A\mathbf{i} + B\mathbf{j} + C\mathbf{k}) \cdot [(x - x_0)\mathbf{i} + (y - y_0)\mathbf{j} + (z - z_0)\mathbf{k}] = 0.$$

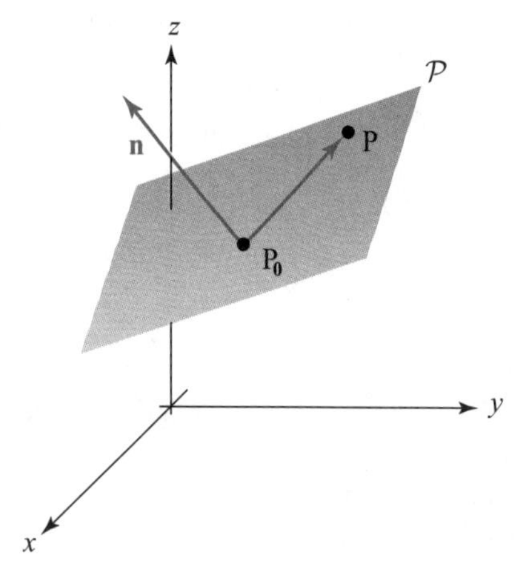

Figure 1.3.6 The points P of the plane through P_0 and perpendicular to $\mathbf{n}$ satisfy the equation $\overrightarrow{\mathrm{P}_0\mathrm{P}} \cdot \mathbf{n} = 0$.

Thus,

$$A(x - x_0) + B(y - y_0) + C(z - z_0) = 0.$$

Equation of a Plane in Space The equation of the plane $\mathcal{P}$ through (x_0, y_0, z_0) that has a normal vector $\mathbf{n} = A\mathbf{i} + B\mathbf{j} + C\mathbf{k}$ is

$$A(x - x_0) + B(y - y_0) + C(z - z_0) = 0;$$

that is, $(x, y, z) \in \mathcal{P}$ if and only if

$$Ax + By + Cz + D = 0$$

where $D = -Ax_0 - By_0 - Cz_0$.

The four numbers A, B, C, and D are not determined uniquely by the plane $\mathcal{P}$. To see this, note that (x, y, z) satisfies the equation $Ax + By + Cz + D = 0$ if and only if it also satisfies the relation

$$(\lambda A)x + (\lambda B)y + (\lambda C)z + (\lambda D) = 0$$

for any constant $\lambda \neq 0$. Furthermore, if A, B, C, D and A', B', C', D' determine the same plane $\mathcal{P}$, then $A = \lambda A'$, $B = \lambda B'$, $C = \lambda C'$, $D = \lambda D'$ for a scalar λ. Consequently, A, B, C, D are ***determined by*** $\mathcal{P}$ ***up to a scalar multiple.***

EXAMPLE 10 Determine an equation for the plane that is perpendicular to the vector $\mathbf{i} + \mathbf{j} + \mathbf{k}$ and contains the point $(1, 0, 0)$.

SOLUTION Using the general form $A(x - x_0) + B(y - y_0) + C(z - z_0) = 0$, the plane is $1(x - 1) + 1(y - 0) + 1(z - 0) = 0$; that is, $x + y + z = 1$. ▲

EXAMPLE 11 Find an equation for the plane containing the three points $(1, 1, 1)$, $(2, 0, 0)$, and $(1, 1, 0)$.

SOLUTION *Method 1.* This is a "brute force" method that you can use if you have forgotten the vector methods. The equation for any plane is of the form $Ax + By + Cz + D = 0$. Because the points $(1, 1, 1)$, $(2, 0, 0)$, and $(1, 1, 0)$ lie in the plane, we have

$$\begin{aligned} A + B + C + D &= 0, \\ 2A \qquad\qquad + D &= 0, \\ A + B \qquad + D &= 0. \end{aligned}$$

Proceeding by elimination, we reduce this system of equations to the form

$$\begin{aligned} 2A + D &= 0 \quad \text{(second equation)} \\ 2B + D &= 0 \quad (2 \times \text{third} - \text{second}), \\ C &= 0 \quad (\text{first} - \text{third}). \end{aligned}$$

Because the numbers A, B, C, and D are determined only up to a scalar multiple, we can fix the value of one of them, say $A = 1$, and then the others will be determined uniquely. We get $A = 1$, $D = -2$, $B = 1$, $C = 0$. Thus, an equation of the plane that contains the given points is $x + y - 2 = 0$.

Method 2. Let $\mathrm{P} = (1, 1, 1)$, $\mathrm{Q} = (2, 0, 0)$, $\mathrm{R} = (1, 1, 0)$. Any vector normal to the plane must be orthogonal to the vectors $\overrightarrow{\mathrm{QP}}$ and $\overrightarrow{\mathrm{RP}}$, which are parallel to the plane, because their endpoints lie on the plane. Thus, $\mathbf{n} = \overrightarrow{\mathrm{QP}} \times \overrightarrow{\mathrm{RP}}$ is normal to the plane. Computing the cross product, we have

$$\mathbf{n} = \begin{vmatrix} \mathbf{i} & \mathbf{j} & \mathbf{k} \\ -1 & 1 & 1 \\ 0 & 0 & 1 \end{vmatrix} = \mathbf{i} + \mathbf{j}.$$

Because the point (2, 0, 0) lies on the plane, we conclude that the equation is given by $(x - 2) + (y - 0) + 0 \cdot (z - 0) = 0$; that is, $x + y - 2 = 0$. ▲

Two planes are called ***parallel*** when their normal vectors are parallel. Thus, the planes $A_1x + B_1y + C_1z + D_1 = 0$ and $A_2x + B_2y + C_2z + D_2 = 0$ are parallel when $\mathbf{n}_1 = A_1\mathbf{i} + B_1\mathbf{j} + C_1\mathbf{k}$ and $\mathbf{n}_2 = A_2\mathbf{i} + B_2\mathbf{j} + C_2\mathbf{k}$ are parallel; that is, $\mathbf{n}_1 = \sigma\mathbf{n}_2$ for a constant σ. For example, the planes

$$x - 2y + z = 0 \quad \text{and} \quad -2x + 4y - 2z = 10$$

are parallel, but the planes

$$x - 2y + z = 0 \quad \text{and} \quad 2x - 2y + z = 10$$

are not parallel.

Distance: Point to Plane

Let us now determine the distance from a point $\mathrm{E} = (x_1, y_1, z_1)$ to the plane $\mathcal{P}$ described by the equation $A(x - x_0) + B(y - y_0) + C(z - z_0) = Ax + By + Cz + D = 0$. To do so, consider the unit normal vector

$$\mathbf{n} = \frac{A\mathbf{i} + B\mathbf{j} + C\mathbf{k}}{\sqrt{A^2 + B^2 + C^2}},$$

which is a unit vector normal to the plane. Drop a perpendicular from E to the plane and construct the triangle REQ shown in Figure 1.3.7. The distance $d = \|\overrightarrow{\mathrm{EQ}}\|$ is the length of the projection of $\mathbf{v} = \overrightarrow{\mathrm{RE}}$ (the vector from R to E) onto $\mathbf{n}$;

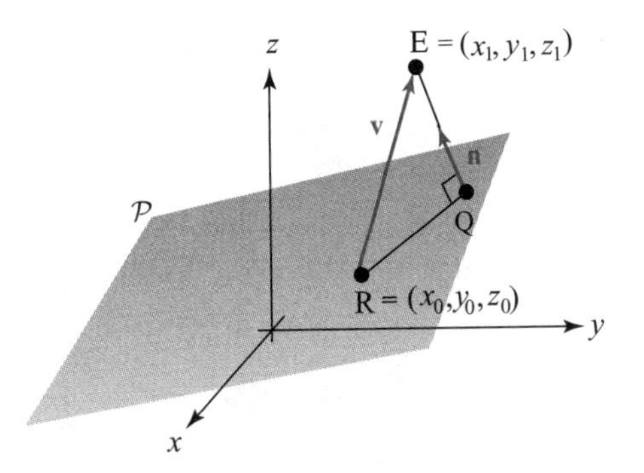

Figure 1.3.7 The geometry for determining the distance from the point E to plane $\mathcal{P}$.

thus,

$$\begin{aligned}\text{Distance} &= |\mathbf{v}\cdot\mathbf{n}| = |[(x_1-x_0)\mathbf{i}+(y_1-y_0)\mathbf{j}+(z_1-z_0)\mathbf{k}]\cdot\mathbf{n}| \\ &= \frac{|A(x_1-x_0)+B(y_1-y_0)+C(z_1-z_0)|}{\sqrt{A^2+B^2+C^2}}.\end{aligned}$$

If the plane is given in the form $Ax+By+Cz+D=0$, then for any point (x_0, y_0, z_0) on it, $D=-(Ax_0+By_0+Cz_0)$. Substitution into the previous formula gives the following:

Distance from a Point to a Plane The distance from (x_1, y_1, z_1) to the plane $Ax+By+Cz+D=0$ is

$$\text{Distance} = \frac{|Ax_1+By_1+Cz_1+D|}{\sqrt{A^2+B^2+C^2}}.$$

EXAMPLE 12 Find the distance from $Q=(2,0,-1)$ to the plane $3x-2y+8z+1=0$.

SOLUTION We substitute into the formula in the preceding box the values $x_1=2$ $y_1=0$, $z_1=-1$ (the point) and $A=3$, $B=-2$, $C=8$, $D=1$ (the plane) to give

$$\text{Distance} = \frac{|3\cdot 2+(-2)\cdot 0+8(-1)+1|}{\sqrt{3^2+(-2)^2+8^2}} = \frac{|-1|}{\sqrt{77}} = \frac{1}{\sqrt{77}}. \quad \blacktriangle$$

Historical Note

The Origins of the Vector, Scalar, Dot, and Cross Products

QUADRATIC EQUATIONS, CUBIC EQUATIONS, AND IMAGINARY NUMBERS. We know from Babylonian clay tablets that this great civilization possessed the quadratic formula, enabling them (in verbal form) to solve quadratic equations. Because the concept of negative numbers had to wait until the sixteenth century to see the light of day, the Babylonians did not consider either negative (or imaginary) solutions.

With the Renaissance and the rediscovery of ancient learning, Italian mathematicians began to wonder about the solutions of cubic equations, $x^3 + ax^2 + bx + c = 0$, where a, b, and c are positive numbers.

Around 1500, Scipione del Ferro, a professor in Bologna (the oldest European university) was able to solve cubics of the form $x^3 + bx = c$, but kept his discovery secret. Before his death, he passed his formula to his successor, Antonio Fior, who for a while also kept the formula to himself. It remained a secret until a brilliant, self-taught mathematician named Nicolo Fontana, also known as Tartaglia (the stammerer), appeared on the scene. Tartaglia claimed he could solve the cubic, and Fior felt he needed to protect the priority of del Ferro, and so in response challenged Tartaglia to a public competition.

We are told that Tartaglia was able to solve all of the thirty cubic equations posed by Fior. Amazingly, some scholars believe that Tartaglia discovered the formula for solutions to $x^3 + cx = d$ only days before the contest was to take place.

The greatest mathematician of the sixteenth century, Gerolamo Cardano (1501–1576)—a Renaissance scholar, mathematician, physician, and fortuneteller—gave the first published solution of the general cubic. Although born of modest means, he (like Tartaglia) rose, through effort and natural brilliance, to great fame. Cardano is the author of the first book on games of chance (marking the beginning of modern probability theory) and also of *Ars Magna* (the Great Art), which marks the beginning of modern algebra. It was in this book that Cardano published the solution to the *general* cubic $x^3 + ax^2 + bx + c = 0$. How did he get it?

While working on his algebra book, and aware that Tartaglia was able to solve forms of cubic, Cardano, in 1539, wrote to Tartaglia asking for a meeting. After some cajoling, Tartaglia agreed. It was at this meeting that, in exchange for a pledge of secrecy (and we know how these generally go), Tartaglia revealed his solution, from which Cardano was able to derive a solution to the general equation, which then appeared in *Ars Magna*. Feeling betrayed, Tartaglia led a scathing attack on Cardano, leading to a small soap opera.

What is important for us, at the moment, is that as a consequence of the method of solution, something very strange occurred. Consider the cubic $x^3 - 15x = 4$. Its only positive root is 4. However, the Tartaglia–Cardano solution formula yields

$$x = \sqrt[3]{2 + \sqrt{-121}} + \sqrt[3]{2 - \sqrt{-121}} \tag{1}$$

as the positive root. Thus, this number must be equal to 4. Yet this must be *nonsense*, because inside the cube root we are taking the square root of a negative number—at the time, an absolute impossibility. This was a real shock. Over 100 years later, in 1702, when Leibniz, codiscoverer of calculus, showed the great Dutch scientist Christian Huygens the formula

$$\sqrt{6} = \sqrt{1 + \sqrt{-3}} + \sqrt{1 - \sqrt{-3}} \tag{2}$$

Huygens was completely flabbergasted, and remarked that this equality "defies all human understanding." [Try, informally, to verify both formulas (1) and (2) for yourself.]

Whether nonsense or not, Tartaglia and Cardano's formula forced mathematicians to confront square roots of negative numbers (or *imaginary numbers*, as they are called today). This historical incident is another example that negates the (widespread) view that mathematics is "made up" by mathematicians. As is often the case, *it is the mathematics itself that speaks to us.*

THE MATURING OF COMPLEX NUMBERS. For well over two centuries, numbers like $i = \sqrt{-1}$ were looked at with great suspicion. The square root of any negative number can be written in terms of i; for example, $\sqrt{-a} = \sqrt{(a)(-1)} = \sqrt{a}\sqrt{-1}$. In the middle of the eighteenth century, the Swiss mathematician Leonhard Euler connected the universal cosmic numbers e and π with the imaginary number i. Whatever i was or meant, it necessarily follows that

$$e^{\pi i} = -1,$$

that is, e "raised to the power πi equals -1. Thus, these cosmic numbers, reflecting perhaps some deeper mystery, are in fact connected to each other by a very simple formula.

At the beginning of the nineteenth century, the German mathematician Karl Friedrich Gauss was able to prove the *fundamental theorem of algebra*, which says that any nth-degree polynomial has n roots (some or all of which may be imaginary; that is, the roots have the form $a + bi$, where, as earlier, $i = \sqrt{-1}$ and where a and b are real numbers).

By the middle of the nineteenth century, the French mathematician Augusten-Louis Cauchy and the German mathematician Bernhard Riemann had developed the differential calculus for functions of one complex variable. An example of such a function is $F(z) = z^n$, where

$z = a + bi$. In this case, the usual formula for the derivative, $F'(z) = nz^{n-1}$, still holds. However, by introducing imaginary numbers, Cauchy was able to evaluate "real integrals" that heretofore could not be evaluated. For example, it is possible to show that

$$\int_0^{\infty} \frac{\sin x}{x}\, dx = \frac{\pi}{2}$$

and that

$$\int_0^{\pi} \log \sin x\, dx = -\pi \log 2,$$

These were stunning results.

In summary, the solution of the cubic equation, the fundamental theorem of algebra, and the evaluation of real integrals proved how valuable it was to consider imaginary numbers $a + bi$, even though they were not (at least not yet) on *terra firma*. Did they really exist or were they simply phantoms of our *imagination*, and thus truly *imaginary*?

HAMILTON'S DEFINITION OF COMPLEX NUMBERS. Many mathematicians after Cardano made important contributions to imaginary (or complex) numbers, including Argand, Wessel, and Gauss—all of whom represented them geometrically. However, the modern, intellectually rigorous definition of a complex number is due to the great Irish mathematician William Rowan Hamilton (see Figure 1.3.8). After Newton, who created the vector concept through his invention of the notion of force, Hamilton was, beyond any doubt, the most important and singular figure in

Figure 1.3.8 Sir William Rowan Hamilton (1805–1865).

the development of vector calculus. It was Hamilton who gave us the terms *vector* and *scalar quantity*.

William Rowan Hamilton was born in Dublin, Ireland, at midnight on August 3, 1805. In 1823, he entered Trinity College, Dublin. His university career, by any standard, was phenomenal. By his third year, Trinity offered him a professorship, the Andrew's Chair of Astronomy, and the State named him Royal Astronomer of Ireland. These honors were based on his theoretical prediction (in 1824) of two entirely new and unexpected optical phenomena, namely, internal and external conical refraction.

By 1827 he had become interested in imaginary numbers. He wrote that "the symbol $\sqrt{-1}$ is absurd, and denotes an impossible extraction . . ." He set out to put the idea of a complex number on a firm logical foundation. His solution was to define a complex number $a + bi$ as a point (a, b) in the plane $\mathbb{R}^2$, much as we do today. Thus, the imaginary number bi for Hamilton was simply the point $(0, b)$ on the y axis. The difference between complex numbers and the Cartesian plane was that Hamilton followed the proforma multiplication of complex numbers:

$$(a + bi)(c + di) = (ac - bd) + (ad + bc)i,$$

and defined a new multiplication on the complex plane:

$$(a, b) \cdot (c, d) = (ac - bd, ad + bc).$$

Thus, $i = \sqrt{-1}$ just disappears into the point $(0, 1)$, and the mystery and confusion over complex numbers disappears along with it.

From Complex Numbers to Quaternions. From Hamilton's interpretation, *complex numbers are nothing more than the extension of real numbers into a new dimension, two dimensions*. Hamilton, however, also did fundamental work in mechanics, and he knew well that two dimensions were too limiting for the space analysis necessary for understanding the physics of the three-dimensional world. Therefore, Hamilton set out to find a triplet system; that is, an acceptable[1] multiplication scheme on points (a,b,c) in $\mathbb{R}^3$, or, as it were, on vectors $a\mathbf{i} + b\mathbf{j} + c\mathbf{k}$.

By 1843, Hamilton realized that his quest was hopeless. But then, on October 16, 1843, Hamilton discovered that what he could not achieve for $\mathbb{R}^3$ he could achieve for $\mathbb{R}^4$; he discovered *quaternions*, an entirely new number system.

[1]For him, "acceptable" meant that the associative law of multiplication would hold.

Let us revisit the important historical moment in Hamilton's own words:

> But on the 16th day of the same month—which happened to be a Monday, and a Council Day of the Royal Irish Academy—I was walking in to attend and preside, and your mother was walking with me, along the Royal Canal, to which she had perhaps driven; and although she talked with me now and then, yet another *under-current* of thought was going on in my mind, which gave at last a *result*, whereof it is not too much to say that I felt at *once* the importance. An *electric* circuit seemed to *close*; and a spark flashed forth, the herald (as I *foresaw, immediately*) of many long years to come of definitely directed thought and work, by *myself* if spared, and at all events on the parts of *others*, if I should even be allowed to live long enough to distinctly communicate the discovery. Nor could I resist the impulse—unphilosophical as it may have been—to cut with a knife on a stone of Brougham Bridge, as we passed it, the fundamental formula with the symbols i, j, k; namely
>
> $$\mathbf{i}^2 = \mathbf{j}^2 = \mathbf{k}^2 = \mathbf{ijk} = -1,$$
>
> which contains the *Solution of the Problem*, but of course, as an inscription, has long since moldered away.[2]

Hamilton had realized that the multiplication he had been searching for could be introduced on 4-tuples (a, b, c, d), which he had denoted by

$$a + b\mathbf{i} + c\mathbf{j} + d\mathbf{k}.$$

The a was called the *scalar part* and $b\mathbf{i} + c\mathbf{j} + d\mathbf{k}$ was called the *vector part*, which in reality, as with complex numbers, meant the point (a, b, c, d) in $\mathbb{R}^4$. The multiplication table he introduced was

$$\mathbf{ij} = \mathbf{k} = -\mathbf{ji}$$
$$\mathbf{ki} = \mathbf{j} = -\mathbf{ki}$$
$$\mathbf{jk} = \mathbf{i} = -\mathbf{kj}$$
$$\mathbf{i}^2 = \mathbf{j}^2 = \mathbf{k}^2 = \mathbf{ijk} = -1.$$

Hamilton continued to *passionately* believe in his quaternions until the end of his life. Unfortunately, historical development went in another direction.

---

[2] *North British Review* **14** (1858): 57.

The first step away from the quaternions was in fact taken by a firm believer in the importance of quaternions, namely, Peter Guthrie Tait, who was born in 1831 near Edinburgh, Scotland. In 1860, Tait was appointed to the Chair of Natural Philosophy at Edinburgh University, where he remained until his death in 1901. In 1867, he wrote his *Elementary Treatises on Quaternions*, a text stressing physical applications. His third chapter was most significant. It was here that Tait looked at the quaternionic product of two vectors:

$$\mathbf{v} = a\mathbf{i} + b\mathbf{j} + c\mathbf{k} \qquad \text{and} \qquad \mathbf{w} = a'\mathbf{i} + b'\mathbf{j} + c'\mathbf{k}.$$

Then the product $\mathbf{vw}$, as defined by Hamilton, yields:

$$\begin{aligned}(a\mathbf{i} + b\mathbf{j} + c\mathbf{k})&(a'\mathbf{i} + b'\mathbf{j} + c'\mathbf{k}) \\ &= -(aa' + bb' + cc') + (bc' - cb')\mathbf{i} + (ac' - ca')\mathbf{j} + (ab' - ba')\mathbf{k}\end{aligned}$$

or, in modern form:

$$\mathbf{vw} = -(\mathbf{v} \cdot \mathbf{w}) + \mathbf{v} \times \mathbf{w},$$

where $\cdot$ is the modern dot or inner product of vectors and $\times$ is the cross product. Tait discovered the formulas

$$\mathbf{v} \cdot \mathbf{w} = \|\mathbf{v}\|\|\mathbf{w}\| \cos\theta \qquad \text{and} \qquad \|\mathbf{v} \times \mathbf{w}\| = \|\mathbf{v}\|\|\mathbf{w}\| \sin\theta,$$

where θ is the angle formed by $\mathbf{v}$ and $\mathbf{w}$. Moreover, he showed that $\mathbf{v} \times \mathbf{w}$ was orthogonal to $\mathbf{v}$ and $\mathbf{w}$, therefore giving a *geometric* interpretation of the quaternionic product of two vectors.

This began the move away from the study of quaternions and back to Newton's vectors, with the quaternionic product eventually being replaced by two separate products, the inner product and the cross product.

By the way, you might wonder why Hamilton did not at first discover the cross product, since it is a product on $\mathbb{R}^3$. The reason is that it did not have a fundamental property that he required—namely, it was not associative:[3]

$$0 = (\mathbf{i} \times \mathbf{i}) \times \mathbf{k} \neq \mathbf{i} \times (\mathbf{i} \times \mathbf{k}) = -\mathbf{k}$$

---

[3] Interestingly, if one is willing to continue to live with nonassociativity, there is also a vector product with most of the properties of the cross product in $\mathbb{R}^7$; this involves yet another number system called the *octonians*, which exists in $\mathbb{R}^8$. The nonexistence of a cross product in other dimensions is a result that goes beyond the scope of this text. For further information, see the *American Mathematical Monthly*, **74** (1967), pp. 188–194, and **90** (1983), p. 697, as well as J. Baez, "The Octonians," *Bulletin of the American Mathematical Society*, **39** (2002), pp. 145–206. One can show that systems like the quaternions and octonians occur only in dimension 1 (the reals $\mathbb{R}$), dimension 2 (the complex numbers), dimension 4 (the quaternions), and dimension 8 (the octonians). On the other hand, the "right" way to extend the cross product is to introduce the notion of *differential forms*, which exists in *any* dimension. We discuss their construction in Section 8.6.

The Move Away from Quaternions. The scientists ultimately responsible for the demise of quaternions were James Clerk Maxwell (see Figure 1.3.9), Oliver Heaviside, and Josiah Willard Gibbs, a founder of statistical mechanics. In the 1860s, Maxwell wrote down his monumental equations of electricity and magnetism. No vector notation was used (it did not exist). Instead, Maxwell wrote out his equations in what we would now call "component form." Around 1870, Tait began to correspond with Maxwell, piquing his interest in quaternions.

Figure 1.3.9 James Clerk Maxwell (1831–1879).

In 1873, Maxwell published his epic work, *Treatise on Electricity and Magnetism*. Here (as we shall do in Chapter 8), Maxwell wrote down the equations of the electromagnetic field using quaternions, thus motivating physicists and mathematicians alike to take a closer look at them. From this manuscript many have concluded that Maxwell was a supporter of the "quaternionic approach" to physics. The truth, however, is that Maxwell was reluctant to use quaternions. It was Maxwell, in fact, who began the process of separating the *vector* part of a product of two quaternions (the cross product) from its *scalar* part (the dot product).

It is known that Maxwell was troubled by the fact that the scalar part of the "square"of a vector ($\mathbf{vv}$) was always negative ($-\mathbf{v}\cdot\mathbf{v}$), which in the case of a velocity vector could be interpreted as negative kinetic energy—an unacceptable idea!

It was Heaviside and Gibbs who made the final push away from quaternions. Heaviside, an independent researcher interested in electricity and magnetism, and Gibbs, a professor of mathematical physics at Yale, almost simultaneously—and independently—created our modern system of vector analysis, which we have just started to study.

In 1879, Gibbs taught a course at Yale in vector analysis with applications to electricity and magnetism. This treatise was clearly motivated by the advent of Maxwell's equations, which we will be studying in Chapter 8. In 1884, he published his *Elements of Vector Analysis,* a book in which all the properties of the dot and cross products are fully developed. Knowing that much of what Gibbs wrote was in fact due to Tait, Gibbs's contemporaries did not view his book as highly original. However, it is one of the sources from which modern vector analysis has come into existence.

Heaviside was also largely motivated by Maxwell's brilliant work. His great *Electromagnetic Theory* was published in three volumes. Volume I (1893) contained the first extensive treatment of modern vector analysis.

We all owe a great debt to E. B. Wilson's 1901 book *Vector Analysis: A Textbook for the Use of Students of Mathematics and Physics Founded upon the Lectures of J. Willard Gibbs.* Wilson was reluctant to take Gibbs's course, because he had just completed a full-year course in quaternions at Harvard under J. M. Pierce, a champion of quaternionic methods; but he was forced by a dean to add the course to his program, and he did so in 1899. Wilson was later asked by the editor of the Yale Bicentennial Series to write a book based on Gibbs's lectures. For a picture of Gibbs and for additional historical comments on divergence and curl, see the Historical Note in Section 4.4.

EXERCISES

1. Verify that interchanging the first two rows of the 3×3 determinant

$$\begin{vmatrix} 1 & 2 & 1 \\ 3 & 0 & 1 \\ 2 & 0 & 2 \end{vmatrix}$$

changes the sign of the determinant.

2. Evaluate the determinants

(a) $\begin{vmatrix} 2 & -1 & 0 \\ 4 & 3 & 2 \\ 3 & 0 & 1 \end{vmatrix}$

(b) $\begin{vmatrix} 36 & 18 & 17 \\ 45 & 24 & 20 \\ 3 & 5 & -2 \end{vmatrix}$

(c) $\begin{vmatrix} 1 & 4 & 9 \\ 4 & 9 & 16 \\ 9 & 16 & 25 \end{vmatrix}$

(d) $\begin{vmatrix} 2 & 3 & 5 \\ 7 & 11 & 13 \\ 17 & 19 & 23 \end{vmatrix}$

3. Compute $\mathbf{a} \times \mathbf{b}$, where $\mathbf{a} = \mathbf{i} - 2\mathbf{j} + \mathbf{k}$, $\mathbf{b} = 2\mathbf{i} + \mathbf{j} + \mathbf{k}$.

4. Compute $\mathbf{a} \cdot (\mathbf{b} \times \mathbf{c})$, where $\mathbf{a}$ and $\mathbf{b}$ are as in Exercise 3 and $\mathbf{c} = 3\mathbf{i} - \mathbf{j} + 2\mathbf{k}$.

5. Find the area of the parallelogram with sides $\mathbf{a}$ and $\mathbf{b}$ given in Exercise 3.

6. A triangle has vertices $(0, 0, 0)$, $(1, 1, 1)$, and $(0, -2, 3)$. Find its area.

7. What is the volume of the parallelepiped with sides $2\mathbf{i} + \mathbf{j} - \mathbf{k}$, $5\mathbf{i} - 3\mathbf{k}$, and $\mathbf{i} - 2\mathbf{j} + \mathbf{k}$?

8. What is the volume of the parallelepiped with sides $\mathbf{i}$, $3\mathbf{j} - \mathbf{k}$, and $4\mathbf{i} + 2\mathbf{j} - \mathbf{k}$?

In Exercises 9 to 12, describe all unit vectors orthogonal to both of the given vectors.

9. $\mathbf{i}, \mathbf{j}$

10. $-5\mathbf{i} + 9\mathbf{j} - 4\mathbf{k}$, $7\mathbf{i} + 8\mathbf{j} + 9\mathbf{k}$

11. $-5\mathbf{i} + 9\mathbf{j} - 4\mathbf{k}$, $7\mathbf{i} + 8\mathbf{j} + 9\mathbf{k}$, $\mathbf{0}$

12. $2\mathbf{i} - 4\mathbf{j} + 3\mathbf{k}$, $-4\mathbf{i} + 8\mathbf{j} - 6\mathbf{k}$

13. Compute $\mathbf{u} + \mathbf{v}$, $\mathbf{u} \cdot \mathbf{v}$, $\|\mathbf{u}\|$, $\|\mathbf{v}\|$, and $\mathbf{u} \times \mathbf{v}$, where $\mathbf{u} = \mathbf{i} - 2\mathbf{j} + \mathbf{k}$, $\mathbf{v} = 2\mathbf{i} - \mathbf{j} + 2\mathbf{k}$.

14. Repeat Exercise 13 for $\mathbf{u} = 3\mathbf{i} + \mathbf{j} - \mathbf{k}$, $\mathbf{v} = -6\mathbf{i} - 2\mathbf{j} - 2\mathbf{k}$.

15. Find an equation for the plane that

(a) is perpendicular to $\mathbf{v} = (1, 1, 1)$ and passes through $(1, 0, 0)$.
(b) is perpendicular to $\mathbf{v} = (1, 2, 3)$ and passes through $(1, 1, 1)$.
(c) is perpendicular to the line $\mathbf{l}(t) = (5, 0, 2)t + (3, -1, 1)$ and passes through $(5, -1, 0)$.
(d) is perpendicular to the line $\mathbf{l}(t) = (-1, -2, 3)t + (0, 7, 1)$ and passes through $(2, 4, -1)$.

16. Find an equation for the plane that passes through

(a) $(0, 0, 0)$, $(2, 0, -1)$, and $(0, 4, -3)$.
(b) $(1, 2, 0)$, $(0, 1, -2)$, and $(4, 0, 1)$.
(c) $(2, -1, 3)$, $(0, 0, 5)$, and $(5, 7, -1)$.

17. (a) Show that two parallel planes are either identical or they never intersect.
(b) How do two nonparallel planes intersect?

18. Find the intersection of the planes $x + 2y + z = 0$ and $x - 3y - z = 0$.

19. Find the intersection of the planes $x + (y - 1) + z = 0$ and $-x + (y + 1) - z = 0$.

20. Find the intersection of the two planes with equations $3(x - 1) + 2y + (z + 1) = 0$ and $(x - 1) + 4y - (z + 1) = 0$.

21. (a) Prove the two triple-vector-product identities

$$(\mathbf{a} \times \mathbf{b}) \times \mathbf{c} = (\mathbf{a} \cdot \mathbf{c})\mathbf{b} - (\mathbf{b} \cdot \mathbf{c})\mathbf{a} \quad \text{and} \quad \mathbf{a} \times (\mathbf{b} \times \mathbf{c}) = (\mathbf{a} \cdot \mathbf{c})\mathbf{b} - (\mathbf{a} \cdot \mathbf{b})\mathbf{c}.$$

(b) Prove $(\mathbf{u} \times \mathbf{v}) \times \mathbf{w} = \mathbf{u} \times (\mathbf{v} \times \mathbf{w})$ if and only if $(\mathbf{u} \times \mathbf{w}) \times \mathbf{v} = \mathbf{0}$.

(c) Also prove that $(\mathbf{u} \times \mathbf{v}) \times \mathbf{w} + (\mathbf{v} \times \mathbf{w}) \times \mathbf{u} + (\mathbf{w} \times \mathbf{u}) \times \mathbf{v} = \mathbf{0}$ (called the ***Jacobi identity***).

22. (a) Prove, without recourse to geometry, that

$$\begin{aligned}\mathbf{u} \cdot (\mathbf{v} \times \mathbf{w}) &= \mathbf{v} \cdot (\mathbf{w} \times \mathbf{u}) = \mathbf{w} \cdot (\mathbf{u} \times \mathbf{v}) = -\mathbf{u} \cdot (\mathbf{w} \times \mathbf{v}) \\ &= -\mathbf{w} \cdot (\mathbf{v} \times \mathbf{u}) = -\mathbf{v} \cdot (\mathbf{u} \times \mathbf{w}).\end{aligned}$$

(b) Use part (a) and Exercise 21(a) to prove that

$$(\mathbf{u} \times \mathbf{v}) \cdot (\mathbf{u}' \times \mathbf{v}') = (\mathbf{u} \cdot \mathbf{u}')(\mathbf{v} \cdot \mathbf{v}') - (\mathbf{u} \cdot \mathbf{v}')(\mathbf{u}' \cdot \mathbf{v}) = \begin{vmatrix} \mathbf{u} \cdot \mathbf{u}' & \mathbf{u} \cdot \mathbf{v}' \\ \mathbf{u}' \cdot \mathbf{v} & \mathbf{v} \cdot \mathbf{v}' \end{vmatrix}.$$

23. Verify Cramer's rule.

24. Find an equation for the plane that passes through the point $(2, -1, 3)$ and is perpendicular to the line $\mathbf{v} = (1, -2, 2) + t(3, -2, 4)$.

25. Find an equation for the plane that passes through the point $(1, 2, -3)$ and is perpendicular to the line $\mathbf{v} = (0, -2, 1) + t(1, -2, 3)$.

26. Find the equation of the line that passes through the point $(1, -2, -3)$ and is perpendicular to the plane $3x - y - 2z + 4 = 0$.

27. Find an equation for the plane containing the two (parallel) lines

$$\mathbf{v}_1 = (0, 1, -2) + t(2, 3, -1) \qquad \text{and} \qquad \mathbf{v}_2 = (2, -1, 0) + t(2, 3, -1).$$

28. Find the distance from the point $(2, 1, -1)$ to the plane $x - 2y + 2z + 5 = 0$.

29. Find an equation for the plane that contains the line $\mathbf{v} = (-1, 1, 2) + t(3, 2, 4)$ and is perpendicular to the plane $2x + y - 3z + 4 = 0$.

30. Find an equation for the plane that passes through $(3, 2, -1)$ and $(1, -1, 2)$ and that is parallel to the line $\mathbf{v} = (1, -1, 0) + t(3, 2, -2)$.

31. Redo Exercises 19 and 20 of Section 1.1 using the dot product and what you know about normals to planes.

32. Given vectors $\mathbf{a}$ and $\mathbf{b}$, do the equations $\mathbf{x} \times \mathbf{a} = \mathbf{b}$ and $\mathbf{x} \cdot \mathbf{a} = \|\mathbf{a}\|$ determine a unique vector $\mathbf{x}$? Argue both geometrically and analytically.

33. Determine the distance from the plane $12x + 13y + 5z + 2 = 0$ to the point $(1, 1, -5)$.

34. Find the distance to the point $(6, 1, 0)$ from the plane through the origin that is perpendicular to $\mathbf{i} - 2\mathbf{j} + \mathbf{k}$.

35. (a) In mechanics, the ***moment*** ***M*** *of a force* **F** *about a point O* is defined to be the magnitude of **F** times the perpendicular distance d from O to the line of action of **F**. The ***vector moment*** **M** is the vector of magnitude M whose direction is perpendicular to the plane of O and $\mathbf{F}$, determined by the right-hand rule. Show that $\mathbf{M} = \mathbf{R} \times \mathbf{F}$, where $\mathbf{R}$ is any vector from O to the line of action of $\mathbf{F}$. (See Figure 1.3.10.)

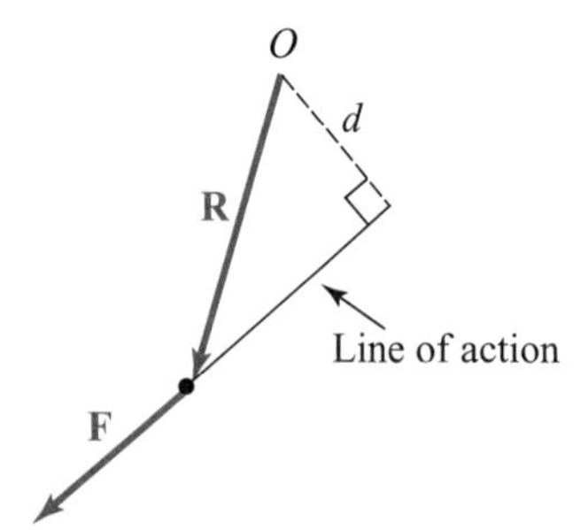

Figure 1.3.10 Moment of a force.

(b) Find the moment of the force vector $\mathbf{F} = \mathbf{i} - \mathbf{j} + 2\mathbf{k}$ newtons about the origin if the line of action is $x = 1 + t$, $y = 1 - t$, $z = 2t$.

36. Show that the plane that passes through the three points $A = (a_1, a_2, a_3)$, $B = (b_1, b_2, b_3)$, and $C = (c_1, c_2, c_3)$ consists of the points $\mathrm{P} = (x, y, z)$ given by

$$\begin{vmatrix} a_1 - x & a_2 - y & a_3 - z \\ b_1 - x & b_2 - y & b_3 - z \\ c_1 - x & c_2 - y & c_3 - z \end{vmatrix} = 0.$$

(HINT: Write the determinant as a triple product.)

37. Two media with indices of refraction n_1 and n_2 are separated by a plane surface perpendicular to the unit vector $\mathbf{N}$. Let $\mathbf{a}$ and $\mathbf{b}$ be unit vectors along the incident and refracted rays, respectively, their directions being those of the light rays. Show that $n_1(\mathbf{N} \times \mathbf{a}) = n_2(\mathbf{N} \times \mathbf{b})$ by using *Snell's law*, $\sin\theta_1 / \sin\theta_2 = n_2/n_1$, where θ_1 and θ_2 are the angles of incidence and refraction, respectively. (See Figure 1.3.11.)

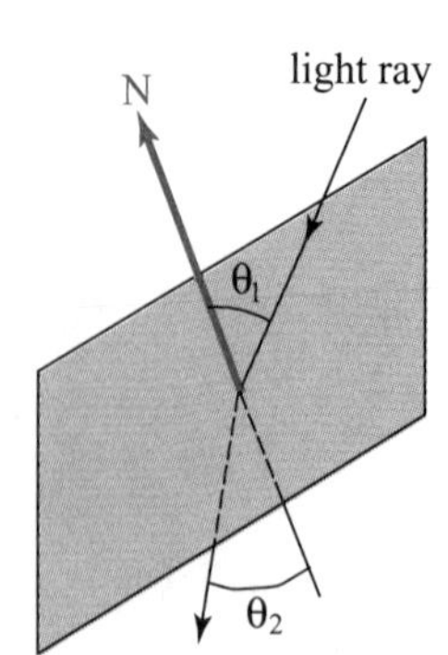

Figure 1.3.11 Snell's law.

38. Justify the steps in the following computation:

$$\begin{vmatrix} 1 & 2 & 3 \\ 4 & 5 & 6 \\ 7 & 8 & 10 \end{vmatrix} = \begin{vmatrix} 1 & 2 & 3 \\ 0 & -3 & -6 \\ 7 & 8 & 10 \end{vmatrix} = \begin{vmatrix} 1 & 2 & 3 \\ 0 & -3 & -6 \\ 0 & -6 & -11 \end{vmatrix} = \begin{vmatrix} -3 & -6 \\ -6 & -11 \end{vmatrix} = 33 - 36 = -3.$$

39. Show that adding a multiple of the first row of a matrix to the second row leaves the determinant unchanged; that is,

$$\begin{vmatrix} a_1 & b_1 & c_1 \\ a_2 + \lambda a_1 & b_2 + \lambda b_1 & c_2 + \lambda c_1 \\ a_3 & b_3 & c_3 \end{vmatrix} = \begin{vmatrix} a_1 & b_1 & c_1 \\ a_2 & b_2 & c_2 \\ a_3 & b_3 & c_3 \end{vmatrix}.$$

[In fact, adding a multiple of any row (column) of a matrix to another row (column) leaves the determinant unchanged.]

1.4 Cylindrical and Spherical Coordinates

A standard way to represent a point in the plane $\mathbb{R}^2$ is by means of rectangular coordinates (x, y). However, as the reader has probably learned in elementary calculus, polar coordinates in the plane can be extremely useful. As portrayed in Figure 1.4.1, the coordinates (r, θ) are related to (x, y) by the formulas

$$x = r\cos\theta \qquad \text{and} \qquad y = r\sin\theta,$$

where we usually take $r \geq 0$ and $0 \leq \theta < 2\pi$.

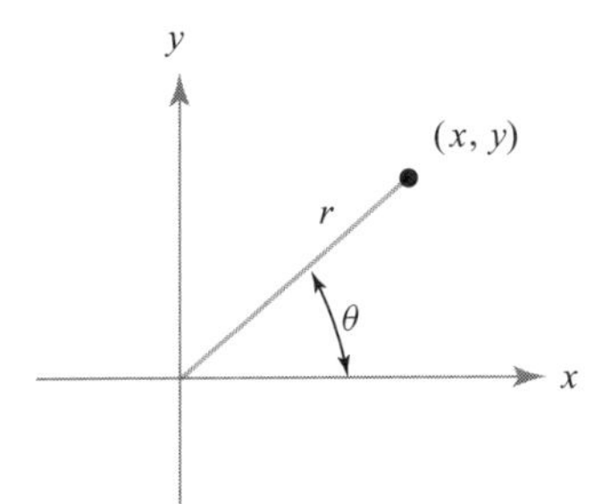

Figure 1.4.1 The polar coordinates of (x, y) are (r, θ).

Readers not familiar with polar coordinates are advised to study the relevant section of their calculus texts. We now set forth two ways of representing points in space other than by using rectangular Cartesian coordinates (x, y, z). These alternative coordinate systems are particularly well suited for certain types of problems, such as the evaluation of integrals using a change of variables.

Historical Note

In 1671, Isaac Newton wrote a manuscript entitled *The Method of Fuxions and Infinite Series*, which contains many uses of coordinate geometry to sketch the solutions of equations. In particular, he introduces the polar coordinate system, among various other coordinate systems.

In 1691, Jacob Bernoulli published a paper also containing polar coordinates. Because Newton's manuscript was not published until after his death in 1727, credit for the discovery of polar coordinates is usually attributed to Bernoulli.

Cylindrical Coordinates

DEFINITION The ***cylindrical coordinates*** (r, θ, z) of a point (x, y, z) are defined by (see Figure 1.4.2)

$$x = r\cos\theta, \qquad y = r\sin\theta, \qquad z = z. \tag{1}$$

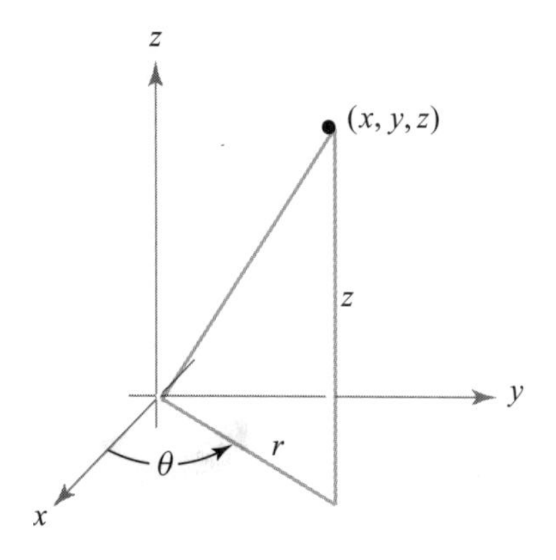

Figure 1.4.2 Representing a point (x, y, z) in terms of its cylindrical coordinates r, θ, and z.

To express r, θ, and z in terms of x, y, and z, and to ensure that θ lies between 0 and 2π, we can write

$$r = \sqrt{x^2 + y^2}, \qquad \theta = \begin{cases} \tan^{-1}(y/x) & \text{if } x > 0 \text{ and } y \geq 0 \\ \pi + \tan^{-1}(y/x) & \text{if } x < 0 \\ 2\pi + \tan^{-1}(y/x) & \text{if } x > 0 \text{ and } y < 0, \end{cases} \qquad z = z,$$

where $\tan^{-1}(y/x)$ is taken to lie between $-\pi/2$ and $\pi/2$. The requirement that $0 \leq \theta < 2\pi$ uniquely determines θ and $r \geq 0$ for a given x and y. If $x = 0$, then $\theta = \pi/2$ for $y > 0$ and $3\pi/2$ for $y < 0$. If $x = y = 0$, θ is undefined.

In other words, for any point (x, y, z), we represent the first and second coordinates in terms of polar coordinates and leave the third coordinate unchanged. Formula (1) shows that, given (r, θ, z), the triple (x, y, z) is completely determined, and vice versa, if we restrict θ to the interval $[0, 2\pi)$ (sometimes the range $(-\pi, \pi]$ is convenient) and require that $r > 0$.

To see why we use the term *cylindrical coordinates*, note that if the conditions $0 \leq \theta < 2\pi$, $-\infty < z < \infty$ hold and if $r = a$ is some positive constant, then the locus of these points is a cylinder of radius a (see Figure 1.4.3).

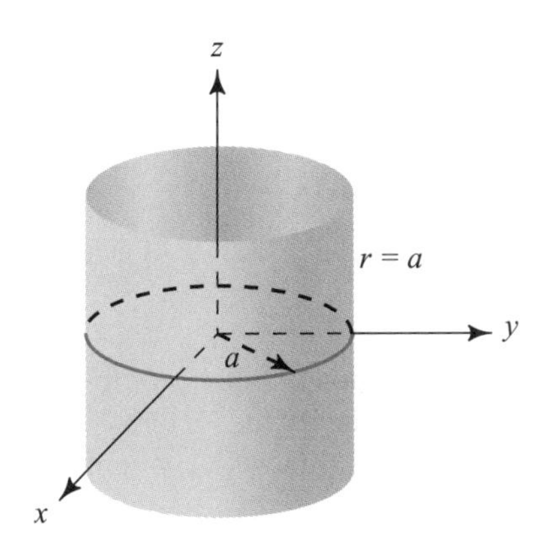

Figure 1.4.3 The graph of the points whose cylindrical coordinates satisfy $r = a$ is a cylinder.

EXAMPLE 1 (a) Find and plot the cylindrical coordinates of (6, 6, 8). (b) If a point has cylindrical coordinates $(8, 2\pi/3, -3)$, what are its Cartesian coordinates? Plot.

SOLUTION For part (a), we have $r = \sqrt{6^2 + 6^2} = 6\sqrt{2}$ and $\theta = \tan^{-1}(6/6) = \tan^{-1}(1) = \pi/4$. Thus, the cylindrical coordinates are $(6\sqrt{2}, \pi/4, 8)$. This is point P in Figure 1.4.4. For part (b), note that $2\pi/3 = \pi/2 + \pi/6$ and compute

$$x = r\cos\theta = 8\cos\frac{2\pi}{3} = -\frac{8}{2} = -4$$

and

$$y = r\sin\theta = 8\sin\frac{2\pi}{3} = 8\frac{\sqrt{3}}{2} = 4\sqrt{3}.$$

Thus, the Cartesian coordinates are $(-4, 4\sqrt{3}, -3)$. This is point Q in the figure. ▲

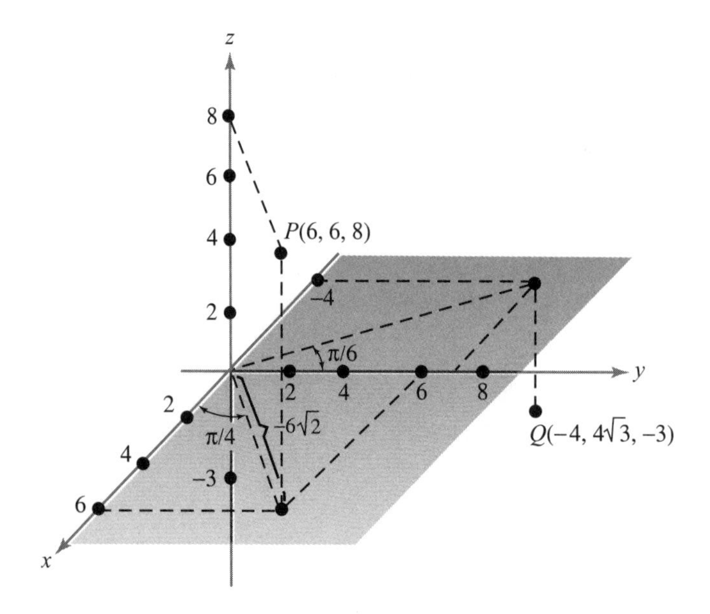

Figure 1.4.4 Some examples of the conversion between Cartesian and cylindrical coordinates.

Spherical Coordinates

Cylindrical coordinates are not the only possible generalization of polar coordinates to three dimensions. Recall that in two dimensions the magnitude of the vector $x\mathbf{i} + y\mathbf{j}$ (that is, $\sqrt{x^2 + y^2}$) is the r in the polar coordinate system. For cylindrical coordinates, the length of the vector $x\mathbf{i} + y\mathbf{j} + z\mathbf{k}$, namely,

$$\rho = \sqrt{x^2 + y^2 + z^2},$$

is not one of the coordinates of that system—instead, we used the magnitude $r = \sqrt{x^2 + y^2}$, the angle θ, and the "height" z.

We now modify this by introducing the *spherical coordinate* system, which *does use* ρ as a coordinate. Spherical coordinates are often useful for problems that possess spherical symmetry (symmetry about a point), whereas cylindrical coordinates can be applied when cylindrical symmetry (symmetry about a line) is involved.

Given a point $(x, y, z) \in \mathbb{R}^3$, let

$$\rho = \sqrt{x^2 + y^2 + z^2}$$

and represent x and y by polar coordinates in the xy plane:

$$x = r\cos\theta, \qquad y = r\sin\theta \tag{2}$$

where $r = \sqrt{x^2 + y^2}$ and θ is determined by formula (1) [see the expression for θ following formula (1)]. The coordinate z is given by

$$z = \rho\cos\phi,$$

where ϕ is the angle (chosen to lie between 0 and π, inclusive) that the radius vector $\mathbf{v} = x\mathbf{i} + y\mathbf{j} + z\mathbf{k}$ makes with the positive z axis, in the plane containing the vector $\mathbf{v}$ and the z axis (see Figure 1.4.5). Using the dot product, we can express ϕ as follows:

$$\cos\phi = \frac{\mathbf{v}\cdot\mathbf{k}}{\|\mathbf{v}\|}, \qquad \text{that is,} \qquad \phi = \cos^{-1}\left(\frac{\mathbf{v}\cdot\mathbf{k}}{\|\mathbf{v}\|}\right).$$

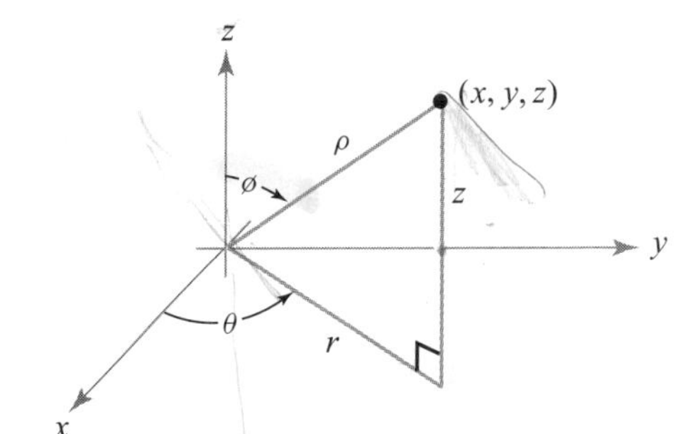

Figure 1.4.5 Spherical coordinates (ρ, θ, ϕ); the graph of points satisfying $\rho = a$ is a sphere.

We take as our coordinates the quantities ρ, θ, ϕ. Because

$$r = \rho\sin\phi,$$

we can use formula (2) to find x, y, z in terms of the spherical coordinates ρ, θ, ϕ.

DEFINITION The ***spherical coordinates*** of (x, y, z) is the triple (ρ, θ, ϕ), defined as follows:

$$x = \rho\sin\phi\cos\theta, \qquad y = \rho\sin\phi\sin\theta, \qquad z = \rho\cos\phi \tag{3}$$

where

$$\rho \geq 0, \qquad 0 \leq \theta < 2\pi, \qquad 0 \leq \phi \leq \pi.$$

Historical Note

In 1773, Joseph Louis Lagrange was working on Newton's gravitational theory as it applied to ellipsoids of revolution. In attempting to calculate the total gravitational attraction of such an ellipsoid, he encountered an integral that was difficult to evaluate. Motivated by this application, he introduced spherical coordinates, which allowed him to calculate the integral. We will be discussing the method of changing coordinates as it applies to multiple integrals in Section 6.2, and applications to gravitation in Section 6.3, where we show how the inverse square law of gravity allowed Newton to consider spherical masses as point masses.

Spherical coordinates are also closely connected to navigation by latitude and longitude. To see the connection, first note that the sphere of radius a centered at the origin is described by a very simple equation in spherical coordinates, namely, $\rho = a$. Fixing the radius a, the spherical coordinates θ and ϕ are similar to the geographic coordinates of longitude and latitude if we take the earth's axis to be the z axis. There are differences, though: The geographical longitude is $|\theta|$ and is called east or west longitude, according to whether θ is a positive or negative measure from the Greenwich meridian; the geographical latitude is $|\pi/2 - \phi|$ and is called north or south latitude, according to whether $\pi/2 - \phi$ is positive or negative.

EXAMPLE 2

(a) Find the spherical coordinates of the Cartesian point $(1, -1, 1)$ and plot.

(b) Find the Cartesian coordinates of the spherical coordinate point $(3, \pi/6, \pi/4)$ and plot.

(c) Let a point have Cartesian coordinates $(2, -3, 6)$. Find its spherical coordinates and plot.

(d) Let a point have spherical coordinates $(1, -\pi/2, \pi/4)$. Find its Cartesian coordinates and plot.

SOLUTION

(a) $\rho = \sqrt{x^2 + y^2 + z^2} = \sqrt{1^2 + (-1)^2 + 1^2} = \sqrt{3},$

$$\theta = 2\pi + \tan^{-1}\left(\frac{y}{x}\right) = 2\pi + \tan^{-1}\left(\frac{-1}{1}\right) = 2\pi - \frac{\pi}{4} = \frac{7\pi}{4}$$

$$\phi = \cos^{-1}\left(\frac{z}{\rho}\right) = \cos^{-1}\left(\frac{1}{\sqrt{3}}\right) \approx 0.955 \approx 54.74^\circ.$$

See Figure 1.4.6(a) and the formula for θ following formula (1).

(b) $x = \rho \sin\phi \cos\theta = 3 \sin\left(\frac{\pi}{4}\right) \cos\left(\frac{\pi}{6}\right) = 3\left(\frac{1}{\sqrt{2}}\right)\frac{\sqrt{3}}{2} = \frac{3\sqrt{3}}{2\sqrt{2}},$

$$y = \rho \sin\phi \sin\theta = 3 \sin\left(\frac{\pi}{4}\right) \sin\left(\frac{\pi}{6}\right) = 3\left(\frac{1}{\sqrt{2}}\right)\left(\frac{1}{2}\right) = \frac{3}{2\sqrt{2}},$$

$$z = \rho \cos\phi = 3 \cos\left(\frac{\pi}{4}\right) = \frac{3}{\sqrt{2}} = \frac{3\sqrt{2}}{2}.$$

See Figure 1.4.6(b).

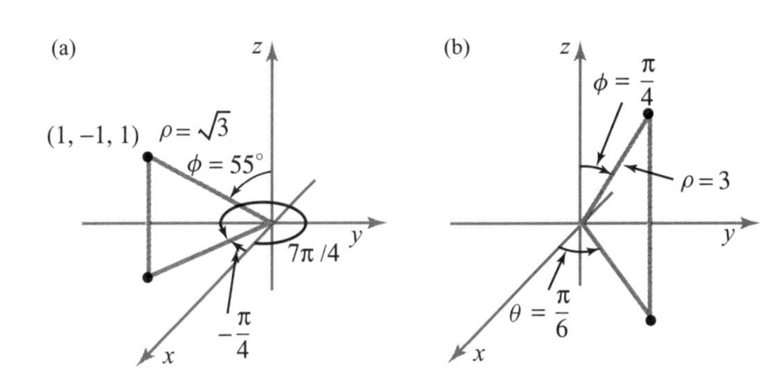

Figure 1.4.6 Finding (a) the spherical coordinates of the point (1, –1, 1), and (b) the Cartesian coordinates of $(3, \pi/6, \pi/4)$.

(c) $$\rho = \sqrt{x^2 + y^2 + z^2} = \sqrt{2^2 + (-3)^2 + 6^2} = \sqrt{49} = 7,$$

$$\theta = 2\pi + \tan^{-1}\left(\frac{y}{x}\right) = 2\pi + \tan^{-1}\left(\frac{-3}{2}\right) \approx 5.3004 \text{ radians} \approx 303.69^\circ,$$

$$\phi = \cos^{-1}\left(\frac{z}{\rho}\right) = \cos^{-1}\left(\frac{6}{7}\right) \approx 0.541 \approx 31.0^\circ.$$

See Figure 1.4.7(a).

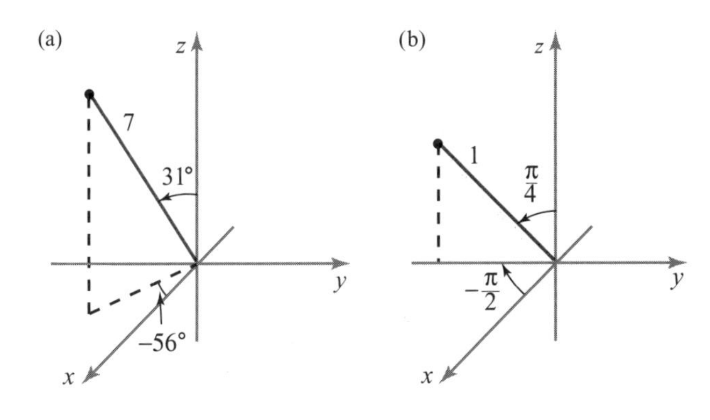

Figure 1.4.7 Finding (a) the spherical coordinates of the point (2, –3, 6), and (b) the Cartesian coordinates of $(1, -\pi/2, \pi/4)$.

(d) $$x = \rho \sin\phi \cos\theta = 1 \sin\left(\frac{\pi}{4}\right) \cos\left(-\frac{\pi}{2}\right) = \left(\frac{\sqrt{2}}{2}\right) \cdot 0 = 0,$$

$$y = \rho \sin\phi \sin\theta = 1 \sin\left(\frac{\pi}{4}\right) \sin\left(-\frac{\pi}{2}\right) = \left(\frac{\sqrt{2}}{2}\right)(-1) = -\frac{\sqrt{2}}{2},$$

$$z = \rho \cos\phi = 1 \cos\left(\frac{\pi}{4}\right) = \frac{\sqrt{2}}{2}.$$

See Figure 1.4.7(b) ▲

EXAMPLE 3 Express (a) the surface $xz = 1$ and (b) the surface $x^2 + y^2 - z^2 = 1$ in spherical coordinates.

SOLUTION From formula (3), $x = \rho \sin\phi \cos\theta$, and $z = \rho \cos\phi$, and so the surface $xz = 1$ in (a) consists of all (ρ, θ, ϕ) such that

$$\rho^2 \sin\phi \cos\theta \cos\phi = 1, \qquad \text{that is,} \qquad \rho^2 \sin 2\phi \cos\theta = 2.$$

For part (b), we can write

$$x^2 + y^2 - z^2 = x^2 + y^2 + z^2 - 2z^2 = \rho^2 - 2\rho^2 \cos^2\phi,$$

so that the surface is $\rho^2(1 - 2\cos^2\phi) = 1$; that is, $-\rho^2 \cos(2\phi) = 1$. ▲

Associated with cylindrical and spherical coordinates are unit vectors that are the counterparts of **i**, **j**, and **k** for rectangular coordinates. They are shown in Figure 1.4.8. For example, $\mathbf{e}_r$ is the unit vector parallel to the xy plane and in the radial direction, so that $\mathbf{e}_r = (\cos\theta)\mathbf{i} + (\sin\theta)\mathbf{j}$. Similarly, in spherical coordinates, $\mathbf{e}_\phi$ is the unit vector tangent to the curve parametrized by the variable ϕ with the variables ρ and θ held fixed. We shall use these unit vectors later when we use cylindrical and spherical coordinates in vector calculations.

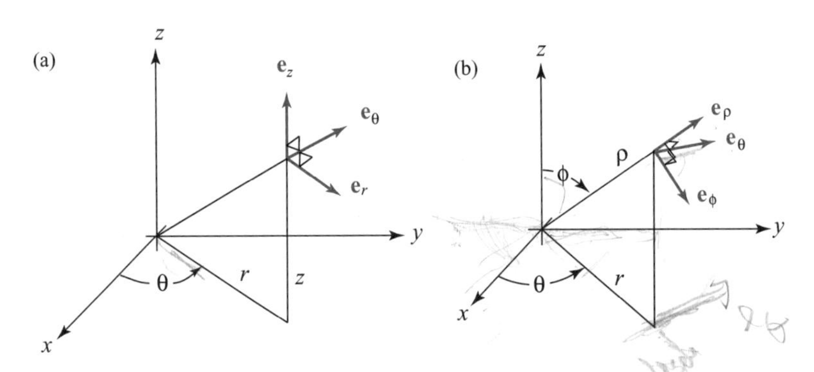

Figure 1.4.8 (a) Orthonormal vectors $\mathbf{e}_r$, $\mathbf{e}_\theta$, and $\mathbf{e}_z$ associated with cylindrical coordinates. The vector $\mathbf{e}_r$ is parallel to the line labeled r. (b) Orthonormal vectors $\mathbf{e}_\rho$, $\mathbf{e}_\theta$, and $\mathbf{e}_\phi$ associated with spherical coordinates.

EXERCISES

1. (a) The following points are given in cylindrical coordinates; express each in rectangular coordinates and spherical coordinates: $(1, 45^\circ, 1)$, $(2, \pi/2, -4)$, $(0, 45^\circ, 10)$, $(3, \pi/6, 4)$, $(1, \pi/6, 0)$, and $(2, 3\pi/4, -2)$. (Only the first point is solved in the Study Guide.)

(b) Change each of the following points from rectangular coordinates to spherical coordinates and to cylindrical coordinates: $(2, 1, -2)$, $(0, 3, 4)$, $(\sqrt{2}, 1, 1)$, $(-2\sqrt{3}, -2, 3)$. (Only the first point is solved in the Study Guide.)

2. Describe the geometric meaning of the following mappings in cylindrical coordinates:

(a) $(r, \theta, z) \mapsto (r, \theta, -z)$
(b) $(r, \theta, z) \mapsto (r, \theta + \pi, -z)$
(c) $(r, \theta, z) \mapsto (-r, \theta - \pi/4, z)$

3. Describe the geometric meaning of the following mappings in spherical coordinates:

(a) $(\rho, \theta, \phi) \mapsto (\rho, \theta + \pi, \phi)$
(b) $(\rho, \theta, \phi) \mapsto (\rho, \theta, \pi - \phi)$
(c) $(\rho, \theta, \phi) \mapsto (2\rho, \theta + \pi/2, \phi)$

4. (a) Describe the surfaces $r =$ constant, $\theta =$ constant, and $z =$ constant in the cylindrical coordinate system.

(b) Describe the surfaces $\rho =$ constant, $\theta =$ constant, and $\phi =$ constant in the spherical coordinate system.

5. Show that to represent each point in $\mathbb{R}^3$ by spherical coordinates it is necessary to take only values of θ between 0 and 2π, values of ϕ between 0 and π, and values of $\rho \geq 0$. Are coordinates unique if we allow $\rho \leq 0$?

6. Using cylindrical coordinates and the orthonormal (orthogonal normalized) vectors $\mathbf{e}_r$, $\mathbf{e}_\theta$, and $\mathbf{e}_z$ (see Figure 1.4.8),

(a) express each of $\mathbf{e}_r$, $\mathbf{e}_\theta$, and $\mathbf{e}_z$ in terms of $\mathbf{i}$, $\mathbf{j}$, $\mathbf{k}$ and (x, y, z); and
(b) calculate $\mathbf{e}_\theta \times \mathbf{j}$ both analytically, using part (a), and geometrically.

7. Using spherical coordinates and the orthonormal (orthogonal normalized) vectors $\mathbf{e}_\rho$, $\mathbf{e}_\theta$, and $\mathbf{e}_\phi$ (see Figure 1.4.8(b)),

(a) express each of $\mathbf{e}_\rho$, $\mathbf{e}_\theta$, and $\mathbf{e}_\phi$ in terms of $\mathbf{i}$, $\mathbf{j}$, $\mathbf{k}$ and (x, y, z); and
(b) calculate $\mathbf{e}_\theta \times \mathbf{j}$ and $\mathbf{e}_\phi \times \mathbf{j}$ both analytically and geometrically.

8. Express the plane $z = x$ in (a) cylindrical, and (b) spherical coordinates.

9. Show that in spherical coordinates:

(a) ρ is the length of $x\mathbf{i} + y\mathbf{j} + z\mathbf{k}$.
(b) $\phi = \cos^{-1}(\mathbf{v} \cdot \mathbf{k}/\|\mathbf{v}\|)$, where $\mathbf{v} = x\mathbf{i} + y\mathbf{j} + z\mathbf{k}$.
(c) $\theta = \cos^{-1}(\mathbf{u} \cdot \mathbf{i}/\|\mathbf{u}\|)$, where $\mathbf{u} = x\mathbf{i} + y\mathbf{j}$.

10. Two surfaces are described in spherical coordinates by the two equations $\rho = f(\theta, \phi)$ and $\rho = -2f(\theta, \phi)$, where $f(\theta, \phi)$ is a function of two variables. How is the second surface obtained geometrically from the first?

11. A circular membrane in space lies over the region $x^2 + y^2 \leq a^2$. The maximum z component of points in the membrane is b. Assume that (x, y, z) is a point on the membrane. Show that the corresponding point (r, θ, z) in cylindrical coordinates satisfies the conditions $0 \leq r \leq a$, $0 \leq \theta \leq 2\pi$, $|z| \leq b$.

12. A tank in the shape of a right-circular cylinder of radius 10 ft and height 16 ft is half filled and lying on its side. Describe the air space inside the tank by suitably chosen cylindrical coordinates.

13. A vibrometer is to be designed that withstands the heating effects of its spherical enclosure of diameter d, which is buried to a depth $d/3$ in the earth, the upper portion being heated by the sun (assume the surface is flat). Heat conduction analysis requires a description of the buried portion of the enclosure in spherical coordinates. Find it.

14. An oil filter cartridge is a porous right-circular cylinder inside which oil diffuses from the axis to the outer curved surface. Describe the cartridge in cylindrical coordinates, if the diameter of the filter is 4.5 inches, the height is 5.6 inches, and the center of the cartridge is drilled (all the way through) from the top to admit a $\frac{5}{8}$-inch-diameter bolt.

15. Describe the surface given in spherical coordinates by $\rho = \cos 2\theta$.

1.5 n-Dimensional Euclidean Space

Vectors in n-space

In Sections 1.1 and 1.2 we studied the spaces $\mathbb{R} = \mathbb{R}^1$, $\mathbb{R}^2$, and $\mathbb{R}^3$ and gave geometric interpretations to them. For example, a point (x, y, z) in $\mathbb{R}^3$ can be thought of as a geometric object, namely, the directed line segment or vector emanating from the origin and ending at the point (x, y, z). We can therefore think of $\mathbb{R}^3$ in either of two ways:

(i) Algebraically, as a set of triples (x, y, z) where x, y, and z are real numbers

(ii) Geometrically, as a set of directed line segments

These two ways of looking at $\mathbb{R}^3$ are equivalent. For generalization it is easier to use definition (i). Specifically, we can define $\mathbb{R}^n$, where n is a positive integer (possibly greater than 3), to be the set of all ordered n-tuples $(x_1, x_2, \dots, x_n)$, where the x_i are real numbers. For instance, $(1, \sqrt{5}, 2, \sqrt{3}) \in \mathbb{R}^4$.

The set $\mathbb{R}^n$ so defined is known as ***Euclidean n-space***, and its elements, which we write as $\mathbf{x} = (x_1, x_2, \dots, x_n)$, are known as ***vectors*** or ***n-vectors***. By setting $n = 1, 2$, or 3, we recover the line, the plane, and three-dimensional space, respectively.

We launch our study of Euclidean n-space by introducing several algebraic operations. These are analogous to those introduced in Section 1.1 for $\mathbb{R}^2$ and $\mathbb{R}^3$. The first two, addition and scalar multiplication, are defined as follows:

(i) $(x_1, x_2, \dots, x_n) + (y_1, y_2, \dots, y_n) = (x_1 + y_1, x_2 + y_2, \dots, x_n + y_n)$;

and

(ii) for any real number α,

$$\alpha(x_1, x_2, \dots, x_n) = (\alpha x_1, \alpha x_2, \dots, \alpha x_n).$$

The geometric significance of these operations for $\mathbb{R}^2$ and $\mathbb{R}^3$ was discussed in Section 1.1.

The n vectors

$$\mathbf{e}_1 = (1, 0, 0, \dots, 0), \mathbf{e}_2 = (0, 1, 0, \dots, 0), \dots, \mathbf{e}_n = (0, 0, \dots, 0, 1)$$

are called the ***standard basis vectors*** of $\mathbb{R}^n$, and they generalize the three mutually orthogonal unit vectors $\mathbf{i}, \mathbf{j}, \mathbf{k}$ of $\mathbb{R}^3$. The vector $\mathbf{x} = (x_1, x_2, \dots, x_n)$ can then be written as $\mathbf{x} = x_1\mathbf{e}_1 + x_2\mathbf{e}_2 + \cdots + x_n\mathbf{e}_n$.

For two vectors $\mathbf{x} = (x_1, x_2, x_3)$ and $\mathbf{y} = (y_1, y_2, y_3)$ in $\mathbb{R}^3$, we defined the *dot* or *inner product* $\mathbf{x}\cdot\mathbf{y}$ to be the real number $\mathbf{x}\cdot\mathbf{y} = x_1y_1 + x_2y_2 + x_3y_3$. This definition easily extends to $\mathbb{R}^n$; specifically, for $\mathbf{x} = (x_1, x_2, \dots, x_n)$, $\mathbf{y} = (y_1, y_2, \dots, y_n)$, we define the ***inner product*** of $\mathbf{x}$ and $\mathbf{y}$ to be $\mathbf{x}\cdot\mathbf{y} = x_1y_1 + x_2y_2 + \cdots + x_ny_n$. In $\mathbb{R}^n$, the notation $\langle\mathbf{x}, \mathbf{y}\rangle$ is often used in place of $\mathbf{x}\cdot\mathbf{y}$ for the inner product.

Continuing the analogy with $\mathbb{R}^3$, we are led to define the notion of the ***length*** or ***norm*** of a vector $\mathbf{x}$ by the formula

$$\text{Length of } \mathbf{x} = \|\mathbf{x}\| = \sqrt{\mathbf{x}\cdot\mathbf{x}} = \sqrt{x_1^2 + x_2^2 + \cdots + x_n^2}.$$

If $\mathbf{x}$ and $\mathbf{y}$ are two vectors in the plane ($\mathbb{R}^2$) or in space ($\mathbb{R}^3$), then we know that the angle θ between them is given by the formula

$$\cos\theta = \frac{\mathbf{x}\cdot\mathbf{y}}{\|\mathbf{x}\|\|\mathbf{y}\|}.$$

The right side of this equation can be defined in $\mathbb{R}^n$ as well as in $\mathbb{R}^2$ or $\mathbb{R}^3$. It still represents the cosine of the angle between $\mathbf{x}$ and $\mathbf{y}$; this angle is geometrically well defined, because $\mathbf{x}$ and $\mathbf{y}$ lie in a two-dimensional subspace of $\mathbb{R}^n$ (the plane determined by $\mathbf{x}$ and $\mathbf{y}$) and our usual geometry ideas apply to such planes.

It will be useful to have available some algebraic properties of the inner product. These are summarized in the next theorem [compare with properties (i), (ii), (iii), and (iv) of Section 1.2].

THEOREM 3 For $\mathbf{x}, \mathbf{y}, \mathbf{z} \in \mathbb{R}^n$ and α, β, real numbers, we have

(i) $(\alpha\mathbf{x} + \beta\mathbf{y})\cdot\mathbf{z} = \alpha(\mathbf{x}\cdot\mathbf{z}) + \beta(\mathbf{y}\cdot\mathbf{z})$.

(ii) $\mathbf{x}\cdot\mathbf{y} = \mathbf{y}\cdot\mathbf{x}$.

(iii) $\mathbf{x}\cdot\mathbf{x} \geq 0$.

(iv) $\mathbf{x}\cdot\mathbf{x} = 0$ if and only if $\mathbf{x} = \mathbf{0}$.

PROOF Each of the four assertions can be proved by a simple computation. For example, to prove property (i) we write

$$\begin{aligned}(\alpha\mathbf{x} + \beta\mathbf{y})\cdot\mathbf{z} &= (\alpha x_1 + \beta y_1, \alpha x_2 + \beta y_2, \dots, \alpha x_n + \beta y_n)\cdot(z_1, z_2, \dots, z_n)\\ &= (\alpha x_1 + \beta y_1)z_1 + (\alpha x_2 + \beta y_2)z_2 + \cdots + (\alpha x_n + \beta y_n)z_n\\ &= a x_1 z_1 + \beta y_1 z_1 + \alpha x_2 z_2 + \beta y_2 z_2 + \cdots + \alpha x_n z_n + \beta y_n z_n\\ &= \alpha(\mathbf{x}\cdot\mathbf{z}) + \beta(\mathbf{y}\cdot\mathbf{z}).\end{aligned}$$

The other proofs are similar. ■

In Section 1.2, we proved an interesting property of dot products, called the Cauchy–Schwarz inequality.[4] For $\mathbb{R}^2$ our proof required the use of the law of cosines. For $\mathbb{R}^n$ we could also use this method, by confining our attention to a plane in $\mathbb{R}^n$. However, we can also give a direct, completely algebraic proof.

THEOREM 4: Cauchy–Schwarz Inequality in $\mathbb{R}^n$ Let $\mathbf{x}, \mathbf{y}$ be vectors in $\mathbb{R}^n$. Then

$$|\mathbf{x} \cdot \mathbf{y}| \leq \|\mathbf{x}\|\|\mathbf{y}\|.$$

PROOF Let $a = \mathbf{y} \cdot \mathbf{y}$ and $b = -\mathbf{x} \cdot \mathbf{y}$. If $a = 0$, the theorem is clearly valid, because then $\mathbf{y} = \mathbf{0}$ and both sides of the inequality reduce to 0. Thus, we may suppose $a \neq 0$. By Theorem 3 we have

$$\begin{aligned} 0 \leq (a\mathbf{x} + b\mathbf{y}) \cdot (a\mathbf{x} + b\mathbf{y}) &= a^2\mathbf{x} \cdot \mathbf{x} + 2ab\mathbf{x} \cdot \mathbf{y} + b^2\mathbf{y} \cdot \mathbf{y} \\ &= (\mathbf{y} \cdot \mathbf{y})^2\mathbf{x} \cdot \mathbf{x} - (\mathbf{y} \cdot \mathbf{y})(\mathbf{x} \cdot \mathbf{y})^2. \end{aligned}$$

Dividing by $\mathbf{y} \cdot \mathbf{y}$ gives $0 \leq (\mathbf{y} \cdot \mathbf{y})(\mathbf{x} \cdot \mathbf{x}) - (\mathbf{x} \cdot \mathbf{y})^2$, that is, $(\mathbf{x} \cdot \mathbf{y})^2 \leq (\mathbf{x} \cdot \mathbf{x})(\mathbf{y} \cdot \mathbf{y}) = \|\mathbf{x}\|^2\|\mathbf{y}\|^2$. Taking square roots on both sides of this inequality yields the desired result. ■

There is a useful consequence of the Cauchy–Schwarz inequality in terms of lengths. The triangle inequality is geometrically clear in $\mathbb{R}^3$ and was discussed in Section 1.2. The *analytic* proof of the triangle inequality that we gave in Section 1.2 works exactly the same in $\mathbb{R}^n$ and proves the following:

COROLLARY: Triangle Inequality in $\mathbb{R}^n$ Let $\mathbf{x}, \mathbf{y}$ be vectors in $\mathbb{R}^n$. Then

$$\|\mathbf{x} + \mathbf{y}\| \leq \|\mathbf{x}\| + \|\mathbf{y}\|.$$

If Theorem 4 and its corollary are written out algebraically, they become the following useful inequalities:

$$\left|\sum_{i=1}^{n} x_i y_i\right| \leq \left(\sum_{i=1}^{n} x_i^2\right)^{1/2} \left(\sum_{i=1}^{n} y_i^2\right)^{1/2};$$

$$\left(\sum_{i=1}^{n} (x_i + y_i)^2\right)^{1/2} \leq \left(\sum_{i=1}^{n} x_i^2\right)^{1/2} + \left(\sum_{i=1}^{n} y_i^2\right)^{1/2}.$$

[4] Sometimes called the Cauchy–Bunyakovskii–Schwarz inequality, or simply the CBS inequality, because it was independently discovered in special cases by the French mathematician Cauchy, the Russian mathematician Bunyakovskii, and the German mathematician Schwarz.

EXAMPLE 1 Let $\mathbf{x} = (1, 2, 0, -1)$ and $\mathbf{y} = (-1, 1, 1, 0)$. Verify Theorem 4 and its corollary in this case.

SOLUTION

$$\begin{aligned}
\|\mathbf{x}\| &= \sqrt{1^2 + 2^2 + 0^2 + (-1)^2} = \sqrt{6} \\
\|\mathbf{y}\| &= \sqrt{(-1)^2 + 1^2 + 1^2 + 0^2} = \sqrt{3} \\
\mathbf{x} \cdot \mathbf{y} &= 1(-1) + 2 \cdot 1 + 0 \cdot 1 + (-1)0 = 1 \\
\mathbf{x} + \mathbf{y} &= (0, 3, 1, -1) \\
\|\mathbf{x} + \mathbf{y}\| &= \sqrt{0^2 + 3^2 + 1^2 + (-1)^2} = \sqrt{11}.
\end{aligned}$$

We compute $\mathbf{x} \cdot \mathbf{y} = 1 \le 4.24 \approx \sqrt{6}\sqrt{3} = \|\mathbf{x}\|\,\|\mathbf{y}\|$, which verifies Theorem 4. Similarly, we can check its corollary:

$$\begin{aligned}
\|\mathbf{x} + \mathbf{y}\| &= \sqrt{11} \approx 3.32 \\
&\le 4.18 = 2.45 + 1.73 \approx \sqrt{6} + \sqrt{3} = \|\mathbf{x}\| + \|\mathbf{y}\|. \quad \blacktriangle
\end{aligned}$$

By analogy with $\mathbb{R}^3$, we can define the notion of distance in $\mathbb{R}^n$; namely, if $\mathbf{x}$ and $\mathbf{y}$ are points in $\mathbb{R}^n$, the ***distance between* x *and* y** is defined to be $\|\mathbf{x} - \mathbf{y}\|$, or the length of the vector $\mathbf{x} - \mathbf{y}$. We do not attempt to define the cross product on $\mathbb{R}^n$ except for $n = 3$.

General Matrices

Generalizing 2×2 and 3×3 matrices (see Section 1.3), we can consider $m \times n$ matrices, which are arrays of mn numbers:

$$A = \begin{bmatrix} a_{11} & a_{12} & \cdots & a_{1n} \\ a_{21} & a_{22} & \cdots & a_{2n} \\ \vdots & \vdots & & \vdots \\ a_{m1} & a_{m2} & \cdots & a_{mn} \end{bmatrix}.$$

We shall also write A as $[a_{ij}]$. We define addition and multiplication by a scalar componentwise, just as we did for vectors. Given two $m \times n$ matrices A and B, we can add them to obtain a new $m \times n$ matrix $C = A + B$, whose ijth entry c_{ij} is the sum of a_{ij} and b_{ij}. It is clear that $A + B = B + A$.

EXAMPLE 2

(a) $\begin{bmatrix} 2 & 1 & 0 \\ 3 & 4 & 1 \end{bmatrix} + \begin{bmatrix} -1 & 1 & 3 \\ 0 & 0 & 7 \end{bmatrix} = \begin{bmatrix} 1 & 2 & 3 \\ 3 & 4 & 8 \end{bmatrix}.$

(b) $[1 \quad 2] + [0 \quad -1] = [1 \quad 1]$.

(c) $\begin{bmatrix} 2 & 1 \\ 1 & 2 \end{bmatrix} - \begin{bmatrix} 1 & 0 \\ 0 & 1 \end{bmatrix} = \begin{bmatrix} 1 & 1 \\ 1 & 1 \end{bmatrix}$. ▲

Given a scalar λ and an $m \times n$ matrix A, we can multiply A by λ to obtain a new $m \times n$ matrix $\lambda A = C$, whose ijth entry c_{ij} is the product λa_{ij}.

EXAMPLE 3

$$3\begin{bmatrix} 1 & -1 & 2 \\ 0 & 1 & 5 \\ 1 & 0 & 3 \end{bmatrix} = \begin{bmatrix} 3 & -3 & 6 \\ 0 & 3 & 15 \\ 3 & 0 & 9 \end{bmatrix}. \quad ▲$$

Next we turn to matrix multiplication. If $A = [a_{ij}]$, $B = [b_{ij}]$ are $n \times n$ matrices, then the product $AB = C$ has entries given by

$$c_{ij} = \sum_{k=1}^{n} a_{ik} b_{kj},$$

which is the dot product of the ith row of A and the jth column of B:

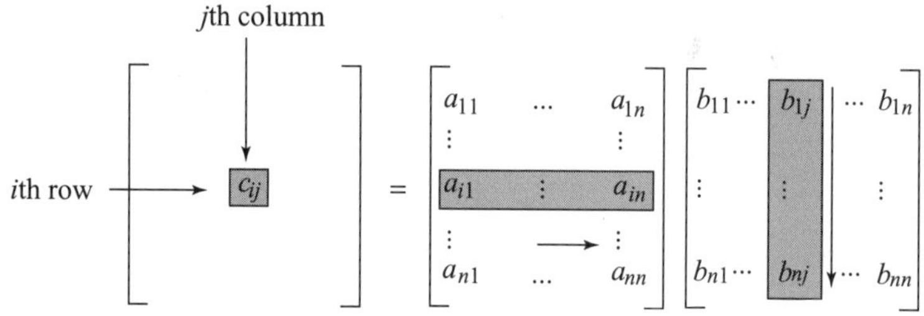

jth column

ith row

$$\begin{bmatrix} & & \\ & c_{ij} & \\ & & \end{bmatrix} = \begin{bmatrix} a_{11} & \cdots & a_{1n} \\ \vdots & & \vdots \\ a_{i1} & \cdots & a_{in} \\ \vdots & & \vdots \\ a_{n1} & \cdots & a_{nn} \end{bmatrix} \begin{bmatrix} b_{11} \cdots & b_{1j} & \cdots b_{1n} \\ \vdots & \vdots & \vdots \\ b_{n1} \cdots & b_{nj} & \cdots b_{nn} \end{bmatrix}$$

EXAMPLE 4 Let

$$A = \begin{bmatrix} 1 & 0 & 3 \\ 2 & 1 & 0 \\ 1 & 0 & 0 \end{bmatrix} \quad \text{and} \quad B = \begin{bmatrix} 0 & 1 & 0 \\ 1 & 0 & 0 \\ 0 & 1 & 1 \end{bmatrix}.$$

Then

$$AB = \begin{bmatrix} 0 & 4 & 3 \\ 1 & 2 & 0 \\ 0 & 1 & 0 \end{bmatrix} \quad \text{and} \quad BA = \begin{bmatrix} 2 & 1 & 0 \\ 1 & 0 & 3 \\ 3 & 1 & 0 \end{bmatrix}.$$

Observe that $AB \neq BA$. ▲

Similarly, we can multiply an $m \times n$ matrix (m rows, n columns) by an $n \times p$ matrix (n rows, p columns) to obtain an $m \times p$ matrix (m rows, p columns) by the same rule. Note that for AB to be defined, the number of *columns* of A must equal the number of *rows* of B.

EXAMPLE 5 Let

$$A = \begin{bmatrix} 2 & 0 & 1 \\ 1 & 1 & 2 \end{bmatrix} \quad \text{and} \quad B = \begin{bmatrix} 1 & 0 & 2 \\ 0 & 2 & 1 \\ 1 & 1 & 1 \end{bmatrix}.$$

Then

$$AB = \begin{bmatrix} 3 & 1 & 5 \\ 3 & 4 & 5 \end{bmatrix},$$

and BA is not defined. ▲

EXAMPLE 6 Let

$$A = \begin{bmatrix} 1 \\ 2 \\ 1 \\ 3 \end{bmatrix} \quad \text{and} \quad B = [2 \quad 2 \quad 1 \quad 2].$$

Then

$$AB = \begin{bmatrix} 2 & 2 & 1 & 2 \\ 4 & 4 & 2 & 4 \\ 2 & 2 & 1 & 2 \\ 6 & 6 & 3 & 6 \end{bmatrix} \quad \text{and} \quad BA = [13]. \quad ▲$$

Any $m \times n$ matrix A determines a mapping of $\mathbb{R}^n$ to $\mathbb{R}^m$ defined as follows: Let $\mathbf{x} = (x_1, \dots, x_n) \in \mathbb{R}^n$; consider the $n \times 1$ column matrix associated with $\mathbf{x}$, which we shall *temporarily* denote $\mathbf{x}^T$

$$\mathbf{x}^T = \begin{bmatrix} x_1 \\ \vdots \\ x_n \end{bmatrix}$$

and multiply A by $\mathbf{x}^T$ (considered to be an $n \times 1$ matrix) to get a new $m \times 1$ matrix:

$$A\mathbf{x}^T = \begin{bmatrix} a_{11} & \cdots & a_{1n} \\ \vdots & & \vdots \\ a_{m1} & \cdots & a_{mn} \end{bmatrix} \begin{bmatrix} x_1 \\ \vdots \\ x_n \end{bmatrix} = \begin{bmatrix} y_1 \\ \vdots \\ y_m \end{bmatrix} = \mathbf{y}^T,$$

corresponding to the vector $\mathbf{y} = (y_1, \ldots, y_m)$.[5] Thus, although it may cause some confusion, we will write $\mathbf{x} = (x_1, \ldots, x_n)$ and $\mathbf{y} = (y_1, \ldots, y_m)$ as column matrices

$$\mathbf{x} = \begin{bmatrix} x_1 \\ \vdots \\ x_n \end{bmatrix}, \qquad \mathbf{y} = \begin{bmatrix} y_1 \\ \vdots \\ y_m \end{bmatrix}$$

when dealing with matrix multiplication; that is, we will *identify* these two forms of writing vectors. Thus, we will delete the T on $\mathbf{x}^T$ and view $\mathbf{x}^T$ and $\mathbf{x}$ as the same.

Thus, $A\mathbf{x} = \mathbf{y}$ will "really" mean the following: Write $\mathbf{x}$ as a column matrix, multiply it by A, and let $\mathbf{y}$ be the vector whose components are those of the resulting column matrix. The rule $\mathbf{x} \mapsto A\mathbf{x}$ therefore defines a mapping of $\mathbb{R}^n$ to $\mathbb{R}^m$. This mapping is linear; that is, it satisfies

$$A(\mathbf{x} + \mathbf{y}) = A\mathbf{x} + A\mathbf{y}$$

$$A(\alpha\mathbf{x}) = \alpha(A\mathbf{x}), \qquad \alpha \text{ a scalar,}$$

as may be easily verified. One learns in a linear algebra course that, conversely, any linear transformation of $\mathbb{R}^n$ to $\mathbb{R}^m$ is representable in this way by an $m \times n$ matrix.

If $A = [a_{ij}]$ is an $m \times n$ matrix and $\mathbf{e}_j$ is the jth standard basis vector of $\mathbb{R}^n$, then $A\mathbf{e}_j$ is a vector in $\mathbb{R}^m$ with components the same as the jth column of A. That is, the ith component of $A\mathbf{e}_j$ is a_{ij}. In symbols, $(A\mathbf{e}_j)_i = a_{ij}$.

EXAMPLE 7 If

$$A = \begin{bmatrix} 1 & 0 & 3 \\ -1 & 0 & 1 \\ 2 & 1 & 2 \\ -1 & 2 & 1 \end{bmatrix},$$

then $\mathbf{x} \mapsto A\mathbf{x}$ of $\mathbb{R}^3$ to $\mathbb{R}^4$ is the mapping defined by

$$\begin{bmatrix} x_1 \\ x_2 \\ x_3 \end{bmatrix} \mapsto \begin{bmatrix} x_1 + 3x_3 \\ -x_1 + x_3 \\ 2x_1 + x_2 + 2x_3 \\ -x_1 + 2x_2 + x_3 \end{bmatrix}. \quad \blacktriangle$$

[5] To use a matrix A to get a mapping from vectors $\mathbf{x} = (x_1, \ldots, x_n)$ to vectors $\mathbf{y} = (y_1, \ldots, y_n)$ according to the equation $A\mathbf{x}^T = \mathbf{y}^T$, we write the vectors in the column form $\mathbf{x}^T$ instead of the row form $(x_1, \ldots, x_n)$. This sudden switch from writing $\mathbf{x}$ as a row to writing $\mathbf{x}$ as a column is necessitated by standard conventions on matrix multiplication.

EXAMPLE 8 The following illustrates what happens to a specific point when mapped by a 4×3 matrix:

$$A\mathbf{e}_2 = \begin{bmatrix} 4 & 2 & 9 \\ 3 & 5 & 4 \\ 1 & 2 & 3 \\ 0 & 1 & 2 \end{bmatrix} \begin{bmatrix} 0 \\ 1 \\ 0 \end{bmatrix} = \begin{bmatrix} 2 \\ 5 \\ 2 \\ 1 \end{bmatrix} = \text{2nd column of } A. \quad \blacktriangle$$

Properties of Matrices

Matrix multiplication is not, in general, ***commutative***: If A and B are $n \times n$ matrices, then generally

$$AB \neq BA,$$

as Examples 4, 5, and 6 show.

An $n \times n$ matrix is said to be ***invertible*** if there is an $n \times n$ matrix B such that

$$AB = BA = I_n,$$

where

$$I_n = \begin{bmatrix} 1 & 0 & 0 & \cdots & 0 \\ 0 & 1 & 0 & \cdots & 0 \\ 0 & 0 & 1 & \cdots & 0 \\ \vdots & \vdots & \vdots & & \vdots \\ 0 & 0 & 0 & \cdots & 1 \end{bmatrix}$$

is the $n \times n$ identity matrix: I_n has the property that $I_nC = CI_n = C$ for any $n \times n$ matrix C. We denote B by A^{-1} and call A^{-1} the ***inverse*** of A. The inverse, when it exists, is unique.

EXAMPLE 9 If

$$A = \begin{bmatrix} 2 & 4 & 0 \\ 0 & 2 & 1 \\ 3 & 0 & 2 \end{bmatrix}, \qquad \text{then} \qquad A^{-1} = \frac{1}{20}\begin{bmatrix} 4 & -8 & 4 \\ 3 & 4 & -2 \\ -6 & 12 & 4 \end{bmatrix},$$

because $AA^{-1} = I_3 = A^{-1}A$, as may be checked by matrix multiplication. ▲

Methods of computing inverses are learned in linear algebra; we won't require these methods in this book. If A is invertible, the equation $A\mathbf{x} = \mathbf{y}$ can be solved for the vector $\mathbf{x}$ by multiplying both sides by A^{-1} to obtain[6] $\mathbf{x} = A^{-1}\mathbf{y}$.

---

[6]In fact, Cramer's rule from Section 1.3 provides one way to invert matrices. Numerically more efficient methods based on elimination methods are learned in linear algebra or computer science.

In Section 1.3, we defined the determinant of a 3×3 matrix. This can be generalized by induction to $n \times n$ determinants. We illustrate here how to write the determinant of a 4×4 matrix in terms of the determinants of 3×3 matrices:

$$\begin{vmatrix} a_{11} & a_{12} & a_{13} & a_{14} \\ a_{21} & a_{22} & a_{23} & a_{24} \\ a_{31} & a_{32} & a_{33} & a_{34} \\ a_{41} & a_{42} & a_{43} & a_{44} \end{vmatrix} = a_{11} \begin{vmatrix} a_{22} & a_{23} & a_{24} \\ a_{32} & a_{33} & a_{34} \\ a_{42} & a_{43} & a_{44} \end{vmatrix} - a_{12} \begin{vmatrix} a_{21} & a_{23} & a_{24} \\ a_{31} & a_{33} & a_{34} \\ a_{41} & a_{43} & a_{44} \end{vmatrix}$$

$$+ a_{13} \begin{vmatrix} a_{21} & a_{22} & a_{24} \\ a_{31} & a_{32} & a_{34} \\ a_{41} & a_{42} & a_{44} \end{vmatrix} - a_{14} \begin{vmatrix} a_{21} & a_{22} & a_{23} \\ a_{31} & a_{32} & a_{33} \\ a_{41} & a_{42} & a_{43} \end{vmatrix}$$

[see formula (2) of Section 1.3; the signs alternate $+, -, +, -$].

The basic properties of 3×3 determinants reviewed in Section 1.3 remain valid for $n \times n$ determinants. In particular, we note the fact that if A is an $n \times n$ matrix and B is the matrix formed by adding a scalar multiple of one row (or column) of A to another row (or, respectively, column) of A, then the determinant of A is equal to the determinant of B (see Example 10).

A basic theorem of linear algebra states that an $n \times n$ matrix A is invertible if and only if the determinant of A is not zero. Another basic property is that the determinant is multiplicative: $\det(AB) = (\det A)(\det B)$. In this text, we shall not make use of many details of linear algebra, and so we shall leave these assertions unproved.

EXAMPLE 10 Let

$$A = \begin{bmatrix} 1 & 0 & 1 & 0 \\ 1 & 1 & 1 & 1 \\ 2 & 1 & 0 & 1 \\ 1 & 1 & 0 & 2 \end{bmatrix}.$$

Find det A. Does A have an inverse?

SOLUTION Adding $(-1) \times$ first column to the third column, we get

$$\det A = \begin{vmatrix} 1 & 0 & 0 & 0 \\ 1 & 1 & 0 & 1 \\ 2 & 1 & -2 & 1 \\ 1 & 1 & -1 & 2 \end{vmatrix} = 1 \begin{vmatrix} 1 & 0 & 1 \\ 1 & -2 & 1 \\ 1 & -1 & 2 \end{vmatrix}.$$

Adding $(-1) \times$ first column to the third column of this 3×3 determinant gives

$$\det A = \begin{vmatrix} 1 & 0 & 0 \\ 1 & -2 & 0 \\ 1 & -1 & 1 \end{vmatrix} = \begin{vmatrix} -2 & 0 \\ -1 & 1 \end{vmatrix} = -2.$$

Thus, $\det A = -2 \neq 0$, and so A has an inverse. ▲

If we have three matrices A, B, and C such that the products AB and BC are defined, then the products $(AB)C$ and $A(BC)$ are defined and are in fact equal (that is, matrix multiplication is *associative*). We call this the *triple product* of matrices and denote it by ABC.

EXAMPLE 11 Let

$$A = \begin{bmatrix} 3 \\ 5 \end{bmatrix}, \qquad B = [1 \quad 1], \qquad \text{and} \qquad C = \begin{bmatrix} 1 \\ 2 \end{bmatrix}.$$

Then

$$ABC = A(BC) = \begin{bmatrix} 3 \\ 5 \end{bmatrix} [3] = \begin{bmatrix} 9 \\ 15 \end{bmatrix}. \quad \blacktriangle$$

EXAMPLE 12

$$\begin{bmatrix} 2 & 0 \\ 0 & 1 \end{bmatrix} \begin{bmatrix} 1 & 1 \\ 1 & 1 \end{bmatrix} \begin{bmatrix} 0 & -1 \\ 1 & 1 \end{bmatrix} = \begin{bmatrix} 2 & 0 \\ 0 & 1 \end{bmatrix} \begin{bmatrix} 1 & 0 \\ 1 & 0 \end{bmatrix} = \begin{bmatrix} 2 & 0 \\ 1 & 0 \end{bmatrix}. \quad \blacktriangle$$

Historical Note

The founder of modern (coordinate) geometry was René Descartes (see Figure 1.5.1), a great physicist, philosopher, and mathematician, as well as a founder of modern biology.

Born in Touraine, France, in 1596, Descartes had a fascinating life. After studying law, he settled in Paris, where he developed an interest in mathematics. In 1628, he moved to Holland, where he wrote his only mathematical work, *La Geometrie*, one of the origins of modern coordinate geometry.

Descartes had been highly critical of the geometry of the ancient Greeks, with all their undefined terms and with their proofs requiring ever newer and more ingenious approaches. For Descartes, this geometry was so tied to geometrical figures "that it can exercise the understanding only on condition of greatly fatiguing the imagination." He undertook to exploit, in geometry, the use of algebra, which had recently been developed. The result was *La Geometrie*, which made possible analytic or computational methods in geometry.

Remember that the Greeks were, like Descartes, philosophers as well as mathematicians and physicists. Their answer to the question of the meaning of space was "Euclidean geometry." Descartes had therefore succeeded in "algebrizing"the Greek model of space.

Gottfried Wilhelm Leibniz, cofounder (with Isaac Newton) of calculus, was also interested in "space analysis," but he did not think that Descartes'

Figure 1.5.1 René Descartes (1596–1650).

algebra went far enough. Leibniz called for a direct method of space analysis (*analysis situs*) that could be interpreted as a call for the development of vector analysis.

On September 8, 1679, Leibniz outlined his ideas in a letter to Christian Huygens:

> I am still not satisfied with algebra, because it does not give the shortest methods or the most beautiful constructions in geometry. This is why I believe that, so far as geometry is concerned, we need still another analysis which is distinctly geometrical or linear and which will express situation (situs) directly as algebra expresses magnitude directly. And I believe that I have found the way and that we can represent figures and even machines and movements by characters, as algebra represents numbers or magnitudes. I am sending you an essay which seems to me to be important.

In the essay, Leibniz described his ideas in greater detail:

> I have discovered certain elements of a new characteristic which is entirely different from algebra and which will have great advantages in representing to the mind, exactly and in a way faithful to its nature, even without figures, everything which depends on sense perception. Algebra is the characteristic for undetermined numbers of magnitudes only, but it does not express situation, angles, and

motion directly. Hence it is often difficult to analyze the properties of a figure by calculation, and still more difficult to find very convenient geometrical demonstrations and constructions, even when the algebraic calculation is completed. But this new characteristic, which follows the visual figures, cannot fail to give the solution, the construction, and the geometric demonstration all at the same time, and in a natural way and in one analysis, that is, through determined procedure.

Leibniz's ideas influenced Hamilton and others. In the middle of the nineteenth century, Bolyai and Lobachevsky developed their "non-Euclidean" geometry, and Gauss studied and developed a theory of curved surfaces in three-dimensional space. Gauss developed two measures of curvature, the *mean curvature* and the *Gauss curvature*. For example, soap bubbles and soap films have constant mean curvature, whereas only soap bubbles have constant Gauss curvature. We discuss these ideas further in Section 7.7.

Bernhard Riemann (see Figure 1.5.2), possibly the greatest mathematical genius of all time, gave an inaugural address in 1854 before the faculty of Göttingen University, entitled "On the Hypotheses Which Lie at the Foundation of Geometry." It was this monumental work that would lay the foundation, 50 years later, of Einstein's general theory of relativity. Riemann, like Leibniz and the early Greeks, was interested in space, especially its metric (or distance) properties.

Figure 1.5.2 Bernhard Riemann (1826–1866).

Riemann called for the study of n-dimensional spaces and surfaces. He showed how to measure the curvature of three-, four-, ..., n-dimensional surfaces and (incredibly) showed that in order to be called "curved," a surface need not be "curving" inside anything else; curvature was simply a consequence of the intrinsic "metric properties of space." Once Riemann demonstrated that mathematical models permitted us to think of spaces of any dimension, the question naturally arose as to why our space is three-dimensional and not four-, five-, or more dimensional. Surprisingly, no one has yet put forth a convincing explanation why, at the moment of creation, space became three-dimensional.

Around 1910, Albert Einstein realized that gravity could be explained as a consequence of the curvature of a four-dimensional space–time (matter and energy curve space and time), and, thanks to Riemann, Einstein's space–time need not be enclosed in an ambient universe. Exactly how matter and energy curve space–time is the essence of Einstein's field equations in general relativity. In Section 7.7, we will discuss the ideas of curvature in greater depth and will indicate some of the ideas behind general relativity. The idea of n-dimensions also began to creep into mathematics from another, very different direction—from matrices.

The definition of a matrix, as an isolated abstract object, is due to the English mathematician Arthur Cayley. Cayley was born in 1821, and in 1863 was appointed Sedlesian Professor of Mathematics at Cambridge University. Around 1855, one year after Riemann's inaugural address, Cayley, in an effort to simplify notation for his study of linear equations (as we saw in Section 1.5), introduced the abstract idea of a matrix of m columns and n rows. Naturally, a $1 \times n$ matrix could also be viewed as a vector in an "n-dimensional space."

After this concept took hold, mathematicians realized that they lost little in working in general dimensions, and the subject of modern linear algebra was off and running. Again, physics was to be a major impetus. Modern, abstract, linear algebra, including abstract vector spaces, began to turn up in textbooks after the appearance in 1918 of Hermann Weyl's *Space–Time–Matter*.

EXERCISES

1. Calculate the dot product of $\mathbf{x} = (1, -1, 0, 2) \in \mathbb{R}^4$ and $\mathbf{y} = (1, 2, 3, 4) \in \mathbb{R}^4$.

2. In $\mathbb{R}^n$ show that

(a) $2\|\mathbf{x}\|^2 + 2\|\mathbf{y}\|^2 = \|\mathbf{x} + \mathbf{y}\|^2 + \|\mathbf{x} - \mathbf{y}\|^2$ (This is known as the ***parallelogram law***.)
(b) $\|\mathbf{x} - \mathbf{y}\| \|\mathbf{x} + \mathbf{y}\| \le \|\mathbf{x}\|^2 + \|\mathbf{y}\|^2$
(c) $4\langle \mathbf{x}, \mathbf{y} \rangle = \|\mathbf{x} + \mathbf{y}\|^2 - \|\mathbf{x} - \mathbf{y}\|^2$ (This is called the ***polarization identity***.)

Interpret these results geometrically in terms of the parallelogram formed by $\mathbf{x}$ and $\mathbf{y}$.

Verify the Cauchy–Schwarz inequality and the triangle inequality for the vectors in Exercises 3 to 6.

3. $\mathbf{x} = (2, 0, -1), \mathbf{y} = (4, 0, -2)$

4. $\mathbf{x} = (1, 0, 2, 6), \mathbf{y} = (3, 8, 4, 1)$

5. $\mathbf{x} = (1, -1, 1, -1, 1), \mathbf{y} = (3, 0, 0, 0, 2)$

6. $\mathbf{x} = (1, 0, 0, 1), \mathbf{y} = (-1, 0, 0, 1)$

7. Compute AB, $\det A$, $\det B$, $\det(AB)$, and $\det(A + B)$ for

$$A = \begin{bmatrix} 1 & -1 & 0 \\ 0 & 3 & 2 \\ 3 & 1 & 1 \end{bmatrix} \quad \text{and} \quad B = \begin{bmatrix} -2 & 0 & 2 \\ -1 & 1 & -1 \\ 1 & 4 & 3 \end{bmatrix}.$$

8. Compute AB, $\det A$, $\det B$, $\det(AB)$, and $\det(A + B)$ for

$$A = \begin{bmatrix} 3 & 0 & 1 \\ 1 & 2 & -1 \\ 1 & 0 & 1 \end{bmatrix} \quad \text{and} \quad B = \begin{bmatrix} 1 & 0 & -1 \\ 2 & 0 & 1 \\ 0 & 1 & 0 \end{bmatrix}.$$

9. Use induction on k to prove that if $\mathbf{x}_1, \dots, \mathbf{x}_k \in \mathbb{R}^n$, then

$$\|\mathbf{x}_1 + \cdots + \mathbf{x}_k\| \leq \|\mathbf{x}_1\| + \cdots + \|\mathbf{x}_k\|.$$

10. Prove using algebra, the ***identity of Lagrange***: For real numbers $x_1, \dots, x_n$ and $y_1, \dots, y_n$.

$$\left(\sum_{i=1}^{n} x_i y_i\right)^2 = \left(\sum_{i=1}^{n} x_i^2\right)\left(\sum_{i=1}^{n} y_i^2\right) - \sum_{i<j} (x_i y_j - x_j y_i)^2.$$

Use this to give another proof of the Cauchy–Schwarz inequality in $\mathbb{R}^n$,

11. Prove that if A is an $n \times n$ matrix, then

(a) $\det(\lambda A) = \lambda^n \det A$; and

(b) if B is a matrix obtained from A by multiplying any row or column by a scalar λ, then $\det B = \lambda \det A$.

In Exercises 12 to 14, A, B, and C denote $n \times n$ matrices.

12. Is $\det(A + B) = \det A + \det B$? Give a proof or counterexample.

13. Does $(A + B)(A - B) = A^2 - B^2$?

14. Assuming the law $\det(AB) = (\det A)(\det B)$, prove that $\det(ABC) = (\det A)(\det B)(\det C)$.

15. (This exercise assumes a knowledge of integration of continuous functions of one variable.) Note that the proof of the Cauchy–Schwarz inequality (Theorem 4) depends only on the properties of the inner product listed in Theorem 1. Use this observation to establish the following inequality for continuous functions $f, g\colon [0, 1] \to \mathbb{R}$:

$$\left| \int_0^1 f(x)g(x)\, dx \right| \le \sqrt{\int_0^1 [f(x)]^2\, dx}\, \sqrt{\int_0^1 [g(x)]^2\, dx}.$$

Do this by

(a) verifying that the space of continuous functions from [0, 1] to $\mathbb{R}$ forms a vector space; that is, we may think of functions f, g abstractly as "vectors" that can be added to each other and multiplied by scalars.

(b) introducing the inner product of functions

$$f \cdot g = \int_0^1 f(x)g(x)\, dx$$

and verifying that it satisfies conditions (i) to (iv) of Theorem 3.

16. Define the transpose A^T of an $n \times n$ matrix A as follows: the ijth element of A^T is a_{ji} where a_{ij} is the ijth entry of A. Show that A^T is characterized by the following property: For all $\mathbf{x}, \mathbf{y}$ in $\mathbb{R}^n$,

$$(A^T\mathbf{x}) \cdot \mathbf{y} = \mathbf{x} \cdot (A\mathbf{y}).$$

17. Verify that the inverse of

$$\begin{bmatrix} a & b \\ c & d \end{bmatrix} \quad \text{is} \quad \frac{1}{ad - bc}\begin{bmatrix} d & -b \\ -c & a \end{bmatrix}.$$

18. Use your answer in Exercise 17 to show that the solution of the system

$$ax + by = e$$

$$cx + dy = f$$

is

$$\begin{bmatrix} x \\ y \end{bmatrix} = \frac{1}{ad - bc}\begin{bmatrix} d & -b \\ -c & a \end{bmatrix}\begin{bmatrix} e \\ f \end{bmatrix}.$$

19. Assuming the law $\det(AB) = (\det A)(\det B)$, verify that $(\det A)(\det A^{-1}) = 1$ and conclude that if A has an inverse, then $\det A \neq 0$.

REVIEW EXERCISES FOR CHAPTER 1

1. Let $\mathbf{v} = 3\mathbf{i} + 4\mathbf{j} + 5\mathbf{k}$ and $\mathbf{w} = \mathbf{i} - \mathbf{j} + \mathbf{k}$. Compute $\mathbf{v} + \mathbf{w}$, $3\mathbf{v}$, $6\mathbf{v} + 8\mathbf{w}$, $-2\mathbf{v}$, $\mathbf{v} \cdot \mathbf{w}$, $\mathbf{v} \times \mathbf{w}$. Interpret each operation geometrically by graphing the vectors.

2. Repeat Exercise 1 with $\mathbf{v} = 2\mathbf{j} + \mathbf{k}$ and $\mathbf{w} = -\mathbf{i} - \mathbf{k}$.

3. (a) Find the equation of the line through $(-1, 2, -1)$ in the direction of $\mathbf{j}$.
(b) Find the equation of the line passing through $(0, 2, -1)$ and $(-3, 1, 0)$.
(c) Find the equation for the plane perpendicular to the vector $(-2, 1, 2)$ and passing through the point $(-1, 1, 3)$.

4. (a) Find the equation of the line through $(0, 1, 0)$ in the direction of $3\mathbf{i} + \mathbf{k}$.
(b) Find the equation of the line passing through $(0, 1, 1)$ and $(0, 1, 0)$.
(c) Find an equation for the plane perpendicular to the vector $(-1, 1, -1)$ and passing through the point $(1, 1, 1)$.

5. Compute $\mathbf{v} \cdot \mathbf{w}$ for the following sets of vectors:

(a) $\mathbf{v} = -\mathbf{i} + \mathbf{j}; \mathbf{w} = \mathbf{k}$.
(b) $\mathbf{v} = \mathbf{i} + 2\mathbf{j} - \mathbf{k}; \mathbf{w} = 3\mathbf{i} + \mathbf{j}$.
(c) $\mathbf{v} = -2\mathbf{i} - \mathbf{j} + \mathbf{k}; \mathbf{w} = 3\mathbf{i} + 2\mathbf{j} - 2\mathbf{k}$.

6. Compute $\mathbf{v} \times \mathbf{w}$ for the vectors in Exercise 5. [Only part (b) is solved in the Study Guide.]

7. Find the cosine of the angle between the vectors in Exercise 5. [Only part (b) is solved in the Study Guide.]

8. Find the area of the parallelogram spanned by the vectors in Exercise 5. [Only part (b) is solved in the Study Guide.]

9. Use vector notation to describe the triangle in space whose vertices are the origin and the endpoints of vectors **a** and **b**.

10. Show that three vectors **a, b, c** lie in the same plane through the origin if and only if there are three scalars α, β, γ, not all zero, such that $\alpha\mathbf{a} + \beta\mathbf{b} + \gamma\mathbf{c} = \mathbf{0}$.

11. For real numbers $a_1, a_2, a_3, b_1, b_2, b_3$, show that

$$(a_1b_1 + a_2b_2 + a_3b_3)^2 \leq (a_1^2 + a_2^2 + a_3^2)(b_1^2 + b_2^2 + b_3^2).$$

12. Let **u**, **v**, **w** be unit vectors that are orthogonal to each other. If $\mathbf{a} = \alpha\mathbf{u} + \beta\mathbf{v} + \gamma\mathbf{w}$, show that

$$\alpha = \mathbf{a} \cdot \mathbf{u}, \qquad \beta = \mathbf{a} \cdot \mathbf{v}, \qquad \gamma = \mathbf{a} \cdot \mathbf{w}.$$

Interpret the results geometrically.

13. Let **a**, **b** be two vectors in the plane, $\mathbf{a} = (a_1, a_2)$, $\mathbf{b} = (b_1, b_2)$, and let λ be a real number. Show that the area of the parallelogram determined by **a** and $\mathbf{b} + \lambda\mathbf{a}$ is the same as that determined by **a** and **b**. Sketch. Relate this result to a known property of determinants.

14. Find the volume of the parallelepiped determined by the vertices $(0, 1, 0)$, $(1, 1, 1)$, $(0, 2, 0)$, $(3, 1, 2)$.

15. Given nonzero vectors **a** and **b** in $\mathbb{R}^3$, show that the vector $\mathbf{v} = \|\mathbf{a}\|\mathbf{b} + \|\mathbf{b}\|\mathbf{a}$ bisects the angle between **a** and **b**.

16. Use vector methods to prove that the distance from the point (x_1, y_1) to the line $ax + by = c$ is

$$\frac{|ax_1 + by_1 - c|}{\sqrt{a^2 + b^2}}.$$

17. Verify that the direction of $\mathbf{b} \times \mathbf{c}$ is given by the right-hand rule, by choosing $\mathbf{b}, \mathbf{c}$ to be two of the vectors $\mathbf{i}$, $\mathbf{j}$, and $\mathbf{k}$.

18. (a) Suppose $\mathbf{a} \cdot \mathbf{b} = \mathbf{a}' \cdot \mathbf{b}$ for all $\mathbf{b}$. Show that $\mathbf{a} = \mathbf{a}'$.
(b) Suppose $\mathbf{a} \times \mathbf{b} = \mathbf{a}' \times \mathbf{b}$ for all $\mathbf{b}$. Is it true that $\mathbf{a} = \mathbf{a}'$?

19. (a) Using vector methods, show that the distance between two nonparallel lines l_1 and l_2 is given by

$$d = \frac{|(\mathbf{v}_2 - \mathbf{v}_1) \cdot (\mathbf{a}_1 \times \mathbf{a}_2)|}{\|\mathbf{a}_1 \times \mathbf{a}_2\|},$$

where $\mathbf{v}_1, \mathbf{v}_2$ are any two points on l_1 and l_2, respectively, and $\mathbf{a}_1$ and $\mathbf{a}_2$ are the directions of l_1 and l_2. [Hint: Consider the plane through l_2 that is parallel to l_1. Show that the vector $(\mathbf{a}_1 \times \mathbf{a}_2)/\|\mathbf{a}_1 \times \mathbf{a}_2\|$ is a unit normal for this plane; now project $\mathbf{v}_2 - \mathbf{v}_1$ onto this normal direction.]

(b) Find the distance between the line l_1 determined by the points $(-1, -1, 1)$ and $(0, 0, 0)$ and the line l_2 determined by the points $(0, -2, 0)$ and $(2, 0, 5)$.

20. Show that two planes given by the equations $Ax + By + Cz + D_1 = 0$ and $Ax + By + Cz + D_2 = 0$ are parallel, and that the distance between them is

$$\frac{|D_1 - D_2|}{\sqrt{A^2 + B^2 + C^2}}.$$

21. (a) Prove that the area of the triangle in the plane with vertices $(x_1, y_1), (x_2, y_2), (x_3, y_3)$ is the absolute value of

$$\frac{1}{2}\begin{vmatrix} 1 & 1 & 1 \\ x_1 & x_2 & x_3 \\ y_1 & y_2 & y_3 \end{vmatrix}.$$

(b) Find the area of the triangle with vertices $(1, 2), (0, 1), (-1, 1)$.

22. Convert the following points from Cartesian to cylindrical and spherical coordinates and plot:

(a) $(0, 3, 4)$
(b) $(-\sqrt{2}, 1, 0)$
(c) $(0, 0, 0)$
(d) $(-1, 0, 1)$
(e) $(-2\sqrt{3}, -2, 3)$

23. Convert the following points from cylindrical to Cartesian and spherical coordinates and plot:

(a) $(1, \pi/4, 1)$
(b) $(3, \pi/6, -4)$

(c) $(0, \pi/4, 1)$
(d) $(2, -\pi/2, 1)$
(e) $(-2, -\pi/2, 1)$

24. Convert the following points from spherical to Cartesian and cylindrical coordinates and plot:

(a) $(1, \pi/2, \pi)$
(b) $(2, -\pi/2, \pi/6)$
(c) $(0, \pi/8, \pi/35)$
(d) $(2, -\pi/2, -\pi)$
(e) $(-1, \pi, \pi/6)$

25. Rewrite the equation $z = x^2 - y^2$ using cylindrical and spherical coordinates.

26. Using spherical coordinates, show that

$$\phi = \cos^{-1}\left(\frac{\mathbf{u} \cdot \mathbf{k}}{\|\mathbf{u}\|}\right)$$

where $\mathbf{u} = x\mathbf{i} + y\mathbf{j} + z\mathbf{k}$. Interpret geometrically.

27. Verify the Cauchy–Schwarz and triangle inequalities for

$$\mathbf{x} = (3, 2, 1, 0) \qquad \text{and} \qquad \mathbf{y} = (1, 1, 1, 2).$$

28. Multiply the matrices

$$A = \begin{bmatrix} 3 & 0 & 1 \\ 2 & 0 & 1 \\ 1 & 0 & 1 \end{bmatrix} \qquad \text{and} \qquad B = \begin{bmatrix} 1 & 0 & 1 \\ 1 & 1 & 1 \\ 0 & 0 & 1 \end{bmatrix}.$$

Does $AB = BA$?

29. (a) Show that for two $n \times n$ matrices A and B, and $\mathbf{x} \in \mathbb{R}^n$,

$$(AB)\mathbf{x} = A(B\mathbf{x}).$$

(b) What does the equality in part (a) imply about the relationship between the composition of the mappings $\mathbf{x} \mapsto B\mathbf{x}$, $\mathbf{y} \mapsto A\mathbf{y}$, and matrix multiplication?

30. Find the volume of the parallelepiped spanned by the vectors

$$(1, 0, 1), \quad (1, 1, 1), \quad \text{and} \quad (-3, 2, 0).$$

31. (For students with some knowledge of linear algebra.) Verify that a linear mapping T of $\mathbb{R}^n$ to $\mathbb{R}^n$ is determined by an $n \times n$ matrix.

32. Find an equation for the plane that contains $(3, -1, 2)$ and the line with equation $\mathbf{v} = (2, -1, 0) + t(2, 3, 0)$.

33. The work W done in moving an object from $(0, 0)$ to $(7, 2)$ subject to a constant force $\mathbf{F}$ is $W = \mathbf{F} \cdot \mathbf{r}$, where $\mathbf{r}$ is the vector with its head at $(7, 2)$ and tail at $(0, 0)$. The units are feet and pounds.

(a) Suppose the force $\mathbf{F} = 10 \cos \theta \mathbf{i} + 10 \sin \theta \mathbf{j}$. Find W in terms of θ.

(b) Suppose the force $\mathbf{F}$ has magnitude of 6 lb and makes an angle of $\pi/6$ rad with the horizontal, pointing right. Find W in foot-pounds.

34. If a particle with mass m moves with velocity $\mathbf{v}$, its *momentum* is $\mathbf{p} = m\mathbf{v}$. In a game of marbles, a marble with mass 2 grams (g) is shot with velocity 2 meters per second (m/s), hits two marbles with mass 1 g each, and comes to a dead halt. One of the marbles flies off with a velocity of 3 m/s at an angle of 45° to the incident direction of the larger marble as in Figure 1.R.1. Assuming that the total momentum before and after the collision is the same (according to the law of conservation of momentum), at what angle and speed does the second marble move?

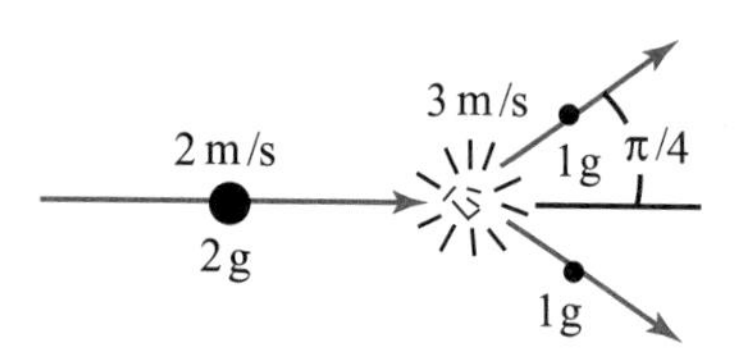

Figure 1.R.1 Momentum and marbles.

35. Show that for all x, y, z,

$$\begin{vmatrix} x+2 & y & z \\ z & y+1 & 10 \\ 5 & 5 & 2 \end{vmatrix} = -\begin{vmatrix} y & x+2 & z \\ 1 & z-x-2 & 10-z \\ 5 & 5 & 2 \end{vmatrix}.$$

36. Show that

$$\begin{vmatrix} 1 & x & x^2 \\ 1 & y & y^2 \\ 1 & z & z^2 \end{vmatrix} \neq 0$$

if x, y, and z are all different.

37. Show that

$$\begin{vmatrix} 66 & 628 & 246 \\ 88 & 435 & 24 \\ 2 & -1 & 1 \end{vmatrix} = \begin{vmatrix} 68 & 627 & 247 \\ 86 & 436 & 23 \\ 2 & -1 & 1 \end{vmatrix}.$$

38. Show that

$$\begin{vmatrix} n & n+1 & n+2 \\ n+3 & n+4 & n+5 \\ n+6 & n+7 & n+8 \end{vmatrix}$$

has the same value no matter what n is. What is this value?

39. The volume of a tetrahedron with concurrent edges $\mathbf{a}$, $\mathbf{b}$, $\mathbf{c}$ is given by $V = \frac{1}{6}\mathbf{a} \cdot (\mathbf{b} \times \mathbf{c})$.

(a) Express the volume as a determinant.
(b) Evaluate V when $\mathbf{a} = \mathbf{i} + \mathbf{j} + \mathbf{k}$, $\mathbf{b} = \mathbf{i} - \mathbf{j} + \mathbf{k}$, $\mathbf{c} = \mathbf{i} + \mathbf{j}$.

Use the following definition for Exercises 40 and 41: Let $\mathbf{r}_1, \ldots, \mathbf{r}_n$ be vectors in $\mathbb{R}^3$ from 0 *to the masses $m_1, \ldots, m_n$. The* ***center of mass*** *is the vector*

$$\mathbf{c} = \frac{\sum_{i=1}^{n} m_i \mathbf{r}_i}{\sum_{i=1}^{n} m_i}.$$

40. A tetrahedron sits in xyz coordinates with one vertex at $(0, 0, 0)$, and the three edges concurrent at $(0, 0, 0)$ are coincident with the vectors $\mathbf{a}$, $\mathbf{b}$, $\mathbf{c}$.

(a) Draw a figure and label the heads of the vectors $\mathbf{a}$, $\mathbf{b}$, $\mathbf{c}$.
(b) Find the center of mass of each of the four triangular faces of the tetrahedron if a unit mass is placed at each vertex.

41. Show that for any vector $\mathbf{r}$, the center of mass of a system satisfies

$$\sum_{i=1}^{n} m_i \|\mathbf{r} - \mathbf{r}_i\|^2 = \sum_{i=1}^{n} m_i \|\mathbf{r}_i - \mathbf{c}\|^2 + m\|\mathbf{r} - \mathbf{c}\|^2,$$

where $m = \sum_{i=1}^{n} m_i$ is the total mass of the system.

In Exercises 42 to 47, find a unit vector that has the given property.

42. Parallel to the line $x = 3t + 1$, $y = 16t - 2$, $z = -(t + 2)$.

43. Orthogonal to the plane $x - 6y + z = 12$.

44. Parallel to both the planes $8x + y + z = 1$ and $x - y - z = 0$.

45. Orthogonal to $\mathbf{i} + 2\mathbf{j} - \mathbf{k}$ and to $\mathbf{k}$.

46. Orthogonal to the line $x = 2t - 1$, $y = -t - 1$, $z = t + 2$, and the vector $\mathbf{i} - \mathbf{j}$.

47. At an angle of 30° to $\mathbf{i}$ and making equal angles with $\mathbf{j}$ and $\mathbf{k}$.

2

Differentiation

I turn away with fright and horror from the lamentable evil of functions which do not have derivatives.

Charles Hermite,
in a letter to Thomas Jan Stieltjes

This chapter extends the principles of differential calculus for functions of one variable to functions of several variables. We begin in Section 2.1 with the geometry of real-valued functions and study the graphs of these functions as an aid in visualizing them. Section 2.2 gives some basic definitions relating to limits and continuity. This subject is treated briefly, because it requires time and mathematical maturity to develop fully and is therefore best left to a more advanced course. Fortunately, a complete understanding of all the subtleties of the limit concept is not necessary for our purposes; the student who has difficulty with this section should bear this in mind. However, we hasten to add that the notion of a limit is central to the definition of the derivative, but not to the computation of most derivatives in specific problems, as we already know from one-variable calculus. Sections 2.3 and 2.5 deal with the definition of the derivative, and establish some basic rules of calculus: namely, how to differentiate a sum, product, quotient, or composition. In Section 2.6, we study directional derivatives and tangent planes, relating these ideas to those in Section 2.1. Finally, the Internet supplement gives some of the technical proofs.

In generalizing calculus from one dimension to several, it is often convenient to use the language of matrix algebra. What we shall need has been summarized in Section 1.5.

2.1 The Geometry of Real-Valued Functions

We launch our investigation of real-valued functions by developing methods for visualizing them. In particular, we introduce the notions of a graph, a level curve, and a level surface of such functions.

Functions and Mappings

Let f be a function whose domain is a subset A of $\mathbb{R}^n$ and with a range contained in $\mathbb{R}^m$. By this we mean that to each $\mathbf{x} = (x_1, \ldots, x_n) \in A$, f assigns a value $f(\mathbf{x})$, an m-tuple in $\mathbb{R}^m$. Such functions f are called ***vector-valued functions***[1] if $m > 1$, and ***scalar-valued functions*** if $m = 1$. For example, the scalar-valued function $f(x, y, z) = (x^2 + y^2 + z^2)^{-3/2}$ maps the set A of $(x, y, z) \neq (0, 0, 0)$ in $\mathbb{R}^3$ ($n = 3$ in this case) to $\mathbb{R}$ ($m = 1$). To denote f we sometimes write

$$f\colon (x, y, z) \mapsto (x^2 + y^2 + z^2)^{-3/2}.$$

Note that in $\mathbb{R}^3$ we often use the notation (x, y, z) instead of (x_1, x_2, x_3). In general, the notation $\mathbf{x} \mapsto f(\mathbf{x})$ is useful for indicating the value to which a point $\mathbf{x} \in \mathbb{R}^n$ is sent. We write $f\colon A \subset \mathbb{R}^n \to \mathbb{R}^m$ to signify that A is the domain of f (a subset of $\mathbb{R}^n$) and the range is contained in $\mathbb{R}^m$. We also use the expression f *maps* A *into* $\mathbb{R}^m$. Such functions f are called ***functions of several variables*** if $A \subset \mathbb{R}^n$, $n > 1$.

As another example we can take the vector-valued function $g\colon \mathbb{R}^6 \to \mathbb{R}^2$ defined by the rule

$$g(\mathbf{x}) = g(x_1, x_2, x_3, x_4, x_5, x_6) = \left(x_1x_2x_3x_4x_5x_6, \sqrt{x_1^2 + x_6^2}\right).$$

The first coordinate of the value of g at $\mathbf{x}$ is the product of the coordinates of $\mathbf{x}$.

Functions from $\mathbb{R}^n$ to $\mathbb{R}^m$ are not just mathematical abstractions, they arise naturally in problems studied in all the sciences. For example, to specify the temperature T in a region A of space requires a function $T\colon A \subset \mathbb{R}^3 \to \mathbb{R}$ ($n = 3, m = 1$); thus, $T(x, y, z)$ is the temperature at the point (x, y, z). To specify the velocity of a fluid moving in space requires a map $\mathbf{V}\colon \mathbb{R}^4 \to \mathbb{R}^3$, where $\mathbf{V}(x, y, z, t)$ is the velocity vector of the fluid at the point (x, y, z) in space at time t (see Figure 2.1.1). To

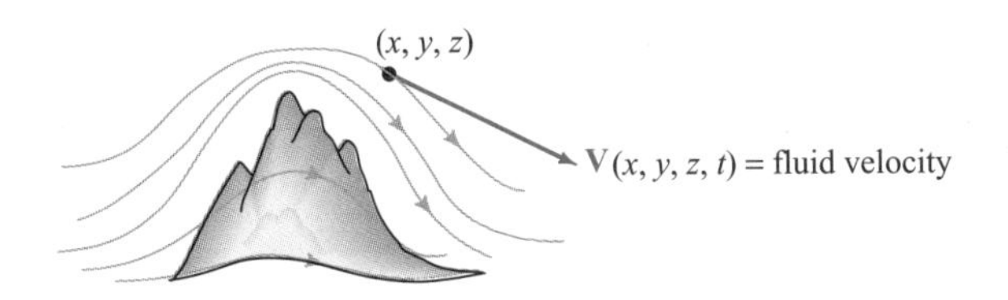

Figure 2.1.1 A fluid in motion defines a vector field $\mathbf{V}$ by specifying the velocity of the fluid particles at each point in space and time.

[1] Some mathematicians would write such an f in boldface, using the notation $\mathbf{f}(\mathbf{x})$, because the function is vector-valued. We did not do so, as a matter of personal taste. We use boldface primarily for mappings that are vector fields, introduced later. The notion of function was developed over many centuries, with the definition extended to cover more cases as they arose. For example, in 1667 James Gregory defined a function as "a quantity obtained from other quantities by a succession of algebraic operations or by any other operation imaginable." In 1755 Euler gave the following definition: "If some quantities depend on others in such a way as to undergo variation when the latter are varied then the former are called functions of the latter."

specify the reaction rate of a solution consisting of six reacting chemicals A, B, C, D, E, F in proportions x, y, z, w, u, v requires a map $\sigma\colon U \subset \mathbb{R}^6 \to \mathbb{R}$, where $\sigma(x, y, z, w, u, v)$ gives the rate when the chemicals are in the indicated proportions. To specify the cardiac vector (the vector giving the magnitude and direction of electric current flow in the heart) at time t requires a map $\mathbf{c}\colon \mathbb{R} \to \mathbb{R}^3$, $t \mapsto \mathbf{c}(t)$.

When $f\colon U \subset \mathbb{R}^n \to \mathbb{R}$, we say that f is a ***real-valued function of n variables with domain** U*. The reason we say "n variables" is simply that we regard the coordinates of a point $\mathbf{x} = (x_1, \dots, x_n) \in U$ as n variables, and $f(\mathbf{x}) = f(x_1, \dots, x_n)$ depends on these variables. We say "real-valued" because $f(x_1, \dots, x_n)$ is a real number. A good deal of our work will be with real-valued functions, so we give them special attention.

Graphs of Functions

For $f\colon U \subset \mathbb{R} \to \mathbb{R}$ ($n = 1$), the ***graph*** of f is the subset of $\mathbb{R}^2$ consisting of all points $(x, f(x))$ in the plane, for x in U. This subset can be thought of as a curve in $\mathbb{R}^2$. In symbols, we write this as

$$\operatorname{graph} f = \{(x, f(x)) \in \mathbb{R}^2 \mid x \in U\},$$

where the curly braces mean "the set of all" and the vertical bar is read "such that." Drawing the graph of a function of one variable is a useful device to help visualize how the function actually behaves. (See Figure 2.1.2.) It will be helpful to generalize

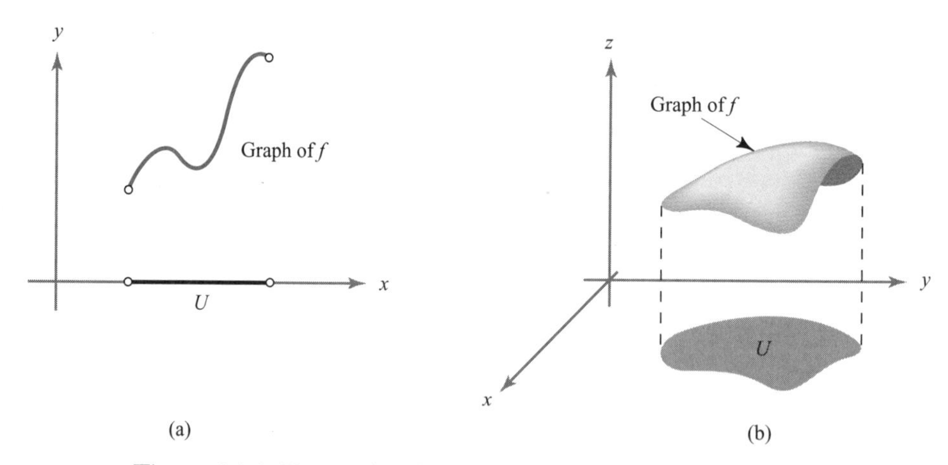

Figure 2.1.2 The graphs of (a) a function of one variable, and (b) a function of two variables.

the idea of a graph to functions of several variables. This leads to the following definition:

DEFINITION: Graph of a Function Let $f: U \subset \mathbb{R}^n \to \mathbb{R}$. Define the ***graph*** of f to be the subset of $\mathbb{R}^{n+1}$ consisting of all the points

$$(x_1, \ldots, x_n, f(x_1, \ldots, x_n))$$

in $\mathbb{R}^{n+1}$ for $(x_1, \ldots, x_n)$ in U. In symbols,

$$\text{graph } f = \{(x_1, \ldots, x_n, f(x_1, \ldots, x_n)) \in \mathbb{R}^{n+1} \mid (x_1, \ldots, x_n) \in U\}.$$

For the case $n = 1$ the graph is a curve in $\mathbb{R}^2$, while for $n = 2$ it is a surface in $\mathbb{R}^3$ (see Figure 2.1.2). For $n = 3$ it is difficult to visualize the graph, because, since we are humans living in a three-dimensional world, it is hard for us to envisage sets in $\mathbb{R}^4$. To help overcome this handicap, we introduce the idea of a level set.

Level Sets, Curves, and Surfaces

Suppose $f(x, y, z) = x^2 + y^2 + z^2$. A ***level set*** is a subset of $\mathbb{R}^3$ on which f is constant; for instance, the set where $x^2 + y^2 + z^2 = 1$ is a level set for f. This we can visualize: It is just a sphere of radius 1 in $\mathbb{R}^3$. Formally, a level set is the set of (x, y, z) such that $f(x, y, z) = c$, where c is a constant. The behavior or structure of a function is determined in part by the shape of its level sets; consequently, understanding these sets aids us in understanding the function in question. Level sets are also useful for understanding functions of two variables $f(x, y)$, in which case we speak of ***level curves*** or ***level contours***.

The idea is similar to that used to prepare contour maps, where one draws lines to represent constant altitudes; walking along such a line would mean walking on a level path. In the case of a hill rising from the xy plane, a graph of all the level curves gives us a good idea of the function $h(x, y)$, which represents the height of the hill at point (x, y) (see Figure 2.1.3).

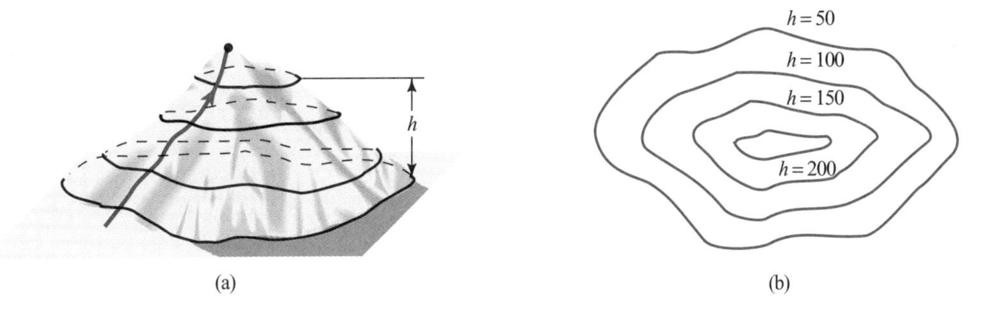

Figure 2.1.3 Level contours of a function are defined in the same manner as contour lines for a topographical map.

EXAMPLE 1 The constant function $f\colon \mathbb{R}^2 \to \mathbb{R}, (x, y) \mapsto 2$, that is, the function $f(x, y) = 2$, has as its graph the horizontal plane $z = 2$ in $\mathbb{R}^3$. The level curve of value c is empty if $c \neq 2$, and is the whole xy plane if $c = 2$. ▲

EXAMPLE 2 The function $f\colon \mathbb{R}^2 \to \mathbb{R}$, defined by $f(x, y) = x + y + 2$ has as its graph the inclined plane $z = x + y + 2$. This plane intersects the xy plane ($z = 0$) in the line $y = -x - 2$ and the z axis at the point $(0, 0, 2)$. For any value $c \in \mathbb{R}$, the level curve of value c is the straight line $y = -x + (c - 2)$; or in symbols, the set

$$L_c = \{(x, y) \mid y = -x + (c - 2)\} \subset \mathbb{R}^2.$$

We indicate a few of the level curves of the function in Figure 2.1.4. This is a contour map of the function f.

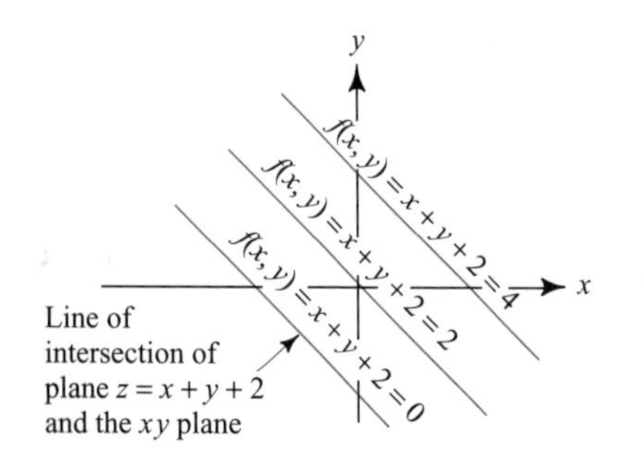

Figure 2.1.4 The level curves of $f(x, y) = x + y + 2$ show the sets on which f takes a given value.

From level curves labeled with the value or "height" of the function, the shape of the graph may be inferred by mentally elevating each level curve to the appropriate height, without stretching, tilting, or sliding it. If this procedure is visualized for all level curves, L_c, that is, for all values $c \in \mathbb{R}$, they will assemble to give the entire graph of f, as indicated by the shaded plane in Figure 2.1.5. If the graph is visualized

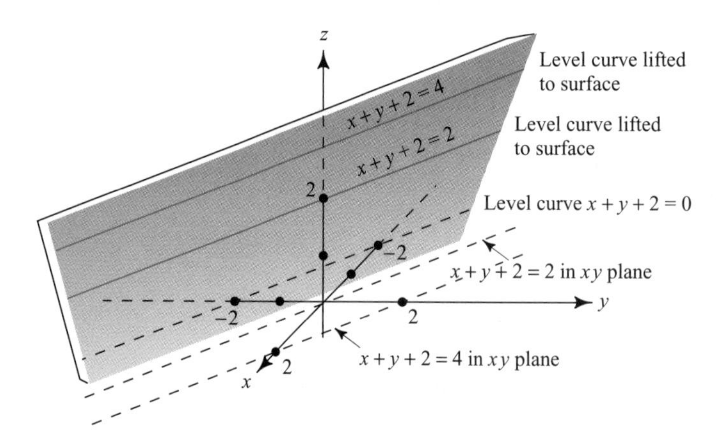

Figure 2.1.5 The relationship of level curves of Figure 2.1.4 to the graph of the function $f(x, y) = x + y + 2$, which is the plane $z = x + y + 2$.

using a finite number of level curves, a contour model is produced. If f is a smooth function, its graph will be a smooth surface, and so the contour model, mentally smoothed over, gives a good impression of the graph. ▲

DEFINITION: Level Curves and Surfaces Let $f: U \subset \mathbb{R}^n \to \mathbb{R}$ and let $c \in \mathbb{R}$. Then the ***level set of value*** c is defined to be the set of those points $\mathbf{x} \in U$ at which $f(\mathbf{x}) = c$. If $n = 2$, we speak of a ***level curve*** (of value c); and if $n = 3$, we speak of a ***level surface***. In symbols, the level set of value c is written

$$\{\mathbf{x} \in U \mid f(\mathbf{x}) = c\} \subset \mathbb{R}^n.$$

Note that the level set is always in the domain space.

EXAMPLE 3 Describe the graph of the quadratic function

$$f: \mathbb{R}^2 \to \mathbb{R}, (x, y) \mapsto x^2 + y^2.$$

SOLUTION The graph is the ***paraboloid of revolution*** $z = x^2 + y^2$, oriented upward from the origin, around the z axis. The level curve of value c is empty for $c < 0$; for $c > 0$ the level curve of value c is the set $\{(x, y) \mid x^2 + y^2 = c\}$, a circle of radius $\sqrt{c}$ centered at the origin. Thus, raised to height c above the xy plane, the level set is a circle of radius $\sqrt{c}$, indicating a parabolic shape (see Figures 2.1.6 and 2.1.7). ▲

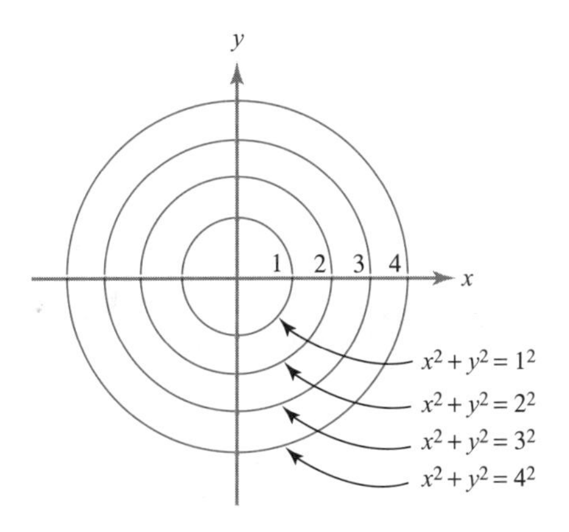

Figure 2.1.6 Some level curves for the function $f(x, y) = x^2 + y^2$.

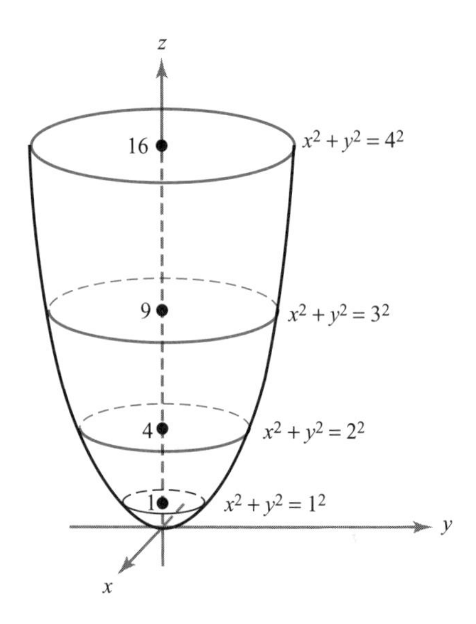

Figure 2.1.7 Level curves in Figure 2.1.6 raised to the graph.

The Method of Sections

By a ***section*** of the graph of f we mean the intersection of the graph and a (vertical) plane. For example, if P_1 is the xz plane in $\mathbb{R}^3$, defined by $y = 0$, then the section of f in Example 3 is the set

$$P_1 \cap \text{graph } f = \{(x, y, z) \mid y = 0, z = x^2\},$$

which is a parabola in the xz plane. Similarly, if P_2 denotes the yz plane, defined by $x = 0$, then the section

$$P_2 \cap \text{graph } f = \{(x, y, z) \mid x = 0, z = y^2\}$$

is a parabola in the yz plane (see Figure 2.1.8). It is usually helpful to compute at least one section to complement the information given by the level sets.

EXAMPLE 4 The graph of the quadratic function

$$f: \mathbb{R}^2 \to \mathbb{R}, (x, y) \mapsto x^2 - y^2$$

is called a ***hyperbolic paraboloid***, or ***saddle***, centered at the origin. Sketch the graph.

SOLUTION To visualize this surface, we first draw the level curves. To determine the level curves, we solve the equation $x^2 - y^2 = c$. Consider the values $c = 0, \pm 1, \pm 4$. For $c = 0$, we have $y^2 = x^2$, or $y = \pm x$, so that this level set consists of two straight lines through the origin. For $c = 1$, the level curve is $x^2 - y^2 = 1$, or $y = \pm\sqrt{x^2 - 1}$, which is a hyperbola that passes vertically through the x axis at the points $(\pm 1, 0)$ (see Figure 2.1.9). Similarly, for $c = 4$, the level curve is defined by $y = \pm\sqrt{x^2 - 4}$, the hyperbola passing vertically through the x axis at $(\pm 2, 0)$. For

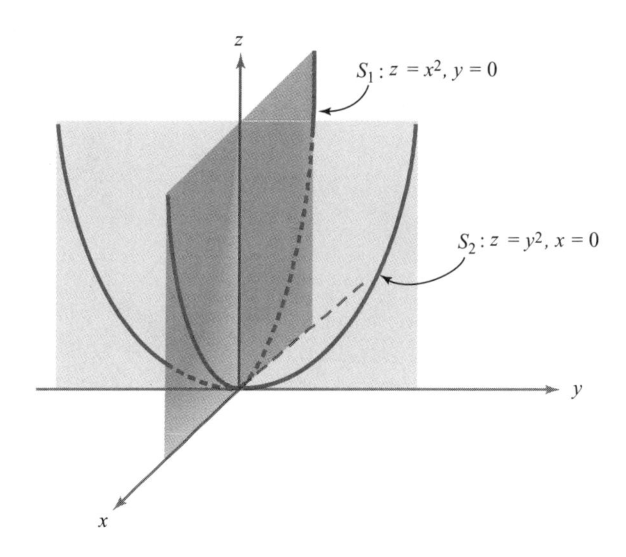

Figure 2.1.8 Two sections of the graph of $f(x, y) = x^2 + y^2$.

$c = -1$, we obtain the curve $x^2 - y^2 = -1$, that is, $x = \pm\sqrt{y^2 - 1}$, the hyperbola passing horizontally through the y axis at $(0, \pm 1)$. And for $c = -4$, the hyperbola through $(0, \pm 2)$ is obtained. These level curves are shown in Figure 2.1.9. Because it is not easy to visualize the graph of f from these data alone, we shall compute two sections, as in the previous example. For the section in the xz plane, we have

$$P_1 \cap \text{graph of } f = \{(x, y, z) \mid y = 0, z = x^2\},$$

which is a parabola opening upward; and for the yz plane,

$$P_2 \cap \text{graph } f = \{(x, y, z) \mid x = 0, z = -y^2\},$$

which is a parabola opening downward. The graph may now be visualized by lifting the level curves to the appropriate heights and smoothing out the resulting surface. Their

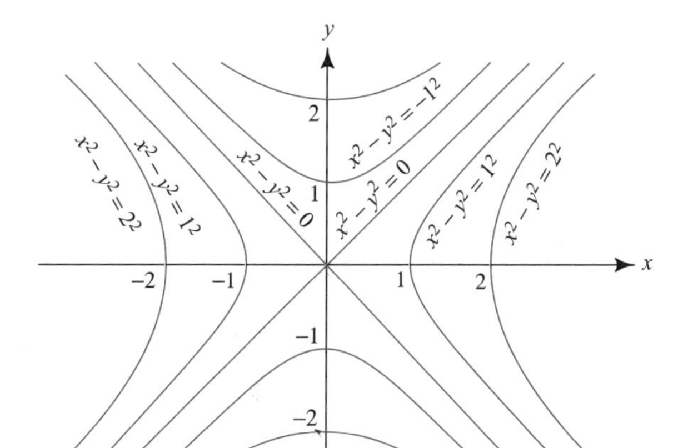

Figure 2.1.9 Level curves for the function $f(x, y) = x^2 - y^2$.

placement is aided by computing the parabolic sections. This procedure generates the hyperbolic saddle indicated in Figure 2.1.10. Compare this with the computer-generated graphs in Figure 2.1.11 (note that the orientation of the axes has been changed). ▲

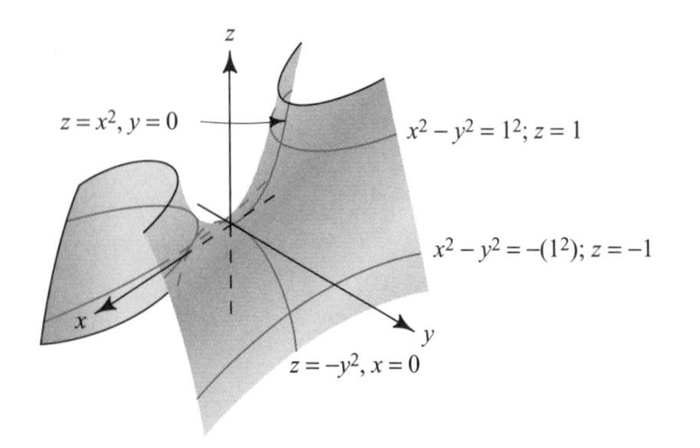

Figure 2.1.10 Some level curves on the graph of $f(x, y) = x^2 - y^2$.

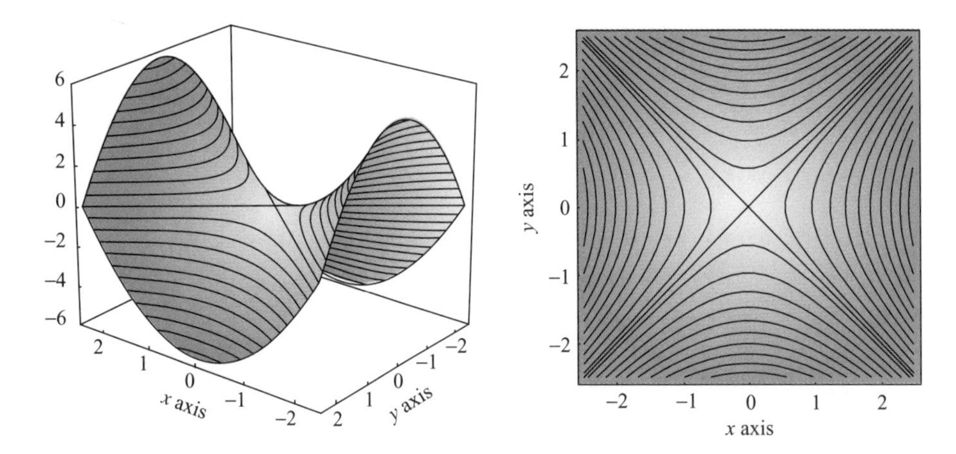

Figure 2.1.11 The graph of $z = x^2 - y^2$ and its level curves.

EXAMPLE 5 Describe the level sets of the function

$$f\colon \mathbb{R}^3 \to \mathbb{R}, (x, y, z) \mapsto x^2 + y^2 + z^2.$$

SOLUTION This is the three-dimensional analogue of Example 3. In this context, level sets are surfaces in the three-dimensional domain $\mathbb{R}^3$. The graph, in $\mathbb{R}^4$, cannot be visualized directly, but sections can nevertheless be computed.

The level set with value c is the set

$$L_c = \{(x, y, z) \mid x^2 + y^2 + z^2 = c\},$$

which is the sphere centered at the origin with radius $\sqrt{c}$ for $c > 0$, is a single point at the origin for $c = 0$, and is empty for $c < 0$. The level sets for $c = 0, 1, 4$, and 9 are indicated in Figure 2.1.12. ▲

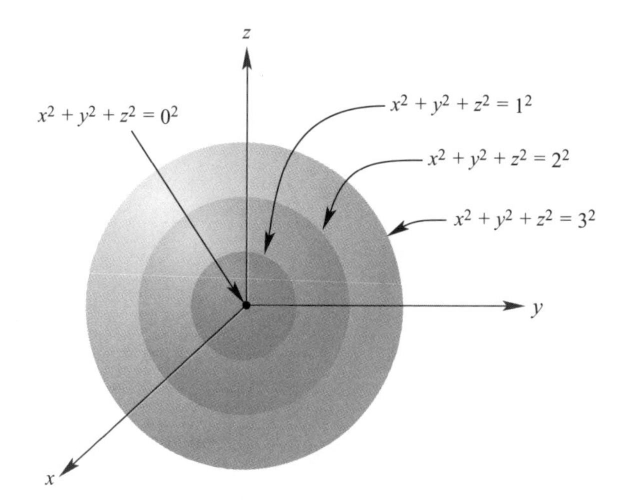

Figure 2.1.12 Some level surfaces for $f(x, y, z) = x^2 + y^2 + z^2$.

EXAMPLE 6 Describe the graph of the function $f\colon \mathbb{R}^3 \to \mathbb{R}$ defined by $f(x, y, z) = x^2 + y^2 - z^2$, which is the three-dimensional analogue of Example 4, and is also called a ***saddle***.

SOLUTION Formally, the graph of f is a subset of four-dimensional space. If we denote points in this space by (x, y, z, t), then the graph is given by

$$\{(x, y, z, t) \mid t = x^2 + y^2 - z^2\}.$$

The level surfaces of f are defined by

$$L_c = \{(x, y, z) \mid x^2 + y^2 - z^2 = c\}.$$

For $c = 0$, this is the cone $z = \pm\sqrt{x^2 + y^2}$ centered on the z axis. For c negative, say, $c = -a^2$, we obtain $z = \pm\sqrt{x^2 + y^2 + a^2}$, which is a hyperboloid of two sheets around the z axis, passing through the z axis at the points $(0, 0, \pm a)$. For c positive, say, $c = b^2$, the level surface is the ***single-sheeted hyperboloid of revolution*** around the z axis defined by $z = \pm\sqrt{x^2 + y^2 - b^2}$, which intersects the xy plane in the circle of radius $|b|$. These level surfaces are sketched in Figure 2.1.13.

Another view of the graph may be obtained from a section. For example, the subspace $S_{y=0} = \{(x, y, z, t) \mid y = 0\}$ intersects the graph in the section

$$S_{y=0} \cap \text{graph } f = \{(x, y, z, t) \mid y = 0, t = x^2 - z^2\},$$

that is, the set of points of the form $(x, 0, z, x^2 - z^2)$, which may be considered to be a surface in xzt space (see Figure 2.1.14). ▲

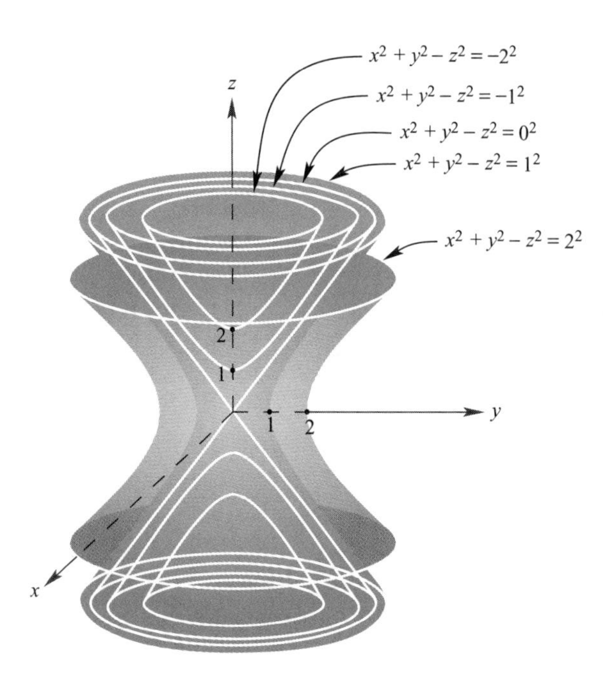

Figure 2.1.13 Some level surfaces of the function $f(x, y, z) = x^2 + y^2 - z^2$.

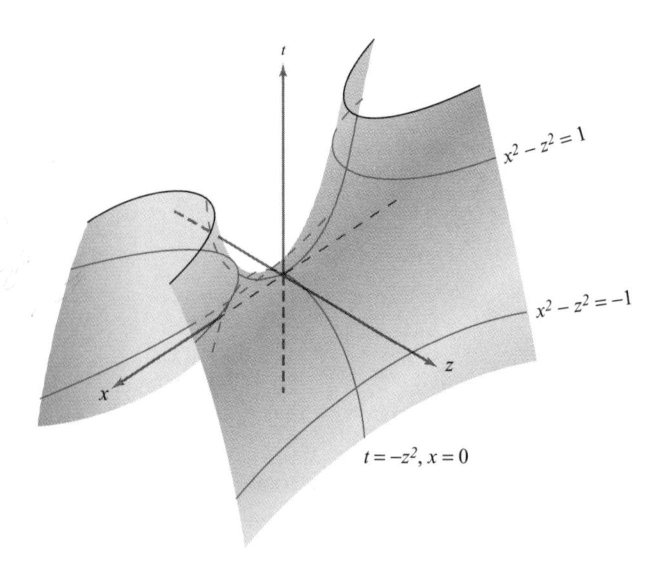

Figure 2.1.14 The $y = 0$ section of the graph of $f(x, y, z) = x^2 + y^2 - z^2$.

We have seen how the methods of sections and level sets can be used to understand the behavior of a function and its graph; these techniques can be quite useful to people who desire comprehensive visualization of complicated data. There are many computer programs available to do this, and we show the results of one such program in Figure 2.1.15.

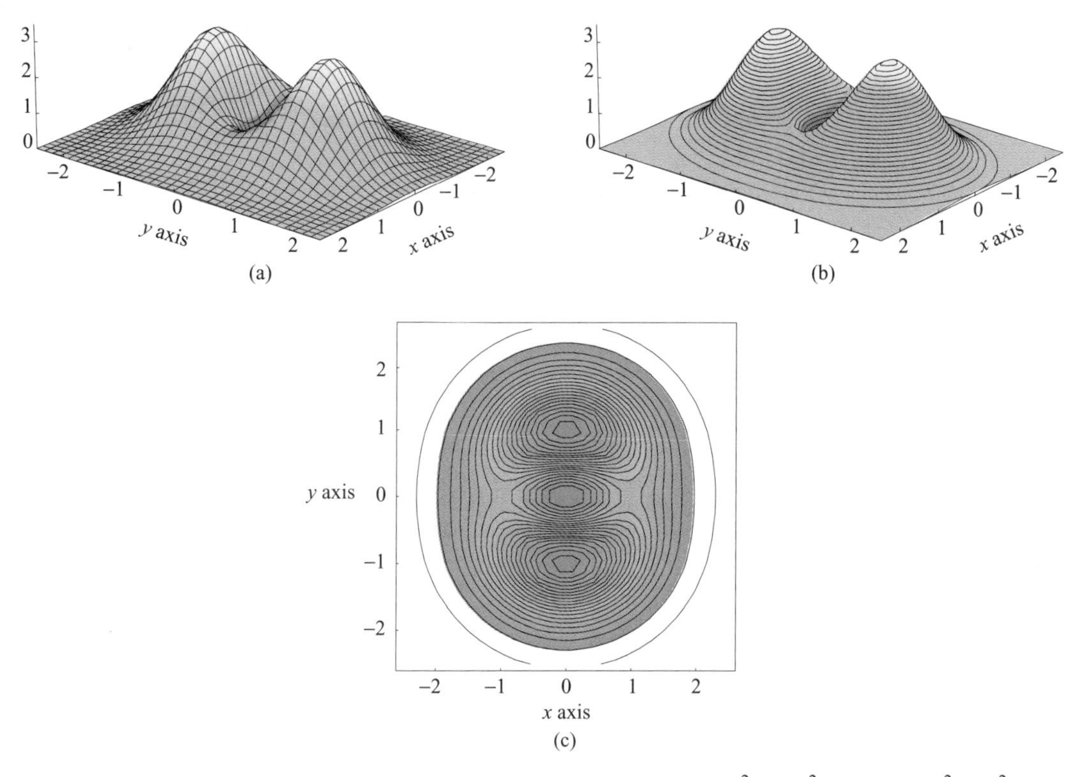

Figure 2.1.15 Computer-generated graph of $z = (x^2 + 3y^2)\exp(1 - x^2 - y^2)$ represented in three ways: (a) by sections, (b) by level curves on a graph, and (c) by level curves in the xy plane.

EXERCISES

1. Sketch the level curves and graphs of the following functions:

(a) $f\colon \mathbb{R}^2 \to \mathbb{R},\ (x, y) \mapsto x - y + 2$

(b) $f\colon \mathbb{R}^2 \to \mathbb{R},\ (x, y) \mapsto x^2 + 4y^2$

(c) $f\colon \mathbb{R}^2 \to \mathbb{R},\ (x, y) \mapsto -xy$

2. Describe the behavior, as c varies, of the level curve $f(x, y) = c$ for each of these functions:

(a) $f(x, y) = x^2 + y^2 + 1$ (b) $f(x, y) = 1 - x^2 - y^2$ (c) $f(x, y) = x^3 - x$

3. For the functions in Examples 2, 3, and 4, compute the section of the graph defined by the plane

$$S_\theta = \{(x, y, z) \mid y = x \tan\theta\}$$

for a given *constant* θ. Do this by expressing z as a function of r, where $x = r\cos\theta$, $y = r\sin\theta$. Determine which of these functions f have the property that the shape of the section $S_\theta \cap \text{graph } f$ is independent of θ. (The solution for Example 3 only is in the Study Guide.)

In Exercises 4 to 10, draw the level curves (in the xy plane) for the given function f and specified values of c. Sketch the graph of $z = f(x, y)$.

4. $f(x, y) = 4 - 3x + 2y$, $c = 0, 1, 2, 3, -1, -2, -3$

5. $f(x, y) = (100 - x^2 - y^2)^{1/2}$, $c = 0, 2, 4, 6, 8, 10$

6. $f(x, y) = (x^2 + y^2)^{1/2}$, $c = 0, 1, 2, 3, 4, 5$

7. $f(x, y) = x^2 + y^2$, $c = 0, 1, 2, 3, 4, 5$

8. $f(x, y) = 3x - 7y$, $c = 0, 1, 2, 3, -1, -2, -3$

9. $f(x, y) = x^2 + xy$, $c = 0, 1, 2, 3, -1, -2, -3$

10. $f(x, y) = x/y$, $c = 0, 1, 2, 3, -1, -2, -3$

In Exercises 11 to 13, sketch or describe the level surfaces and a section of the graph of each function.

11. $f\colon \mathbb{R}^3 \to \mathbb{R}, (x, y, z) \mapsto -x^2 - y^2 - z^2$

12. $f\colon \mathbb{R}^3 \to \mathbb{R}, (x, y, z) \mapsto 4x^2 + y^2 + 9z^2$

13. $f\colon \mathbb{R}^3 \to \mathbb{R}, (x, y, z) \mapsto x^2 + y^2$

In Exercises 14 to 18, describe the graph of each function by computing some level sets and sections.

14. $f\colon \mathbb{R}^3 \to \mathbb{R}, (x, y, z) \mapsto xy$

15. $f\colon \mathbb{R}^3 \to \mathbb{R}, (x, y, z) \mapsto xy + yz$

16. $f\colon \mathbb{R}^3 \to \mathbb{R}, (x, y, z) \mapsto xy + z^2$

17. $f\colon \mathbb{R}^2 \to \mathbb{R}, (x, y) \mapsto |y|$

18. $f\colon \mathbb{R}^2 \to \mathbb{R}, (x, y) \mapsto \max(|x|, |y|)$

Sketch or describe the surfaces in $\mathbb{R}^3$ of the equations presented in Exercises 19 to 31.

19. $4x^2 + y^2 = 16$

20. $x + 2z = 4$

21. $z^2 = y^2 + 4$

22. $x^2 + y^2 - 2x = 0$

23. $\dfrac{x}{4} = \dfrac{y^2}{4} + \dfrac{z^2}{9}$

24. $\dfrac{y^2}{9} + \dfrac{z^2}{4} = 1 + \dfrac{x^2}{16}$

25. $z = x^2$

26. $y^2 + z^2 = 4$

27. $z = \frac{y^2}{4} - \frac{x^2}{9}$

28. $y^2 = x^2 + z^2$

29. $4x^2 - 3y^2 + 2z^2 = 0$

30. $\frac{x^2}{9} + \frac{y^2}{12} + \frac{z^2}{9} = 1$

31. $x^2 + y^2 + z^2 + 4x - by + 9z - b = 0$, where b is a constant

32. Using polar coordinates, describe the level curves of the function defined by

$$f(x, y) = 2xy/(x^2 + y^2) \text{ if } (x, y) \neq (0, 0) \text{ and } f(0, 0) = 0.$$

33. Let $f\colon \mathbb{R}^2\backslash\{\mathbf{0}\} \to \mathbb{R}$ be given in polar coordinates by $f(r, \theta) = (\cos 2\theta)/r^2$. Sketch a few level curves in the xy plane. Here, $\mathbb{R}^2\backslash\{\mathbf{0}\} = \{\mathbf{x} \in \mathbb{R}^2 \mid \mathbf{x} \neq \mathbf{0}\}$.

34. Show that in Figure 2.1.15, the level "curve" $z = 3$ consists of two points.

2.2 Limits and Continuity

This section develops the concepts of open sets, limits, and continuity; open sets are needed to understand limits, and limits are in turn needed to understand continuity and differentiability.

As in elementary calculus, it is not necessary to completely master the limit concept in order to work problems in differentiation. For this reason, instructors may treat the following material with varying degrees of rigor. The student should consult with the instructor about the depth of understanding required.

Open Sets

We begin formulating the concept of an open set by defining an open disk. Let $\mathbf{x}_0 \in \mathbb{R}^n$ and let r be a positive real number. The ***open disk*** (or ***open ball***) of radius r and center $\mathbf{x}_0$ is defined to be the set of all points $\mathbf{x}$ such that $\|\mathbf{x} - \mathbf{x}_0\| < r$. This set is denoted $D_r(\mathbf{x}_0)$, and is the set of points $\mathbf{x}$ in $\mathbb{R}^n$ whose distance from $\mathbf{x}_0$ is less than r. Notice that we include only those $\mathbf{x}$ for which *strict* inequality holds. The disk $D_r(\mathbf{x}_0)$ is illustrated in Figure 2.2.1 for $n = 1, 2, 3$. For the case $n = 1$ and $x_0 \in \mathbb{R}$, the open disk $D_r(x_0)$ is the open interval $(x_0 - r, x_0 + r)$, which consists of all numbers $x \in \mathbb{R}$ *strictly* between $x_0 - r$ and $x_0 + r$. For the case $n = 2$, $\mathbf{x}_0 \in \mathbb{R}^2$, $D_r(\mathbf{x}_0)$ is the "inside" of the disk of radius r centered at $\mathbf{x}_0$. For the case $n = 3$, $\mathbf{x}_0 \in \mathbb{R}^3$, $D_r(\mathbf{x}_0)$ is the part strictly "inside" of the ball of radius r centered at $\mathbf{x}_0$.

DEFINITION: Open Sets Let $U \subset \mathbb{R}^n$ (that is, let U be a subset of $\mathbb{R}^n$). We call U an ***open set*** when for every point $\mathbf{x}_0$ in U there exists some $r > 0$ such that $D_r(\mathbf{x}_0)$ is contained within U; symbolically, we write $D_r(\mathbf{x}_0) \subset U$ (see Figure 2.2.2).

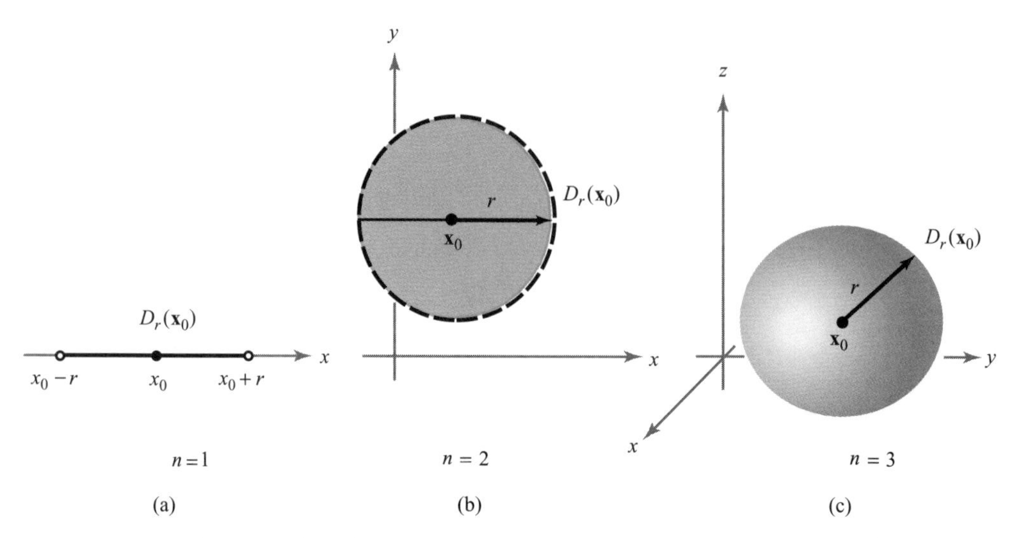

Figure 2.2.1 What disks $D_r(\mathbf{x}_0)$ look like in (a) one, (b) two, and (c) three dimensions.

The number $r > 0$ can depend on the point $\mathbf{x}_0$, and generally r will shrink as $\mathbf{x}_0$ gets closer to the "edge" of U. Intuitively speaking, a set U is open when the "boundary" points of U do not lie in U. In Figure 2.2.2, the dashed line is *not* included in U.

We establish the convention that the empty set $\emptyset$ (the set consisting of no elements) is open.

We have defined an open disk and an open set. From our choice of terms it would seem that an open disk should also be an open set. A little thought shows that this fact requires some proof. The following theorem does this.

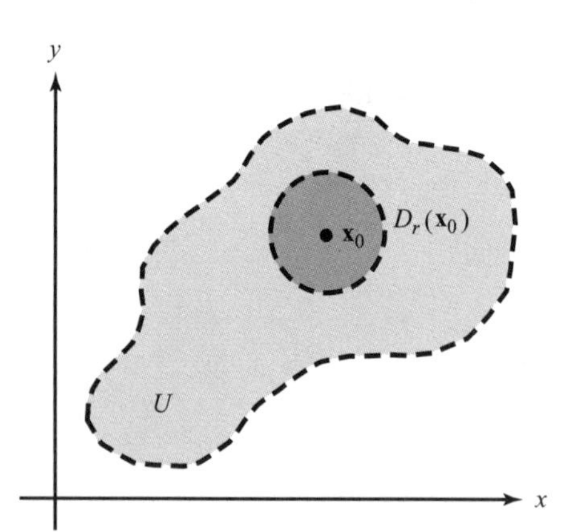

Figure 2.2.2 An open set U is one that completely encloses some disk $D_r(\mathbf{x}_0)$ about each of its points $\mathbf{x}_0$.

THEOREM 1 For each $\mathbf{x}_0 \in \mathbb{R}^n$ and $r > 0$, $D_r(\mathbf{x}_0)$ is an open set.

PROOF Let $\mathbf{x} \in D_r(\mathbf{x}_0)$, that is, let $\|\mathbf{x} - \mathbf{x}_0\| < r$. According to the definition of an open set, we must find an $s > 0$ such that $D_s(\mathbf{x}) \subset D_r(\mathbf{x}_0)$. Referring to Figure 2.2.3, we see that $s = r - \|\mathbf{x} - \mathbf{x}_0\|$ is a reasonable choice; note that $s > 0$, but that s becomes smaller if $\mathbf{x}$ is nearer the edge of $D_r(\mathbf{x}_0)$.

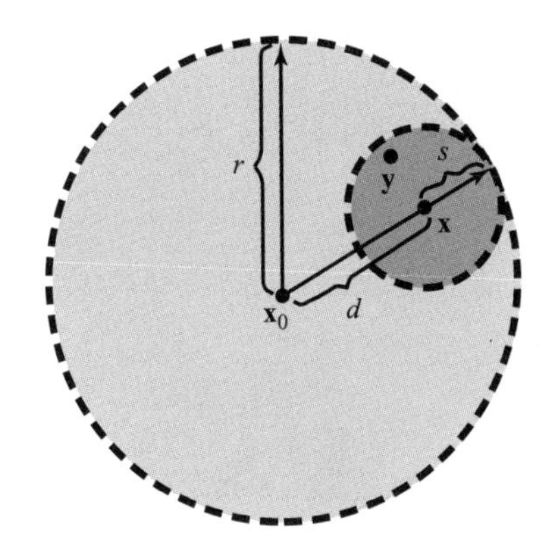

Figure 2.2.3 The geometry of the proof that an open disk is an open set.

To prove that $D_s(\mathbf{x}) \subset D_r(\mathbf{x}_0)$, let $\mathbf{y} \in D_s(\mathbf{x})$; that is, let $\|\mathbf{y} - \mathbf{x}\| < s$. We want to prove that $\mathbf{y} \in D_r(\mathbf{x}_0)$ as well. Proving this, in view of the definition of an r-disk, entails showing that $\|\mathbf{y} - \mathbf{x}_0\| < r$. This is done by using the triangle inequality for vectors in $\mathbb{R}^n$:

$$\|\mathbf{y} - \mathbf{x}_0\| = \|(\mathbf{y} - \mathbf{x}) + (\mathbf{x} - \mathbf{x}_0)\| \leq \|\mathbf{y} - \mathbf{x}\| + \|\mathbf{x} - \mathbf{x}_0\| < s + \|\mathbf{x} - \mathbf{x}_0\| = r.$$

Hence, $\|\mathbf{y} - \mathbf{x}_0\| < r$. ■

The following example illustrates some techniques that are useful in establishing the openness of sets.

EXAMPLE 1 Prove that $A = \{(x, y) \in \mathbb{R}^2 \mid x > 0\}$ is an open set.

SOLUTION The set is pictured in Figure 2.2.4.

Intuitively, this set is open, because no points on the "boundary," $x = 0$, are contained in the set. Such an argument will often suffice after one becomes accustomed to the concept of openness. At first, however, we should give details. To prove that A is open, we show that for every point $(x, y) \in A$ there exists an $r > 0$ such that $D_r(x, y) \subset A$. If $(x, y) \in A$, then $x > 0$. Choose $r = x$. If $(x_1, y_1) \in D_r(x, y)$, we have

$$|x_1 - x| = \sqrt{(x_1 - x)^2} \leq \sqrt{(x_1 - x)^2 + (y_1 - y)^2} < r = x,$$

and so $x_1 - x < x$ and $x - x_1 < x$. The latter inequality implies $x_1 > 0$, that is, $(x_1, y_1) \in A$. Hence $D_r(x, y) \subset A$, and therefore A is open (see Figure 2.2.5). ▲

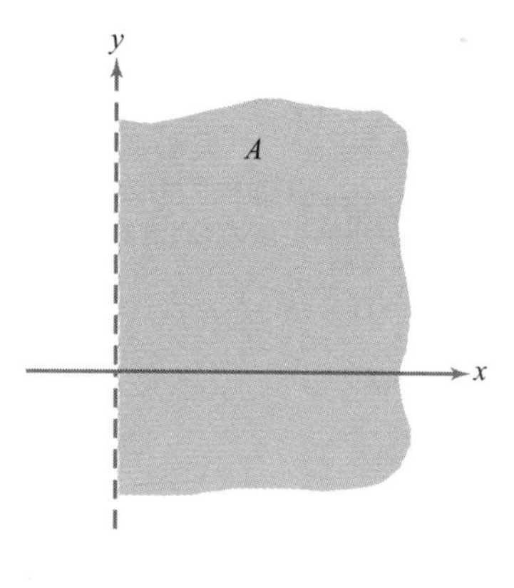

Figure 2.2.4 Show that A is an open set.

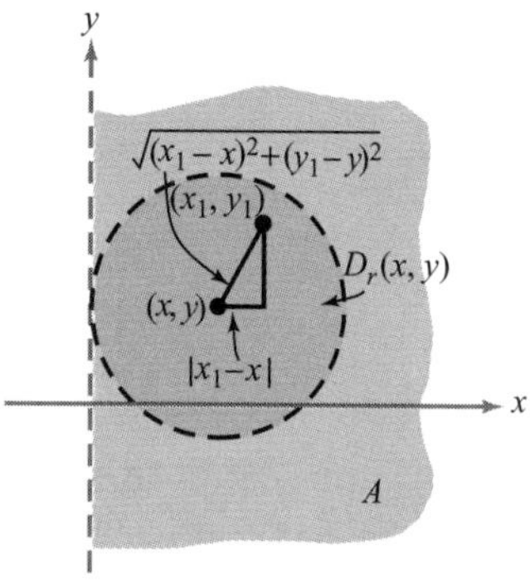

Figure 2.2.5 The construction of a disk about a point in A that is completely enclosed in A.

It is useful to have a special name for an open set containing a given point $\mathbf{x}$, because this idea arises often in the study of limits and continuity. Thus, by a ***neighborhood*** of $\mathbf{x} \in \mathbb{R}^n$ we merely mean an open set U containing the point $\mathbf{x}$. For example, $D_r(\mathbf{x}_0)$ is a neighborhood of $\mathbf{x}_0$ for any $r > 0$. The set A in Example 1 is a neighborhood of the point $\mathbf{x}_0 = (3, -10)$.

Boundary

Let us formally introduce the concept of a boundary point, which we alluded to in Example 1.

> DEFINITION: Boundary Points Let $A \subset \mathbb{R}^n$. A point $\mathbf{x} \in \mathbb{R}^n$ is called a ***boundary point*** of A if every neighborhood of $\mathbf{x}$ contains at least one point in A and at least one point not in A.

In this definition, $\mathbf{x}$ itself may or may not be in A; if $\mathbf{x} \in A$, then $\mathbf{x}$ is a boundary point if every neighborhood of $\mathbf{x}$ contains at least one point *not* in A (it already contains a point of A, namely, $\mathbf{x}$). Similarly, if $\mathbf{x}$ is not in A, it is a boundary point if every neighborhood of $\mathbf{x}$ contains at least one point of A.

We shall be particularly interested in boundary points of open sets. By the definition of an open set, no point of an open set A can be a boundary point of A. Thus, *a point* $\mathbf{x}$ *is a boundary point of an open set* A *if and only if* $\mathbf{x}$ *is not in* A *and every neighborhood of* $\mathbf{x}$ *has a nonempty intersection with* A.

This expresses in precise terms the intuitive idea that a boundary point of A is a point just on the "edge" of A. In many examples it is perfectly clear what the boundary points are.

EXAMPLE 2 (a) Let $A = (a, b)$ in $\mathbb{R}$. Then the boundary points of A consist of the points a and b. A consideration of Figure 2.2.6 and the definition will make this clear. [The reader will be asked to prove this in Exercise 20(c).]

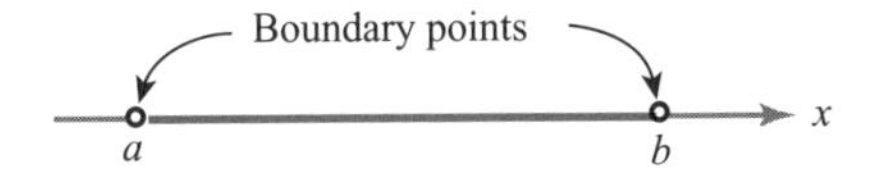

Figure 2.2.6 The boundary points of the interval (a, b).

(b) Let $A = D_r(x_0, y_0)$ be an r-disk about (x_0, y_0) in the plane. The boundary consists of points (x, y) with $(x - x_0)^2 + (y - y_0)^2 = r^2$ (Figure 2.2.7).

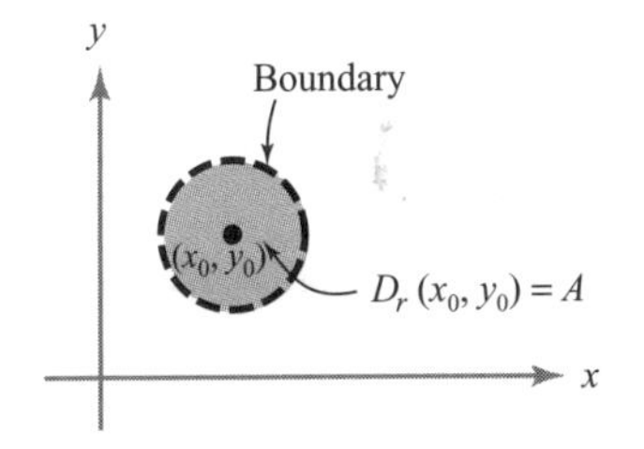

Figure 2.2.7 The boundary of A consists of points on the edge of A.

(c) Let $A = \{(x, y) \in \mathbb{R}^2 \mid x > 0\}$. Then the boundary of A consists of all points on the y axis (the student should draw a figure).

(d) Let A be $D_r(\mathbf{x}_0)$ minus the point $\mathbf{x}_0$ (a "punctured" disk about $\mathbf{x}_0$). Then $\mathbf{x}_0$ is a boundary point of A. ▲

Limits

We now turn our attention to the concept of a limit. *Throughout the following discussions the domain of definition of the function* f *will be an open set* A. We are interested in finding the limit of f as $\mathbf{x} \in A$ approaches either a point of A or a boundary point of A.

The reader should appreciate the fact that the limit concept is a basic and useful tool for the analysis of functions; it enables us to study derivatives, and hence maxima and minima, asymptotes, improper integrals, and other important features of functions, as well as being useful for infinite series and sequences. We will present a theory of limits for functions of several variables that includes the theory for functions of one variable as a special case.

In one-variable calculus, the student has encountered the notion of $\underset{x \to x_0}{\text{limit}} f(x) = l$ for a function $f\colon A \subset \mathbb{R} \to \mathbb{R}$ from a subset A of the real numbers to the real numbers. Intuitively, this means that as x gets closer and closer to x_0, the values $f(x)$ get closer and closer to (the limiting value) l. To put this intuitive idea on a firm, mathematical foundation, either the "epsilon (ε) and delta (δ) method" or the "neighborhood method" is usually introduced. The same is true for functions of several variables. In what follows we develop the neighborhood approach to limits. The epsilon-delta approach is left for optional study at the end of this section.

DEFINITION: Limit Let $f\colon A \subset \mathbb{R}^n \to \mathbb{R}^m$, where A is an open set. Let $\mathbf{x}_0$ be in A or be a boundary point of A, and let N be a neighborhood of $\mathbf{b} \in \mathbb{R}^m$. We say f is ***eventually in** N **as** $\mathbf{x}$ **approaches*** $\mathbf{x}_0$ if there exists a neighborhood U of $\mathbf{x}_0$ such that $\mathbf{x} \neq \mathbf{x}_0$, $\mathbf{x} \in U$, and $\mathbf{x} \in A$ imply $f(\mathbf{x}) \in N$. [The geometric meaning of this assertion is illustrated in Figure 2.2.8; note that $\mathbf{x}_0$ need not be in the set A, so that $f(\mathbf{x}_0)$ is not necessarily defined.] We say $f(\mathbf{x})$ ***approaches*** $\mathbf{b}$ as $\mathbf{x}$ approaches $\mathbf{x}_0$, or, in symbols,

$$\underset{\mathbf{x} \to \mathbf{x}_0}{\text{limit}}\, f(\mathbf{x}) = \mathbf{b} \quad \text{or} \quad f(\mathbf{x}) \to \mathbf{b} \quad \text{as} \quad \mathbf{x} \to \mathbf{x}_0,$$

when, given *any* neighborhood N of $\mathbf{b}$, f is eventually in N as $\mathbf{x}$ approaches $\mathbf{x}_0$ [that is, "$f(\mathbf{x})$ is close to $\mathbf{b}$ if $\mathbf{x}$ is close to $\mathbf{x}_0$"]. It may be that as $\mathbf{x}$ approaches $\mathbf{x}_0$, the values $f(\mathbf{x})$ do not get close to any particular number. In this case, we say that $\underset{\mathbf{x} \to \mathbf{x}_0}{\text{limit}} f(\mathbf{x})$ ***does not exist***.

Henceforth, whenever we consider the notion $\underset{\mathbf{x} \to \mathbf{x}_0}{\text{limit}} f(\mathbf{x})$, we shall always assume that $\mathbf{x}_0$ either belongs to some open set on which f is defined or is on the boundary of such a set.

One reason we insist on $\mathbf{x} \neq \mathbf{x}_0$ in the definition of limit will become clear if we remember from one-variable calculus that we want to be able to define the derivative $f'(x_0)$ of a function f at a point x_0 by

$$f'(x_0) = \underset{x \to x_0}{\text{limit}} \frac{f(x) - f(x_0)}{x - x_0},$$

and this expression is not defined at $x = x_0$.

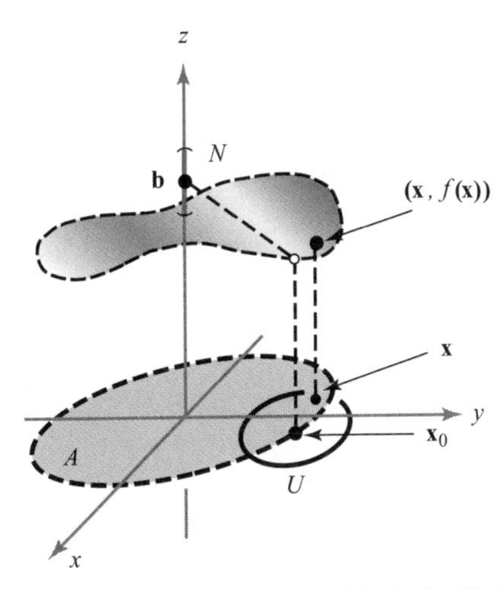

Figure 2.2.8 Limits in terms of neighborhoods; if $\mathbf{x}$ is in U, then $f(\mathbf{x})$ will be in N. (The little open circle denotes that the point does not lie on the graph.) In the figure, $f\colon A = \{(x, y) \mid x^2 + y^2 < 1\} \to \mathbb{R}$. (The dashed line is not in the graph of f.)

EXAMPLE 3 (a) This example illustrates a limit that does not exist. Consider the function $f\colon \mathbb{R} \to \mathbb{R}$ defined by

$$f(x) = \begin{cases} 1 & \text{if } x > 0 \\ -1 & \text{if } x \leq 0. \end{cases}$$

The $\lim_{x \to 0} f(x)$ does not exist, since there are points x_1 arbitrarily close to 0 with $f(x_1) = 1$ and also points x_2 arbitrarily close to 0 with $f(x_2) = -1$; that is, there is no single number that f is close to when x is close to 0 (see Figure 2.2.9). If f is restricted to the domain $(0, 1)$ or $(-1, 0)$, then the limit does exist. Can you say why?

(b) This example illustrates a function whose limit does exist, but whose limiting value does not equal its value at the limiting point. Define $f\colon \mathbb{R} \to \mathbb{R}$ by

$$f(x) = \begin{cases} 0 & \text{if } x \neq 0 \\ 1 & \text{if } x = 0. \end{cases}$$

It is true that $\lim_{x \to 0} f(x) = 0$, since for any neighborhood U of 0, $x \in U$ and $x \neq 0$ implies that $f(x) = 0$. One sees from the graph in Figure 2.2.10 that f approaches 0 as $x \to 0$; we do not care that f happens to take on some other value at 0. ▲

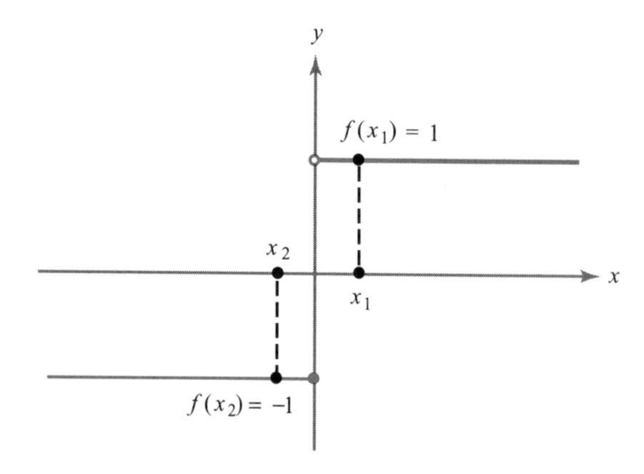

Figure 2.2.9 The limit of this function as $x \to 0$ does not exist.

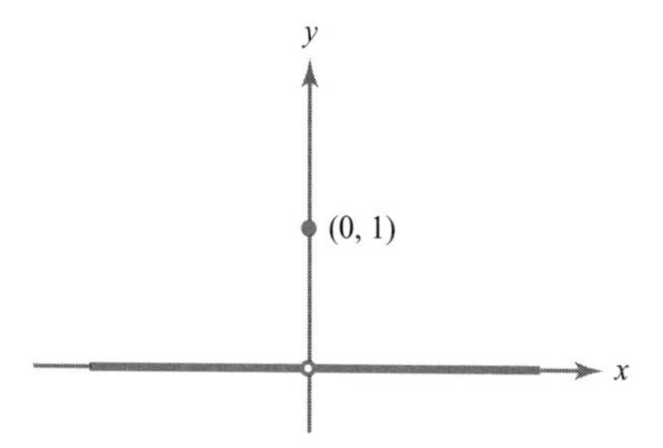

Figure 2.2.10 The limit of this function as $x \to 0$ is zero.

EXAMPLE 4 Use the definition to verify that the "obvious" $\lim_{\mathbf{x}\to\mathbf{x}_0} \mathbf{x} = \mathbf{x}_0$ holds, where $\mathbf{x}$ and $\mathbf{x}_0 \in \mathbb{R}^n$.

SOLUTION Let f be the function defined by $f(\mathbf{x}) = \mathbf{x}$, and let N be any neighborhood of $\mathbf{x}_0$. We must show that $f(\mathbf{x})$ is eventually in N as $\mathbf{x} \to \mathbf{x}_0$. According to the definition, we must find a neighborhood U of $\mathbf{x}_0$ with the property that if $\mathbf{x} \neq \mathbf{x}_0$ and $\mathbf{x} \in U$, then $f(\mathbf{x}) \in N$. Pick $U = N$. If $\mathbf{x} \in U$, then $\mathbf{x} \in N$; because $\mathbf{x} = f(\mathbf{x})$, it follows that $f(\mathbf{x}) \in N$. Thus, we have shown that $\lim_{\mathbf{x}\to\mathbf{x}_0} \mathbf{x} = \mathbf{x}_0$. In a similar way, we have

$$\lim_{(x,y)\to(x_0,y_0)} x = x_0, \qquad \text{etc.} \quad \blacktriangle$$

In what follows, the student may assume, without proof, the validity of limits from one-variable calculus. For example, $\lim_{x\to 1} \sqrt{x} = \sqrt{1} = 1$ and $\lim_{\theta\to 0} \sin\theta = \sin 0 = 0$ may be used.

EXAMPLE 5 (This example demonstrates another case in which the limit cannot simply be "read off" from the function.) Find $\lim_{x\to 1} g(x)$ where

$$g\colon x \mapsto \frac{x-1}{\sqrt{x}-1}.$$

SOLUTION This function is graphed in Figure 2.2.11(a).

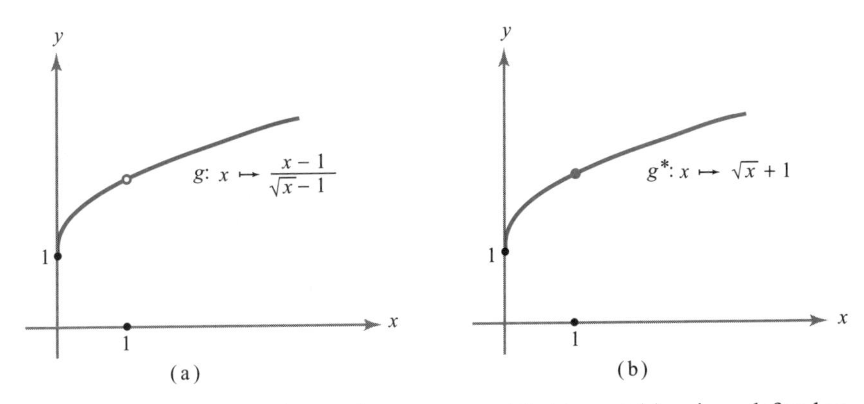

Figure 2.2.11 These graphs are the same except that in part (a), g is undefined at $x = 1$, whereas in part (b), g^* is defined for all $x \geq 0$.

We see that $g(1)$ is not defined, because division by zero is not defined. However, if we multiply the numerator and denominator of $g(x)$ by $\sqrt{x} + 1$, we find that for all x in the domain of g we have

$$g(x) = \frac{x-1}{\sqrt{x}-1} = \sqrt{x} + 1, \qquad x \neq 1.$$

The expression $g^*(x) = \sqrt{x} + 1$ is defined and takes the value 2 at $x = 1$; from one-variable calculus, $g^*(x) \to 2$ as $x \to 1$. But because $g^*(x) = g(x)$ for all $x \geq 0$, $x \neq 1$, we must have as well that $g(x) \to 2$ as $x \to 1$. ▲

We will consider other examples in two variables shortly.

Properties of Limits

To properly speak of *the* limit, we should establish that f can have *at most one* limit as $\mathbf{x} \to \mathbf{x}_0$. This is intuitively clear and we now state it formally. (See the Internet supplement for the proof.)

THEOREM 2: Uniqueness of Limits

$$\text{If} \quad \lim_{\mathbf{x} \to \mathbf{x}_0} f(\mathbf{x}) = \mathbf{b}_1 \quad \text{and} \quad \lim_{\mathbf{x} \to \mathbf{x}_0} f(\mathbf{x}) = \mathbf{b}_2, \quad \text{then} \quad \mathbf{b}_1 = \mathbf{b}_2.$$

To carry out practical computations with limits, we require some rules for limits, for example, that the limit of a sum is the sum of the limits. These rules are summarized in the following theorem (see the Internet supplement for Chapter 2 for the proof).

THEOREM 3: Properties of Limits Let $f\colon A \subset \mathbb{R}^n \to \mathbb{R}^m$, $g\colon A \subset \mathbb{R}^n \to \mathbb{R}^m$, $\mathbf{x}_0$ be in A or be a boundary point of A, $\mathbf{b} \in \mathbb{R}^m$, and $c \in \mathbb{R}$; then

(i) If $\lim_{\mathbf{x}\to\mathbf{x}_0} f(\mathbf{x}) = \mathbf{b}$, then $\lim_{\mathbf{x}\to\mathbf{x}_0} cf(\mathbf{x}) = c\mathbf{b}$, where $cf\colon A \to \mathbb{R}^m$ is defined by $\mathbf{x} \mapsto c(f(\mathbf{x}))$.

(ii) If $\lim_{\mathbf{x}\to\mathbf{x}_0} f(\mathbf{x}) = \mathbf{b}_1$ and $\lim_{\mathbf{x}\to\mathbf{x}_0} g(\mathbf{x}) = \mathbf{b}_2$, then $\lim_{\mathbf{x}\to\mathbf{x}_0} (f+g)(\mathbf{x}) = \mathbf{b}_1 + \mathbf{b}_2$, where $(f+g)\colon A \to \mathbb{R}^m$ is defined by $\mathbf{x} \mapsto f(\mathbf{x}) + g(\mathbf{x})$.

(iii) If $m = 1$, $\lim_{\mathbf{x}\to\mathbf{x}_0} f(\mathbf{x}) = b_1$, and $\lim_{\mathbf{x}\to\mathbf{x}_0} g(\mathbf{x}) = b_2$, then $\lim_{\mathbf{x}\to\mathbf{x}_0} (fg)(\mathbf{x}) = b_1 b_2$, where $(fg)\colon A \to \mathbb{R}$ is defined by $\mathbf{x} \mapsto f(\mathbf{x})g(\mathbf{x})$.

(iv) If $m = 1$, $\lim_{\mathbf{x}\to\mathbf{x}_0} f(\mathbf{x}) = b \neq 0$, and $f(\mathbf{x}) \neq 0$ for all $\mathbf{x} \in A$, then $\lim_{\mathbf{x}\to\mathbf{x}_0} 1/f(\mathbf{x}) = 1/b$, where $1/f\colon A \to \mathbb{R}$ is defined by $\mathbf{x} \mapsto 1/f(\mathbf{x})$.

(v) If $f(\mathbf{x}) = (f_1(\mathbf{x}), \dots, f_m(\mathbf{x}))$ where $f_i\colon A \to \mathbb{R}$, $i = 1, \dots, m$, are the component functions of f, then $\lim_{\mathbf{x}\to\mathbf{x}_0} f(\mathbf{x}) = \mathbf{b} = (b_1, \dots, b_m)$ if and only if $\lim_{\mathbf{x}\to\mathbf{x}_0} f_i(\mathbf{x}) = b_i$ for each $i = 1, \dots, m$.

These results ought to be intuitively clear. For instance, rule (ii) says that if $f(\mathbf{x})$ is close to $\mathbf{b}_1$ and $g(\mathbf{x})$ is close to $\mathbf{b}_2$ when $\mathbf{x}$ is close to $\mathbf{x}_0$, then $f(\mathbf{x}) + g(\mathbf{x})$ is close to $\mathbf{b}_1 + \mathbf{b}_2$ when $\mathbf{x}$ is close to $\mathbf{x}_0$. The following example illustrates how this works.

EXAMPLE 6 Let $f\colon \mathbb{R}^2 \to \mathbb{R}$, $(x, y) \mapsto x^2 + y^2 + 2$. Compute the limit

$$\lim_{(x,y)\to(0,1)} f(x, y).$$

SOLUTION Here f is the sum of the three functions $(x, y) \mapsto x^2$, $(x, y) \mapsto y^2$, and $(x, y) \mapsto 2$. The limit of a sum is the sum of the limits, and the limit of a product is the product of the limits (Theorem 3). Hence, using the fact that $\lim_{(x,y)\to(x_0,y_0)} x = x_0$ (Example 4), we obtain

$$\lim_{(x,y)\to(x_0,y_0)} x^2 = \left(\lim_{(x,y)\to(x_0,y_0)} x\right)\left(\lim_{(x,y)\to(x_0,y_0)} x\right) = x_0^2$$

and, using the same reasoning, $\lim_{(x,y)\to(x_0,y_0)} y^2 = y_0^2$. Consequently,

$$\lim_{(x,y)\to(0,1)} f(x, y) = 0^2 + 1^2 + 2 = 3. \quad \blacktriangle$$

Continuous Functions

In single-variable calculus we learned that the idea of a continuous function is based on the intuitive notion of a function whose graph is an unbroken curve, that is, a curve that has no *jumps*, or the kind of curve that would be traced by a particle in motion or by a moving pencil point that is not lifted from the paper.

To perform a detailed analysis of functions, we need concepts more precise than this rather vague notion. An example may clarify these ideas. Consider the specific function $f\colon \mathbb{R} \to \mathbb{R}$ defined by $f(x) = -1$ if $x \le 0$ and $f(x) = 1$ if $x > 0$. The graph of f is shown in Figure 2.2.12(a). [The little open circle denotes the fact that the point $(0, 1)$ does *not* lie on the graph of f]. Clearly, the graph of f is broken at $x = 0$. Consider also the function $g\colon x \mapsto x^2$. This function is pictured in Figure 2.2.12(b). The graph of g is not broken at any point.

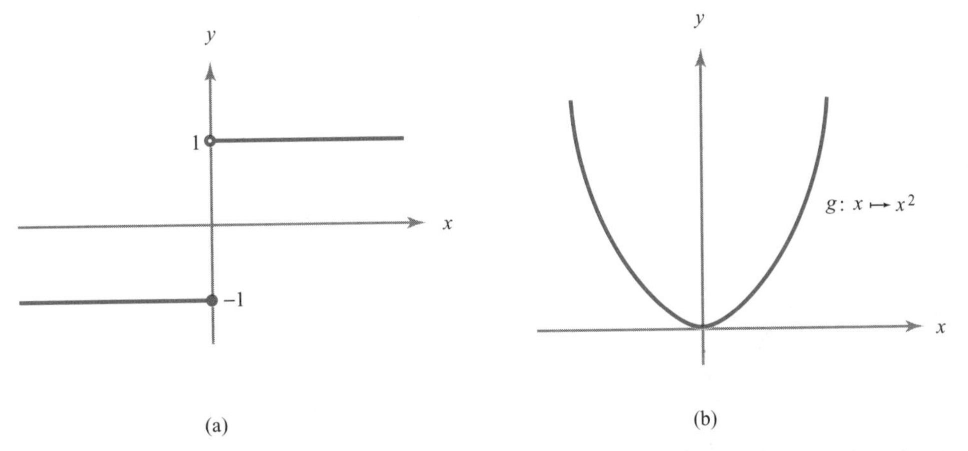

Figure 2.2.12 The function f in part (a) is not continuous, because its value jumps as x crosses 0, whereas the function g in part (b) is continuous.

If one examines examples of functions like f, whose graphs are broken at some point x_0, and functions like g, whose graphs are not broken, one sees that the principal difference between them is that for a function like g, the values of $g(x)$ get closer to $g(x_0)$ as x gets closer and closer to x_0. The same idea works for functions of several variables. But the notion of closer and closer does not suffice as a mathematical definition; thus, we shall formulate these concepts precisely in terms of limits.

Because the condition $\lim_{\mathbf{x} \to \mathbf{x}_0} f(\mathbf{x}) = f(\mathbf{x}_0)$ means that $f(\mathbf{x})$ is close to $f(\mathbf{x}_0)$ when $\mathbf{x}$ is close to $\mathbf{x}_0$, we see that this limit condition does indeed correspond to the requirement that the graph of f be unbroken (see Figure 2.2.13, where we illustrate the case $f\colon \mathbb{R} \to \mathbb{R}$). The case of several variables is easiest to visualize if we deal with real-valued functions, say $f\colon \mathbb{R}^2 \to \mathbb{R}$. In this case, we can visualize f by drawing its graph, which consists of all points (x, y, z) in $\mathbb{R}^3$ with $z = f(x, y)$. The continuity of f thus means that its graph has no "breaks" in it (see Figure 2.2.14).

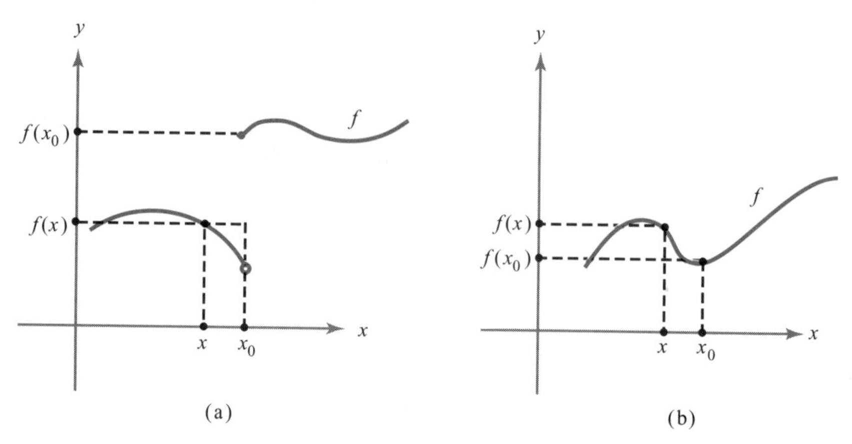

Figure 2.2.13 (a) Discontinuous function for which $\text{limit}_{x \to x_0} f(x)$ does not exist. (b) Continuous function for which this limit exists and equals $f(x_0)$.

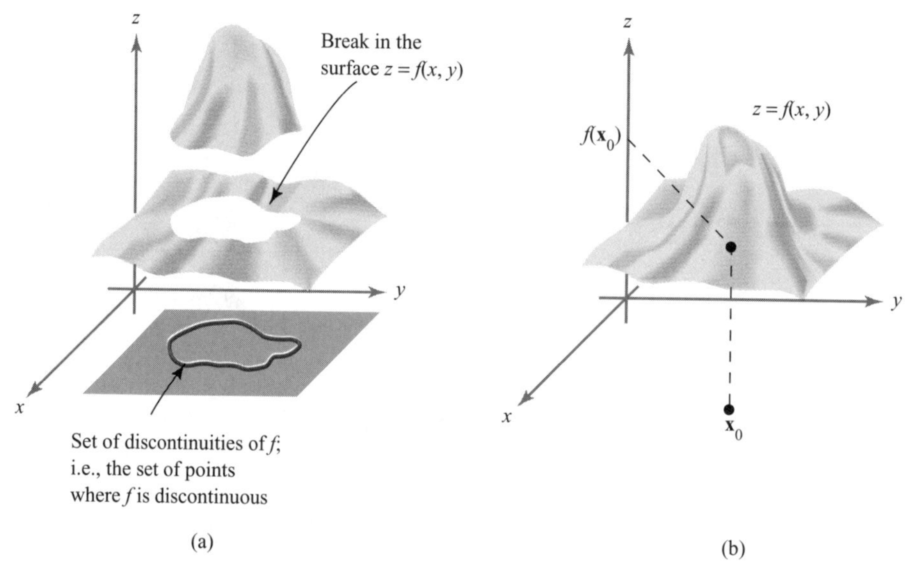

Figure 2.2.14 (a) A discontinuous function of two variables. (b) A continuous function.

DEFINITION: Continuity Let $f \colon A \subset \mathbb{R}^n \to \mathbb{R}^m$ be a given function with domain A. Let $\mathbf{x}_0 \in A$. We say f is ***continuous*** at $\mathbf{x}_0$ if and only if

$$\lim_{\mathbf{x} \to \mathbf{x}_0} f(\mathbf{x}) = f(\mathbf{x}_0).$$

If we just say that f is ***continuous***, we shall mean that f is continuous at each point $\mathbf{x}_0$ of A. If f is not continuous at $\mathbf{x}_0$, we say f is ***discontinuous*** at $\mathbf{x}_0$. If f is discontinuous at some point in its domain, we say f is ***discontinuous***.

EXAMPLE 7 Any polynomial $p(x) = a_0 + a_1 x + \cdots + a_n x^n$ is continuous from $\mathbb{R}$ to $\mathbb{R}$. Indeed, from Theorem 3 and Example 4,

$$\begin{aligned}\operatorname*{limit}_{x\to x_0} \left(a_0 + a_1 x + \cdots + a_n x^n\right) &= \operatorname*{limit}_{x\to x_0} a_0 + \operatorname*{limit}_{x\to x_0} a_1 x + \cdots + \operatorname*{limit}_{x\to x_0} a_n x^n \\ &= a_0 + a_1 x_0 + \cdots + a_n x_0^n,\end{aligned}$$

because the limit of a product is the product of the limits, which gives

$$\operatorname*{limit}_{x\to x_0} x^n = \left(\operatorname*{limit}_{x\to x_0} x\right)^n = x_0^n. \quad \blacktriangle$$

EXAMPLE 8 Let $f\colon \mathbb{R}^2 \to \mathbb{R}$, $f(x, y) = xy$. Then f is continuous, because, by the limit theorems and Example 4,

$$\operatorname*{limit}_{(x,y)\to(x_0,y_0)} xy = \left(\operatorname*{limit}_{(x,y)\to(x_0,y_0)} x\right)\left(\operatorname*{limit}_{(x,y)\to(x_0,y_0)} y\right) = x_0 y_0. \quad \blacktriangle$$

One can see by the same method that any polynomial $p(x, y)$ [for example, $p(x, y) = 3x^2 - 6xy^2 + y^3$] in x and y is continuous.

EXAMPLE 9 The function $f\colon \mathbb{R}^2 \to \mathbb{R}$ defined by

$$f(x, y) = \begin{cases} 1 & \text{if } x \le 0 \text{ or } y \le 0 \\ 0 & \text{otherwise} \end{cases}$$

is not continuous at (0, 0) or at any point on the positive x axis or positive y axis. Indeed, if $(x_0, y_0) = \mathbf{u}$ is such a point (i.e., $x_0 = 0$ and $y_0 \ge 0$, or $y_0 = 0$ and $x_0 \ge 0$) and $\delta > 0$, there are points $(x, y) \in D_\delta(\mathbf{u})$, a neighborhood of $\mathbf{u}$, with $f(x, y) = 1$ and other points $(x, y) \in D_\delta(\mathbf{u})$ with $f(x, y) = 0$. Thus, it is *not* true that $f(x, y) \to f(x_0, y_0) = 1$ as $(x, y) \to (x_0, y_0)$. ▲

To prove that specific functions are continuous, we can avail ourselves of the limit theorems (see Theorem 3 and Example 7). If we transcribe those results in terms of continuity, we are led to the following:

THEOREM 4: Properties of Continuous Functions Suppose that $f\colon A \subset \mathbb{R}^n \to \mathbb{R}^m$, $g\colon A \subset \mathbb{R}^n \to \mathbb{R}^m$, and let c be a real number.

(i) If f is continuous at $\mathbf{x}_0$, so is cf, where $(cf)(\mathbf{x}) = c[f(\mathbf{x})]$.

(ii) If f and g are continuous at $\mathbf{x}_0$, so is $f + g$, where the sum of f and g is defined by $(f + g)(\mathbf{x}) = f(\mathbf{x}) + g(\mathbf{x})$.

(iii) If f and g are continuous at $\mathbf{x}_0$ and $m = 1$, then the product function fg defined by $(fg)(\mathbf{x}) = f(\mathbf{x})g(\mathbf{x})$ is continuous at $\mathbf{x}_0$.

(iv) If $f\colon A \subset \mathbb{R}^n \to \mathbb{R}$ is continuous at $\mathbf{x}_0$ and nowhere zero on A, then the quotient $1/f$ is continuous at $\mathbf{x}_0$, where $(1/f)(\mathbf{x}) = 1/f(\mathbf{x})$.

(v) If $f\colon A \subset \mathbb{R}^n \to \mathbb{R}^m$ and $f(\mathbf{x}) = (f_1(\mathbf{x}), \ldots, f_m(\mathbf{x}))$, then f is continuous at $\mathbf{x}_0$ if and only if each of the real-valued functions $f_1, \ldots, f_m$ is continuous at $\mathbf{x}_0$.

A variant of (iv) is often used: If $f(\mathbf{x}_0) \neq 0$ and f is continuous, then $f(\mathbf{x}) \neq 0$ in a neighborhood of $\mathbf{x}_0$ and so $1/f$ is defined in that neighborhood, and $1/f$ is continuous at $\mathbf{x}_0$.

EXAMPLE 10 Let $f\colon \mathbb{R}^2 \to \mathbb{R}^2, (x, y) \mapsto (x^2y, (y + x^3)/(1 + x^2))$. Show that f is continuous.

SOLUTION To see this, it is sufficient, by property (v) of Theorem 4, to show that each component is continuous. As we have mentioned, any polynomial in two variables is continuous; thus, the map $(x, y) \mapsto x^2y$ is continuous. Because $1 + x^2$ is continuous and nonzero, by property (iv), we know that $1/(1 + x^2)$ is continuous; hence, $(y + x^3)/(1 + x^2)$ is a product of continuous functions, and by (iii) is continuous. ▲

Similar reasoning applies to examples like the function $\mathbf{c}\colon \mathbb{R} \to \mathbb{R}^3$ given by $\mathbf{c}(t) = (t^2, 1, t^3/(1 + t^2))$ to show they are continuous as well.

Composition

Next we discuss *composition*, another basic operation that can be performed on functions. If g maps A to B and f maps B to C, the ***composition of g with f***, or of f on g, denoted by $f \circ g$, maps A to C by sending $\mathbf{x} \mapsto f(g(\mathbf{x}))$ (see Figure 2.2.15). For example, $\sin(x^2)$ is the composition of $x \mapsto x^2$ with $y \mapsto \sin y$.

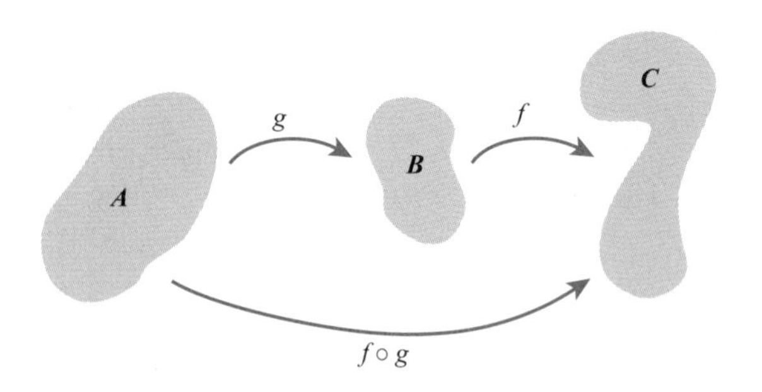

Figure 2.2.15 The composition of f on g.

THEOREM 5: Continuity of Compositions Let $g\colon A \subset \mathbb{R}^n \to \mathbb{R}^m$ and let $f\colon B \subset \mathbb{R}^m \to \mathbb{R}^p$. Suppose $g(A) \subset B$, so that $f \circ g$ is defined on A. If g is continuous at $\mathbf{x}_0 \in A$ and f is continuous at $\mathbf{y}_0 = g(\mathbf{x}_0)$, then $f \circ g$ is continuous at $\mathbf{x}_0$.

The intuition behind this is easy; the formal proof in the Internet supplement follows a similar pattern. Intuitively, we must show that as $\mathbf{x}$ gets close to $\mathbf{x}_0$, $f(g(\mathbf{x}))$ gets close to $f(g(\mathbf{x}_0))$. But as $\mathbf{x}$ gets close to $\mathbf{x}_0$, $g(\mathbf{x})$ gets close to $g(\mathbf{x}_0)$ (by continuity of g at $\mathbf{x}_0$); and as $g(\mathbf{x})$ gets close to $g(\mathbf{x}_0)$, $f(g(\mathbf{x}))$ gets close to $f(g(\mathbf{x}_0))$ [by continuity of f at $g(\mathbf{x}_0)$].

EXAMPLE 11 Let $f(x, y, z) = (x^2 + y^2 + z^2)^{30} + \sin z^3$. Show that f is continuous.

SOLUTION Here we can write f as a sum of the two functions $(x^2 + y^2 + z^2)^{30}$ and $\sin z^3$, so it suffices to show that each is continuous. The first is the composite of $(x, y, z) \mapsto (x^2 + y^2 + z^2)$ with $u \mapsto u^{30}$, and the second is the composite of $(x, y, z) \mapsto z^3$ with $u \mapsto \sin u$, and so we have continuity by Theorem 5. ▲

Limits in Terms of ε's and δ's

We now state a theorem (proved in the Internet supplement for Chapter 2) giving a useful formulation of the notion of limit in terms of epsilons and deltas that is often taken as the *definition* of limit. This is, in fact, another way of making precise the intuitive statement that "$f(\mathbf{x})$ is close to $\mathbf{b}$ when $\mathbf{x}$ is close to $\mathbf{x}_0$." To help understand this formulation, the reader should consider it with respect to each of the examples already presented.

THEOREM 6 Let $f\colon A \subset \mathbb{R}^n \to \mathbb{R}^m$ and let $\mathbf{x}_0$ be in A or be a boundary point of A. Then $\lim_{\mathbf{x}\to\mathbf{x}_0} f(\mathbf{x}) = \mathbf{b}$ if and only if for every number $\varepsilon > 0$ there is a $\delta > 0$ such that for any $\mathbf{x} \in A$ satisfying $0 < \|\mathbf{x} - \mathbf{x}_0\| < \delta$, we have $\|f(\mathbf{x}) - \mathbf{b}\| < \varepsilon$ (see Figure 2.2.16).

To illustrate the methodology of the epsilon-delta technique in Theorem 6, we consider the following examples.

EXAMPLE 12 Show that $\lim_{(x,y)\to(0,0)} x = 0$ using the ε-δ method.

SOLUTION Note that if $\delta > 0$, $\|(x, y) - (0, 0)\| = \sqrt{x^2 + y^2} < \delta$ implies $|x - 0| = |x| = \sqrt{x^2} \le \sqrt{x^2 + y^2} < \delta$. Thus, if $\|(x, y) - (0, 0)\| < \delta$, then $|x - 0|$ is also less than δ. Given $\varepsilon > 0$, we are required to find a $\delta > 0$ (generally depending on ε) with the property that $0 < \|(x, y) - (0, 0)\| < \delta$ implies $|x - 0| < \varepsilon$. What are we to pick as our δ? From the preceding calculation, we see that if we choose $\delta = \varepsilon$,

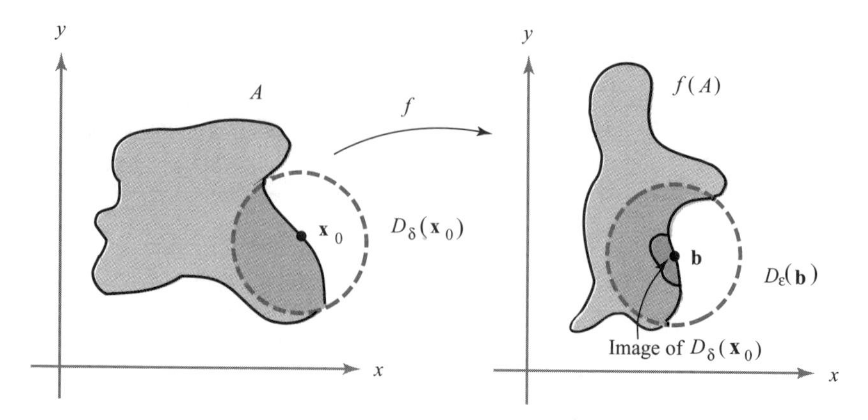

Figure 2.2.16 The geometry of the ε-δ definition of limit.

then $\|(x, y) - (0, 0)\| < \delta$ implies $|x - 0| < \varepsilon$. This shows that $\lim_{(x,y)\to(0,0)} x = 0$. Given $\varepsilon > 0$, we could have also chosen $\delta = \varepsilon/2$ or $\varepsilon/3$, but it suffices to find just one δ satisfying the requirements of the definition of a limit. ▲

EXAMPLE 13 Consider the function

$$f(x, y) = \frac{\sin(x^2 + y^2)}{x^2 + y^2}.$$

Even though f is not defined at (0, 0), determine whether $f(x, y)$ approaches some number as (x, y) approaches (0, 0).

SOLUTION From one-variable calculus or L'Hôpital's rule we know that

$$\lim_{\alpha\to 0} \frac{\sin \alpha}{\alpha} = 1.$$

Thus, it is reasonable to guess that

$$\lim_{\mathbf{v}\to(0,0)} f(\mathbf{v}) = \lim_{\mathbf{v}\to(0,0)} \frac{\sin \|\mathbf{v}\|^2}{\|\mathbf{v}\|^2} = 1.$$

Indeed, because $\lim_{\alpha\to 0}(\sin \alpha)/\alpha = 1$, given $\varepsilon > 0$ we are able to find a $\delta > 0$, with $0 < \delta < 1$, such that $0 < |\alpha| < \delta$ implies that $|(\sin \alpha)/\alpha - 1| < \varepsilon$. If $0 < \|\mathbf{v}\| < \delta$, then $0 < \|\mathbf{v}\|^2 < \delta^2 < \delta$, and therefore

$$|f(\mathbf{v}) - 1| = \left| \frac{\sin \|\mathbf{v}\|^2}{\|\mathbf{v}\|^2} - 1 \right| < \varepsilon.$$

Thus, $\lim_{\mathbf{v}\to(0,0)} f(\mathbf{v}) = 1$. If we plot $[\sin(x^2 + y^2)]/(x^2 + y^2)$ on a computer, we get a graph that is indeed well behaved near (0, 0) (Figure 2.2.17). ▲

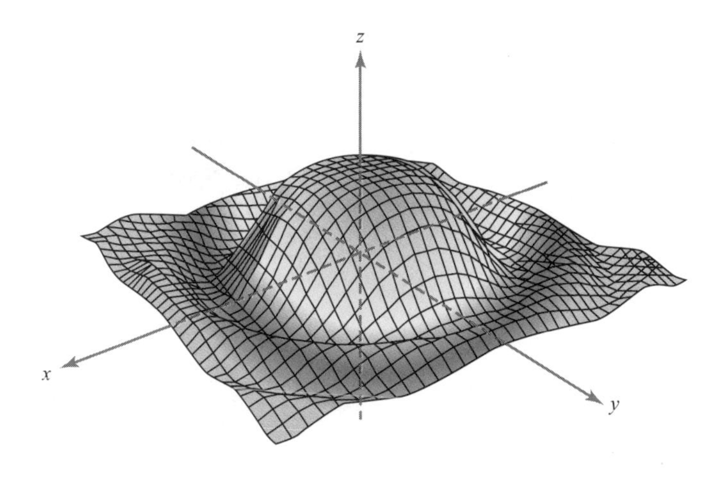

Figure 2.2.17 Graph of the function $f(x, y) = [\sin(x^2 + y^2)]/(x^2 + y^2)$.

EXAMPLE 14 Show that

$$\lim_{(x,y)\to(0,0)} \frac{x^2}{\sqrt{x^2 + y^2}} = 0.$$

SOLUTION We must show that $x^2/\sqrt{x^2 + y^2}$ is small when (x, y) is close to the origin. To do this, we use the following inequality:

$$\begin{aligned} 0 \le \frac{x^2}{\sqrt{x^2 + y^2}} &\le \frac{x^2 + y^2}{\sqrt{x^2 + y^2}} \qquad (\text{because } y^2 \ge 0) \\ &= \sqrt{x^2 + y^2}. \end{aligned}$$

Given $\varepsilon > 0$, choose $\delta = \varepsilon$. Then $\|(x, y) - (0, 0)\| = \|(x, y)\| = \sqrt{x^2 + y^2}$, and so $\|(x, y) - (0, 0)\| < \delta$ implies that

$$\left| \frac{x^2}{\sqrt{x^2 + y^2}} - 0 \right| = \frac{x^2}{\sqrt{x^2 + y^2}} \le \sqrt{x^2 + y^2} = \|(x, y) - (0, 0)\| < \delta = \varepsilon.$$

Thus, the conditions of Theorem 6 have been fulfilled and the limit is verified. ▲

EXAMPLE 15 (a) Does

$$\lim_{(x,y)\to(0,0)} x^2/(x^2 + y^2)$$

exist? [See Figure 2.2.18(a).]

(b) Prove that [see Figure 2.2.18(b)]

$$\lim_{(x,y)\to(0,0)} \frac{2x^2 y}{x^2 + y^2} = 0.$$

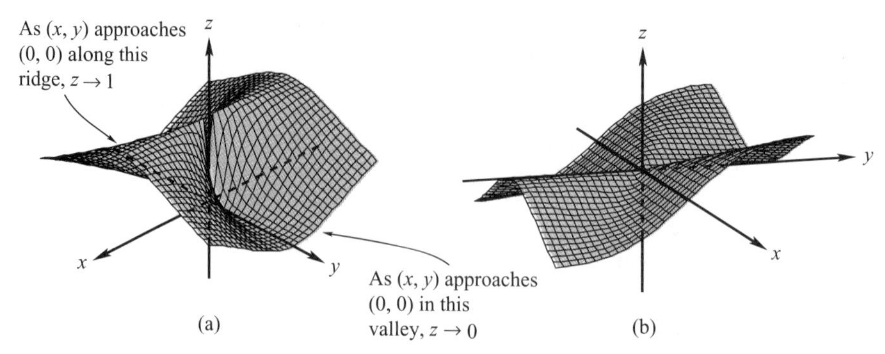

Figure 2.2.18 (a) The function $z = x^2/(x^2 + y^2)$ has no limit at (0, 0). (b) The function $z = (2x^2y)/(x^2 + y^2)$ has limit 0 at (0, 0).

SOLUTION (a) If the limit exists, $x^2/(x^2 + y^2)$ should approach a definite value, say a, as (x, y) gets near (0, 0). In particular, if (x, y) approaches zero along any given path, then $x^2/(x^2 + y^2)$ should approach the limiting value a. If (x, y) approaches (0, 0) along the line $y = 0$, the limiting value is clearly 1 (just set $y = 0$ in the preceding expression to get $x^2/x^2 = 1$). If (x, y) approaches (0, 0) along the line $x = 0$, the limiting value is

$$\lim_{y \to 0} \frac{0^2}{0^2 + y^2} = 0 \neq 1.$$

Hence, $\displaystyle\operatorname*{limit}_{(x,y)\to(0,0)} x^2/(x^2 + y^2)$ does not exist.

(b) Note that

$$\left|\frac{2x^2y}{x^2 + y^2}\right| \leq \left|\frac{2x^2y}{x^2}\right| = 2|y|.$$

Thus, given $\varepsilon > 0$, choose $\delta = \varepsilon/2$; then $0 < \|(x, y) - (0, 0)\| = \sqrt{x^2 + y^2} < \delta$ implies $|y| < \delta$, and thus

$$\left|\frac{2x^2y}{x^2 + y^2} - 0\right| < 2\delta = \varepsilon. \quad \blacktriangle$$

Using the ε-δ notation, we are led to the following reformulation of the definition of continuity.

THEOREM 7 Let $f\colon A \subset \mathbb{R}^n \to \mathbb{R}^m$ be given. Then f is continuous at $\mathbf{x}_0 \in A$ if and only if for every number $\varepsilon > 0$ there is a number $\delta > 0$ such that

$$\mathbf{x} \in A \qquad \text{and} \qquad \|\mathbf{x} - \mathbf{x}_0\| < \delta \qquad \text{implies} \qquad \|f(\mathbf{x}) - f(\mathbf{x}_0)\| < \varepsilon.$$

The proof is almost immediate. Notice that in Theorem 6 we insisted that $0 < \|\mathbf{x} - \mathbf{x}_0\|$, that is, $\mathbf{x} \neq \mathbf{x}_0$. That is *not* imposed here; indeed, the conclusion of Theorem 7 is certainly valid when $\mathbf{x} = \mathbf{x}_0$, and so there is no need to exclude this case. Here we do care about the value of f at $\mathbf{x}_0$; we want f at nearby points to be close to *this* value.

EXERCISES

In the following exercises the reader may assume that the exponential, sine, and cosine functions are continuous and may freely use techniques from one-variable calculus, such as L'Hôpital's rule.

Show that the subsets of the plane in Exercises 1–4 are open:

1. $A = \{(x, y) \mid -1 < x < 1, -1 < y < 1\}$

2. $B = \{(x, y) \mid y > 0\}$

3. $C = \{(x, y) \mid 2 < x^2 + y^2 < 4\}$

4. $D = \{(x, y) \mid x \neq 0 \text{ and } y \neq 0\}$

5. Compute the limits:

(a) $\displaystyle\operatorname*{limit}_{(x,y)\to(0,1)} x^3 y$

(b) $\displaystyle\operatorname*{limit}_{x\to 0} \frac{\cos x - 1}{x^2}$

(c) $\displaystyle\operatorname*{limit}_{h\to 0} \frac{e^h - 1}{h}$.

6. Compute the following limits:

(a) $\displaystyle\operatorname*{limit}_{(x,y)\to(0,1)} e^x y$

(b) $\displaystyle\operatorname*{limit}_{x\to 0} \frac{\sin^2 x}{x}$

(c) $\displaystyle\operatorname*{limit}_{x\to 0} \frac{\sin^2 x}{x^2}$

7. Compute the following limits:

(a) $\displaystyle\operatorname*{limit}_{x\to 3}(x^2 - 3x + 5)$

(b) $\displaystyle\operatorname*{limit}_{x\to 0} \sin x$

(c) $\displaystyle\operatorname*{limit}_{h\to 0} \frac{(x+h)^2 - x^2}{h}$

8. Compute the following limits if they exist:

(a) $\displaystyle\operatorname*{limit}_{(x,y)\to(0,0)} \frac{(x+y)^2 - (x-y)^2}{xy}$

(b) $\displaystyle\operatorname*{limit}_{(x,y)\to(0,0)} \frac{\sin xy}{y}$

(c) $\displaystyle\operatorname*{limit}_{(x,y)\to(0,0)} \frac{x^3 - y^3}{x^2 + y^2}$

9. Compute the following limits if they exist:

(a) $\displaystyle\operatorname*{limit}_{(x,y)\to(0,0)} \frac{e^{xy} - 1}{y}$

(b) $\displaystyle\operatorname*{limit}_{(x,y)\to(0,0)} \frac{\cos(xy) - 1}{x^2 y^2}$

(c) $\displaystyle\operatorname*{limit}_{(x,y)\to(0,0)} \frac{xy}{x^2 + y^2 + 2}$

10. Compute the following limits, if they exist:

(a) $\displaystyle \underset{(x,y)\to(0,0)}{\text{limit}} \frac{e^{xy}}{x+1}$

(b) $\displaystyle \underset{(x,y)\to(0,0)}{\text{limit}} \frac{\cos x - 1 - (x^2/2)}{x^4+y^4}$

(c) $\displaystyle \underset{(x,y)\to(0,0)}{\text{limit}} \frac{(x-y)^2}{x^2+y^2}$

11. Compute the following limits if they exist:

(a) $\displaystyle \underset{(x,y)\to(0,0)}{\text{limit}} \frac{\sin xy}{xy}$

(b) $\displaystyle \underset{(x,y,z)\to(0,0,0)}{\text{limit}} \frac{\sin(xyz)}{xyz}$

(c) $\displaystyle \underset{(x,y,z)\to(0,0,0)}{\text{limit}} f(x,y,z)$, where $f(x,y,z)=(x^2+3y^2)/(x+1)$.

12. Compute the following limits if they exist:

(a) $\displaystyle \underset{x\to 0}{\text{limit}} \frac{\sin 2x - 2x}{x^3}$

(b) $\displaystyle \underset{(x,y)\to(0,0)}{\text{limit}} \frac{\sin 2x - 2x + y}{x^3+y}$

(c) $\displaystyle \underset{(x,y,z)\to(0,0,0)}{\text{limit}} \frac{2x^2y\cos z}{x^2+y^2}$

13. Compute $\underset{\mathbf{x}\to\mathbf{x}_0}{\text{limit}} f(\mathbf{x})$, if it exists, for the following cases:

(a) $f\colon \mathbb{R}\to\mathbb{R},\ x\mapsto |x|,\ x_0=1$
(b) $f\colon \mathbb{R}^n\to\mathbb{R},\ \mathbf{x}\mapsto \|\mathbf{x}\|$, arbitrary $\mathbf{x}_0$
(c) $f\colon \mathbb{R}\to\mathbb{R}^2,\ x\mapsto (x^2,e^x),\ x_0=1$
(d) $f\colon \mathbb{R}^2\backslash\{(0,0)\}\to\mathbb{R}^2,\ (x,y)\mapsto(\sin(x-y), e^{x(y+1)}-x-1)/\|(x,y)\|,\ \mathbf{x}_0=(0,0)$.

14. Let $A\subset\mathbb{R}^2$ be the open unit disk $D_1(0,0)$ with the point $\mathbf{x}_0=(1,0)$ added, and let $f\colon A\to\mathbb{R},\ \mathbf{x}\mapsto f(\mathbf{x})$ be the constant function $f(\mathbf{x})=1$. Show that $\underset{\mathbf{x}\to\mathbf{x}_0}{\text{limit}} f(\mathbf{x})=1$.

15. If $f\colon\mathbb{R}^n\to\mathbb{R}$ and $g\colon\mathbb{R}^n\to\mathbb{R}$ are continuous, show that the functions

$$f^2g\colon \mathbb{R}^n\to\mathbb{R},\ \mathbf{x}\mapsto [f(\mathbf{x})]^2g(\mathbf{x})$$

and

$$f^2+g\colon \mathbb{R}^n\to\mathbb{R},\ x\mapsto [f(\mathbf{x})]^2+g(\mathbf{x})$$

are continuous.

16. (a) Show that $f\colon\mathbb{R}\to\mathbb{R},\ x\mapsto(1-x)^8+\cos(1+x^3)$ is continuous.
(b) Show that the map $f\colon\mathbb{R}\to\mathbb{R},\ x\mapsto x^2e^x/(2-\sin x)$ is continuous.

17. (a) Can $[\sin(x+y)]/(x+y)$ be made continuous by suitably defining it at $(0,0)$?
(b) Can $xy/(x^2+y^2)$ be made continuous by suitably defining it at $(0,0)$?
(c) Prove that $f\colon\mathbb{R}^2\to\mathbb{R},\ (x,y)\mapsto ye^x+\sin x+(xy)^4$ is continuous.

18. Using either ε's and δ's or spherical coordinates, show that

$$\underset{(x,y,z)\to(0,0,0)}{\text{limit}} \frac{xyz}{x^2+y^2+z^2}=0.$$

19. Use the ε-δ formulation of limits to prove that $x^2 \to 4$ as $x \to 2$. Give another proof using Theorem 3.

20. (a) Prove that for $\mathbf{x} \in \mathbb{R}^n$ and $s < t$, $D_s(\mathbf{x}) \subset D_t(\mathbf{x})$.
(b) Prove that if U and V are neighborhoods of $\mathbf{x} \in \mathbb{R}^n$, then so are $U \cap V$ and $U \cup V$.
(c) Prove that the boundary points of an open interval $(a, b) \subset \mathbb{R}$ are the points a and b.

21. Suppose $\mathbf{x}$ and $\mathbf{y}$ are in $\mathbb{R}^n$ and $\mathbf{x} \neq \mathbf{y}$. Show that there is a continuous function $f\colon \mathbb{R}^n \to \mathbb{R}$ with $f(\mathbf{x}) = 1$, $f(\mathbf{y}) = 0$, and $0 \leq f(\mathbf{z}) \leq 1$ for every $\mathbf{z}$ in $\mathbb{R}^n$.

22. Let $f\colon A \subset \mathbb{R}^n \to \mathbb{R}$ be given and let $\mathbf{x}_0$ be a boundary point of A. We say that $\operatorname{limit}_{\mathbf{x}\to\mathbf{x}_0} f(\mathbf{x}) = \infty$ if for every $N > 0$ there is a $\delta > 0$ such that $0 < \|\mathbf{x} - \mathbf{x}_0\| < \delta$ and $\mathbf{x} \in A$ implies $f(\mathbf{x}) > N$.

(a) Prove that $\operatorname{limit}_{x\to 1} (x-1)^{-2} = \infty$.
(b) Prove that $\operatorname{limit}_{x\to 0} 1/|x| = \infty$. Is it true that $\operatorname{limit}_{x\to 0} 1/x = \infty$?
(c) Prove that $\operatorname{limit}_{(x,y)\to(0,0)} 1/(x^2+y^2) = \infty$.

23. Let $b \in \mathbb{R}$ and $f\colon \mathbb{R}\backslash[b] \to \mathbb{R}$ be a function. We write $\operatorname{limit}_{x\to b-} f(x) = L$ and say that L is the ***left-hand limit*** of f at b, if for every $\varepsilon > 0$, there is a $\delta > 0$ such that $x < b$ and $0 < |x - b| < \delta$ implies $|f(x) - L| < \varepsilon$.

(a) Formulate a definition of ***right-hand limit***, or $\operatorname{limit}_{x\to b+} f(x)$.
(b) Find $\operatorname{limit}_{x\to 0-} 1/(1+e^{1/x})$ and $\operatorname{limit}_{x\to 0+} 1/(1+e^{1/x})$.
(c) Sketch the graph of $1/(1+e^{1/x})$.

24. Show that f is continuous at $\mathbf{x}_0$ if and only if

$$\operatorname{limit}_{\mathbf{x}\to\mathbf{x}_0} \|f(\mathbf{x}) - f(\mathbf{x}_0)\| = 0.$$

25. Let $f\colon A \subset \mathbb{R}^n \to \mathbb{R}^m$ satisfy $\|f(\mathbf{x}) - f(\mathbf{y})\| \leq K\|\mathbf{x} - \mathbf{y}\|^\alpha$ for all $\mathbf{x}$ and $\mathbf{y}$ in A for positive constants K and α. Show that f is continuous. (Such functions are called ***Hölder-continuous*** or, if $\alpha = 1$, ***Lipschitz-continuous***.)

26. Show that $f\colon \mathbb{R}^n \to \mathbb{R}^m$ is continuous at all points if and only if the inverse image of every open set is open.

27. (a) Find a specific number $\delta > 0$ such that if $|a| < \delta$, then $|a^3 + 3a^2 + a| < 1/100$.
(b) Find a specific number $\delta > 0$ such that if $x^2 + y^2 < \delta^2$, then

$$|x^2 + y^2 + 3xy + 180xy^5| < 1/10{,}000.$$

2.3 Differentiation

In Section 2.1 we considered a few methods for graphing functions. By these methods alone it may be impossible to compute enough information to grasp even the general features of a complicated function. From elementary calculus we know that the idea of the derivative can greatly aid us in this task; for example, it enables us to locate maxima and minima and to compute rates of change. The derivative also has many applications beyond this, as the student surely has discovered in elementary calculus.

Intuitively, we know from our work in Section 2.2 that a continuous function is one that has no "breaks" in its graph. A differentiable function from $\mathbb{R}^2$ to $\mathbb{R}$ ought to be such that not only are there no breaks in its graph, but there is a well-defined plane tangent to the graph at each point. Thus, there must not be any sharp folds, corners, or peaks in the graph (see Figure 2.3.1). In other words, the graph must be *smooth*.

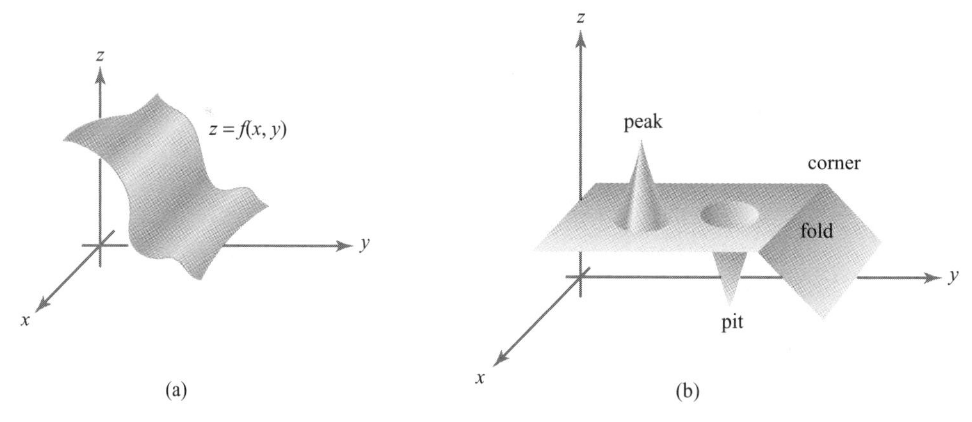

Figure 2.3.1 (a) A smooth graph and (b) a nonsmooth one.

Partial Derivatives

To make these ideas precise, we need a sound definition of what we mean by the phrase "$f(x_1, \dots, x_n)$ is differentiable at $\mathbf{x} = (x_1, \dots, x_n)$." Actually, this definition is not quite as simple as one might think. Toward this end, however, let us introduce the notion of the *partial derivative*. This notion relies only on our knowledge of one-variable calculus. (A quick review of the definition of the derivative in a one-variable calculus text might be advisable at this point.)

DEFINITION: Partial Derivatives Let $U \subset \mathbb{R}^n$ be an open set and suppose $f\colon U \subset \mathbb{R}^n \to \mathbb{R}$ is a real-valued function. Then $\partial f/\partial x_1, \dots, \partial f/\partial x_n$, the ***partial derivatives*** of f with respect to the first, second, ..., nth variable, are the real-valued functions of n variables, which, at the point $(x_1, \dots, x_n) = \mathbf{x}$, are defined by

$$\begin{aligned}\frac{\partial f}{\partial x_j}(x_1, \dots, x_n) &= \lim_{h\to 0} \frac{f(x_1, x_2, \dots, x_j + h, \dots, x_n) - f(x_1, \dots, x_n)}{h}\\ &= \lim_{h\to 0} \frac{f(\mathbf{x} + h\mathbf{e}_j) - f(\mathbf{x})}{h}\end{aligned}$$

if the limits exist, where $1 \leq j \leq n$ and $\mathbf{e}_j$ is the jth standard basis vector defined by $\mathbf{e}_j = (0, \dots, 1, \dots, 0)$, with 1 in the jth slot (see Section 1.5). The domain of the function $\partial f/\partial x_j$ is the set of $\mathbf{x} \in \mathbb{R}^n$ for which the limit exists.

In other words, $\partial f/\partial x_j$ is just the derivative of f with respect to the variable x_j, with the other variables held fixed. If $f\colon \mathbb{R}^3 \to \mathbb{R}$, we shall often use the notation $\partial f/\partial x, \partial f/\partial y, \partial f/\partial z$ in place of $\partial f/\partial x_1, \partial f/\partial x_2, \partial f/\partial x_3$. If $f\colon U \subset \mathbb{R}^n \to \mathbb{R}^m$, then we can write

$$f(x_1, \dots, x_n) = (f_1(x_1, \dots, x_n), \dots, f_m(x_1, \dots, x_n)),$$

so that we can speak of the partial derivatives of each component; for example, $\partial f_m/\partial x_n$ is the partial derivative of the mth *component* with respect to x_n, the nth variable.

EXAMPLE 1 If $f(x, y) = x^2y + y^3$, find $\partial f/\partial x$ and $\partial f/\partial y$.

SOLUTION To find $\partial f/\partial x$ we hold y constant (think of it as some number, say 1) and differentiate only with respect to x; this yields

$$\frac{\partial f}{\partial x} = \frac{d(x^2y + y^3)}{dx} = 2xy.$$

Similarly, to find $\partial f/\partial y$ we hold x constant and differentiate only with respect to y:

$$\frac{\partial f}{\partial y} = \frac{d(x^2y + y^3)}{dy} = x^2 + 3y^2. \quad \blacktriangle$$

To indicate that a partial derivative is to be evaluated at a particular point, for example, at (x_0, y_0), we write

$$\frac{\partial f}{\partial x}(x_0, y_0) \quad \text{or} \quad \left.\frac{\partial f}{\partial x}\right|_{x=x_0, y=y_0} \quad \text{or} \quad \left.\frac{\partial f}{\partial x}\right|_{(x_0, y_0)}.$$

When we write $z = f(x, y)$ for the dependent variable, we sometimes write $\partial z/\partial x$ for $\partial f/\partial x$. Strictly speaking, this is an abuse of notation, but it is common practice to use these two notations interchangeably.

EXAMPLE 2 If $z = \cos xy + x \cos y = f(x, y)$, find the two partial derivatives $(\partial z/\partial x)(x_0, y_0)$ and $(\partial z/\partial y)(x_0, y_0)$.

SOLUTION First we fix y_0 and differentiate with respect to x, giving

$$\begin{aligned}\frac{\partial z}{\partial x}(x_0, y_0) &= \left.\frac{d(\cos xy_0 + x \cos y_0)}{dx}\right|_{x=x_0}\\ &= (-y_0 \sin xy_0 + \cos y_0)|_{x=x_0}\\ &= -y_0 \sin x_0y_0 + \cos y_0.\end{aligned}$$

Similarly, we fix x_0 and differentiate with respect to y to obtain

$$\begin{aligned}\frac{\partial z}{\partial y}(x_0, y_0) &= \left.\frac{d(\cos x_0 y + x_0 \cos y)}{dy}\right|_{y=y_0} \\ &= (-x_0 \sin x_0 y - x_0 \sin y)|_{y=y_0} \\ &= -x_0 \sin x_0 y_0 - x_0 \sin y_0. \quad \blacktriangle\end{aligned}$$

EXAMPLE 3 Find $\partial f/\partial x$ if $f(x, y) = xy/\sqrt{x^2 + y^2}$.

SOLUTION By the quotient rule,

$$\frac{\partial f}{\partial x} = \frac{y\sqrt{x^2 + y^2} - xy(x/\sqrt{x^2 + y^2})}{x^2 + y^2} = \frac{y(x^2 + y^2) - x^2 y}{(x^2 + y^2)^{3/2}} = \frac{y^3}{(x^2 + y^2)^{3/2}}. \quad \blacktriangle$$

A definition of differentiability that requires only the existence of partial derivatives turns out to be insufficient. Many standard results, such as the chain rule for functions of several variables would not follow, as Example 4 shows. Below, we shall see how to rectify this situation.

EXAMPLE 4 Let $f(x, y) = x^{1/3} y^{1/3}$. By definition,

$$\frac{\partial f}{\partial x}(0, 0) = \lim_{h \to 0} \frac{f(h, 0) - f(0, 0)}{h} = \lim_{h \to 0} \frac{0 - 0}{h} = 0,$$

and, similarly, $(\partial f/\partial y)(0, 0) = 0$ (these are not indeterminate forms!). It is necessary to use the original definition of partial derivatives, because the functions $x^{1/3}$ and $y^{1/3}$ are not themselves differentiable at 0. Suppose we restrict f to the line $y = x$ to get $f(x, x) = x^{2/3}$ (see Figure 2.3.2). We can view the substitution $y = x$ as the composition $f \circ g$ of the function g: $\mathbb{R} \to \mathbb{R}^2$, defined by $g(x) = (x, x)$, and f: $\mathbb{R}^2 \to \mathbb{R}$, defined by $f(x, y) = x^{1/3} y^{1/3}$.

Thus, the composite $f \circ g$ is given by $(f \circ g)(x) = x^{2/3}$. Each component of g is differentiable in x, and f has partial derivatives at $(0, 0)$, but *$f \circ g$ is not differentiable* at $x = 0$, in the sense of one-variable calculus. In other words, *the composition of f with g is not differentiable* in contrast to the calculus of functions of one variable, where the composition of differentiable functions *is* differentiable. Below, we shall give a definition of differentiability that has the pleasant consequence that the composition of differentiable functions *is* differentiable.

There is another reason for being dissatisfied with the mere existence of partial derivatives of $f(x, y) = x^{1/3} y^{1/3}$: There is no plane tangent, in any reasonable sense, to the graph at $(0, 0)$. The xy plane is tangent to the graph along the x and y axes because f has slope zero at $(0, 0)$ along these axes; that is, $\partial f/\partial x = 0$ and $\partial f/\partial y = 0$

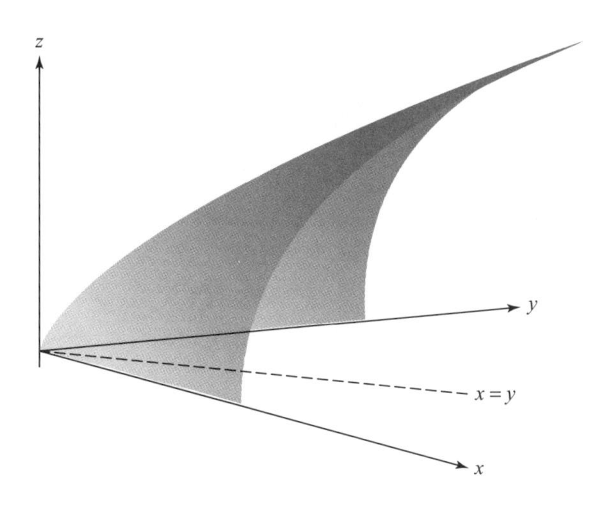

Figure 2.3.2 The portion of the graph of $x^{1/3}y^{1/3}$ in the first quadrant.

at (0, 0). Thus, if there is a tangent plane, it must be the xy plane. However, as is evident from Figure 2.3.2, the xy plane is not tangent to the graph in other directions, because the graph has a severe crinkle, and so the xy plane cannot be said to be tangent to the graph of f. ▲

The Linear Approximation

To "motivate" our definition of differentiability, let us compute what the equation of the plane tangent to the graph of $f: \mathbb{R}^2 \to \mathbb{R}, (x, y) \mapsto f(x, y)$ at (x_0, y_0) ought to be if f is smooth enough. In $\mathbb{R}^3$, a nonvertical plane has an equation of the form

$$z = ax + by + c.$$

If it is to be the plane tangent to the graph of f, the slopes along the x and y axes must be equal to $\partial f/\partial x$ and $\partial f/\partial y$, the rates of change of f with respect to x and y. Thus, $a = \partial f/\partial x, b = \partial f/\partial y$ [evaluated at (x_0, y_0)]. Finally, we may determine the constant c from the fact that $z = f(x_0, y_0)$ when $x = x_0, y = y_0$. Thus, we get the ***linear approximation***:

$$z = f(x_0, y_0) + \left[\frac{\partial f}{\partial x}(x_0, y_0)\right](x - x_0) + \left[\frac{\partial f}{\partial y}(x_0, y_0)\right](y - y_0), \tag{1}$$

which should be the equation of the plane tangent to the graph of f at (x_0, y_0), if f is "smooth enough" (see Figure 2.3.3).

Our definition of differentiability will mean in effect that the plane defined by the linear approximation (1) is a "good" approximation of f near (x_0, y_0). To get an idea of what one might mean by a good approximation, let us return for a moment to

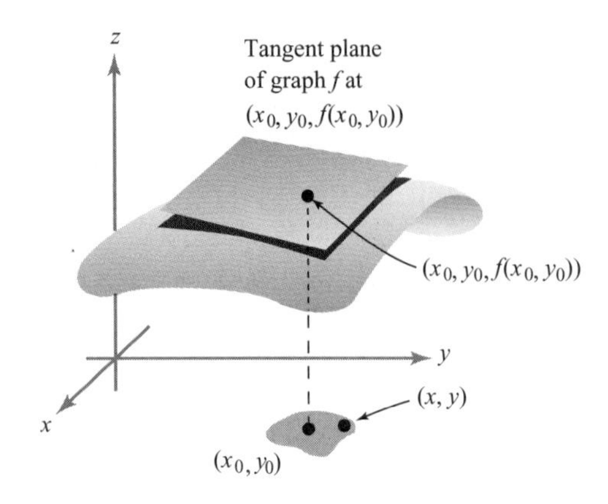

Figure 2.3.3 For points (x, y) near (x_0, y_0), the graph of the tangent plane is close to the graph of f.

one-variable calculus. If f is differentiable at a point x_0, then we know that

$$\lim_{\Delta x \to 0} \frac{f(x_0 + \Delta x) - f(x_0)}{\Delta x} = f'(x_0).$$

Let $x = x_0 + \Delta x$ and rewrite this as

$$\lim_{x \to x_0} \frac{f(x) - f(x_0)}{x - x_0} = f'(x_0).$$

Using the trivial limit $\lim_{x \to x_0} f'(x_0) = f'(x_0)$, we can rewrite the preceding equation as

$$\lim_{x \to x_0} \frac{f(x) - f(x_0)}{x - x_0} = \lim_{x \to x_0} f'(x_0);$$

that is,

$$\lim_{x \to x_0} \left[\frac{f(x) - f(x_0)}{x - x_0} - f'(x_0) \right] = 0;$$

that is,

$$\lim_{x \to x_0} \frac{f(x) - f(x_0) - f'(x_0)(x - x_0)}{x - x_0} = 0.$$

Thus, the tangent line l through $(x_0, f(x_0))$ with slope $f'(x_0)$ is close to f in the sense that the difference between $f(x)$ and $l(x) = f(x_0) + f'(x_0)(x - x_0)$, the equation of the tangent line goes to zero *even* when divided by $x - x_0$ as x goes to x_0. This is the notion of a "good approximation" that we will adapt to functions of several variables, with the tangent line replaced by the tangent plane [see equation (1), given earlier].

Differentiability for Functions of Two Variables

Using the linear approximation, we are ready to define the notion of differentiability.

DEFINITION: Differentiable: Two Variables Let $f\colon \mathbb{R}^2 \to \mathbb{R}$. We say f is ***differentiable*** at (x_0, y_0), if $\partial f/\partial x$ and $\partial f/\partial y$ exist at (x_0, y_0) and if

$$\frac{f(x, y) - f(x_0, y_0) - \left[\dfrac{\partial f}{\partial x}(x_0, y_0)\right](x - x_0) - \left[\dfrac{\partial f}{\partial y}(x_0, y_0)\right](y - y_0)}{\|(x, y) - (x_0, y_0)\|} \to 0 \tag{2}$$

as $(x, y) \to (x_0, y_0)$. This equation expresses what we mean by saying that

$$f(x_0, y_0) + \left[\frac{\partial f}{\partial x}(x_0, y_0)\right](x - x_0) + \left[\frac{\partial f}{\partial y}(x_0, y_0)\right](y - y_0)$$

is a ***good approximation*** to the function f.

It is not always easy to use this definition to see whether f is differentiable, but it will be easy to use another criterion, given shortly in Theorem 9.

Tangent Plane

We have used the informal notion of the plane tangent to the graph of a function to motivate our definition of differentiability. Now we are ready to adopt a formal definition of the tangent plane.

DEFINITION: Tangent Plane Let $f\colon \mathbb{R}^2 \to \mathbb{R}$ be differentiable at $\mathbf{x}_0 = (x_0, y_0)$. The plane in $\mathbb{R}^3$ defined by the equation

$$z = f(x_0, y_0) + \left[\frac{\partial f}{\partial x}(x_0, y_0)\right](x - x_0) + \left[\frac{\partial f}{\partial y}(x_0, y_0)\right](y - y_0),$$

is called the ***tangent plane*** of the graph of f at the point (x_0, y_0).

EXAMPLE 5 Compute the plane tangent to the graph of $z = x^2 + y^4 + e^{xy}$ at the point $(1, 0, 2)$.

SOLUTION Use formula (1), with $x_0 = 1$, $y_0 = 0$, and $z_0 = f(x_0, y_0) = 2$. The partial derivatives are

$$\frac{\partial z}{\partial x} = 2x + ye^{xy} \qquad \text{and} \qquad \frac{\partial z}{\partial y} = 4y^3 + xe^{xy}.$$

At (1, 0, 2), these partial derivatives are 2 and 1, respectively. Thus, by formula (1), the tangent plane is

$$z = 2(x-1) + 1(y-0) + 2, \qquad \text{that is,} \qquad z = 2x + y. \quad \blacktriangle$$

Let us write $\mathbf{D}f(x_0, y_0)$ for the row matrix

$$\left[\frac{\partial f}{\partial x}(x_0, y_0) \quad \frac{\partial f}{\partial y}(x_0, y_0)\right],$$

so that the definition of differentiability asserts that

$$f(x_0, y_0) + \mathbf{D}f(x_0, y_0)\begin{bmatrix} x - x_0 \\ y - y_0 \end{bmatrix}$$

$$= f(x_0, y_0) + \left[\frac{\partial f}{\partial x}(x_0, y_0)\right](x - x_0) + \left[\frac{\partial f}{\partial y}(x_0, y_0)\right](y - y_0) \qquad (3)$$

is our good approximation to f near (x_0, y_0). As earlier, "good" is taken in the sense that expression (3) differs from $f(x, y)$ by something small times $\sqrt{(x - x_0)^2 + (y - y_0)^2}$. We say that expression (3) is the ***best linear approximation*** to f near (x_0, y_0).

Differentiability: The General Case

Now we are ready to give a definition of differentiability for maps f of $\mathbb{R}^n$ to $\mathbb{R}^m$, using the preceding discussion as motivation. The derivative $\mathbf{D}f(\mathbf{x}_0)$ of $f = (f_1, \ldots, f_m)$ at a point $\mathbf{x}_0$ is a matrix $\mathbf{T}$ whose elements are $t_{ij} = \partial f_i / \partial x_j$ evaluated at $\mathbf{x}_0$.[2]

DEFINITION: Differentiable, n Variables, m Functions Let U be an open set in $\mathbb{R}^n$ and let $f\colon U \subset \mathbb{R}^n \to \mathbb{R}^m$ be a given function. We say that f is ***differentiable*** at $\mathbf{x}_0 \in U$ if the partial derivatives of f exist at $\mathbf{x}_0$ and if

$$\lim_{\mathbf{x} \to \mathbf{x}_0} \frac{\| f(\mathbf{x}) - f(\mathbf{x}_0) - \mathbf{T}(\mathbf{x} - \mathbf{x}_0) \|}{\| \mathbf{x} - \mathbf{x}_0 \|} = 0, \qquad (4)$$

where $\mathbf{T} = \mathbf{D}f(\mathbf{x}_0)$ is the $m \times n$ matrix with matrix elements $\partial f_i / \partial x_j$ evaluated at $\mathbf{x}_0$ and $\mathbf{T}(\mathbf{x} - \mathbf{x}_0)$ means the product of $\mathbf{T}$ with $\mathbf{x} - \mathbf{x}_0$ (regarded as a column matrix). We call $\mathbf{T}$ the ***derivative*** of f at $\mathbf{x}_0$.

---

[2] It turns out that we need to postulate the existence of only *some* matrix giving the best linear approximation near $\mathbf{x}_0 \in \mathbb{R}^n$, because in fact this matrix is *necessarily* the matrix whose ijth entry is $\partial f_i / \partial x_j$ (see the Internet supplement for Chapter 2).

We shall always denote the derivative $\mathbf{T}$ of f at $\mathbf{x}_0$ by $\mathbf{D}f(\mathbf{x}_0)$, although in some books it is denoted $df(\mathbf{x}_0)$ and referred to as the ***differential*** of f. In the case where $m = 1$, the matrix $\mathbf{T}$ is just the row matrix

$$\left[\frac{\partial f}{\partial x_1}(\mathbf{x}_0) \quad \cdots \quad \frac{\partial f}{\partial x_n}(\mathbf{x}_0)\right].$$

(Sometimes, when there is danger of confusion, we separate the entries by commas.) Furthermore, setting $n = 2$ and putting the result back into equation (4), we see that conditions (2) and (4) do agree. Thus, if we let $\mathbf{h} = \mathbf{x} - \mathbf{x}_0$, a real-valued function f of n variables is differentiable at a point $\mathbf{x}_0$ if

$$\operatorname*{limit}_{\mathbf{h}\to 0} \frac{1}{\|\mathbf{h}\|}\left| f(\mathbf{x}_0 + \mathbf{h}) - f(\mathbf{x}_0) - \sum_{j=1}^{n} \frac{\partial f}{\partial x_j}(\mathbf{x}_0)h_j \right| = 0,$$

because

$$\mathbf{Th} = \sum_{j=1}^{n} h_j \frac{\partial f}{\partial x_j}(\mathbf{x}_0).$$

For the general case of f mapping a subset of $\mathbb{R}^n$ to $\mathbb{R}^m$, the derivative is the $m \times n$ matrix given by

$$\mathbf{D}f(\mathbf{x}_0) = \begin{bmatrix} \dfrac{\partial f_1}{\partial x_1} & \cdots & \dfrac{\partial f_1}{\partial x_n} \\ \vdots & & \vdots \\ \dfrac{\partial f_m}{\partial x_1} & \cdots & \dfrac{\partial f_m}{\partial x_n} \end{bmatrix},$$

where $\partial f_i/\partial x_j$ is evaluated at $\mathbf{x}_0$. The matrix $\mathbf{D}f(\mathbf{x}_0)$ is, appropriately, called the ***matrix of partial derivatives of f at $\mathbf{x}_0$***.

EXAMPLE 6 Calculate the matrices of partial derivatives for these functions.

(a) $f(x, y) = (e^{x+y} + y, y^2 x)$

(b) $f(x, y) = (x^2 + \cos y, ye^x)$

(c) $f(x, y, z) = (ze^x, -ye^z)$

SOLUTION

(a) Here $f\colon \mathbb{R}^2 \to \mathbb{R}^2$ is defined by $f_1(x, y) = e^{x+y} + y$ and $f_2(x, y) = y^2 x$. Hence, $\mathbf{D}f(x, y)$ is the 2×2 matrix

$$\mathbf{D}f(x, y) = \begin{bmatrix} e^{x+y} & e^{x+y} + 1 \\ y^2 & 2xy \end{bmatrix}.$$

(b) We have

$$\mathbf{D}f(x, y) = \begin{bmatrix} 2x & -\sin y \\ ye^x & e^x \end{bmatrix}.$$

(c) In this case,

$$\mathbf{D}f(x, y, z) = \begin{bmatrix} ze^x & 0 & e^x \\ 0 & -e^z & -ye^z \end{bmatrix}. \quad \blacktriangle$$

Gradients

For real-valued functions we use special terminology for the derivative.

> DEFINITION: Gradient Consider the special case $f\colon U \subset \mathbb{R}^n \to \mathbb{R}$. Here $\mathbf{D}f(\mathbf{x})$ is a $1 \times n$ matrix:
>
> $$\mathbf{D}f(\mathbf{x}) = \begin{bmatrix} \frac{\partial f}{\partial x_1} & \cdots & \frac{\partial f}{\partial x_n} \end{bmatrix}.$$
>
> We can form the corresponding vector $(\partial f/\partial x_1, \ldots, \partial f/\partial x_n)$, called the ***gradient*** of f and denoted by ∇f or grad f.

From the definition, we see that for $f\colon \mathbb{R}^3 \to \mathbb{R}$,

$$\nabla f = \frac{\partial f}{\partial x}\mathbf{i} + \frac{\partial f}{\partial y}\mathbf{j} + \frac{\partial f}{\partial z}\mathbf{k},$$

while for $f\colon \mathbb{R}^2 \to \mathbb{R}$,

$$\nabla f = \frac{\partial f}{\partial x}\mathbf{i} + \frac{\partial f}{\partial y}\mathbf{j}.$$

The geometric significance of the gradient will be discussed in Section 2.6. In terms of inner products, we can write the derivative of f as

$$\mathbf{D}f(\mathbf{x})(\mathbf{h}) = \nabla f(\mathbf{x}) \cdot \mathbf{h}.$$

EXAMPLE 7 Let $f\colon \mathbb{R}^3 \to \mathbb{R}$, $f(x, y, z) = xe^y$. Then

$$\nabla f = \left(\frac{\partial f}{\partial x}, \frac{\partial f}{\partial y}, \frac{\partial f}{\partial z}\right) = (e^y, xe^y, 0). \quad \blacktriangle$$

EXAMPLE 8 If $f\colon \mathbb{R}^2 \to \mathbb{R}$ is given by $(x, y) \mapsto e^{xy} + \sin xy$, then

$$\begin{aligned}\nabla f(x, y) &= (ye^{xy} + y\cos xy)\mathbf{i} + (xe^{xy} + x\cos xy)\mathbf{j}\\ &= (e^{xy} + \cos xy)(y\mathbf{i} + x\mathbf{j}). \quad \blacktriangle\end{aligned}$$

In one-variable calculus it is shown that if f is differentiable, then f is continuous. We will state in Theorem 8 that this is also true for differentiable functions of several variables. As we know, there are plenty of functions of one variable that are continuous but not differentiable, such as $f(x) = |x|$. Before stating the result, let us give an example of a function of two variables whose *partial derivatives exist at a point, but which is not continuous at that point.*

EXAMPLE 9 Let $f\colon \mathbb{R}^2 \to \mathbb{R}$ be defined by

$$f(x, y) = \begin{cases} 1 & \text{if } x = 0 \text{ or if } y = 0\\ 0 & \text{otherwise.}\end{cases}$$

Because f is constant on the x and y axes, where it equals 1,

$$\frac{\partial f}{\partial x}(0, 0) = 0 \qquad \text{and} \qquad \frac{\partial f}{\partial y}(0, 0) = 0.$$

But f is not continuous at $(0, 0)$, because $\lim\limits_{(x,y)\to(0,0)} f(x, y)$ does not exist. ▲

Some Basic Theorems

The first of these basic theorems relates differentiability and continuity.

THEOREM 8 Let $f\colon U \subset \mathbb{R}^n \to \mathbb{R}^m$ be differentiable at $\mathbf{x}_0 \in U$. Then f is continuous at $\mathbf{x}_0$.

This result is very reasonable, because "differentiability" means that there is enough smoothness to have a tangent plane, which is stronger than just being continuous. Consult the Internet supplement for Chapter 2 for the formal proof.

As we have seen, it is usually easy to tell when the partial derivatives of a function exist using what we know from one-variable calculus. However, the definition of differentiability looks somewhat complicated, and the required approximation condition in equation (4) may seem, and sometimes is, difficult to verify. Fortunately, there is a simple criterion, given in the following theorem, that tells us when a function is differentiable.

THEOREM 9 Let $f\colon U \subset \mathbb{R}^n \to \mathbb{R}^m$. Suppose the partial derivatives $\partial f_i/\partial x_j$ of f all exist and are continuous in a neighborhood of a point $\mathbf{x} \in U$. Then f is differentiable at $\mathbf{x}$.

We give the proof in the Internet supplement for Chapter 2. Notice the following hierarchy:

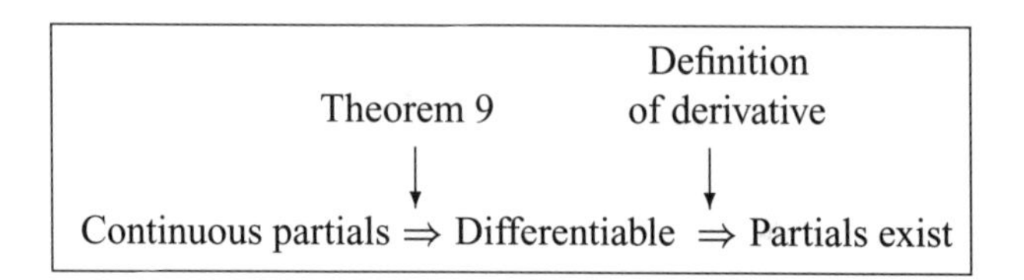

Each converse statement, obtained by reversing an implication, is invalid. [For a counterexample to the converse of the first implication, use $f(x) = x^2 \sin(1/x)$, $f(0) = 0$; for the second, see Example 1 in the Internet supplement for Chapter 2 or use Example 4 in this section.]

A function whose partial derivatives exist and are continuous is said to be of ***class*** C^1. Thus, Theorem 9 says that *any C^1 function is differentiable.*

EXAMPLE 10 Let

$$f(x, y) = \frac{\cos x + e^{xy}}{x^2 + y^2}.$$

Show that f is differentiable at all points $(x, y) \neq (0, 0)$.

SOLUTION Observe that the partial derivatives

$$\frac{\partial f}{\partial x} = \frac{(x^2 + y^2)(ye^{xy} - \sin x) - 2x(\cos x + e^{xy})}{(x^2 + y^2)^2}$$

$$\frac{\partial f}{\partial y} = \frac{(x^2 + y^2)xe^{xy} - 2y(\cos x + e^{xy})}{(x^2 + y^2)^2}$$

are continuous except when $x = 0$ and $y = 0$ (by the results in Section 2.2). Thus, f is differentiable by Theorem 9. ▲

In the Internet supplement we show that $f(x, y) = xy/\sqrt{x^2 + y^2}$ [with $f(0, 0) = 0$] is continuous, has partial derivatives at $(0, 0)$, yet is *not* differentiable there. See Figure 2.3.4. By Theorem 9, its partial derivatives cannot be continuous at $(0, 0)$.

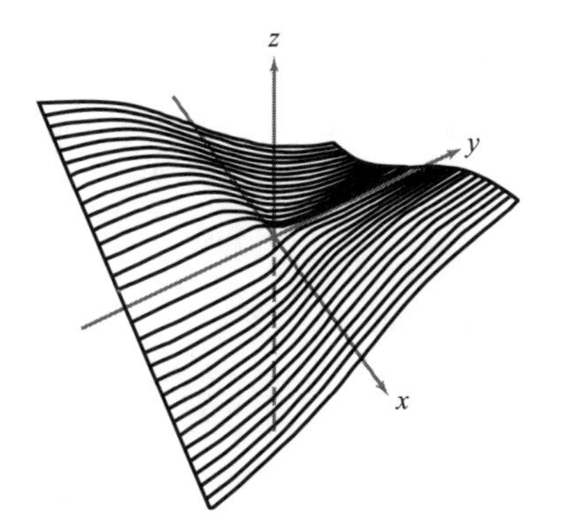

Figure 2.3.4 This function is not differentiable at (0, 0), because it is "crinkled."

EXERCISES

1. Find $\partial f/\partial x, \partial f/\partial y$ if

(a) $f(x, y) = xy$
(b) $f(x, y) = e^{xy}$
(c) $f(x, y) = x \cos x \cos y$
(d) $f(x, y) = (x^2 + y^2) \log (x^2 + y^2)$

2. Evaluate the partial derivatives $\partial z/\partial x, \partial z/\partial y$ for the given function at the indicated points.

(a) $z = \sqrt{a^2 - x^2 - y^2}$; $(0, 0), (a/2, a/2)$
(b) $z = \log \sqrt{1 + xy}$; $(1, 2), (0, 0)$
(c) $z = e^{ax} \cos (bx + y)$; $(2\pi/b, 0)$

3. In each case following, find the partial derivatives $\partial w/\partial x, \partial w/\partial y$.

(a) $w = xe^{x^2+y^2}$
(b) $w = \dfrac{x^2 + y^2}{x^2 - y^2}$
(c) $w = e^{xy} \log (x^2 + y^2)$
(d) $w = x/y$
(e) $w = \cos (ye^{xy}) \sin x$

4. Show that each of the following functions is differentiable at each point in its domain. Decide which of the functions are C^1.

(a) $f(x, y) = \dfrac{2xy}{(x^2 + y^2)^2}$
(b) $f(x, y) = \dfrac{x}{y} + \dfrac{y}{x}$
(c) $f(r, \theta) = \frac{1}{2} r \sin 2\theta, r > 0$
(d) $f(x, y) = \dfrac{xy}{\sqrt{x^2 + y^2}}$
(e) $f(x, y) = \dfrac{x^2 y}{x^4 + y^2}$

5. Find the equation of the plane tangent to the surface $z = x^2 + y^3$ at $(3, 1, 10)$.

6. Using the respective functions in Exercise 1, compute the plane tangent to the graphs at the indicated points.

(a) $(0, 0)$ (b) $(0, 1)$ (c) $(0, \pi)$ (d) $(0, 1)$

7. Compute the matrix of partial derivatives of the following functions:

(a) $f\colon \mathbb{R}^2 \to \mathbb{R}^2$, $f(x, y) = (x, y)$
(b) $f\colon \mathbb{R}^2 \to \mathbb{R}^3$, $f(x, y) = (xe^y + \cos y, x, x + e^y)$
(c) $f\colon \mathbb{R}^3 \to \mathbb{R}^2$, $f(x, y, z) = (x + e^z + y, yx^2)$
(d) $f\colon \mathbb{R}^2 \to \mathbb{R}^3$, $f(x, y) = (xye^{xy}, x \sin y, 5xy^2)$

8. Compute the matrix of partial derivatives of

(a) $f(x, y) = (e^x, \sin xy)$
(b) $f(x, y, z) = (x - y, y + z)$
(c) $f(x, y) = (x + y, x - y, xy)$
(d) $f(x, y, z) = (x + z, y - 5z, x - y)$

9. Where does the plane tangent to $z = e^{x-y}$ at $(1, 1, 1)$ meet the z axis?

10. Why should the graphs of $f(x, y) = x^2 + y^2$ and $g(x, y) = -x^2 - y^2 + xy^3$ be called "tangent" at $(0, 0)$?

11. Let $f(x, y) = e^{xy}$. Show that $x(\partial f/\partial x) = y(\partial f/\partial y)$.

12. Use the linear approximation to approximate a suitable function $f(x, y)$ and thereby estimate the following:

(a) $(0.99e^{0.02})^8$
(b) $(0.99)^3 + (2.01)^3 - 6(0.99)(2.01)$
(c) $\sqrt{(4.01)^2 + (3.98)^2 + (2.02)^2}$

13. Compute the gradients of the following functions:

(a) $f(x, y, z) = x \exp(-x^2 - y^2 - z^2)$ (Note that $\exp u = e^u$.)
(b) $f(x, y, z) = \dfrac{xyz}{x^2 + y^2 + z^2}$
(c) $f(x, y, z) = z^2 e^x \cos y$

14. Compute the tangent plane at $(1, 0, 1)$ for each of the functions in Exercise 13. [The solution to part (c) only is in the Study Guide.]

15. Find the equation of the tangent plane to $z = x^2 + 2y^3$ at $(1, 1, 3)$.

16. Calculate $\nabla h(1, 1, 1)$ if $h(x, y, z) = (x + z)e^{x-y}$.

17. Let $f(x, y, z) = x^2 + y^2 - z^2$. Calculate $\nabla f(0, 0, 1)$.

18. Evaluate the gradient of $f(x, y, z) = \log(x^2 + y^2 + z^2)$ at $(1, 0, 1)$.

19. Describe all Hölder-continuous functions with $\alpha > 1$ (see Exercise 25, Section 2.2). (HINT: What is the derivative of such a function?)

20. Suppose $f: \mathbb{R}^n \to \mathbb{R}^m$ is a linear map. What is the derivative of f?

2.4 Introduction to Paths and Curves

In this section, we introduce some of the basic geometry and computational methods for paths in the plane and space. This will be an important ingredient for the chain rule treated in the next section. We will return to paths with additional topics in Chapter 4.

Paths and Curves

One often thinks of a curve as a line drawn on paper, such as a straight line, a circle, or a sine curve. It is useful to think of a curve C mathematically as the set of values of a function that maps an interval of real numbers into the plane or space. We shall call such a map a ***path***. We usually denote a path by $\mathbf{c}$. The image C of the path then corresponds to the curve we see on paper (see Figure 2.4.1). Often we write t for the independent variable and imagine it to be *time*, so that $\mathbf{c}(t)$ is the position at time t of a moving particle, which ***traces out*** a curve as t varies. We also say $\mathbf{c}$ ***parametrizes*** C. Strictly speaking, we should distinguish between $\mathbf{c}(t)$ as a *point* in space and as a *vector* based at the origin.

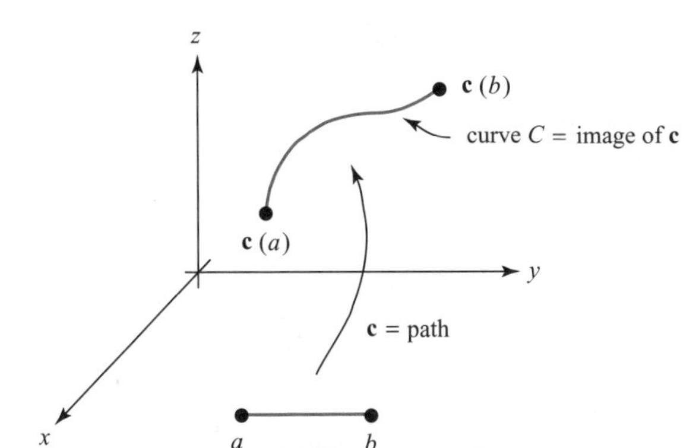

Figure 2.4.1 The map $\mathbf{c}$ is the path; its image C is the curve we "see."

EXAMPLE 1 The straight line L in $\mathbb{R}^3$ through the point (x_0, y_0, z_0) in the direction of vector $\mathbf{v}$ is the image of the path

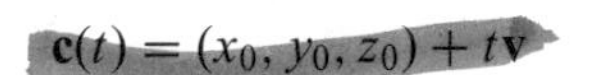

$$\mathbf{c}(t) = (x_0, y_0, z_0) + t\mathbf{v}$$

for $t \in \mathbb{R}$ (see Figure 2.4.2). Thus, our notion of curve includes straight lines as special cases. ▲

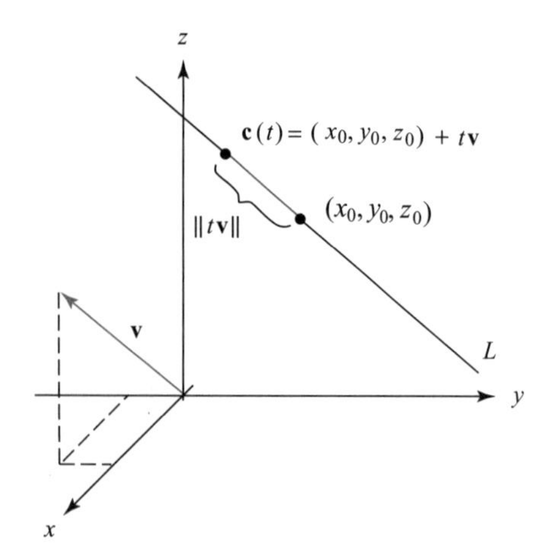

Figure 2.4.2 L is the straight line in space through (x_0, y_0, z_0) and in direction $\mathbf{v}$; its equation is $\mathbf{c}(t) = (x_0, y_0, z_0) + t\mathbf{v}$.

EXAMPLE 2 The unit circle C: $x^2 + y^2 = 1$ in the plane is the image of the path

$$\mathbf{c}: \mathbb{R} \to \mathbb{R}^2, \qquad \mathbf{c}(t) = (\cos t, \sin t), \qquad 0 \le t \le 2\pi,$$

(see Figure 2.4.3). The unit circle is also the image of the path $\tilde{\mathbf{c}}(t) = (\cos 2t, \sin 2t)$, $0 \le t \le \pi$. Thus, different paths may parametrize the same curve. ▲

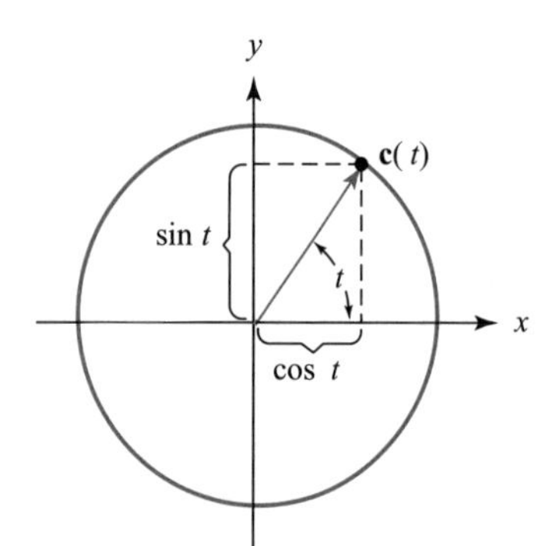

Figure 2.4.3 $\mathbf{c}(t) = (\cos t, \sin t)$ is a path whose image C is the unit circle.

Paths and Curves A ***path*** in $\mathbb{R}^n$ is a map $\mathbf{c}: [a, b] \to \mathbb{R}^n$; it is a ***path in the plane*** if $n = 2$ and a ***path in space*** if $n = 3$. The collection C of points $\mathbf{c}(t)$ as t varies in $[a, b]$ is called a ***curve***, and $\mathbf{c}(a)$ and $\mathbf{c}(b)$ are its ***endpoints***. The path $\mathbf{c}$ is said to ***parametrize*** the curve C. We also say $\mathbf{c}(t)$ ***traces out*** C as t varies.

If $\mathbf{c}$ is a path in $\mathbb{R}^3$, we can write $\mathbf{c}(t) = (x(t), y(t), z(t))$ and we call $x(t)$, $y(t)$, and $z(t)$ the ***component functions*** of $\mathbf{c}$. We form component functions similarly in $\mathbb{R}^2$ or, generally, in $\mathbb{R}^n$.

EXAMPLE 3 The path $\mathbf{c}(t) = (t, t^2)$ traces out a parabolic arc. This curve coincides with the graph $f(x) = x^2$ (see Figure 2.4.4). ▲

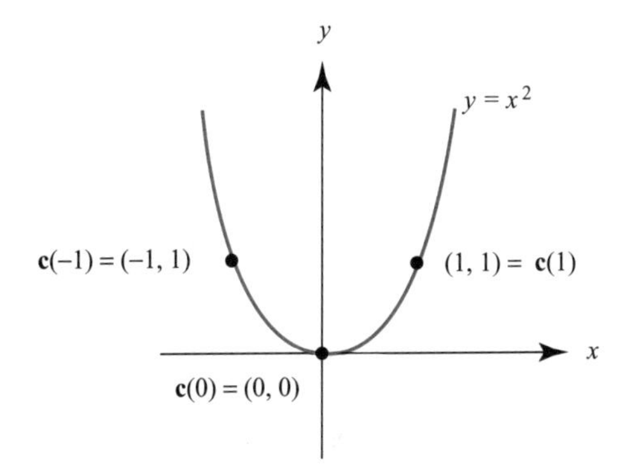

Figure 2.4.4 The image of $\mathbf{c}(t) = (t, t^2)$ is the parabola $y = x^2$.

EXAMPLE 4 A wheel of radius R rolls to the right along a straight line at speed v. Use vector methods to find the path $\mathbf{c}(t)$ of the point on the wheel that initially lies at a distance r below the center.

SOLUTION We place the wheel in the xy plane with its center initially at $(0, R)$, so that the position of the center at time t is given by the path $\mathbf{C}(t) = (vt, R)$. (Refer to Figure 2.4.5.)

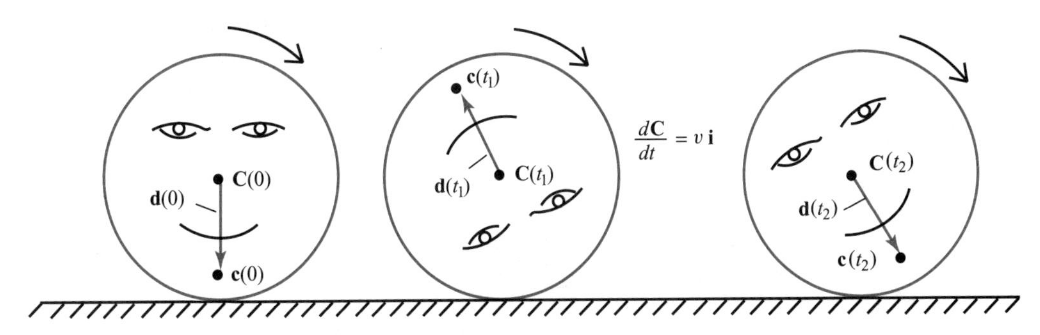

Figure 2.4.5 The vector $\mathbf{d}(t)$ points from the wheel's center, $\mathbf{C}(t)$, to the position $\mathbf{c}(t)$ of a point on the wheel and rotates in the clockwise direction while the wheel moves to the right.

The position of the point $\mathbf{c}(t)$ *relative to the center* is given by the vector $\mathbf{d}(t) = \mathbf{c}(t) - \mathbf{C}(t)$ that has the initial value $-r\mathbf{j}$ and rotates in the *clockwise* direction. The rate of rotation is such that the wheel makes a full rotation after the center has moved a distance $2\pi R$ (equal to the circumference of the wheel). This takes a time $2\pi R/v$, so the angular velocity $d\theta/dt$ of the wheel is v/R. Because the rotation is clockwise, the vector function $\mathbf{d}(t)$ is of the form

$$\mathbf{d}(t) = r\left(\cos\left[-\frac{v}{R}t + \theta\right]\mathbf{i} + \sin\left[-\frac{v}{R}t + \theta\right]\mathbf{j}\right)$$

for some initial angle θ. Because $\mathbf{d}(0) = -r\mathbf{j}$, we have $\cos\theta = 0$ and $\sin\theta = -1$, so $\theta = -\pi/2$, and hence

$$\mathbf{d}(t) = r\left(\cos\left[-\frac{v}{R}t - \frac{\pi}{2}\right]\mathbf{i} + \sin\left[-\frac{v}{R}t - \frac{\pi}{2}\right]\mathbf{j}\right).$$

Using $\cos(\varphi - \pi/2) = \sin\varphi$ and $\sin(\varphi - \pi/2) = -\cos\varphi$, along with $\cos(-\varphi) = \cos\varphi$ and $\sin(-\varphi) = -\sin\varphi$, we get

$$\mathbf{d}(t) = r\left(-\sin\frac{vt}{R}\mathbf{i} - \cos\frac{vt}{R}\mathbf{j}\right).$$

Finally, the path $\mathbf{c}(t)$ is given by adding the components of the vector function $\mathbf{d}(t)$ to the coordinates of the path $\mathbf{C}(t)$; the result is

$$\mathbf{c}(t) = \left(vt - r\sin\frac{vt}{R},\ R - r\cos\frac{vt}{R}\right).$$

In the special case $v = R = r = 1$, we get $\mathbf{c}(t) = (t - \sin t, 1 - \cos t)$. The image curve C of this path $\mathbf{c}$ is shown in Figure 2.4.6; it is called a ***cycloid***. ▲

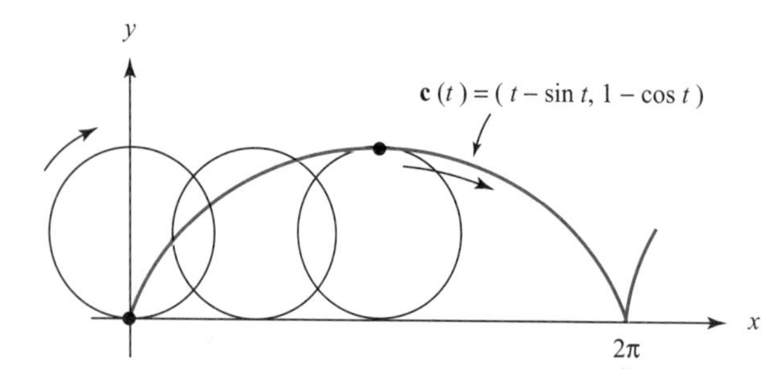

Figure 2.4.6 The curve traced by a point moving on the rim of a rolling circle is called a cycloid.

The preceding example considered the path of a point not necessarily on the rim of a wheel rolling along a straight line. When the wheel rolls on a circle, the resulting curve is called an ***epicycle***. These are the epicycles discussed in the Ptolemaic theory in the introduction. If the wheel is outside the circle and the point is on the rim, the curve is called an ***epicycloid***, and when the wheel is inside the circle it is a ***hypocycloid***. An example of the latter is shown in Figure 2.4.7.

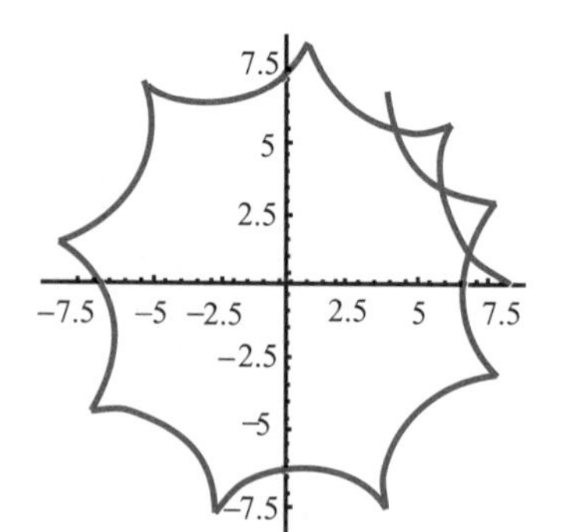

Figure 2.4.7 An example of a hypocycloid.

Historical Note

The French mathematician Blaise Pascal studied the cycloid in 1649 as a way of distracting himself at a time when he was suffering from a painful toothache. When the pain disappeared, he took it as a sign that God was not displeased with his thoughts. Pascal's results stimulated other mathematicians to investigate this curve, and subsequently numerous remarkable properties were found. One of these was discovered by the Dutchman Christian Huygens, who used it in the construction of a "perfect" pendulum clock.

Velocity and Tangents to Paths

If we think of $\mathbf{c}(t)$ as the curve traced out by a particle and t as time, it is reasonable to define the velocity vector as follows.

DEFINITION: Velocity Vector If $\mathbf{c}$ is a path and it is differentiable, we say $\mathbf{c}$ is a ***differentiable path***. The ***velocity*** of $\mathbf{c}$ at time t is defined by[3]

$$\mathbf{c}'(t) = \lim_{h \to 0} \frac{\mathbf{c}(t+h) - \mathbf{c}(t)}{h}.$$

We normally draw the vector $\mathbf{c}'(t)$ with its tail at the point $\mathbf{c}(t)$. The ***speed*** of the path $\mathbf{c}(t)$ is $s = \|\mathbf{c}'(t)\|$, the length of the velocity vector. If $\mathbf{c}(t) = (x(t), y(t))$ in $\mathbb{R}^2$, then

$$\mathbf{c}'(t) = (x'(t), y'(t)) = x'(t)\mathbf{i} + y'(t)\mathbf{j}$$

and if $\mathbf{c}(t) = (x(t), y(t), z(t))$ in $\mathbb{R}^3$, then

$$\mathbf{c}'(t) = (x'(t), y'(t), z'(t)) = x'(t)\mathbf{i} + y'(t)\mathbf{j} + z'(t)\mathbf{k}.$$

Here, $x'(t)$ is the one-variable derivative dx/dt. If we accept limits of vectors interpreted componentwise, the formulas for the velocity vector follow from the definition of the derivative. However, the limit can be interpreted in the sense of vectors as well. In Figure 2.4.8, we see that $[\mathbf{c}(t+h) - \mathbf{c}(t)]/h$ approaches the tangent to the path as $h \to 0$.

---

[3]If t lies at the endpoint of an interval, one should, as in one-variable calculus, take right- or left-handed limits.

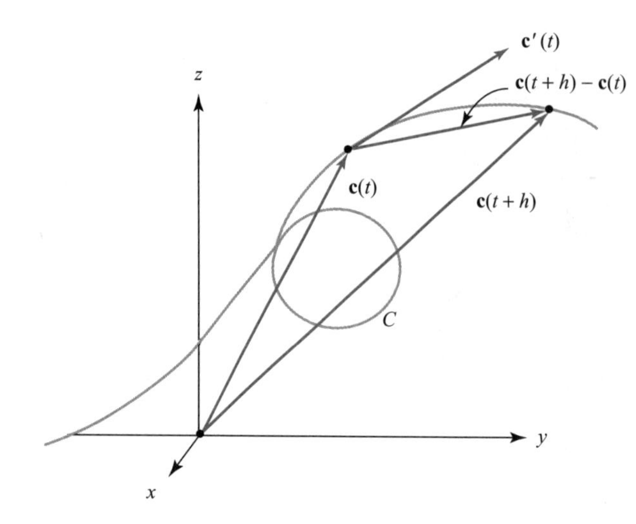

Figure 2.4.8 The vector $\mathbf{c}'(t)$ is tangent to the path $\mathbf{c}(t)$.

> **Tangent Vector** The velocity $\mathbf{c}'(t)$ is a vector ***tangent*** to the path $\mathbf{c}(t)$ at time t. If C is a curve traced out by $\mathbf{c}$ and if $\mathbf{c}'(t)$ is not equal to $\mathbf{0}$, then $\mathbf{c}'(t)$ is a vector tangent to the curve C at the point $\mathbf{c}(t)$.

If we think of the derivative $\mathbf{D}\mathbf{c}(t)$ as a matrix, it will be a column vector with the entries $x'(t)$, $y'(t)$, and $z'(t)$. Thus, the derivative here is consistent with our earlier notion.

EXAMPLE 5 Compute the tangent vector to the path $\mathbf{c}(t) = (t, t^2, e^t)$ at $t = 0$.

SOLUTION Here $\mathbf{c}'(t) = (1, 2t, e^t)$, and so at $t = 0$ we obtain the tangent vector $(1, 0, 1)$. ▲

EXAMPLE 6 Describe the path $\mathbf{c}(t) = (\cos t, \sin t, t)$. Find the velocity vector at the point on the image curve where $t = \pi/2$.

SOLUTION For a given t, the point $(\cos t, \sin t, 0)$ lies on the circle $x^2 + y^2 = 1$ in the xy plane. Therefore, the point $(\cos t, \sin t, t)$ lies t units above the point $(\cos t, \sin t, 0)$ if t is positive and $-t$ units below $(\cos t, \sin t, 0)$ if t is negative. As t increases, $(\cos t, \sin t, t)$ wraps around the cylinder $x^2 + y^2 = 1$ with the z coordinate increasing. The curve this traces out is called a ***helix***, which is depicted in Figure 2.4.9. At $t = \pi/2$, $\mathbf{c}'(\pi/2) = (-\sin \pi/2, \cos \pi/2, 1) = (-1, 0, 1) = -\mathbf{i} + \mathbf{k}$. ▲

EXAMPLE 7 The cycloidal path of a particle on the edge of a wheel of radius R with velocity v is given by $\mathbf{c}(t) = (vt - R\sin(vt/R), R - R\cos(vt/R))$. (See Example 4.) Find the velocity $\mathbf{c}'(t)$ of the particle as a function of t. When is the velocity zero? Is the velocity vector ever vertical?

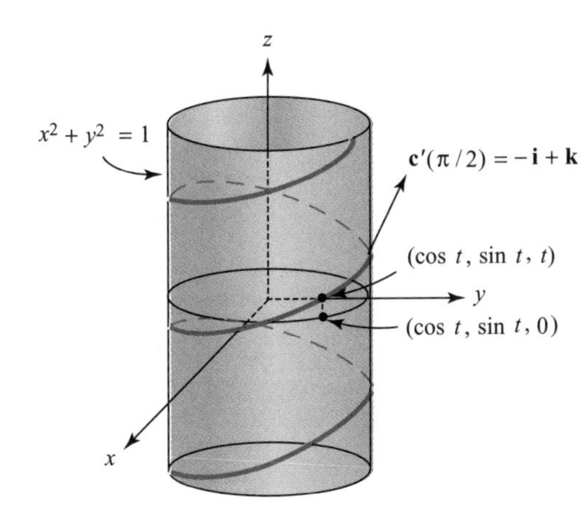

Figure 2.4.9 The helix $\mathbf{c}(t) = (\cos t, \sin t, t)$ wraps around the cylinder $x^2 + y^2 = 1$.

SOLUTION To find the velocity, we differentiate:

$$\mathbf{c}'(t) = \left(\frac{d}{dt}\left(vt - R\sin\frac{vt}{R}\right), \frac{d}{dt}\left(R - R\cos\frac{vt}{R}\right)\right)$$
$$= \left(v - v\cos\frac{vt}{R}, v\sin\frac{vt}{R}\right).$$

In vector notation, $\mathbf{c}'(t) = (v - v\cos(vt/R))\mathbf{i} + (v\sin(vt/R))\mathbf{j}$. The component in the direction of $\mathbf{i}$ is $v(1 - \cos(vt/R))$, which is zero whenever vt/R is an integer multiple of 2π. For such values of t, $\sin(vt/R)$ is zero as well, so the only times at which the velocity is zero are when $t = 2\pi nR/v$ for some integer n. At such times, $\mathbf{c}(t) = (2\pi nR, 0)$, so the moving point is touching the ground. These moments occur at time intervals of $2\pi R/v$ (more frequently for small wheels, as well as for rapidly rolling ones).

The velocity vector is never vertical, because the horizontal component vanishes only when the vertical one does as well. ▲

Figure 2.4.10 shows some velocity vectors superimposed on the cycloidal path of Figure 2.4.6.

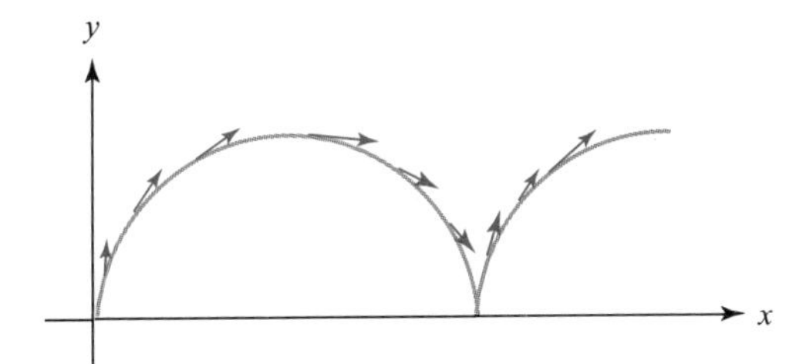

Figure 2.4.10 Velocity vectors for the curve traced out by a point on the rim of a rolling wheel.

Tangent Line

The tangent line to a path at a point is the line through the point in the direction of the tangent vector. Using the point-direction form of the equation of a line, we obtain the parametric equation for the tangent line.

> **Tangent Line to a Path** If $\mathbf{c}(t)$ is a path, and if $\mathbf{c}'(t_0) \neq \mathbf{0}$, the equation of its ***tangent line*** at the point $\mathbf{c}(t_0)$ is
>
> $$\boldsymbol{l}(t) = \mathbf{c}(t_0) + (t - t_0)\mathbf{c}'(t_0).$$
>
> If C is the curve traced out by $\mathbf{c}$, then the line traced out by $\boldsymbol{l}$ is the tangent line to the curve C at $\mathbf{c}(t_0)$.

Notice that we have written the equation in such a way that $\boldsymbol{l}$ goes through the point $\mathbf{c}(t_0)$ at $t = t_0$ (rather than $t = 0$). See Figure 2.4.11.

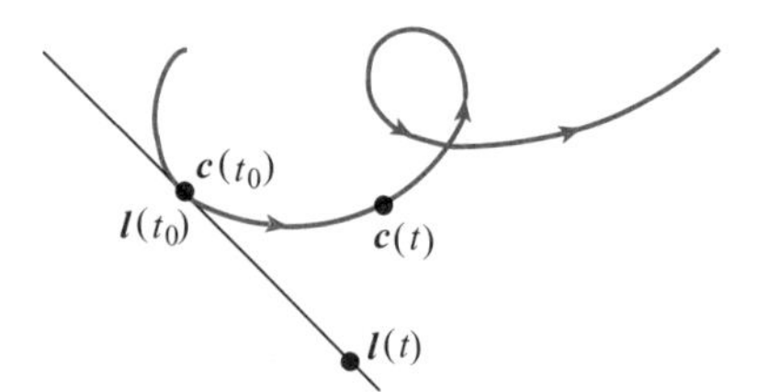

Figure 2.4.11 The tangent line to a path.

EXAMPLE 8 A path in $\mathbb{R}^3$ goes through the point $(3, 6, 5)$ at $t = 0$ with tangent vector $\mathbf{i} - \mathbf{j}$. Find the equation of the tangent line.

SOLUTION The equation of the tangent line is

$$\boldsymbol{l}(t) = (3, 6, 5) + t(\mathbf{i} - \mathbf{j}) = (3, 6, 5) + t(1, -1, 0) = (3 + t, 6 - t, 5).$$

In (x, y, z) coordinates, the tangent line is $x = 3 + t$, $y = 6 - t$, $z = 5$. ▲

Physically, we can interpret motion along the tangent line as the path that a particle on a curve would follow if it were set free at a certain moment.

EXAMPLE 9 Suppose that a particle follows the path $\mathbf{c}(t) = (e^t, e^{-t}, \cos t)$ until it flies off on a tangent at $t = 1$. Where is it at $t = 3$?

SOLUTION The velocity vector is $(e^t, -e^{-t}, -\sin t)$, which at $t = 1$ is the vector $(e, -1/e, -\sin 1)$. The particle is at $(e, 1/e, \cos 1)$ at $t = 1$. The equation of the tangent line is $\boldsymbol{l}(t) = (e, 1/e, \cos 1) + (t - 1)(e, -1/e, -\sin 1)$. At $t = 3$, the position on this line is

$$\boldsymbol{l}(3) = \left(e, \frac{1}{e}, \cos 1\right) + 2\left(e, -\frac{1}{e}, -\sin 1\right) = \left(3e, -\frac{1}{e}, \cos 1 - 2\sin 1\right)$$
$$\cong (8.155, -0.368, -1.143). \quad \blacktriangle$$

EXERCISES

Sketch the curves that are the images of the paths in Exercises 1 to 4.

1. $x = \sin t$, $y = 4\cos t$, where $0 \leq t \leq 2\pi$

2. $x = 2\sin t$, $y = 4\cos t$, where $0 \leq t \leq 2\pi$

3. $\mathbf{c}(t) = (2t - 1, t + 2, t)$

4. $\mathbf{c}(t) = (-t, 2t, 1/t)$, where $1 \leq t \leq 3$

In Exercises 5 to 8, determine the velocity vector of the given path.

5. $\mathbf{c}(t) = 6t\mathbf{i} + 3t^2\mathbf{j} + t^3\mathbf{k}$

6. $\mathbf{c}(t) = (\sin 3t)\mathbf{i} + (\cos 3t)\mathbf{j} + 2t^{3/2}\mathbf{k}$

7. $\mathbf{r}(t) = (\cos^2 t, 3t - t^3, t)$

8. $\mathbf{r}(t) = (4e^t, 6t^4, \cos t)$

In Exercises 9 to 12, compute the tangent vector to the given path.

9. $\mathbf{c}(t) = (e^t, \cos t)$

10. $\mathbf{c}(t) = (3t^2, t^3)$

11. $\mathbf{c}(t) = (t\sin t, 4t)$

12. $\mathbf{c}(t) = (t^2, e^2)$

13. When is the velocity vector of a point on the rim of a rolling wheel *horizontal*? What is the speed at this point?

14. If the position of a particle in space is $(6t, 3t^2, t^3)$ at time t, what is its velocity vector at $t = 0$?

In Exercises 15 and 16, determine the equation of the tangent line to the given path at the specified value of t.

15. $(\sin 3t, \cos 3t, 2t^{5/2}); t = 1$

16. $(\cos^2 t, 3t - t^3, t); t = 0$

In Exercises 17 to 20, suppose that a particle following the given path $\mathbf{c}(t)$ flies off on a tangent at $t = t_0$. Compute the position of the particle at the given time t_1.

17. $\mathbf{c}(t) = (t^2, t^3 - 4t, 0)$, where $t_0 = 2, t_1 = 3$

18. $\mathbf{c}(t) = (e^t, e^{-t}, \cos t)$, where $t_0 = 1, t_1 = 2$

19. $\mathbf{c}(t) = (4e^t, 6t^4, \cos t)$, where $t_0 = 0, t_1 = 1$

20. $\mathbf{c}(t) = (\sin e^t, t, 4 - t^3)$, where $t_0 = 1, t_1 = 2$

2.5 Properties of the Derivative

In elementary calculus, we learn how to differentiate sums, products, quotients, and composite functions. We now generalize these ideas to functions of several variables, paying particular attention to the differentiation of composite functions. The rule for differentiating composites, called the *chain rule*, takes on a more profound form for functions of several variables than for those of one variable.

If f is a real-valued function of one variable, written as $z = f(y)$, and y is a function of x, written $y = g(x)$, then z becomes a function of x through substitution, namely, $z = f(g(x))$, and we have the familiar chain rule:

$$\frac{dz}{dx} = \frac{dz}{dy}\frac{dy}{dx} = f'(g(x))g'(x).$$

If f is a real-valued function of three variables u, v, and w, written in the form $z = f(u, v, w)$, and the variables u, v, w are each functions of x, $u = g(x)$, $v = h(x)$, and $w = k(x)$, then by substituting $g(x)$, $h(x)$, and $k(x)$ for u, v, and w, we obtain z as a function of x: $z = f(g(x), h(x), k(x))$. The chain rule in this case reads:

$$\frac{dz}{dx} = \frac{\partial z}{\partial u}\frac{du}{dx} + \frac{\partial z}{\partial v}\frac{dv}{dx} + \frac{\partial z}{\partial w}\frac{dw}{dx}.$$

One of the goals of this section is to explain such formulas in detail.

Sums, Products, Quotients

These rules work just as they do in one-variable calculus.

THEOREM 10: Sums, Products, Quotients

(i) **Constant Multiple Rule**. Let $f: U \subset \mathbb{R}^n \to \mathbb{R}^m$ be differentiable at $\mathbf{x}_0$ and let c be a real number. Then $h(\mathbf{x}) = cf(\mathbf{x})$ is differentiable at $\mathbf{x}_0$ and

$$\mathbf{D}h(\mathbf{x}_0) = c\mathbf{D}f(\mathbf{x}_0) \qquad \text{(equality of matrices).}$$

(ii) **Sum Rule**. Let $f: U \subset \mathbb{R}^n \to \mathbb{R}^m$ and $g: U \subset \mathbb{R}^n \to \mathbb{R}^m$ be differentiable at $\mathbf{x}_0$. Then $h(\mathbf{x}) = f(\mathbf{x}) + g(\mathbf{x})$ is differentiable at $\mathbf{x}_0$ and

$$\mathbf{D}h(\mathbf{x}_0) = \mathbf{D}f(\mathbf{x}_0) + \mathbf{D}g(\mathbf{x}_0) \qquad \text{(sum of matrices).}$$

(iii) **Product Rule**. Let $f: U \subset \mathbb{R}^n \to \mathbb{R}$ and $g: U \subset \mathbb{R}^n \to \mathbb{R}$ be differentiable at $\mathbf{x}_0$ and let $h(\mathbf{x}) = g(\mathbf{x})f(\mathbf{x})$. Then $h: U \subset \mathbb{R}^n \to \mathbb{R}$ is differentiable at $\mathbf{x}_0$ and

$$\mathbf{D}h(\mathbf{x}_0) = g(\mathbf{x}_0)\mathbf{D}f(\mathbf{x}_0) + f(\mathbf{x}_0)\mathbf{D}g(\mathbf{x}_0).$$

(Note that each side of this equation is a $1 \times n$ matrix; a more general product rule is presented in Exercise 29 at the end of this section.)

(iv) **Quotient Rule**. With the same hypotheses as in rule (iii), let $h(\mathbf{x}) = f(\mathbf{x})/g(\mathbf{x})$ and suppose g is never zero on U. Then h is differentiable at $\mathbf{x}_0$ and

$$\mathbf{D}h(\mathbf{x}_0) = \frac{g(\mathbf{x}_0)\mathbf{D}f(\mathbf{x}_0) - f(\mathbf{x}_0)\mathbf{D}g(\mathbf{x}_0)}{[g(\mathbf{x}_0)]^2}.$$

PROOF The proofs of rules (i) through (iv) proceed almost exactly as in the one-variable case with a slight difference in notation. We shall prove rules (i) and (ii), leaving the proofs of rules (iii) and (iv) as Exercise 25.

(i) To show that $\mathbf{D}h(\mathbf{x}_0) = c\mathbf{D}f(\mathbf{x}_0)$, we must show that

$$\operatorname*{limit}_{\mathbf{x}\to\mathbf{x}_0} \frac{\|h(\mathbf{x}) - h(\mathbf{x}_0) - c\mathbf{D}f(\mathbf{x}_0)(\mathbf{x} - \mathbf{x}_0)\|}{\|\mathbf{x} - \mathbf{x}_0\|} = 0,$$

that is, that

$$\operatorname*{limit}_{\mathbf{x}\to\mathbf{x}_0} \frac{\|cf(\mathbf{x}) - cf(\mathbf{x}_0) - c\mathbf{D}f(\mathbf{x}_0)(\mathbf{x} - \mathbf{x}_0)\|}{\|\mathbf{x} - \mathbf{x}_0\|} = 0,$$

[see equation (4) of Section 2.3]. This is certainly true, since f is differentiable and the constant c can be factored out [see Theorem 3(i), Section 2.2].

(ii) By the triangle inequality, we may write

$$\frac{\|h(\mathbf{x}) - h(\mathbf{x}_0) - [\mathbf{D}f(\mathbf{x}_0) + \mathbf{D}g(\mathbf{x}_0)](\mathbf{x} - \mathbf{x}_0)\|}{\|\mathbf{x} - \mathbf{x}_0\|}$$

$$= \frac{\|f(\mathbf{x}) - f(\mathbf{x}_0) - [\mathbf{D}f(\mathbf{x}_0)](\mathbf{x} - \mathbf{x}_0) + g(\mathbf{x}) - g(\mathbf{x}_0) - [\mathbf{D}g(\mathbf{x}_0)](\mathbf{x} - \mathbf{x}_0)\|}{\|\mathbf{x} - \mathbf{x}_0\|}$$

$$\leq \frac{\|f(\mathbf{x}) - f(\mathbf{x}_0) - [\mathbf{D}f(\mathbf{x}_0)](\mathbf{x} - \mathbf{x}_0)\|}{\|\mathbf{x} - \mathbf{x}_0\|} + \frac{\|g(\mathbf{x}) - g(\mathbf{x}_0) - [\mathbf{D}g(\mathbf{x}_0)](\mathbf{x} - \mathbf{x}_0)\|}{\|\mathbf{x} - \mathbf{x}_0\|},$$

and each term approaches 0 as $\mathbf{x} \to \mathbf{x}_0$. Hence, rule (ii) holds. ■

EXAMPLE 1 Verify the formula for $\mathbf{D}h$ in rule (iv) of Theorem 10 with

$$f(x, y, z) = x^2 + y^2 + z^2 \text{ and } g(x, y, z) = x^2 + 1.$$

SOLUTION Here

$$h(x, y, z) = \frac{x^2 + y^2 + z^2}{x^2 + 1},$$

so that by direct differentiation

$$\mathbf{D}h(x, y, z) = \left[\frac{\partial h}{\partial x}, \frac{\partial h}{\partial y}, \frac{\partial h}{\partial z}\right] = \left[\frac{(x^2 + 1)2x - (x^2 + y^2 + z^2)2x}{(x^2 + 1)^2}, \frac{2y}{x^2 + 1}, \frac{2z}{x^2 + 1}\right]$$

$$= \left[\frac{2x(1 - y^2 - z^2)}{(x^2 + 1)^2}, \frac{2y}{x^2 + 1}, \frac{2z}{x^2 + 1}\right].$$

By rule (iv), we get

$$\mathbf{D}h = \frac{g\mathbf{D}f - f\mathbf{D}g}{g^2} = \frac{(x^2 + 1)[2x, 2y, 2z] - (x^2 + y^2 + z^2)[2x, 0, 0]}{(x^2 + 1)^2},$$

which is the same as what we obtained directly. ▲

Chain Rule

As we mentioned earlier, it is in the differentiation of composite functions that we meet apparently substantial alterations of the formula from one-variable calculus. However, if we use the **D** notation, that is, matrix notation for derivatives, the chain rule for functions of several variables looks similar to the one-variable rule.

THEOREM 11: Chain Rule Let $U \subset \mathbb{R}^n$ and $V \subset \mathbb{R}^m$ be open sets. Let $g: U \subset \mathbb{R}^n \to \mathbb{R}^m$ and $f: V \subset \mathbb{R}^m \to \mathbb{R}^p$ be given functions such that g maps U into V, so that $f \circ g$ is defined. Suppose g is differentiable at $\mathbf{x}_0$ and f is differentiable at $\mathbf{y}_0 = g(\mathbf{x}_0)$. Then $f \circ g$ is differentiable at $\mathbf{x}_0$ and

$$\mathbf{D}(f \circ g)(\mathbf{x}_0) = \mathbf{D}f(\mathbf{y}_0)\mathbf{D}g(\mathbf{x}_0). \tag{1}$$

The right-hand side is the matrix product of $\mathbf{D}f(\mathbf{y}_0)$ with $\mathbf{D}g(\mathbf{x}_0)$.

We shall now give a proof of the chain rule *under the additional assumption that the partial derivatives of f are continuous*, building up to the general case by developing two special cases that are themselves important. (The complete proof of Theorem 11 without the additional assumption of continuity is given in the Internet supplement for Chapter 2.)

First Special Case of the Chain Rule

Suppose $\mathbf{c}: \mathbb{R} \to \mathbb{R}^3$ is a differentiable path and $f: \mathbb{R}^3 \to \mathbb{R}$. Let $h(t) = f(\mathbf{c}(t)) = f(x(t), y(t), z(t))$, where $\mathbf{c}(t) = (x(t), y(t), z(t))$. Then

$$\frac{dh}{dt} = \frac{\partial f}{\partial x}\frac{dx}{dt} + \frac{\partial f}{\partial y}\frac{dy}{dt} + \frac{\partial f}{\partial z}\frac{dz}{dt}. \tag{2}$$

That is,

$$\frac{dh}{dt} = \nabla f(\mathbf{c}(t)) \cdot \mathbf{c}'(t),$$

where $\mathbf{c}'(t) = (x'(t), y'(t), z'(t))$.

This is the special case of Theorem 11 in which we take $\mathbf{c} = g$ and f to be real-valued, and $m = 3$. Notice that

$$\nabla f(\mathbf{c}(t)) \cdot \mathbf{c}'(t) = \mathbf{D}f(\mathbf{c}(t))\mathbf{D}\mathbf{c}(t),$$

where the product on the left-hand side is the dot product of vectors, while the product on the right-hand side is matrix multiplication, and where we regard $\mathbf{D}f(\mathbf{c}(t))$ as a *row* matrix and $\mathbf{D}\mathbf{c}(t)$ as *a column* matrix. The vectors $\nabla f(\mathbf{c}(t))$ and $\mathbf{c}'(t)$ have the same components as their matrix equivalents; the notational change indicates the switch from matrices to vectors.

PROOF OF EQUATION (2). By definition,

$$\frac{dh}{dt}(t_0) = \underset{t \to t_0}{\text{limit}} \frac{h(t) - h(t_0)}{t - t_0}.$$

Adding and subtracting two terms, we write

$$\begin{aligned}\frac{h(t)-h(t_0)}{t-t_0} &= \frac{f(x(t),y(t),z(t))-f(x(t_0),y(t_0),z(t_0))}{t-t_0}\\ &= \frac{f(x(t),y(t),z(t))-f(x(t_0),y(t),z(t))}{t-t_0}\\ &\quad+\frac{f(x(t_0),y(t),z(t))-f(x(t_0),y(t_0),z(t))}{t-t_0}\\ &\quad+\frac{f(x(t_0),y(t_0),z(t))-f(x(t_0),y(t_0),z(t_0))}{t-t_0}.\end{aligned}$$

Now we invoke the *mean-value theorem* from one-variable calculus, which states: *If* $g\colon [a,b]\to\mathbb{R}$ *is continuous and is differentiable on the open interval* (a,b), *then there is a point* c *in* (a,b) *such that* $g(b)-g(a)=g'(c)(b-a)$. Applying this to f as a function of x, we can assert that for some c between x and x_0,

$$f(x,y,z)-f(x_0,y,z)=\left[\frac{\partial f}{\partial x}(c,y,z)\right](x-x_0).$$

In this way, we find that

$$\begin{aligned}\frac{h(t)-h(t_0)}{t-t_0} &= \left[\frac{\partial f}{\partial x}(c,y(t),z(t))\right]\frac{x(t)-x(t_0)}{t-t_0}+\left[\frac{\partial f}{\partial y}(x(t_0),d,z(t))\right]\frac{y(t)-y(t_0)}{t-t_0}\\ &\quad+\left[\frac{\partial f}{\partial z}(x(t_0),y(t_0),e)\right]\frac{z(t)-z(t_0)}{t-t_0},\end{aligned}$$

where c, d, and e lie between $x(t)$ and $x(t_0)$, between $y(t)$ and $y(t_0)$, and between $z(t)$ and $z(t_0)$, respectively. Taking the limit $t\to t_0$, using the continuity of the partials $\partial f/\partial x$, $\partial f/\partial y$, $\partial f/\partial z$, and the fact that c, d, and e converge to $x(t_0)$, $y(t_0)$, and $z(t_0)$, respectively, we obtain formula (2). ■

Second Special Case of the Chain Rule

Let $f\colon \mathbb{R}^3\to\mathbb{R}$ and let $g\colon \mathbb{R}^3\to\mathbb{R}^3$. Write

$$g(x,y,z)=(u(x,y,z),v(x,y,z),w(x,y,z))$$

and define $h\colon \mathbb{R}^3\to\mathbb{R}$ by setting

$$h(x,y,z)=f(u(x,y,z),v(x,y,z),w(x,y,z)).$$

In this case, the chain rule states that

$$\begin{bmatrix}\frac{\partial h}{\partial x} & \frac{\partial h}{\partial y} & \frac{\partial h}{\partial z}\end{bmatrix}=\begin{bmatrix}\frac{\partial f}{\partial u} & \frac{\partial f}{\partial v} & \frac{\partial f}{\partial w}\end{bmatrix}\begin{bmatrix}\frac{\partial u}{\partial x} & \frac{\partial u}{\partial y} & \frac{\partial u}{\partial z}\\ \frac{\partial v}{\partial x} & \frac{\partial v}{\partial y} & \frac{\partial v}{\partial z}\\ \frac{\partial w}{\partial x} & \frac{\partial w}{\partial y} & \frac{\partial w}{\partial z}\end{bmatrix}. \quad (3)$$

In this special case, we have taken $n = m = 3$ and $p = 1$ for concreteness, and $U = \mathbb{R}^3$ and $V = \mathbb{R}^3$ for simplicity, and have written out the matrix product $[\mathbf{D}f(\mathbf{y}_0)][\mathbf{D}g(\mathbf{x}_0)]$ explicitly (with the arguments $\mathbf{x}_0$ and $\mathbf{y}_0$ suppressed in the matrices).

PROOF OF THE SECOND SPECIAL CASE OF THE CHAIN RULE. By definition, $\partial h/\partial x$ is obtained by differentiating h with respect to x, holding y and z fixed. But then $(u(x, y, z), v(x, y, z), w(x, y, z))$ may be regarded as a vector function of the single variable x. The first special case applies to this situation and, after the variables are renamed, gives

$$\frac{\partial h}{\partial x} = \frac{\partial f}{\partial u}\frac{\partial u}{\partial x} + \frac{\partial f}{\partial v}\frac{\partial v}{\partial x} + \frac{\partial f}{\partial w}\frac{\partial w}{\partial x}. \tag{3'}$$

Similarly,

$$\frac{\partial h}{\partial y} = \frac{\partial f}{\partial u}\frac{\partial u}{\partial y} + \frac{\partial f}{\partial v}\frac{\partial v}{\partial y} + \frac{\partial f}{\partial w}\frac{\partial w}{\partial y} \tag{3''}$$

and

$$\frac{\partial h}{\partial z} = \frac{\partial f}{\partial u}\frac{\partial u}{\partial z} + \frac{\partial f}{\partial v}\frac{\partial v}{\partial z} + \frac{\partial f}{\partial w}\frac{\partial w}{\partial z}. \tag{3'''}$$

These equations are exactly what would be obtained by multiplying out the matrices in equation (3). ∎

PROOF OF THEOREM 11. The general case in equation (1) may be proved in two steps. First, equation (2) is generalized to m variables; that is, for $f(x_1, \ldots, x_m)$ and $\mathbf{c}(t) = (x_1(t), \ldots, x_m(t))$, one has

$$\frac{dh}{dt} = \sum_{i=1}^{m} \frac{\partial f}{\partial x_i}\frac{dx_i}{dt},$$

where $h(t) = f(x_1(t), \ldots, x_m(t))$. Second, the result obtained in the first step is used to obtain the formula

$$\frac{\partial h_j}{\partial x_i} = \sum_{k=1}^{m} \frac{\partial f_j}{\partial y_k}\frac{\partial y_k}{\partial x_i},$$

where $f = (f_1, \ldots, f_p)$ is a vector function of arguments $y_1, \ldots, y_m$; $g(x_1, \ldots, x_n) = (y_1(x_1, \ldots, x_n), \ldots, y_m(x_1, \ldots, x_n))$; and $h_j(x_1, \ldots, x_n) = f_j(y_1(x_1, \ldots, x_n), \ldots, y_m(x_1, \ldots, x_n))$. (Using the letter y for both functions and arguments is an abuse of notation, but it can help one remember the formula.) This formula is equivalent to formula (1) after the matrices are multiplied out. ∎

The pattern of the chain rule will become clear once the student has worked some additional examples. For instance,

$$\frac{\partial}{\partial x} f(u(x, y), v(x, y), w(x, y), z(x, y)) = \frac{\partial f}{\partial u}\frac{\partial u}{\partial x} + \frac{\partial f}{\partial v}\frac{\partial v}{\partial x} + \frac{\partial f}{\partial w}\frac{\partial w}{\partial x} + \frac{\partial f}{\partial z}\frac{\partial z}{\partial x},$$

with a similar formula for $\partial f/\partial y$.

The chain rule can help us understand the relationship between the geometry of a mapping $f: \mathbb{R}^2 \to \mathbb{R}^2$ and the geometry of curves in $\mathbb{R}^2$. (Similar statements may be made about $\mathbb{R}^3$ or, generally, $\mathbb{R}^n$.) If $\mathbf{c}(t)$ is a path in the plane, then as we saw in Section 2.4, $\mathbf{c}'(t)$ represents the tangent (or velocity) vector of the path $\mathbf{c}(t)$, and this tangent (or velocity) vector is thought of as beginning at $\mathbf{c}(t)$. Now let $\mathbf{p}(t) = f(\mathbf{c}(t))$, where $f: \mathbb{R}^2 \to \mathbb{R}^2$. The path $\mathbf{p}$ represents the image of the path $\mathbf{c}(t)$ under the mapping f. The tangent vector to $\mathbf{p}$ is given by the chain rule:

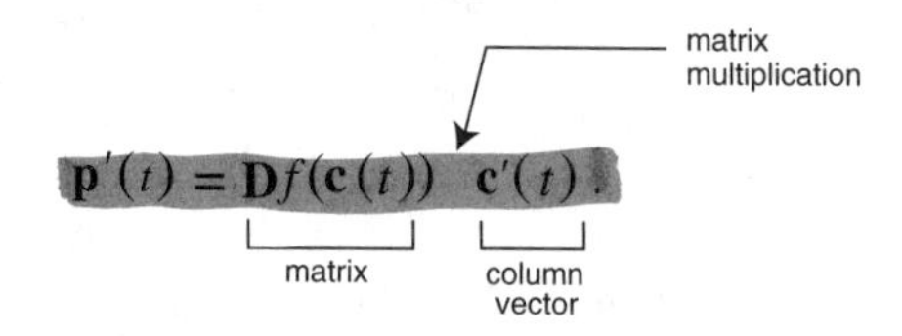

In other words, *the derivative matrix of f maps the tangent (or velocity) vector of a path* $\mathbf{c}$ *to the tangent (or velocity) vector of the corresponding image path* $\mathbf{p}$ (see Figure 2.5.1). Thus, points are mapped by f, while tangent vectors to curves are mapped by the derivative of f, evaluated at the base point of the tangent vector in the domain.

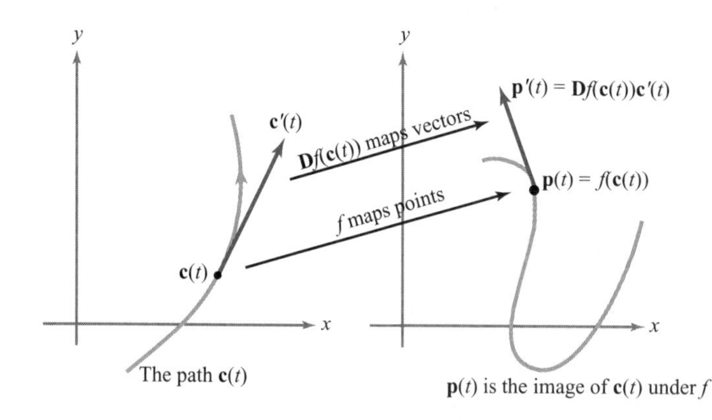

Figure 2.5.1 Tangent vectors are mapped by the derivative matrix.

EXAMPLE 2 Verify the chain rule in the form of formula (3′) for

$$f(u, v, w) = u^2 + v^2 - w,$$

where

$$u(x, y, z) = x^2 y, \qquad v(x, y, z) = y^2, \qquad w(x, y, z) = e^{-xz}.$$

SOLUTION Here

$$\begin{aligned} h(x, y, z) &= f(u(x, y, z), v(x, y, z), w(x, y, z)) \\ &= (x^2y)^2 + y^4 - e^{-xz} = x^4y^2 + y^4 - e^{-xz}. \end{aligned}$$

Thus, differentiating directly,

$$\frac{\partial h}{\partial x} = 4x^3y^2 + ze^{-xz}.$$

On the other hand, using the chain rule,

$$\begin{aligned} \frac{\partial h}{\partial x} &= \frac{\partial f}{\partial u}\frac{\partial u}{\partial x} + \frac{\partial f}{\partial v}\frac{\partial v}{\partial x} + \frac{\partial f}{\partial w}\frac{\partial w}{\partial x} = 2u(2xy) + 2v \cdot 0 + (-1)(-ze^{-xz}) \\ &= (2x^2y)(2xy) + ze^{-xz}, \end{aligned}$$

which is the same as the preceding equation. ▲

EXAMPLE 3 Given $g(x, y) = (x^2 + 1, y^2)$ and $f(u, v) = (u + v, u, v^2)$, compute the derivative of $f \circ g$ at the point $(x, y) = (1, 1)$ using the chain rule.

SOLUTION The matrices of partial derivatives are

$$\mathbf{D}f(u, v) = \begin{bmatrix} \frac{\partial f_1}{\partial u} & \frac{\partial f_1}{\partial v} \\ \frac{\partial f_2}{\partial u} & \frac{\partial f_2}{\partial v} \\ \frac{\partial f_3}{\partial u} & \frac{\partial f_3}{\partial v} \end{bmatrix} = \begin{bmatrix} 1 & 1 \\ 1 & 0 \\ 0 & 2v \end{bmatrix} \quad \text{and} \quad \mathbf{D}g(x, y) = \begin{bmatrix} 2x & 0 \\ 0 & 2y \end{bmatrix}.$$

When $(x, y) = (1, 1)$, note that $g(x, y) = (u, v) = (2, 1)$. Hence,

$$\mathbf{D}(f \circ g)(1, 1) = \mathbf{D}f(2, 1)\mathbf{D}g(1, 1) = \begin{bmatrix} 1 & 1 \\ 1 & 0 \\ 0 & 2 \end{bmatrix}\begin{bmatrix} 2 & 0 \\ 0 & 2 \end{bmatrix} = \begin{bmatrix} 2 & 2 \\ 2 & 0 \\ 0 & 4 \end{bmatrix}$$

is the required derivative. ▲

EXAMPLE 4 Let $f(x, y)$ be given and make the substitution $x = r\cos\theta$, $y = r\sin\theta$ (polar coordinates). Write a formula for $\partial f/\partial\theta$.

SOLUTION By the chain rule,

$$\frac{\partial f}{\partial \theta} = \frac{\partial f}{\partial x}\frac{\partial x}{\partial \theta} + \frac{\partial f}{\partial y}\frac{\partial y}{\partial \theta}$$

that is,

$$\frac{\partial f}{\partial \theta} = -r\sin\theta\frac{\partial f}{\partial x} + r\cos\theta\frac{\partial f}{\partial y}. \quad \blacktriangle$$

EXAMPLE 5 Let $f(x, y) = (\cos y + x^2, e^{x+y})$ and $g(u, v) = (e^{u^2}, u - \sin v)$. (a) Write a formula for $f \circ g$. (b) Calculate $\mathbf{D}(f \circ g)(0, 0)$ using the chain rule.

SOLUTION (a) We have

$$\begin{aligned}(f \circ g)(u, v) &= f(e^{u^2}, u - \sin v)\\ &= \left(\cos(u - \sin v) + e^{2u^2}, e^{e^{u^2} + u - \sin v}\right).\end{aligned}$$

(b) By the chain rule,

$$\mathbf{D}(f \circ g)(0, 0) = [\mathbf{D}f(g(0, 0))][\mathbf{D}g(0, 0)] = [\mathbf{D}f(1, 0)][\mathbf{D}g(0, 0)].$$

Now

$$\mathbf{D}g(0, 0) = \begin{bmatrix} 2ue^{u^2} & 0 \\ 1 & -\cos v \end{bmatrix}_{(u,v)=(0,0)} = \begin{bmatrix} 0 & 0 \\ 1 & -1 \end{bmatrix}$$

and

$$\mathbf{D}f(1, 0) = \begin{bmatrix} 2x & -\sin y \\ e^{x+y} & e^{x+y} \end{bmatrix}_{(x,y)=(1,0)} = \begin{bmatrix} 2 & 0 \\ e & e \end{bmatrix}.$$

[Remember that $\mathbf{D}f$ is evaluated at $g(0, 0)$, not at $(0, 0)$!] Thus,

$$\mathbf{D}(f \circ g)(0, 0) = \begin{bmatrix} 2 & 0 \\ e & e \end{bmatrix}\begin{bmatrix} 0 & 0 \\ 1 & -1 \end{bmatrix} = \begin{bmatrix} 0 & 0 \\ e & -e \end{bmatrix}. \quad \blacktriangle$$

EXAMPLE 6 Let $f\colon U \subset \mathbb{R}^n \to \mathbb{R}^m$ be differentiable, with $f = (f_1, \ldots, f_m)$, and let $g(\mathbf{x}) = \sin[f(\mathbf{x}) \cdot f(\mathbf{x})]$. Compute $\mathbf{D}g(\mathbf{x})$.

SOLUTION By the chain rule, $\mathbf{D}g(\mathbf{x}) = \cos[f(\mathbf{x}) \cdot f(\mathbf{x})]\mathbf{D}h(\mathbf{x})$, where $h(\mathbf{x}) = [f(\mathbf{x}) \cdot f(\mathbf{x})] = f_1^2(\mathbf{x}) + \cdots + f_m^2(\mathbf{x})$. Then

$$\begin{aligned}\mathbf{D}h(x) &= \begin{bmatrix} \dfrac{\partial h}{\partial x_1} & \cdots & \dfrac{\partial h}{\partial x_n} \end{bmatrix}\\ &= \begin{bmatrix} 2f_1\dfrac{\partial f_1}{\partial x_1} + \cdots + 2f_m\dfrac{\partial f_m}{\partial x_1} & \cdots & 2f_1\dfrac{\partial f_1}{\partial x_n} + \cdots + 2f_m\dfrac{\partial f_m}{\partial x_n} \end{bmatrix},\end{aligned}$$

which can be written $2f(\mathbf{x})\mathbf{D}f(\mathbf{x})$, where we regard f as a row matrix,

$$f = [f_1 \quad \cdots \quad f_m] \qquad \text{and} \qquad \mathbf{D}f = \begin{bmatrix} \dfrac{\partial f_1}{\partial x_1} & \cdots & \dfrac{\partial f_1}{\partial x_n} \\ \vdots & & \vdots \\ \dfrac{\partial f_m}{\partial x_1} & \cdots & \dfrac{\partial f_m}{\partial x_n} \end{bmatrix}.$$

Thus, $\mathbf{D}g(\mathbf{x}) = 2[\cos(f(\mathbf{x}) \cdot f(\mathbf{x}))]f(\mathbf{x})\mathbf{D}f(\mathbf{x})$. ▲

EXERCISES

1. If $f\colon U \subset \mathbb{R}^n \to \mathbb{R}$ is differentiable, prove that $\mathbf{x} \mapsto f^2(\mathbf{x}) + 2f(\mathbf{x})$ is differentiable as well, and compute its derivative in terms of $\mathbf{D}f(\mathbf{x})$.

2. Prove that the following functions are differentiable, and find their derivatives at an arbitrary point:

(a) $f\colon \mathbb{R}^2 \to \mathbb{R}, (x, y) \mapsto 2$
(b) $f\colon \mathbb{R}^2 \to \mathbb{R}, (x, y) \mapsto x + y$
(c) $f\colon \mathbb{R}^2 \to \mathbb{R}, (x, y) \mapsto 2 + x + y$
(d) $f\colon \mathbb{R}^2 \to \mathbb{R}, (x, y) \mapsto x^2 + y^2$
(e) $f\colon \mathbb{R}^2 \to \mathbb{R}, (x, y) \mapsto e^{xy}$
(f) $f\colon U \to \mathbb{R}, (x, y) \mapsto \sqrt{1 - x^2 - y^2}$, where $U = \{(x, y) \mid x^2 + y^2 < 1\}$
(g) $f\colon \mathbb{R}^2 \to \mathbb{R}, (x, y) \mapsto x^4 - y^4$

3. Write out the chain rule for each of the following functions and justify your answer in each case using Theorem 11.

(a) $\partial h/\partial x$ where $h(x, y) = f(x, u(x, y))$
(b) dh/dx where $h(x) = f(x, u(x), v(x))$
(c) $\partial h/\partial x$ where $h(x, y, z) = f(u(x, y, z), v(x, y), w(x))$

4. Verify the chain rule for $\partial h/\partial x$, where $h(x, y) = f(u(x, y), v(x, y))$ and

$$f(u, v) = \frac{u^2 + v^2}{u^2 - v^2}, \qquad u(x, y) = e^{-x-y}, \qquad v(x, y) = e^{xy}.$$

5. Verify the first special case of the chain rule for the composition $f \circ \mathbf{c}$ in each of the cases:

(a) $f(x, y) = xy$, $\mathbf{c}(t) = (e^t, \cos t)$
(b) $f(x, y) = e^{xy}$, $\mathbf{c}(t) = (3t^2, t^3)$
(c) $f(x, y) = (x^2 + y^2)\log\sqrt{x^2 + y^2}$, $\mathbf{c}(t) = (e^t, e^{-t})$
(d) $f(x, y) = x\exp(x^2 + y^2)$, $\mathbf{c}(t) = (t, -t)$

6. What is the velocity vector for each path $\mathbf{c}(t)$ in Exercise 5? [The solution to part (b) only is in the Study Guide to this text.]

7. Let $f\colon \mathbb{R}^3 \to \mathbb{R}$ and $g\colon \mathbb{R}^3 \to \mathbb{R}$ be differentiable. Prove that

$$\nabla(fg) = f\nabla g + g\nabla f.$$

8. Let $f\colon \mathbb{R}^3 \to \mathbb{R}$ be differentiable. Making the substitution

$$x = \rho\cos\theta\sin\phi, \qquad y = \rho\sin\theta\sin\phi, \qquad z = \rho\cos\phi$$

(spherical coordinates) into $f(x, y, z)$, compute $\partial f/\partial\rho$, $\partial f/\partial\theta$, and $\partial f/\partial\phi$ in terms of $\partial f/\partial x$, $\partial f/\partial y$, and $\partial f/\partial z$.

9. Let $f(u, v) = (\tan(u-1) - e^v, u^2 - v^2)$ and $g(x, y) = (e^{x-y}, x - y)$. Calculate $f \circ g$ and $\mathbf{D}(f \circ g)(1, 1)$.

10. Let $f(u, v, w) = (e^{u-w}, \cos(v + u) + \sin(u + v + w))$ and $g(x, y) = (e^x, \cos(y - x), e^{-y})$. Calculate $f \circ g$ and $\mathbf{D}(f \circ g)(0, 0)$.

11. Find $(\partial/\partial s)(f \circ T)(1, 0)$, where $f(u, v) = \cos u \sin v$ and $T\colon \mathbb{R}^2 \to \mathbb{R}^2$ is defined by $T(s, t) = (\cos(t^2 s), \log\sqrt{1 + s^2})$.

12. Suppose that the temperature at the point (x, y, z) in space is $T(x, y, z) = x^2 + y^2 + z^2$. Let a particle follow the right-circular helix $\boldsymbol{\sigma}(t) = (\cos t, \sin t, t)$ and let $T(t)$ be its temperature at time t.

(a) What is $T'(t)$?

(b) Find an approximate value for the temperature at $t = (\pi/2) + 0.01$.

13. Suppose that a duck is swimming in the circle $x = \cos t$, $y = \sin t$ and that the water temperature is given by the formula $T = x^2 e^y - xy^3$. Find dT/dt, the rate of change in temperature the duck might feel: (a) by the chain rule; (b) by expressing T in terms of t and differentiating.

14. Let $f\colon \mathbb{R}^n \to \mathbb{R}^m$ be a linear mapping so that (by Exercise 20, Section 2.3) $\mathbf{D}f(\mathbf{x})$ is the matrix of f. Check the validity of the chain rule directly for linear mappings.

15. Let $f\colon \mathbb{R}^2 \to \mathbb{R}^2; (x, y) \mapsto (e^{x+y}, e^{x-y})$. Let $\mathbf{c}(t)$ be a path with $\mathbf{c}(0) = (0, 0)$ and $\mathbf{c}'(0) = (1, 1)$. What is the tangent vector to the image of $\mathbf{c}(t)$ under f at $t = 0$?

16. Let $f(x, y) = 1/\sqrt{x^2 + y^2}$. Compute $\nabla f(x, y)$.

17. (a) Let $y(x)$ be defined implicitly by $G(x, y(x)) = 0$, where G is a given function of two variables. Prove that if $y(x)$ and G are differentiable, then

$$\frac{dy}{dx} = -\frac{\partial G/\partial x}{\partial G/\partial y} \quad \text{if} \quad \frac{\partial G}{\partial y} \neq 0.$$

(b) Obtain a formula analogous to that in part (a) if y_1, y_2 are defined implicitly by

$$G_1(x, y_1(x), y_2(x)) = 0,$$

$$G_2(x, y_1(x), y_2(x)) = 0.$$

(c) Let y be defined implicitly by

$$x^2 + y^3 + e^y = 0.$$

Compute dy/dx in terms of x and y.

18. Thermodynamics texts[4] use the relationship

$$\left(\frac{\partial y}{\partial x}\right)\left(\frac{\partial z}{\partial y}\right)\left(\frac{\partial x}{\partial z}\right) = -1.$$

Explain the meaning of this equation and prove that it is true. [HINT: Start with a relationship $F(x, y, z) = 0$ that defines $x = f(y, z)$, $y = g(x, z)$, and $z = h(x, y)$ and differentiate implicitly.]

19. Dieterici's equation of state for a gas is

$$P(V - b)e^{a/RVT} = RT,$$

where a, b, and R are constants. Regard volume V as a function of temperature T and pressure P and prove that

$$\frac{\partial V}{\partial T} = \left(R + \frac{a}{TV}\right) \bigg/ \left(\frac{RT}{V - b} - \frac{a}{V^2}\right).$$

20. This exercise gives another example of the fact that the chain rule is not applicable if f is not differentiable. Consider the function

$$f(x, y) = \begin{cases} \dfrac{xy^2}{x^2 + y^2} & (x, y) \neq (0, 0) \\ 0 & (x, y) = (0, 0). \end{cases}$$

Show that

(a) $\partial f/\partial x$ and $\partial f/\partial y$ exist at $(0, 0)$.

(b) If $\mathbf{g}(t) = (at, bt)$ for constants a and b, then $f \circ \mathbf{g}$ is differentiable and $(f \circ \mathbf{g})'(0) = ab^2/(a^2 + b^2)$, but $\nabla f(0, 0) \cdot \mathbf{g}'(0) = 0$.

21. Prove that if $f\colon U \subset \mathbb{R}^n \to \mathbb{R}$ is differentiable at $\mathbf{x}_0 \in U$, there is a neighborhood V of $\mathbf{0} \in \mathbb{R}^n$ and a function $R_1\colon V \to \mathbb{R}$ such that for all $\mathbf{h} \in V$, we have $\mathbf{x}_0 + \mathbf{h} \in U$,

$$f(\mathbf{x}_0 + \mathbf{h}) = f(\mathbf{x}_0) + [\mathbf{D}f(\mathbf{x}_0)]\mathbf{h} + R_1(\mathbf{h})$$

[4] See S. M. Binder, "Mathematical Methods in Elementary Thermodynamics," *J. Chem. Educ.* 43 (1966): 85–92. A proper understanding of partial differentiation can be of significant use in applications; for example, see M. Feinberg, "Constitutive Equation for Ideal Gas Mixtures and Ideal Solutions as Consequences of Simple Postulates," *Chem. Eng. Sci.* 32 (1977): 75–78.

and

$$\frac{R_1(\mathbf{h})}{\|\mathbf{h}\|} \to 0 \quad \text{as} \quad \mathbf{h} \to \mathbf{0}.$$

22. Suppose $\mathbf{x}_0 \in \mathbb{R}^n$ and $0 \le r_1 < r_2$. Show that there is a C^1 function $f: \mathbb{R}^n \to \mathbb{R}$ such that $f(\mathbf{x}) = 0$ for $\|\mathbf{x} - \mathbf{x}_0\| \ge r_2$; $0 < f(\mathbf{x}) < 1$ for $r_1 < \|\mathbf{x} - \mathbf{x}_0\| < r_2$; and $f(\mathbf{x}) = 1$ for $\|\mathbf{x} - \mathbf{x}_0\| \le r_1$. [HINT: Apply a cubic polynomial with $g(r_1^2) = 1$ and $g(r_2^2) = g'(r_2^2) = g'(r_1^2) = 0$ to $\|\mathbf{x} - \mathbf{x}_0\|^2$ when $r_1 < \|\mathbf{x} - \mathbf{x}_0\| < r_2$.]

23. Find a C^1 mapping $f: \mathbb{R}^3 \to \mathbb{R}^3$ that takes the vector $\mathbf{i} + \mathbf{j} + \mathbf{k}$ emanating from the origin to $\mathbf{i} - \mathbf{j}$ emanating from $(1, 1, 0)$ and takes $\mathbf{k}$ emanating from $(1, 1, 0)$ to $\mathbf{k} - \mathbf{i}$ emanating from the origin.

24. What is wrong with the following argument? Suppose $w = f(x, y, z)$ and $z = g(x, y)$. By the chain rule,

$$\frac{\partial w}{\partial x} = \frac{\partial w}{\partial x}\frac{\partial x}{\partial x} + \frac{\partial w}{\partial y}\frac{\partial y}{\partial x} + \frac{\partial w}{\partial z}\frac{\partial z}{\partial x} = \frac{\partial w}{\partial x} + \frac{\partial w}{\partial z}\frac{\partial z}{\partial x}.$$

Hence, $0 = (\partial w/\partial z)(\partial z/\partial x)$, and so $\partial w/\partial z = 0$ or $\partial z/\partial x = 0$, which is, in general, absurd.

25. Prove rules (iii) and (iv) of Theorem 10. (HINT: Use the same addition and subtraction tricks as in the one-variable case and Theorem 8.)

26. Show that $h: \mathbb{R}^n \to \mathbb{R}^m$ is differentiable if and only if each of the m components $h_i: \mathbb{R}^n \to \mathbb{R}$ is differentiable. (HINT: Use the coordinate projection function and the chain rule for one implication and consider

$$\left[\frac{\|h(\mathbf{x}) - h(\mathbf{x}_0) - \mathbf{D}h(\mathbf{x}_0)(\mathbf{x} - \mathbf{x}_0)\|}{\|\mathbf{x} - \mathbf{x}_0\|}\right]^2 = \frac{\sum_{i=1}^m [h_i(\mathbf{x}) - h_i(\mathbf{x}_0)\mathbf{D}h_i(\mathbf{x}_0)(\mathbf{x} - \mathbf{x}_0)]^2}{\|\mathbf{x} - \mathbf{x}_0\|^2}$$

to obtain the other.)

27. Use the chain rule and differentiation under the integral sign, namely,

$$\frac{d}{dx}\int_a^b f(x, y)\,dy = \int_a^b \frac{\partial f}{\partial x}(x, y)\,dy,$$

to show that

$$\frac{d}{dx}\int_0^x f(x, y)\,dy = f(x, x) + \int_0^x \frac{\partial f}{\partial x}(x, y)\,dy.$$

28. For what integers $p > 0$ is

$$f(x) = \begin{cases} x^p \sin(1/x) & x \ne 0 \\ 0 & x = 0 \end{cases}$$

differentiable? For what p is the derivative continuous?

29. Suppose $f: \mathbb{R}^n \to \mathbb{R}$ and $g: \mathbb{R}^n \to \mathbb{R}^m$ are differentiable. Show that the product function $h(\mathbf{x}) = f(\mathbf{x})g(\mathbf{x})$ from $\mathbb{R}^n$ to $\mathbb{R}^m$ is differentiable and that if $\mathbf{x}_0$ and $\mathbf{y}$ are in $\mathbb{R}^n$, then $[\mathbf{D}h(\mathbf{x}_0)]\mathbf{y} = f(\mathbf{x}_0)\{[\mathbf{D}g(\mathbf{x}_0)]\mathbf{y}\} + \{[\mathbf{D}f(\mathbf{x}_0)]\mathbf{y}\}g(\mathbf{x}_0)$.

2.6 Gradients and Directional Derivatives

In Section 2.1 we studied the graphs of real-valued functions. Now we take up this study again, using the methods of calculus. Specifically, gradients will be used to obtain a formula for the plane tangent to a level surface.

Gradients in $\mathbb{R}^3$

Let us recall the definition.

DEFINITION: The Gradient If $f: U \subset \mathbb{R}^3 \to \mathbb{R}$ is differentiable, the ***gradient*** of f at (x, y, z) is the vector in space given by

$$\nabla f = \left(\frac{\partial f}{\partial x}, \frac{\partial f}{\partial y}, \frac{\partial f}{\partial z}\right).$$

This vector is also denoted $\nabla f(x, y, z)$. Thus, ∇f is just the matrix of the derivative $\mathbf{D}f$, written as a vector.

EXAMPLE 1 Let $f(x, y, z) = \sqrt{x^2 + y^2 + z^2} = r$, the distance from $\mathbf{0}$ to (x, y, z). Then

$$\begin{aligned}\nabla f(x, y, z) &= \left(\frac{\partial f}{\partial x}, \frac{\partial f}{\partial y}, \frac{\partial f}{\partial z}\right)\\ &= \left(\frac{x}{\sqrt{x^2+y^2+z^2}}, \frac{y}{\sqrt{x^2+y^2+z^2}}, \frac{z}{\sqrt{x^2+y^2+z^2}}\right) = \frac{\mathbf{r}}{r},\end{aligned}$$

where $\mathbf{r}$ is the point (x, y, z). Thus, ∇f is the unit vector in the direction of (x, y, z). ▲

EXAMPLE 2 If $f(x, y, z) = xy + z$, then

$$\nabla f(x, y, z) = \left(\frac{\partial f}{\partial x}, \frac{\partial f}{\partial y}, \frac{\partial f}{\partial z}\right) = (y, x, 1). \quad ▲$$

Suppose $f: \mathbb{R}^3 \to \mathbb{R}$ is a real-valued function. Let $\mathbf{v}$ and $\mathbf{x} \in \mathbb{R}^3$ be fixed vectors and consider the function from $\mathbb{R}$ to $\mathbb{R}$ defined by $t \mapsto f(\mathbf{x} + t\mathbf{v})$. The set of points of the form $\mathbf{x} + t\mathbf{v}$, $t \in \mathbb{R}$, is the line L through the point $\mathbf{x}$ parallel to the vector $\mathbf{v}$ (see Figure 2.6.1).

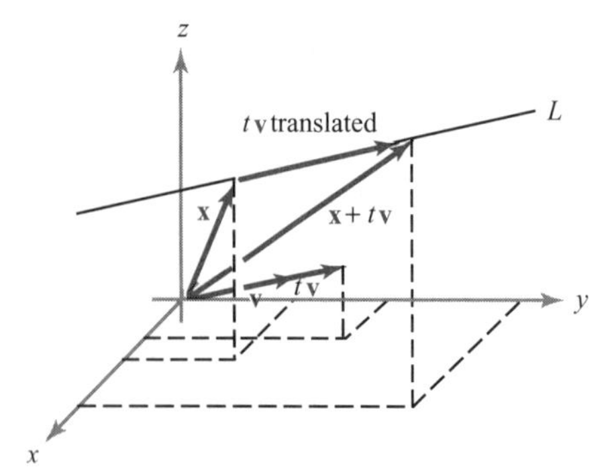

Figure 2.6.1 The equation of L is $\mathbf{l}(t) = \mathbf{x} + t\mathbf{v}$.

Directional Derivatives

The function $t \mapsto f(\mathbf{x} + t\mathbf{v})$ represents the function f restricted to the line L. For example, if a bird flies along this line with velocity $\mathbf{v}$ so that $\mathbf{x} + t\mathbf{v}$ is its position at time t, and if f represents the temperature as a function of position, then $f(\mathbf{x} + t\mathbf{v})$ is the temperature at time t. We may ask: How fast are the values of f changing along the line L at the point $\mathbf{x}$? Because the rate of change of a function is given by a derivative, we could say that the answer to this question is the value of the derivative of this function of t at $t = 0$ (when $t = 0$, $\mathbf{x} + t\mathbf{v}$ reduces to $\mathbf{x}$). This would be the derivative of f at the point $\mathbf{x}$ in the direction of L, that is, of $\mathbf{v}$. We can formalize this concept as follows.

DEFINITION: Directional Derivatives If $f: \mathbb{R}^3 \to \mathbb{R}$, the ***directional derivative*** of f at $\mathbf{x}$ along the vector $\mathbf{v}$ is given by

$$\left.\frac{d}{dt} f(\mathbf{x} + t\mathbf{v})\right|_{t=0}$$

if this exists.

In the definition of a directional derivative, we normally choose $\mathbf{v}$ to be a *unit* vector. In this case we are moving in the direction $\mathbf{v}$ with unit speed and we refer to $\frac{d}{dt} f(\mathbf{x} + t\mathbf{v})|_{t=0}$ as the ***directional derivative of f in the direction*** $\mathbf{v}$.

We now elaborate on why a *unit vector* is chosen in the definition of the directional derivative. Suppose that f measures the temperature in degrees and that we are interested in how fast the temperature changes as we move in a particular direction. If we are measuring distance in meters, then the rate of change of temperature will be measured in degrees per meter. Suppose, for simplicity, that the temperature is changing at a constant rate—say, two degrees per meter—as we move in a given direction **v** starting at **x**. Thus, when we go one meter ahead, the temperature changes by two degrees. That is,

$$f(\mathbf{x}+\mathbf{v}) - f(\mathbf{x}) = 2$$

Such a relation is going to hold only when **v** is a unit vector, reflecting the fact that we are going ahead by *one* meter. More generally, the definition of the directional derivative is going to truly measure only the rate of change of f with respect to distance along a line in a given direction if **v** is a unit vector.

From the definition, we can see that the directional derivative can also be defined by the formula

$$\operatorname*{limit}_{h\to 0} \frac{f(\mathbf{x}+h\mathbf{v}) - f(\mathbf{x})}{h}.$$

THEOREM 12 If $f\colon \mathbb{R}^3 \to \mathbb{R}$ is differentiable, then all directional derivatives exist. The directional derivative at **x** in the direction **v** is given by

$$\mathbf{D}f(\mathbf{x})\mathbf{v} = \operatorname{grad} f(\mathbf{x}) \cdot \mathbf{v} = \nabla f(\mathbf{x}) \cdot \mathbf{v} = \left[\frac{\partial f}{\partial x}(\mathbf{x})\right] v_1 + \left[\frac{\partial f}{\partial y}(\mathbf{x})\right] v_2 + \left[\frac{\partial f}{\partial z}(\mathbf{x})\right] v_3,$$

where $\mathbf{v} = (v_1, v_2, v_3)$.

PROOF Let $\mathbf{c}(t) = \mathbf{x} + t\mathbf{v}$, so that $f(\mathbf{x}+t\mathbf{v}) = f(\mathbf{c}(t))$. By the first special case of the chain rule, $(d/dt)f(\mathbf{c}(t)) = \nabla f(\mathbf{c}(t)) \cdot \mathbf{c}'(t)$. However, $\mathbf{c}(0) = \mathbf{x}$ and $\mathbf{c}'(0) = \mathbf{v}$, and so

$$\left.\frac{d}{dt} f(\mathbf{x}+t\mathbf{v})\right|_{t=0} = \nabla f(\mathbf{x}) \cdot \mathbf{v},$$

as we were required to prove. ■

Notice that one does not have to use straight lines when computing the rate of change of f in a specific direction **v**. Indeed, for a general path $\mathbf{c}(t)$ with $\mathbf{c}(0) = \mathbf{x}$ and $\mathbf{c}'(0) = \mathbf{v}$, we have from the chain rule,

$$\left.\frac{d}{dt} f(\mathbf{c}(t))\right|_{t=0} = \left.\nabla f(\mathbf{c}(t)) \cdot \mathbf{c}'(t)\right|_{t=0} = \nabla f(\mathbf{x}) \cdot \mathbf{v}.$$

EXAMPLE 3 Let $f(x, y, z) = x^2 e^{-yz}$. Compute the rate of change of f in the direction of the unit vector

$$\mathbf{v} = \left(\frac{1}{\sqrt{3}}, \frac{1}{\sqrt{3}}, \frac{1}{\sqrt{3}}\right) \qquad \text{at the point} \qquad (1, 0, 0).$$

SOLUTION The required rate of change is, using Theorem 12,

$$\nabla f \cdot \mathbf{v} = (2xe^{-yz}, -x^2ze^{-yz}, -x^2ye^{-yz}) \cdot \left(\frac{1}{\sqrt{3}}, \frac{1}{\sqrt{3}}, \frac{1}{\sqrt{3}}\right),$$

which, at the point (1, 0, 0), becomes

$$(2, 0, 0) \cdot \left(\frac{1}{\sqrt{3}}, \frac{1}{\sqrt{3}}, \frac{1}{\sqrt{3}}\right) = \frac{2}{\sqrt{3}}. \quad ▲$$

Directions of Fastest Increase

From Theorem 12 we can also obtain the geometrical significance of the gradient:

THEOREM 13 Assume $\nabla f(\mathbf{x}) \neq \mathbf{0}$. Then $\nabla f(\mathbf{x})$ points in the direction along which f is increasing the fastest.

PROOF If $\mathbf{n}$ is a unit vector, the rate of change of f in direction $\mathbf{n}$ is given by $\nabla f(\mathbf{x}) \cdot \mathbf{n} = \|\nabla f(\mathbf{x})\| \cos\theta$, where θ is the angle between $\mathbf{n}$ and $\nabla f(\mathbf{x})$. This is maximum when $\theta = 0$; that is, when $\mathbf{n}$ and ∇f are parallel. [If $\nabla f(\mathbf{x}) = \mathbf{0}$ this rate of change is 0 for any $\mathbf{n}$.] ■

In other words, if one wishes to move in a direction in which f will increase most quickly, one should proceed in the direction $\nabla f(\mathbf{x})$. Analogously, if one wishes to move in a direction in which f *decreases* the fastest, one should proceed in the direction $-\nabla f(\mathbf{x})$.

EXAMPLE 4 In what direction from (0, 1) does $f(x, y) = x^2 - y^2$ increase the fastest?

SOLUTION The gradient is

$$\nabla f = 2x\mathbf{i} - 2y\mathbf{j},$$

and so at (0, 1) this is

$$\nabla f|_{(0,1)} = -2\mathbf{j}.$$

By Theorem 13, f increases fastest in the direction $-\mathbf{j}$. (Can you see why this answer is consistent with Figure 2.1.9?) ▲

Gradients and Tangent Planes to Level Sets

Now we find the relationship between the gradient of a function f and its level surfaces. The gradient points in the direction in which the values of f change most rapidly, whereas a level surface lies in the directions in which they do not change at all. If f is reasonably well behaved, the gradient and the level surface will be perpendicular.

THEOREM 14: The Gradient is Normal to Level Surfaces Let $f: \mathbb{R}^3 \to \mathbb{R}$ be a C^1 map and let (x_0, y_0, z_0) lie on the level surface S defined by $f(x, y, z) = k$, for k a constant. Then $\nabla f(x_0, y_0, z_0)$ is normal to the level surface in the following sense: If $\mathbf{v}$ is the tangent vector at $t = 0$ of a path $\mathbf{c}(t)$ in S with $\mathbf{c}(0) = (x_0, y_0, z_0)$, then $\nabla f(x_0, y_0, z_0) \cdot \mathbf{v} = 0$ (see Figure 2.6.2).

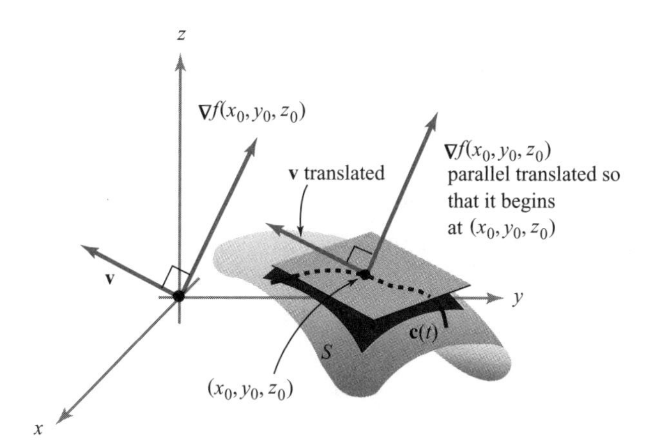

Figure 2.6.2 Geometric significance of the gradient: ∇f is orthogonal to the surface S on which f is constant.

PROOF Let $\mathbf{c}(t)$ lie in S; then $f(\mathbf{c}(t)) = k$. Let $\mathbf{v}$ be as in the hypothesis; then $\mathbf{v} = \mathbf{c}'(0)$. Hence, the fact that $f(\mathbf{c}(t))$ is constant in t, and the chain rule give

$$0 = \frac{d}{dt} f(\mathbf{c}(t))\bigg|_{t=0} = \nabla f(\mathbf{c}(0)) \cdot \mathbf{v}. \quad \blacksquare$$

If we study the conclusion of Theorem 14, we see that it is reasonable to *define* the plane tangent to S as the orthogonal plane to the gradient.

DEFINITION: Tangent Planes to Level Surfaces Let S be the surface consisting of those (x, y, z) such that $f(x, y, z) = k$, for k a constant. The ***tangent plane*** of S at a point (x_0, y_0, z_0) of S is defined by the equation

$$\nabla f(x_0, y_0, z_0) \cdot (x - x_0, y - y_0, z - z_0) = 0 \tag{1}$$

if $\nabla f(x_0, y_0, z_0) \neq \mathbf{0}$. That is, the tangent plane is the set of points (x, y, z) that satisfy equation (1).

This extends the definition we gave earlier for the tangent plane of the graph of a function (see Exercise 11 at the end of this section).

EXAMPLE 5 Compute the equation of the plane tangent to the surface defined by $3xy + z^2 = 4$ at $(1, 1, 1)$.

SOLUTION Here $f(x, y, z) = 3xy + z^2$ and $\nabla f = (3y, 3x, 2z)$, which at $(1, 1, 1)$ is the vector $(3, 3, 2)$. Thus, the tangent plane is

$$(3, 3, 2) \cdot (x - 1, y - 1, z - 1) = 0;$$

that is,

$$3x + 3y + 2z = 8. \quad \blacktriangle$$

In Theorem 14 and the definition following it, we could just as well have worked in two dimensions as in three. Thus, if we have $f\colon \mathbb{R}^2 \to \mathbb{R}$ and consider a level *curve*

$$C = \{(x, y) \mid f(x, y) = k\},$$

then $\nabla f(x_0, y_0)$ is perpendicular to C for any point (x_0, y_0) on C. Likewise, the tangent line to C at (x_0, y_0) has the equation

$$\nabla f(x_0, y_0) \cdot (x - x_0, y - y_0) = 0 \tag{2}$$

if $\nabla f(x_0, y_0) \neq \mathbf{0}$; that is, the tangent line is the set of points (x, y) that satisfy equation (2) (see Figure 2.6.3).

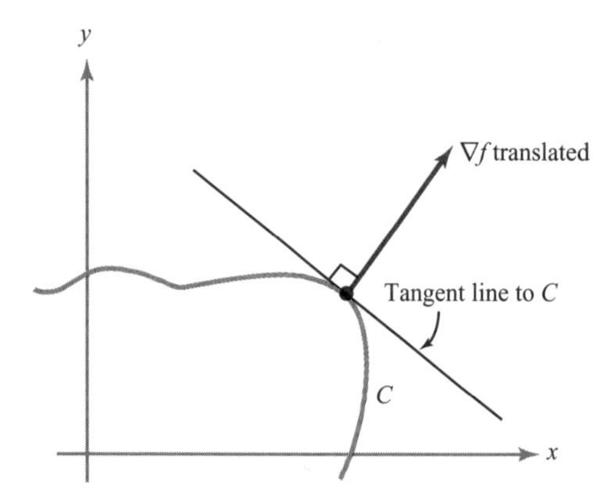

Figure 2.6.3 In the plane, the gradient ∇f is orthogonal to the curve f = constant.

The Gradient Vector Field

We often speak of ∇f as a ***gradient vector field***. The word "field" means that ∇f assigns a vector to each point in the domain of f. In Figure 2.6.4 we describe the gradient ∇f not by drawing its graph, which, if $f\colon \mathbb{R}^3 \to \mathbb{R}$, would be a subset of

$\mathbb{R}^6$, that is, the set of tuples $(\mathbf{x}, \nabla f(\mathbf{x}))$, but by representing $\nabla f(\mathrm{P})$, for each point P, as a vector emanating from the point P rather than from the origin. Like a graph, this pictorial method of depicting ∇f contains the point P and the value $\nabla f(\mathrm{P})$ in the same picture.

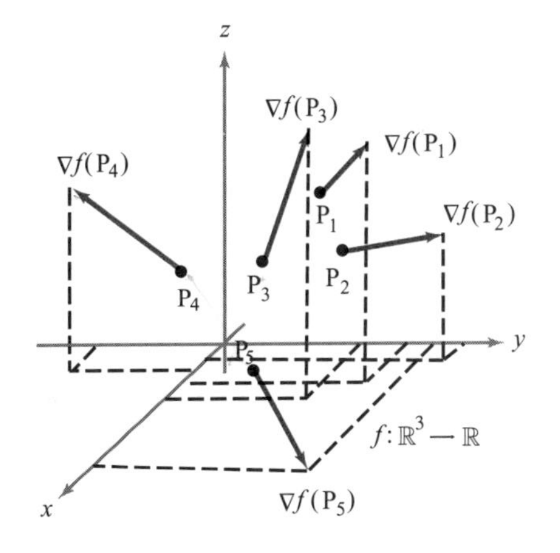

Figure 2.6.4 The gradient ∇f of a function $f\colon \mathbb{R}^3 \to \mathbb{R}$ is a vector field on $\mathbb{R}^3$; at each point P_i, $\nabla f(\mathrm{P}_i)$ is a vector emanating from P_i.

The gradient vector field has important geometric significance. It shows the direction in which f is increasing the fastest and the direction that is orthogonal to the level surfaces (or curves in the plane) of f. That it does both of these at once is quite plausible. To see this, imagine a hill as shown in Figure 2.6.5(a). Let h be the height function, a function of two variables. If we draw level curves of h, these are just level contours of the hill. We could imagine them as level paths on the hill [see Figure 2.6.5(b)]. One thing should be obvious to anyone who has gone for a hike: To get to the top of the hill the fastest, we should walk perpendicular to level contours.[5] This is consistent with Theorems 13 and 14, which state that the direction of fastest increase (the gradient) is orthogonal to the level curves.

EXAMPLE 6 The gravitational force on a unit mass m at (x, y, z) produced by a mass M at the origin in $\mathbb{R}^3$ is, according to Newton's law of gravitation, given by

$$\mathbf{F} = -\frac{GmM}{r^2}\mathbf{n},$$

where G is a constant; $r = \|\mathbf{r}\| = \sqrt{x^2 + y^2 + z^2}$, which is the distance of (x, y, z) from the origin; and $\mathbf{n} = \mathbf{r}/r$, the unit vector in the direction of $\mathbf{r} = x\mathbf{i} + y\mathbf{j} + z\mathbf{k}$, which is the position vector from the origin to (x, y, z).

Note that $\mathbf{F} = \nabla(GmM/r) = -\nabla V$, that is, $\mathbf{F}$ is the negative of the gradient of the gravitational potential $V = -GmM/r$. This can be verified as in Example 1.

[5]This discussion assumes that one walks at the same speed in all directions. Of course, hikers know that this is not necessarily realistic.

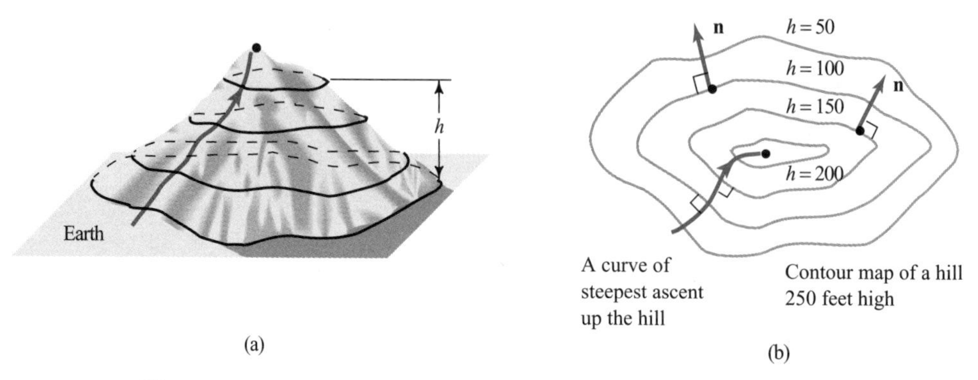

Figure 2.6.5 A physical illustration of the two facts (a) ∇f is the direction of fastest increase of f, and (b) ∇f is orthogonal to the level curves.

Notice that $\mathbf{F}$ is directed inward toward the origin. Also, the level surfaces of V are spheres. The gradient vector field $\mathbf{F}$ is normal to these spheres, which confirms the result of Theorem 14. ▲

EXAMPLE 7 Find a unit vector normal to the surface S given by $z = x^2y^2 + y + 1$ at the point $(0, 0, 1)$.

SOLUTION Let $f(x, y, z) = x^2y^2 + y + 1 - z$, and consider the level surface defined by $f(x, y, z) = 0$. Because this is the set of points (x, y, z) with $z = x^2y^2 + y + 1$, we see that this level set coincides with the surface S. The gradient is given by

$$\nabla f(x, y, z) = \frac{\partial f}{\partial x}\mathbf{i} + \frac{\partial f}{\partial y}\mathbf{j} + \frac{\partial f}{\partial z}\mathbf{k} = 2xy^2\mathbf{i} + (2x^2y + 1)\mathbf{j} - \mathbf{k},$$

and so

$$\nabla f(0, 0, 1) = \mathbf{j} - \mathbf{k}.$$

This vector is perpendicular to S at $(0, 0, 1)$, and so to find a unit normal $\mathbf{n}$ we divide this vector by its length to obtain

$$\mathbf{n} = \frac{\nabla f(0, 0, 1)}{\|\nabla f(0, 0, 1)\|} = \frac{1}{\sqrt{2}}(\mathbf{j} - \mathbf{k}). \quad ▲$$

EXAMPLE 8 Consider two conductors, one charged positively and the other negatively. Between them, an electric potential is set up. This potential is a function $\phi\colon \mathbb{R}^3 \to \mathbb{R}$ (an example of a *scalar* field). The electric field is given by $\mathbf{E} = -\nabla\phi$. From Theorem 14 we know that $\mathbf{E}$ is perpendicular to level surfaces of ϕ. These level surfaces are called ***equipotential surfaces***, because the potential is constant on them (see Figure 2.6.6). ▲

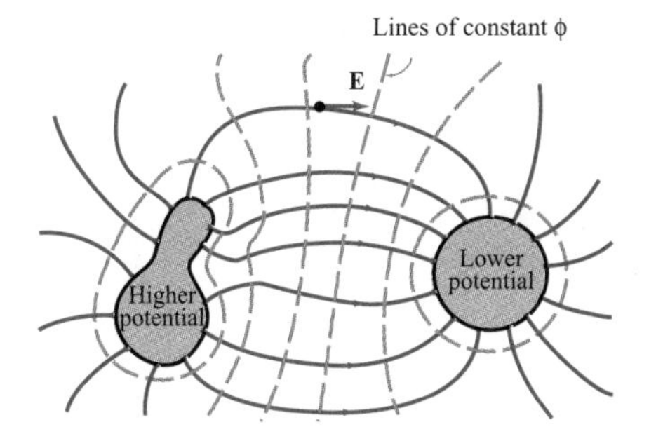

Figure 2.6.6 Equipotential surfaces (the dotted lines) are orthogonal to the electric force field **E**.

EXERCISES

1. Show that the directional derivative of $f(x, y, z) = z^2x + y^3$ at $(1, 1, 2)$ in the direction $(1/\sqrt{5})\mathbf{i} + (2/\sqrt{5})\mathbf{j}$ is $2\sqrt{5}$.

2. Compute the directional derivatives of the following functions at the indicated points in the given directions:

(a) $f(x, y) = x + 2xy - 3y^2$, $(x_0, y_0) = (1, 2)$, $\mathbf{v} = \frac{3}{5}\mathbf{i} + \frac{4}{5}\mathbf{j}$
(b) $f(x, y) = \log\sqrt{x^2 + y^2}$, $(x_0, y_0) = (1, 0)$, $\mathbf{v} = (1/\sqrt{5})(2\mathbf{i} + \mathbf{j})$
(c) $f(x, y) = e^x \cos(\pi y)$, $(x_0, y_0) = (0, -1)$, $\mathbf{v} = -(1/\sqrt{5})\mathbf{i} + (2/\sqrt{5})\mathbf{j}$
(d) $f(x, y) = xy^2 + x^3y$, $(x_0, y_0) = (4, -2)$, $\mathbf{v} = (1/\sqrt{10})\mathbf{i} + (3/\sqrt{10})\mathbf{j}$

3. Compute the directional derivatives of the following functions along unit vectors at the indicated points in directions *parallel* to the given vector:

(a) $f(x, y) = x^y$, $(x_0, y_0) = (e, e)$, $\mathbf{d} = 5\mathbf{i} + 12\mathbf{j}$
(b) $f(x, y, z) = e^x + yz$, $(x_0, y_0, z_0) = (1, 1, 1)$, $\mathbf{d} = (1, -1, 1)$
(c) $f(x, y, z) = xyz$, $(x_0, y_0, z_0) = (1, 0, 1)$, $\mathbf{d} = (1, 0, -1)$

4. Find the planes tangent to the following surfaces at the indicated points:

(a) $x^2 + 2y^2 + 3xz = 10$, at the point $(1, 2, \frac{1}{3})$
(b) $y^2 - x^2 = 3$, at the point $(1, 2, 8)$
(c) $xyz = 1$, at the point $(1, 1, 1)$

5. Find the equation for the plane tangent to each surface $z = f(x, y)$ at the indicated point:

(a) $z = x^3 + y^3 - 6xy$, at the point $(1, 2, -3)$
(b) $z = (\cos x)(\cos y)$, at the point $(0, \pi/2, 0)$
(c) $z = (\cos x)(\sin y)$, at the point $(0, \pi/2, 1)$

6. Compute the gradient ∇f for each of the following functions:

(a) $f(x, y, z) = 1/\sqrt{x^2 + y^2 + z^2}$
(b) $f(x, y, z) = xy + yz + xz$
(c) $f(x, y, z) = \dfrac{1}{x^2 + y^2 + z^2}$

7. For the functions in Exercise 6, what is the direction of fastest increase at $(1, 1, 1)$? [The solution to part (c) only is in the Study Guide to this text.]

8. Show that a unit normal to the surface $x^3y^3 + y - z + 2 = 0$ at $(0, 0, 2)$ is given by $\mathbf{n} = (1/\sqrt{2})(\mathbf{j} - \mathbf{k})$.

9. Find a unit normal to the surface $\cos(xy) = e^z - 2$ at $(1, \pi, 0)$.

10. Verify Theorems 13 and 14 for $f(x, y, z) = x^2 + y^2 + z^2$.

11. Show that the definition following Theorem 14 yields, as a special case, the formula for the plane tangent to the graph of $f(x, y)$ by regarding the graph as a level surface of $F(x, y, z) = f(x, y) - z$ (see Section 2.3).

12. Let $f(x, y) = -(1 - x^2 - y^2)^{1/2}$ for (x, y) such that $x^2 + y^2 < 1$. Show that the plane tangent to the graph of f at $(x_0, y_0, f(x_0, y_0))$ is orthogonal to the vector with components $(x_0, y_0, f(x_0, y_0))$. Interpret this geometrically.

13. For the following functions $f: \mathbb{R}^3 \to \mathbb{R}$ and $\mathbf{g}: \mathbb{R} \to \mathbb{R}^3$, find ∇f and $\mathbf{g}'$ and evaluate $(f \circ \mathbf{g})'(1)$.

(a) $f(x, y, z) = xz + yz + xy$, $\mathbf{g}(t) = (e^t, \cos t, \sin t)$
(b) $f(x, y, z) = e^{xyz}$, $\mathbf{g}(t) = (6t, 3t^2, t^3)$
(c) $f(x, y, z) = (x^2 + y^2 + z^2) \log \sqrt{x^2 + y^2 + z^2}$, $\mathbf{g}(t) = (e^t, e^{-t}, t)$

14. Compute the directional derivative of f in the given directions $\mathbf{v}$ at the given points P.

(a) $f(x, y, z) = xy^2 + y^2z^3 + z^3x$, $\mathrm{P} = (4, -2, -1)$, $\mathbf{v} = 1/\sqrt{14}(\mathbf{i} + 3\mathbf{j} + 2\mathbf{k})$
(b) $f(x, y, z) = x^{yz}$, $\mathrm{P} = (e, e, 0)$, $\mathbf{v} = \frac{12}{13}\mathbf{i} + \frac{3}{13}\mathbf{j} + \frac{4}{13}\mathbf{k}$

15. Let $\mathbf{r} = x\mathbf{i} + y\mathbf{j} + z\mathbf{k}$ and $r = \|\mathbf{r}\|$. Prove that

$$\nabla\left(\frac{1}{r}\right) = -\frac{\mathbf{r}}{r^3}.$$

16. Captain Ralph is in trouble near the sunny side of Mercury. The temperature of the ship's hull when he is at location (x, y, z) will be given by $T(x, y, z) = e^{-x^2-2y^2-3z^2}$, where x, y, and z are measured in meters. He is currently at $(1, 1, 1)$.

(a) In what direction should he proceed in order to decrease the temperature most rapidly?
(b) If the ship travels at e^8 meters per second, how fast will be the temperature decrease if he proceeds in that direction?
(c) Unfortunately, the metal of the hull will crack if cooled at a rate greater than $\sqrt{14}e^2$ degrees per second. Describe the set of possible directions in which he may proceed to bring the temperature down at no more than that rate.

17. A function $f: \mathbb{R}^2 \to \mathbb{R}$ is said to be *independent of the second variable* if there is a function $g: \mathbb{R} \to \mathbb{R}$ such that $f(x, y) = g(x)$ for all x in $\mathbb{R}$. In this case, calculate ∇f in terms of g'.

18. Let f and g be functions from $\mathbb{R}^3$ to $\mathbb{R}$. Suppose f is differentiable and $\nabla f(\mathbf{x}) = g(\mathbf{x})\mathbf{x}$. Show that spheres centered at the origin are contained in the level sets for f; that is, f is constant on such spheres.

19. A function $f: \mathbb{R}^n \to \mathbb{R}$ is called an *even* function if $f(\mathbf{x}) = f(-\mathbf{x})$ for every $\mathbf{x}$ in $\mathbb{R}^n$. If f is differentiable and even, find $\mathbf{D}f$ at the origin.

20. Suppose that a mountain has the shape of an elliptic paraboloid $z = c - ax^2 - by^2$, where a, b, and c are positive constants, x and y are the east-west and north-south map coordinates, and z is the altitude above sea level (x, y, z are all measured in meters). At the point $(1, 1)$, in what direction is the altitude increasing most rapidly? If a marble were released at $(1, 1)$, in what direction would it begin to roll?

21. An engineer wishes to build a railroad up the mountain of Exercise 20. Straight up the mountain is much too steep for the power of the engines. At the point $(1, 1)$, in what directions may the track be laid so that it will be climbing with a 3% grade—that is, an angle whose tangent is 0.03? (There are two possibilities.) Make a sketch of the situation indicating the two possible directions for a 3% grade at $(1, 1)$.

22. In electrostatics, the force $\mathbf{P}$ of attraction between two particles of opposite charge is given by $\mathbf{P} = k(\mathbf{r}/\|\mathbf{r}\|^3)$ (*Coulomb's law*), where k is a constant and $\mathbf{r} = x\mathbf{i} + y\mathbf{j} + z\mathbf{k}$. Show that $\mathbf{P}$ is the gradient of $f = -k/\|\mathbf{r}\|$.

23. The electrostatic potential V due to two infinite parallel filaments with linear charge densities λ and $-\lambda$ is $V = (\lambda/2\pi\varepsilon_0)\ln(r_2/r_1)$, where $r_1^2 = (x - x_0)^2 + y^2$ and $r_2^2 = (x + x_0)^2 + y^2$. We think of the filaments as being in the z direction, passing through the xy plane at $(-x_0, 0)$ and $(x_0, 0)$. Find $\nabla V(x, y)$.

24. For each of the following, find the maximum and minimum values attained by the function f along the path $\mathbf{c}(t)$:

(a) $f(x, y) = xy$; $\mathbf{c}(t) = (\cos t, \sin t)$; $0 \le t \le 2\pi$.
(b) $f(x, y) = x^2 + y^2$; $\mathbf{c}(t) = (\cos t, 2\sin t)$; $0 \le t \le 2\pi$.

25. Suppose that a particle is ejected from the surface $x^2 + y^2 - z^2 = -1$ at the point $(1, 1, \sqrt{3})$ along the normal directed toward the xy plane to the surface at time $t = 0$ with a speed of 10 units per second. When and where does it cross the xy plane?

26. Let $f: \mathbb{R}^3 \to \mathbb{R}$ and regard $\mathbf{D}f(x, y, z)$ as a linear map of $\mathbb{R}^3$ to $\mathbb{R}$. Show that the kernel (that is, the set of vectors mapped to zero) of $\mathbf{D}f$ is the plane in $\mathbb{R}^3$ orthogonal to ∇f.

REVIEW EXERCISES FOR CHAPTER 2

1. Describe the graphs of:

(a) $f(x, y) = 3x^2 + y^2$ (b) $f(x, y) = xy + 3x$

2. Describe some appropriate level surfaces and sections of the graphs of:

(a) $f(x, y, z) = 2x^2 + y^2 + z^2$

(b) $f(x, y, z) = x^2$
(c) $f(x, y, z) = xyz$

3. Compute the derivative $\mathbf{D}f(\mathbf{x})$ of each of the following functions:

(a) $f(x, y) = (x^2 y, e^{-xy})$
(b) $f(x) = (x, x)$
(c) $f(x, y, z) = e^x + e^y + e^z$
(d) $f(x, y, z) = (x, y, z)$

4. Suppose $f(x, y) = f(y, x)$ for all (x, y). Prove that

$$(\partial f/\partial x)(a, b) = (\partial f/\partial y)(b, a).$$

5. Let $f(x, y) = (1 - x^2 - y^2)^{1/2}$. Show that the plane tangent to the graph of f at $(x_0, y_0, f(x_0, y_0))$ is orthogonal to the vector $(x_0, y_0, f(x_0, y_0))$. Interpret geometrically.

6. Let $F(u, v)$ and $u = h(x, y, z)$, $v = k(x, y, z)$ be given (differentiable) real-valued functions and let $f(x, y, z)$ be defined by $f(x, y, z) = F(h(x, y, z), k(x, y, z))$. Write a formula for the gradient of f in terms of the partial derivatives of F, h, and k.

7. Find an equation for the tangent plane of the graph of f at the point $(x_0, y_0, f(x_0, y_0))$ for:

(a) $f\colon \mathbb{R}^2 \to \mathbb{R}, (x, y) \mapsto x - y + 2,\quad (x_0, y_0) = (1, 1)$
(b) $f\colon \mathbb{R}^2 \to \mathbb{R}, (x, y) \mapsto x^2 + 4y^2,\quad (x_0, y_0) = (2, -1)$
(c) $f\colon \mathbb{R}^2 \to \mathbb{R}, (x, y) \mapsto xy,\quad (x_0, y_0) = (-1, -1)$
(d) $f(x, y) = \log(x + y) + x\cos y + \arctan(x + y),\quad (x_0, y_0) = (1, 0)$
(e) $f(x, y) = \sqrt{x^2 + y^2},\quad (x_0, y_0) = (1, 1)$
(f) $f(x, y) = xy,\quad (x_0, y_0) = (2, 1)$

8. Compute an equation for the tangent planes of the following surfaces at the indicated points.

(a) $x^2 + y^2 + z^2 = 3,\quad (1, 1, 1)$
(b) $x^3 - 2y^3 + z^3 = 0,\quad (1, 1, 1)$
(c) $(\cos x)(\cos y)e^z = 0,\quad (\pi/2, 1, 0)$
(d) $e^{xyz} = 1,\quad (1, 1, 0)$

9. Draw some level curves for the following functions:

(a) $f(x, y) = 1/xy$
(b) $f(x, y) = x^2 - xy - y^2$

10. Consider a temperature function $T(x, y) = x \sin y$. Plot a few level curves. Compute ∇T and explain its meaning.

11. Find the following limits if they exist:

(a) $\displaystyle \lim_{(x,y)\to(0,0)} \frac{\cos xy - 1}{x}$ (b) $\displaystyle \lim_{(x,y)\to(0,0)} \sqrt{|(x + y)/(x - y)|},\ x \neq y$

12. Compute the first partial derivatives and gradients of the following functions:

(a) $f(x, y, z) = xe^z + y \cos x$
(b) $f(x, y, z) = (x + y + z)^{10}$
(c) $f(x, y, z) = (x^2 + y)/z$

13. Compute $\dfrac{\partial}{\partial x}[x \exp(1 + x^2 + y^2)]$

14. Let $y(x)$ be a differentiable function defined implicitly by $F(x, y(x)) = 0$. From Exercise 17(a), Section 2.5, we know that

$$\frac{dy}{dx} = -\frac{\partial F/\partial x}{\partial F/\partial y}.$$

Consider the surface $z = F(x, y)$, and suppose F is increasing as a function of x and as a function of y; that is, $\partial F/\partial x > 0$ and $\partial F/\partial y > 0$. By considering the graph and the plane $z = 0$, show that for z fixed at $z = 0$, y should *decrease* as x increases and x should *decrease* as y increases. Does this agree with the minus sign in the formula for dy/dx?

15. (a) Consider the graph of a function $f(x, y)$ [Figure 2.R.1(a)]. Let (x_0, y_0) lie on a level curve C, so $\nabla f(x_0, y_0)$ is perpendicular to this curve. Show that the tangent plane of the graph is the plane that (i) contains the line perpendicular to $\nabla f(x_0, y_0)$ and lying in the horizontal plane $z = f(x_0, y_0)$, and (ii) has slope $\|\nabla f(x_0, y_0)\|$ relative to the xy plane. (By the *slope* of a plane P relative to the xy plane we mean the tangent of the angle θ, $0 \leq \theta \leq \pi$, between the upward-pointing normal $\mathbf{p}$ to P and the unit vector $\mathbf{k}$.)

(b) Use this method to show that the tangent plane of the graph of $f(x, y) = (x + \cos y)x^2$ at $(1, 0, 2)$ is as sketched in Figure 2.R.1(b).

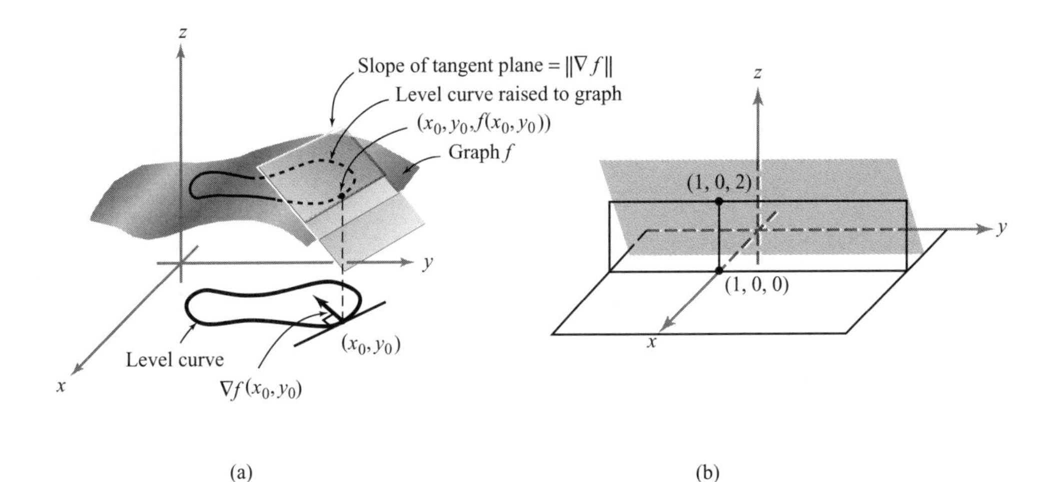

Figure 2.R.1 (a) The relationship between the gradient of a function and the tangent plane to the graph [Exercise 15(a)]. A specific instance of the plane in (b) for Exercise 15(b).

16. Find the plane tangent to the surface $z = x^2 + y^2$ at the point $(1, -2, 5)$. Explain the geometric significance, for this surface, of the gradient of $f(x, y) = x^2 + y^2$ (see Exercise 15).

17. In which direction is the directional derivative of $f(x, y) = (x^2 - y^2)/(x^2 + y^2)$ at $(1, 1)$ equal to zero?

18. Find the directional derivative of the given function at the given point and in the direction of the given vector.

(a) $f(x, y, z) = e^x \cos(yz)$, $\quad p_0 = (0, 0, 0)$, $\quad \mathbf{v} = (2, 1, -2)$
(b) $f(x, y, z) = xy + yz + zx$, $\quad p_0 = (1, 1, 2)$, $\quad \mathbf{v} = (10, -1, 2)$

19. Find the tangent plane and normal to the hyperboloid $x^2 + y^2 - z^2 = 18$ at $(3, 5, -4)$.

20. Let $(x(t), y(t))$ be a path in the plane, $0 \le t \le 1$, and let $f(x, y)$ be a C^1 function of two variables. Assume that $(dx/dt)f_x + (dy/dt)f_y \le 0$. Show that $f(x(1), y(1)) \le f(x(0), y(0))$.

21. A bug finds itself in a toxic environment. The toxicity level is given by $T(x, y) = 2x^2 - 4y^2$. The bug is at $(-1, 2)$. In what direction should it move to lower the toxicity the fastest?

22. Find the direction in which the function $w = x^2 + xy$ increases most rapidly at the point $(-1, 1)$. What is the magnitude of ∇w at this point? Interpret this magnitude geometrically.

23. Let f be defined on an open set S in $\mathbb{R}^n$. We say that f is ***homogeneous of degree*** p over S if $f(\lambda \mathbf{x}) = \lambda^p f(\mathbf{x})$ for every real λ and for every $\mathbf{x}$ in S for which $\lambda \mathbf{x} \in S$.

(a) If such a function is differentiable at $\mathbf{x}$, show that $\mathbf{x} \cdot \nabla f(\mathbf{x}) = pf(\mathbf{x})$. This is known as ***Euler's theorem*** for homogeneous functions. [HINT: For fixed $\mathbf{x}$, define $g(\lambda) = f(\lambda \mathbf{x})$ and compute $g'(1)$.]

(b) Find p and check Euler's theorem for the function $f(x, y, z) = x - 2y - \sqrt{xz}$, on the region where $xz > 0$.

24. If $z = [f(x - y)]/y$ (where f is differentiable and $y \neq 0$), show that the identity $z + y(\partial z/\partial x) + y(\partial z/\partial y) = 0$ holds.

25. Given $z = f((x + y)/(x - y))$ for f a C^1 function, show that

$$x \frac{\partial z}{\partial x} + y \frac{\partial z}{\partial y} = 0.$$

26. Let f have partial derivatives $\partial f(\mathbf{x})/\partial x_i$, where $i = 1, 2, \ldots, n$, at each point $\mathbf{x}$ of an open set U in $\mathbb{R}^n$. If f has a local maximum or a local minimum at the point $\mathbf{x}_0$ in U, show that $\partial f(\mathbf{x}_0)/\partial x_i = 0$ for each i.

27. Consider the functions defined in $\mathbb{R}^2$ by the following formulas:

$$\begin{array}{lll}\text{(i) } f(x,y) = xy/(x^2+y^2) & \text{if } (x,y) \neq (0,0), & f(0,0)=0 \\ \text{(ii) } f(x,y) = x^2y^2/(x^2+y^4) & \text{if } (x,y) \neq (0,0), & f(0,0)=0\end{array}$$

(a) In each case, show that the partial derivatives $\partial f(x,y)/\partial x$ and $\partial f(x,y)/\partial y$ exist for every (x,y) in $\mathbb{R}^2$, and evaluate these derivatives explicitly in terms of x and y.

(b) Explain why the functions described in (i) and (ii) are or are not differentiable at $(0,0)$.

28. Compute the gradient vector $\nabla f(x,y)$ at all points (x,y) in $\mathbb{R}^2$ for each of the following functions:

$$\begin{array}{lll}\text{(a) } f(x,y) = x^2y^2\log(x^2+y^2) & \text{if } (x,y) \neq (0,0), & f(0,0)=0 \\ \text{(b) } f(x,y) = xy\sin[1/(x^2+y^2)] & \text{if } (x,y) \neq (0,0), & f(0,0)=0\end{array}$$

29. Find the directional derivatives of the following functions at the point $(1,1)$ in the direction $(\mathbf{i}+\mathbf{j})/\sqrt{2}$:

(a) $f(x,y) = x\tan^{-1}(x/y)$
(b) $f(x,y) = \cos(\sqrt{x^2+y^2})$
(c) $f(x,y) = \exp(-x^2-y^2)$

30. (a) Let $\mathbf{u} = \mathbf{i} - 2\mathbf{j} + 2\mathbf{k}$ and $\mathbf{v} = 2\mathbf{i} + \mathbf{j} - 3\mathbf{k}$. Find: $\|\mathbf{u}\|$, $\mathbf{u}\cdot\mathbf{v}$, $\mathbf{u}\times\mathbf{v}$, and a vector in the same direction as $\mathbf{u}$, but of unit length.

(b) Find the rate of change of $e^{xy}\sin(xyz)$ in the direction $\mathbf{u}$ at $(0,1,1)$.

31. Let $h(x,y) = 2e^{-x^2} + e^{-3y^2}$ denote the height on a mountain at position (x,y). In what direction from $(1,0)$ should one begin walking in order to climb the fastest?

32. Compute an equation for the plane tangent to the graph of

$$f(x,y) = \frac{e^x}{x^2+y^2}$$

at $x=1$, $y=2$.

33. (a) Give a careful statement of the general form of the chain rule.

(b) Let $f(x,y) = x^2+y$ and $\mathbf{h}(u) = (\sin 3u, \cos 8u)$. Let $g(u) = f(\mathbf{h}(u))$. Compute dg/du at $u=0$ both directly *and* by using the chain rule.

34. (a) Sketch the level curves of $f(x,y) = -x^2-9y^2$ for $c = 0, -1, -10$.

(b) On your sketch, draw in ∇f at $(1,1)$. Discuss.

35. At time $t=0$, a particle is ejected from the surface $x^2+2y^2+3z^2=6$ at the point $(1,1,1)$ in a direction normal to the surface at a speed of 10 units per second. At what time does it cross the sphere $x^2+y^2+z^2=103$?

36. At what point(s) on the surface in Exercise 35 is the normal vector parallel to the line $x = y = z$?

37. Compute $\partial z/\partial x$ and $\partial z/\partial y$ if

$$z = \frac{u^2 + v^2}{u^2 - v^2}, \qquad u = e^{-x-y}, \qquad v = e^{xy}$$

(a) by substitution and direct calculation, and (b) by the chain rule.

38. Compute the partial derivatives as in Exercise 37 if $z = uv$, $u = x + y$, and $v = x - y$.

39. What is wrong with the following argument? Suppose that $w = f(x, y)$ and $y = x^2$. By the chain rule,

$$\frac{\partial w}{\partial x} = \frac{\partial w}{\partial x}\frac{\partial x}{\partial x} + \frac{\partial w}{\partial y}\frac{\partial y}{\partial x} = \frac{\partial w}{\partial x} + 2x\frac{\partial w}{\partial y}.$$

Hence, $0 = 2x(\partial w/\partial y)$, and so $\partial w/\partial y = 0$. Choose an explicit example to really see that this is incorrect.

40. A boat is sailing northeast at 20 km/h. Assuming that the temperature drops at a rate of 0.2°C/km in the northerly direction and 0.3°C/km in the easterly direction, what is the time rate of change of temperature as observed on the boat?

41. Use the chain rule to find a formula for $(d/dt)\exp[f(t)g(t)]$.

42. Use the chain rule to find a formula for $(d/dt)(f(t)^{g(t)})$.

43. Verify the chain rule for the function $f(x, y, z) = [\ln(1 + x^2 + 2z^2)]/(1 + y^2)$ and the path $\mathbf{c}(t) = (t, 1 - t^2, \cos t)$.

44. Verify the chain rule for the function $f(x, y) = x^2/(2 + \cos y)$ and the path $x = e^t$, $y = e^{-t}$.

45. Suppose that $u(x, t)$ satisfies the differential equation $u_t + uu_x = 0$ and that x, as a function $x = f(t)$ of t, satisfies $dx/dt = u(x, t)$. Prove that $u(f(t), t)$ is constant in t.

46. The displacement at time t and horizontal position on a line x of a certain violin string is given by $u = \sin(x - 6t) + \sin(x + 6t)$. Calculate the velocity of the string at $x = 1$ when $t = \frac{1}{3}$.

47. The ***ideal gas law*** $PV = nRT$ involves a constant R, the number n of moles of the gas, the volume V, the Kelvin temperature T, and the pressure P.

(a) Show that each of n, P, T, V is a function of the remaining variables, and determine explicitly the defining equations.
(b) Calculate $\partial V/\partial T$, $\partial T/\partial P$, $\partial P/\partial V$ and show that their product equals -1.

48. The ***potential temperature*** θ is defined in terms of temperature T and pressure p by

$$\theta = T\left(\frac{1000}{p}\right)^{0.286}.$$

The temperature and pressure may be thought of as functions of position (x, y, z) in the atmosphere and also of time t.

(a) Find formulas for $\partial\theta/\partial x$, $\partial\theta/\partial y$, $\partial\theta/\partial z$, $\partial\theta/\partial t$ in terms of partial derivatives of T and p.

(b) The condition $\partial\theta/\partial z < 0$ is regarded as an unstable atmosphere, for it leads to large vertical excursions of air parcels from a single upward or downward impetus. Meteorologists use the formula

$$\frac{\partial\theta}{\partial z} = \frac{\theta}{T}\left(\frac{\partial T}{\partial z} + \frac{g}{C_p}\right)$$

where $g = 32.2$ and C_p is a positive constant. How does the temperature change in the upward direction for an unstable atmosphere?

49. The specific volume V, pressure P, and temperature T of a van der Waals gas are related by $P = RT/(V - \beta) - \alpha/V^2$, where α, β, and R are constants.

(a) Explain why any two of V, P, and T can be considered independent variables that determine the third variable.

(b) Find $\partial T/\partial P$, $\partial P/\partial V$, $\partial V/\partial T$. Identify which variables are constant, and interpret each partial derivative physically.

(c) Verify that $(\partial T/\partial P)(\partial P/\partial V)(\partial V/\partial T) = -1$(not $+1$!).

50. The height h of the Hawaiian volcano Mauna Loa is (roughly) described by the function $h(x, y) = 2.59 - 0.00024y^2 - 0.00065x^2$, where h is the height above sea level in miles and x and y measure east-west and north-south distances in miles from the top of the mountain. At $(x, y) = (-2, -4)$:

(a) How fast is the height increasing in the direction $(1, 1)$ (that is, northeastward)? Express your answer in miles of height per mile of horizontal distance traveled.

(b) In what direction is the steepest upward path?

51. (a) In what direction is the directional derivative of $f(x, y) = (x^2 - y^2)/(x^2 + y^2)$ at $(1, 1)$ equal to zero?

(b) How about at an arbitrary point (x_0, y_0) in the first quadrant?

(c) Describe the level curves of f. In particular, discuss them in terms of the result of part (b).

52. (a) Show that the curve $x^2 - y^2 = c$, for *any* value of c, satisfies the differential equation $dy/dx = x/y$.

(b) Draw in a few of the curves $x^2 - y^2 = c$, say for $c = \pm 1$. At several points (x, y) along each of these curves, draw a short segment of slope x/y; check that these segments appear to be tangent to the curve. What happens when $y = 0$? What happens when $c = 0$?

53. Suppose that f is a differentiable function of one variable and that a function $u = g(x, y)$ is defined by

$$u = g(x, y) = xyf\left(\frac{x + y}{xy}\right).$$

Show that u satisfies a (partial) differential equation of the form

$$x^2 \frac{\partial u}{\partial x} - y^2 \frac{\partial u}{\partial y} = G(x, y)u$$

and find the function $G(x, y)$.

54. (a) Let F be a function of one variable and f a function of two variables. Show that the gradient vector of $g(x, y) = F(f(x, y))$ is parallel to the gradient vector of $f(x, y)$.

(b) Let $f(x, y)$ and $g(x, y)$ be functions such that $\nabla f = \lambda \nabla g$ for some function $\lambda(x, y)$. What is the relation between the level curves of f and g? Explain why there might be a function F such that $g(x, y) = F(f(x, y))$.

3

Higher-Order Derivatives: Maxima and Minima

Leonhard Euler (by Emanuel Handman) (1707–1783).

All that is superfluous displeases God and Nature.
All that displeases God and Nature is evil.

Dante Alighieri, circa 1300

. . . namely, because the shape of the whole universe is most perfect, and, in fact, designed by the wisest creator, nothing in all of the world will occur in which no maximum or minimum rule is somehow shining forth.

Leonhard Euler

In one-variable calculus, to test a function $f(x)$ for a local maximum or minimum, one often uses the second derivative. We look for critical points x_0, that is, points x_0 for which $f'(x_0) = 0$, and at each such point we check the sign of the second derivative $f''(x_0)$. If $f''(x_0) < 0$, $f(x_0)$ is a local maximum of f; if $f''(x_0) > 0$, $f(x_0)$ is a local minimum of f; if $f''(x_0) = 0$, the test fails.

This chapter extends these methods to real-valued functions of several variables. We begin in Section 3.1 with a discussion of iterated and higher-order partial derivatives, and in Section 3.2 we discuss the multivariable form of Taylor's theorem; this is then used in Section 3.3 to derive tests for maxima, minima, and saddle points. As with functions of one variable, such methods help one to visualize the shape of a graph.

In Section 3.4, we study the problem of maximizing a real-valued function subject to supplementary conditions, also referred to as constraints. For example, we might wish to maximize $f(x, y, z)$ among those (x, y, z) constrained to lie on the unit sphere, $x^2 + y^2 + z^2 = 1$. Section 3.5 discusses a technical theorem (the implicit function theorem) useful for studying constraints. It will also be useful later in our study of surfaces.

3.1 Iterated Partial Derivatives

The preceding chapter developed considerable information concerning the derivative of a map and investigated the geometry associated with the derivative of real-valued functions by making use of the gradient. In this section, we proceed to study higher-order derivatives, with the goal of proving the equality of the "mixed second partial derivatives" of a function. We begin by defining the necessary terms.

Let $f: \mathbb{R}^3 \to \mathbb{R}$ be of class C^1. Recall that this means that $\partial f/\partial x, \partial f/\partial y$, and $\partial f/\partial z$ exist and are continuous. If these derivatives, in turn, have continuous partial derivatives, we say that f is of ***class*** C^2, or is ***twice continuously differentiable***. Likewise, if we say f is of class C^3, we mean f has continuous iterated partial derivatives of third order, and so on. Here are a few examples of how second-order derivatives are written:

$$\frac{\partial^2 f}{\partial x^2} = \frac{\partial}{\partial x}\left(\frac{\partial f}{\partial x}\right), \qquad \frac{\partial^2 f}{\partial x\, \partial y} = \frac{\partial}{\partial x}\left(\frac{\partial f}{\partial y}\right), \qquad \frac{\partial^2 f}{\partial z\, \partial y} = \frac{\partial}{\partial z}\left(\frac{\partial f}{\partial y}\right), \qquad \text{etc.}$$

The process can, of course, be repeated for third-order derivatives, and so on. If f is a function of only x and y and $\partial f/\partial x, \partial f/\partial y$ are continuously differentiable, then by taking second partial derivatives we get the four functions

$$\frac{\partial^2 f}{\partial x^2}, \qquad \frac{\partial^2 f}{\partial y^2}, \qquad \frac{\partial^2 f}{\partial x\, \partial y}, \qquad \text{and} \qquad \frac{\partial^2 f}{\partial y\, \partial x}.$$

All of these are called ***iterated partial derivatives***, while $\partial^2 f/\partial x\, \partial y$ and $\partial^2 f/\partial y\, \partial x$ are called ***mixed partial derivatives***.

EXAMPLE 1 Find all second partial derivatives of $f(x, y) = xy + (x + 2y)^2$.

SOLUTION The first partials are

$$\frac{\partial f}{\partial x} = y + 2(x + 2y), \qquad \frac{\partial f}{\partial y} = x + 4(x + 2y).$$

Now differentiate each of these expressions with respect to x and y:

$$\frac{\partial^2 f}{\partial x^2} = 2, \qquad \frac{\partial^2 f}{\partial y^2} = 8$$

$$\frac{\partial^2 f}{\partial x\, \partial y} = 5, \qquad \frac{\partial^2 f}{\partial y\, \partial x} = 5. \quad \blacktriangle$$

EXAMPLE 2 Find all second partial derivatives of $f(x, y) = \sin x \sin^2 y$.

SOLUTION We proceed just as in Example 1:

$$\frac{\partial f}{\partial x} = \cos x \sin^2 y, \qquad \frac{\partial f}{\partial y} = 2 \sin x \sin y \cos y = \sin x \sin 2y;$$

$$\frac{\partial^2 f}{\partial x^2} - = -\sin x \sin^2 y, \qquad \frac{\partial^2 f}{\partial y^2} = 2 \sin x \cos 2y;$$

$$\frac{\partial^2 f}{\partial x\, \partial y} = \cos x \sin 2y, \qquad \frac{\partial^2 f}{\partial y\, \partial x} = 2 \cos x \sin y \cos y = \cos x \sin 2y. \quad \blacktriangle$$

EXAMPLE 3 Let $f(x, y, z) = e^{xy} + z \cos x$. Then

$$\frac{\partial f}{\partial x} = ye^{xy} - z \sin x, \qquad \frac{\partial f}{\partial y} = xe^{xy}, \qquad \frac{\partial f}{\partial z} = \cos x,$$

$$\frac{\partial^2 f}{\partial z\, \partial x} = -\sin x, \qquad \frac{\partial^2 f}{\partial x\, \partial z} = -\sin x, \qquad \text{etc.} \quad \blacktriangle$$

The Mixed Partials are Equal

In all these examples note that the pairs of mixed partial derivatives, such as $\partial^2 f/\partial x\, \partial y$ and $\partial^2 f/\partial y\, \partial x$, or $\partial^2 f/\partial z\, \partial x$ and $\partial^2 f/\partial x\, \partial z$, are equal. It is a basic and perhaps surprising fact that *this is always the case for C^2 functions*. We shall prove this in the next theorem for functions $f(x, y)$ of two variables, but the proof can be readily extended to functions of n variables.

THEOREM 1: Equality of Mixed Partials If $f(x, y)$ is of class C^2 (is twice continuously differentiable), then the mixed partial derivatives are equal; that is,

$$\frac{\partial^2 f}{\partial x\, \partial y} = \frac{\partial^2 f}{\partial y\, \partial x}.$$

PROOF Consider the following expression (see Figure 3.1.1):

$$S(\Delta x, \Delta y) = f(x_0 + \Delta x, y_0 + \Delta y) - f(x_0 + \Delta x, y_0) \\ - f(x_0, y_0 + \Delta y) + f(x_0, y_0).$$

Holding y_0 and Δy fixed, define

$$g(x) = f(x, y_0 + \Delta y) - f(x, y_0),$$

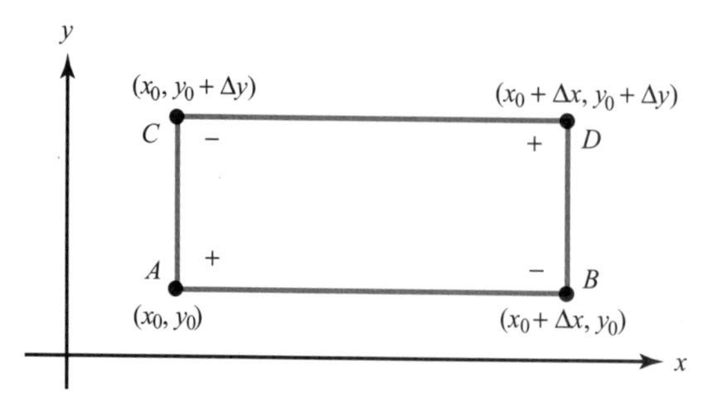

Figure 3.1.1 The algebra behind the equality of mixed partials: writing the difference of differences in two ways.

so that $S(\Delta x, \Delta y) = g(x_0 + \Delta x) - g(x_0)$, *which expresses S as a difference of differences*. By the mean-value theorem for functions of one variable, $g(x_0 + \Delta x) - g(x_0)$ equals $g'(\bar{x})\Delta x$ for some $\bar{x}$ between x_0 and $x_0 + \Delta x$. Hence,

$$S(\Delta x, \Delta y) = \left[\frac{\partial f}{\partial x}(\bar{x}, y_0 + \Delta y) - \frac{\partial f}{\partial x}(\bar{x}, y_0)\right]\Delta x.$$

Applying the mean-value theorem again, there is a $\bar{y}$ between y_0 and $y_0 + \Delta y$ such that

$$S(\Delta x, \Delta y) = \frac{\partial^2 f}{\partial y\, \partial x}(\bar{x}, \bar{y})\, \Delta x\, \Delta y.$$

Because $\partial^2 f/\partial y\, \partial x$ is continuous, it follows that

$$\frac{\partial^2 f}{\partial y\, \partial x}(x_0, y_0) = \lim_{(\Delta x, \Delta y) \to (0,0)} \frac{1}{\Delta x\, \Delta y}[S(\Delta x, \Delta y)].$$

Noting that S is symmetric in Δx and Δy, one shows in a similar way that $\partial^2 f/\partial x\, \partial y$ is given by the *same limit formula*, which proves the result. ■

The equality of mixed partial derivatives is one of the most important results of multivariable calculus. It will reappear on several occasions later in the book, when we study vector identities.

In the next historical note, we will discuss the role of partial derivatives in the formulation of many of the basic equations governing physical phenomena. One of the giants in this era was Leonhard Euler (1707–1783), who developed the equations of fluid mechanics that bear his name—the Euler equations. It was in connection with the needs of this development that he discovered, around 1734, the equality of mixed partial derivatives. Euler was about 27 years old at the time.

In Exercise 11 we ask the reader to deduce from Theorem 1 that for a C^3 function of x, y, and z,

$$\frac{\partial^3 f}{\partial x\, \partial y\, \partial z} = \frac{\partial^3 f}{\partial z\, \partial y\, \partial x} = \frac{\partial^3 f}{\partial y\, \partial z\, \partial x}, \qquad \text{etc.}$$

In other words, we can compute iterated partial derivatives in any order we please.

EXAMPLE 4 Verify the equality of the mixed second partial derivatives for the function

$$f(x, y) = xe^y + yx^2.$$

SOLUTION Here

$$\frac{\partial f}{\partial x} = e^y + 2xy, \qquad \frac{\partial f}{\partial y} = xe^y + x^2,$$

$$\frac{\partial^2 f}{\partial y\, \partial x} = e^y + 2x, \qquad \frac{\partial^2 f}{\partial x\, \partial y} = e^y + 2x,$$

and so we have

$$\frac{\partial^2 f}{\partial y\, \partial x} = \frac{\partial^2 f}{\partial x\, \partial y}. \quad ▲$$

Sometimes the notation f_x, f_y, f_z is used for the partial derivatives: $f_x = \partial f/\partial x$, and so on. With this notation, we write $f_{xy} = (f_x)_y$, and so equality of the mixed partials is denoted by $f_{xy} = f_{yx}$. Notice that $f_{xy} = \partial^2 f/\partial y\, \partial x$, so the order of x and y is reversed in the two notations; fortunately, the equality of mixed partials makes this potential ambiguity irrelevant. The following example illustrates this subscript notation.

EXAMPLE 5 Let

$$z = f(x, y) = e^x \sin xy$$

and write $x = g(s, t)$, $y = h(s, t)$ for certain functions g and h. Let

$$k(s, t) = f(g(s, t), h(s, t)).$$

Calculate k_{st}.

SOLUTION By the chain rule,

$$k_s = f_x g_s + f_y h_s = (e^x \sin xy + ye^x \cos xy)g_s + (xe^x \cos xy)h_s.$$

Differentiating in t using the product rules gives

$$k_{st} = (f_x)_t g_s + f_x(g_s)_t + (f_y)_t h_s + f_y(h_s)_t.$$

Applying the chain rule again to $(f_x)_t$ and $(f_y)_t$ gives

$$(f_x)_t = f_{xx}g_t + f_{xy}h_t \qquad \text{and} \qquad (f_y)_t = f_{yx}g_t + f_{yy}h_t,$$

and so k_{st} becomes

$$\begin{aligned} k_{st} &= (f_{xx}g_t + f_{xy}h_t)g_s + f_x g_{st} + (f_{yx}g_t + f_{yy}h_t)h_s + f_y h_{st} \\ &= f_{xx}g_t g_s + f_{xy}(h_t g_s + h_s g_t) + f_{yy}h_t h_s + f_x g_{st} + f_y h_{st}. \end{aligned}$$

Notice that this last formula is symmetric in (s, t), verifying the equality $k_{st} = k_{ts}$. Computing f_{xx}, f_{xy}, and f_{yy}, we get

$$\begin{aligned} k_{st} = {} & (e^x \sin xy + 2ye^x \cos xy - y^2 e^x \sin xy)g_t g_s \\ & + (xe^x \cos xy + e^x \cos xy - xye^x \sin xy)(h_t g_s + h_s g_t) \\ & - (x^2 e^x \sin xy)h_t h_s + (e^x \sin xy + ye^x \cos xy)g_{st} + (xe^x \cos xy)h_{st}, \end{aligned}$$

in which it is understood that $x = g(s, t)$ and $y = h(s, t)$. ▲

Historical Note

Some Partial Differential Equations

> Philosophy [nature] is written in that great book which ever is before our eyes—I mean the universe—but we cannot understand it if we do not first learn the language and grasp the symbols in which it is written. The book is written in mathematical language, and the symbols are triangles, circles and other geometrical figures, without whose help it is impossible to comprehend a single word of it; without which one wanders in vain through a dark labyrinth.
>
> GALILEO

This quotation illustrates the Greek belief, again popular in the time of Galileo, that much of nature could be described using mathematics. In the latter part of the seventeenth century this thinking was dramatically reinforced when Newton used his law of gravitation to derive Kepler's three laws of celestial motion (see Section 4.1) to explain the tides, and to show that the earth was flattened at the poles. The impact of this philosophy on mathematics was substantial, and many mathematicians sought to "mathematize" nature. The extent to which mathematics pervades the physical sciences today (and, to an increasing amount, economics and the social and life sciences) is testament to the success of these endeavors. Correspondingly, the attempts to mathematize nature have often led to new mathematical discoveries.

Many of the laws of nature were described in terms of either ordinary differential equations (ODEs, equations involving the derivatives of functions of one variable alone, such as the laws of planetary motion) or partial differential equations (PDEs), that is, equations involving partial derivatives of functions. To give the reader some historical perspective and offer motivation for studying partial derivatives, we present a brief description of three of the most famous partial differential equations: the heat equation, the potential equation (or Laplace's equation), and the wave equation. (Further information on some PDEs is given in Section 8.5.)

The Heat Equation. In the early part of the nineteenth century the French mathematician Joseph Fourier (1768–1830) took up the study of heat. Heat flow had obvious applications to both industrial and scientific problems: A better understanding of it would, for example, make possible more efficient smelting of metals and would enable scientists to determine the temperature of a body given the temperature at its boundary, and to approximate the temperature of the earth's interior.

Let a homogeneous body $\mathcal{B} \subset \mathbb{R}^3$ (Figure 3.1.2) be represented by some region in 3-space. Let $T(x, y, z, t)$ denote the temperature of the body at the

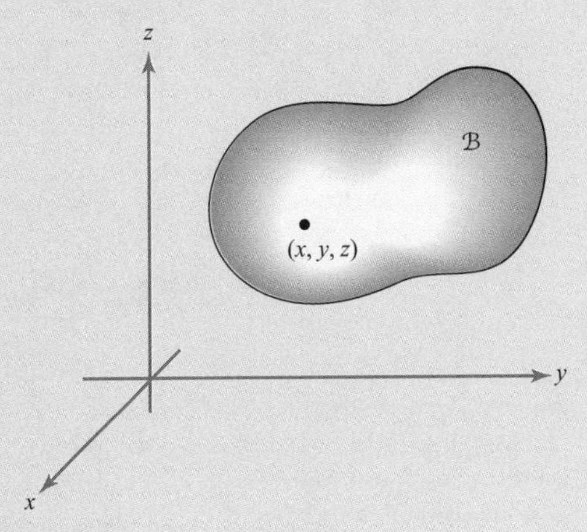

Figure 3.1.2 A homogeneous body in space.

point (x, y, z) at time t. Fourier proved, on the basis of physical principles (described in Section 8.5), that T must satisfy the partial differential equation called the ***heat equation***,

$$k\left(\frac{\partial^2 T}{\partial x^2} + \frac{\partial^2 T}{\partial y^2} + \frac{\partial^2 T}{\partial z^2}\right) = \frac{\partial T}{\partial t}, \tag{1}$$

where k is a constant whose value depends on the conductivity of the material comprising the body.

Fourier used this equation to solve problems in heat conduction. In fact, his investigations into the solutions of equation (1) led him to the discovery of *Fourier series*.

The Potential Equation. Consider the gravitational potential V (often called Newton's potential) of a mass m at a point (x, y, z) caused by a point mass M situated at the origin. This potential is given by $V = -GmM/r$, where $r = \sqrt{x^2 + y^2 + z^2}$. The potential V satisfies the equation

$$\frac{\partial^2 V}{\partial x^2} + \frac{\partial^2 V}{\partial y^2} + \frac{\partial^2 V}{\partial z^2} = 0 \tag{2}$$

everywhere except at the origin, as we will check in the next chapter (see also Exercise 23). This equation is known as ***Laplace's equation***. Pierre-Simon de Laplace (1749–1827) had worked on the gravitational attraction of nonpoint masses and was the first to consider equation (2) with regard to gravitational attraction. He gave arguments (later shown to be incorrect) that equation (2) held for any body and any point whether inside or outside that body. However, Laplace was not the first person to write down equation (2). The potential equation appeared for the first time in one of Euler's major papers in 1752, "Principles of the Motions of Fluids," in which he derived the potential equation with regard to the motion of (incompressible) fluids. Euler remarked that he had no idea how to solve equation (2). Poisson later showed that if (x, y, z) lies inside an attracting body, then V satisfies the equation

$$\frac{\partial^2 V}{\partial x^2} + \frac{\partial^2 V}{\partial y^2} + \frac{\partial^2 V}{\partial z^2} = -4\pi\rho, \tag{3}$$

where ρ is the mass density of the attracting body. Equation (3) is now called ***Poisson's equation***. Poisson was also the first to point out the importance of this equation for problems involving electric fields. Notice that if the temperature T is constant in time, then the heat equation (1) reduces to Laplace's equation (2).

Laplace's and Poisson's equations are fundamental to many fields besides fluid mechanics, gravitational fields, and electrostatic fields. For example, they are useful for studying soap films and liquid crystals (see *The Parsimonious Universe: Shape and Form in the Natural World* by S. Hildebrandt and A. Tromba, Springer-Verlag, New York/Berlin, 1995).

THE WAVE EQUATION. The linear wave equation in space has the form

$$\frac{\partial^2 f}{\partial x^2} + \frac{\partial^2 f}{\partial y^2} + \frac{\partial^2 f}{\partial z^2} = c^2 \frac{\partial^2 f}{\partial t^2}. \tag{4}$$

The one-dimensional wave equation

$$\frac{\partial^2 f}{\partial x^2} = c^2 \frac{\partial^2 f}{\partial t^2} \tag{4'}$$

was derived in about 1727 by Johann II Bernoulli and several years later by Jean Le Rond d'Alembert in the study of how to determine the motion of a vibrating string (such as a violin string). Equation (4) became useful in the study of both vibrating bodies and elasticity. As we shall see when we consider Maxwell's equations for electromagnetism in Section 8.5, this equation also arises in the study of the propagation of electromagnetic radiation and sound waves.

EXAMPLE 6 The partial differential equation $u_t + u_{xxx} + uu_x = 0$, called the ***Korteweg–de Vries equation*** (or KdV equation, for short), describes the motion of water waves in a shallow channel.

(a) Show that for any positive constant c, the function

$$u(x, t) = 3c\,\mathrm{sech}^2\left[\tfrac{1}{2}(x - ct)\sqrt{c}\right]$$

is a solution of the Korteweg–de Vries equation. (This solution represents a traveling "hump" of water in the channel and is called a ***soliton***.)[1]

(b) How do the shape and speed of the soliton depend on c?

SOLUTION (a) We compute u_t, u_x, u_{xx}, and u_{xxx} using the chain rule and the differentiation formula $(d/dx)\,\mathrm{sech}\,x = -\,\mathrm{sech}\,x \tanh x$ from one-variable calculus.

[1] Solitons were first observed by J. Scott Russell around 1840 in barge canals near Edinburgh. He reported his results in *Trans. R. Soc. Edinburgh* **14** (1840): 47–109.

Letting $\alpha = (x - ct)\sqrt{c}/2$,

$$u_t = 6c \operatorname{sech} \alpha \frac{\partial}{\partial t} \operatorname{sech} \alpha = -6c \operatorname{sech}^2 \alpha \tanh \alpha \frac{\partial \alpha}{\partial t}$$
$$= 3c^{5/2} \operatorname{sech}^2 \alpha \tanh \alpha = c^{3/2} u \tanh \alpha.$$

Also,

$$u_x = -6c \operatorname{sech}^2 \alpha \, \tanh \alpha \frac{\partial \alpha}{\partial x}$$
$$= -3c^{3/2} \operatorname{sech}^2 \alpha \tanh \alpha = -\sqrt{c}\, u \tanh \alpha,$$

and so $u_t + cu_x = 0$ and

$$u_{xx} = -\sqrt{c}\left[u_x \tanh \alpha + u(\operatorname{sech}^2 \alpha)\frac{\sqrt{c}}{2}\right] = -\sqrt{c}\,(\tanh \alpha) u_x - \frac{u^2}{6}$$
$$= c(\tanh^2 \alpha)u - \frac{u^2}{6} = c(1 - \operatorname{sech}^2 \alpha)u - \frac{u^2}{6}$$
$$= cu - \frac{u^2}{3} - \frac{u^2}{6} = cu - \frac{u^2}{2}.$$

Thus,

$$u_{xxx} = cu_x - uu_x; \qquad \text{that is,} \qquad u_{xxx} + uu_x = cu_x.$$

Hence,

$$u_t + u_{xxx} + uu_x = u_t + cu_x = 0.$$

(b) The speed of the soliton is c, because $u(x + ct, t) = u(x, 0)$. The soliton is higher and thinner when c is larger. Its shape at time $t = 10$ is shown in Figure 3.1.3. ▲

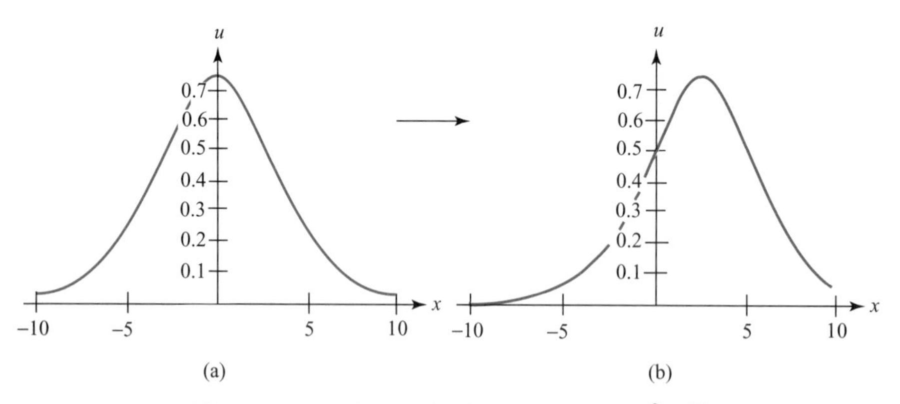

Figure 3.1.3 The graph of $u(x, t) = 3 \operatorname{sech}^2(\sqrt{c}(x - ct)/2)$ for $c = \frac{1}{4}$ at times (a) $t = 0$ and (b) $t = 10$.

EXERCISES

In Exercises 1 to 6, compute the second partial derivatives $\partial^2 f/\partial x^2$, $\partial^2 f/\partial x\,\partial y$, $\partial^2 f/\partial y\,\partial x$, $\partial^2 f/\partial y^2$ for each of the following functions. Verify Theorem 1 in each case.

1. $f(x, y) = 2xy/(x^2 + y^2)^2$, on the region where $(x, y) \neq (0, 0)$

2. $f(x, y, z) = e^z + (1/x) + xe^{-y}$, on the region where $x \neq 0$

3. $f(x, y) = \cos(xy^2)$

4. $f(x, y) = e^{-xy^2} + y^3x^4$

5. $f(x, y) = 1/(\cos^2 x + e^{-y})$

6. $f(x, y) = \log(x - y)$

7. Find $\partial^2 z/\partial x^2$, $\partial^2 z/\partial x\,\partial y$, $\partial^2 z/\partial y\,\partial x$, and $\partial^2 z/\partial y^2$ for

(a) $z = 3x^2 + 2y^2$
(b) $z = (2x^2 + 7x^2y)/3xy$, on the region where $x \neq 0$ and $y \neq 0$

8. Find all the second partial derivatives of

(a) $z = \sin(x^2 - 3xy)$
(b) $z = x^2y^2e^{2xy}$

9. Find f_{xy}, f_{yz}, f_{zx}, and f_{xyz} for

$$f(x, y, z) = x^2y + xy^2 + yz^2.$$

10. Let $z = x^4y^3 - x^8 + y^4$.

(a) Compute $\partial^3 z/\partial y\,\partial x\,\partial x$, $\partial^3 z/\partial x\,\partial y\,\partial x$, and $\partial^3 z/\partial x\,\partial x\,\partial y$ (also denoted $\partial^3 z/\partial x^2\partial y$).
(b) Compute $\partial^3 z/\partial x\,\partial y\,\partial y$, $\partial^3 z/\partial y\,\partial x\,\partial y$, and $\partial^3 z/\partial y\,\partial y\,\partial x$ (also denoted $\partial^3 z/\partial y^2\partial x$).

11. Use Theorem 1 to show that if $f(x, y, z)$ is of class C^3, then

$$\frac{\partial^3 f}{\partial x\,\partial y\,\partial z} = \frac{\partial^3 f}{\partial y\,\partial z\,\partial x}.$$

12. Verify that

$$\frac{\partial^3 f}{\partial x\,\partial y\,\partial z} = \frac{\partial^3 f}{\partial z\,\partial y\,\partial x}$$

for $f(x, y, z) = ze^{xy} + yz^3x^2$.

13. Verify that $f_{xzw} = f_{zwx}$ for $f(x, y, z, w) = e^{xyz}\sin(xw)$.

14. If $f(x, y, z, w)$ is of class C^3, show that $f_{xzw} = f_{zwx}$.

15. Evaluate all first and second partial derivatives of the following functions:

(a) $f(x, y) = x \arctan(x/y)$
(b) $f(x, y) = \cos\sqrt{x^2 + y^2}$
(c) $f(x, y) = \exp(-x^2 - y^2)$

16. Let $w = f(x, y)$ be a function of two variables and let $x = u + v$, $y = u - v$. Show that

$$\frac{\partial^2 w}{\partial u\,\partial v} = \frac{\partial^2 w}{\partial x^2} - \frac{\partial^2 w}{\partial y^2}.$$

17. Let $f\colon \mathbb{R}^2 \to \mathbb{R}$ be a C^2 function and let $\mathbf{c}(t)$ be a C^2 curve in $\mathbb{R}^2$. Write a formula for the second derivative $(d^2/dt^2)((f \circ \mathbf{c})(t))$ using the chain rule twice.

18. Let $f(x, y, z) = e^{xz} \tan(yz)$ and let $x = g(s, t)$, $y = h(s, t)$, $z = k(s, t)$, and define the function $m(s, t) = f(g(s, t), h(s, t), k(s, t))$. Find a formula for m_{st} using the chain rule and verify that your answer is symmetric in s and t.

19. A function $u = f(x, y)$ with continuous second partial derivatives satisfying Laplace's equation

$$\frac{\partial^2 u}{\partial x^2} + \frac{\partial^2 u}{\partial y^2} = 0$$

is called a ***harmonic function***. Show that the function $u(x, y) = x^3 - 3xy^2$ is harmonic.

20. Which of the following functions are harmonic? (See Exercise 19.)

(a) $f(x, y) = x^2 - y^2$
(b) $f(x, y) = x^2 + y^2$
(c) $f(x, y) = xy$
(d) $f(x, y) = y^3 + 3x^2 y$
(e) $f(x, y) = \sin x \cosh y$
(f) $f(x, y) = e^x \sin y$

21. Let f and g be C^2 functions of one variable. Set $\phi = f(x - t) + g(x + t)$.

(a) Prove that ϕ satisfies the wave equation: $\partial^2\phi/\partial t^2 = \partial^2\phi/\partial x^2$.
(b) Sketch the graph of ϕ against t and x if $f(x) = x^2$ and $g(x) = 0$.

22. (a) Show that function $g(x, t) = 2 + e^{-t} \sin x$ satisfies the heat equation: $g_t = g_{xx}$. [Here $g(x, t)$ represents the temperature in a metal rod at position x and time t.]

(b) Sketch the graph of g for $t \geq 0$. (HINT: Look at sections by the planes $t = 0$, $t = 1$, and $t = 2$.)

(c) What happens to $g(x, t)$ as $t \to \infty$? Interpret this limit in terms of the behavior of heat in the rod.

23. Show that Newton's potential $V = -GmM/r$ satisfies Laplace's equation

$$\frac{\partial^2 V}{\partial x^2} + \frac{\partial^2 V}{\partial y^2} + \frac{\partial^2 V}{\partial z^2} = 0 \qquad \text{for} \qquad (x, y, z) \neq (0, 0, 0).$$

24. Let

$$f(x, y) = \begin{cases} xy(x^2 - y^2)/(x^2 + y^2), & (x, y) \neq (0, 0) \\ 0, & (x, y) = (0, 0) \end{cases}$$

(see Figure 3.1.4).

(a) If $(x, y) \neq (0, 0)$, calculate $\partial f/\partial x$ and $\partial f/\partial y$.
(b) Show that $(\partial f/\partial x)(0, 0) = 0 = (\partial f/\partial y)(0, 0)$.
(c) Show that $(\partial^2 f/\partial x\, \partial y)(0, 0) = 1$, $(\partial^2 f/\partial y\, \partial x)(0, 0) = -1$.
(d) What went wrong? Why are the mixed partials not equal?

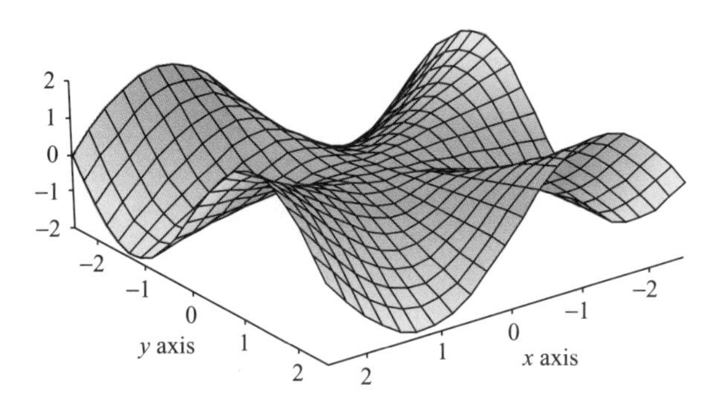

Figure 3.1.4 The graph of the function in Exercise 24.

3.2 Taylor's Theorem

When we introduced the derivative in Chapter 2, we saw that the *linear approximation* of a function played an essential role for a geometric reason—finding the equation of a tangent plane—as well as an analytic reason—finding approximate values of functions. Taylor's theorem deals with the important issue of finding *quadratic and higher-order approximations*.

Taylor's theorem is a central tool for finding accurate numerical approximations of functions, and as such plays an important role in many areas of applied and computational mathematics. We shall use it in the next section to develop the second derivative test for maxima and minima of functions of several variables.

The strategy used to prove Taylor's theorem is to reduce it to the one-variable case by probing a function of many variables along lines of the form $\mathbf{l}(t) = \mathbf{x}_0 + t\mathbf{h}$

emanating from a point $\mathbf{x}_0$ and heading in the direction $\mathbf{h}$. Thus, it will be useful for us to begin by reviewing Taylor's theorem from one-variable calculus.

Single-Variable Taylor Theorem

When recalling a theorem from an earlier course, it is helpful to ask these basic questions: What is the main point of the theorem? What are the key ideas in the proof? Can I understand the result better the second time around?

The main point of the single-variable Taylor theorem is to find approximations of a function near a given point that are accurate to a higher order than the linear approximation. The key idea in the proof is to use the *fundamental theorem of calculus*, followed by *integration by parts*. In fact, just by recalling these basic ideas, one can reconstruct the entire proof. Thinking in this way will help organize all the pieces that need to come together to develop a mastery of Taylor approximations of functions of one and several variables.

For a smooth function $f\colon \mathbb{R} \to \mathbb{R}$ of one variable, Taylor's theorem asserts that:

$$f(x_0 + h) = f(x_0) + f'(x_0) \cdot h + \frac{f''(x_0)}{2} h^2 + \cdots + \frac{f^{(k)}(x_0)}{k!} h^k + R_k(x_0, h), \quad (1)$$

where

$$R_k(x_0, h) = \int_{x_0}^{x_0+h} \frac{(x_0 + h - \tau)^k}{k!} f^{k+1}(\tau)\, d\tau$$

is the remainder. For small h, this remainder is small to order k in the sense that

$$\lim_{h \to 0} \frac{R_k(x_0, h)}{h^k} = 0. \quad (2)$$

In other words, $R_k(x_0, h)$ is small compared to the already small quantity h^k.

The preceding is the formal statement of Taylor's theorem. What about the proof? As promised, we begin with the fundamental theorem of calculus, written in the form:

$$f(x_0 + h) = f(x_0) + \int_{x_0}^{x_0+h} f'(\tau)\, d\tau.$$

Next, we write $d\tau = -d(x_0 + h - \tau)$ and integrate parts[2] to give:

$$f(x_0 + h) = f(x_0) + f'(x_0)h + \int_{x_0}^{x_0+h} f''(\tau)(x_0 + h - \tau)\, d\tau,$$

which is the first-order Taylor formula. Integrating by parts again:

$$\begin{aligned}
&\int_{x_0}^{x_0+h} f''(\tau)(x_0 + h - \tau)\, d\tau \\
&\quad = -\frac{1}{2}\int_{x_0}^{x_0+h} f''(\tau)\, d(x_0 + h - \tau)^2 \\
&\quad = \frac{1}{2} f''(x_0)h^2 + \frac{1}{2}\int_{x_0}^{x_0+h} f'''(\tau)(x_0 + h - \tau)^2\, d\tau,
\end{aligned}$$

which, when substituted into the preceding formula, gives the ***second-order Taylor formula***:

$$f(x_0 + h) = f(x_0) + f'(x_0)h + \frac{1}{2} f''(x_0)h^2 + \frac{1}{2}\int_{x_0}^{x_0+h} f'''(\tau)(x_0 + h - \tau)^2\, d\tau.$$

This is Taylor's theorem for $k = 2$.

Taylor's theorem for general k proceeds by repeated integration by parts. The statement (2) that $R_k(x_0, h)/h^k \to 0$ as $h \to 0$ is seen as follows. For τ in the interval $[x_0, x_0 + h]$, we have $|x_0 + h - \tau| \le |h|$ and $f^{k+1}(\tau)$, being continuous, is bounded; say, $|f^{k+1}(\tau)| \le M$. Then:

$$|R_k(x_0, h)| = \left| \int_{x_0}^{x_0+h} \frac{(x_0 + h - \tau)^k}{k!} f^{k+1}(\tau)\, d\tau \right| \le \frac{|h|^{k+1}}{k!} M$$

and, in particular, $|R_k(x_0, h)/h^k| \le |h|\, M/k! \to 0$ as $h \to 0$.

---

[2] Recall that integration by parts (the product rule for the derivative read backwards) reads as:

$$\int_a^b u\, dv = uv|_a^b - \int_a^b v\, du.$$

Here we choose $u = f'(\tau)$ and $v = x_0 + h - \tau$.

Taylor's Theorem for Many Variables

Our next goal in this section is to prove an analogous theorem that is valid for functions of several variables. We already know a first-order version, that is, when $k = 1$. Indeed, if $f\colon \mathbb{R}^n \to \mathbb{R}$ is differentiable at $\mathbf{x}_0$ and we define

$$R_1(\mathbf{x}_0, \mathbf{h}) = f(\mathbf{x}_0 + \mathbf{h}) - f(\mathbf{x}_0) - [\mathbf{D}f(\mathbf{x}_0)](\mathbf{h}),$$

so that

$$f(\mathbf{x}_0 + \mathbf{h}) = f(\mathbf{x}_0) + [\mathbf{D}f(\mathbf{x}_0)](\mathbf{h}) + R_1(\mathbf{x}_0, \mathbf{h}),$$

then by the definition of differentiability,

$$\frac{|R_1(\mathbf{x}_0, \mathbf{h})|}{\|\mathbf{h}\|} \to 0 \qquad \text{as} \qquad \mathbf{h} \to 0,$$

that is, $R_1(\mathbf{x}_0, \mathbf{h})$ vanishes to first order at $\mathbf{x}_0$. In summary, we have:

THEOREM 2: First-Order Taylor Formula Let $f\colon U \subset \mathbb{R}^n \to \mathbb{R}$ be differentiable at $\mathbf{x}_0 \in U$. Then

$$f(\mathbf{x}_0 + \mathbf{h}) = f(\mathbf{x}_0) + \sum_{i=1}^{n} h_i \frac{\partial f}{\partial x_i}(\mathbf{x}_0) + R_1(\mathbf{x}_0, \mathbf{h}),$$

where $R_1(\mathbf{x}_0, \mathbf{h})/\|\mathbf{h}\| \to 0$ as $\mathbf{h} \to \mathbf{0}$ in $\mathbb{R}^n$.

The second-order version is as follows:

THEOREM 3: Second-Order Taylor Formula Let $f\colon U \subset \mathbb{R}^n \to \mathbb{R}$ have continuous partial derivatives of third order.[3] Then we may write

$$f(\mathbf{x}_0 + \mathbf{h}) = f(\mathbf{x}_0) + \sum_{i=1}^{n} h_i \frac{\partial f}{\partial x_i}(\mathbf{x}_0) + \frac{1}{2} \sum_{i,j=1}^{n} h_i h_j \frac{\partial^2 f}{\partial x_i\, \partial x_j}(\mathbf{x}_0) + R_2(\mathbf{x}_0, \mathbf{h}),$$

where $R_2(\mathbf{x}_0, \mathbf{h})/\|\mathbf{h}\|^2 \to 0$ as $\mathbf{h} \to \mathbf{0}$ and the second sum is over all i's and j's between 1 and n (so there are n^2 terms).

[3] For the statement of the theorem as given here, f actually needs only to be of class C^2, but for a convenient form of the remainder we assume f is of class C^3.

Notice that this result can be written in matrix form as

$$f(\mathbf{x}_0+\mathbf{h}) = f(\mathbf{x}_0) + \left(\frac{\partial f}{\partial x_1},\ldots,\frac{\partial f}{\partial x_n}\right)\begin{pmatrix} h_1 \\ \vdots \\ h_n \end{pmatrix}$$

$$+\frac{1}{2}(h_1,\ldots,h_n)\begin{pmatrix} \dfrac{\partial^2 f}{\partial x_1\,\partial x_1} & \dfrac{\partial^2 f}{\partial x_1\,\partial x_2} & \cdots & \dfrac{\partial^2 f}{\partial x_1\,\partial x_n} \\ \dfrac{\partial^2 f}{\partial x_2\,\partial x_1} & \dfrac{\partial^2 f}{\partial x_2\,\partial x_2} & \cdots & \dfrac{\partial^2 f}{\partial x_2\,\partial x_n} \\ \vdots & & & \\ \dfrac{\partial^2 f}{\partial x_n\,\partial x_1} & \dfrac{\partial^2 f}{\partial x_n\,\partial x_2} & \cdots & \dfrac{\partial^2 f}{\partial x_n\,\partial x_n} \end{pmatrix}\begin{pmatrix} h_1 \\ h_2 \\ \vdots \\ h_n \end{pmatrix},$$

$$+ R_2(\mathbf{x}_0,\mathbf{h}),$$

where the derivatives of f are evaluated at $\mathbf{x}_0$.

In the course of the proof of the Theorem 3, we shall obtain a useful explicit formula for the remainder, as in the single-variable theorem.

PROOF OF THEOREM 3 Let $g(t) = f(\mathbf{x}_0 + t\mathbf{h})$ with $\mathbf{x}_0$ and $\mathbf{h}$ fixed, which is a C^3 function of t. Now apply the single-variable Taylor theorem (1) to g, with $k = 2$ to obtain

$$\left.\begin{aligned} g(1) &= g(0) + g'(0) + \frac{g''(0)}{2!} + R_2, \\ \text{where}\quad & \\ R_2 &= \int_0^1 \frac{(t-1)^2}{2!} g'''(t)\,dt. \end{aligned}\right\} \tag{3}$$

By the chain rule,

$$g'(t) = \sum_{i=1}^{n} \frac{\partial f}{\partial x_i}(\mathbf{x}_0 + t\mathbf{h})\mathbf{h}_i; \qquad g''(t) = \sum_{i,j=1}^{n} \frac{\partial^2 f}{\partial x_i\,\partial x_j}(\mathbf{x}_0 + t\mathbf{h})h_i h_j,$$

and

$$g'''(t) = \sum_{i,j,k=1}^{n} \frac{\partial^3 f}{\partial x_i\,\partial x_j\,\partial x_k}(\mathbf{x}_0 + t\mathbf{h})h_i h_j h_k.$$

Writing $R_2 = R_2(\mathbf{x}_0, \mathbf{h})$ we have thus proved:

$$\left.\begin{aligned} f(\mathbf{x}_0+\mathbf{h}) &= f(\mathbf{x}_0) + \sum_{i=1}^{n} h_i \frac{\partial f}{\partial x_i}(\mathbf{x}_0) + \frac{1}{2}\sum_{i,j=1}^{n} h_i h_j \frac{\partial^2 f}{\partial x_i\,\partial x_j}(\mathbf{x}_0) + R_2(\mathbf{x}_0, \mathbf{h}), \\ &\text{where} \\ R_2(\mathbf{x}_0, \mathbf{h}) &= \sum_{i,j,k=1}^{n} \int_0^1 \frac{(t-1)^2}{2} \frac{\partial^3 f}{\partial x_i\,\partial x_j\,\partial x_k}(\mathbf{x}_0 + t\mathbf{h}) h_i h_j h_k\, dt. \end{aligned}\right\} \quad (4)$$

The integrand is a continuous function of t and is therefore bounded by a positive constant C on a small neighborhood of $\mathbf{x}_0$ (because it has to be close to its value at $\mathbf{x}_0$). Also note that $|h_i| \le \|\mathbf{h}\|$, for $\|\mathbf{h}\|$ small, and so

$$|R_2(\mathbf{x}_0, \mathbf{h})| \le \|\mathbf{h}\|^3 C.$$

In particular,

$$\frac{|R_2(\mathbf{x}_0, \mathbf{h})|}{\|\mathbf{h}\|^2} \le \|\mathbf{h}\| C \to 0 \qquad \text{as} \qquad \mathbf{h} \to \mathbf{0},$$

as required by the theorem.

The proof of Theorem 2 follows analogously from the Taylor formula (1) with $k = 1$. A similar argument for R_1 shows that $|R_1(\mathbf{x}_0, \mathbf{h})|/\|\mathbf{h}\| \to 0$ as $\mathbf{h} \to \mathbf{0}$, although this also follows directly from the definition of differentiability. ■

Forms of the Remainder In Theorem 2,

$$R_1(\mathbf{x}_0, \mathbf{h}) = \sum_{i,j=1}^{n} \int_0^1 (1-t) \frac{\partial^2 f}{\partial x_i \partial x_j}(\mathbf{x}_0 + t\mathbf{h}) h_i h_j\, dt = \sum_{i,j=1}^{n} \frac{1}{2} \frac{\partial^2 f}{\partial x_i \partial x_j}(\mathbf{c}_{ij}) h_i h_j, \quad (5)$$

where $\mathbf{c}_{ij}$ lies somewhere on the line joining $\mathbf{x}_0$ to $\mathbf{x}_0 + \mathbf{h}$.

In Theorem 3,

$$\begin{aligned} R_2(\mathbf{x}_0, \mathbf{h}) &= \sum_{i,j,k=1}^{n} \int_0^1 \frac{(t-1)^2}{2} \frac{\partial^3 f}{\partial x_i\,\partial x_j\,\partial x_k}(\mathbf{x}_0 + t\mathbf{h}) h_i h_j h_k\, dt \\ &= \sum_{i,j,k=1}^{n} \frac{1}{3!} \frac{\partial^3 f}{\partial x_i\,\partial x_j\,\partial x_k}(\mathbf{c}_{ijk}) h_i h_j h_k, \end{aligned} \quad (5')$$

where $\mathbf{c}_{ijk}$ lies somewhere on the line joining $\mathbf{x}_0$ to $\mathbf{x}_0 + \mathbf{h}$.

The formulas involving $\mathbf{c}_{ij}$ and $\mathbf{c}_{ijk}$ (called Lagrange's form of the remainder) are obtained by making use of the *second mean-value theorem for integrals*. This states that

$$\int_a^b h(t)g(t)\,dt = h(c)\int_a^b g(t)\,dt,$$

provided h and g are continuous and $g \geq 0$ on $[a, b]$; *here c is some number between* a and b.[4] This is applied in formula (5) for the explicit form of the remainder with $h(t) = (\partial^2 f/\partial x_i \partial x_j)(\mathbf{x}_0 + t\mathbf{h})$ and $g(t) = 1 - t$.

The third-order Taylor formula is

$$f(\mathbf{x}_0 + \mathbf{h}) = f(\mathbf{x}_0) + \sum_{i=1}^n h_i \frac{\partial f}{\partial x_i}(\mathbf{x}_0) + \frac{1}{2}\sum_{i,j=1}^n h_i h_j \frac{\partial^2 f}{\partial x_i\,\partial x_j}(\mathbf{x}_0)$$
$$+ \frac{1}{3!}\sum_{i,j,k=1}^n h_i h_j h_k \frac{\partial^3 f}{\partial x_i\,\partial x_j\,\partial x_k}(\mathbf{x}_0) + R_3(\mathbf{x}_0, \mathbf{h}),$$

where $R_3(\mathbf{x}_0, \mathbf{h})/\|\mathbf{h}\|^3 \to 0$ as $\mathbf{h} \to \mathbf{0}$, and so on. The general formula can be proved by induction, using the method of proof already given.

EXAMPLE 1 Compute the second-order Taylor formula for the function $f(x, y) = \sin(x + 2y)$, about the point $\mathbf{x}_0 = (0, 0)$.

SOLUTION Notice that

$$f(0, 0) = 0,$$
$$\frac{\partial f}{\partial x}(0, 0) = \cos(0 + 2\cdot 0) = 1, \qquad \frac{\partial f}{\partial y}(0, 0) = 2\cos(0 + 2\cdot 0) = 2,$$
$$\frac{\partial^2 f}{\partial x^2}(0, 0) = 0, \qquad \frac{\partial^2 f}{\partial y^2}(0, 0) = 0, \qquad \frac{\partial^2 f}{\partial x\,\partial y}(0, 0) = 0.$$

Thus,

$$f(\mathbf{h}) = f(h_1, h_2) = h_1 + 2h_2 + R_2(\mathbf{0}, \mathbf{h}),$$

---

[4]*Proof* If $g = 0$, the result is clear, so we can suppose $g \neq 0$; thus, we can assume $\int_a^b g(t)\,dt > 0$. Let M and m be the maximum and minimum values of h, achieved at t_M and t_m, respectively. Because $g(t) \geq 0$,

$$m\int_a^b g(t)\,dt \leq \int_a^b h(t)g(t)\,dt \leq M\int_a^b g(t)\,dt.$$

Thus, $\left(\int_a^b h(t)g(t)\,dt\right)\Big/\left(\int_a^b g(t)\,dt\right)$ lies between $m = h(t_m)$ and $M = h(t_M)$ and therefore, by the intermediate-value theorem, equals $h(c)$ for some intermediate c. ■

where

$$\frac{R_2(\mathbf{0}, \mathbf{h})}{\|\mathbf{h}\|^2} \to 0 \qquad \text{as} \qquad \mathbf{h} \to \mathbf{0}. \quad \blacktriangle$$

EXAMPLE 2 Compute the second-order Taylor formula for $f(x, y) = e^x \cos y$ about the point $x_0 = 0,\ y_0 = 0$.

SOLUTION Here

$$f(0,0) = 1, \qquad \frac{\partial f}{\partial x}(0,0) = 1, \qquad \frac{\partial f}{\partial y}(0,0) = 0,$$

$$\frac{\partial^2 f}{\partial x^2}(0,0) = 1, \qquad \frac{\partial^2 f}{\partial y^2}(0,0) = -1, \qquad \frac{\partial^2 f}{\partial x\, \partial y}(0,0) = 0,$$

and so

$$f(\mathbf{h}) = f(h_1, h_2) = 1 + h_1 + \tfrac{1}{2}h_1^2 - \tfrac{1}{2}h_2^2 + R_2(\mathbf{0}, \mathbf{h}),$$

where

$$\frac{R_2(\mathbf{0}, \mathbf{h})}{\|\mathbf{h}\|^2} \to 0 \qquad \text{as} \qquad \mathbf{h} \to \mathbf{0}. \quad \blacktriangle$$

In the case of functions of one variable, one can expand $f(x)$ in an infinite power series, called the ***Taylor series***:

$$f(x_0 + h) = f(x_0) + f'(x_0)h + \frac{f''(x_0)h^2}{2} + \cdots + \frac{f^{(k)}(x_0)h^k}{k!} + \cdots,$$

provided one can show that $R_k(x_0, h) \to 0$ as $k \to \infty$. Similarly, for functions of several variables the preceding terms are replaced by the corresponding ones involving partial derivatives, as we have seen in Theorem 3. Again, one can represent such a function by its Taylor series provided one can show that $R_k \to 0$ as $k \to \infty$. This point is examined further in Exercise 7.

EXAMPLE 3 Find the first- and second-order Taylor approximations to $f(x, y) = \sin(xy)$ at the point $(x_0, y_0) = (1, \pi/2)$.

SOLUTION Here

$$f(x_0, y_0) = \sin(x_0 y_0) = \sin(\pi/2) = 1$$

$$f_x(x_0, y_0) = y_0 \cos(x_0 y_0) = \frac{\pi}{2} \cos(\pi/2) = 0$$

$$f_y(x_0, y_0) = x_0 \cos(x_0 y_0) = \cos(\pi/2) = 0$$

$$f_{xx}(x_0, y_0) = -y_0^2 \sin(x_0 y_0) = -\frac{\pi^2}{4}\sin(\pi/2) = -\pi^2/4$$

$$f_{xy}(x_0, y_0) = \cos(x_0 y_0) - x_0 y_0 \sin(x_0 y_0) = -\frac{\pi}{2}\sin(\pi/2) = -\pi/2$$

$$f_{yy}(x_0, y_0) = -x_0^2 \sin(x_0 y_0) = -\sin(\pi/2) = -1.$$

Thus, the linear (first-order) approximation is

$$\begin{aligned} l(x, y) &= f(x_0, y_0) + f_x(x_0, y_0)(x - x_0) + f_y(x_0, y_0)(y - y_0) \\ &= 1 + 0 + 0 = 1, \end{aligned}$$

and the second-order (or quadratic) approximation is

$$\begin{aligned} g(x, y) &= 1 + 0 + 0 + \frac{1}{2}\left(-\frac{\pi^2}{4}\right)(x-1)^2 + \left(-\frac{\pi}{2}\right)(x-1)\left(y - \frac{\pi}{2}\right) \\ &\quad + \frac{1}{2}(-1)\left(y - \frac{\pi}{2}\right)^2 \\ &= 1 - \frac{\pi^2}{8}(x-1)^2 - \frac{\pi}{2}(x-1)\left(y - \frac{\pi}{2}\right) - \frac{1}{2}\left(y - \frac{\pi}{2}\right)^2. \end{aligned}$$

See Figure 3.2.1. ▲

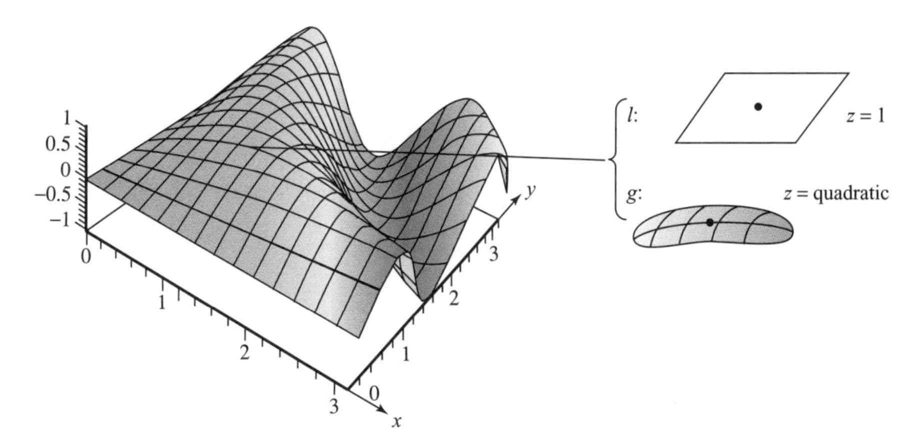

Figure 3.2.1 The linear and quadratic approximations to $z = \sin(xy)$ near $(1, \pi/2)$.

EXAMPLE 4 Find linear and quadratic approximations to the expression $(3.98-1)^2/(5.97-3)^2$. Compare with the exact value.

SOLUTION Let $f(x, y) = (x-1)^2/(y-3)^2$. The desired expression is close to $f(4, 6) = 1$. To find the approximations, we differentiate:

$$f_x = \frac{2(x-1)}{(y-3)^2}, \qquad f_y = \frac{-2(x-1)^2}{(y-3)^3}$$

$$f_{xy} = f_{yx} = \frac{-4(x-1)}{(y-3)^3}, \qquad f_{xx} = \frac{2}{(y-3)^2}, \qquad f_{yy} = \frac{6(x-1)^2}{(y-3)^4}.$$

At the point of approximation, we have

$$f_x(4, 6) = \frac{2}{3}, \qquad f_y = -\frac{2}{3}, \qquad f_{xy} = f_{yx} = -\frac{4}{9}, \qquad f_{xx} = \frac{2}{9}, \qquad f_{yy} = \frac{2}{3}.$$

The linear approximation is then

$$1 + \frac{2}{3}(-0.02) - \frac{2}{3}(-0.03) = 1.00666.$$

The quadratic approximation is

$$1 + \frac{2}{3}(-0.02) - \frac{2}{3}(-0.03) + \frac{2}{9}\frac{(-0.02)^2}{2} - \frac{4}{9}(-0.02)(-0.03) + \frac{2}{3}\frac{(-0.03)^2}{2}$$
$$= 1.00674.$$

The "exact" value using a calculator is 1.00675. ▲

EXERCISES

In each of Exercises 1 to 6, determine the second-order Taylor formula for the given function about the given point (x_0, y_0).

1. $f(x, y) = (x+y)^2$, where $x_0 = 0, y_0 = 0$

2. $f(x, y) = 1/(x^2 + y^2 + 1)$, where $x_0 = 0, y_0 = 0$

3. $f(x, y) = e^{x+y}$, where $x_0 = 0, y_0 = 0$

4. $f(x, y) = e^{-x^2-y^2}\cos(xy)$, where $x_0 = 0, y_0 = 0$

5. $f(x, y) = \sin(xy) + \cos(xy)$, where $x_0 = 0, y_0 = 0$

6. $f(x, y) = e^{(x-1)^2}\cos y$, where $x_0 = 1, y_0 = 0$

7. (Challenging) A function $f\colon \mathbb{R} \to \mathbb{R}$ is called an ***analytic*** function provided

$$f(x+h) = f(x) + f'(x)h + \cdots + \frac{f^{(k)}(x)}{k!}h^k + \cdots$$

[i.e., the series on the right-hand side converges and equals $f(x+h)$].

(a) Suppose f satisfies the following condition: On any closed interval $[a, b]$, there is a constant M such that for all $k = 1, 2, 3, \ldots, |f^{(k)}(x)| \le M^k$ for all $x \in [a, b]$. Prove that f is analytic.

(b) Let $f(x) = \begin{cases} e^{-1/x} & x > 0 \\ 0 & x \le 0. \end{cases}$.

Show that f is a C^∞ function, but f is not analytic.

(c) Give a definition of analytic functions from $\mathbb{R}^n$ to $\mathbb{R}$. Generalize the proof of part (a) to this class of functions.

(d) Develop $f(x, y) = e^{x+y}$ in a power series about $x_0 = 0$, $y_0 = 0$.

3.3 Extrema of Real-Valued Functions

— Historical Note —

As we saw in the book's Historical introduction, the early Greeks sought to mathematize nature and to find, as in the geometric Ptolemaic model of planetary motion, mathematical laws governing the universe. With the revival of Greek learning during the Renaissance, this point of view again took hold and the search for these laws recommenced. In particular, the question was raised as to whether there was *one* law, one mathematical principle that governed and superseded all others, a principle that the Creator used in His Grand Design of the Universe.

MAUPERTUIS'S PRINCIPLE. In 1744, the French scientist Pierre-Louis de Maupertuis (see Figure 3.3.1) put forth his grand scheme of the world. The "metaphysical principle" of Maupertuis is the assumption that nature always operates with the greatest possible economy. In short, physical laws are a consequence of a principle of "economy of means"; nature always acts in such a way as to minimize some quantity that Maupertuis called the *action*. Action was nothing more than the expenditure of energy over time, or energy × time. In applications, the type of energy changes with each case. For example, physical systems often try to "rearrange themselves" to have a minimum energy—such as a ball rolling from a mountain peak to a valley, or the primordial irregular Earth assuming a more nearly spherical shape. As another example, the spherical shape of soap bubbles is connected with the fact that spheres are the surfaces of least area containing a fixed volume.

We state Maupertuis's principle formally as: *Nature always minimizes action.* Maupertuis saw in this principle an expression of the wisdom of the Supreme Being, of God, according to which everything in nature is

Figure 3.3.1 Pierre-Louis de Maupertuis (1698–1759).

performed in the most economical way. He wrote:

> What satisfaction for the human spirit that, in contemplating these laws which contain the principle of motion and of rest for all bodies in the universe, he finds the proof of existence of Him who governs the world.

Maupertuis indeed believed that he had discovered God's fundamental law, the very secret of Creation itself, but he was actually not the first person to pose this principle.

In 1707, Leibniz wrote down the principle of least action in a letter to Johann Bernoulli, which became lost until 1913, when it was discovered in Germany's Gotha library. For Leibniz, this principle was a natural outgrowth of his great philosophical treatise *The Theodicy*, in which he argues that God may indeed think of all possible worlds, but would want to create only the best among them; and hence our world is necessarily the *best of all possible worlds*.

Action, as defined by Leibniz, was motivated by the following reasoning, used in his letter. Think of a hiker walking along a road, and consider how to describe his action. If he travels 2 kilometers in 1 hour, you would say that he has carried out twice as much action as he would if he traveled 2 kilometers in 2 hours. However, you would also say that he carries out twice as much action in traveling 2 kilometers in 2 hours as he would in traveling 1 kilometer in 1 hour. Altogether then, our hiker, by walking 2 kilometers in 1 hour, carries out 4 times as much action as he would in traveling 1 kilometer in 1 hour.

Using this intuitive idea, Maupertuis defined action as the product of distance, velocity, and mass:

$$\text{Action} = \text{Mass} \times \text{Distance} \times \text{Velocity}.$$

Mass is included in this definition to account for the hiker's backpack. Moreover, according to Leibniz, the kinetic energy E is given by the formula:

$$E = \frac{1}{2} \times \text{Mass} \times (\text{Velocity})^2.$$

So action has the same physical dimension as

$$\text{Energy} \times \text{Time},$$

because velocity is distance divided by time.

Principle of Least Action. In the 250 years after Maupertuis formulated his principle, this *principle of least action* has been found to be a "theoretical basis" for Newton's law of gravity, Maxwell's equations for electromagnetism, Schrödinger's equation of quantum mechanics, and Einstein's field equation in general relativity.

Max Planck (see Figure 3.3.2), one of the greatest scientists of the modern era and the discoverer of the "quantization" of nature, was also a profound believer in the mathematical design of the universe. On June 29, 1922, on "Leibniz Day" in Berlin, Germany, just a few years after World War I, with all its terrible carnage, Planck delivered an address honoring this great scholar.

What follows are some excerpts from Planck's remarks:

> *The Theodicy* culminates with the statement that whatever occurs in our world, in the large as in the small, in nature as in spiritual life, is once and for all regulated by divine reason, and in such a way that our world is the best among possible worlds.
>
> Would Leibniz reaffirm this statement even today, in view of the misery of the present time, in view of the bitter failure of many efforts not immediately aimed at material gain, in view of the

Figure 3.3.2 Max Planck (1858–1947).

undeniable fact that the imagined general harmony of people today seems to be further away from its realization than ever? No doubt, we should have to answer this question in the affirmative, even if we did not know that Leibniz never ceased to earnestly occupy himself until his last years despite an adverse fate and disappointments of all kinds, and we shall hardly err in assuming that it was exactly the *Theodicy* that gave him support and comfort in the most sorrowful days of his life. This once again is a touching example of the old truth that our most profound and most sacred principles are firmly rooted in our innermost soul, independent of experiences in the outer world.

Modern science, in particular under the influence of the development of the notion of causality, has moved far away from Leibniz's teleological point of view. Science has abandoned the assumption of a special, anticipating reason, and it considers each event in the natural and spiritual world, at least in principle, as reducible to prior states. But still we notice a fact, particularly in the most exact science, which, at least in this context, is most surprising. Present-day physics, as far as it is theoretically organized, is completely governed by a system of space–time differential equations which state that each process in nature is totally determined by the events which occur in its immediate temporal and spatial neighborhood. This entire rich system of differential equations, though they differ in detail, since they refer to mechanical, electric, magnetic, and thermal processes, is now completely contained in a single dictum—*the principle of least action*. This, in short, states that, of all possible processes, the only ones that actually occur are those that involve minimum expenditure of action. As we can see, only a short step is required to recognize in the preference for the smallest quantity of action the ruling of divine reason, and thus to discover a part of Leibniz's teleological ordering of the universe.[5]

In present-day physics the principle of least action plays a relatively minor role. It does not quite fit into the framework of present theories. Of course, admittedly it is a correct statement; yet usually it serves not as the foundation of the theory, but as a true but dispensable appendix, because present theoretical physics is entirely tailored to the principle of infinitesimal local effects, and sees extensions to larger spaces and times as an unnecessary and uneconomical complication of the method of treatment. Hence, physics is inclined to view the principle of least action more as a formal and accidental curiosity than as a pillar of physical knowledge.

There is much more to the story of the least action principle, which we will revisit in Section 4.1.

[5] For more information and history, consult S. Hildebrandt and A. J. Tromba, *The Parsimonious Universe: Shape and Form in the Natural World*, Springer-Verlag, New York/Berlin, 1995.

Maxima and Minima for Functions of n-Variables

As the previous remarks show, for Leibniz, Euler, and Maupertuis, and for much of modern science as well, all in nature is a consequence of some maximum or minimum principle. To make such grand schemes—as well as some that are more down to earth—effective, one must first learn the techniques of how to find maxima and minima of functions of n variables.

Extreme Points

Among the most basic geometric features of the graph of a function are its extreme points, at which the function attains its greatest and least values. In this section, we derive a method for determining these points. In fact, the method locates local extrema as well. These are points at which the function attains a maximum or minimum value relative only to nearby points. Let us begin by defining our terms.

DEFINITION If $f: U \subset \mathbb{R}^n \to \mathbb{R}$ is a given scalar function, a point $\mathbf{x}_0 \in U$ is called a ***local minimum*** of f if there is a neighborhood V of $\mathbf{x}_0$ such that for all points $\mathbf{x}$ in V, $f(\mathbf{x}) \geq f(\mathbf{x}_0)$. (See Figure 3.3.3.) Similarly, $\mathbf{x}_0 \in U$ is a ***local maximum*** if there is a neighborhood V of $\mathbf{x}_0$ such that $f(\mathbf{x}) \leq f(\mathbf{x}_0)$ for all $\mathbf{x} \in V$. The point

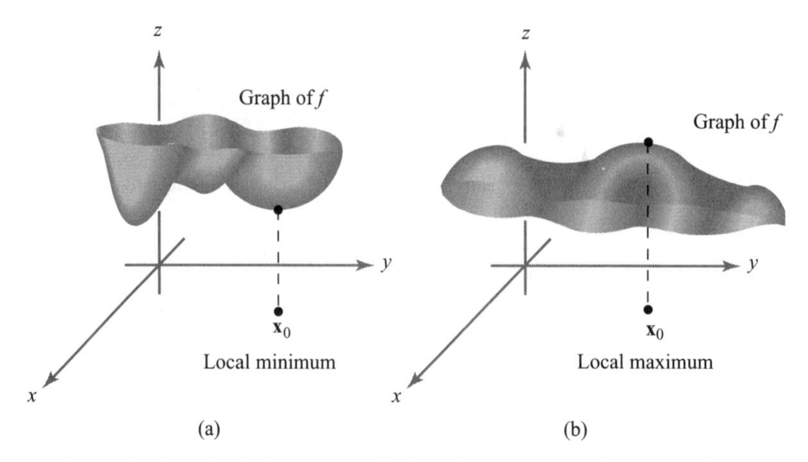

Figure 3.3.3 (a) Local minimum and (b) local maximum points for a function of two variables.

$\mathbf{x}_0 \in U$ is said to be a ***local***, or ***relative***, ***extremum*** if it is either a local minimum or a local maximum. A point $\mathbf{x}_0$ is a ***critical point*** of f if either f is *not differentiable at* $\mathbf{x}_0$, or if it is, $\mathbf{D}f(\mathbf{x}_0) = \mathbf{0}$. A critical point that is not a local extremum is called a ***saddle point***.[6]

First Derivative Test for Local Extrema

The location of extrema is based on the following fact, which should be familiar from one-variable calculus (the case $n = 1$): *Every extremum is a critical point.*

> THEOREM 4: First Derivative Test for Local Extrema If $U \subset \mathbb{R}^n$ is open, the function $f\colon U \subset \mathbb{R}^n \to \mathbb{R}$ is differentiable, and $\mathbf{x}_0 \in U$ is a local extremum, then $\mathbf{D}f(\mathbf{x}_0) = \mathbf{0}$; that is, $\mathbf{x}_0$ is a critical point of f.

PROOF Suppose that f achieves a local maximum at $\mathbf{x}_0$. Then for any $\mathbf{h} \in \mathbb{R}^n$, the function $g(t) = f(\mathbf{x}_0 + t\mathbf{h})$ has a local maximum at $t = 0$. Thus, from one-variable calculus $g'(0) = 0$.[7] On the other hand, by the chain rule,

$$g'(0) = [\mathbf{D}f(\mathbf{x}_0)]\mathbf{h}.$$

Thus, $[\mathbf{D}f(\mathbf{x}_0)]\mathbf{h} = 0$ for every $\mathbf{h}$, and so $\mathbf{D}f(\mathbf{x}_0) = \mathbf{0}$. The case in which f achieves a local minimum at $\mathbf{x}_0$ is entirely analogous. ■

---

[6] The term "saddle point" is sometimes not used this generally; we shall discuss saddle points further in the subsequent development.

[7] Recall the proof from one-variable calculus: Because $g(0)$ is a local maximum, $g(t) \le g(0)$ for small $t > 0$, so $g(t) - g(0) \le 0$, and hence $g'(0) = \operatorname{limit}_{t\to 0^+} (g(t) - g(0))/t \le 0$, where $\operatorname{limit}_{t\to 0^+}$ means the limit as $t \to 0$, $t > 0$. For small $t < 0$, we similarly have $g'(0) = \operatorname{limit}_{t\to 0^-} (g(t) - g(0))/t \ge 0$. Therefore, $g'(0) = 0$.

If we remember that $\mathbf{D}f(\mathbf{x}_0) = \mathbf{0}$ means that all the components of $\mathbf{D}f(\mathbf{x}_0)$ are zero, we can rephrase the result of Theorem 4: If $\mathbf{x}_0$ is a local extremum, then

$$\frac{\partial f}{\partial x_i}(\mathbf{x}_0) = 0, \qquad i = 1, \ldots, n;$$

that is, each partial derivative is zero at $\mathbf{x}_0$. In other words, $\nabla f(\mathbf{x}_0) = \mathbf{0}$, where ∇f is the gradient of f.

If we seek to find the extrema or local extrema of a function, then Theorem 4 states that we should look among the critical points. Sometimes these can be tested by inspection, but usually we use tests (to be developed below) analogous to the second-derivative test in one-variable calculus.

EXAMPLE 1 Find the maxima and minima of the function $f\colon \mathbb{R}^2 \to \mathbb{R}$, defined by $f(x, y) = x^2 + y^2$. (Ignore the fact that this example can be done by inspection.)

SOLUTION We first identify the critical points of f by solving the two equations $\partial f/\partial x = 0$ and $\partial f/\partial y = 0$, for x and y. But

$$\frac{\partial f}{\partial x} = 2x \qquad \text{and} \qquad \frac{\partial f}{\partial y} = 2y,$$

so the only critical point is the origin (0, 0), where the value of the function is zero. Because $f(x, y) \geq 0$, this point is a relative minimum—in fact, an absolute, or global, minimum—of f. Because (0, 0) is the only critical point, there are no maxima. ▲

EXAMPLE 2 Consider the function $f\colon \mathbb{R}^2 \to \mathbb{R}, (x, y) \mapsto x^2 - y^2$. Ignoring for the moment that this function has a saddle and no extrema, apply the method of Theorem 4 for the location of extrema.

SOLUTION As in Example 1, we find that f has only one critical point, at the origin, and the value of f there is zero. Examining values of f directly for points near the origin, we see that $f(x, 0) \geq f(0, 0)$ and $f(0, y) \leq f(0, 0)$, with strict inequalities when $x \neq 0$ and $y \neq 0$. Because x or y can be taken arbitrarily small, the origin cannot be either a relative minimum or a relative maximum (so it is a saddle point). Therefore, this function can have no relative extrema (see Figure 3.3.4). ▲

EXAMPLE 3 Find all the critical points of $z = x^2y + y^2x$.

SOLUTION Differentiating, we obtain

$$\frac{\partial z}{\partial x} = 2xy + y^2, \qquad \frac{\partial z}{\partial y} = 2xy + x^2.$$

Equating the partial derivatives to zero yields

$$2xy + y^2 = 0, \qquad 2xy + x^2 = 0.$$

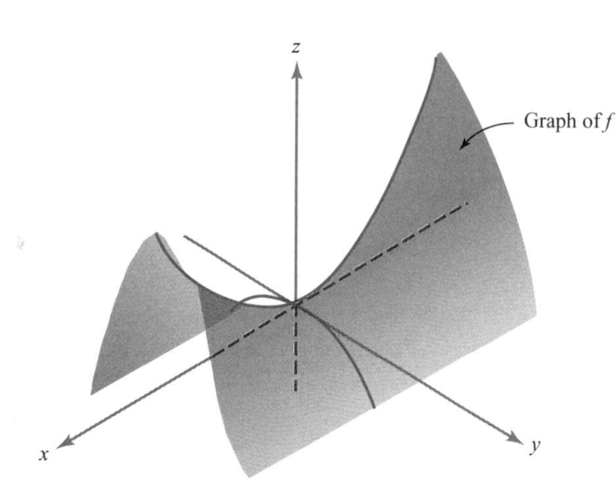

Figure 3.3.4 A function of two variables with a saddle point.

Subtracting, we obtain $x^2 = y^2$. Thus, $x = \pm y$. Substituting $x = +y$ in the first of the two preceding equations, we find that

$$2y^2 + y^2 = 3y^2 = 0,$$

so that $y = 0$ and thus $x = 0$. If $x = -y$, then

$$-2y^2 + y^2 = -y^2 = 0,$$

so $y = 0$ and therefore $x = 0$. Hence, the only critical point is $(0, 0)$. For $x = y$, $z = 2x^3$, which is both positive and negative for x near zero. Thus, $(0, 0)$ is not a relative extremum. ▲

EXAMPLE 4 Refer to Figure 3.3.5, a computer-drawn graph of the function $z = 2(x^2 + y^2)\, e^{-x^2-y^2}$. Where are the critical points?

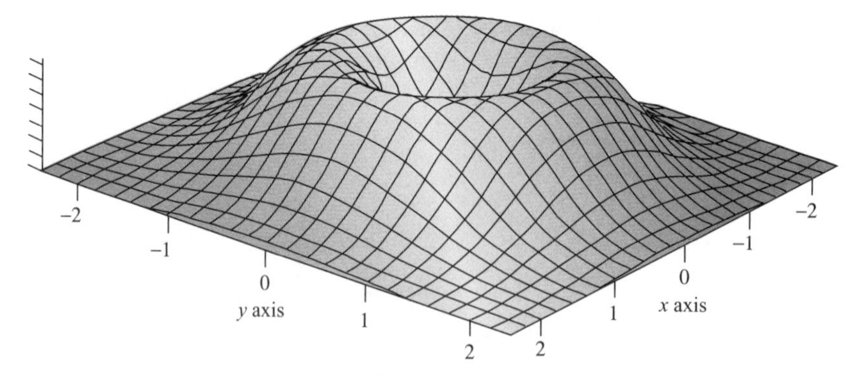

Figure 3.3.5 The volcano: $z = 2(x^2 + y^2)\exp(-x^2 - y^2)$.

SOLUTION Because $z = 2(x^2 + y^2)e^{-x^2-y^2}$, we have

$$\begin{aligned}\frac{\partial z}{\partial x} &= 4x(e^{-x^2-y^2}) + 2(x^2 + y^2)e^{-x^2-y^2}(-2x)\\ &= e^{-x^2-y^2}[4x - 4x(x^2 + y^2)]\\ &= 4x(e^{-x^2-y^2})(1 - x^2 - y^2)\end{aligned}$$

and

$$\frac{\partial z}{\partial y} = 4y(e^{-x^2-y^2})(1 - x^2 - y^2).$$

These both vanish when $x = y = 0$ or when $x^2 + y^2 = 1$. This is consistent with the figure: Points on the crater's rim are maxima and the origin is a minimum. ▲

Second Derivative Test for Local Extrema

The remainder of this section is devoted to deriving a criterion, depending on the second derivative, for a critical point to be a relative extremum. In the special case $n = 1$, our criterion will reduce to the familiar condition from one-variable calculus: $f''(x_0) > 0$ for a minimum and $f''(x_0) < 0$ for a maximum. But in the general context, the second derivative is a fairly complicated mathematical object. To state our criterion, we will introduce a version of the second derivative called the Hessian, which in turn is related to quadratic functions. ***Quadratic functions*** are functions $g\colon \mathbb{R}^n \to \mathbb{R}$ that have the form

$$g(h_1, \ldots, h_n) = \sum_{i,j=1}^{n} a_{ij}h_ih_j$$

for a $n \times n$ matrix $[a_{ij}]$. In terms of matrix multiplication, we can write

$$g(h_1, \ldots, h_n) = [h_1 \cdots h_n] \begin{bmatrix} a_{11} & a_{12} & \cdots & a_{1n} \\ \vdots & \vdots & & \vdots \\ a_{n1} & a_{n2} & \cdots & a_{nn} \end{bmatrix} \begin{bmatrix} h_1 \\ \vdots \\ h_n \end{bmatrix}.$$

For example, if $n = 3$,

$$g(h_1, h_2, h_3) = h_1^2 - 2h_1h_2 + h_3^2 = [h_1\, h_2\, h_3] \begin{bmatrix} 1 & -1 & 0 \\ -1 & 0 & 0 \\ 0 & 0 & 1 \end{bmatrix} \begin{bmatrix} h_1 \\ h_2 \\ h_3 \end{bmatrix}$$

is a quadratic function.

We can, if we wish, assume that $[a_{ij}]$ is symmetric; in fact, g is unchanged if we replace $[a_{ij}]$ by the symmetric matrix $[b_{ij}]$, where $b_{ij} = \frac{1}{2}(a_{ij} + a_{ji})$, because

$h_i h_j = h_j h_i$ and the sum is over all i and j. The *quadratic* nature of g is reflected in the identity

$$g(\lambda h_1, \dots, \lambda h_n) = \lambda^2 g(h_1, \dots, h_n),$$

which follows from the definition.

Now we are ready to define Hessian functions (named after Ludwig Otto Hesse, who introduced them in 1844).

DEFINITION Suppose that $f\colon U \subset \mathbb{R}^n \to \mathbb{R}$ has second-order continuous derivatives $(\partial^2 f/\partial x_i\, \partial x_j)(\mathbf{x}_0)$, for $i, j = 1, \dots, n$, at a point $\mathbf{x}_0 \in U$. The ***Hessian of f at*** $\mathbf{x}_0$ is the quadratic function defined by

$$\begin{aligned} Hf(\mathbf{x}_0)(\mathbf{h}) &= \frac{1}{2} \sum_{i,j=1}^{n} \frac{\partial^2 f}{\partial x_i\, \partial x_j}(\mathbf{x}_0) h_i h_j \\ &= \frac{1}{2}(h_1, \dots, h_n) \begin{pmatrix} \dfrac{\partial^2 f}{\partial x_1\, \partial x_1} & \cdots & \dfrac{\partial^2 f}{\partial x_1\, \partial x_n} \\ \vdots & & \\ \dfrac{\partial^2 f}{\partial x_n\, \partial x_1} & \cdots & \dfrac{\partial^2 f}{\partial x_n\, \partial x_n} \end{pmatrix} \begin{pmatrix} h_1 \\ \vdots \\ h_n \end{pmatrix}. \end{aligned}$$

Notice that, by equality of mixed partials, the second derivative matrix is symmetric.

This function is usually used at critical points $\mathbf{x}_0 \in U$. In this case, $\mathbf{D}f(\mathbf{x}_0) = \mathbf{0}$, so the Taylor formula (see Theorem 2, Section 3.2) may be written in the form

$$f(\mathbf{x}_0 + \mathbf{h}) = f(\mathbf{x}_0) + Hf(\mathbf{x}_0)(\mathbf{h}) + R_2(\mathbf{x}_0, \mathbf{h}).$$

Thus, *at a critical point the Hessian equals the first nonconstant term in the Taylor series of f.*

A quadratic function $g\colon \mathbb{R}^n \to \mathbb{R}$ is called ***positive-definite*** if $g(\mathbf{h}) \geq 0$ for all $\mathbf{h} \in \mathbb{R}^n$ and $g(\mathbf{h}) = 0$ only for $\mathbf{h} = \mathbf{0}$. Similarly, g is ***negative-definite*** if $g(\mathbf{h}) \leq 0$ and $g(\mathbf{h}) = 0$ for $\mathbf{h} = \mathbf{0}$ only. Note that if $n = 1$, $Hf(x_0)(h) = \frac{1}{2} f''(x_0)h^2$, which is positive-definite if and only if $f''(x_0) > 0$.

THEOREM 5: Second Derivative Test for Local Extrema If $f\colon U \subset \mathbb{R}^n \to \mathbb{R}$ is of class C^3, $\mathbf{x}_0 \in U$ is a critical point of f, and the Hessian $Hf(\mathbf{x}_0)$ is positive-definite, then $\mathbf{x}_0$ is a relative minimum of f. Similarly, if $Hf(\mathbf{x}_0)$ is negative-definite, then $\mathbf{x}_0$ is a relative maximum.

Actually, we shall prove that the extrema given by this criterion are *strict*. A relative maximum $\mathbf{x}_0$ is said to be ***strict*** if $f(\mathbf{x}) < f(\mathbf{x}_0)$ for nearby $\mathbf{x} \neq \mathbf{x}_0$. A strict relative minimum is defined similarly. Also, the theorem is valid even if f is only C^2 but we have assumed C^3 for simplicity.

The proof of Theorem 5 requires Taylor's theorem and the following result from linear algebra.

LEMMA 1 If $B = [b_{ij}]$ is an $n \times n$ real matrix, and if the associated quadratic function

$$H\colon \mathbb{R}^n \to \mathbb{R}, (h_1, \ldots, h_n) \mapsto \frac{1}{2} \sum_{i,j=1}^{n} b_{ij} h_i h_j$$

is positive-definite, then there is a constant $M > 0$ such that for all $\mathbf{h} \in \mathbb{R}^n$,

$$H(\mathbf{h}) \geq M\|\mathbf{h}\|^2.$$

PROOF For $\|\mathbf{h}\| = 1$, set $g(\mathbf{h}) = H(\mathbf{h})$. Then g is a continuous function of $\mathbf{h}$ for $\|\mathbf{h}\| = 1$ and so achieves a minimum value, say M.[8] Because H is quadratic, we have

$$H(\mathbf{h}) = H\left(\frac{\mathbf{h}}{\|\mathbf{h}\|}\|\mathbf{h}\|\right) = H\left(\frac{\mathbf{h}}{\|\mathbf{h}\|}\right)\|\mathbf{h}\|^2 = g\left(\frac{\mathbf{h}}{\|\mathbf{h}\|}\right)\|\mathbf{h}\|^2 \geq M\|\mathbf{h}\|^2$$

for any $\mathbf{h} \neq \mathbf{0}$. (The result is obviously valid if $\mathbf{h} = \mathbf{0}$). ■

Note that the quadratic function associated with the symmetric matrix $\frac{1}{2}(\partial^2 f/\partial x_i\, \partial x_j)$ is exactly the Hessian.

PROOF OF THEOREM 5 Recall that if $f\colon U \subset \mathbb{R}^n \to \mathbb{R}$ is of class C^3 and $\mathbf{x}_0 \in U$ is a critical point, Taylor's theorem may be expressed in the form

$$f(\mathbf{x}_0 + \mathbf{h}) - f(\mathbf{x}_0) = Hf(\mathbf{x}_0)(\mathbf{h}) + R_2(\mathbf{x}_0, \mathbf{h}),$$

where $(R_2(\mathbf{x}_0, \mathbf{h}))/\|\mathbf{h}\|^2 \to 0$ as $\mathbf{h} \to \mathbf{0}$.

Because $Hf(\mathbf{x}_0)$ is positive-definite, lemma 1 assures us of a constant $M > 0$ such that for all $\mathbf{h} \in \mathbb{R}^n$

$$Hf(\mathbf{x}_0)(\mathbf{h}) \geq M\|\mathbf{h}\|^2.$$

[8]Here we are using, without proof, a theorem analogous to a theorem in calculus that states that every continuous function on an interval $[a, b]$ achieves a maximum and a minimum; see Theorem 7.

Because $R_2(\mathbf{x}_0, \mathbf{h})/\|\mathbf{h}\|^2 \to 0$ as $\mathbf{h} \to \mathbf{0}$, there is a $\delta > 0$ such that for $0 < \|\mathbf{h}\| < \delta$

$$|R_2(\mathbf{x}_0, \mathbf{h})| < M\|\mathbf{h}\|^2.$$

Thus, $0 < Hf(\mathbf{x}_0)(\mathbf{h}) + R_2(\mathbf{x}_0, \mathbf{h}) = f(\mathbf{x}_0 + \mathbf{h}) - f(\mathbf{x}_0)$ for $0 < \|\mathbf{h}\| < \delta$, so that $\mathbf{x}_0$ is a relative minimum, in fact, a strict relative minimum.

The proof in the negative-definite case is similar, or else follows by applying the preceding to $-f$, and is left as an exercise. ■

EXAMPLE 5 Consider again the function $f: \mathbb{R}^2 \to \mathbb{R}, (x, y) \mapsto x^2 + y^2$. Then $(0, 0)$ is a critical point, and f is already in the form of Taylor's theorem:

$$f((0,0) + (h_1, h_2)) = f(0,0) + (h_1^2 + h_2^2) + 0.$$

We can see directly that the Hessian at $(0, 0)$ is

$$Hf(\mathbf{0})(\mathbf{h}) = h_1^2 + h_2^2,$$

which is clearly positive-definite. Thus, $(0, 0)$ is a relative minimum. This simple case can, of course, be done without calculus. Indeed, it is clear that $f(x, y) > 0$ for all $(x, y) \neq (0, 0)$. ▲

For functions of two variables $f(x, y)$, the Hessian may be written as follows:

$$Hf(x,y)(\mathbf{h}) = \tfrac{1}{2}[h_1, h_2] \begin{bmatrix} \dfrac{\partial^2 f}{\partial x^2} & \dfrac{\partial^2 f}{\partial y\, \partial x} \\ \dfrac{\partial^2 f}{\partial x\, \partial y} & \dfrac{\partial^2 f}{\partial y^2} \end{bmatrix} \begin{bmatrix} h_1 \\ h_2 \end{bmatrix}.$$

Now we shall give a useful criterion for when a quadratic function defined by such a 2×2 matrix is positive-definite. This will be useful in conjunction with Theorem 5.

LEMMA 2 Let

$$B = \begin{bmatrix} a & b \\ b & c \end{bmatrix} \quad \text{and} \quad H(\mathbf{h}) = \tfrac{1}{2}[h_1, h_2] B \begin{bmatrix} h_1 \\ h_2 \end{bmatrix}.$$

Then $H(\mathbf{h})$ is positive-definite if and only if $a > 0$ and $\det B = ac - b^2 > 0$.

PROOF We have

$$H(\mathbf{h}) = \tfrac{1}{2}[h_1, h_2] \begin{bmatrix} ah_1 + bh_2 \\ bh_1 + ch_2 \end{bmatrix} = \tfrac{1}{2}(ah_1^2 + 2bh_1h_2 + ch_2^2).$$

Let us complete the square, writing

$$H(\mathbf{h}) = \tfrac{1}{2}a\left(h_1 + \frac{b}{a}h_2\right)^2 + \frac{1}{2}\left(c - \frac{b^2}{a}\right)h_2^2.$$

Suppose H is positive-definite. Setting $h_2 = 0$, we see that $a > 0$. Setting $h_1 = -(b/a)h_2$, we get $c - b^2/a > 0$ or $ac - b^2 > 0$. Conversely, if $a > 0$ and $c - b^2/a > 0$, $H(\mathbf{h})$ is a sum of squares, so that $H(\mathbf{h}) \geq 0$. If $H(\mathbf{h}) = 0$, then each square must be zero. This implies that both h_1 and h_2 must be zero, so that $H(\mathbf{h})$ is positive-definite. ■

Similarly, one can see that $H(\mathbf{h})$ is negative-definite if and only if $a < 0$ and $ac - b^2 > 0$. We note that an alternative formulation is that $H(\mathbf{h})$ is positive- (respectively, negative-) definite if $a + c = \text{trace } B > 0$ (respectively, < 0) and $\det B > 0$.

Determinant Test for Positive Definiteness

There are similar criteria for an $n \times n$ symmetric matrix B. Consider the n square submatrices along the diagonal (see Figure 3.3.6). B is positive-definite (that is, the quadratic function associated with B is positive-definite) if and only if the determinants of these diagonal submatrices are all greater than zero. For negative-definite B, the signs should be alternately < 0 and > 0. We shall not prove this general case here.[9] In case the determinants of the diagonal submatrices are all nonzero, but the matrix is not positive- or negative-definite, the critical point is of ***saddle type***; in this case, one can show that the point is neither a maximum nor a minimum in the manner of Example 2.

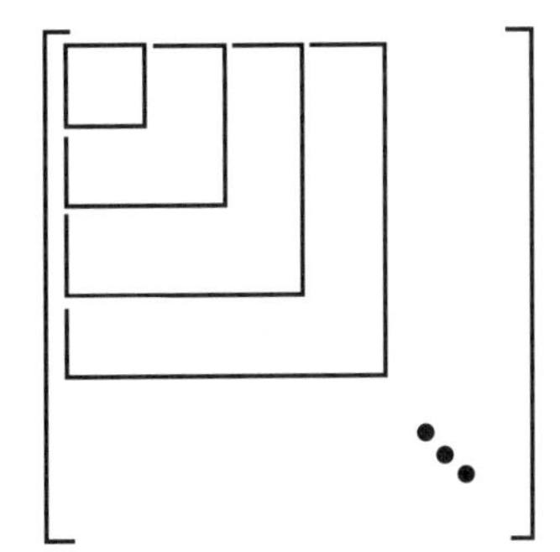

Figure 3.3.6 "Diagonal" submatrices are used in the criterion for positive definiteness; they must all have determinant > 0.

[9]This is proved in, for example, K. Hoffman and R. Kunze, *Linear Algebra*, Prentice Hall, Englewood Cliffs, N.J., 1961, pp. 249–251. For students with sufficient background in linear algebra, it should be noted that B is positive-definite when all of its eigenvalues (which are necessarily real, because B is symmetric) are positive.

Second Derivative Test

Lemma 2 and Theorem 5 imply the following result:

THEOREM 6: Second Derivative Maximum-Minimum Test for Functions of Two Variables Let $f(x, y)$ be of class C^3 on an open set U in $\mathbb{R}^2$. A point (x_0, y_0) is a (strict) local minimum of f provided the following three conditions hold:

(i) $\dfrac{\partial f}{\partial x}(x_0, y_0) = \dfrac{\partial f}{\partial y}(x_0, y_0) = 0$

(ii) $\dfrac{\partial^2 f}{\partial x^2}(x_0, y_0) > 0$

(iii) $D = \left(\dfrac{\partial^2 f}{\partial x^2}\right)\left(\dfrac{\partial^2 f}{\partial y^2}\right) - \left(\dfrac{\partial^2 f}{\partial x\, \partial y}\right)^2 > 0$ at (x_0, y_0)

(D is called the ***discriminant*** of the Hessian.) If in (ii) we have <0 instead of >0 and condition (iii) is unchanged, then we have a (strict) local maximum.

EXAMPLE 6 Classify the critical points of the function $f\colon \mathbb{R}^2 \to \mathbb{R}$, defined by $(x, y) \mapsto x^2 - 2xy + 2y^2$.

SOLUTION As in Example 5, we find that $f(0, 0) = 0$, the origin is the only critical point, and the Hessian is

$$Hf(\mathbf{0})(\mathbf{h}) = h_1^2 - 2h_1h_2 + 2h_2^2 = (h_1 - h_2)^2 + h_2^2,$$

which is clearly positive-definite. Thus, f has a relative minimum at $(0, 0)$. Alternatively, we can apply Theorem 6. At $(0, 0)$, $\partial^2 f/\partial x^2 = 2$, $\partial^2 f/\partial y^2 = 4$, and $\partial^2 f/\partial x\, \partial y = -2$. Conditions (i), (ii), and (iii) hold, so f has a relative minimum at $(0, 0)$ ▲

If $D < 0$ in Theorem 6, then we have a saddle point. In fact, one can prove that $f(x, y)$ is larger than $f(x_0, y_0)$ as we move away from (x_0, y_0) in some direction and smaller in the orthogonal direction (see Exercise 26). The general appearance is thus similar to that shown in Figure 3.3.4. The appearance of the graph near (x_0, y_0) in the case $D = 0$ must be determined by further analysis.

We summarize the procedure for dealing with functions of two variables: After all critical points have been found and their associated Hessians computed, some of these Hessians may be positive-definite, indicating relative minima; some may be negative-definite, indicating relative maxima; and some may be neither positive- nor negative-definite, indicating saddle points. The shape of the graph at a saddle point where $D < 0$ is like that in Figure 3.3.4. Critical points for which $D \neq 0$ are called ***nondegenerate critical points***. Such points are maxima, minima, or saddle points. The remaining critical points, where $D = 0$, may be tested directly, with level sets and sections or by some other method. Such critical points are said to be ***degenerate***; the methods developed in this chapter fail to provide a picture of the behavior of a function near such points, so we examine them case by case.

EXAMPLE 7 Locate the relative maxima, minima, and saddle points of the function

$$f(x, y) = \log(x^2 + y^2 + 1).$$

SOLUTION We must first locate the critical points of this function; therefore, according to Theorem 3, we calculate

$$\nabla f(x, y) = \frac{2x}{x^2 + y^2 + 1}\mathbf{i} + \frac{2y}{x^2 + y^2 + 1}\mathbf{j}.$$

Thus, $\nabla f(x, y) = \mathbf{0}$ if and only if $(x, y) = (0, 0)$, and so the only critical point of f is $(0, 0)$. Now we must determine whether this is a maximum, a minimum, or a saddle point. The second partial derivatives are

$$\frac{\partial^2 f}{\partial x^2} = \frac{2(x^2 + y^2 + 1) - (2x)(2x)}{(x^2 + y^2 + 1)^2},$$

$$\frac{\partial^2 f}{\partial y^2} = \frac{2(x^2 + y^2 + 1) - (2y)(2y)}{(x^2 + y^2 + 1)^2},$$

and

$$\frac{\partial^2 f}{\partial x\, \partial y} = \frac{-2x(2y)}{(x^2 + y^2 + 1)^2}.$$

Therefore,

$$\frac{\partial^2 f}{\partial x^2}(0, 0) = 2 = \frac{\partial^2 f}{\partial y^2}(0, 0) \quad \text{and} \quad \frac{\partial^2 f}{\partial x\, \partial y}(0, 0) = 0,$$

which yields

$$D = 2 \cdot 2 = 4 > 0.$$

Because $(\partial^2 f/\partial x^2)(0, 0) > 0$, we conclude by Theorem 6 that $(0, 0)$ is a local minimum. (Can you show this just from the fact that $\log t$ is an increasing function of $t > 0$?) ▲

EXAMPLE 8 The graph of the function $g(x, y) = 1/xy$ is a surface S in $\mathbb{R}^3$. Find the points on S that are closest to the origin $(0, 0, 0)$. (See Figure 3.3.7.)

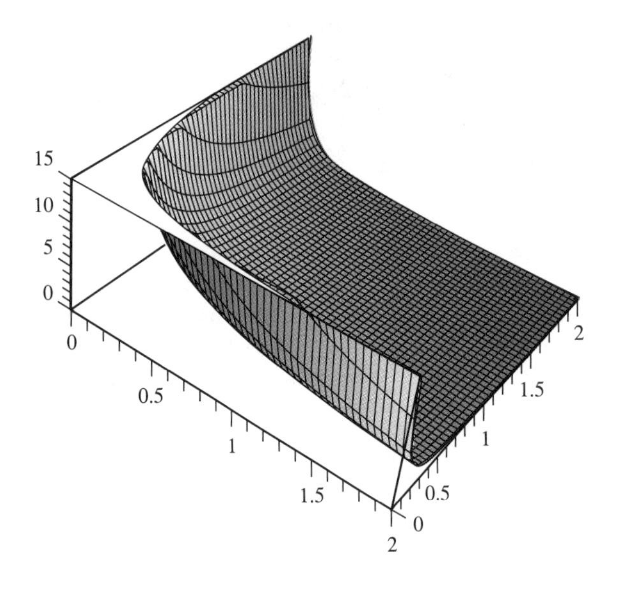

Figure 3.3.7 The surface $z = 1/xy$ in the first quadrant. (There are similar figures in the other quadrants, but notice that $z < 0$ in the second and fourth quadrants.)

SOLUTION Each point on S is of the form $(x, y, 1/xy)$. The distance from this point to the origin is

$$d(x, y) = \sqrt{x^2 + y^2 + \frac{1}{x^2 y^2}}.$$

It is easier to work with the square of d, so let $f(x, y) = x^2 + y^2 + (1/x^2 y^2)$, which will have the same minimum point. Notice that $f(x, y)$ becomes very large as x and y get larger and larger; $f(x, y)$ also becomes very large as (x, y) approaches the x or y axis where f is not defined, so f must attain a minimum at some critical point. The critical points are determined by:

$$\frac{\partial f}{\partial x} = 2x - \frac{2}{x^3 y^2} = 0,$$

$$\frac{\partial f}{\partial y} = 2y - \frac{2}{y^3 x^2} = 0,$$

that is, $x^4 y^2 - 1 = 0$, and $x^2 y^4 - 1 = 0$. From the first equation we get $y^2 = 1/x^4$, and, substituting this into the second equation, we obtain

$$\frac{x^2}{x^8} = 1 = \frac{1}{x^6}.$$

Thus, $x = \pm 1$ and $y = \pm 1$, and it therefore follows that f has four critical points, namely, $(1, 1)$, $(1, -1)$, $(-1, 1)$, and $(-1, -1)$. Note that f has the value 3 for all these points, so they are all minima. Therefore, the points on the surface closest to the point $(0, 0, 0)$ are $(1, 1, 1)$, $(1, -1, -1)$, $(-1, 1, -1)$, and $(-1, -1, 1)$ and the minimum distance is $\sqrt{3}$. Is this consistent with the graph in Figure 3.3.7? ▲

EXAMPLE 9 Analyze the behavior of $z = x^5y + xy^5 + xy$ at its critical points.

SOLUTION The first partial derivatives are

$$\frac{\partial z}{\partial x} = 5x^4y + y^5 + y = y(5x^4 + y^4 + 1)$$

and

$$\frac{\partial z}{\partial y} = x(5y^4 + x^4 + 1).$$

The terms $5x^4 + y^4 + 1$ and $5y^4 + x^4 + 1$ are always greater than or equal to 1, and so it follows that the only critical point is $(0, 0)$.

The second partial derivatives are

$$\frac{\partial^2 z}{\partial x^2} = 20x^3y, \qquad \frac{\partial^2 z}{\partial y^2} = 20xy^3,$$

and

$$\frac{\partial^2 z}{\partial x\, \partial y} = 5x^4 + 5y^4 + 1.$$

Thus, at $(0, 0)$, $D = -1$, and so $(0, 0)$ is a nondegenerate saddle point and the graph of z near $(0, 0)$ looks like the graph in Figure 3.3.4. ▲

Global Maxima and Minima

We end this section with a discussion of the theory of *absolute*, or *global*, maxima and minima of functions of several variables. Unfortunately, the location of absolute maxima and minima for functions on $\mathbb{R}^n$ is, in general, a more difficult problem than for functions of one variable.

DEFINITION Suppose $f\colon A \to \mathbb{R}$ is a function defined on a set A in $\mathbb{R}^2$ or $\mathbb{R}^3$. A point $\mathbf{x}_0 \in A$ is said to be an ***absolute maximum*** (or ***absolute minimum***) point of f if $f(\mathbf{x}) \le f(\mathbf{x}_0)$ [or $f(\mathbf{x}) \ge f(\mathbf{x}_0)$] for all $\mathbf{x} \in A$.

In one-variable calculus, one learns—but often does not prove—that every continuous function on a closed interval I assumes its absolute maximum (or minimum) value at some point $\mathbf{x}_0$ in I. A generalization of this theoretical fact also holds in $\mathbb{R}^n$.

Such theorems guarantee that the maxima or minima one is seeking actually exist; therefore, the search for them is not in vain.

DEFINITION A set $D \in \mathbb{R}^n$ is said to be ***bounded*** if there is a number $M > 0$ such that $\|\mathbf{x}\| < M$ for all $\mathbf{x} \in D$. A set is ***closed*** if it contains all its boundary points.

Thus, a set is bounded if it can be strictly contained in some (large) ball. The appropriate generalization of the one-variable theorem on maxima and minima is the following result, stated without proof.

THEOREM 7: Global Existence Theorem for Maxima and Minima

Let D be closed and bounded in $\mathbb{R}^n$ and let $f\colon D \to \mathbb{R}$ be continuous. Then f assumes its absolute maximum and minimum values at some points $\mathbf{x}_0$ and $\mathbf{x}_1$ of D.

Simply stated, $\mathbf{x}_0$ and $\mathbf{x}_1$ are points where f assumes its largest and smallest values. As in one-variable calculus, these points need not be uniquely determined.

Suppose now that $D = U \cup \partial U$, where U is open and ∂U is its boundary. If $D \subset \mathbb{R}^2$, we suppose that ∂U is a piecewise smooth curve; that is, D is a region bounded by a collection of smooth curves—for example, a square or the sets depicted in Figure 3.3.8.

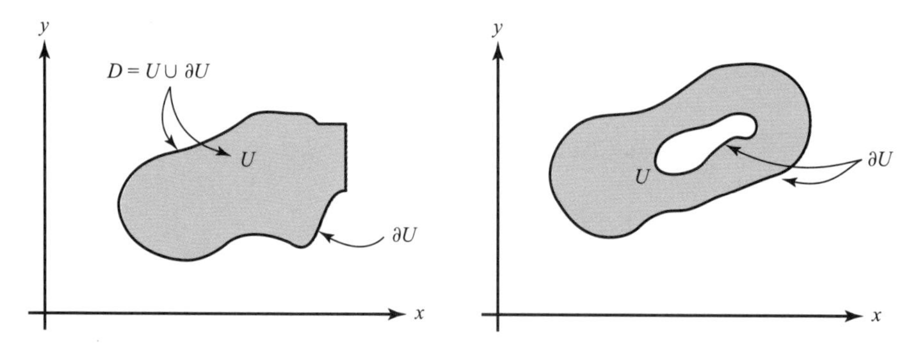

Figure 3.3.8 $D = U \cup \partial U$: Two examples of regions whose boundary is a piecewise smooth curve.

If $\mathbf{x}_0$ and $\mathbf{x}_1$ are in U, we know from Theorem 4 that they are critical points of f. If they are in ∂U, and ∂U is a smooth curve (i.e., the image of a smooth path $\mathbf{c}$ with $\mathbf{c}' \neq 0$), then they are maximum or minimum points of f viewed as a function on ∂U. These observations provide a method of finding the absolute maximum and minimum values of f on a region D.

Strategy for Finding the Absolute Maxima and Minima on a Region with Boundary Let f be a continuous function of two variables defined on a closed and bounded region D in $\mathbb{R}^2$, which is bounded by a smooth closed curve. To find the absolute maximum and minimum of f on D:

(i) Locate all critical points for f in U.

(ii) Find the maximum and minimum of f viewed as a function only on ∂U.

(iii) Compute the value of f at all of these critical points.

(iv) Compare all these values and select the largest and the smallest.

If D is a region bounded by a collection of smooth curves (such as a square), then one follows a similar procedure, but including in step (iii) the points where the curves meet (such as the corners of the square).

All the steps except step (ii) should now be familiar to the student. To carry out step (ii) in the plane, one way is to find a smooth parametrization of ∂U; that is, we find a path $\mathbf{c}: I \to \partial U$, where I is some interval, which is onto ∂U. Second, we consider the function of one variable $t \mapsto f(\mathbf{c}(t))$, where $t \in I$, and locate the maximum and minimum points $t_0, t_1 \in I$ (remember to check the endpoints!). Then $\mathbf{c}(t_0)$, $\mathbf{c}(t_1)$ will be maximum and minimum *points* for f as a function on ∂U. Another method for dealing with step (ii) is the Lagrange multiplier method, to be presented in the next section.

EXAMPLE 10 Find the maximum and minimum values of the function $f(x, y) = x^2 + y^2 - x - y + 1$ in the disk D defined by $x^2 + y^2 \leq 1$.

SOLUTION (i) To find the critical points we set $\partial f/\partial x = \partial f/\partial y = 0$. Thus, $2x - 1 = 0$, $2y - 1 = 0$, and hence $(x, y) = (\frac{1}{2}, \frac{1}{2})$ is the only critical point in the open disk $U = \{(x, y) \mid x^2 + y^2 < 1\}$.

(ii) The boundary ∂U can be parametrized by $\mathbf{c}(t) = (\sin t, \cos t)$, $0 \leq t \leq 2\pi$. Thus,

$$f(\mathbf{c}(t)) = \sin^2 t + \cos^2 t - \sin t - \cos t + 1 = 2 - \sin t - \cos t = g(t).$$

To find the maximum and minimum of f on ∂U, it suffices to locate the maximum and minimum of g. Now $g'(t) = 0$ only when

$$\sin t = \cos t, \qquad \text{that is, when} \qquad t = \frac{\pi}{4}, \frac{5\pi}{4}.$$

Thus, the candidates for the maximum and minimum for f on ∂U are the points $\mathbf{c}(\pi/4)$, $\mathbf{c}(5\pi/4)$ and the endpoints $\mathbf{c}(0) = \mathbf{c}(2\pi)$.

(iii) The values of f at the critical points are: $f(\frac{1}{2}, \frac{1}{2}) = \frac{1}{2}$ from step (i) and, from step (ii),

$$f\left(\mathbf{c}\left(\frac{\pi}{4}\right)\right) = f\left(\frac{\sqrt{2}}{2}, \frac{\sqrt{2}}{2}\right) = 2 - \sqrt{2},$$

$$f\left(\mathbf{c}\left(\frac{5\pi}{4}\right)\right) = f\left(-\frac{\sqrt{2}}{2}, -\frac{\sqrt{2}}{2}\right) = 2 + \sqrt{2},$$

and

$$f(\mathbf{c}(0)) = f(\mathbf{c}(2\pi)) = f(0, 1) = 1.$$

(iv) Comparing all the values $\frac{1}{2}, 2 - \sqrt{2}, 2 + \sqrt{2}, 1$, it is clear that the absolute minimum occurs at $(1/2, 1/2)$ and the absolute maximum occurs at $(-\sqrt{2}/2, -\sqrt{2}/2)$. ▲

In Section 3.4, we shall consider a generalization of the strategy for finding the absolute maximum and minimum to regions D in $\mathbb{R}^n$.

EXERCISES

In Exercises 1 to 16, find the critical points of the given function and then determine whether they are local maxima, local minima, or saddle points.

1. $f(x, y) = x^2 - y^2 + xy$

2. $f(x, y) = x^2 + y^2 - xy$

3. $f(x, y) = x^2 + y^2 + 2xy$

4. $f(x, y) = x^2 + y^2 + 3xy$

5. $f(x, y) = e^{1+x^2-y^2}$

6. $f(x, y) = x^2 - 3xy + 5x - 2y + 6y^2 + 8$

7. $f(x, y) = 3x^2 + 2xy + 2x + y^2 + y + 4$

8. $f(x, y) = \sin(x^2 + y^2)$ [consider only the critical point $(0, 0)$]

9. $f(x, y) = \cos(x^2 + y^2)$ [consider only the three critical points $(0, 0)$, $(\sqrt{\pi/2}, \sqrt{\pi/2})$, and $(0, \sqrt{\pi})$]

10. $f(x, y) = y + x \sin y$

11. $f(x, y) = e^x \cos y$

12. $f(x, y) = (x - y)(xy - 1)$

13. $f(x, y) = xy + \dfrac{1}{x} + \dfrac{1}{y}$

14. $f(x, y) = \log(2 + \sin xy)$

15. $f(x, y) = x \sin y$

16. $f(x, y) = (x + y)(xy + 1)$

17. Find the local maxima and minima for $z = (x^2 + 3y^2)e^{1-x^2-y^2}$. (See Figure 2.1.15.)

18. Let $f(x, y) = x^2 + y^2 + kxy$. If you imagine the graph changing as k increases, at what values of k does the shape of the graph change qualitatively?

19. An examination of the function $f: \mathbb{R}^2 \to \mathbb{R}, (x, y) \mapsto (y - 3x^2)(y - x^2)$ will give an idea of the difficulty of finding conditions that guarantee that a critical point is a relative extremum when Theorem 6 fails.[10] Show that

(a) the origin is a critical point of f;
(b) f has a relative minimum at $(0, 0)$ on every straight line through $(0, 0)$; that is, if $g(t) = (at, bt)$, then $f \circ g: \mathbb{R} \to \mathbb{R}$ has a relative minimum at 0, for every choice of a and b;
(c) the origin is not a relative minimum of f.

20. Let $f(x, y) = Ax^2 + E$ where A and E are constants. What are the critical points of f? Are they local maxima or local minima?

21. Let $f(x, y) = x^2 - 2xy + y^2$. Here $D = 0$. Can you say whether the critical points are local minima, local maxima, or saddle points?

22. Find the point on the plane $2x - y + 2z = 20$ nearest the origin.

23. Show that a rectangular box of given volume has minimum surface area when the box is a cube.

24. Show that the rectangular parallelepiped with fixed surface area and maximum volume is a cube.

25. Write the number 120 as a sum of three numbers so that the sum of the products taken two at a time is a maximum.

26. Show that if (x_0, y_0) is a critical point of a quadratic function $f(x, y)$ and $D < 0$, then there are points (x, y) near (x_0, y_0) at which $f(x, y) > f(x_0, y_0)$ and, similarly, points for which $f(x, y) < f(x_0, y_0)$.

27. Determine the nature of the critical points of the function

$$f(x, y, z) = x^2 + y^2 + z^2 + xy.$$

28. Let n be an integer greater than 2 and set $f(x, y) = ax^n + cy^n$, where $ac \neq 0$. Determine the nature of the critical points of f.

---

[10] This interesting phenomenon was first pointed out by the famous mathematician Giuseppe Peano (1858–1932). Another curious "pathology" is given in Exercise 41.

29. Determine the nature of the critical points of $f(x, y) = x^3 + y^2 - 6xy + 6x + 3y$.

30. Find the absolute maximum and minimum values of the function $f(x, y) = (x^2 + y^2)^4$ on the disk $x^2 + y^2 \leq 1$. (You do not *have* to use calculus.)

31. Repeat Exercise 30 for the function $f(x, y) = x^2 + xy + y^2$.

32. A curve C in space is defined *implicitly* on the cylinder $x^2 + y^2 = 1$ by the additional equation $x^2 - xy + y^2 - z^2 = 1$. Find the point or points on C closest to the origin.

33. Find the absolute maximum and minimum values for $f(x, y) = \sin x + \cos y$ on the rectangle R defined by $0 \leq x \leq 2\pi, 0 \leq y \leq 2\pi$.

34. Find the absolute maximum and minimum values for the function $f(x, y) = xy$ on the rectangle R defined by $-1 \leq x \leq 1, -1 \leq y \leq 1$.

35. Determine the nature of the critical points of $f(x, y) = xy + 1/x + 8/y$.

In Exercises 36 through 40, D denotes the unit disk.

36. Let u be a C^2 function on D which is "strictly subharmonic"; that is, the following inequality holds: $\nabla^2 u = (\partial^2 u/\partial x^2) + (\partial^2 u/\partial y^2) > 0$. Show that u cannot have a maximum point in $D \backslash \partial D$ (the set of points in D, but not in ∂D).

37. Let u be a harmonic function on D; that is, $\nabla^2 u = 0$ on $D \backslash \partial D$ and be continuous on D. Show that if u achieves its maximum value in $D \backslash \partial D$, it also achieves it on ∂D. This is sometimes called the "weak maximum principle" for harmonic functions. [Hint: Consider $\nabla^2(u + \varepsilon e^x)$, $\varepsilon > 0$. You can use the following fact, which is proved in more advanced texts: Given a sequence $\{\mathbf{p}_n\}$, $n = 1, 2, \ldots,$ of points in a closed bounded set A in $\mathbb{R}^2$ or $\mathbb{R}^3$, there exists a point $\mathbf{q}$ such that every neighborhood of $\mathbf{q}$ contains at least one member of $\{\mathbf{p}_n\}$.]

38. Define the notion of a strict superharmonic function u on D by mimicking Exercise 36. Show that u cannot have a minimum in $D \backslash \partial D$.

39. Let u be harmonic in D as in Exercise 37. Show that if u achieves its minimum value in $D \backslash \partial D$, it also achieves it on ∂D. This is sometimes called the "weak minimum principle" for harmonic functions.

40. Let $\phi\colon \partial D \to \mathbb{R}$ be continuous and let T be a solution on D to $\nabla^2 T = 0$, continuous on D and $T = \phi$ on ∂D.

(a) Use Exercises 36 to 39 to show that such a solution, if it exists, must be unique.

(b) Suppose that $T(x, y)$ represents a temperature function that is independent of time, with ϕ representing the temperature of a circular plate at its boundary. Can you give a physical interpretation of the principle stated in part (a)?

41. (a) Let f be a C^1 function on the real line $\mathbb{R}$. Suppose that f has exactly one critical point x_0 that is a strict local minimum for f. Show that x_0 is also an absolute minimum for f, that is, that $f(x) \geq f(x_0)$ for all x.

(b) The next example shows that the conclusion of part (a) does not hold for functions of more than one variable. Let $f\colon \mathbb{R}^2 \to \mathbb{R}$ be defined by

$$f(x, y) = -y^4 - e^{-x^2} + 2y^2\sqrt{e^x + e^{-x^2}}.$$

(i) Show that (0, 0) is the only critical point for f and that it is a local minimum.
(ii) Argue informally that f has no absolute minimum.

42. Suppose that a pentagon is composed of a rectangle topped by an isosceles triangle (see Figure 3.3.9). If the length of the perimeter is fixed, find the maximum possible area.

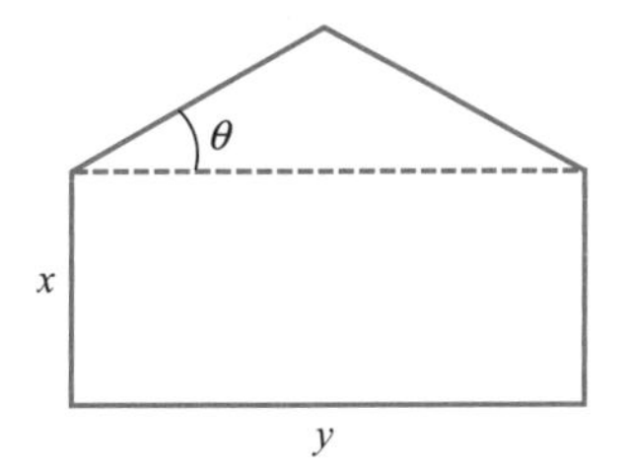

Figure 3.3.9 Maximize the area for fixed perimeter.

3.4 Constrained Extrema and Lagrange Multipliers

Often one is required to maximize or minimize a function subject to certain *constraints* or *side conditions*. For example, we might need to maximize $f(x, y)$ subject to the condition that $x^2 + y^2 = 1$; that is, that (x, y) lie on the unit circle. More generally, we might need to maximize or minimize $f(x, y)$ subject to the side condition that (x, y) also satisfies an equation $g(x, y) = c$ where g is some function and c equals a constant [in the preceding example, $g(x, y) = x^2 + y^2$, and $c = 1$]. The set of such (x, y) is a level curve for g.

The purpose of this section is to develop some methods for handling this sort of problem. In Figure 3.4.1 we picture a graph of a function $f(x, y)$. In this picture,

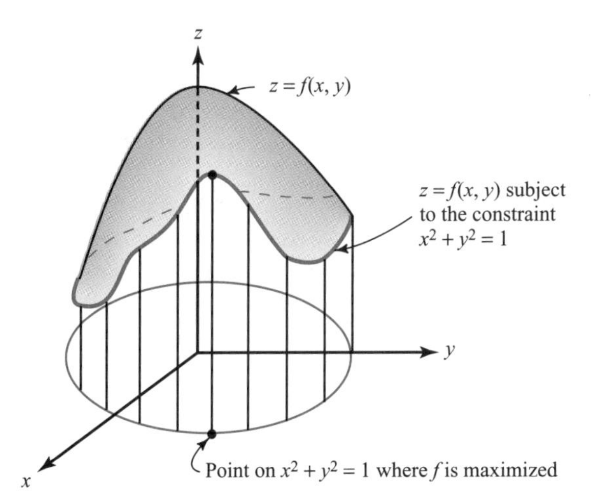

Figure 3.4.1 The geometric meaning of maximizing f subject to the constraint $x^2 + y^2 = 1$.

the maximum of f might be at (0, 0). However, suppose we are not interested in this maximum but only the maximum of $f(x, y)$ when (x, y) belongs to the unit circle; that is, when $x^2 + y^2 = 1$. The cylinder over $x^2 + y^2 = 1$ intersects the graph of $z = f(x, y)$ in a curve that lies on this graph. The problem of maximizing or minimizing $f(x, y)$ subject to the constraint $x^2 + y^2 = 1$ amounts to finding the point on this curve where z is the greatest or the least.

The Lagrange Multiplier Method

In general, let $f: U \subset \mathbb{R}^n \to \mathbb{R}$ and $g: U \subset \mathbb{R}^n \to \mathbb{R}$ be given C^1 functions, and let S be the level set for g with value c [recall that this is the set of points $\mathbf{x} \in \mathbb{R}^n$ with $g(\mathbf{x}) = c$].

When f is restricted to S we again have the notion of local maxima or local minima of f (local extrema), and an absolute maximum (largest value) or absolute minimum (smallest value) must be a local extremum. The following method provides a necessary condition for a constrained extremum:

THEOREM 8: The Method of Lagrange Multipliers Suppose that $f: U \subset \mathbb{R}^n \to \mathbb{R}$ and $g: U \subset \mathbb{R}^n \to \mathbb{R}$ are given C^1 real-valued functions. Let $\mathbf{x}_0 \in U$ and $g(\mathbf{x}_0) = c$, and let S be the level set for g with value c (recall that this is the set of points $\mathbf{x} \in \mathbb{R}^n$ satisfying $g(\mathbf{x}) = c$). Assume $\nabla g(\mathbf{x}_0) \neq \mathbf{0}$.

If $f|S$, which denotes "f restricted to S," has a local maximum or minimum on S at $\mathbf{x}_0$, then there is a real number λ such that

$$\nabla f(\mathbf{x}_0) = \lambda \nabla g(\mathbf{x}_0). \tag{1}$$

In general, a point $\mathbf{x}_0$ where equation (1) holds is said to be a ***critical point*** of $f|S$.

PROOF We have not developed enough techniques to give a complete proof, but we can provide the essential points. (The additional technicalities needed are discussed in Section 3.5 and in the Internet supplement.)

In Section 2.6 we learned that for $n = 3$ the tangent space or tangent plane of S at $\mathbf{x}_0$ is the space orthogonal to $\nabla g(\mathbf{x}_0)$. For arbitrary n we can give the same definition for the tangent space of S at $\mathbf{x}_0$. This definition can be motivated by considering tangents to paths $\mathbf{c}(t)$ that lie in S, as follows: If $\mathbf{c}(t)$ is a path in S and $\mathbf{c}(0) = \mathbf{x}_0$, then $\mathbf{c}'(0)$ is a tangent vector to S at $\mathbf{x}_0$, but

$$\frac{d}{dt} g(\mathbf{c}(t)) = \frac{d}{dt} c = 0,$$

and on the other hand, by the chain rule,

$$\left.\frac{d}{dt}g(\mathbf{c}(t))\right|_{t=0} = \nabla g(\mathbf{x}_0)\cdot \mathbf{c}'(0),$$

so that $\nabla g(\mathbf{x}_0)\cdot \mathbf{c}'(0) = 0$; that is, $\mathbf{c}'(0)$ is orthogonal to $\nabla g(\mathbf{x}_0)$.

If $f|S$ has a maximum at $\mathbf{x}_0$, then $f(\mathbf{c}(t))$ has a maximum at $t = 0$. By one-variable calculus, $df(\mathbf{c}(t))/dt|_{t=0} = 0$. Hence, by the chain rule,

$$0 = \left.\frac{d}{dt}f(\mathbf{c}(t))\right|_{t=0} = \nabla f(\mathbf{x}_0)\cdot \mathbf{c}'(0).$$

Thus, $\nabla f(\mathbf{x}_0)$ is perpendicular to the tangent of every curve in S and so is perpendicular to the whole tangent space to S at $\mathbf{x}_0$. Because the space perpendicular to this tangent space is a line, $\nabla f(\mathbf{x}_0)$ and $\nabla g(\mathbf{x}_0)$ are parallel. Because $\nabla g(\mathbf{x}_0) \neq \mathbf{0}$, it follows that $\nabla f(\mathbf{x}_0)$ is a multiple of $\nabla g(\mathbf{x}_0)$, which is the conclusion of the theorem. ■

Let us extract some geometry from this proof.

THEOREM 9 If f, when constrained to a surface S, has a maximum or minimum at $\mathbf{x}_0$, then $\nabla f(\mathbf{x}_0)$ is perpendicular to S at $\mathbf{x}_0$ (see Figure 3.4.2).

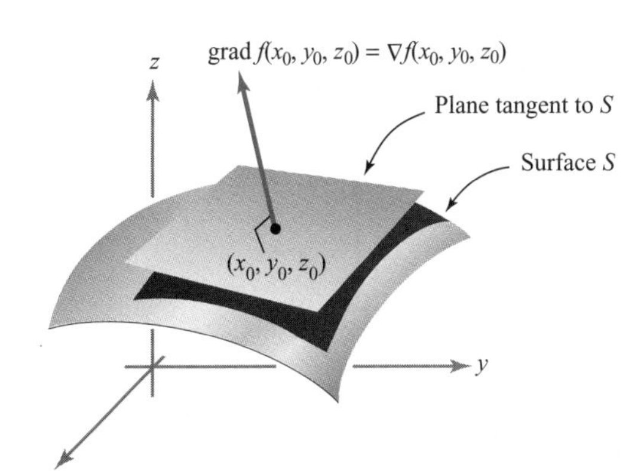

Figure 3.4.2 The geometry of constrained extrema.

These results tell us that in order to find the constrained extrema of f, we must look among those points $\mathbf{x}_0$ satisfying the conclusions of these two theorems. We shall give several illustrations of how to use each.

When the method of Theorem 8 is used, we look for a point $\mathbf{x}_0$ and a constant λ, called a ***Lagrange multiplier***, such that $\nabla f(\mathbf{x}_0) = \lambda\nabla g(\mathbf{x}_0)$. This method is more analytic in nature than the geometric method of theorem 9. Surprisingly, Euler introduced these multipliers in 1744, some 40 years before Lagrange!

Equation (1) says that the partial derivatives of f are proportional to those of g. Finding such points $\mathbf{x}_0$ at which this occurs means solving the simultaneous equations

$$\left.\begin{aligned}
\frac{\partial f}{\partial x_1}(x_1,\dots,x_n) &= \lambda\frac{\partial g}{\partial x_1}(x_1,\dots,x_n)\\
\frac{\partial f}{\partial x_2}(x_1,\dots,x_n) &= \lambda\frac{\partial g}{\partial x_2}(x_1,\dots,x_n)\\
&\vdots\\
\frac{\partial f}{\partial x_n}(x_1,\dots,x_n) &= \lambda\frac{\partial g}{\partial x_n}(x_1,\dots,x_n)\\
g(x_1,\dots,x_n) &= c
\end{aligned}\right\} \tag{2}$$

for $x_1, \dots, x_n$ and λ.

Another way of looking at these equations is as follows: Think of λ as an additional variable and form the auxiliary function

$$h(x_1,\dots,x_n,\lambda) = f(x_1,\dots,x_n) - \lambda[g(x_1,\dots,x_n) - c].$$

The Lagrange multiplier theorem says that to find the extreme points of $f|S$, we should examine the critical points of h. These are found by solving the equations

$$\left.\begin{aligned}
0 &= \frac{\partial h}{\partial x_1} = \frac{\partial f}{\partial x_1} - \lambda\frac{\partial g}{\partial x_1}\\
&\vdots\\
0 &= \frac{\partial h}{\partial x_n} = \frac{\partial f}{\partial x_n} - \lambda\frac{\partial g}{\partial x_n}\\
0 &= \frac{\partial h}{\partial \lambda} = g(x_1,\dots,x_n) - c
\end{aligned}\right\}, \tag{3}$$

which are the same as equations (2) above.

Second derivative tests for maxima and minima analogous to those in Section 3.3 will be given in Theorem 10 later in this section. However, in many problems it is possible to distinguish between maxima and minima by direct observation or by geometric means. Because this is often simpler, we consider examples of the latter type first.

EXAMPLE 1 Let $S \subset \mathbb{R}^2$ be the line through $(-1, 0)$ inclined at 45°, and let $f\colon \mathbb{R}^2 \to \mathbb{R}, (x, y) \mapsto x^2 + y^2$. Find the extrema of $f|S$.

SOLUTION Here $S = \{(x, y) \mid y - x - 1 = 0\}$, and therefore we set $g(x, y) = y - x - 1$ and $c = 0$. We have $\nabla g(x, y) = -\mathbf{i} + \mathbf{j} \neq \mathbf{0}$. The relative extrema of $f|S$ must be found among the points at which ∇f is orthogonal to S, that is, inclined at $-45°$. But $\nabla f(x, y) = (2x, 2y)$, which has the desired slope only when $x = -y$, or when (x, y) lies on the line L through the origin inclined at $-45°$. This can occur in the set S only for the single point at which L and S intersect (see Figure 3.4.3). Reference to the level curves of f indicates that this point, $(-1/2, 1/2)$, is a relative minimum of $f|S$ (but not of f).

Notice that in this problem, f on S has a minimum but no maximum. ▲

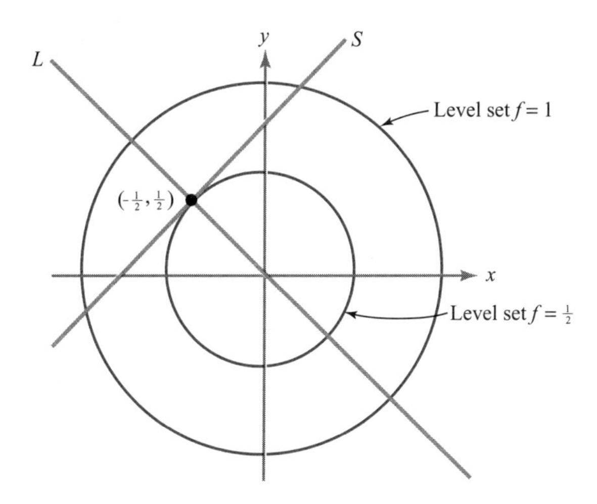

Figure 3.4.3 The geometry associated with finding the extrema of $f(x, y) = x^2 + y^2$ restricted to $S = \{(x, y) \mid y - x - 1 = 0\}$.

EXAMPLE 2 Let $f\colon \mathbb{R}^2 \to \mathbb{R}$, $(x, y) \mapsto x^2 - y^2$, and let S be the circle of radius 1 around the origin. Find the extrema of $f|S$.

SOLUTION The set S is the level curve for g with value 1, where $g\colon \mathbb{R}^2 \to \mathbb{R}$, $(x, y) \mapsto x^2 + y^2$. Because both of these functions have been studied in previous examples, we know their level curves; these are shown in Figure 3.4.4. In two dimensions, the condition that $\nabla f = \lambda \nabla g$ at $\mathbf{x}_0$, that is, that ∇f and ∇g are parallel at $\mathbf{x}_0$ is the same as the level curves being tangent at $\mathbf{x}_0$ (why?). Thus, the extreme points of $f|S$ are $(0, \pm 1)$ and $(\pm 1, 0)$. Evaluating f, we find $(0, \pm 1)$ are minima and $(\pm 1, 0)$ are maxima.

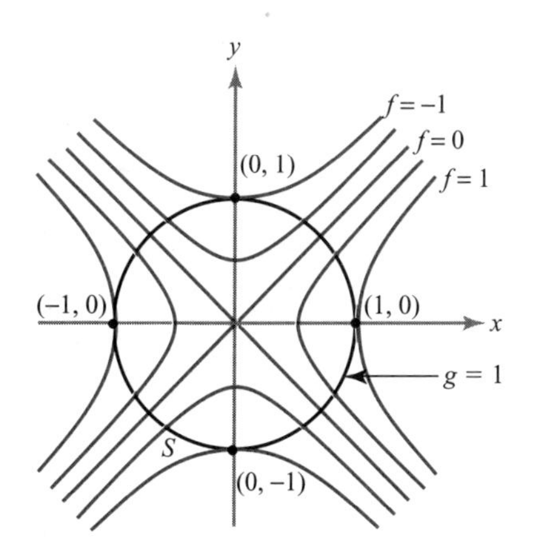

Figure 3.4.4 The geometry associated with the problem of finding the extrema of $x^2 - y^2$ on $S = \{(x, y) \mid x^2 + y^2 = 1\}$.

Let us also do this problem analytically by the method of Lagrange multipliers. Clearly,

$$\nabla f(x, y) = \left(\frac{\partial f}{\partial x}, \frac{\partial f}{\partial y}\right) = (2x, -2y) \qquad \text{and} \qquad \nabla g(x, y) = (2x, 2y).$$

Note that $\nabla g(x, y) \neq \mathbf{0}$ if $x^2 + y^2 = 1$. Thus, according to the Lagrange multiplier theorem, we must find a λ such that

$$(2x, -2y) = \lambda(2x, 2y) \qquad \text{and} \qquad (x, y) \in S, \qquad \text{i.e., } x^2 + y^2 = 1.$$

These conditions yield three equations, which can be solved for the three unknowns x, y, and λ. From $2x = \lambda 2x$ we conclude that either $x = 0$ or $\lambda = 1$. If $x = 0$, then $y = \pm 1$ and $-2y = \lambda 2y$ implies $\lambda = -1$. If $\lambda = 1$, then $y = 0$ and $x = \pm 1$. Thus, we get the points $(0, \pm 1)$ and $(\pm 1, 0)$, as before. As we have mentioned, this method only locates potential extrema; whether they are maxima, minima, or neither must be determined by other means, such as geometric arguments or the second derivative test given below.[11] ▲

EXAMPLE 3 Maximize the function $f(x, y, z) = x + z$ subject to the constraint $x^2 + y^2 + z^2 = 1$.

SOLUTION By Theorem 7 we know that the function f restricted to the unit sphere $x^2 + y^2 + z^2 = 1$ has a maximum (and also a minimum). To find the maximum, we again use the Lagrange multiplier theorem. We seek λ and (x, y, z) such that

$$1 = 2x\lambda, \qquad 0 = 2y\lambda, \qquad \text{and} \qquad 1 = 2z\lambda,$$

and

$$x^2 + y^2 + z^2 = 1.$$

From the first or the third equation, we see that $\lambda \neq 0$. Thus, from the second equation, we get $y = 0$. From the first and third equations, $x = z$, and so from the fourth, $x = \pm 1/\sqrt{2} = z$. Hence, our points are $(1/\sqrt{2}, 0, 1/\sqrt{2})$ and $(-1/\sqrt{2}, 0, -1/\sqrt{2})$. Comparing the values of f at these points, we can see that the first point yields the maximum of f (restricted to the constraint) and the second the minimum. ▲

EXAMPLE 4 Assume that among all rectangular boxes with fixed surface area of 10 square meters there is a box of largest possible volume. Find its dimensions.

SOLUTION If x, y, and z are the lengths of the sides, $x \geq 0$, $y \geq 0$, $z \geq 0$, respectively, and the volume is $f(x, y, z) = xyz$. The constraint is $2(xy + xz + yz) = 10$;

[11] In these examples, $\nabla g(\mathbf{x}_0) \neq \mathbf{0}$ on the surface S, as required by the Lagrange multiplier theorem. If $\nabla g(\mathbf{x}_0)$ were zero for some $\mathbf{x}_0$ on S, then it would have to be included among the possible extrema.

This case may be proved by generalizing the method used to prove the Lagrange multiplier theorem. Let us give an example of how this more general formulation is used.

EXAMPLE 5 Find the extreme points of $f(x, y, z) = x + y + z$ subject to the two conditions $x^2 + y^2 = 2$ and $x + z = 1$.

SOLUTION Here there are two constraints:

$$g_1(x, y, z) = x^2 + y^2 - 2 = 0 \qquad \text{and} \qquad g_2(x, y, z) = x + z - 1 = 0.$$

Thus, we must find x, y, z, λ_1, and λ_2 such that

$$\nabla f(x, y, z) = \lambda_1 \nabla g_1(x, y, z) + \lambda_2 \nabla g_2(x, y, z),$$
$$g_1(x, y, z) = 0, \qquad \text{and} \qquad g_2(x, y, z) = 0.$$

Computing the gradients and equating components, we get

$$1 = \lambda_1 \cdot 2x + \lambda_2 \cdot 1,$$
$$1 = \lambda_1 \cdot 2y + \lambda_2 \cdot 0,$$
$$1 = \lambda_1 \cdot 0 + \lambda_2 \cdot 1,$$
$$x^2 + y^2 = 2, \qquad \text{and} \qquad x + z = 1.$$

These are five equations for x, y, z, λ_1, and λ_2. From the third equation, $\lambda_2 = 1$, and so $2x\lambda_1 = 0$, $2y\lambda_1 = 1$. Because the second implies $\lambda_1 \neq 0$, we have $x = 0$. Thus, $y = \pm\sqrt{2}$ and $z = 1$. Hence, the possible extrema are $(0, \pm\sqrt{2}, 1)$. By inspection, $(0, \sqrt{2}, 1)$ gives a relative maximum, and $(0, -\sqrt{2}, 1)$ a relative minimum.

The condition $x^2 + y^2 = 2$ implies that x and y must be bounded. The condition $x + z = 1$ implies that z is also bounded. If follows that the constraint set S is closed and bounded. By Theorem 7 it follows that f has a maximum and minimum on S that must therefore occur at $(0, \sqrt{2}, 1)$ and $(0, -\sqrt{2}, 1)$, respectively. ▲

The method of Lagrange multipliers provides us with another tool to locate the absolute maxima and minima of differentiable functions on bounded regions in $\mathbb{R}^2$ (see the strategy for finding absolute maximum and minimum in Section 3.3).

EXAMPLE 6 Find the absolute maximum of $f(x, y) = xy$ on the unit disk D, where D is the set of points (x, y) with $x^2 + y^2 \leq 1$.

SOLUTION By Theorem 7 of Section 3.3, we know the absolute maximum exists. First, we find all the critical points of f in U, the set of points (x, y) with

$x^2 + y^2 < 1$. Because

$$\frac{\partial f}{\partial x} = y \quad \text{and} \quad \frac{\partial f}{\partial y} = x,$$

$(0, 0)$ is the only critical point of f in U. Now consider f on the unit circle, the level curve $g(x, y) = 1$, where $g(x, y) = x^2 + y^2$. To locate the maximum and minimum of f on C, we write down the Lagrange multiplier equations: $\nabla f(x, y) = (y, x) = \lambda \nabla g(x, y) = \lambda(2x, 2y)$ and $x^2 + y^2 = 1$. Rewriting these in component form, we get

$$\begin{aligned} y &= 2\lambda x, \\ x &= 2\lambda y, \\ x^2 + y^2 &= 1. \end{aligned}$$

Thus,

$$y = 4\lambda^2 y,$$

or $\lambda = \pm 1/2$ and $y = \pm x$, which means that $x^2 + x^2 = 2x^2 = 1$ or $x = \pm 1/\sqrt{2}$, $y = \pm 1/\sqrt{2}$. On C we compute four candidates for the absolute maximum and minimum, namely,

$$\left(-\frac{1}{\sqrt{2}}, -\frac{1}{\sqrt{2}}\right), \quad \left(-\frac{1}{\sqrt{2}}, \frac{1}{\sqrt{2}}\right), \quad \left(\frac{1}{\sqrt{2}}, \frac{1}{\sqrt{2}}\right), \quad \left(\frac{1}{\sqrt{2}}, -\frac{1}{\sqrt{2}}\right).$$

The value of f at both $(-1/\sqrt{2}, -1/\sqrt{2})$ and $(1/\sqrt{2}, 1/\sqrt{2})$ is $1/2$. The value of f at $(-1/\sqrt{2}, 1/\sqrt{2})$ and $(1/\sqrt{2}, -1/\sqrt{2})$ is $-1/2$, and the value of f at $(0, 0)$ is 0. Therefore, the absolute maximum of f is $1/2$ and the absolute minimum is $-1/2$, both occurring on C. At $(0, 0)$, $\partial^2 f/\partial x^2 = 0$, $\partial^2 f/\partial y^2 = 0$ and $\partial^2 f/\partial x\, \partial y = 1$, so the discriminant is -1 and thus $(0, 0)$ is a saddle point. ▲

EXAMPLE 7 Find the absolute maximum and minimum of $f(x, y) = \frac{1}{2}x^2 + \frac{1}{2}y^2$ in the elliptical region D defined by $\frac{1}{2}x^2 + y^2 \le 1$.

SOLUTION Again by Theorem 7, Section 3.3, the absolute maximum exists. We first locate the critical points of f in U, the set of points (x, y) with $\frac{1}{2}x^2 + y^2 < 1$. Because

$$\frac{\partial f}{\partial x} = x, \quad \frac{\partial f}{\partial y} = y,$$

the only critical point is the origin $(0, 0)$.

We now find the maximum and minimum of f on C, the boundary of U, which is the level curve $g(x, y) = 1$, where $g(x, y) = \frac{1}{2}x^2 + y^2$. The Lagrange multiplier equations are

$$\nabla f(x, y) = (x, y) = \lambda \nabla g(x, y) = \lambda(x, 2y)$$

and $(x^2/2) + y^2 = 1$. In other words,

$$x = \lambda x$$
$$y = 2\lambda y$$
$$\frac{x^2}{2} + y^2 = 1.$$

If $x = 0$, then $y = \pm 1$ and $\lambda = \frac{1}{2}$. If $y = 0$, then $x = \pm\sqrt{2}$ and $\lambda = 1$. If $x \neq 0$ and $y \neq 0$, we get both $\lambda = 1$ and $1/2$, which is impossible. Thus, the candidates for the maxima and minima of f on C are $(0, \pm 1)$, $(\pm\sqrt{2}, 0)$ and for f inside D, the candidate is $(0, 0)$. The value of f at $(0, \pm 1)$ is $1/2$, at $(\pm\sqrt{2}, 0)$ it is 1, and at $(0, 0)$ it is 0. Thus, the absolute minimum of f occurs at $(0, 0)$ and is 0. The absolute maximum of f on D is thus 1 and occurs at the points $(\pm\sqrt{2}, 0)$. ▲

Global Maxima and Minima

The method of Lagrange multipliers enhances our techniques for finding global maxima and minima. In this respect, the following is useful.

DEFINITION Let U be an open region in $\mathbb{R}^n$ with boundary ∂U. We say that ∂U is ***smooth*** if ∂U is the level set of a smooth function g whose gradient ∇g never vanishes (i.e., $\nabla g \neq \mathbf{0}$). Then we have the following strategy.

Lagrange Multiplier Strategy for Finding Absolute Maxima and Minima on Regions with Boundary Let f be a differentiable function on a closed and bounded region $D = U \cup \partial U$, U open in $\mathbb{R}^n$, with smooth boundary ∂U.

To find the absolute maximum and minimum of f on D:

(i) Locate all critical points of f in U.

(ii) Use the method of Lagrange multiplier to locate all the critical points of $f|\partial U$.

(iii) Compute the values of f at all these critical points.

(iv) Select the largest and the smallest.

EXAMPLE 8 Find the absolute maximum and minimum of the function $f(x, y, z) = x + y + z$ on the set $D = \{(x, y, z) \mid x^2 + y^2 + z^2 \leq 1\}$.

SOLUTION As in the previous examples, we know the absolute maximum and minimum exists. Now $D = U \cup \partial U$, where

$$U = \{(x, y, z) \mid x^2 + y^2 + z^2 < 1\}$$

and

$$\partial U = \{(x, y, z) \mid x^2 + y^2 + z^2 = 1\}.$$

The gradient of f is $\nabla f = (1, 1, 1)$, and so f has no critical points in U. Therefore, the maximum and minimum values of f must occur on ∂U.

Let $g(x, y, z) = x^2 + y^2 + z^2$. Then ∂U is the level set $g(x, y, z) = 1$. By the method of Lagrange multipliers, the maximum and minimum must occur at a critical point of $f|\partial U$, that is, at a point $\mathbf{x}_0$ where $\nabla f(\mathbf{x}_0) = \lambda \nabla g(\mathbf{x}_0)$ for some scalar λ.

Thus,

$$(1, 1, 1) = \lambda(2x, 2y, 2z); \qquad \text{that is,} \qquad x = \frac{\lambda}{2}, \qquad y = \frac{\lambda}{2}, \qquad z = \frac{\lambda}{2}.$$

Because $x^2 + y^2 + z^2 = 1$, we obtain $\lambda = \pm 2/\sqrt{3}$ and so $\mathbf{x}_0 = \pm(1/\sqrt{3}, 1/\sqrt{3}, 1/\sqrt{3})$. Clearly, $-(1/\sqrt{3}, 1/\sqrt{3}, 1/\sqrt{3})$ is the point where f assumes its absolute minimum (namely, $-\sqrt{3}$) and $(1/\sqrt{3}, 1/\sqrt{3}, 1\sqrt{3})$, the point where f assumes its maximum value $\sqrt{3}$. ▲

Two Additional Applications

We now present two further applications of the mathematical techniques developed in this section to geometry and to economics. We shall begin wth a geometric example.

EXAMPLE 9 Suppose we have a curve defined by the equation

$$\phi(x, y) = Ax^2 + 2Bxy + Cy^2 - 1 = 0.$$

Find the maximum and minimum distance of the curve to the origin. (These are the lengths of the ***semimajor*** and the ***semiminor*** axis of this quadric.)

SOLUTION The problem is equivalent to finding the extreme values of $f(x, y) = x^2 + y^2$ subject to the constraining condition $\phi(x, y) = 0$. Using the Lagrange multiplier method, we have the following equations:

$$2x + \lambda(2Ax + 2By) = 0 \tag{6}$$

$$2y + \lambda(2Bx + 2Cy) = 0 \tag{7}$$

$$Ax^2 + 2Bxy + Cy^2 = 1. \tag{8}$$

Adding x times equation (6) to y times equation (7), we obtain

$$2(x^2 + y^2) + 2\lambda(Ax^2 + 2Bxy + Cy^2) = 0.$$

By equation (8), it follows that $x^2 + y^2 + \lambda = 0$. Let $t = -1/\lambda = 1/(x^2 + y^2)$ [the case $\lambda = 0$ is impossible, because (0, 0) is not on the curve $\phi(x, y) = 0$]. Then equations (6) and (7) can be written as follows:

$$\begin{aligned} 2(A - t)x + 2By &= 0 \\ 2Bx + 2(C - t)y &= 0. \end{aligned} \tag{9}$$

If these two equations are to have a nontrivial solution [remember that $(x, y) = (0, 0)$ is not on our curve and so is not a solution], it follows from a theorem of linear algebra that their determinant vanishes:[13]

$$\begin{vmatrix} A - t & B \\ B & C - t \end{vmatrix} = 0.$$

Because this equation is quadratic in t, there are two solutions, which we shall call t_1 and t_2. Because $-\lambda = x^2 + y^2$, we have $\sqrt{x^2 + y^2} = \sqrt{-\lambda}$. Now $\sqrt{x^2 + y^2}$ is the distance from the point (x, y) to the origin. Therefore, if (x_1, y_1) and (x_2, y_2) denote the nontrivial solutions to equation (9) corresponding to t_1 and t_2, and if t_1 and t_2 are positive, we get $\sqrt{x_2^2 + y_2^2} = 1/\sqrt{t_2}$ and $\sqrt{x_1^2 + y_1^2} = 1/\sqrt{t_1}$. Consequently, if $t_1 > t_2$, the lengths of the semiminor and semimajor axes are $1/\sqrt{t_1}$ and $1/\sqrt{t_2}$, respectively. If the curve is an ellipse, both t_1 and t_2 are, in fact, real and positive. What happens with a hyperbola or a parabola? ▲

Finally, we discuss an application to economics.

EXAMPLE 10 Suppose that the output of a manufacturing firm is a quantity Q of a certain product, where Q is a function $f(K, L)$, where K is the amount of capital equipment (or investment) and L is the amount of labor used. If the price of labor is p, the price of capital is q, and the firm can spend no more than B dollars, how can we find the amount of capital and labor to maximize the output Q?

SOLUTION We would expect that if the amount of capital or labor is increased, then the output Q should also increase; that is,

$$\frac{\partial Q}{\partial K} \geq 0 \quad \text{and} \quad \frac{\partial Q}{\partial L} \geq 0.$$

[13] The matrix of coefficients of the equations cannot have an inverse, because this would imply that the solution is zero. Recall that a matrix that does not have an inverse has determinant zero.

We also expect that as more labor is added to a given amount of capital equipment, we get less additional output for our effort; that is,

$$\frac{\partial^2 Q}{\partial L^2} < 0.$$

Similarly,

$$\frac{\partial^2 Q}{\partial K^2} < 0.$$

With these assumptions on Q, it is reasonable to expect the level curves of output (called ***isoquants***) $Q(K, L) = c$ to look something like the curves sketched in Figure 3.4.5, with $c_1 < c_2 < c_3$.

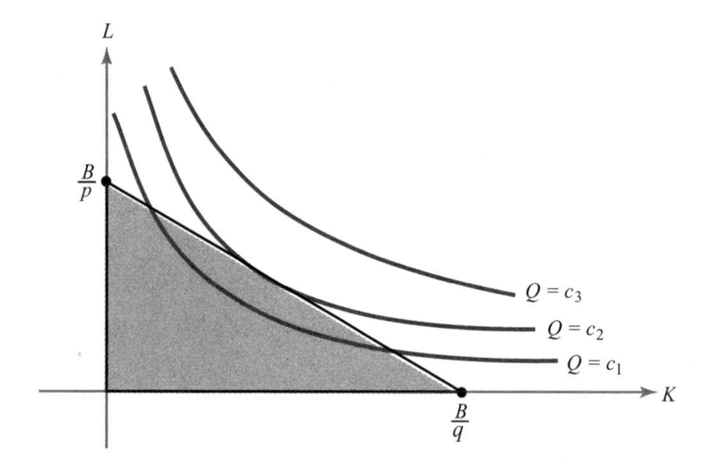

Figure 3.4.5 What is the largest value of Q in the shaded triangle?

We can interpret the convexity of the isoquants as follows: As one moves to the right along a given isoquant, it takes more and more capital to replace a unit of labor and still produce the same output. The budget constraint means that we must stay inside the triangle bounded by the axes and the line $pL + qK = B$. Geometrically, it is clear that we produce the most by spending all our money in such a way as to pick the isoquant that just touches, but does not cross, the budget line.

Because the maximum point lies on the boundary of our domain, we apply the method of Lagrange multipliers to find the maximum. To maximize $Q = f(K, L)$ subject to the constraint $pL + qK = B$, we look for critical points of the auxiliary function,

$$h(K, L, \lambda) = f(K, L) - \lambda(pL + qK - B).$$

Thus, we want

$$\frac{\partial Q}{\partial K} = \lambda q, \qquad \frac{\partial Q}{\partial L} = \lambda p, \qquad \text{and} \qquad pL + qK = B.$$

These are the conditions we must meet in order to maximize output. (The reader is asked to work out a specific case in Exercise 31.) ▲

In the preceding example, λ represents something interesting. Let $k = qK$ and $l = pL$, so that k is the dollar value of the capital used and l is the dollar value of the labor used. Then the first two equations become

$$\frac{\partial Q}{\partial k} = \frac{1}{q}\frac{\partial Q}{\partial K} = \lambda = \frac{1}{p}\frac{\partial Q}{\partial L} = \frac{\partial Q}{\partial l}.$$

Thus, at the optimum production point the marginal change in output per dollar's worth of additional capital investment is equal to the marginal change of output per dollar's worth of additional labor, and λ is this common value. At the optimum point, the exchange of a dollar's worth of capital for a dollar's worth of labor does not change the output. Away from the optimum point the marginal outputs are different, and one exchange or the other will increase the output.

A Second Derivative Test for Constrained Extrema

In Section 3.3, we developed a second derivative test for extrema of functions of several variables by looking at the second-degree term in the Taylor series of f. If the Hessian matrix of second partial derivatives is either positive-definite or negative-definite at a critical point of f, this point is a relative minimum or maximum, respectively.

The question naturally arises as to whether there is a second derivative test for maximum and minimum problems *in the presence of constraints*. The answer is yes and the test involves a matrix called a *bordered Hessian*. We will first discuss the test and how to apply it for the case of a function $f(x, y)$ of two variables subject to the constraint $g(x, y) = c$.

THEOREM 10 Let $f\colon U \subset \mathbb{R}^2 \to \mathbb{R}$ and $g\colon U \subset \mathbb{R}^2 \to \mathbb{R}$ be smooth (at least C^2) functions. Let $\mathbf{v}_0 \in U$, $g(\mathbf{v}_0) = c$, and S be the level curve for g with value c. Assume that $\nabla g(\mathbf{v}_0) \neq \mathbf{0}$ and that there is a real number λ such that $\nabla f(\mathbf{v}_0) = \lambda \nabla g(\mathbf{v}_0)$. Form the auxiliary function $h = f - \lambda g$ and the ***bordered Hessian*** determinant

$$|\bar{H}| = \begin{vmatrix} 0 & -\dfrac{\partial g}{\partial x} & -\dfrac{\partial g}{\partial y} \\ -\dfrac{\partial g}{\partial x} & \dfrac{\partial^2 h}{\partial x^2} & \dfrac{\partial^2 h}{\partial x\,\partial y} \\ -\dfrac{\partial g}{\partial y} & \dfrac{\partial^2 h}{\partial x\,\partial y} & \dfrac{\partial^2 h}{\partial y^2} \end{vmatrix} \quad \text{evaluated at } \mathbf{v}_0.$$

(i) If $|\bar{H}| > 0$, then $\mathbf{v}_0$ is a local maximum point for $f|S$.

(ii) If $|\bar{H}| < 0$, then $\mathbf{v}_0$ is a local minimum point for $f|S$.

(iii) If $|\bar{H}| = 0$, the test is inconclusive and $\mathbf{v}_0$ may be a minimum, a maximum, or neither.

This theorem is proved in the Internet supplement for this section.

EXAMPLE 11 Find extreme points of $f(x, y) = (x - y)^n$ subject to the constraint $x^2 + y^2 = 1$, where $n \geq 1$.

SOLUTION We set the first derivatives of the auxiliary function h defined by $h(x, y, \lambda) = (x - y)^n - \lambda(x^2 + y^2 - 1)$ equal to 0:

$$\begin{aligned} n(x-y)^{n-1} - 2\lambda x &= 0 \\ -n(x-y)^{n-1} - 2\lambda y &= 0 \\ -(x^2 + y^2 - 1) &= 0. \end{aligned}$$

From the first two equations we see that $\lambda(x + y) = 0$. If $\lambda = 0$, then $x = y = \pm\sqrt{2}/2$. If $\lambda \neq 0$, then $x = -y$. The four critical points are represented in Figure 3.4.6 and the corresponding values of $f(x, y)$ are listed below:

(A) $x = \sqrt{2}/2$ $\quad y = \sqrt{2}/2$ $\quad \lambda = 0$ $\quad f(x, y) = 0$

(B) $x = \sqrt{2}/2$ $\quad y = -\sqrt{2}/2$ $\quad \lambda = n(\sqrt{2})^{n-2}$ $\quad f(x, y) = (\sqrt{2})^n$

(C) $x = -\sqrt{2}/2$ $\quad y = -\sqrt{2}/2$ $\quad \lambda = 0$ $\quad f(x, y) = 0$

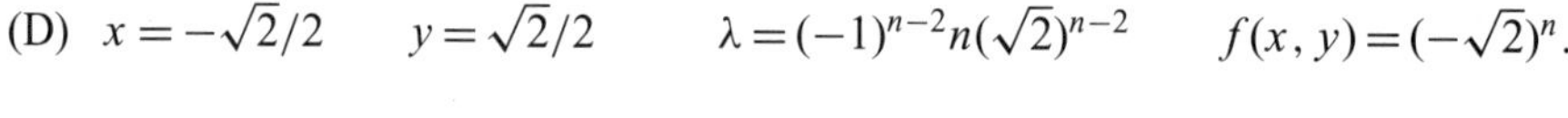

(D) $x = -\sqrt{2}/2$ $\quad y = \sqrt{2}/2$ $\quad \lambda = (-1)^{n-2} n(\sqrt{2})^{n-2}$ $\quad f(x, y) = (-\sqrt{2})^n$.

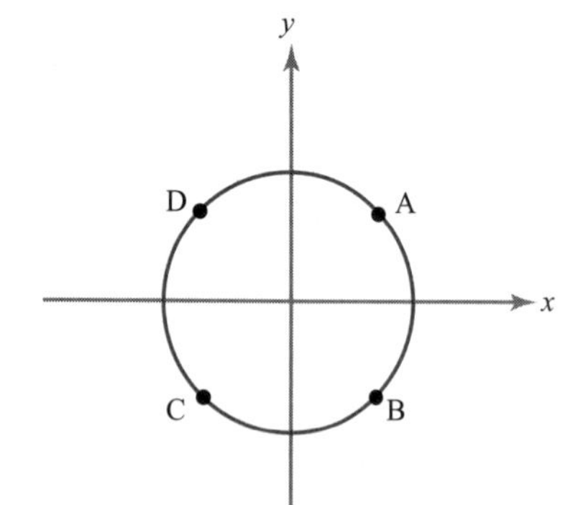

Figure 3.4.6 The four critical points in Example 11.

By inspection, we see that if n is even, then A and C are minimum points and B and D are maxima. If n is odd, then B is a maximum point, D is a minimum, and A and C are neither. Let us see whether Theorem 10 is consistent with these observations.

The bordered Hessian determinant is

$$|\bar{H}| = \begin{vmatrix} 0 & -2x & -2y \\ -2x & n(n-1)(x-y)^{n-2} - 2\lambda & -n(n-1)(x-y)^{n-2} \\ -2y & -n(n-1)(x-y)^{n-2} & n(n-1)(x-y)^{n-2} - 2\lambda \end{vmatrix}$$

$$= -4n(n-1)(x-y)^{n-2}(x+y)^2 + 8\lambda(x^2 - y^2).$$

If $n = 1$ or if $n \geq 3$, $|\bar{H}| = 0$ at A, B, C, and D. If $n = 2$, then $|\bar{H}| = 0$ at B and D and -16 at A and C. Thus, the second-derivative test picks up the minima at A and C, but is inconclusive in testing the maxima at B and D for $n = 2$. It is also inconclusive for all other values of n. ▲

Just as in the unconstrained case, there is also a second-derivative test for functions of more than two variables. If we are to find extreme points for $f(x_1, \dots, x_n)$ subject to a single constraint $g(x_1, \dots, x_n) = c$, we first form the bordered Hessian for the auxiliary function $h(x_1, \dots, x_n) = f(x_1, \dots, x_n) - \lambda(g(x_1, \dots, x_n) - c)$ as follows:

$$\begin{vmatrix} 0 & \dfrac{-\partial g}{\partial x_1} & \dfrac{-\partial g}{\partial x_2} & \cdots & \dfrac{-\partial g}{\partial x_n} \\ \dfrac{-\partial g}{\partial x_1} & \dfrac{\partial^2 h}{\partial x_1^2} & \dfrac{\partial^2 h}{\partial x_1\,\partial x_2} & \cdots & \dfrac{\partial^2 h}{\partial x_1\,\partial x_n} \\ \dfrac{-\partial g}{\partial x_2} & \dfrac{\partial^2 h}{\partial x_1\,\partial x_2} & \dfrac{\partial^2 h}{\partial x_2^2} & \cdots & \dfrac{\partial^2 h}{\partial x_2\,\partial x_n} \\ \vdots & \vdots & \vdots & & \vdots \\ \dfrac{-\partial g}{\partial x_n} & \dfrac{\partial^2 h}{\partial x_1\,\partial x_n} & \dfrac{\partial^2 h}{\partial x_2\,\partial x_n} & \cdots & \dfrac{\partial^2 h}{\partial x_n^2} \end{vmatrix}.$$

Second, we examine the determinants of the diagonal submatrices of order ≥ 3 at the critical points of h. If they are all negative, that is, if

$$\begin{vmatrix} 0 & -\dfrac{\partial g}{\partial x_1} & -\dfrac{\partial g}{\partial x_2} \\ -\dfrac{\partial g}{\partial x_1} & \dfrac{\partial^2 h}{\partial x_1^2} & \dfrac{\partial^2 h}{\partial x_1\,\partial x_2} \\ -\dfrac{\partial g}{\partial x_2} & \dfrac{\partial^2 h}{\partial x_1\,\partial x_2} & \dfrac{\partial^2 h}{\partial x_2^2} \end{vmatrix} < 0, \quad \begin{vmatrix} 0 & -\dfrac{\partial g}{\partial x_1} & -\dfrac{\partial g}{\partial x_2} & -\dfrac{\partial g}{\partial x_3} \\ -\dfrac{\partial g}{\partial x_1} & \dfrac{\partial^2 h}{\partial x_1^2} & \dfrac{\partial^2 h}{\partial x_1\,\partial x_2} & \dfrac{\partial^2 h}{\partial x_1\,\partial x_3} \\ -\dfrac{\partial g}{\partial x_2} & \dfrac{\partial^2 h}{\partial x_1\,\partial x_2} & \dfrac{\partial^2 h}{\partial x_2^2} & \dfrac{\partial^2 h}{\partial x_2\,\partial x_3} \\ -\dfrac{\partial g}{\partial x_3} & \dfrac{\partial^2 h}{\partial x_1\,\partial x_3} & \dfrac{\partial^2 h}{\partial x_2\,\partial x_3} & \dfrac{\partial^2 h}{\partial x_3^2} \end{vmatrix} < 0, \dots,$$

then we are at a local minimum of $f|S$. If they start out with a positive 3×3 subdeterminant and alternate in sign (that is, >0, <0, >0, $<0, \dots$), then we are at a local maximum. If they are all nonzero and do not fit one of these patterns, then the point is neither a maximum nor a minimum (it is said to be of the saddle type).[14]

EXAMPLE 12 Study the local extreme points of $f(x, y, z) = xyz$ on the surface of the unit sphere $x^2 + y^2 + z^2 = 1$ using the second-derivative test.

SOLUTION Setting the partial derivatives of the auxiliary function $h(x, y, z, \lambda) = xyz - \lambda(x^2 + y^2 + z^2 - 1)$ equal to zero gives

$$\begin{aligned} yz &= 2\lambda x \\ xz &= 2\lambda y \\ xy &= 2\lambda z \\ x^2 + y^2 + z^2 &= 1. \end{aligned}$$

Thus, $3xyz = 2\lambda(x^2 + y^2 + z^2) = 2\lambda$. If $\lambda = 0$, the solutions are $(x, y, z, \lambda) = (\pm 1, 0, 0, 0)$, $(0, \pm 1, 0, 0)$, and $(0, 0, \pm 1, 0)$. If $\lambda \neq 0$, then we have $2\lambda = 3xyz = 6\lambda z^2$ and so $z^2 = \frac{1}{3}$. Similarly, $x^2 = y^2 = \frac{1}{3}$. Thus, the solutions are given by $\lambda = \frac{3}{2}xyz = \pm\sqrt{3}/6$. The critical points of h and the corresponding values of f are given in Table 3.1. From it, we see that points E, F, G, and K are minima. Points D, H, I, and J are maxima. To see whether this is in accord with the second-derivative

Table 3.1 The critical points A, B, ..., J, K of h and corresponding values of f

	x	y	z	λ	$f(x, y, z)$
±A	± 1	0	0	0	0
±B	0	± 1	0	0	0
±C	0	0	± 1	0	0
D	$\sqrt{3}/3$	$\sqrt{3}/3$	$\sqrt{3}/3$	$\sqrt{3}/6$	$\sqrt{3}/9$
E	$-\sqrt{3}/3$	$\sqrt{3}/3$	$\sqrt{3}/3$	$-\sqrt{3}/6$	$-\sqrt{3}/9$
F	$\sqrt{3}/3$	$-\sqrt{3}/3$	$\sqrt{3}/3$	$-\sqrt{3}/6$	$-\sqrt{3}/9$
G	$\sqrt{3}/3$	$\sqrt{3}/3$	$-\sqrt{3}/3$	$-\sqrt{3}/6$	$-\sqrt{3}/9$
H	$\sqrt{3}/3$	$-\sqrt{3}/3$	$-\sqrt{3}/3$	$\sqrt{3}/6$	$\sqrt{3}/9$
I	$-\sqrt{3}/3$	$\sqrt{3}/3$	$-\sqrt{3}/3$	$\sqrt{3}/6$	$\sqrt{3}/9$
J	$-\sqrt{3}/3$	$-\sqrt{3}/3$	$\sqrt{3}/3$	$\sqrt{3}/6$	$\sqrt{3}/9$
K	$-\sqrt{3}/3$	$-\sqrt{3}/3$	$-\sqrt{3}/3$	$-\sqrt{3}/6$	$-\sqrt{3}/9$

[14] For a detailed discussion, see C. Caratheodory, *Calculus of Variations and Partial Differential Equations*, Holden-Day, San Francisco, 1965; Y. Murata, *Mathematics for Stability and Optimization of Economic Systems*, Academic Press, New York, 1977, pp. 263–271; or D. Spring, *Am. Math. Mon.* 92 (1985): 631–643.

test, we need to consider two determinants. First, we look at the following:

$$|\bar{H}_2| = \begin{vmatrix} 0 & -\partial g/\partial x & -\partial g/\partial y \\ -\partial g/\partial x & \partial^2 h/\partial x^2 & \partial^2/\partial x\,\partial y \\ -\partial g/\partial y & \partial^2 h/\partial x\,\partial y & \partial^2 h/\partial y^2 \end{vmatrix} = \begin{vmatrix} 0 & -2x & -2y \\ -2x & -2\lambda & z \\ -2y & z & -2\lambda \end{vmatrix}$$
$$= 8\lambda x^2 + 8\lambda y^2 + 8xyz = 8\lambda(x^2 + y^2 + 2z^2).$$

Observe that sign $(|\bar{H}_2|) =$ sign $\lambda =$ sign (xyz), where the sign of a number is 1 if that number is positive, or is -1 if that number is negative. Second, we consider

$$|\bar{H}_3| = \begin{vmatrix} 0 & -\partial g/\partial x & -\partial g/\partial y & -\partial g/\partial z \\ -\partial g/\partial x & \partial^2 h/\partial x^2 & \partial^2 h/\partial x\,\partial y & \partial^2 h/\partial x\,\partial z \\ -\partial g/\partial y & \partial^2 h/\partial x\,\partial y & \partial^2 h/\partial y^2 & \partial^2 h/\partial y\,\partial z \\ -\partial g/\partial z & \partial^2 h/\partial x\,\partial z & \partial^2 h/\partial y\,\partial z & \partial^2 h/\partial z^2 \end{vmatrix}$$
$$= \begin{vmatrix} 0 & -2x & -2y & -2z \\ -2x & -2\lambda & z & y \\ -2y & z & -2\lambda & x \\ -2z & y & x & -2\lambda \end{vmatrix},$$

which works out to be $+4$ at points $\pm$A, $\pm$B, and $\pm$C and $-\frac{16}{3}$ at the other eight points. At E, F, G, and K, we have $|\bar{H}_2| < 0$ *and* $|\bar{H}_3| < 0$, and so the test indicates these are local minima. At D, H, I, and J we have $|\bar{H}_2| > 0$ and $|\bar{H}_3| < 0$, and so the test says these are local maxima. Finally, the second-derivative test shows that $\pm$A, $\pm$B, and $\pm$C are saddle points. ▲

EXERCISES

In Exercises 1 to 5 find the extrema of f subject to the stated constraints.

1. $f(x, y, z) = x - y + z$, subject to $x^2 + y^2 + z^2 = 2$

2. $f(x, y) = x - y$, subject to $x^2 - y^2 = 2$

3. $f(x, y) = x$, subject to $x^2 + 2y^2 = 3$

4. $f(x, y, z) = x + y + z$, subject to $x^2 - y^2 = 1, 2x + z = 1$

5. $f(x, y) = 3x + 2y$, subject to $2x^2 + 3y^2 = 3$

Find the relative extrema of $f|S$ in Exercises 6 to 9.

6. $f\colon \mathbb{R}^2 \to \mathbb{R}, (x, y) \mapsto x^2 + y^2, S = \{(x, 2) \mid x \in \mathbb{R}\}$

7. $f\colon \mathbb{R}^2 \to \mathbb{R}, (x, y) \mapsto x^2 + y^2, S = \{(x, y) \mid y \geq 2\}$

8. $f\colon \mathbb{R}^2 \to \mathbb{R}, (x, y) \mapsto x^2 - y^2, S = \{(x, \cos x) \mid x \in \mathbb{R}\}$

9. $f\colon \mathbb{R}^3 \to \mathbb{R}, (x, y, z) \mapsto x^2 + y^2 + z^2, S = \{(x, y, z) \mid z \geq 2 + x^2 + y^2\}$

10. Use the method of Lagrange multipliers to find the absolute maximum and minimum values of $f(x, y) = x^2 + y^2 - x - y + 1$ on the unit disk (see Example 10 of Section 3.3).

11. Consider the function $f(x, y) = x^2 + xy + y^2$ defined on the unit disk, namely, $D = \{(x, y) \mid x^2 + y^2 \leq 1\}$. Use the method of Lagrange multipliers to locate the maximum and minimum points for f on the unit circle. Use this to determine the absolute maximum and minimum values for f on D.

12. A rectangular box with no top is to have a surface area of 16 m^2. Find the dimensions that maximize its volume.

13. Design a cylindrical can (with a lid) to contain 1 liter ($= 1000$ cm^3) of water, using the minimum amount of metal.

14. Show that solutions of equations (4) and (5) are in one-to-one correspondence with the critical points of

$$\begin{aligned} h(x_1, \ldots, x_n, \lambda_1, \ldots, \lambda_k) = f(x_1, \ldots, x_n) &- \lambda_1[g_1(x_1, \ldots, x_n) - c_1] \\ &- \cdots - \lambda_k[g_k(x_1, \ldots, x_n) - c_k]. \end{aligned}$$

15. Find the absolute maximum and minimum for the function $f(x, y, z) = x + y - z$ on the ball $B = \{(x, y, z) \mid x^2 + y^2 + z^2 \leq 1\}$.

16. Repeat Exercise 15 for $f(x, y, z) = x + yz$.

17. A rectangular mirror with area A square feet is to have trim along the edges. If the trim along the horizontal edges costs p cents per foot and that for the vertical edges costs q cents per foot, find the dimensions that will minimize the total cost.

18. An irrigation canal in Arizona has concrete sides and bottom with trapezoidal cross section of area $A = y(x + y \tan\theta)$ and wetted perimeter $P = x + 2y/\cos\theta$, where $x =$ bottom width, $y =$ water depth, $\theta =$ side inclination, measured from vertical. The best design for a fixed inclination θ is found by solving $P =$ minimum subject to the condition $A =$ constant. Show that $y^2 = (A\cos\theta)/(2 - \sin\theta)$.

19. Apply the second-derivative test to study the nature of the extrema in Exercises 1 and 5.

20. A light ray travels from point A to point B crossing a boundary between two media (see Figure 3.4.7). In the first medium its speed is v_1, and in the second it is v_2. Show that the trip is made in minimum time when *Snell's law* holds:

$$\frac{\sin\theta_1}{\sin\theta_2} = \frac{v_1}{v_2}.$$

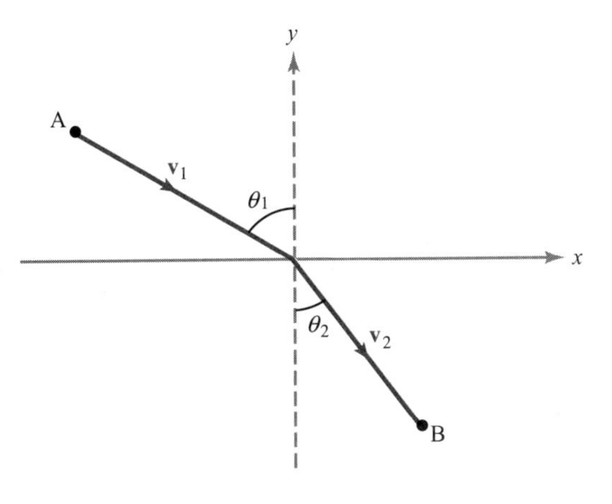

Figure 3.4.7 Snell's law of refraction.

21. A parcel delivery service requires that the dimensions of a rectangular box be such that the length plus twice the width plus twice the height be no more than 108 inches ($l + 2w + 2h \leq 108$). What is the volume of the largest-volume box the company will deliver?

22. Let P be a point on a surface S in $\mathbb{R}^3$ defined by the equation $f(x, y, z) = 1$, where f is of class C^1. Suppose that P is a point where the distance from the origin to S is maximized. Show that the vector emanating from the origin and ending at P is perpendicular to S.

23. Let A be a nonzero symmetric 3×3 matrix. Thus, its entries satisfy $a_{ij} = a_{ji}$. Consider the function $f(\mathbf{x}) = \frac{1}{2}(A\mathbf{x}) \cdot \mathbf{x}$.

(a) What is ∇f?

(b) Consider the restriction of f to the unit sphere $S = \{(x, y, z) \mid x^2 + y^2 + z^2 = 1\}$ in $\mathbb{R}^3$. By Theorem 7 we know that f must have a maximum and a minimum on S. Show that there must be an $\mathbf{x} \in S$ and a $\lambda \neq 0$ such that $A\mathbf{x} = \lambda\mathbf{x}$. (The vector $\mathbf{x}$ is called an ***eigenvector***, while the scalar λ is called an ***eigenvalue***.)

(c) What are the maxima and minima for f on $B = \{(x, y, z) \mid x^2 + y^2 + z^2 \leq 1\}$?

24. Suppose that A in the function f defined in Exercise 23 is not necessarily symmetric.

(a) What is ∇f?

(b) Can one conclude the existence of an eigenvector and eigenvalues as in Exercise 23?

25. (a) Find the critical points of $x + y^2$ subject to the constraint $2x^2 + y^2 = 1$.

(b) Use the bordered Hessian to classify the critical points.

26. Answer the question posed in the last line of Example 9.

27. Try to find the extrema of $xy + yz$ among points satisfying $xz = 1$.

28. A company's production function is $Q(x, y) = xy$. The cost of production is $C(x, y) = 2x + 3y$. If this company can spend $C(x, y) = 10$, what is the maximum quantity that can be produced?

29. Find the point on the curve $(\cos t, \sin t, \sin(t/2))$ that is farthest from the origin.

30. A firm uses wool and cotton fiber to produce cloth. The amount of cloth produced is given by $Q(x, y) = xy - x - y + 1$, where x is the number of pounds of wool, y the number of pounds of cotton, $x > 1$, and $y > 1$. If wool costs p dollars per pound, and cotton q dollars per pound and the firm can spend B dollars on material, what should the ratio of cotton and wool be to produce the most cloth?

31. Carry out the analysis of Example 10 for the production function $Q(K, L) = AK^{\alpha}L^{1-\alpha}$, where A and α are positive constants and $0 < \alpha < 1$. This is called a ***Cobb–Douglas production function*** and is sometimes used as a simple model for the national economy. Q is then the aggregate output of the economy for a given input of capital and labor.

3.5 The Implicit Function Theorem

In this section, we state two versions of the *implicit function theorem*, arguably the most important theorem in all of mathematical analysis. The entire theoretical basis of the idea of a surface as well as the method of Lagrange multipliers depends on it. Moreover, it is a cornerstone of several fields of mathematics, such as differential topology and geometry.

The One-Variable Implicit Function Theorem

In one-variable calculus, we learn the importance of the inversion process. For example, $x = \ln y$ is the inverse of $y = e^x$, and $x = \sin^{-1} y$ is the inverse of $y = \sin x$. The inversion process is also important for functions of several variables; for example, the switch between Cartesian and polar coordinates in the plane involves inverting two functions of two variables.

Recall from one-variable calculus that if $y = f(x)$ is a C^1 function and $f'(x_0) \neq 0$, then locally near x_0 we can solve for x to give the inverse function: $x = f^{-1}(y)$. We learn that $(f^{-1})'(y) = 1/f'(x)$; that is, $dx/dy = 1/(dy/dx)$. That $y = f(x)$ can be inverted is plausible because $f'(x_0) \neq 0$ means that the slope of $y = f(x)$ is nonzero, so that the graph is rising or falling near x_0. Thus, if we reflect the graph across the line $y = x$, it is still a graph *near* (x_0, y_0) where $y_0 = f(x_0)$. For example, in Figure 3.5.1, we can invert $y = f(x)$ in the shaded box, so in this range, $x = f^{-1}(y)$ is defined.

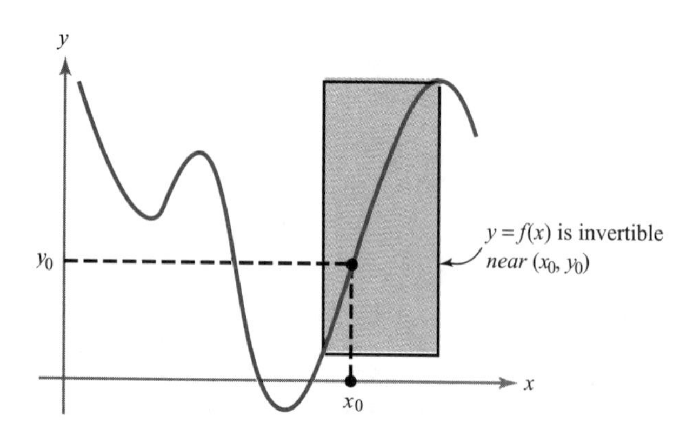

Figure 3.5.1 If $f'(x_0) \neq 0$, then $y = f(x)$ is locally invertible.

A Special Result

We next turn to the situation for real-valued functions of variables $x_1, \dots, x_n$ and z.

THEOREM 11: Special Implicit Function Theorem Suppose that $F\colon \mathbb{R}^{n+1} \to \mathbb{R}$ has continuous partial derivatives. Denoting points in $\mathbb{R}^{n+1}$ by $(\mathbf{x}, z)$, where $\mathbf{x} \in \mathbb{R}^n$ and $z \in \mathbb{R}$, assume that $(\mathbf{x}_0, z_0)$ satisfies

$$F(\mathbf{x}_0, z_0) = 0 \qquad \text{and} \qquad \frac{\partial F}{\partial z}(\mathbf{x}_0, z_0) \neq 0.$$

Then there is a ball U containing $\mathbf{x}_0$ in $\mathbb{R}^n$ and a neighborhood V of z_0 in $\mathbb{R}$ such that there is a unique function $z = g(\mathbf{x})$ defined for $\mathbf{x}$ in U and z in V that satisfies

$$F(\mathbf{x}, g(\mathbf{x})) = 0.$$

Moreover, if $\mathbf{x}$ in U and z in V satisfy $F(\mathbf{x}, z) = 0$, then $z = g(\mathbf{x})$. Finally, $z = g(\mathbf{x})$ is continuously differentiable, with the derivative given by

$$\mathbf{D}g(\mathbf{x}) = -\frac{1}{\dfrac{\partial F}{\partial z}(\mathbf{x}, z)} \mathbf{D}_{\mathbf{x}} F(\mathbf{x}, z)\Bigg|_{z=g(\mathbf{x})},$$

where $\mathbf{D}_{\mathbf{x}} F$ denotes the (partial) derivative of F with respect to the variable $\mathbf{x}$, that is, we have $\mathbf{D}_{\mathbf{x}} F = [\partial F/\partial x_1, \dots, \partial F/\partial x_n]$; in other words,

$$\frac{\partial g}{\partial x_i} = -\frac{\partial F/\partial x_i}{\partial F/\partial z}, \qquad i = 1, \dots, n. \tag{1}$$

A proof of this theorem is given in the Internet supplement.

Once it is known that $z = g(\mathbf{x})$ exists and is differentiable, formula (1) may be checked by implicit differentiation; to see this, note that the chain rule applied to $F(\mathbf{x}, g(\mathbf{x})) = 0$ gives

$$\mathbf{D}_{\mathbf{x}} F(\mathbf{x}, g(\mathbf{x})) + \left[\frac{\partial F}{\partial z}(\mathbf{x}, g(\mathbf{x}))\right][\mathbf{D}g(\mathbf{x})] = 0,$$

which is equivalent to formula (1).

EXAMPLE 1 In the special implicit function theorem, it is important to recognize the necessity of taking sufficiently small neighborhoods U and V. For example, consider the equation

$$x^2 + z^2 - 1 = 0,$$

that is, $F(x, z) = x^2 + z^2 - 1$, with $n = 1$. Here $(\partial F/\partial z)(x, z) = 2z$, and so the special implicit function theorem applies to a point (x_0, z_0) satisfying $x_0^2 + z_0^2 - 1 = 0$ and $z_0 \neq 0$. Thus, near such points, z is a unique function of x. This function is

$z = -\sqrt{1-x^2}$ if $z_0 > 0$ and $z = -\sqrt{1-x^2}$ if $z_0 < 0$. Note that z is defined for $|x| < 1$ only (U must not be too big) and z is unique only if it is near z_0 (V must not be too big). These facts and the nonexistence of $\partial z/\partial x$ at $z_0 = 0$ are, of course, clear from the fact that $x^2 + z^2 = 1$ defines a circle in the xz plane (Figure 3.5.2). ▲

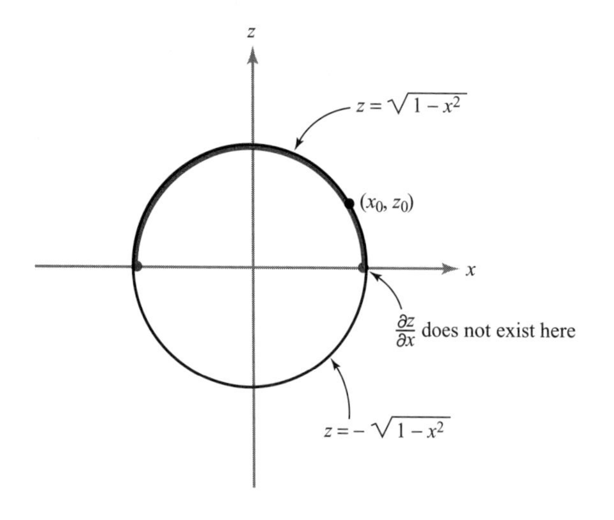

Figure 3.5.2 It is necessary to take small neighborhoods in the implicit function theorem.

The Implicit Function Theorem and Surfaces

Let us apply Theorem 11 to the study of surfaces. We are concerned with the level set of a function $g\colon U \subset \mathbb{R}^n \to \mathbb{R}$, that is, with the surface S consisting of the set of $\mathbf{x}$ satisfying $g(\mathbf{x}) = c_0$, where $c_0 = g(\mathbf{x}_0)$ and where $\mathbf{x}_0$ is given. Let us take $n = 3$ for concreteness. Thus, we are dealing with the level surface of a function $g(x, y, z)$ through a given point (x_0, y_0, z_0). As in the Lagrange multiplier theorem, assume that $\nabla g(x_0, y_0, z_0) \neq \mathbf{0}$. This means that at least one of the partial derivatives of g is nonzero. For definiteness, suppose that $(\partial g/\partial z)(x_0, y_0, z_0) \neq 0$. By applying Theorem 11 to the function $(x, y, z) \mapsto g(x, y, z) - c_0$, we know there is a unique function $z = k(x, y)$ satisfying $g(x, y, k(x, y)) = c_0$ for (x, y) near (x_0, y_0) and z near z_0. Thus, near z_0 the surface S is the graph of the function k. Because k is continuously differentiable, this surface has a tangent plane at (x_0, y_0, z_0) given by

$$z = z_0 + \left[\frac{\partial k}{\partial x}(x_0, y_0)\right](x - x_0) + \left[\frac{\partial k}{\partial y}(x_0, y_0)\right](y - y_0). \tag{2}$$

But by formula (1),

$$\frac{\partial k}{\partial x}(x_0, y_0) = -\frac{\dfrac{\partial g}{\partial x}(x_0, y_0, z_0)}{\dfrac{\partial g}{\partial z}(x_0, y_0, z_0)} \quad \text{and} \quad \frac{\partial k}{\partial y}(x_0, y_0) = -\frac{\dfrac{\partial g}{\partial y}(x_0, y_0, z_0)}{\dfrac{\partial g}{\partial z}(x_0, y_0, z_0)}.$$

Substituting these two equations into the equation for the tangent plane gives this equivalent description:

$$0 = (z - z_0)\frac{\partial g}{\partial z}(x_0, y_0, z_0) + (x - x_0)\frac{\partial g}{\partial x}(x_0, y_0, z_0) + (y - y_0)\frac{\partial g}{\partial y}(x_0, y_0, z_0);$$

that is,

$$(x - x_0, y - y_0, z - z_0) \cdot \nabla g(x_0, y_0, z_0) = 0.$$

Thus, the tangent plane to the level surface of g is the orthogonal complement to $\nabla g(x_0, y_0, z_0)$ through the point (x_0, y_0, z_0). This agrees with our characterization of tangent planes to level sets from Chapter 2.

We are now ready to complete the proof of the Lagrange multiplier theorem. To do this, we must show that every vector tangent to S at (x_0, y_0, z_0) is tangent to a curve in S. By Theorem 11, we need only show this for a graph of the form $z = k(x, y)$. However, if $\mathbf{v} = (x - x_0, y - y_0, z - z_0)$ is tangent to the graph [that is, if it satisfies equation (2)], then $\mathbf{v}$ is tangent to the path in S given by

$$\mathbf{c}(t) = (x_0 + t(x - x_0), y_0 + t(y - y_0), k(x_0 + t(x - x_0), y_0 + t(y - y_0)))$$

at $t = 0$. This can be checked by using the chain rule. (See Figure 3.5.3.)

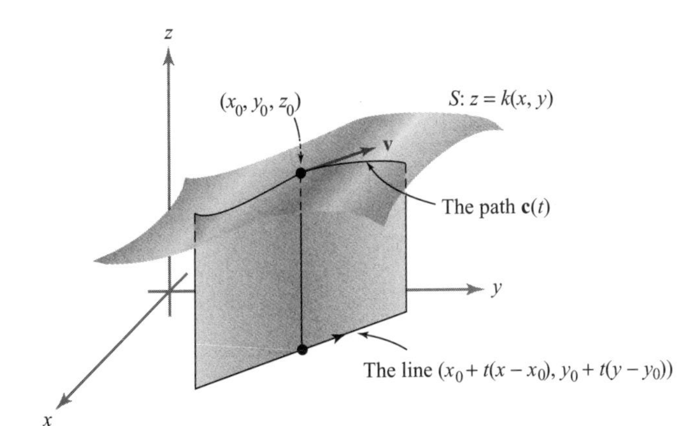

Figure 3.5.3 The construction of a path $\mathbf{c}(t)$ in the surface S whose tangent vector is $\mathbf{v}$.

EXAMPLE 2 Near what points may the surface

$$x^3 + 3y^2 + 8xz^2 - 3z^3y = 1$$

be represented as a graph of a differentiable function $z = k(x, y)$?

SOLUTION Here we take $F(x, y, z) = x^3 + 3y^2 + 8xz^2 - 3z^3y - 1$ and attempt to solve $F(x, y, z) = 0$ for z as a function of (x, y). By Theorem 11, this may be done near a point (x_0, y_0, z_0) if $(\partial F/\partial z)(x_0, y_0, z_0) \neq 0$, that is, if

$$z_0(16x_0 - 9z_0y_0) \neq 0,$$

which means, in turn,

$$z_0 \neq 0 \quad \text{and} \quad 16x_0 \neq 9z_0y_0. \quad \blacktriangle$$

General Implicit Function Theorem

Next we shall state, without proof, the *general implicit function theorem.*[15] Instead of attempting to solve one equation for one variable, we attempt to solve m equations for m variables $z_1, \dots, z_m$:

$$\begin{aligned} F_1(x_1, \dots, x_n, z_1, \dots, z_m) &= 0 \\ F_2(x_1, \dots, x_n, z_1, \dots, z_m) &= 0 \\ \vdots \qquad\qquad & \vdots \\ F_m(x_1, \dots, x_n, z_1, \dots, z_m) &= 0. \end{aligned} \tag{3}$$

In Theorem 11 we had the condition $\partial F/\partial z \neq 0$. The condition appropriate to the general implicit function theorem is that $\Delta \neq 0$,[16] where Δ is the determinant of the $m \times m$ matrix

$$\begin{bmatrix} \dfrac{\partial F_1}{\partial z_1} & \cdots & \dfrac{\partial F_1}{\partial z_m} \\ \vdots & & \vdots \\ \dfrac{\partial F_m}{\partial z_1} & \cdots & \dfrac{\partial F_m}{\partial z_m} \end{bmatrix}$$

[15] For three different proofs of the general case, consult:

(a) E. Goursat, *A Course in Mathematical Analysis*, I, Dover, New York, 1959, p. 45. (This proof derives the general theorem by successive application of Theorem 11.)

(b) T. M. Apostol, *Mathematical Analysis*, 2d ed., Addison-Wesley, Reading, Mass., 1974.

(c) J. E. Marsden and M. Hoffman, *Elementary Classical Analysis*, 2d ed., Freeman, New York, 1993.

Of these sources, the last two use more sophisticated ideas that are usually not covered until a junior-level course in analysis. The first, however, is easily understood by the reader who has some knowledge of linear algebra.

[16] For students who have had linear algebra: The condition $\Delta \neq 0$ has a simple interpretation in the case that F *is linear;* namely, $\Delta \neq 0$ is equivalent to the rank of F being equal to m, which in turn is equivalent to the fact that the solution space of $F = 0$ is m-dimensional.

evaluated at the point $(\mathbf{x}_0, \mathbf{z}_0)$; in the neighborhood of such a point, we can uniquely solve for $\mathbf{z}$ in terms of $\mathbf{x}$.

THEOREM 12: General Implicit Function Theorem If $\Delta \neq 0$, then near the point $(\mathbf{x}_0, \mathbf{z}_0)$, equation (3) defines unique (smooth) functions

$$z_i = k_i(x_1, \ldots, x_n) \qquad (i = 1, \ldots, m).$$

Their derivatives may be computed by implicit differentiation.

EXAMPLE 3 Show that near the point $(x, y, u, v) = (1, 1, 1, 1)$, we can solve

$$\begin{aligned} xu + yvu^2 &= 2 \\ xu^3 + y^2v^4 &= 2 \end{aligned}$$

uniquely for u and v as functions of x and y. Compute $\partial u/\partial x$ at the point $(1, 1)$.

SOLUTION To check solvability, we form the equations

$$\begin{aligned} F_1(x, y, u, v) &= xu + yvu^2 - 2 \\ F_2(x, y, u, v) &= xu^3 + y^2v^4 - 2 \end{aligned}$$

and the determinant

$$\begin{aligned} \Delta &= \begin{vmatrix} \dfrac{\partial F_1}{\partial u} & \dfrac{\partial F_1}{\partial v} \\ \dfrac{\partial F_2}{\partial u} & \dfrac{\partial F_2}{\partial v} \end{vmatrix} \qquad \text{at} \qquad (1, 1, 1, 1) \\ &= \begin{vmatrix} x + 2yuv & yu^2 \\ 3u^2x & 4y^2v^3 \end{vmatrix} \qquad \text{at} \qquad (1, 1, 1, 1) \\ &= \begin{vmatrix} 3 & 1 \\ 3 & 4 \end{vmatrix} = 9. \end{aligned}$$

Because $\Delta \neq 0$, solvability is assured by the general implicit function theorem. To find $\partial u/\partial x$, we implicitly differentiate the given equations in x using the chain rule:

$$x\frac{\partial u}{\partial x} + u + y\frac{\partial v}{\partial x}u^2 + 2yvu\frac{\partial u}{\partial x} = 0$$

$$3xu^2\frac{\partial u}{\partial x} + u^3 + 4y^2v^3\frac{\partial v}{\partial x} = 0.$$

Setting $(x, y, u, v) = (1, 1, 1, 1)$ gives

$$3\frac{\partial u}{\partial x} + \frac{\partial v}{\partial x} = -1$$

$$3\frac{\partial u}{\partial x} + 4\frac{\partial v}{\partial x} = -1.$$

Solving for $\partial u/\partial x$ by multiplying the first equation by 4 and subtracting gives $\partial u/\partial x = -\frac{1}{3}$. ▲

Inverse Function Theorem

A special case of the general implicit function theorem is the *inverse function theorem*. Here we attempt to solve the n equations

$$\left.\begin{aligned} f_1(x_1, \dots, x_n) &= y_1 \\ &\cdots \\ f_n(x_1, \dots, x_n) &= y_n \end{aligned}\right\} \tag{4}$$

for $x_1, \dots, x_n$ as functions of $y_1, \dots, y_n$; that is, we are trying to invert the equations of system (4). This is analogous to forming the inverses of functions like $\sin x = y$ and $e^x = y$, with which the reader should be familiar from elementary calculus. Now, however, we are concerned with functions of several variables. The question of solvability is answered by the general implicit function theorem applied to the functions $y_i - f_i(x_1, \dots, x_n)$ with the unknowns $x_1, \dots, x_n$ (called $z_1, \dots, z_n$ earlier). The condition for solvability in a neighborhood of a point $\mathbf{x}_0$ is $\Delta \neq 0$, where Δ is the determinant of the matrix $\mathbf{D}f(\mathbf{x}_0)$, and $f = (f_1, \dots, f_n)$. The quantity Δ is denoted by $\partial(f_1, \dots, f_n)/\partial(x_1, \dots, x_n)$, or $\partial(y_1, \dots, y_n)/\partial(x_1, \dots, x_n)$ or $J(f)(\mathbf{x}_0)$ and is called the ***Jacobian determinant*** of f. Explicitly,

$$\left.\frac{\partial(f_1, \dots, f_n)}{\partial(x_1, \dots, x_n)}\right|_{\mathbf{x}=\mathbf{x}_0} = J(f)(\mathbf{x}_0) = \begin{vmatrix} \frac{\partial f_1}{\partial x_1}(\mathbf{x}_0) & \cdots & \frac{\partial f_1}{\partial x_n}(\mathbf{x}_0) \\ \vdots & & \vdots \\ \frac{\partial f_n}{\partial x_1}(\mathbf{x}_0) & \cdots & \frac{\partial f_n}{\partial x_n}(\mathbf{x}_0) \end{vmatrix}. \tag{5}$$

The reader should note that in the case when f is linear, for example $f(x) = Ax$, where A is an $n \times n$ matrix, the condition $\Delta \neq 0$ is equivalent to the fact that the determinant of A, $\det A \neq 0$, and from Section 1.5 we know that A, and therefore f, has an inverse.

The Jacobian determinant will play an important role in our work on integration (see Chapter 5). The following theorem summarizes this discussion:

THEOREM 13: Inverse Function Theorem Let $U \subset \mathbb{R}^n$ be open and let $f_1: U \to \mathbb{R}, \ldots, f_n: U \to \mathbb{R}$ have continuous partial derivatives. Consider the equations (4) near a given solution $\mathbf{x}_0, \mathbf{y}_0$. If $J(f)(\mathbf{x}_0)$ [defined by equation (5)] is nonzero, then equation (4) can be solved uniquely as $\mathbf{x} = g(\mathbf{y})$ for $\mathbf{x}$ near $\mathbf{x}_0$ and $\mathbf{y}$ near $\mathbf{y}_0$. Moreover, the function g has continuous partial derivatives.

EXAMPLE 4 Consider the equations

$$\frac{x^4 + y^4}{x} = u, \qquad \sin x + \cos y = v.$$

Near which points (x, y) can we solve for x, y in terms of u, v?

SOLUTION Here the functions are $u = f_1(x, y) = (x^4 + y^4)/x$ and $v = f_2(x, y) = \sin x + \cos y$. We want to know the points near which we can solve for x, y as functions of u and v. According to the inverse function theorem, we must first compute the Jacobian determinant $\partial(f_1, f_2)/\partial(x, y)$. We take the domain of $f = (f_1, f_2)$ to be $U = \{(x, y) \in \mathbb{R}^2 \mid x \neq 0\}$. Now

$$\frac{\partial(f_1, f_2)}{\partial(x, y)} = \begin{vmatrix} \dfrac{\partial f_1}{\partial x} & \dfrac{\partial f_1}{\partial y} \\ \dfrac{\partial f_2}{\partial x} & \dfrac{\partial f_2}{\partial y} \end{vmatrix} = \begin{vmatrix} \dfrac{3x^4 - y^4}{x^2} & \dfrac{4y^3}{x} \\ \cos x & -\sin y \end{vmatrix} = \frac{\sin y}{x^2}(y^4 - 3x^4) - \frac{4y^3}{x}\cos x.$$

Therefore, at points where this does not vanish we can solve for x, y in terms of u and v. In other words, we can solve for x, y near those x, y for which $x \neq 0$ and $(\sin y)(y^4 - 3x^4) \neq 4xy^3 \cos x$. Such conditions generally cannot be solved explicitly. For example, if $x_0 = \pi/2$, $y_0 = \pi/2$, we can solve for x, y near (x_0, y_0) because there, $\partial(f_1, f_2)/\partial(x, y) \neq 0$. ▲

EXERCISES

1. Let $F(x, y) = 0$ define a curve in the xy plane through the point (x_0, y_0), where F is C^1. Assume that $(\partial F/\partial y)(x_0, y_0) \neq 0$. Show that this curve can be locally represented by the graph of a function $y = g(x)$. Show that (i) the line orthogonal to $\nabla F(x_0, y_0)$ agrees with (ii) the tangent line to the graph of $y = g(x)$.

2. Show that $xy + z + 3xz^5 = 4$ is solvable for z as a function of (x, y) near $(1, 0, 1)$. Compute $\partial z/\partial x$ and $\partial z/\partial y$ at $(1, 0)$.

3. (a) Check directly (i.e., without using Theorem 11) where we can solve the equation $F(x, y) = y^2 + y + 3x + 1 = 0$ for y in terms of x.

(b) Check that your answer in part (a) agrees with the answer you expect from the implicit function theorem. Compute dy/dx.

4. Repeat Exercise 3 with $F(x, y) = xy^2 - 2y + x^2 + 2 = 0$.

5. Show that $x^3z^2 - z^3yx = 0$ is solvable for z as a function of (x, y) near $(1, 1, 1)$, but not near the origin. Compute $\partial z/\partial x$ and $\partial z/\partial y$ at $(1, 1)$.

6. Discuss the solvability in the system

$$\begin{aligned} 3x + 2y + z^2 + u + v^2 &= 0 \\ 4x + 3y + z + u^2 + v + w + 2 &= 0 \\ x + z + w + u^2 + 2 &= 0 \end{aligned}$$

for u, v, w in terms of x, y, z near $x = y = z = 0$, $u = v = 0$, and $w = -2$.

7. Discuss the solvability of

$$\begin{aligned} y + x + uv &= 0 \\ uxy + v &= 0 \end{aligned}$$

for u, v in terms of x, y near $x = y = u = v = 0$ and check directly.

8. Investigate whether or not the system

$$\begin{aligned} u(x, y, z) &= x + xyz \\ v(x, y, z) &= y + xy \\ w(x, y, z) &= z + 2x + 3z^2 \end{aligned}$$

can be solved for x, y, z in terms of u, v, w near $(x, y, z) = (0, 0, 0)$.

9. Consider $f(x, y) = ((x^2 - y^2)/(x^2 + y^2), xy/(x^2 + y^2))$. Does this map of $\mathbb{R}^2\backslash(0, 0)$ to $\mathbb{R}^2$ have a local inverse near $(x, y) = (0, 1)$?

10. (a) Define $x\colon \mathbb{R}^2 \to \mathbb{R}$ by $x(r, \theta) = r\cos\theta$ and define $y\colon \mathbb{R}^2 \to \mathbb{R}$ by $y(r, \theta) = r\sin\theta$. Show that

$$\left.\frac{\partial(x, y)}{\partial(r, \theta)}\right|_{(r_0, \theta_0)} = r_0.$$

(b) When can we form a smooth inverse function $(r(x, y), \theta(x, y))$? Check directly and with the inverse function theorem.

(c) Consider the following transformations for spherical coordinates (see Section 1.4):

$$\begin{aligned} x(\rho, \phi, \theta) &= \rho\sin\phi\cos\theta \\ x(\rho, \phi, \theta) &= \rho\sin\phi\sin\theta \\ z(\rho, \phi, \theta) &= \rho\cos\phi. \end{aligned}$$

Show that the Jacobian determinant is given by

$$\frac{\partial(x, y, z)}{\partial(\rho, \phi, \theta)} = \rho^2 \sin\phi.$$

(d) When can we solve for (ρ, ϕ, θ) in terms of (x, y, z)?

11. Let (x_0, y_0, z_0) be a point of the locus defined by $z^2 + xy - a = 0$, $z^2 + x^2 - y^2 - b = 0$, where a and b are constants.

(a) Under what conditions may the part of the locus near (x_0, y_0, z_0) be represented in the form $x = f(z)$, $y = g(z)$?
(b) Compute $f'(z)$ and $g'(z)$.

12. Is it possible to solve the system of equations

$$xy^2 + xzu + yv^2 = 3$$
$$u^3yz + 2xv - u^2v^2 = 2$$

for $u(x, y, z)$, $v(x, y, z)$ near $(x, y, z) = (1, 1, 1)$, $(u, v) = (1, 1)$? Compute $\partial v/\partial y$ at $(x, y, z) = (1, 1, 1)$.

13. The problem of factoring a polynomial $x^n + a_{n-1}x^{n-1} + \cdots + a_0$ into linear factors is, in a sense, an "inverse function" problem. The coefficients a_i may be thought of as functions of the n roots r_j. We would like to find the roots as functions of the coefficients in some region. With $n = 3$, apply the inverse function theorem to this problem and state what it tells you about the possibility of doing this.

REVIEW EXERCISES FOR CHAPTER 3

1. Analyze the behavior of the following functions at the indicated points. [Your answer in part (b) may depend on the constant C.]

(a) $z = x^2 - y^2 + 3xy$, $(x, y) = (0, 0)$
(b) $z = x^2 - y^2 + Cxy$, $(x, y) = (0, 0)$

2. Find and classify the extreme values (if any) of the functions on $\mathbb{R}^2$ defined by the following expressions:

(a) $y^2 - x^3$ (b) $(x - 1)^2 + (x - y)^2$ (c) $x^2 + xy^2 + y^4$

3. (a) Find the minimum distance from the origin in $\mathbb{R}^3$ to the surface $z = \sqrt{x^2 - 1}$.
(b) Repeat part (a) for the surface $z = 6xy + 7$.

4. Find the first few terms in the Taylor expansion of $f(x, y) = e^{xy} \cos x$ about $x = 0$, $y = 0$.

5. Prove that

$$z = \frac{3x^4 - 4x^3 - 12x^2 + 18}{12(1 + 4y^2)}$$

has one local maximum, one local minimum, and one saddle point. (The graph is shown in Figure 3.R.1.)

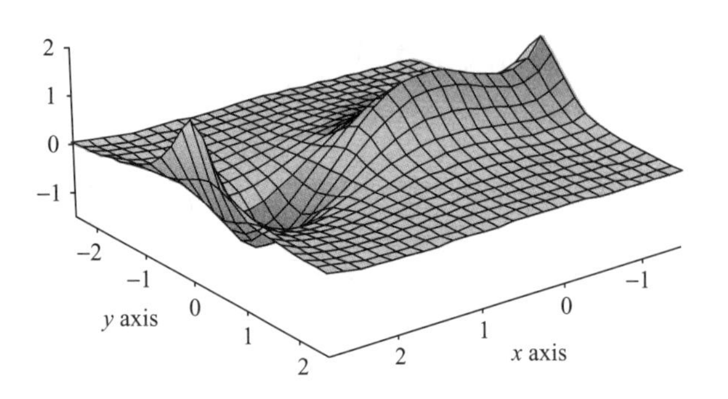

Figure 3.R.1 Graph of $z = (3x^4 - 4x^3 - 12x^2 + 18)/12(1 + 4y^2)$.

6. Find the maxima, minima, and saddles of the function $z = (2 + \cos \pi x)(\sin \pi y)$, which is graphed in Figure 3.R.2.

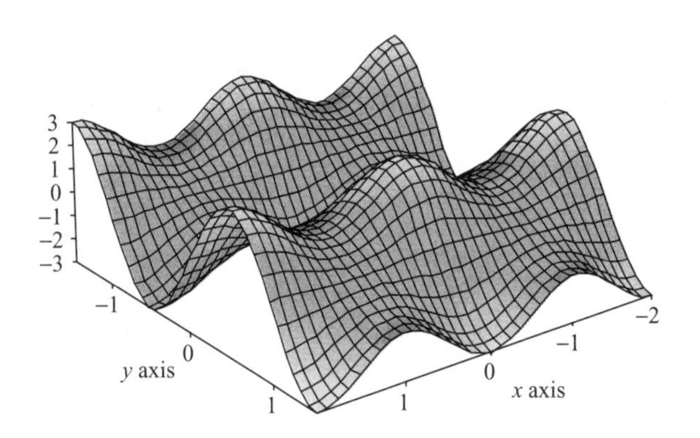

Figure 3.R.2 Graph of $z = (2 + \cos \pi x)\,(\sin \pi y)$.

7. Find and describe the critical points of $f(x, y) = y \sin(\pi x)$. (See Figure 3.R.3.)

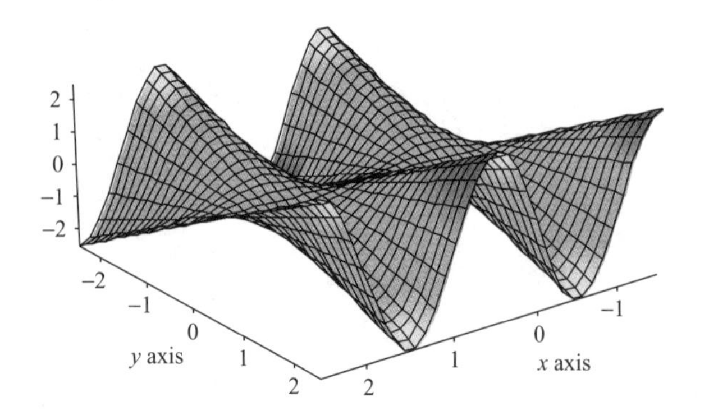

Figure 3.R.3 Graph of $z = y \sin(\pi x)$.

8. A graph of the function $z = \sin(\pi x)/(1 + y^2)$ is shown in Figure 3.R.4. Verify that this function has alternating maxima and minima on the x axis, with no other critical points.

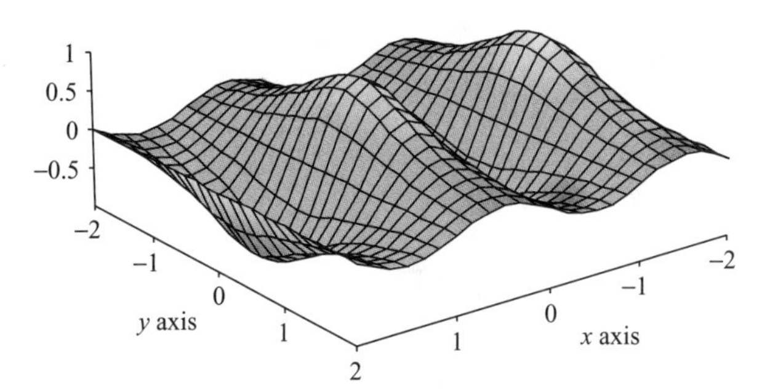

Figure 3.R.4 Graph of $z = \sin(\pi x)/(1 + y^2)$.

In Exercises 9 to 14 find the extrema of the given functions subject to the given constraints.

9. $f(x, y) = x^2 - 2xy + 2y^2$, subject to $x^2 + y^2 = 1$

10. $f(x, y) = xy - y^2$, subject to $x^2 + y^2 = 1$

11. $f(x, y) = \cos(x^2 - y^2)$, subject to $x^2 + y^2 = 1$

12. $f(x, y) = \dfrac{x^2 - y^2}{x^2 + y^2}$, subject to $x + y = 1$

13. $z = xy$, subject to the condition $x + y = 1$.

14. $z = \cos^2 x + \cos^2 y$, subject to the condition $x + y = \pi/4$.

15. Find the points on the surface $z^2 - xy = 1$ nearest to the origin.

16. Use the implicit function theorem to compute dy/dx for

(a) $x/y = 10$ (b) $x^3 - \sin y + y^4 = 4$ (c) $e^{x+y^2} + y^3 = 0$

17. Find the shortest distance from the point $(0, b)$ to the parabola $x^2 - 4y = 0$. Solve this problem using the Lagrange multiplier method and also without using Lagrange's method.

18. Solve the following geometric problems by Lagrange's method.

(a) Find the shortest distance from the point (a_1, a_2, a_3) in $\mathbb{R}^3$ to the plane whose equation is given by $b_1x_1 + b_2x_2 + b_3x_3 + b_0 = 0$, where $(b_1, b_2, b_3) \neq (0, 0, 0)$.

(b) Find the point on the line of intersection of the two planes $a_1x_1 + a_2x_2 + a_3x_3 = 0$ and $b_1x_1 + b_2x_2 + b_3x_3 + b_0 = 0$ that is nearest to the origin.

(c) Show that the volume of the largest rectangular parallelepiped that can be inscribed in the ellipsoid

$$\frac{x^2}{a^2} + \frac{y^2}{b^2} + \frac{z^2}{c^2} = 1$$

is $8abc/3\sqrt{3}$.

19. A particle moves in a potential $V(x, y) = x^3 - y^2 + x^2 + 3xy$. Determine whether $(0, 0)$ is a stable equilibrium point—that is, whether or not $(0, 0)$ is a strict local minimum of V.

20. Study the nature of the function $f(x, y) = x^3 - 3xy^2$ near $(0, 0)$. Show that the point $(0, 0)$ is a degenerate critical point, that is, $D = 0$. This surface is called a *monkey saddle*.

21. Find the maximum of $f(x, y) = xy$ on the curve $(x + 1)^2 + y^2 = 1$.

22. Find the maximum and minimum of $f(x, y) = xy - y + x - 1$ on the set $x^2 + y^2 \leq 2$.

23. The Baraboo, Wisconsin, plant of International Widget Co., Inc., uses aluminium, iron, and magnesium to produce high-quality widgets. The quantity of widgets that may be produced using x tons of aluminum, y tons of iron, and z tons of magnesium is $Q(x, y, z) = xyz$. The cost of raw materials is aluminum, \$6 per ton; iron, \$4 per ton; and magnesium, \$8 per ton. How many tons each of aluminum, iron, and magnesium should be used to manufacture 1000 widgets at the lowest possible cost? (HINT: Find an extreme value for what function subject to what constraint?)

24. Let $f: \mathbb{R} \to \mathbb{R}$ be of class C^1 and let

$$\begin{aligned} u &= f(x) \\ v &= -y + xf(x). \end{aligned}$$

If $f'(x_0) \neq 0$, show that this transformation of $\mathbb{R}^2$ to $\mathbb{R}^2$ is invertible near (x_0, y_0) and its inverse is given by

$$\begin{aligned} x &= f^{-1}(u) \\ y &= -v + uf^{-1}(u). \end{aligned}$$

25. Show that the pair of equations

$$\begin{aligned} x^2 - y^2 - u^3 + v^2 + 4 &= 0 \\ 2xy + y^2 - 2u^2 + 3v^4 + 8 &= 0 \end{aligned}$$

determine functions $u(x, y)$ and $v(x, y)$ defined for (x, y) near $x = 2$ and $y = -1$ such that $u(2, -1) = 2$ and $v(2, -1) = 1$. Compute $\partial u/\partial x$ at $(2, -1)$.

26. Show that there are positive numbers p and q and unique functions u and v from the interval $(-1 - p, -1 + p)$ into the interval $(1 - q, 1 + q)$ satisfying

$$xe^{u(x)} + u(x)e^{v(x)} = 0 = xe^{v(x)} + v(x)e^{u(x)}$$

for all x in the interval $(-1 - p, -1 + p)$ with $u(-1) = 1 = v(-1)$.

27. To work this exercise, the reader should be familiar with the technique of diagonalizing a 2×2 matrix. Let $a(x)$, $b(x)$, and $c(x)$ be three continuous functions defined on $U \cup \partial U$, where U is an open set and ∂U denotes its set of boundary points (see Section 2.2). Use the notation of Lemma 2 in Section 3.3, and assume that for each $x \in U \cup \partial U$ the quadratic form defined by the matrix

$$\begin{bmatrix} a & b \\ b & c \end{bmatrix}$$

is positive-definite. For a C^2 function v on $U \cup \partial U$, we define a differential operator L by $Lv = a(\partial^2 v/\partial x^2) + 2b(\partial^2 v/\partial x \partial y) + c(\partial^2 v/\partial y^2)$. With this positive-definite condition, such an operator is said to be ***elliptic***. A function v is said to be ***strictly subharmonic relative*** to L if $Lv > 0$. Show that a strictly subharmonic function cannot have a maximum point in U.

28. A function v is said to be in the ***kernel*** of the operator L described in Exercise 27 if $Lv = 0$ on $U \cup \partial U$. Arguing as in Exercise 37 of Section 3.3, show that if v achieves its maximum on U it also achieves it on ∂U. This is called the weak maximum principle for elliptic operators.

29. Let L be an elliptic differential operator as in Exercises 27 and 28.

(a) Define the notion of a strict superharmonic function.
(b) Show that such functions cannot achieve a minimum on U.
(c) If v is as in Exercise 28, show that if v achieves its minimum on U it also achieves it on ∂U.

*The following **method of least squares** should be applied to Exercises 30 to 35.*

It sometimes happens that the theory behind an experiment indicates that the experimental data should lie approximately along a straight line of the form $y = mx + b$. The actual results, of course, never match the theory exactly. We are then faced with the problem of finding the straight line that *best fits* some set of experimental data $(x_1, y_1), \ldots, (x_n, y_n)$ as in Figure 3.R.5. If we guess at a straight line $y = mx + b$ to fit the data, each point will deviate vertically from the line by an amount $d_i = y_i - (mx_i + b)$.

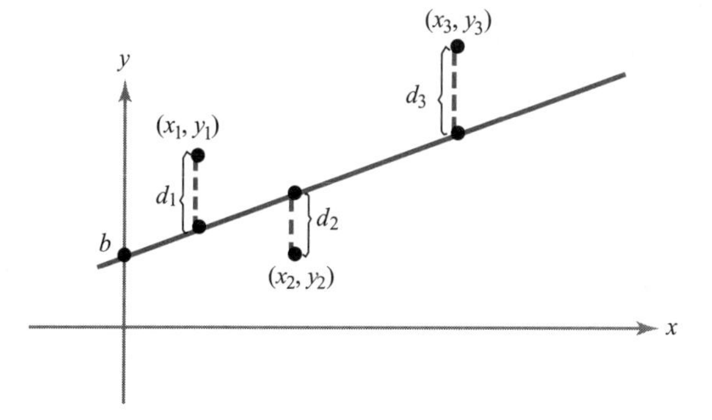

Figure 3.R.5 The method of least squares tries to find a straight line that best approximates a set of data.

We would like to choose m and b in such a way as to make the total effect of these deviations as small as possible. However, because some are negative and some positive, we could get a lot of cancellations and still have a pretty bad fit. This leads us to suspect that a better measure of the total error might be the sum of the *squares* of these deviations. Thus, we are led to the problem of finding the m and b that minimize the function

$$s = f(m, b) = d_1^2 + d_2^2 + \cdots + d_n^2 = \sum_{i=1}^{n} (y_i - mx_i - b)^2,$$

where $x_1, \ldots, x_n$ and $y_1, \ldots, y_n$ are the given data.

30. For each set of three data points, plot the points, write down the function $f(m, b)$ from the preceding equation, find m and b to give the best straight-line fit according to the method of least squares, and plot the straight line.

(a) $(x_1, y_1) = (1, 1)$
$(x_2, y_2) = (2, 3)$
$(x_3, y_3) = (4, 3)$

(b) $(x_1, y_1) = (0, 0)$
$(x_2, y_2) = (1, 2)$
$(x_3, y_3) = (2, 3)$

31. Show that if only two data points (x_1, y_1) and (x_2, y_2) are given, this method produces the line through (x_1, y_1) and (x_2, y_2).

32. Show that the equations for a critical point, $\partial s/\partial b = 0$ and $\partial s/\partial m = 0$, are equivalent to

$$m\left(\sum x_i\right) + nb = \left(\sum y_i\right) \quad \text{and} \quad m\left(\sum x_i^2\right) + b\left(\sum x_i\right) = \left(\sum x_i y_i\right),$$

where all the sums run from $i = 1$ to $i = n$.

33. If $y = mx + b$ is the best-fitting straight line to the data points $(x_1, y_1), \ldots, (x_n, y_n)$ according to the least-square method, show that

$$\sum_{i=1}^{n} (y_i - mx_i - b) = 0;$$

that is, the positive and negative deviations cancel (see Exercise 32).

34. Use the second derivative test to show that the critical point of f is a minimum.

35. Use the method of least squares to find the straight line that best fits the points $(0, 1)$, $(1, 3)$, $(2, 2)$, $(3, 4)$, and $(4, 5)$. Plot the points and line.[17]

---

[17]The method of least squares may be varied and generalized in a number of ways. The basic idea can be applied to equations of more complicated curves than the straight line. For example, this might be done to find the parabola that best fits a given set of data points. These ideas also formed part of the basis for the development of the science of cybernetics by Norbert Wiener. Another version of the data is the following problem of least-square approximation: Given a function f defined and integrable on an interval $[a, b]$, find a polynomial P of degree $\le n$ such that the mean square error

$$\int_a^b |f(x) - P(x)|^2 \, dx$$

is as small as possible.

4

Vector-Valued Functions

... who by vigor of mind almost divine, the motions and figures of the planets, the paths of comets, and the tides of the seas first demonstrated.

Newton's Epitaph

Chapters 2 and 3 focused on *real*-valued functions. This chapter is largely concerned with *vector*-valued functions. We begin in the first section with a continuation of our study of paths, adding applications of Newton's second law. Then we study arc length of paths. Following this, we introduce the divergence and curl of a vector field which, in addition to the gradient, are basic operations in vector *differential* calculus. The basic geometry and calculus of the divergence and curl are studied. The associated *integral* calculus will be given in Chapter 8.

4.1 Acceleration and Newton's Second Law

In Section 2.4, we studied the basic geometry of paths, learning how to sketch curves (the images of paths) and compute tangent lines. We also learned to think of, as the name suggests, a path as the trajectory of a particle and to regard the derivative of the path as its velocity vector. In this section, we continue our study of paths, including additional topics, especially acceleration and Newton's second law.

Differentiation of Paths

Recall that a path in $\mathbb{R}^n$ is a map $\mathbf{c}$ of $\mathbb{R}$ or an interval in $\mathbb{R}$ to $\mathbb{R}^n$. If the path is differentiable, its derivative at each time t is an $n \times 1$ matrix. Specifically, if $x_1(t), \ldots, x_n(t)$ are the component functions of $\mathbf{c}$, the derivative matrix is

$$\mathbf{c}'(t) = \begin{bmatrix} dx_1/dt \\ dx_2/dt \\ \vdots \\ dx_n/dt \end{bmatrix},$$

which can also be written in vector form as

$$(dx_1/dt, \ldots, dx_n/dt) \qquad \text{or as} \qquad (x_1'(t), \ldots, x_n'(t)).$$

Recall from Section 2.4 that $\mathbf{c}'(t)$ is the *tangent vector* to the path at the point $\mathbf{c}(t)$. Also recall that if $\mathbf{c}$ represents the path of a moving particle, then its *velocity vector* is

$$\mathbf{v} = \mathbf{c}'(t),$$

and its *speed* is $s = \|\mathbf{v}\|$.

The differentiation of paths is facilitated by the following rules.

Differentiation Rules Let $\mathbf{b}(t)$ and $\mathbf{c}(t)$ be differentiable paths in $\mathbb{R}^3$ and $p(t)$ and $q(t)$ be differentiable scalar functions:

Sum Rule: $$\frac{d}{dt}[\mathbf{b}(t) + \mathbf{c}(t)] = \mathbf{b}'(t) + \mathbf{c}'(t)$$

Scalar Multiplication Rule: $$\frac{d}{dt}[p(t)\mathbf{c}(t)] = p'(t)\mathbf{c}(t) + p(t)\mathbf{c}'(t)$$

Dot Product Rule: $$\frac{d}{dt}[\mathbf{b}(t) \cdot \mathbf{c}(t)] = \mathbf{b}'(t) \cdot \mathbf{c}(t) + \mathbf{b}(t) \cdot \mathbf{c}'(t)$$

Cross Product Rule: $$\frac{d}{dt}[\mathbf{b}(t) \times \mathbf{c}(t)] = \mathbf{b}'(t) \times \mathbf{c}(t) + \mathbf{b}(t) \times \mathbf{c}'(t)$$

Chain Rule: $$\frac{d}{dt}[\mathbf{c}(q(t))] = q'(t)\mathbf{c}'(q(t)).$$

These rules follow by applying the usual differentiation rules to the components.

EXAMPLE 1 Show that if $\mathbf{c}(t)$ is a vector function such that $\|\mathbf{c}(t)\|$ is constant, then $\mathbf{c}'(t)$ is perpendicular to $\mathbf{c}(t)$ for all t.

SOLUTION Because $\|\mathbf{c}(t)\|$ is constant, so is its square $\|\mathbf{c}(t)\|^2 = \mathbf{c}(t) \cdot \mathbf{c}(t)$. The derivative of this constant is zero, so by the dot product rule,

$$0 = \frac{d}{dt}[\mathbf{c}(t) \cdot \mathbf{c}(t)] = \mathbf{c}'(t) \cdot \mathbf{c}(t) + \mathbf{c}(t) \cdot \mathbf{c}'(t) = 2\mathbf{c}(t) \cdot \mathbf{c}'(t);$$

thus, $\mathbf{c}(t) \cdot \mathbf{c}'(t) = 0$; that is, $\mathbf{c}'(t)$ is perpendicular to $\mathbf{c}(t)$. ▲

For a path describing uniform rectilinear motion, the velocity vector is constant. In general, the velocity vector is a vector function $\mathbf{v} = \mathbf{c}'(t)$ that depends on t. The derivative $\mathbf{a} = d\mathbf{v}/dt = \mathbf{c}''(t)$ is called the ***acceleration*** of the curve. If the curve is

$(x(t), y(t), z(t))$, then the acceleration at time t is given by

$$\mathbf{a}(t) = x''(t)\mathbf{i} + y''(t)\mathbf{j} + z''(t)\mathbf{k}.$$

EXAMPLE 2 A particle moves in such a way that its acceleration is constantly equal to $-\mathbf{k}$. If the position when $t = 0$ is $(0, 0, 1)$ and the velocity at $t = 0$ is $\mathbf{i} + \mathbf{j}$, when and where does the particle fall below the plane $z = 0$? Describe the path traveled by the particle (assume $t \geq 0$).

SOLUTION Let $(x(t), y(t), z(t))$ be the path traced out by the particle, so that the velocity vector is $\mathbf{c}'(t) = x'(t)\mathbf{i} + y'(t)\mathbf{j} + z'(t)\mathbf{k}$. The acceleration $\mathbf{c}''(t)$ is $-\mathbf{k}$, so $x''(t) = 0$, $y''(t) = 0$, and $z''(t) = -1$. It follows that $x'(t)$ and $y'(t)$ are constant functions, and $z'(t)$ is a linear function with slope -1. Because $\mathbf{c}'(0) = \mathbf{i} + \mathbf{j}$, we get $\mathbf{c}'(t) = \mathbf{i} + \mathbf{j} - t\mathbf{k}$. Integrating again and using the initial position $(0, 0, 1)$, we find that $(x(t), y(t), z(t)) = (t, t, 1 - \frac{1}{2}t^2)$. The particle drops below the plane $z = 0$ when $1 - \frac{1}{2}t^2 = 0$; that is, $t = \sqrt{2}$ (because $t \geq 0$). At that instant, the position is $(\sqrt{2}, \sqrt{2}, 0)$. The path traveled by the particle is a parabola in the plane $y = x$ (see Figure 4.1.1), because in this plane the equation is described by $z = 1 - \frac{1}{2}x^2$. ▲

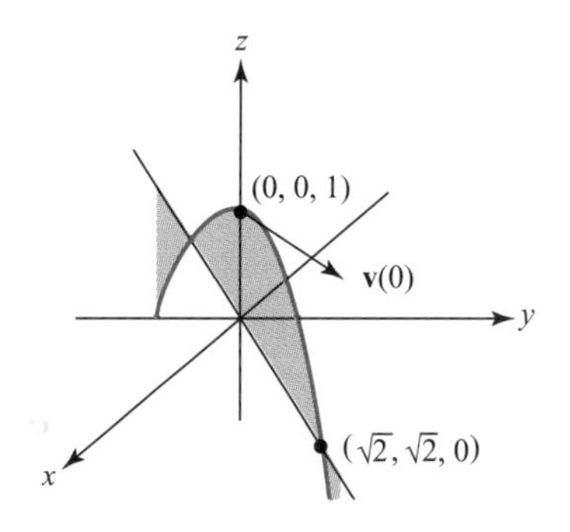

Figure 4.1.1 The path of the particle with initial position $(0, 0, 1)$, initial velocity $\mathbf{i} + \mathbf{j}$, and constant acceleration $-\mathbf{k}$ is a parabola in the plane $y = x$.

The image of a C^1 path is not necessarily "very smooth"; indeed, it may have sharp bends or changes of direction. For instance, the cycloid $\mathbf{c}(t) = (t - \sin t, 1 - \cos t)$ shown in Figure 2.4.6 has cusps at all points where $\mathbf{c}$ touches the x axis (that is, when $1 - \cos t = 0$, which happens when $t = 2\pi n$, $n = 0, \pm 1, \ldots$). Another example is the ***hypocycloid of four cusps***, $\mathbf{c}\colon [0, 2\pi] \to \mathbb{R}^2$, $t \mapsto (\cos^3 t, \sin^3 t)$, which has cusps at four points (Figure 4.1.2). At all such points, however, $\mathbf{c}'(t) = \mathbf{0}$, and the tangent line is not well defined. Evidently, the direction of $\mathbf{c}'(t)$ may change abruptly at points where it slows to rest.

A differentiable path $\mathbf{c}$ is said to be ***regular*** at $t = t_0$ if $\mathbf{c}'(t_0) \neq \mathbf{0}$. If $\mathbf{c}'(t) \neq \mathbf{0}$ for all t, we say that c is a regular path. In this case, the image curve looks smooth.

EXAMPLE 3 A particle moves along a hypocycloid according to the equations

$$x = \cos^3 t, \qquad y = \sin^3 t, \qquad a \leq t \leq b.$$

What are the velocity and speed of the particle?

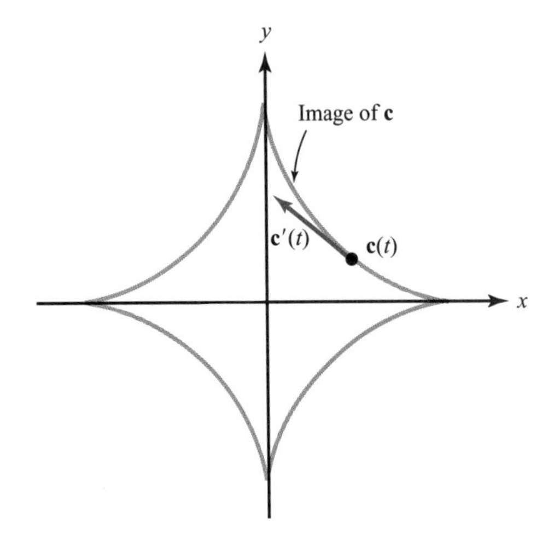

Figure 4.1.2 The image of the smooth path $\mathbf{c}(t) = (\cos^3 t, \sin^3 t)$, a hypocycloid, does not "look smooth."

SOLUTION The velocity vector of the particle is

$$\mathbf{v} = \frac{dx}{dt}\mathbf{i} + \frac{dy}{dt}\mathbf{j} = -(3\sin t\cos^2 t)\mathbf{i} + (3\cos t\sin^2 t)\mathbf{j},$$

and its speed is

$$s = \|\mathbf{v}\| = (9\sin^2 t\cos^4 t + 9\cos^2 t\sin^4 t)^{1/2} = 3\,|\sin t|\;|\cos t|. \quad \blacktriangle$$

Newton's Second Law

If a particle of mass m moves in $\mathbb{R}^3$, the force $\mathbf{F}$ acting on it at the point $\mathbf{c}(t)$ is related to the acceleration $\mathbf{a}(t)$ by ***Newton's second law***:[1]

$$\mathbf{F}(\mathbf{c}(t)) = m\mathbf{a}(t).$$

In particular, if no forces act on a particle, then $\mathbf{a}(t) = \mathbf{0}$, so $\mathbf{c}'(t)$ is constant and the particle follows a straight line.

Acceleration and Newton's Second Law The ***acceleration*** of a path $\mathbf{c}(t)$ is

$$\mathbf{a}(t) = \mathbf{c}''(t).$$

If $\mathbf{F}$ is the force acting and m is the mass of the particle, then

$$\mathbf{F} = m\mathbf{a}.$$

[1]Most scientists acknowledge that $\mathbf{F} = m\mathbf{a}$ is the single most important equation in all of science and engineering.

In the problem of determining the path $\mathbf{c}(t)$ of a particle under the influence of a given force field, $\mathbf{F}$, Newton's law becomes a differential equation (i.e., an equation involving derivatives) for $\mathbf{c}(t)$.

For example, the motion of a planet moving along a path $\mathbf{r}(t)$ around the sun (considered to be located at the origin in $\mathbb{R}^3$) obeys the law

$$m\mathbf{r}'' = -\frac{GmM}{r^3}\mathbf{r},$$

where M is the mass of the sun, m that of the planet, $r = \|\mathbf{r}\|$, and G is the gravitational constant. The relation used in determining the force, $\mathbf{F} = -GmM\mathbf{r}/r^3$, is called ***Newton's law of gravitation*** (see Figure 4.1.3). We shall not make a general study of such equations in this book, but content ourselves with the special case of circular orbits. (More general orbits—the conic sections—are discussed in the Internet supplement.)

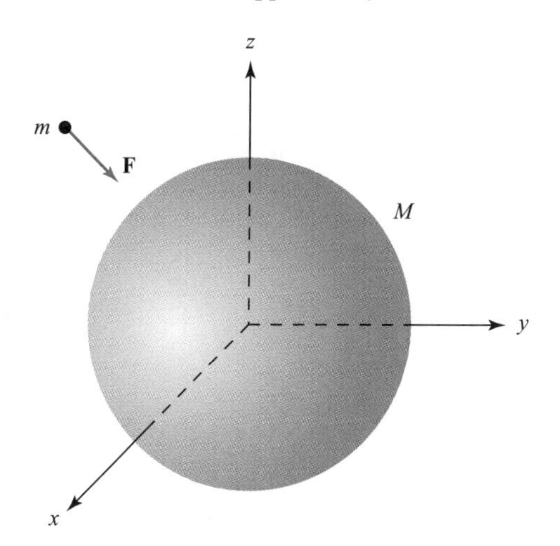

Figure 4.1.3 A mass M attracts a mass m with a force $\mathbf{F}$ given by Newton's law of gravitation: $\mathbf{F} = -GmM\mathbf{r}/r^3$.

Circular Orbits

Consider a particle of mass m moving at constant speed s in a circular path of radius r_0. Supposing that it moves in the xy plane, we can suppress the third component and write its location as

$$\mathbf{r}(t) = \left(r_0 \cos\frac{st}{r_0}, r_0 \sin\frac{st}{r_0}\right).$$

Note that this is a circle of radius r_0 and that its speed is given by $\|\mathbf{r}'(t)\| = s$. The quantity s/r_0 is called the ***frequency*** and is denoted ω. Thus,

$$\mathbf{r}(t) = (r_0 \cos \omega t,\ r_0 \sin \omega t).$$

The acceleration is given by

$$\mathbf{a}(t) = \mathbf{r}''(t) = \left(-\frac{s^2}{r_0}\cos\frac{st}{r_0},\ -\frac{s^2}{r_0}\sin\frac{st}{r_0}\right) = -\frac{s^2}{r_0^2}\mathbf{r}(t) = -\omega^2\mathbf{r}(t).$$

Thus, the acceleration is in a direction opposite to $\mathbf{r}(t)$; that is, it is directed toward the center of the circle (see Figure 4.1.4). This acceleration multiplied by the mass of the particle is called the ***centripetal force***. Even though the speed is constant, the direction of the velocity is continuously changing and therefore the acceleration, which is a rate of change in either speed or direction or both, is nonzero.

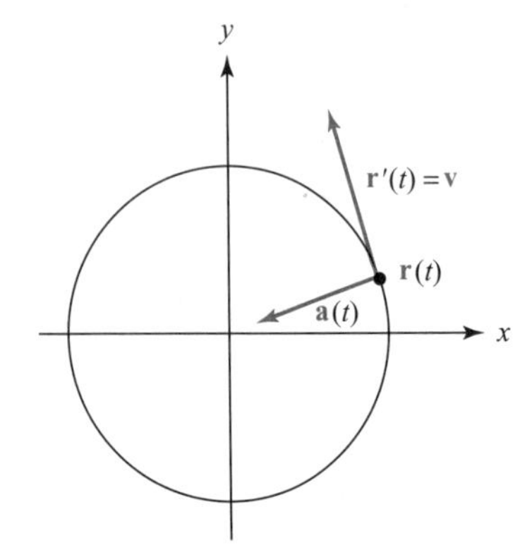

Figure 4.1.4 The position, velocity, and acceleration of a particle in circular motion.

Newton's law helps us discover a relationship between the radius of the orbit of a revolving body and the period, that is, the time it takes for one complete revolution. Consider a satellite of mass m moving with a speed s around a central body with mass M in a *circular* orbit of radius r_0 (distance from the *center* of the spherical central body). By Newton's second law $\mathbf{F} = m\mathbf{a}$, we get

$$-\frac{s^2 m}{r_0^2}\mathbf{r}(t) = -\frac{GmM}{r_0^3}\mathbf{r}(t).$$

The lengths of the vectors on both sides of this equation must be equal. Hence,

$$s^2 = \frac{GM}{r_0}.$$

If T denotes the period, then $s = 2\pi r_0/T$; substituting this value for s in the preceding equation and solving for T, we obtain the following:

Kepler's Law

$$T^2 = r_0^3\frac{(2\pi)^2}{GM}.$$

Thus, the *square of the period is proportional to the cube of the radius*.

We have defined two basic concepts associated with a path; its velocity and its acceleration. Both involve *differential* calculus. The basic concept of the length of a path, which involves *integral* calculus, will be taken up in the next section.

EXAMPLE 4 Suppose that a satellite is to be in a circular orbit about the earth such that it stays fixed in the sky over one point on the equator. What is the radius of such a *geosynchronous* orbit? (The mass of the earth is 5.98×10^{24} kilograms and $G = 6.67 \times 10^{-11}$ in the meter-kilogram-second system of units.)

SOLUTION The period of the satellite should be 1 day, so $T = 60 \times 60 \times 24 = 86{,}400$ seconds. From the formula $T^2 = r_0^3(2\pi)^2/GM$, we get $r_0^3 = T^2GM/(2\pi)^2$, and so

$$r_0^3 = \frac{T^2GM}{(2\pi)^2} = \frac{(86{,}400)^2 \times (6.67 \times 10^{-11}) \times (5.98 \times 10^{24})}{(2\pi)^2} \approx 7.54 \times 10^{22}\,\mathrm{m}^3.$$

Thus, $r_0 = 4.23 \times 10^7\,\mathrm{m} = 42{,}300\,\mathrm{km} \approx 26{,}200\,\mathrm{mi}$. ▲

Supplement to Section 4.1: Planetary Orbits, Hamilton's Principle, and Spacecraft Trajectories

In this section, we have been studying paths in space and Newton's second law. Hopefully, the student realizes that these ideas apply to the real world—the motion of our earth around the sun, for example, is governed by these laws. But there is more to the story, and we will try to convey some of it here.

Historical Note

Kepler, Newton, and Hamilton

As we discussed in the historical introduction, the law of planetary motion stating that the square of the period is proportional to the cube of the radius of an orbit is one of the three that Kepler observed before Newton formulated his laws of motion, known more generally as Newton's mechanics. These mechanics enable one to compute the period of a satellite about the earth or a planet about the sun (when the radius of its orbit is given), and, as we will indicate shortly, trajectories of space missions.

Kepler discovered and used results like this not only for circular orbits but more generally for elliptical orbits. Newton was able to derive Kepler's three celestial laws from his own law of gravitation. The neat mathematical order of the universe that these laws provided had a great impact on eighteenth-century thought.

Newton never wrote down his laws of mechanics as differential equations. This was first done by Euler around 1730. Newton made most of his deductions (at least those in published form) by geometric methods. Euler also showed how Newton's equations followed from Maupertuis's action principle. The clearest version of the action principle in mechanics, now known as ***Hamilton's principle***, is due to William Rowan Hamilton around 1830, who, as we all should now know, happens to also be the father of vector calculus. Hamilton's version of Maupertuis's principle was elegantly presented by Richard Feynman, as we discuss next.

Feynman and Hamilton's Principle

In his legendary Caltech *Lectures on Physics*, Nobel Prize–winning physicist Richard Phillips Feynman (see Figure 4.1.5) included what he called a "Special Lecture" on a topic clearly very close to his heart—one that he first heard about from his New York high school teacher, Mr. Bader. Mr. Bader told his (apparently bored) student Feynman how principles of maxima and minima apply to the trajectories of moving objects and in particular how the action principle of Maupertuis, Leibniz, and Hamilton (discussed in Section 3.3) applies to Newton's mechanics, governed by $\mathbf{F} = m\mathbf{a}$.

Figure 4.1.5 Richard P. Feynman (1918–1988).

Professor Feynman, at the end of his lecture, notes that "a physicist, a student of Mr. Bader, in 1942 showed how this action principle applied to quantum mechanics." That student was Feynman himself, who received the Nobel Prize for his insights, which also included the discovery of *Feynman integrals*. The moral here is *pay attention to your teachers—especially the best ones*!

We include the first part of Feynman's lecture here and more of it in the Internet supplement; see Volume II, Lecture 19, of the *Feynman Lectures on Physics* for the entire lecture.

The Principle of Least Action, by Richard Feynman

When I was in high school, my physics teacher—whose name was Mr. Bader—called me down one day after physics class and said, "You look bored; I want to tell you something interesting." Then he told me

Figure 4.1.6 Feynman lecturing at Caltech.

something which I found absolutely fascinating, and have, since then, always found fascinating. Every time the subject comes up, I work on it. In fact, when I began to prepare this lecture I found myself making more analyses on the thing. Instead of worrying about the lecture, I got involved in a new problem. The subject is this—the principle of least action.

Mr. Bader told me the following: Suppose you have a particle (in a gravitational field, for instance) which starts somewhere and moves to some other point by free motion—you throw it, and it goes up and comes down [see Figure 4.1.7].

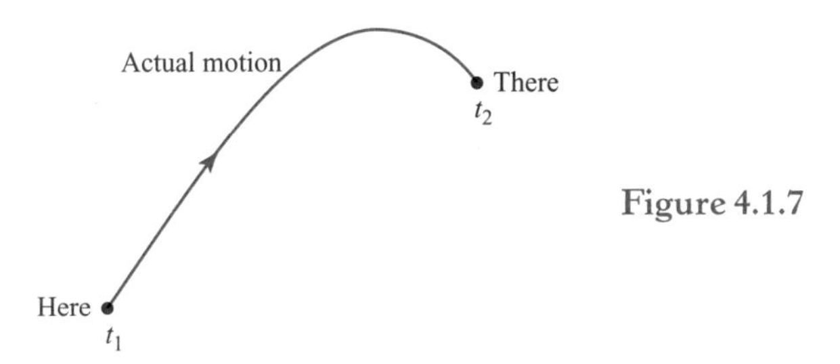

Figure 4.1.7

It goes from the original place to the final place in a certain amount of time. Now, you try a different motion. Suppose that to get from here to there, it went like this [see Figure 4.1.8], but got there in just the same amount of time.

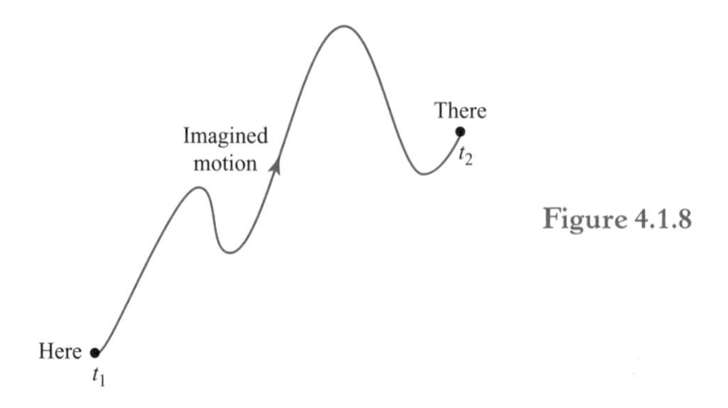

Figure 4.1.8

Then he said this: "If you calculate the kinetic energy at every moment on the path, take away the potential energy, and integrate it over the time during the whole path, you'll find that the number you'll get is bigger than that for the actual motion."

In other words, the laws of Newton could be stated not in the form $F = ma$ but in the form: The average kinetic energy less the average potential energy is as little as possible for the path of an object going from one point to another.

Let me illustrate a little better what this means. If you take the case of the gravitational field, then if the particle has the path $x(t)$ (let's just take one dimension for a moment; we take a trajectory that goes up and down and not sideways), where x is the height above the ground, the kinetic energy is $\frac{1}{2}m(dx/dt)^2$, and the potential energy at any time is mgx. Now I take the kinetic energy minus the potential energy at every moment along the path and integrate that with respect to time from the initial time to the final time. Let's suppose that at the original time t_1 we started at some height and at the end of the time t_2 we are definitely ending at some other place [see Figure 4.1.9].

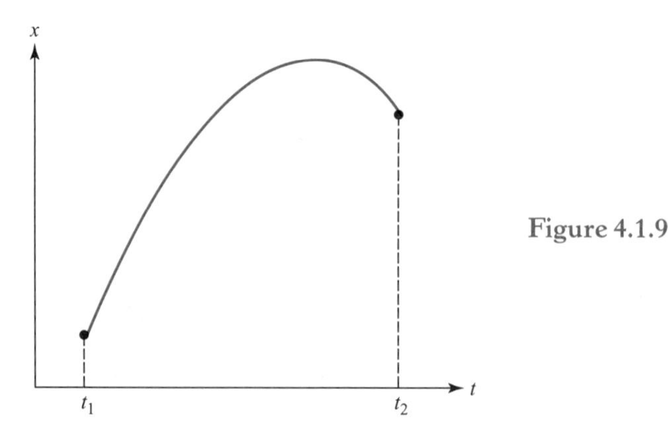

Figure 4.1.9

Then the integral is

$$\int_{t_1}^{t_2} \left[\frac{1}{2} m \left(\frac{dx}{dt} \right)^2 - mgx \right] dt.$$

The actual motion is some kind of curve—it's a parabola if we plot against the time—and gives a certain value for the integral. But we could *imagine* some other motion that went very high and came up and down in some peculiar way [see Figure 4.1.10].

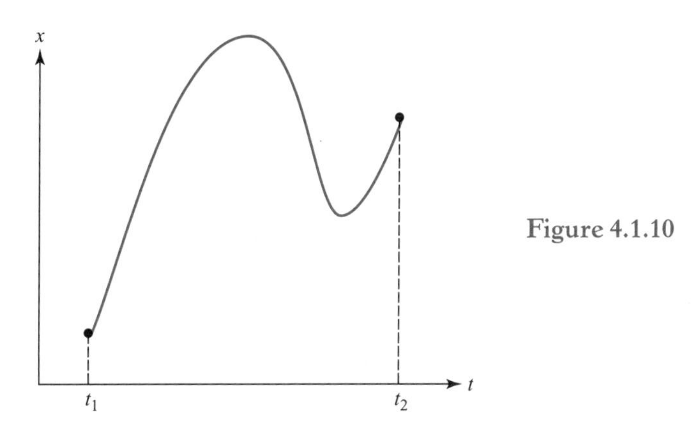

Figure 4.1.10

We can calculate the kinetic energy minus the potential energy and integrate for such a path . . . or for any other path we want. The miracle is that the true path is the one for which that *integral is least.*

Real-Life Trajectories

Interesting paths in $\mathbb{R}^3$ that obey Newton's second law occur in our own solar system and are used by NASA to plan space missions. One such mission, the *Genesis* Discovery Mission, launched from earth August 8, 2001 (and is due to return to earth in September 2004), has a particularly interesting trajectory, as shown in Figure 4.1.11. More information about this trajectory and the mission objectives can be found at http://genesismission.jpl.nasa.gov/.

The points denoted L_1 and L_2 in this figure denote places of balance (discovered by Euler) between the earth and the sun. A motionless spacecraft positioned there will remain there. There are periodic orbits about these points that we have (loosely)

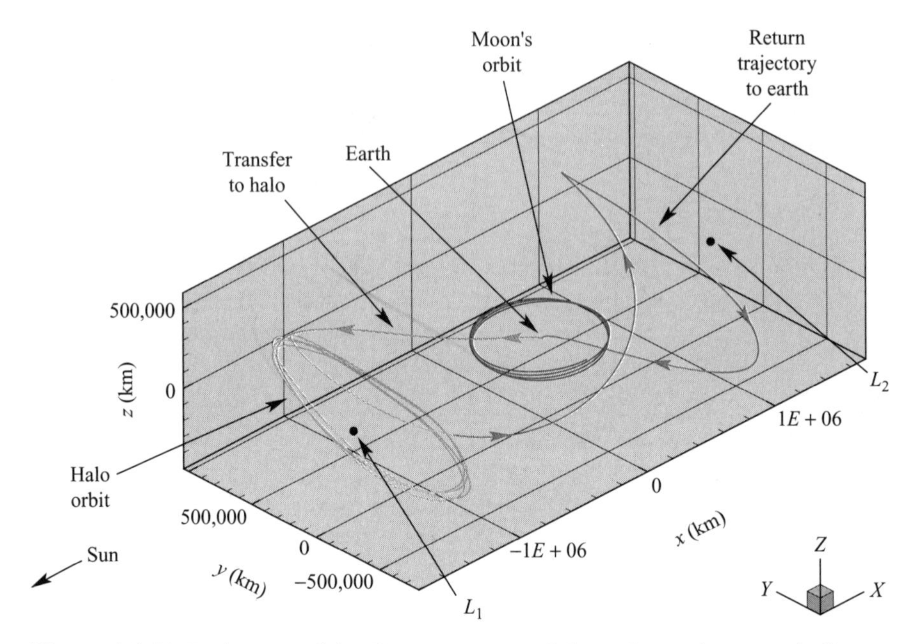

Figure 4.1.11 Trajectory of the *Genesis* spacecraft from the earth to a periodic orbit about a million and a half kilometers from earth and the interesting return trajectory to earth.

called *halo orbits*. The main dynamics of the spacecraft is governed by the pull of both the earth and the sun (and to a very small extent the moon) on the spacecraft. This is thus part of the famous *three-body problem* studied and made famous by Poincaré around 1890.[2]

Emmy Noether and Hamilton's Principle

Emmy Noether (1882–1935) (see Figure 4.1.12) is perhaps best known for her work in algebra, but she made a significant contribution to Hamilton's principle as well.[3] For planetary motion, the angular momentum vector $\mathbf{J} = \mathbf{r}(t) \times m\dot{\mathbf{r}}(t)$ is time-independent (so is a *conserved quantity*), as one can readily see by computing the time derivative of $\mathbf{J}$ and using $\mathbf{F} = m\mathbf{a}$ (see Exercise 20). What Noether discovered was a deep connection between such conserved quantities and symmetries in Hamilton's principle—in the case of angular momentum, this is rotational symmetry. Noether's discoveries have had a profound influence on the study of mechanical systems, from classical to quantum, ever since.

[2]For more information about Poincaré, see F. Diacu and P. Holmes, *Celestial Encounters. The Origins of Chaos and Stability*, Princeton University Press: Princeton, NJ, 1996.

[3]"Invariante Variationsprobleme," *Göttingen Math. Phys.* **2** (1918): 235–257.

Figure 4.1.12 Emmy Noether (1882–1935).

EXERCISES

In Exercises 1 to 4, find the velocity and acceleration vectors and the equation of the tangent line for each of the following curves, at the given value of t.

1. $\mathbf{r}(t) = (\cos t)\mathbf{i} + (\sin 2t)\mathbf{j}$, at $t = 0$

2. $\mathbf{c}(t) = (t\ \sin t, t\ \cos t, \sqrt{3}t)$, at $t = 0$

3. $\mathbf{r}(t) = \sqrt{2}t\mathbf{i} + e^t\mathbf{j} + e^{-t}\mathbf{k}$, at $t = 0$

4. $\mathbf{c}(t) = t\mathbf{i} + t\mathbf{j} + \frac{2}{3}t^{3/2}\mathbf{k}$, at $t = 9$

In Exercises 5 to 8, let $\mathbf{c}_1(t) = e^t\mathbf{i} + (\sin t)\mathbf{j} + t^3\mathbf{k}$ *and* $\mathbf{c}_2(t) = e^{-t}\mathbf{i} + (\cos t)\mathbf{j} - 2t^3\mathbf{k}$. *Find each of the stated derivatives in two different ways to verify the rules in the box preceding Example 1.*

5. $\dfrac{d}{dt}[\mathbf{c}_1(t) + \mathbf{c}_2(t)]$

6. $\dfrac{d}{dt}[\mathbf{c}_1(t) \cdot \mathbf{c}_2(t)]$

7. $\dfrac{d}{dt}[\mathbf{c}_1(t) \times \mathbf{c}_2(t)]$

8. $\dfrac{d}{dt}\{\mathbf{c}_1(t) \cdot [2\mathbf{c}_2(t) + \mathbf{c}_1(t)]\}$

9. If $\mathbf{r}(t) = 6t\mathbf{i} + 3t^2\mathbf{j} + t^3\mathbf{k}$, what force acts on a particle of mass m moving along $\mathbf{r}$ at $t = 0$?

10. Let a particle of mass 1 gram (g) follow the path in Exercise 1, with units in seconds and centimeters. What force acts on it at $t = 0$? (Give the units in your answer.)

11. A body of mass 2 kilograms moves on a circle of radius 3 meters, making one revolution every 5 seconds. Find the centripetal force acting on the body.

12. Find the centripetal force acting on a body of mass 4 kilograms, moving on a circle of radius 10 meters with a frequency of 2 revolutions per second.

13. Show that if the acceleration of an object is always perpendicular to the velocity, then the speed of the object is constant. (HINT: See Example 1.)

14. Show that, at a local maximum or minimum of $\|\mathbf{r}(t)\|$, the vector $\mathbf{r}'(t)$ is perpendicular to $\mathbf{r}(t)$.

15. A satellite is in a circular orbit 500 miles above the surface of the earth. What is the period of the orbit? (You may take the radius of the earth to be 4000 miles, or 6.436×10^6 meters).

16. What is the acceleration of the satellite in Exercise 15? The centripetal force?

17. Find the path $\mathbf{c}$ such that $\mathbf{c}(0) = (0, -5, 1)$ and $\mathbf{c}'(t) = (t, e^t, t^2)$.

18. Let $\mathbf{c}$ be a path in $\mathbb{R}^3$ with zero acceleration. Prove that $\mathbf{c}$ is a straight line or a point.

19. Find paths $\mathbf{c}(t)$ that represent the following curves or trajectories.

(a) $\{(x, y) \mid y = e^x\}$

(b) $\{(x, y) \mid 4x^2 + y^2 = 1\}$

(c) A straight line in $\mathbb{R}^3$ passing through the origin and the point (a, b, c)

(d) $\{(x, y) \mid 9x^2 + 16y^2 = 4\}$

20. Let $\mathbf{c}(t)$ be a path, $\mathbf{v}(t)$ its velocity, and $\mathbf{a}(t)$ the acceleration. Suppose $\mathbf{F}$ is a C^1 mapping of $\mathbb{R}^3$ to $\mathbb{R}^3$, $m > 0$, and $\mathbf{F}(\mathbf{c}(t)) = m\mathbf{a}(t)$ (Newton's second law). Prove that

$$\frac{d}{dt}[m\mathbf{c}(t) \times \mathbf{v}(t)] = \mathbf{c}(t) \times \mathbf{F}(\mathbf{c}(t))$$

(i.e., "rate of change of angular momentum = torque"). What can you conclude if $\mathbf{F}(\mathbf{c}(t))$ is parallel to $\mathbf{c}(t)$? Is this the case in planetary motion?

21. Continue the investigations in Exercise 20 to prove Kepler's law that a planet moving under the influence of gravity about the sun does so in a fixed plane.

4.2 Arc Length

Definition of Arc Length

What is the length of a path $\mathbf{c}(t)$? Because the speed $\|\mathbf{c}'(t)\|$ is the rate of change of distance traveled with respect to time, the distance traveled by a point moving along the curve should be the integral of speed with respect to the time over the interval

$[t_0, t_1]$ of travel time; that is, the length of the path, also called its ***arc length***, is

$$L(\mathbf{c}) = \int_{t_0}^{t_1} \|\mathbf{c}'(t)\| dt.$$

There is the question as to whether or not this formula actually corresponds to the true arc length. For example, suppose we take a curve in space and glue a string tightly to it, cutting the string so it exactly fits the curve. If we then remove the string, straighten it out and measure it with a straight edge, we surely should obtain the length of the curve. That our formula for arc length agrees with such a process is justified in the supplement at the end of this section.

EXAMPLE 1 The arc length of the path $\mathbf{c}(t) = (r \cos t, r \sin t)$, for t lying in the interval $[0, 2\pi]$; that is, for $0 \le t \le 2\pi$, is

$$L(\mathbf{c}) = \int_0^{2\pi} \sqrt{(-r \sin t)^2 + (r \cos t)^2}\, dt = 2\pi r,$$

which is the circumference of a circle of radius r. If we had allowed $0 \le t \le 4\pi$, we would have obtained $4\pi r$, because the path traverses the same circle *twice* (Figure 4.2.1). ▲

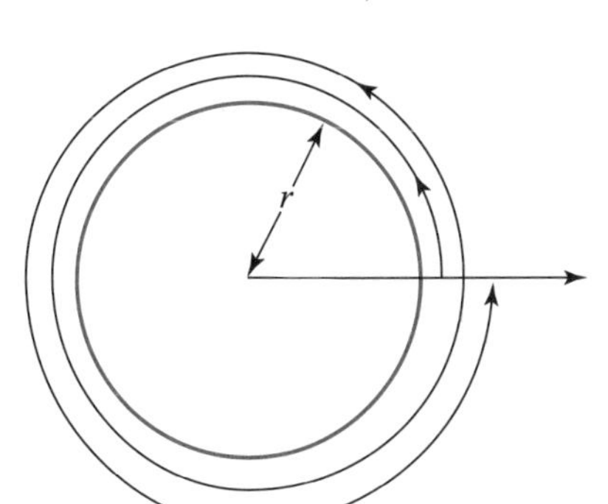

Figure 4.2.1 The arc length of a circle traversed twice is $4\pi r$.

Arc Length The length of the path $\mathbf{c}(t) = (x(t), y(t), z(t))$ for $t_0 \le t \le t_1$, is

$$L(\mathbf{c}) = \int_{t_0}^{t_1} \sqrt{[x'(t)]^2 + [y'(t)]^2 + [z'(t)]^2}\, dt.$$

For planar curves, one omits the $z'(t)$ term, as in Example 1. Here is an example in $\mathbb{R}^3$.

EXAMPLE 2 Find the arc length of $(\cos t, \sin t, t^2)$, $0 \le t \le \pi$.

SOLUTION The path $\mathbf{c}(t) = (\cos t, \sin t, t^2)$ has the velocity vector given by $\mathbf{v} = (-\sin t, \cos t, 2t)$. Because

$$\|\mathbf{v}\| = \sqrt{\sin^2 t + \cos^2 t + 4t^2} = \sqrt{1 + 4t^2} = 2\sqrt{t^2 + \left(\frac{1}{2}\right)^2},$$

the arc length is

$$L(\mathbf{c}) = \int_0^{\pi} 2\sqrt{t^2 + \left(\frac{1}{2}\right)^2}\, dt.$$

This integral may be evaluated using the following formula from the table of integrals:

$$\int \sqrt{x^2 + a^2}\, dx = \frac{1}{2}\left[x\sqrt{x^2 + a^2} + a^2 \log\left(x + \sqrt{x^2 + a^2}\right)\right] + C.$$

Thus,

$$L(\mathbf{c}) = 2 \cdot \frac{1}{2}\left[t\sqrt{t^2 + \left(\frac{1}{2}\right)^2} + \left(\frac{1}{2}\right)^2 \log\left(t + \sqrt{t^2 + \left(\frac{1}{2}\right)^2}\right)\right]\Bigg|_{t=0}^{\pi}$$

$$= \pi\sqrt{\pi^2 + \frac{1}{4}} + \frac{1}{4}\log\left(\pi + \sqrt{\pi^2 + \frac{1}{4}}\right) - \frac{1}{4}\log\left(\sqrt{\frac{1}{4}}\right)$$

$$= \frac{\pi}{2}\sqrt{1 + 4\pi^2} + \frac{1}{4}\log\left(2\pi + \sqrt{1 + 4\pi^2}\right) \approx 10.63.$$

As a check on our answer, we may note that the path $\mathbf{c}$ connects the points $(1, 0, 0)$ and $(-1, 0, \pi^2)$. The distance between these points is $\sqrt{4 + \pi^2} \approx 3.72$, which is less than 10.63, as it should be. ▲

If a curve is made up of a finite number of pieces each of which is C^1 (with bounded derivative), we compute the arc length by adding the lengths of the component pieces. Such curves are called ***piecewise*** C^1. Sometimes we just say "piecewise smooth."

EXAMPLE 3 A billiard ball on a pool table follows the path $\mathbf{c}\colon [-1, 1] \to \mathbb{R}^3$ defined by $\mathbf{c}(t) = (x(t), y(t), z(t)) = (|t|, |t - \frac{1}{2}|, 0)$. Find the distance traveled by the ball.

SOLUTION This path is not smooth, because $x(t) = |t|$ is not differentiable at 0, nor is $y(t) = |t - \frac{1}{2}|$ differentiable at $\frac{1}{2}$. However, if we divide the interval $[-1, 1]$ into the pieces $[-1, 0]$, $[0, \frac{1}{2}]$, and $[\frac{1}{2}, 1]$, we see that $x(t)$ and $y(t)$ have continuous derivatives on each of the intervals $[-1, 0]$, $[0, \frac{1}{2}]$, and $[\frac{1}{2}, 1]$. (See Figure 4.2.2.)

On $[-1, 0]$, $x(t) = -t$, $y(t) = -t + \frac{1}{2}$, and $z(t) = 0$, so $\|\mathbf{c}'(t)\| = \sqrt{2}$. Hence, the arc length of $\mathbf{c}$ between -1 and 0 is $\int_{-1}^{0} \sqrt{2}\, dt = \sqrt{2}$. Similarly, on $[0, \frac{1}{2}]$, $x(t) = t$, $y(t) = -t + \frac{1}{2}$, $z(t) = 0$, and again $\|\mathbf{c}'(t)\| = \sqrt{2}$, so that the arc length of $\mathbf{c}$ between 0

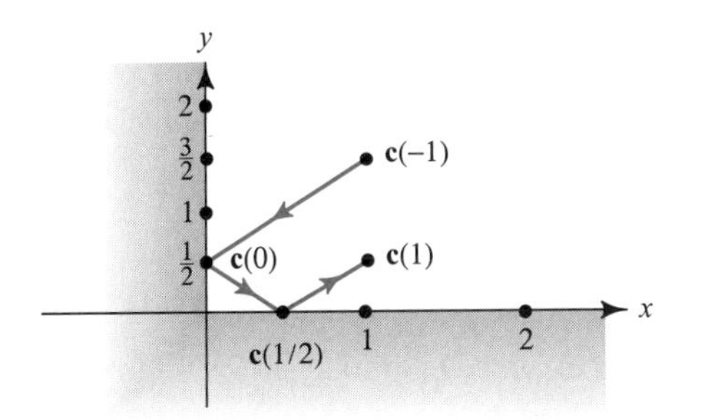

Figure 4.2.2 A piecewise smooth path.

and $\frac{1}{2}$ is $\frac{1}{2}\sqrt{2}$. Finally, on $[\frac{1}{2}, 1]$ we have $x(t) = t$, $y(t) = t - \frac{1}{2}$, $z(t) = 0$, and the arc length of **c** between $\frac{1}{2}$ and 1 is $\frac{1}{2}\sqrt{2}$. Thus, the total arc length of **c** is $2\sqrt{2}$. Of course, one can also compute the answer as the sum of the distances from $\mathbf{c}(-1)$ to $\mathbf{c}(0)$ to $\mathbf{c}(\frac{1}{2})$ to $\mathbf{c}(1)$. ▲

EXAMPLE 4 Consider the point with position function

$$\mathbf{c}(t) = (t - \sin t, 1 - \cos t),$$

which traces out the cycloid discussed in Section 2.4 (see Figure 2.4.6). Find the velocity, the speed, and the length of one arch.

SOLUTION The velocity vector is $\mathbf{c}'(t) = (1 - \cos t, \sin t)$, so the speed of the point $\mathbf{c}(t)$ is

$$\|\mathbf{c}'(t)\| = \sqrt{(1 - \cos t)^2 + \sin^2 t} = \sqrt{2 - 2\cos t}.$$

Hence, $\mathbf{c}(t)$ moves at variable speed although the circle rolls at constant speed. Furthermore, the speed of $\mathbf{c}(t)$ is zero when t is an integral multiple of 2π. At these values of t, the y coordinate of the point $\mathbf{c}(t)$ is zero and so the point lies on the x axis. The arc length of one cycle is

$$\begin{aligned} L(\mathbf{c}) &= \int_0^{2\pi} \sqrt{2 - 2\cos t}\, dt = 2\int_0^{2\pi} \sqrt{\frac{1 - \cos t}{2}}\, dt \\ &= 2\int_0^{2\pi} \sin\frac{t}{2}\, dt \left(\text{because } 1 - \cos t = 2\sin^2\frac{t}{2} \text{ and } \sin\frac{t}{2} \geq 0 \text{ on } [0, 2\pi]\right) \\ &= 4\left(-\cos\frac{t}{2}\right)\Big|_0^{2\pi} = 8. \quad ▲ \end{aligned}$$

The Differential of Arc Length

The arc-length formula suggests that one introduce the following notation, which will be useful in Chapter 7 in our discussion of line integrals.

Arc-Length Differential An ***infinitesimal displacement*** of a particle following a path $\mathbf{c}(t) = x(t)\mathbf{i} + y(t)\mathbf{j} + z(t)\mathbf{k}$ is

$$d\mathbf{s} = dx\mathbf{i} + dy\mathbf{j} + dz\mathbf{k} = \left(\frac{dx}{dt}\mathbf{i} + \frac{dy}{dt}\mathbf{j} + \frac{dz}{dt}\mathbf{k}\right) dt,$$

and its length

$$ds = \sqrt{dx^2 + dy^2 + dz^2} = \sqrt{\left(\frac{dx}{dt}\right)^2 + \left(\frac{dy}{dt}\right)^2 + \left(\frac{dz}{dt}\right)^2}\, dt$$

is the ***differential of arc length***. See Figure 4.2.3.

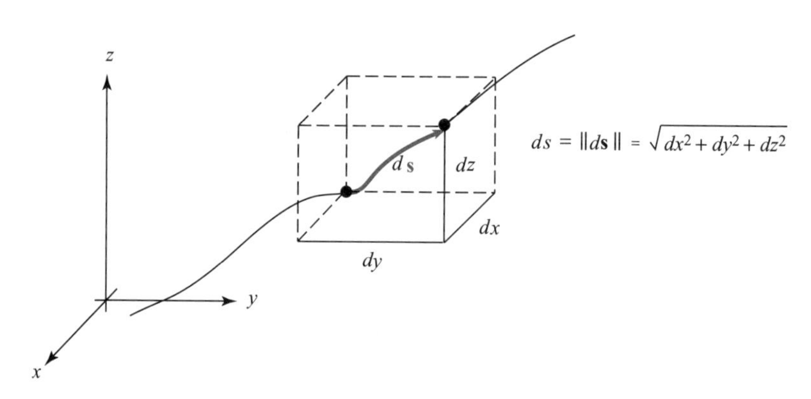

Figure 4.2.3 Differential of arc length.

These formulas help us remember the arc-length formula as

$$\text{arc length} = \int_{t_0}^{t_1} ds.$$

As we have done before with such geometric concepts as length and angle, we can extend the notion of arc length to paths in n-dimensional space.

Arc Length in $\mathbb{R}^n$ Let $\mathbf{c}\colon [t_0, t_1] \to \mathbb{R}^n$ be a piecewise C^1 path. Its ***length*** is defined to be

$$L(\mathbf{c}) = \int_{t_0}^{t_1} \|\mathbf{c}'(t)\|\, dt.$$

The integrand is the square root of the sum of the squares of the coordinate functions of $\mathbf{c}'(t)$: If

$$\mathbf{c}(t) = (x_1(t), x_2(t), \ldots, x_n(t)),$$

then

$$L(\mathbf{c}) = \int_{t_0}^{t_1} \sqrt{(x_1'(t))^2 + (x_2'(t))^2 + \cdots + (x_n'(t))^2}\, dt.$$

EXAMPLE 5 Find the length of the path $\mathbf{c}(t) = (\cos t,\ \sin t,\ \cos 2t, \sin 2t)$ in $\mathbb{R}^4$, defined on the interval from 0 to π.

SOLUTION We have $\mathbf{c}'(t) = (-\sin t,\ \cos t,\ -2\ \sin 2t, 2\ \cos 2t)$, and so

$$\|\mathbf{c}'(t)\| = \sqrt{\sin^2 t + \cos^2 t + 4\ \sin^2 2t + 4\ \cos^2 2t} = \sqrt{1+4} = \sqrt{5},$$

a constant, so the length of the path is

$$\int_0^{\pi} \sqrt{5}\, dt = \sqrt{5}\pi. \quad \blacktriangle$$

It is common to introduce the ***arc-length function*** $s(t)$ associated to a path $\mathbf{c}(t)$ given by

$$s(t) = \int_a^t \|\mathbf{c}'(u)\|\, du,$$

so that (by the fundamental theorem of calculus)

$$s'(t) = \|\mathbf{c}'(t)\|$$

and

$$\int_a^b s'(t)\, dt = s(b) - s(a) = s(b).$$

EXAMPLE 6 Consider the graph of a function of one variable $y = f(x)$ for x in the interval $[a, b]$. We can consider it to be a curve parametrized by $t = x$, namely, $\mathbf{c}(x) = (x, f(x))$ for x ranging from a to b. The arc-length formula gives

$$L(\mathbf{c}) = \int_a^b \sqrt{1 + [f'(x)]^2}\, dx,$$

which agrees with the formula for the length of a graph from one-variable calculus. ▲

Justification for the Arc-Length Formula

The following discussion assumes an acquaintance with the definite integral defined in terms of Riemann sums. If your background in this topic needs reinforcement, the material may be postponed until after Chapter 5.

In $\mathbb{R}^3$ there is another way to justify the arc-length formula based on polygonal approximations. We partition the interval $[a, b]$ into N subintervals of equal length:

$$a = t_0 < t_1 < \cdots < t_N = b;$$

$$t_{i+1} - t_i = \frac{b-a}{N} \qquad \text{for} \qquad 0 \le i \le N-1.$$

We then consider the polygonal line obtained by joining the successive pairs of points $\mathbf{c}(t_i)$, $\mathbf{c}(t_{i+1})$ for $0 \le i \le N-1$. This yields a polygonal approximation to $\mathbf{c}$ as in Figure 4.2.4. By the formula for distance in $\mathbb{R}^3$, it follows that the line segment from $\mathbf{c}(t_i)$ to $\mathbf{c}(t_{i+1})$ has length

$$\|\mathbf{c}(t_{i+1}) - \mathbf{c}(t_i)\| = \sqrt{[x(t_{i+1}) - x(t_i)]^2 + [y(t_{i+1}) - y(t_i)]^2 + [z(t_{i+1}) - z(t_i)]^2},$$

where $\mathbf{c}(t) = (x(t), y(t), z(t))$. Applying the mean-value theorem to $x(t)$, $y(t)$, and $z(t)$ on $[t_i, t_{i+1}]$, we obtain three points t_i^*, t_i^{**}, and t_i^{***} such that

$$x(t_{i+1}) - x(t_i) = x'(t_i^*)(t_{i+1} - t_i),$$
$$y(t_{i+1}) - y(t_i) = y'(t_i^{**})(t_{i+1} - t_i),$$

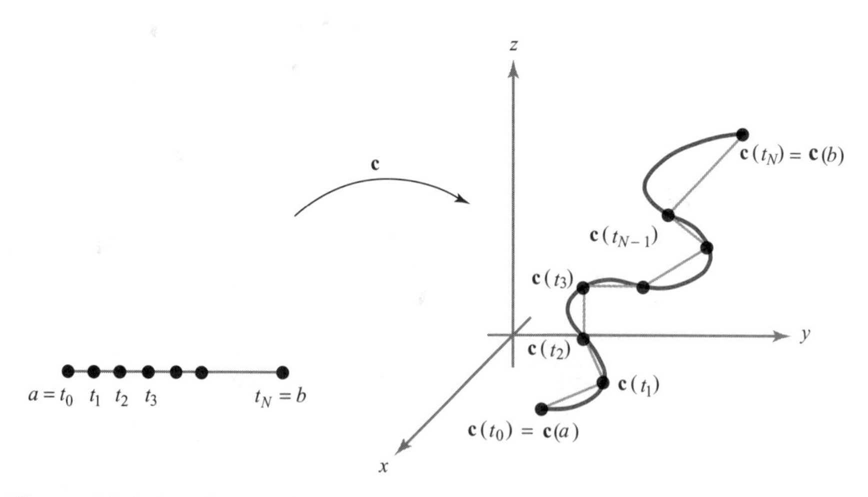

Figure 4.2.4 A path $\mathbf{c}$ may be approximated by a polygonal path obtained by joining each $\mathbf{c}(t_i)$ to $\mathbf{c}(t_{i+1})$ by a straight line.

and

$$z(t_{i+1}) - z(t_i) = z'(t_i^{***})(t_{i+1} - t_i).$$

Thus, the line segment from $\mathbf{c}(t_i)$ to $\mathbf{c}(t_{i+1})$ has length

$$\sqrt{[x'(t_i^*)]^2 + [y'(t_i^{**})]^2 + [z'(t_i^{***})]^2}(t_{i+1} - t_i).$$

Therefore, the length of our approximating polygonal line is

$$S_N = \sum_{i=0}^{N-1} \sqrt{[x'(t_i^*)]^2 + [y'(t_i^{**})]^2 + [z'(t_i^{***})]^2}(t_{i+1} - t_i).$$

As $N \to \infty$, this polygonal line approximates the image of $\mathbf{c}$ more closely. Therefore, we define the arc length of $\mathbf{c}$ as the limit, if it exists, of the sequence S_N as $N \to \infty$. Because the derivatives x', y', and z' are all assumed to be continuous on $[a, b]$, we can conclude that, in fact, the limit does exist and is given by

$$\operatorname*{limit}_{N\to\infty} S_N = \int_a^b \sqrt{[x'(t)]^2 + [y'(t)]^2 + [z'(t)]^2}\, dt.$$

(The theory of integration relates the integral to sums by the formula

$$\int_a^b f(t)\, dt = \operatorname*{limit}_{N\to\infty} \sum_{i=0}^{N-1} f(t_i^*)(t_{i+1} - t_i),$$

where $t_0, \ldots, t_N$ is a partition of $[a, b]$, $t_i^* \in [t_i, t_{i+1}]$ is arbitrary, and f is a continuous function. Here we have *possibly different points* t_i^*, t_i^{**}, and t_i^{***}, and so this formula must be extended slightly.)

EXERCISES

Find the arc length of the given curve on the specified interval in Exercises 1 to 6.[4]

1. $(2\cos t, 2\sin t, t)$, for $0 \le t \le 2\pi$

2. $(1, 3t^2, t^3)$, for $0 \le t \le 1$

3. $(\sin 3t, \cos 3t, 2t^{3/2})$, for $0 \le t \le 1$

---

[4]Several of these problems make use of the formula

$$\int \sqrt{x^2 + a^2}\, dx = \tfrac{1}{2}\left[x\sqrt{x^2 + a^2} + a^2 \log\left(x + \sqrt{x^2 + a^2}\right)\right] + C$$

from the table of integrals in the back of the book.

4. $\left(t+1, \frac{2\sqrt{2}}{3}t^{3/2}+7, \frac{1}{2}t^2\right)$, for $1 \leq t \leq 2$ **5.** (t, t, t^2), for $1 \leq t \leq 2$

6. $(t, t\ \sin t, t\ \cos t)$, for $0 \leq t \leq \pi$

7. Find the length of the path $\mathbf{c}(t)$, defined by $\mathbf{c}(t) = (2\cos t, 2\sin t, t)$, if $0 \leq t \leq 2\pi$ and $\mathbf{c}(t) = (2, t - 2\pi, t)$, if $2\pi \leq t \leq 4\pi$.

8. Let $\mathbf{c}$ be the path $\mathbf{c}(t) = (t, t\ \sin t, t\ \cos t)$. Find the arc length of $\mathbf{c}$ between the two points $(0, 0, 0)$ and $(\pi, 0, -\pi)$.

9. Let $\mathbf{c}$ be the path $\mathbf{c}(t) = (2t, t^2, \log t)$, defined for $t > 0$. Find the arc length of $\mathbf{c}$ between the points $(2, 1, 0)$ and $(4, 4, \log 2)$.

10. The arc-length function $s(t)$ for a given path $\mathbf{c}(t)$, defined by $s(t) = \int_a^t \|\mathbf{c}'(\tau)\|\, d\tau$, represents the distance a particle traversing the trajectory of $\mathbf{c}$ will have traveled by time t if it starts out at time a; that is, it gives the length of $\mathbf{c}$ between $\mathbf{c}(a)$ and $\mathbf{c}(t)$. Find the arc-length functions for the curves $\alpha(t) = (\cosh t,\ \sinh t, t)$ and $\beta(t) = (\cos t,\ \sin t, t)$, with $a = 0$.

11. Let $\mathbf{c}(t)$ be a given path, $a \leq t \leq b$. Let $s = \alpha(t)$ be a new variable, where α is a strictly increasing C^1 function given on $[a, b]$. For each s in $[\alpha(a), \alpha(b)]$ there is a unique t with $\alpha(t) = s$. Define the function $\mathbf{d}\colon [\alpha(a), \alpha(b)] \to \mathbb{R}^3$ by $\mathbf{d}(s) = \mathbf{c}(t)$.

(a) Argue that the image curves of $\mathbf{c}$ and $\mathbf{d}$ are the same.
(b) Show that $\mathbf{c}$ and $\mathbf{d}$ have the same arc length.
(c) Let $s = \alpha(t) = \int_a^t \|\mathbf{c}'(\tau)\|\, d\tau$. Define $\mathbf{d}$ as above by $\mathbf{d}(s) = \mathbf{c}(t)$. Show that

$$\left\| \frac{d}{ds}\mathbf{d}(s) \right\| = 1.$$

The path $s \mapsto \mathbf{d}(s)$ is said to be an ***arc length reparametrization*** of $\mathbf{c}$ (see also Exercise 13).

Exercises 12 to 17 develop some of the classic differential geometry of curves.

12. Let $\mathbf{c}\colon [a, b] \to \mathbb{R}^3$ be an infinitely differentiable path (derivatives of all orders exist). Assume $\mathbf{c}'(t) \neq \mathbf{0}$ for any t. The vector $\mathbf{c}'(t)/\|\mathbf{c}'(t)\| = \mathbf{T}(t)$ is tangent to $\mathbf{c}$ at $\mathbf{c}(t)$, and, because $\|\mathbf{T}(t)\| = 1$, $\mathbf{T}$ is called the ***unit tangent*** to $\mathbf{c}$.

(a) Show that $\mathbf{T}'(t) \cdot \mathbf{T}(t) = 0$. [HINT: Differentiate $\mathbf{T}(t) \cdot \mathbf{T}(t) = 1$.]
(b) Write down a formula for $\mathbf{T}'(t)$ in terms of $\mathbf{c}$.

13. (a) A path $\mathbf{c}(s)$ is said to be ***parametrized by arc length*** or, what is the same thing, to have ***unit speed*** if $\|\mathbf{c}'(s)\| = 1$. For a path parametrized by arc length on $[a, b]$, show that $l(\mathbf{c}) = b - a$.

(b) The ***curvature*** at a point $\mathbf{c}(s)$ on a path is defined by $k = \|\mathbf{T}'(s)\|$ when the path is parametrized by arc length. Show that $k = \|\mathbf{c}''(s)\|$.

(c) If $\mathbf{c}$ is given in terms of some other parameter t and $\mathbf{c}'(t)$ is never $\mathbf{0}$, show that $k = \|\mathbf{c}'(t) \times \mathbf{c}''(t)\|/\|\mathbf{c}'(t)\|^3$.

(d) Calculate the curvature of the helix $\mathbf{c}(t) = (1/\sqrt{2})(\cos t,\ \sin t, t)$. (This $\mathbf{c}$ is a scalar multiple of the right-circular helix.)

14. If $\mathbf{T}'(t) \neq \mathbf{0}$, it follows from Exercise 12 that $\mathbf{N}(t) = \mathbf{T}'(t)/\|\mathbf{T}'(t)\|$ is normal (i.e., perpendicular) to $\mathbf{T}(t)$; $\mathbf{N}$ is called the ***principal normal vector***. Let a third unit vector that is perpendicular to both $\mathbf{T}$ and $\mathbf{N}$ be defined by $\mathbf{B} = \mathbf{T} \times \mathbf{N}$; $\mathbf{B}$ is called the ***binormal vector***. Together, $\mathbf{T}$, $\mathbf{N}$, and $\mathbf{B}$ form a right-handed system of mutually orthogonal vectors that may be thought of as moving along the path (Figure 4.2.5). Show that

(a) $\dfrac{d\mathbf{B}}{dt} \cdot \mathbf{B} = 0$.

(b) $\dfrac{d\mathbf{B}}{dt} \cdot \mathbf{T} = 0$.

(c) $d\mathbf{B}/dt$ is a scalar multiple of $\mathbf{N}$.

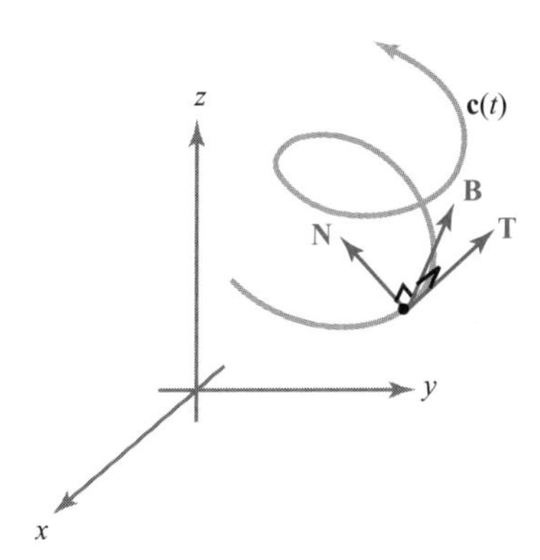

Figure 4.2.5 The tangent **T**, principal normal **N**, and binormal **B**.

15. If $\mathbf{c}(s)$ is parametrized by arc length, we use the result of Exercise 14(c) to define a scalar-valued function τ, called the ***torsion***, by

$$\frac{d\mathbf{B}}{ds} = -\tau\mathbf{N}.$$

(a) Show that $\tau = [\mathbf{c}'(s) \times \mathbf{c}''(s)] \cdot \mathbf{c}'''(s)/\|\mathbf{c}''(s)\|^2$.
(b) Show that if $\mathbf{c}$ is given in terms of some other parameter t,

$$\tau = \frac{[\mathbf{c}'(t) \times \mathbf{c}''(t)] \cdot \mathbf{c}'''(t)}{\|\mathbf{c}'(t) \times \mathbf{c}''(t)\|^2}.$$

Compare with Exercise 13(c).
(c) Compute the torsion of the helix $\mathbf{c}(t) = (1/\sqrt{2})(\cos t,\ \sin t, t)$.

16. Show that if a path lies in a plane, then the torsion is zero. Do this by demonstrating that $\mathbf{B}$ is constant and is a normal vector to the plane in which $\mathbf{c}$ lies. (If the torsion is not zero, it gives a measure of how fast the curve is twisting out of the plane of $\mathbf{T}$ and $\mathbf{N}$.)

17. (a) Use the results of Exercises 13, 14, and 15 to prove the following ***Frenet formulas*** for a unit-speed curve:

$$\frac{d\mathbf{T}}{ds} = k\mathbf{N}; \qquad \frac{d\mathbf{N}}{ds} = -k\mathbf{T} + \tau\mathbf{B}; \qquad \frac{d\mathbf{B}}{ds} = -\tau\mathbf{N}.$$

(b) Reexpress the results of part (a) as

$$\frac{d}{ds}\begin{pmatrix}\mathbf{T}\\ \mathbf{N}\\ \mathbf{B}\end{pmatrix} = \boldsymbol{\omega} \times \begin{pmatrix}\mathbf{T}\\ \mathbf{N}\\ \mathbf{B}\end{pmatrix}$$

for a suitable vector $\boldsymbol{\omega}$.

18. In special relativity, the ***proper time*** of a path $\mathbf{c}\colon [a, b] \to \mathbb{R}^4$ with components given by $\mathbf{c}(\lambda) = (x(\lambda), y(\lambda), z(\lambda), t(\lambda))$ is defined to be the quantity

$$\int_a^b \sqrt{-[x'(\lambda)]^2 - [y'(\lambda)]^2 - [z'(\lambda)]^2 + c^2[t'(\lambda)]^2}\, d\lambda,$$

where c is the velocity of light, a constant. In Figure 4.2.6, show that, using self-explanatory notation, the "twin paradox inequality" holds:

proper time (AB) + proper time (BC) < proper time (AC).

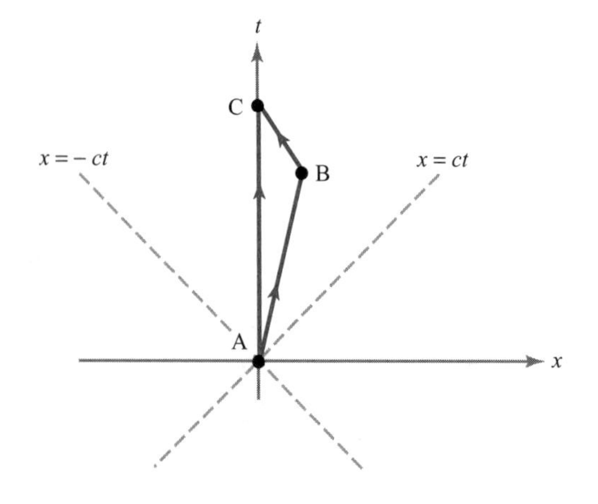

Figure 4.2.6 The relativistic triangle inequality.

19. The early Greeks knew that a straight line was the shortest possible path between two points. Euclid, in his book *Optics*, stated the "principle of the reflection of light"—that is, light traveling in a plane travels in a straight line, and when it is reflected across a mirror, the angle of incidence equals the angle of reflection.

The Greeks could not have had a proof that straight lines provided the shortest path between two points because they, in the first place, had no definition of the length of a path. They saw this property of straight lines as more or less "obvious."

Using the justification of arc length in this section and the triangle inequality of Section 1.5, argue that if $\mathbf{c}_0$ is the straight-line path $c_0(t) = t\mathrm{P} + (1 - t)\mathrm{Q}$ between P and Q in $\mathbb{R}^3$, then

$$L(\mathbf{c}_0) \leq L(\mathbf{c})$$

for any other path $\mathbf{c}$ joining P and Q.

4.3 Vector Fields

The Concept of a Vector Field

In Chapter 2, we introduced a particular kind of vector field, the gradient. In this section we study *general* vector fields, discussing their geometric and physical significance.

> Vector Fields A ***vector field*** in $\mathbb{R}^n$ is a map $\mathbf{F}: A \subset \mathbb{R}^n \to \mathbb{R}^n$ that assigns to each point $\mathbf{x}$ in its domain A a vector $\mathbf{F}(\mathbf{x})$. If $n = 2$, $\mathbf{F}$ is called a ***vector field in the plane***, and if $n = 3$, $\mathbf{F}$ is a ***vector field in space***.

Picture $\mathbf{F}$ as attaching an *arrow* to each point (Figure 4.3.1). By contrast, a map $f: A \subset \mathbb{R}^n \to \mathbb{R}$ that assigns a *number* to each point is a ***scalar field***. A vector field $\mathbf{F}(x, y, z)$ on $\mathbb{R}^3$ has three ***component scalar fields*** F_1, F_2, and F_3, so that

$$\mathbf{F}(x, y, z) = (F_1(x, y, z), F_2(x, y, z), F_3(x, y, z)).$$

Similarly, a vector field on $\mathbb{R}^n$ has n components $F_1, \ldots, F_n$. If each component is a C^k function, we say the vector field $\mathbf{F}$ is of ***class*** C^k. Vector fields will be assumed to be at least of class C^1 unless otherwise noted.

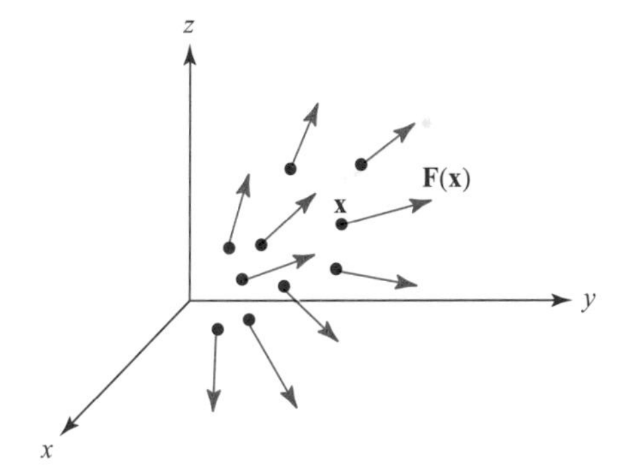

Figure 4.3.1 A vector field $\mathbf{F}$ assigns a vector $\mathbf{F}(\mathbf{x})$ to each point $\mathbf{x}$ of its domain.

In many applications, $\mathbf{F}(\mathbf{x})$ represents a physical vector quantity (force, velocity, etc.) associated with the position $\mathbf{x}$, as in the following examples.

EXAMPLE 1 The flow of water through a pipe is said to be ***steady*** if, at each point inside the pipe, the velocity of the fluid passing through that point does not change with time. (Note that this is quite different from saying that the water in the pipe is not moving.) Attaching to each point the fluid velocity at that point, we obtain the ***velocity field*** $\mathbf{V}$ of the fluid (see Figure 4.3.2). Notice that the length of the arrows (the speed), as well as the direction of flow, may change from point to point. ▲

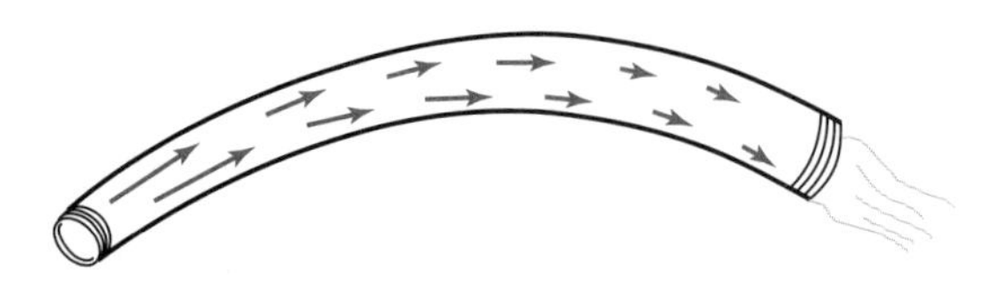

Figure 4.3.2 A vector field describing the velocity of flow in a pipe.

EXAMPLE 2 Some forms of rotary motion (such as the motion of particles on a turntable) can be described by the vector field

$$\mathbf{V}(x, y) = -y\mathbf{i} + x\mathbf{j}.$$

See Figure 4.3.3, in which we have shown instead of $\mathbf{V}$ the shorter vector field $\frac{1}{4}\mathbf{V}$ so that the arrows do not overlap. This is a common convention in drawing pictures of vector fields. ▲

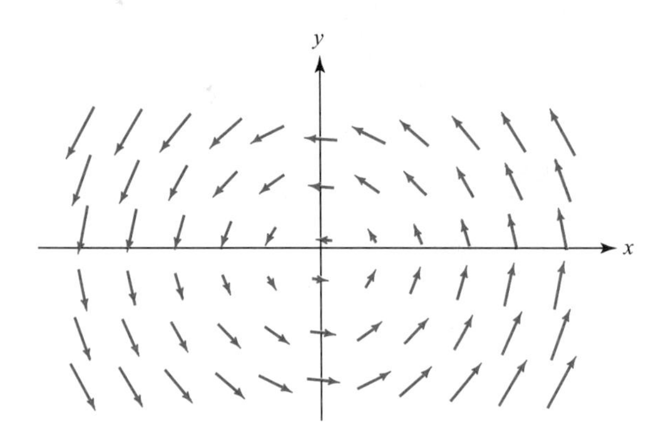

Figure 4.3.3 A rotary vector field.

EXAMPLE 3 In the plane, $\mathbb{R}^2$, let the vector field x be defined by

$$\mathbf{V}(x, y) = \frac{y\mathbf{i}}{x^2 + y^2} - \frac{x\mathbf{j}}{x^2 + y^2} = \left(\frac{y}{x^2 + y^2}, -\frac{x}{x^2 + y^2}\right)$$

(except at the origin, where $\mathbf{V}$ is not defined). This vector field is a good approximation to the planar part of the velocity of water flowing toward a hole in the bottom of a tub (Figure 4.3.4). Notice that the velocity becomes *larger* as you approach the hole. ▲

Gradient Vector Fields

In Section 2.6 we introduced the gradient of a function by

$$\nabla f(x, y, z) = \frac{\partial f}{\partial x}(x, y, z)\mathbf{i} + \frac{\partial f}{\partial y}(x, y, z)\mathbf{j} + \frac{\partial f}{\partial z}(x, y, z)\mathbf{k}.$$

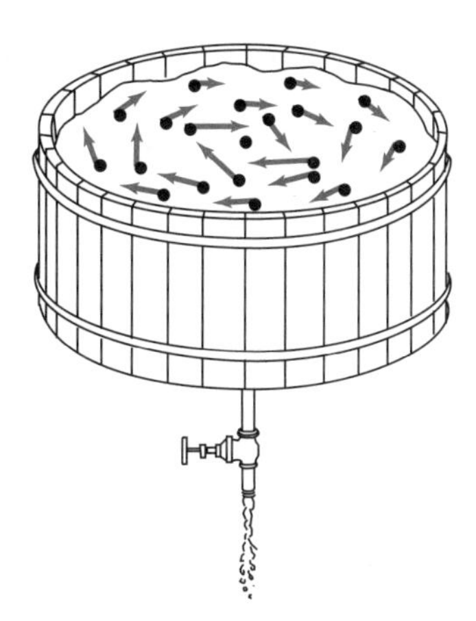

Figure 4.3.4 The vector field describing circular flow in a tub.

Now we want to think of this as an example of a vector field—it assigns a vector to each point (x, y, z). As such, we refer to ∇f as a ***gradient vector field***. Gradient fields come up in a variety of situations, as the next two examples show.

EXAMPLE 4 A piece of material is heated on one side and cooled on another. The temperature at each point within the body is described at a given moment by a scalar field $T(x, y, z)$. The flow of heat may be marked by a field of arrows indicating the direction and magnitude of the flow (Figure 4.3.5). This ***energy*** or ***heat flux vector field*** is given by $\mathbf{J} = -k\nabla T$, where $k > 0$ is a constant called the ***conductivity*** and ∇T is the gradient of the real-valued function T. Level sets of T are called ***isotherms***. Note that the heat flows from hot regions toward cold ones, since $-\nabla T$ points in the direction of decreasing T. ▲

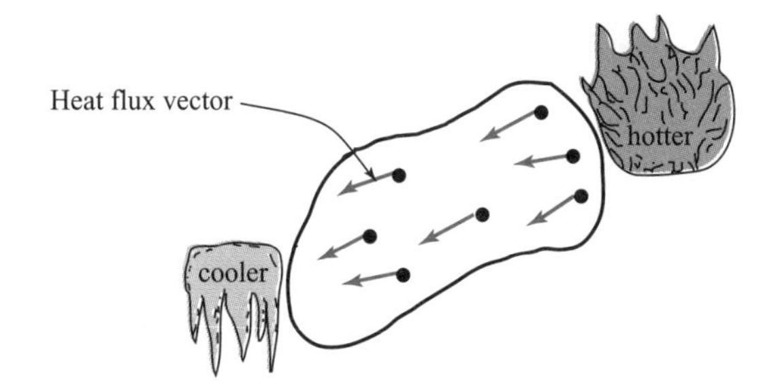

Figure 4.3.5 A vector field describing the direction and magnitude of heat flow.

EXAMPLE 5 The force of attraction of the earth on a mass m can be described by a vector field called the ***gravitational force field***. Place the origin of a coordinate system at the center of the earth (assumed spherical). According to Newton's law of gravity, this field is given by

$$\mathbf{F} = -\frac{mMG}{r^3}\mathbf{r},$$

where $\mathbf{r}(x, y, z) = (x, y, z)$, and $r = \|\mathbf{r}\|$ (see Figure 4.3.6). The domain of this vector field consists of those $\mathbf{r}$ for which $\|\mathbf{r}\|$ is greater than the radius of the earth. As we saw in Example 6, Section 2.6, $\mathbf{F}$ is a gradient field, $\mathbf{F} = -\nabla V$, where

$$V = -\frac{mMG}{r}$$

is the ***gravitational potential***. Note again that $\mathbf{F}$ points in the direction of *decreasing* V. Writing $\mathbf{F}$ in terms of components, we see that

$$\mathbf{F}(x, y, z) = \left(\frac{-mMG}{r^3}x, \frac{-mMG}{r^3}y, \frac{-mMG}{r^3}z\right). \quad \blacktriangle$$

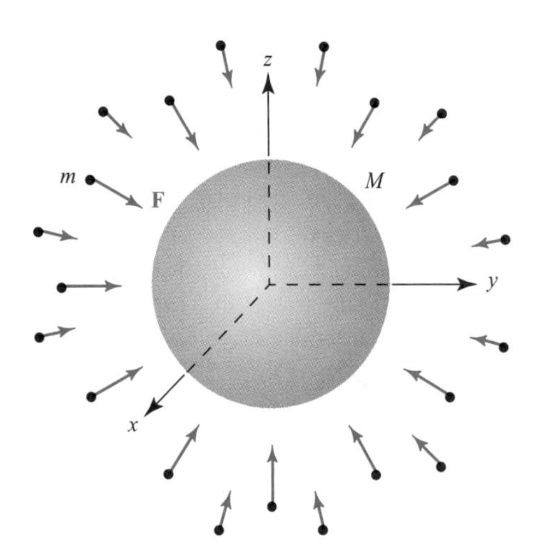

Figure 4.3.6 The vector field $\mathbf{F}$ given by Newton's law of gravitation.

EXAMPLE 6 According to ***Coulomb's law***, the force acting on a charge e at position $\mathbf{r}$ due to a charge Q at the origin is

$$\mathbf{F} = \frac{\varepsilon Qe}{r^3}\mathbf{r} = -\nabla V,$$

where $V = \varepsilon Qe/r$ and ε is a constant that depends on the units used. For $Qe > 0$ (like charges) the force is repulsive [Figure 4.3.7(a)], and for $Qe < 0$ (unlike charges) the force is attractive [Figure 4.3.7(b)]. Because the potential V is constant on the level surfaces of V, they are called ***equipotential surfaces***. Note that the force field is orthogonal to the equipotential surfaces (the force field is radial and the equipotential surfaces are concentric spheres). ▲

The next example shows that not every vector field is a gradient.

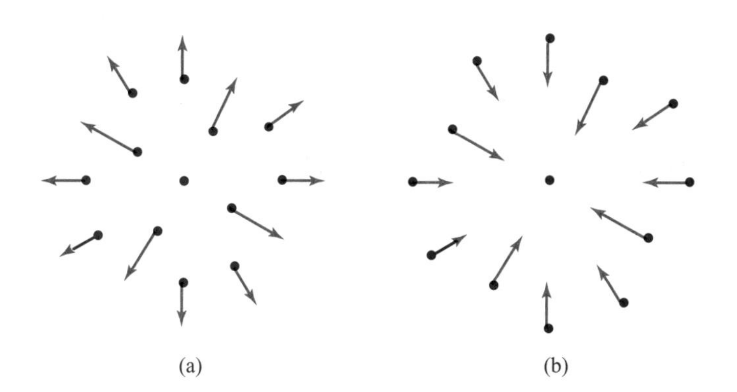

Figure 4.3.7 The vector fields associated with (a) like charges ($Qe > 0$), and (b) unlike charges ($Qe < 0$).

EXAMPLE 7 Show that the vector field $\mathbf{V}$ on $\mathbb{R}^2$ defined by $\mathbf{V}(x, y) = y\mathbf{i} - x\mathbf{j}$ is not a gradient vector field; that is, there is no C^1 function f such that

$$\mathbf{V}(x, y) = \nabla f(x, y) = \frac{\partial f}{\partial x}\mathbf{i} + \frac{\partial f}{\partial y}\mathbf{j}.$$

SOLUTION Suppose that such an f exists. Then $\partial f/\partial x = y$ and $\partial f/\partial y = -x$. Because these are C^1 functions, f itself must have continuous first- *and* second-order partial derivatives. But, $\partial^2 f/\partial x\,\partial y = -1$, and $\partial^2 f/\partial y\,\partial x = 1$, which violates the equality of mixed partials. Thus, $\mathbf{V}$ cannot be a gradient vector field. ▲

Conservation of Energy and Escaping the Earth's Gravitational Field

Consider a particle of mass m moving in a force field $\mathbf{F}$ that is a potential field. That is, assume $\mathbf{F} = -\nabla V$ for a real-valued function V, and that the particle moves according to $\mathbf{F} = m\mathbf{a}$. Thus, if the path is $\mathbf{r}(t)$, then

$$m\ddot{\mathbf{r}}(t) = -\nabla V(\mathbf{r}(t)). \tag{1}$$

A basic fact about such motion is the *conservation of energy*. The energy E of the particle is defined to be the sum of the kinetic and potential energies, defined as

$$E = \frac{1}{2}m\|\dot{\mathbf{r}}(t)\|^2 + V(\mathbf{r}(t)). \tag{2}$$

The principle of *conservation of energy* states that if Newton's second law holds, then E is independent of time; that is, $dE/dt = 0$. The proof of this fact is a simple calculation; we use equation (2), the chain rule, and equation (1):

$$\begin{aligned}\frac{dE}{dt} &= m\dot{\mathbf{r}}\cdot\ddot{\mathbf{r}} + (\nabla V)\cdot\dot{\mathbf{r}}\\ &= \dot{\mathbf{r}}\cdot(-\nabla V + \nabla V) = 0.\end{aligned}$$

Escape Velocity

As an application of conservation of energy, we compute the velocity required for a rocket to escape the earth's gravitational influence. Assume the rocket has mass m and is at a distance R_0 from the center of the earth (or another planet) when its escape velocity v_e has been reached, and that it will coast thereafter. The energy at this time is

$$E_0 = \frac{1}{2}mv_e^2 - \frac{mMG}{R_0}. \tag{3}$$

By conservation of energy, E_0 will equal the energy at a later time, which we write as

$$E_0 = E = \frac{1}{2}mv^2 - \frac{mMG}{R}, \tag{4}$$

where v is the velocity and R is the distance from the center of the earth (or the other planet). What we mean by the term *escape velocity* is that v_e is chosen such that the rocket gets to great distances, at which time it is barely moving. That is, v is close to zero and R is very large. Thus, from equation (4), we see that $E = 0$ and hence $E_0 = 0$; solving $E_0 = 0$ for v_e using equation (3) gives:

$$v_e = \sqrt{\frac{2MG}{R_0}}.$$

Now GM/R_0^2 is exactly g, the acceleration due to gravity at the distance R_0 from the center of the planet. Thus, we can write:

$$v_e = \sqrt{2gR_0}.$$

For the earth, if the escape velocity were to be achieved at the surface of the earth (of course, this is a bit unrealistic), this would give

$$v_e = \sqrt{2 \cdot 9.8 \text{ m/s}^2 \cdot 6{,}371{,}000 \text{ m}} = 11{,}127 \text{ m/s}.$$

However, this is a good approximation to the velocity that a satellite in low earth orbit needs in order to escape the earth's gravitational field.

Flow Lines

An important concept related to general (not necessarily gradient) vector fields is that of a flow line, defined as follows.

Flow Lines If $\mathbf{F}$ is a vector field, a ***flow line*** for $\mathbf{F}$ is a path $\mathbf{c}(t)$ such that

$$\mathbf{c}'(t) = \mathbf{F}(\mathbf{c}(t)).$$

That is, $\mathbf{F}$ yields the velocity field of the path $\mathbf{c}(t)$.

In the context of Example 1, a flow line is the path followed by a small particle suspended in the fluid (Figure 4.3.8). Flow lines are also appropriately called ***streamlines*** or ***integral curves***.

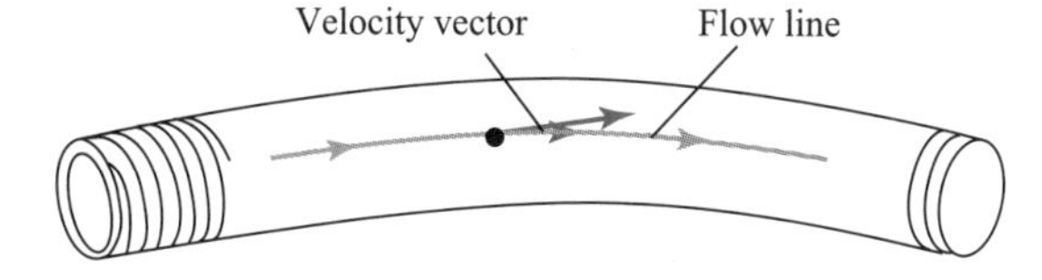

Figure 4.3.8 The velocity vector of a fluid is tangent to a flow line.

Geometrically, a flow line for a given vector filed **F** is a curve that threads its way through the domain of the vector field in such a way that the tangent vector of the curve coincides with the vector field, as in Figure 4.3.9.

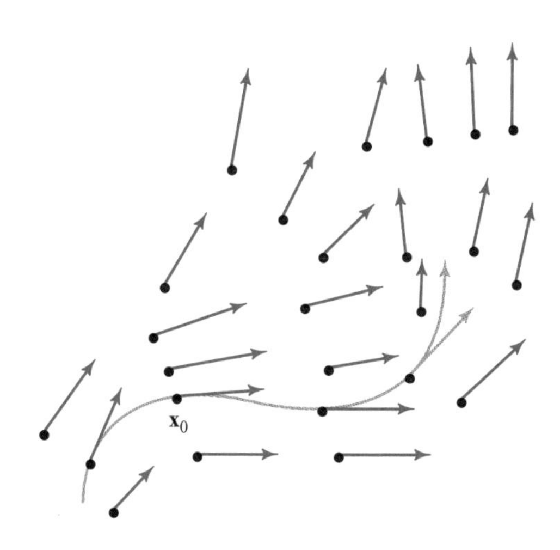

Figure 4.3.9 A flow line threading its way through a vector field in the plane.

A flow line may be viewed as a solution of a system of differential equations. Indeed, we can write the definition $\mathbf{c}'(t) = \mathbf{F}(\mathbf{c}(t))$ as

$$x'(t) = P(x(t), y(t), z(t)),$$
$$y'(t) = Q(x(t), y(t), z(t)),$$
$$z'(t) = R(x(t), y(t), z(t)),$$

where $\mathbf{c}(t) = x(t)\mathbf{i} + y(t)\mathbf{j} + z(t)\mathbf{k}$, and where

$$\mathbf{F} = P\mathbf{i} + Q\mathbf{j} + R\mathbf{k}.$$

One learns about such systems in courses on differential equations, but we are not presuming such a course has been taken.

EXAMPLE 8 Show that the path $\mathbf{c}(t) = (\cos t,\ \sin t)$ is a flow line of the vector field $\mathbf{F}(x, y) = -y\mathbf{i} + x\mathbf{j}$. Can you find others?

SOLUTION We must verify that $\mathbf{c}'(t) = \mathbf{F}(\mathbf{c}(t))$. The left side is $(-\sin t)\mathbf{i} + (\cos t)\mathbf{j}$, while the right side is $\mathbf{F}(\cos t,\ \sin t) = (-\sin t)\mathbf{i} + (\cos t)\mathbf{j}$, so we have a flow line. As suggested by Figure 4.3.3, the other flow lines are also circles. They have the form

$$\mathbf{c}(t) = (r\ \cos(t - t_0), r\ \sin(t - t_0))$$

for constants r and t_0. ▲

In many cases, explicit formulas for flow lines are not available, so one must resort to numerical methods. Figure 4.3.10 shows some output from a program that computes flow lines numerically and plots them on the screen.

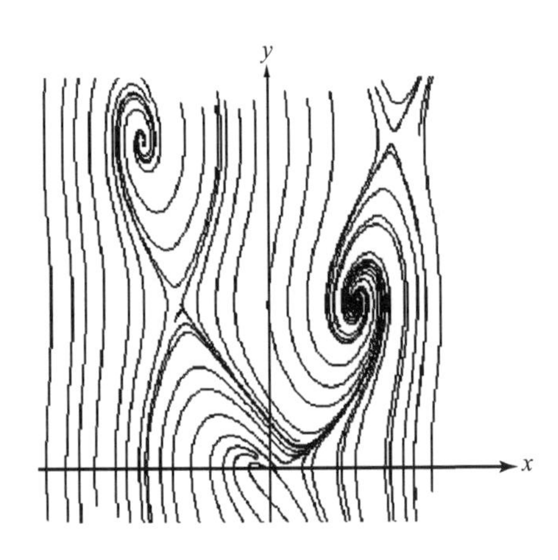

Figure 4.3.10 Computer-generated integral curves of the vector field $\mathbf{F}(x, y) = (\sin y)\mathbf{i} + (x^2 - y)\mathbf{j}$. This figure was created using *3D-XplorMath*, available from Richard Palais' Web site at rsp.math.brandeis.edu/3D-XplorMath.

Historical Note

The Field Concept

The concept of a "field," such as a vector field, has had an enormous impact on the development of conceptual frameworks for physics and engineering. It is truly one of the great breakthrough ideas in the history of human thought. It is the notion that allows one to describe, in a systematic way, influences on objects and between objects that are spatially separated.

The idea of a field began with Newton's concept of the gravitational field. In this case, the gravitational field describes the attractive influence of one body or group of bodies on one another. Similarly, the electric field produced by a charged object or group of objects creates, according to

Coulomb's law, a force on another charged object. Using vector fields to describe these sorts of forces has led to a deeper understanding of attractive and repulsive forces in nature.

However, it was the monumental discovery of the Maxwell field equations, which describe the propagation of electromagnetic energy, that cemented the concept of "field" in scientific thought. This example is particularly interesting because these fields can *propagate*. The contrast between the electromagnetic field that can propagate and the gravitational field that involves instantaneous *action at a distance* has caused great interest among philosophers of science.

Einstein's idea is that gravitation can be described in terms of the metric properties of space–time and that in this theory the associated field can also propagate, just like the electromagnetic field, thus providing profound philosophical evidence that Einstein's version of gravity may be correct. These ideas have also led to modern efforts to detect gravitational waves. For a further discussion of Einstein's work, see Section 7.7.

The idea of a field is also used in engineering to describe elastic systems and interesting microstructural properties of materials. In modern theoretical physics, the field concept is used to describe elementary particles and is central to attempts by modern theoretical physicists to unify gravity with the quantum mechanical physics of elementary particles. It is impossible to imagine a modern theoretical framework that does not incorporate some sort of field concept as a central ingredient.

EXERCISES

In Exercises 1 to 8, sketch the given vector field or a small multiple of it.

1. $\mathbf{F}(x, y) = (2, 2)$

2. $\mathbf{F}(x, y) = (4, 0)$

3. $\mathbf{F}(x, y) = (x, y)$

4. $\mathbf{F}(x, y) = (-x, y)$

5. $\mathbf{F}(x, y) = (2y, x)$

6. $\mathbf{F}(x, y) = (y, -2x)$

7. $\mathbf{F}(x, y) = \left(\dfrac{x}{\sqrt{x^2 + y^2}}, \dfrac{y}{\sqrt{x^2 + y^2}}\right)$

8. $\mathbf{F}(x, y) = \left(\dfrac{y}{\sqrt{x^2 + y^2}}, \dfrac{x}{\sqrt{x^2 + y^2}}\right)$

In Exercises 9 to 12, sketch a few flow lines of the given vector field.

9. $\mathbf{F}(x, y) = (y, -x)$

10. $\mathbf{F}(x, y) = (x, -y)$

11. $\mathbf{F}(x, y) = (x, x^2)$

12. $\mathbf{F}(x, y, z) = (y, -x, 0)$

In Exercises 13 to 16, show that the given curve $\mathbf{c}(t)$ *is a flow line of the given velocity vector field* $\mathbf{F}(x, y, z)$.

13. $\mathbf{c}(t) = (e^{2t}, \log|t|, 1/t), t \neq 0; \mathbf{F}(x, y, z) = (2x, z, -z^2)$

14. $\mathbf{c}(t) = (t^2, 2t - 1, \sqrt{t}), t > 0; \mathbf{F}(x, y, z) = (y + 1, 2, 1/2z)$

15. $\mathbf{c}(t) = (\sin t, \cos t, e^t); \mathbf{F}(x, y, z) = (y, -x, z)$

16. $\mathbf{c}(t) = \left(\frac{1}{t^3}, e^t, \frac{1}{t}\right); \mathbf{F}(x, y, z) = (-3z^4, y, -z^2)$

17. Show that it takes half as much energy to launch a satellite into an orbit just above the earth as it does to escape the earth. (Ignore the rotation of the earth.)

18. Let $\mathbf{c}(t)$ be a flow line of a gradient field $\mathbf{F} = -\nabla V$. Prove that $V(\mathbf{c}(t))$ is a decreasing function of t.

19. Suppose that the isotherms in a region are all concentric spheres centered at the origin. Prove that the energy flux vector field points either toward or away from the origin.

20. Sketch the gradient field $-\nabla V$ for $V(x, y) = (x + y)/(x^2 + y^2)$ and the equipotential surface $V = 1$.

4.4 Divergence and Curl

For each of the divergence and curl operations, we will make use of the ***del operator***, defined by

$$\nabla = \mathbf{i}\frac{\partial}{\partial x} + \mathbf{j}\frac{\partial}{\partial y} + \mathbf{k}\frac{\partial}{\partial z}.$$

For functions of one variable, taking a derivative can be thought of as an operation or process; that is, given a function $y = f(x)$, its derivative is the result of *operating* on y by the derivative *operator* d/dx. Similarly, we can write the gradient as

$$\nabla f = \left(\mathbf{i}\frac{\partial}{\partial x} + \mathbf{j}\frac{\partial}{\partial y}\right)f = \mathbf{i}\frac{\partial f}{\partial x} + \mathbf{j}\frac{\partial f}{\partial y}$$

for functions of two variables, and

$$\nabla f = \left(\mathbf{i}\frac{\partial}{\partial x} + \mathbf{j}\frac{\partial}{\partial y} + \mathbf{k}\frac{\partial}{\partial z}\right) f = \mathbf{i}\frac{\partial f}{\partial x} + \mathbf{j}\frac{\partial f}{\partial y} + \mathbf{k}\frac{\partial f}{\partial z}$$

for three variables. In operational terms, the gradient of f is obtained by taking the ∇ operator and applying it to f.

Definition of Divergence

We define the divergence of a vector field $\mathbf{F}$ by taking the *dot product* of ∇ with $\mathbf{F}$.

Divergence If $\mathbf{F} = F_1\mathbf{i} + F_2\mathbf{j} + F_3\mathbf{k}$, the ***divergence*** of $\mathbf{F}$ is the scalar field

$$\text{div } \mathbf{F} = \nabla \cdot \mathbf{F} = \frac{\partial F_1}{\partial x} + \frac{\partial F_2}{\partial y} + \frac{\partial F_3}{\partial z}.$$

Similarly, if $\mathbf{F} = (F_1, \ldots, F_n)$ is a vector field on $\mathbb{R}^n$, its divergence is

$$\text{div } \mathbf{F} = \sum_{i=1}^{n} \frac{\partial F_i}{\partial x_i} = \frac{\partial F_1}{\partial x_1} + \frac{\partial F_2}{\partial x_2} + \cdots + \frac{\partial F_n}{\partial x_n}.$$

EXAMPLE 1 Compute the divergence of

$$\mathbf{F} = x^2y\mathbf{i} + z\mathbf{j} + xyz\mathbf{k}.$$

SOLUTION

$$\text{div } \mathbf{F} = \frac{\partial}{\partial x}(x^2y) + \frac{\partial}{\partial y}(z) + \frac{\partial}{\partial z}(xyz) = 2xy + 0 + xy = 3xy. \quad \blacktriangle$$

Interpretation

The divergence has an important physical interpretation. If we imagine $\mathbf{F}$ to be the velocity field of a gas (or a fluid), then div $\mathbf{F}$ *represents the rate of expansion per unit volume under the flow of the gas (or fluid)*. If div $\mathbf{F} < 0$, the gas (or fluid) is *compressing*. For a vector field $\mathbf{F}(x, y) = F_1\mathbf{i} + F_2\mathbf{j}$ on the plane, the *divergence*

$$\nabla \cdot \mathbf{F} = \frac{\partial F_1}{\partial x} + \frac{\partial F_2}{\partial y}$$

measures the rate of expansion of area.

This interpretation is explained graphically, as follows. Choose a small region W about a point $\mathbf{x}_0$. For each point $\mathbf{x}$ in W, let $\mathbf{x}(t)$ be the flow line emanating from $\mathbf{x}$. The set of points $\mathbf{x}(t)$ describe how the set W flows after time t (see Figure 4.4.1).

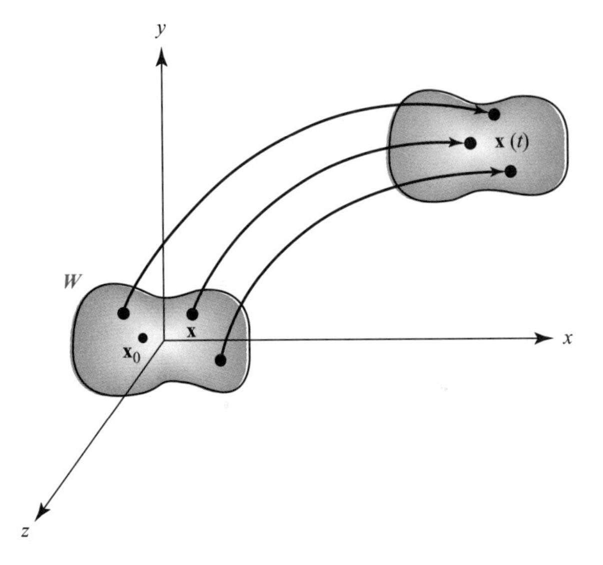

Figure 4.4.1 Flowing a region W along the flow lines of a vector field.

Call the region that results after time t has elapsed $W(t)$, and let $\mathcal{V}(t)$ be its volume (or area in two dimensions). Then the relative rate of change of volume is the divergence; more precisely,

$$\frac{1}{\mathcal{V}(0)}\frac{d}{dt}\mathcal{V}(t)\bigg|_{t=0} \approx \operatorname{div} \mathbf{F}(\mathbf{x}_0),$$

with the approximation being more exact as W shrinks to $\mathbf{x}_0$. A direct proof of this is given in the Internet supplement, but a more natural argument is given in Chapter 8, in the context of the integral theorems of vector calculus.

EXAMPLE 2 Consider the vector field in the plane given by $\mathbf{V}(x, y) = x\mathbf{i}$. Relate the sign of the divergence of $\mathbf{V}$ with the rate of change of areas under the flow.

SOLUTION We think of $\mathbf{V}$ as the velocity field of a fluid in the plane. The vector field $\mathbf{V}$ points to the right for $x > 0$ and to the left if $x < 0$, as we see in Figure 4.4.2. The length of $\mathbf{V}$ gets shorter toward the origin. As the fluid moves, it expands (the area of the shaded rectangle increases), so we expect $\operatorname{div} \mathbf{V} > 0$. Indeed, $\operatorname{div} \mathbf{V} = 1$. ▲

EXAMPLE 3 The flow lines of the vector field $\mathbf{F} = x\mathbf{i} + y\mathbf{j}$ are straight lines directed away from the origin (Figure 4.4.3).

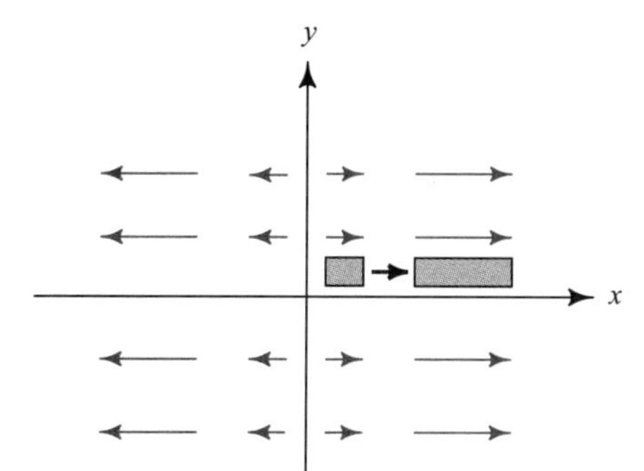

Figure 4.4.2 This fluid is expanding.

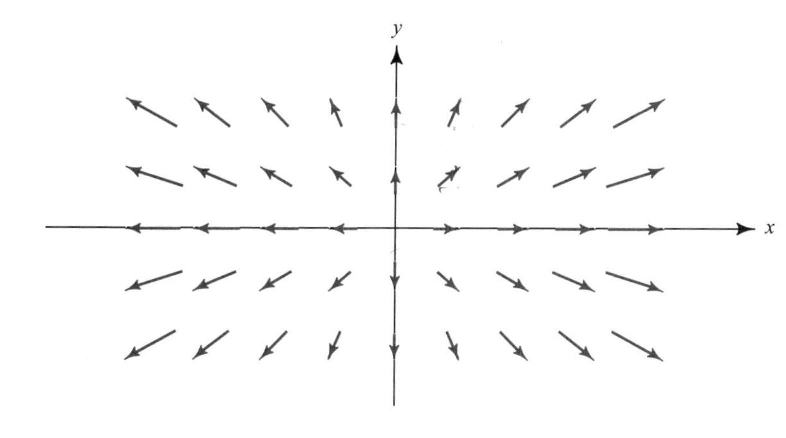

Figure 4.4.3 The vector field $\mathbf{F}(x, y) = x\mathbf{i} + y\mathbf{j}$.

If these flow lines are those of a fluid, the fluid is expanding as it moves out from the origin, so div $\mathbf{F}$ should be positive. In fact,

$$\nabla \cdot \mathbf{F} = \frac{\partial}{\partial x}x + \frac{\partial}{\partial y}y = 2 > 0. \quad \blacktriangle$$

EXAMPLE 4 Consider the vector field $\mathbf{F} = -x\mathbf{i} - y\mathbf{j}$. Here the flow lines point toward the origin instead of away from it (see Figure 4.4.4). Therefore, the fluid is compressing, so we expect (div $\mathbf{F}$) < 0. Calculating, we see that

$$\nabla \cdot \mathbf{F} = \frac{\partial}{\partial x}(-x) + \frac{\partial}{\partial y}(-y) = -1 - 1 = -2 < 0. \quad \blacktriangle$$

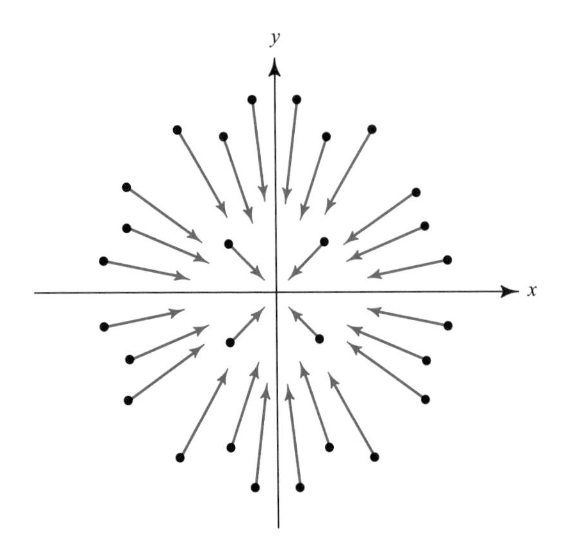

Figure 4.4.4 The vector field $\mathbf{F}(x, y) = -x\mathbf{i} - y\mathbf{j}$.

EXAMPLE 5 As we saw in the last section, the flow lines of $\mathbf{F} = -y\mathbf{i} + x\mathbf{j}$ are concentric circles about the origin, moving counterclockwise (see Figure 4.4.5). From this figure, it appears that the fluid is neither compressing nor expanding. This is confirmed by calculating

$$\nabla \cdot \mathbf{F} = \frac{\partial}{\partial x}(-y) + \frac{\partial}{\partial y}(x) = 0 + 0 = 0. \quad \blacktriangle$$

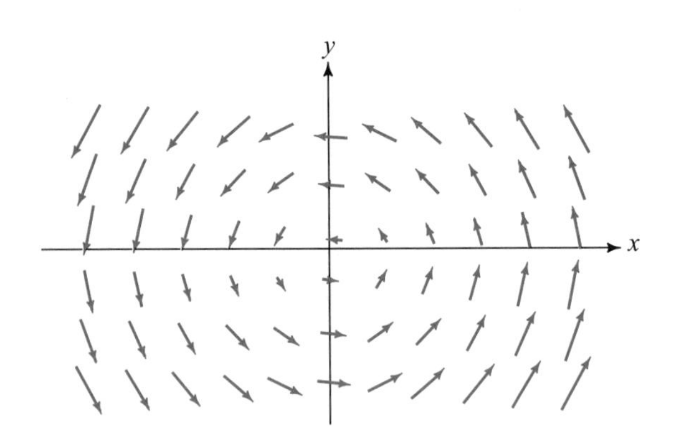

Figure 4.4.5 The vector field $\mathbf{F}(x, y) = -y\mathbf{i} + x\mathbf{j}$ has zero divergence.

EXAMPLE 6 Some flow lines of $\mathbf{F} = x\mathbf{i} - y\mathbf{j}$ are shown in Figure 4.4.6. Here our intuition about expansion or compression is less clear. However, it is true that the shaded regions shown have the same area, and we calculate that

$$\nabla \cdot \mathbf{F} = \frac{\partial}{\partial x}x + \frac{\partial}{\partial y}(-y) = 1 + (-1) = 0. \quad \blacktriangle$$

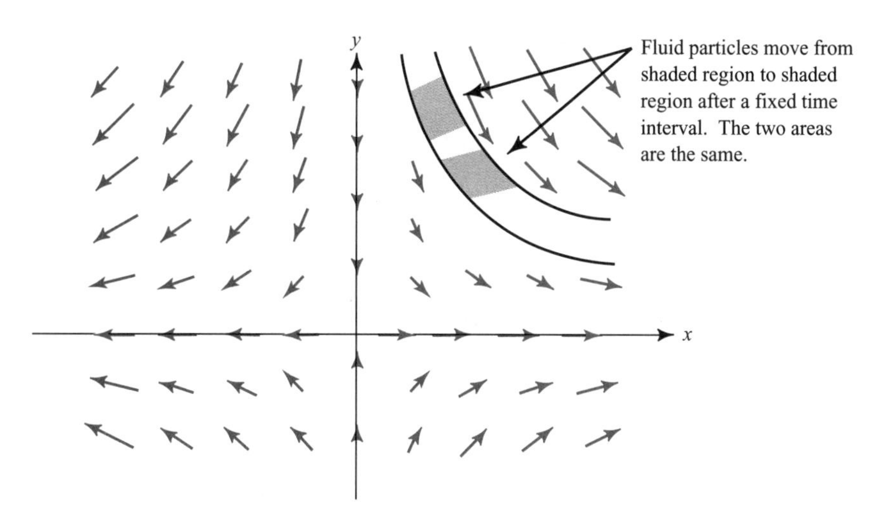

Figure 4.4.6 The vector field $\mathbf{F}(x, y) = x\mathbf{i} - y\mathbf{j}$.

Curl

To calculate the curl, the second basic operation performed on vector fields, we take the *cross product* of ∇ with $\mathbf{F}$.

Curl of a Vector Field If $\mathbf{F} = F_1\mathbf{i} + F_2\mathbf{j} + F_3\mathbf{k}$, the ***curl*** of $\mathbf{F}$ is the *vector field*

$$\operatorname{curl} \mathbf{F} = \nabla \times \mathbf{F} = \begin{vmatrix} \mathbf{i} & \mathbf{j} & \mathbf{k} \\ \dfrac{\partial}{\partial x} & \dfrac{\partial}{\partial y} & \dfrac{\partial}{\partial z} \\ F_1 & F_2 & F_3 \end{vmatrix}$$

$$= \left(\frac{\partial F_3}{\partial y} - \frac{\partial F_2}{\partial z}\right)\mathbf{i} + \left(\frac{\partial F_1}{\partial z} - \frac{\partial F_3}{\partial x}\right)\mathbf{j} + \left(\frac{\partial F_2}{\partial x} - \frac{\partial F_1}{\partial y}\right)\mathbf{k}.$$

If we write $\mathbf{F} = P\mathbf{i} + Q\mathbf{j} + R\mathbf{k}$, which is alternative notation, the same formula for the curl reads

$$\operatorname{curl} \mathbf{F} = \begin{vmatrix} \mathbf{i} & \mathbf{j} & \mathbf{k} \\ \dfrac{\partial}{\partial x} & \dfrac{\partial}{\partial y} & \dfrac{\partial}{\partial z} \\ P & Q & R \end{vmatrix}$$

$$= \left(\frac{\partial R}{\partial y} - \frac{\partial Q}{\partial z}\right)\mathbf{i} - \left(\frac{\partial R}{\partial x} - \frac{\partial P}{\partial z}\right)\mathbf{j} + \left(\frac{\partial Q}{\partial x} - \frac{\partial P}{\partial y}\right)\mathbf{k}.$$

EXAMPLE 7 Let $\mathbf{F}(x, y, z) = x\mathbf{i} + xy\mathbf{j} + \mathbf{k}$. Find $\nabla \times \mathbf{F}$.

SOLUTION We use the preceding formula:

$$\nabla \times \mathbf{F} = \begin{vmatrix} \mathbf{i} & \mathbf{j} & \mathbf{k} \\ \dfrac{\partial}{\partial x} & \dfrac{\partial}{\partial y} & \dfrac{\partial}{\partial z} \\ x & xy & 1 \end{vmatrix} = (0-0)\mathbf{i} - (0-0)\mathbf{j} + (y-0)\mathbf{k}.$$

Thus, $\nabla \times \mathbf{F} = y\mathbf{k}$. ▲

EXAMPLE 8 Find the curl of $xy\mathbf{i} - \sin z\mathbf{j} + \mathbf{k}$.

SOLUTION Letting $\mathbf{F} = xy\mathbf{i} - \sin z\mathbf{j} + \mathbf{k}$,

$$\begin{aligned} \nabla \times \mathbf{F} &= \begin{vmatrix} \mathbf{i} & \mathbf{j} & \mathbf{k} \\ \dfrac{\partial}{\partial x} & \dfrac{\partial}{\partial y} & \dfrac{\partial}{\partial z} \\ xy & -\sin z & 1 \end{vmatrix} \\ &= \begin{vmatrix} \dfrac{\partial}{\partial y} & \dfrac{\partial}{\partial z} \\ -\sin z & 1 \end{vmatrix}\mathbf{i} - \begin{vmatrix} \dfrac{\partial}{\partial x} & \dfrac{\partial}{\partial z} \\ xy & 1 \end{vmatrix}\mathbf{j} + \begin{vmatrix} \dfrac{\partial}{\partial x} & \dfrac{\partial}{\partial y} \\ xy & -\sin z \end{vmatrix}\mathbf{k} \\ &= \cos z\mathbf{i} - x\mathbf{k}. \quad ▲ \end{aligned}$$

Unlike the divergence, which can be defined in $\mathbb{R}^n$ for any n, we define the curl only in three-dimensional space (or for planar vector fields, regarding their third component as zero).

The Curl and Rotations

The physical significance of the curl will be discussed in Chapter 8, when we study Stokes' theorem. However, we can now consider a specific situation, in which the curl is associated with rotations.

EXAMPLE 9 Consider a solid rigid body B rotating about an axis L. The rotational motion of the body can be described by a vector $\boldsymbol{\omega}$ along the axis of rotation, the direction being chosen so that the body rotates about $\boldsymbol{\omega}$, as in Figure 4.4.7. We call $\boldsymbol{\omega}$ the ***angular velocity vector***. The length $\omega = \|\boldsymbol{\omega}\|$ is taken to be the angular speed of the body B, that is, the speed of any point in B divided by its distance from the axis L of rotation. The motion of points in the rotating body is described by the vector field $\mathbf{v}$ whose value at each point is the velocity at that point. To find $\mathbf{v}$, let Q be any point in B and let α be the distance from Q to L.

Figure 4.4.7 shows that $\alpha = \|\mathbf{r}\| \sin\theta$, where $\mathbf{r}$ is the vector whose initial point is the origin and whose terminal point is Q and θ is the angle between $\mathbf{r}$ and the axis L of rotation. The tangential velocity $\mathbf{v}$ of Q is directed counterclockwise along the

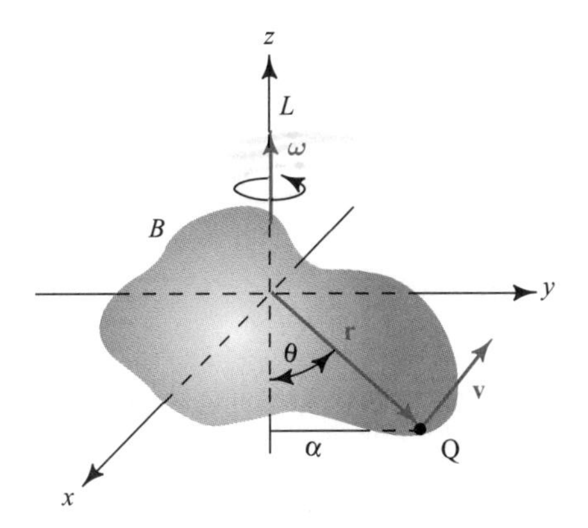

Figure 4.4.7 The velocity $\mathbf{v}$ and angular velocity $\boldsymbol{\omega}$ of a rotating body are related by $\mathbf{v} = \boldsymbol{\omega} \times \mathbf{r}$.

tangent to a circle parallel to the xy plane with radius α and has magnitude

$$\|\mathbf{v}\| = \omega\alpha = \omega\|\mathbf{r}\| \sin\theta = \|\boldsymbol{\omega}\|\,\|\mathbf{r}\| \sin\theta.$$

The direction and magnitude of $\mathbf{v}$ imply that $\mathbf{v} = \boldsymbol{\omega} \times \mathbf{r}$. Selecting a coordinate system in which L is the z axis, we can write $\boldsymbol{\omega} = \omega\mathbf{k}$ and $\mathbf{r} = x\mathbf{i} + y\mathbf{j} + z\mathbf{k}$. Thus,

$$\mathbf{v} = \boldsymbol{\omega} \times \mathbf{r} = -\omega y\mathbf{i} + \omega x\mathbf{j},$$

and so

$$\operatorname{curl}\mathbf{v} = \begin{vmatrix} \mathbf{i} & \mathbf{j} & \mathbf{k} \\ \dfrac{\partial}{\partial x} & \dfrac{\partial}{\partial y} & \dfrac{\partial}{\partial z} \\ -\omega y & \omega x & 0 \end{vmatrix} = 2\,\omega\mathbf{k} = 2\boldsymbol{\omega}.$$

Hence, for the rotation of a rigid body, the curl of the velocity vector field is a vector field whose value is the same at each point. It is directed along the axis of rotation with magnitude *twice* the angular speed. ▲

The Curl and Rotational Flow

If a vector field represents the flow of a *fluid*, then the value of $\nabla \times \mathbf{F}$ at a point is twice the angular velocity vector of a *rigid* body that rotates as the fluid does near that point. In particular, $\nabla \times \mathbf{F} = \mathbf{0}$ at a point P means that the fluid is free from rigid rotations at P; that is, it has no whirlpools. Another justification of this idea depends on Stokes' theorem from Chapter 8. However, we can say informally that curl $\mathbf{F} = \mathbf{0}$ means that if a *small* rigid paddle wheel is placed in the fluid, it will move with the fluid but will not rotate around its own axis. Such a vector field is called ***irrotational***. For example, it has been determined from experiments that fluid draining from a tub is usually irrotational except right at the center, even though the fluid is "rotating" around the drain (see Figure 4.4.8). In Example 10, *the flow lines of the vector field* **V** *are circles about the origin, yet we show that the flow is irrotational.* Thus, the reader should be warned of the possible confusion the word "irrotational" can cause.

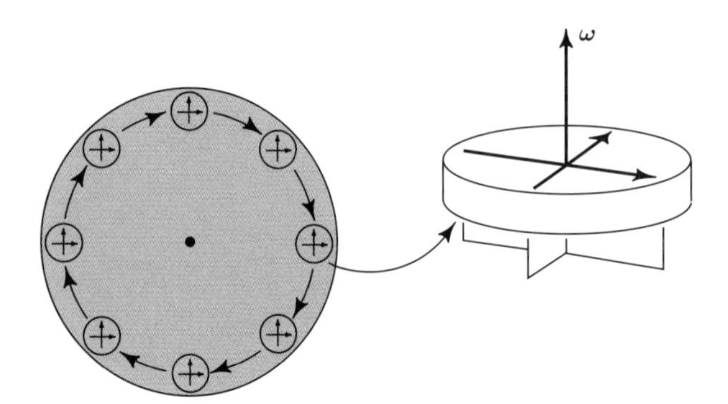

Figure 4.4.8 Looking at a paddle wheel from above a moving fluid. The velocity field $\mathbf{V}(x, y, z) = (y\mathbf{i} - x\mathbf{j})/(x^2 + y^2)$ is irrotational; the paddle wheel does not rotate around its axis $\boldsymbol{\omega}$.

EXAMPLE 10 Verify that the vector field

$$\mathbf{V}(x, y, z) = \frac{y\mathbf{i} - x\mathbf{j}}{x^2 + y^2}$$

is irrotational when $(x, y) \neq (0, 0)$ (i.e., except where **V** is not defined).

SOLUTION The curl is

$$\nabla \times \mathbf{V} = \begin{vmatrix} \mathbf{i} & \mathbf{j} & \mathbf{k} \\ \dfrac{\partial}{\partial x} & \dfrac{\partial}{\partial y} & \dfrac{\partial}{\partial z} \\ \dfrac{y}{x^2 + y^2} & \dfrac{-x}{x^2 + y^2} & 0 \end{vmatrix}$$

$$= 0\mathbf{i} + 0\mathbf{j} + \left[\frac{\partial}{\partial x}\left(\frac{-x}{x^2 + y^2}\right) - \frac{\partial}{\partial y}\left(\frac{y}{x^2 + y^2}\right)\right]\mathbf{k}$$

$$= \left[\frac{-(x^2 + y^2) + 2x^2}{(x^2 + y^2)^2} + \frac{-(x^2 + y^2) + 2y^2}{(x^2 + y^2)^2}\right]\mathbf{k} = \mathbf{0}. \quad \blacktriangle$$

Gradients are Curl Free

The following identity is a basic relation between the gradient and curl, which should be compared with the fact that for any vector **v**, we have $\mathbf{v} \times \mathbf{v} = \mathbf{0}$.

THEOREM 1: Curl of a Gradient For any C^2 function f,

$$\nabla \times (\nabla f) = \mathbf{0}.$$

That is, the curl of any gradient is the zero vector.

PROOF Because $\nabla f = (\partial f/\partial x, \partial f/\partial y, \partial f/\partial z)$ we have, by definition,

$$\nabla \times \nabla f = \begin{vmatrix} \mathbf{i} & \mathbf{j} & \mathbf{k} \\ \dfrac{\partial}{\partial x} & \dfrac{\partial}{\partial y} & \dfrac{\partial}{\partial z} \\ \dfrac{\partial f}{\partial x} & \dfrac{\partial f}{\partial y} & \dfrac{\partial f}{\partial z} \end{vmatrix}$$

$$= \left(\frac{\partial^2 f}{\partial y\,\partial z} - \frac{\partial^2 f}{\partial z\,\partial y}\right)\mathbf{i} + \left(\frac{\partial^2 f}{\partial z\,\partial x} - \frac{\partial^2 f}{\partial x\,\partial z}\right)\mathbf{j} + \left(\frac{\partial^2 f}{\partial x\,\partial y} - \frac{\partial^2 f}{\partial y\,\partial x}\right)\mathbf{k}.$$

Each component is zero because of the equality of mixed partial derivatives. ■

The converse to this theorem (a vector field with zero curl is a gradient, under suitable hypotheses) will be discussed in Chapter 8.

EXAMPLE 11 Let $\mathbf{V}(x, y, z) = y\mathbf{i} - x\mathbf{j}$. Show that $\mathbf{V}$ is not a gradient field.

SOLUTION If $\mathbf{V}$ were a gradient field, then it would satisfy curl $\mathbf{V} = \mathbf{0}$ by Theorem 1. But

$$\text{curl}\,\mathbf{V} = \begin{vmatrix} \mathbf{i} & \mathbf{j} & \mathbf{k} \\ \dfrac{\partial}{\partial x} & \dfrac{\partial}{\partial y} & \dfrac{\partial}{\partial z} \\ y & -x & 0 \end{vmatrix} = -2\mathbf{k} \neq \mathbf{0},$$

so $\mathbf{V}$ cannot be a gradient. ▲

Scalar Curl

There is an operation on vector fields in the plane that is closely related to the curl. If $\mathbf{F} = P(x, y)\mathbf{i} + Q(x, y)\mathbf{j}$ is a vector field in the plane, it can also be regarded as a vector field in space for which the $\mathbf{k}$ component is zero and the other two components are independent of z. The curl of $\mathbf{F}$ then reduces to

$$\nabla \times \mathbf{F} = \left(\frac{\partial Q}{\partial x} - \frac{\partial P}{\partial y}\right)\mathbf{k}$$

and always points in the **k** direction. The function

$$\frac{\partial Q}{\partial x} - \frac{\partial P}{\partial y}$$

of x and y is called the ***scalar curl*** of **F**.

EXAMPLE 12 Find the scalar curl of $\mathbf{V}(x, y) = -y^2\mathbf{i} + x\mathbf{j}$.

SOLUTION The curl is

$$\nabla \times \mathbf{V} = \begin{vmatrix} \mathbf{i} & \mathbf{j} & \mathbf{k} \\ \dfrac{\partial}{\partial x} & \dfrac{\partial}{\partial y} & \dfrac{\partial}{\partial z} \\ -y^2 & x & 0 \end{vmatrix} = (1 + 2y)\,\mathbf{k},$$

so the scalar curl, which is the coefficient of **k**, is $1 + 2y$. ▲

Curls are Divergence Free

A basic relation between the divergence and curl operations is given next.

THEOREM 2: Divergence of a Curl For any C^2 vector field **F**,

$$\text{div curl } \mathbf{F} = \nabla \cdot (\nabla \times \mathbf{F}) = 0.$$

That is, the divergence of any curl is zero.

As with the curl of a gradient, the proof rests on the equality of the mixed partial derivatives. The student should write out the details. A converse will be discussed in Chapter 8.

EXAMPLE 13 Show that the vector field $\mathbf{V}(x, y, z) = x\mathbf{i} + y\mathbf{j} + z\mathbf{k}$ cannot be the curl of some vector field **F**; that is, there is no **F** with $\mathbf{V} = \text{curl } \mathbf{F}$.

SOLUTION If this were so, then div **V** would be zero by Theorem 2. But

$$\text{div } \mathbf{V} = \frac{\partial x}{\partial x} + \frac{\partial y}{\partial y} + \frac{\partial z}{\partial z} = 3 \neq 0,$$

so **V** cannot be curl **F** for any **F**. ▲

Laplacian

The ***Laplace operator*** ∇^2, which operates on functions f, is defined to be the divergence of the gradient:

$$\nabla^2 f = \nabla \cdot (\nabla f) = \frac{\partial^2 f}{\partial x^2} + \frac{\partial^2 f}{\partial y^2} + \frac{\partial^2 f}{\partial z^2}.$$

This operator plays an important role in many physical laws, as we have mentioned in Section 3.1.

EXAMPLE 14 Show that $\nabla^2 f = 0$ for

$$f(x, y, z) = \frac{1}{\sqrt{x^2 + y^2 + z^2}} = \frac{1}{r} \qquad \text{and} \qquad (x, y, z) \neq (0, 0, 0),$$

where $\mathbf{r} = x\mathbf{i} + y\mathbf{j} + z\mathbf{k}$ and $r = \|\mathbf{r}\|$.

SOLUTION The first derivatives are

$$\frac{\partial f}{\partial x} = \frac{-x}{(x^2 + y^2 + z^2)^{3/2}}, \quad \frac{\partial f}{\partial y} = \frac{-y}{(x^2 + y^2 + z^2)^{3/2}}, \quad \frac{\partial f}{\partial z} = \frac{-z}{(x^2 + y^2 + z^2)^{3/2}}.$$

Computing the second derivatives, we find that

$$\begin{aligned}
\frac{\partial^2 f}{\partial x^2} &= \frac{3x^2}{(x^2 + y^2 + z^2)^{5/2}} - \frac{1}{(x^2 + y^2 + z^2)^{3/2}}, \\
\frac{\partial^2 f}{\partial y^2} &= \frac{3y^2}{(x^2 + y^2 + z^2)^{5/2}} - \frac{1}{(x^2 + y^2 + z^2)^{3/2}}, \\
\frac{\partial^2 f}{\partial z^2} &= \frac{3z^2}{(x^2 + y^2 + z^2)^{5/2}} - \frac{1}{(x^2 + y^2 + z^2)^{3/2}}.
\end{aligned}$$

Thus,

$$\begin{aligned}
\frac{\partial^2 f}{\partial x^2} + \frac{\partial^2 f}{\partial y^2} + \frac{\partial^2 f}{\partial z^2} &= \frac{3(x^2 + y^2 + z^2)}{(x^2 + y^2 + z^2)^{5/2}} - \frac{3}{(x^2 + y^2 + z^2)^{3/2}} \\
&= \frac{3}{(x^2 + y^2 + z^2)^{3/2}} - \frac{3}{(x^2 + y^2 + z^2)^{3/2}} = 0. \quad \blacktriangle
\end{aligned}$$

Vector Identities

We now have these basic operations on hand: gradient, divergence, curl, and the Laplace operator. The following box contains some basic general formulas that are useful when computing with vector fields.

Basic Identities of Vector Analysis

1. $\nabla(f+g) = \nabla f + \nabla g$
2. $\nabla(cf) = c\nabla f$, for a constant c
3. $\nabla(fg) = f\nabla g + g\nabla f$
4. $\nabla(f/g) = (g\nabla f - f\nabla g)/g^2$, at points $\mathbf{x}$ where $g(\mathbf{x}) \neq 0$
5. $\operatorname{div}(\mathbf{F}+\mathbf{G}) = \operatorname{div}\mathbf{F} + \operatorname{div}\mathbf{G}$
6. $\operatorname{curl}(\mathbf{F}+\mathbf{G}) = \operatorname{curl}\mathbf{F} + \operatorname{curl}\mathbf{G}$
7. $\operatorname{div}(f\mathbf{F}) = f\operatorname{div}\mathbf{F} + \mathbf{F}\cdot\nabla f$
8. $\operatorname{div}(\mathbf{F}\times\mathbf{G}) = \mathbf{G}\cdot\operatorname{curl}\mathbf{F} - \mathbf{F}\cdot\operatorname{curl}\mathbf{G}$
9. $\operatorname{div}\operatorname{curl}\mathbf{F} = 0$
10. $\operatorname{curl}(f\mathbf{F}) = f\operatorname{curl}\mathbf{F} + \nabla f\times\mathbf{F}$
11. $\operatorname{curl}\nabla f = \mathbf{0}$
12. $\nabla^2(fg) = f\nabla^2 g + g\nabla^2 f + 2(\nabla f\cdot\nabla g)$
13. $\operatorname{div}(\nabla f\times\nabla g) = 0$
14. $\operatorname{div}(f\nabla g - g\nabla f) = f\nabla^2 g - g\nabla^2 f$

EXAMPLE 15 Prove identity 7 in the preceding box.

SOLUTION The vector field $f\mathbf{F}$ has components fF_i, for $i = 1, 2, 3$, and so

$$\operatorname{div}(f\mathbf{F}) = \frac{\partial}{\partial x}(fF_1) + \frac{\partial}{\partial y}(fF_2) + \frac{\partial}{\partial z}(fF_3).$$

However, $(\partial/\partial x)(fF_1) = f\,\partial F_1/\partial x + F_1\,\partial f/\partial x$ by the product rule, with similar expressions for the other terms. Therefore,

$$\begin{aligned}\operatorname{div}(f\mathbf{F}) &= f\left(\frac{\partial F_1}{\partial x} + \frac{\partial F_2}{\partial y} + \frac{\partial F_3}{\partial z}\right) + F_1\frac{\partial f}{\partial x} + F_2\frac{\partial f}{\partial y} + F_3\frac{\partial f}{\partial z}\\ &= f(\nabla\cdot\mathbf{F}) + \mathbf{F}\cdot\nabla f. \quad \blacktriangle\end{aligned}$$

Let us use these identities to redo Example 14.

EXAMPLE 16 Show that for $\mathbf{r} \neq \mathbf{0}$, $\nabla^2(1/r) = 0$.

SOLUTION As in the case of the gravitational potential, $\nabla(1/r) = -\mathbf{r}/r^3$. In general, $\nabla(r^n) = nr^{n-2}\mathbf{r}$ (see Exercise 30). By the identity $\nabla \cdot (f\mathbf{F}) = f\nabla \cdot \mathbf{F} + \nabla f \cdot \mathbf{F}$, we get

$$\nabla \cdot \left(\frac{\mathbf{r}}{r^3}\right) = \frac{1}{r^3}\nabla \cdot \mathbf{r} + \mathbf{r} \cdot \nabla\left(\frac{1}{r^3}\right)$$

$$= \frac{3}{r^3} + \mathbf{r} \cdot \left(\frac{-3\mathbf{r}}{r^5}\right) = \frac{3}{r^3} - \frac{3}{r^3} = 0. \quad \blacktriangle$$

Historical Note

Divergence and Curl

William Rowan Hamilton, in his investigation of quaternions (discussed in Section 1.3) introduced the *del operator*, defined formally as

$$\nabla = \frac{\partial}{\partial x}\mathbf{i} + \frac{\partial}{\partial y}\mathbf{j} + \frac{\partial}{\partial z}\mathbf{k}.$$

Hamilton firmly believed in the significance of this operator. If $f(x, y, z)$ is a scalar function on $\mathbb{R}^3$, then "multiplication" by ∇ gives the gradient of f:

$$\nabla f = \frac{\partial f}{\partial x}\mathbf{i} + \frac{\partial f}{\partial y}\mathbf{j} + \frac{\partial f}{\partial z}\mathbf{k},$$

which, of course, gives the direction of steepest ascent (see Section 2.6). If

$$\mathbf{V}(x,y,z) = V_1(x,y,z)\mathbf{i} + V_2(x,y,z)\mathbf{j} + V_3(x,y,z)\mathbf{k}$$

is a vector field, then the "quaternionic multiplication" of ∇ with $\mathbf{V}$ yields

$$\nabla\mathbf{V} = -\text{div}\,\mathbf{V} + \text{curl}\,\mathbf{V}.$$

Thus, what we now call the divergence of $\mathbf{V}$ is the negative of the scalar part of this product, and curl $\mathbf{V}$ is the vector part (c.f. the quaternion discussion in Section 1.3).

As far as we are aware, Hamilton never gave a physical interpretation of divergence and curl, but he surely believed that, as a consequence of his faith in them, they must have an important physical interpretation. His faith in his mathematical formalism was justified, but a physical explanation of divergence and curl had to wait for James Clerk Maxwell's *Treatise on Electricity and Magnetism*. Here, Maxwell used both the divergence and the curl in his equations for the interaction of electric and magnetic fields (the Maxwell equations are discussed in Chapter 8).

Curiously, Maxwell referred to divergence as *convergence* and to curl as *rotation*, a term still used in the literature. It was Josiah Gibbs (Figure 4.4.9) who renamed convergence and rotation as the more familiar terms we use today—divergence and curl.

Figure 4.4.9 Josiah Willard Gibbs (1839–1903).

Maxwell gave a physical interpretation of the divergence using the Gauss divergence theorem, as we do in Section 8.4. His physical interpretation of the curl as a rotation was rather brief. Gibbs provided a more elementary interpretation of divergence, as we do in this section. In the spirit of Leibniz (who believed in infinitesimal quantities dx, dy, dz), Gibbs imagined placing a small cube of dimensions dx by dy by dz in a fluid. The faces of this cube have areas $dx\,dy$, $dy\,dz$, and $dx\,dz$.

At this point, students may be interested to hear Gibbs through the words of his student E. B. Wilson:

> Consider the amount of fluid which passes through those faces of the cube which are parallel to the YZ plane, i.e., perpendicular to the X axis [see Figure 4.4.10].
>
> The normal to the face whose x coordinate is the lesser, that is, the normal to the left-hand face of the cube is $-\mathbf{i}$. The flux of substance through this face is
>
> $$-\mathbf{i} \cdot \mathbf{V}(x, y, z)\, dy\, dz.$$

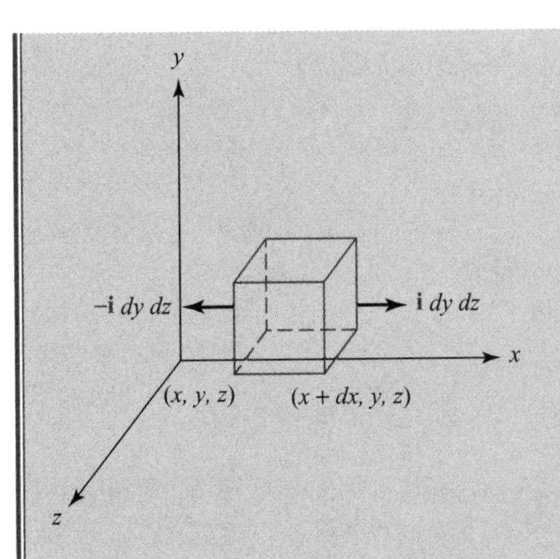

Figure 4.4.10 Cube with faces parallel to the *YZ* plane.

The normal to the opposite face, the face whose x coordinate is greater by the amount dx, is $+\mathbf{i}$, and the flux through it is therefore

$$\begin{aligned}\mathbf{i}\cdot\mathbf{V}(x+dx,y,z)\,dy\,dz &= \mathbf{i}\cdot\left[\mathbf{V}(x,y,z)+\frac{\partial\mathbf{V}}{\partial x}dx\right]dy\,dz\\ &= \mathbf{i}\cdot\mathbf{V}(x,y,z)\,dy\,dz+\mathbf{i}\cdot\frac{\partial\mathbf{V}}{\partial x}dx\,dy\,dz.\end{aligned}$$

The total flux outward from the cube through these two faces is therefore the algebraic sum of these quantities. This is simply

$$\mathbf{i}\cdot\frac{\partial\mathbf{V}}{\partial x}dx\,dy\,dz=\frac{\partial V_1}{\partial x}dx\,dy\,dz.$$

In like manner the fluxes through the other pairs of faces of the cube are

$$\mathbf{j}\cdot\frac{\partial\mathbf{V}}{\partial y}dx\,dy\,dz \qquad\text{and}\qquad \mathbf{k}\cdot\frac{\partial\mathbf{V}}{\partial z}dx\,dy\,dz.$$

The total flux out from the cube is therefore

$$\left(\mathbf{i}\cdot\frac{\partial\mathbf{V}}{\partial x}+\mathbf{j}\cdot\frac{\partial\mathbf{V}}{\partial y}+\mathbf{k}\cdot\frac{\partial\mathbf{V}}{\partial z}\right)dx\,dy\,dz.$$

This is the net quantity of fluid that leaves the cube per unit time. The quotient of this by the volume $dx\,dy\,dz$ of the cube gives the rate of diminution of density. This is

$$\nabla \cdot \mathbf{V} = \mathbf{i} \cdot \frac{\partial \mathbf{V}}{\partial x} + \mathbf{j} \cdot \frac{\partial \mathbf{V}}{\partial y} + \mathbf{k} \cdot \frac{\partial \mathbf{V}}{\partial z} = \frac{\partial V_1}{\partial x} + \frac{\partial V_2}{\partial y} + \frac{\partial V_3}{\partial z}.$$

Because $\nabla \cdot \mathbf{V}$ thus represents the diminution of density or the rate at which matter is leaving a point per unit volume per unit time, it is called the *divergence*. Maxwell employed the term *convergence* to denote the rate at which fluid approaches a point per unit volume per unit time. This is the negative of the divergence. In the case that the fluid is *incompressible*, as much matter must leave the cube as enters it. The total change of contents must therefore be zero. For this reason, the characteristic differential equation that any incompressible fluid must satisfy is

$$\nabla \cdot \mathbf{V} = 0,$$

where $\mathbf{V}$ is the velocity of the fluid. This equation is often known as the *hydrodynamic equation*. It is satisfied by any flow of water, since water is practically incompressible. The great importance of the equation for work in electricity is due to the fact that according to Maxwell's hypothesis, electric displacement obeys the same laws as an incompressible fluid. If, then, $\mathbf{D}$ is the electric displacement,

$$\operatorname{div} \mathbf{D} = \nabla \cdot \mathbf{D} = 0.$$

Gibbs' interpretation of curl was much like the one we gave in Example 9 for the rotation of a rigid body. Wilson remarks that an analysis of the meaning of curl for fluid motion was "rather difficult." It remains a bit elusive, even today, as can be seen from our discussion following Example 9. We provide another interpretation in Chapter 8.

EXERCISES

Find the divergence of the vector fields in Exercises 1 to 4.

1. $\mathbf{V}(x, y, z) = e^{xy}\mathbf{i} - e^{xy}\mathbf{j} + e^{yz}\mathbf{k}$

2. $\mathbf{V}(x, y, z) = yz\mathbf{i} + xz\mathbf{j} + xy\mathbf{k}$

3. $\mathbf{V}(x, y, z) = x\mathbf{i} + (y + \cos x)\mathbf{j} + (z + e^{xy})\mathbf{k}$

4. $\mathbf{V}(x, y, z) = x^2\mathbf{i} + (x + y)^2\mathbf{j} + (x + y + z)^2\mathbf{k}$

5. Figure 4.4.11 shows some flow lines and moving regions for a fluid moving in the plane field velocity field $\mathbf{V}$. Where is $\operatorname{div} \mathbf{V} > 0$, and also where is $\operatorname{div} \mathbf{V} < 0$?

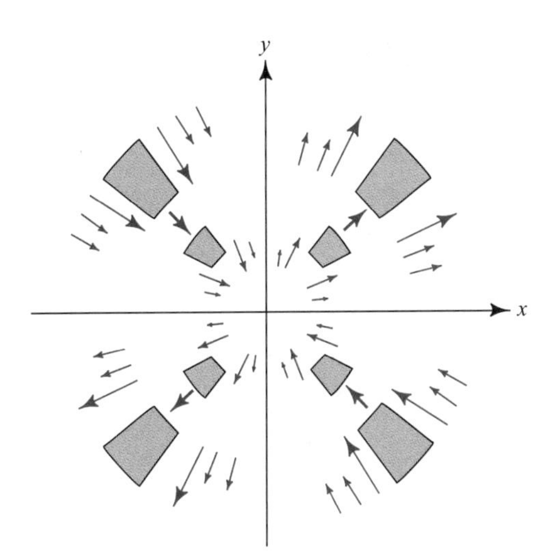

Figure 4.4.11 The flow lines of a fluid moving in the plane.

6. Let $V(x, y, z) = x\mathbf{i}$ be the velocity field of a fluid in space. Relate the sign of the divergence with the rate of change of volume under the flow.

7. Sketch a few flow lines for $\mathbf{F}(x, y) = y\mathbf{i}$. Calculate $\nabla \cdot \mathbf{F}$ and explain why your answer is consistent with your sketch.

8. Sketch a few flow lines for $\mathbf{F}(x, y) = -3x\mathbf{i} - y\mathbf{j}$. Calculate $\nabla \cdot \mathbf{F}$ and explain why your answer is consistent with your sketch.

Calculate the divergence of the vector fields in Exercises 9 to 12.

9. $\mathbf{F}(x, y) = x^3\mathbf{i} - x\sin(xy)\mathbf{j}$

10. $\mathbf{F}(x, y) = y\mathbf{i} - x\mathbf{j}$

11. $\mathbf{F}(x, y) = \sin(xy)\mathbf{i} - \cos(x^2y)\mathbf{j}$

12. $\mathbf{F}(x, y) = xe^y\mathbf{i} - [y/(x + y)]\mathbf{j}$

Compute the curl, $\nabla \times \mathbf{F}$, of the vector fields in Exercises 13 to 16.

13. $\mathbf{F}(x, y, z) = x\mathbf{i} + y\mathbf{j} + z\mathbf{k}$

14. $\mathbf{F}(x, y, z) = yz\mathbf{i} + xz\mathbf{j} + xy\mathbf{k}$

15. $\mathbf{F}(x, y, z) = (x^2 + y^2 + z^2)(3\mathbf{i} + 4\mathbf{j} + 5\mathbf{k})$

16. $\mathbf{F}(x, y, z) = \dfrac{yz\mathbf{i} - xz\mathbf{j} + xy\mathbf{k}}{x^2 + y^2 + z^2}$

Calculate the scalar curl of each of the vector fields in Exercises 17 to 20.

17. $\mathbf{F}(x, y) = \sin x\mathbf{i} + \cos x\mathbf{j}$

18. $\mathbf{F}(x, y) = y\mathbf{i} - x\mathbf{j}$

19. $\mathbf{F}(x, y) = xy\mathbf{i} + (x^2 - y^2)\mathbf{j}$

20. $\mathbf{F}(x, y) = x\mathbf{i} + y\mathbf{j}$

Verify that $\nabla \times (\nabla f) = \mathbf{0}$ *for the functions in Exercises 21 to 24.*

21. $f(x, y, z) = \sqrt{x^2 + y^2 + z^2}$

22. $f(x, y, z) = xy + yz + xz$

23. $f(x, y, z) = 1/(x^2 + y^2 + z^2)$

24. $f(x, y, z) = x^2y^2 + y^2z^2$

25. Show that $\mathbf{F} = y(\cos x)\mathbf{i} + x(\sin y)\mathbf{j}$ is *not* a gradient vector field.

26. Show that $\mathbf{F} = (x^2 + y^2)\mathbf{i} - 2xy\mathbf{j}$ is *not* a gradient field.

27. Prove identity 10 in the list of vector identities.

28. Suppose that $\nabla \cdot \mathbf{F} = 0$ and $\nabla \cdot \mathbf{G} = 0$. Which of the following necessarily have zero divergence?

(a) $\mathbf{F} + \mathbf{G}$ (b) $\mathbf{F} \times \mathbf{G}$

29. Let $\mathbf{F} = 2xz^2\mathbf{i} + \mathbf{j} + y^3zx\mathbf{k}$ and $f = x^2y$. Compute the following quantities:

(a) ∇f (b) $\nabla \times \mathbf{F}$ (c) $\mathbf{F} \times \nabla f$ (d) $\mathbf{F} \cdot (\nabla f)$

30. Let $\mathbf{r}(x, y, z) = (x, y, z)$ and $r = \sqrt{x^2 + y^2 + z^2} = \|\mathbf{r}\|$. Prove the following identities.

(a) $\nabla(1/r) = -\mathbf{r}/r^3, r \neq 0$; and, in general, $\nabla(r^n) = nr^{n-2}\mathbf{r}$ and $\nabla(\log r) = \mathbf{r}/r^2$.
(b) $\nabla^2(1/r) = 0, r \neq 0$; and, in general, $\nabla^2 r^n = n(n+1)r^{n-2}$.
(c) $\nabla \cdot (\mathbf{r}/r^3) = 0$; and, in general, $\nabla \cdot (r^n\mathbf{r}) = (n+3)r^n$.
(d) $\nabla \times \mathbf{r} = \mathbf{0}$; and, in general, $\nabla \times (r^n\mathbf{r}) = \mathbf{0}$.

31. Does $\nabla \times \mathbf{F}$ have to be perpendicular to $\mathbf{F}$?

32. Let $\mathbf{F}(x, y, z) = 3x^2y\mathbf{i} + (x^3 + y^3)\mathbf{j}$.

(a) Verify that curl $\mathbf{F} = \mathbf{0}$.
(b) Find a function f such that $\mathbf{F} = \nabla f$. (Techniques for constructing f in general are given in Chapter 8. The one in this problem should be sought by trial and error.)

33. Show that the real and imaginary parts of each of the following complex functions form the components of an irrotational and incompressible vector field in the plane; here $i = \sqrt{-1}$.

(a) $(x - iy)^2$ (b) $(x - iy)^3$ (c) $e^{x-iy} = e^x(\cos y - i \sin y)$

REVIEW EXERCISES FOR CHAPTER 4

For Exercises 1 to 4, at the indicated point, compute the velocity vector, the acceleration vector, the speed, and the equation of the tangent line.

1. $\mathbf{c}(t) = (t^3 + 1, e^{-t}, \cos(\pi t/2))$, at $t = 1$

2. $\mathbf{c}(t) = (t^2 - 1, \cos(t^2), t^4)$, at $t = \sqrt{\pi}$

3. $\mathbf{c}(t) = (e^t, \sin t, \cos t)$, at $t = 0$

4. $\mathbf{c}(t) = \dfrac{t^2}{1+t^2}\mathbf{i} + t\mathbf{j} + \mathbf{k}$, at $t = 2$

5. Calculate the tangent and acceleration vectors for the helix $\mathbf{c}(t) = (\cos t, \sin t, t)$ at $t = \pi/4$.

6. Calculate the tangent and acceleration vector for the cycloid $\mathbf{c}(t) = (t - \sin t, 1 - \cos t)$ at $t = \pi/4$ and sketch.

7. Let a particle of mass m move on the path $\mathbf{c}(t) = (t^2, \sin t, \cos t)$. Compute the force acting on the particle at $t = 0$.

8. (a) Let $\mathbf{c}(t)$ be a path with $\|\mathbf{c}(t)\| =$ constant; that is, the curve lies on a sphere. Show that $\mathbf{c}'(t)$ is orthogonal to $\mathbf{c}(t)$.

(b) Let $\mathbf{c}$ be a path whose speed is never zero. Show that $\mathbf{c}$ has constant speed if and only if the acceleration vector $\mathbf{c}''$ is always perpendicular to the velocity vector $\mathbf{c}'$.

9. Express the arc length of the curve $x^2 = y^3 = z^5$ between $x = 1$ and $x = 4$ as an integral, using a suitable parametrization.

10. Find the arc length of $\mathbf{c}(t) = t\mathbf{i} + (\log t)\mathbf{j} + 2\sqrt{2}t\mathbf{k}$ for $1 \le t \le 2$.

11. A particle is constrained to move around the unit circle in the xy plane according to the formula $(x, y, z) = (\cos(t^2), \sin(t^2), 0)$, $t \ge 0$.

(a) What are the velocity vector and speed of the particle as functions of t?

(b) At what point on the circle should the particle be released to hit a target at $(2, 0, 0)$? (Be careful about which direction the particle is moving around the circle.)

(c) At what time t should the release take place? (Use the smallest $t > 0$ that will work.)

(d) What are the velocity and speed at the time of release?

(e) At what time is the target hit?

12. A particle of mass m moves under the influence of a force $\mathbf{F} = -k\mathbf{r}$, where k is a constant and $\mathbf{r}(t)$ is the position of the particle at time t.

(a) Write down differential equations for the components of $\mathbf{r}(t)$.
(b) Solve the equations in part (a) subject to the initial conditions $\mathbf{r}(0) = \mathbf{0}$, $\mathbf{r}'(0) = 2\mathbf{j} + \mathbf{k}$.

13. Write the curve described by the equations $x - 1 = 2y + 1 = 3z + 2$ in parametric form.

14. Write the curve $x = y^3 = z^2 + 1$ in parametric form.

15. Show that $\mathbf{c}(t) = (1/(1-t), 0, e^t/(1-t))$ is a flow line of the vector field defined by $\mathbf{F}(x, y, z) = (x^2, 0, z(1+x))$.

16. Let $\mathbf{F}(x, y) = f(x^2 + y^2)[-y\mathbf{i} + x\mathbf{j}]$ for a function f of one variable. What equation must $g(t)$ satisfy for

$$\mathbf{c}(t) = [\cos g(t)]\mathbf{i} + [\sin g(t)]\mathbf{j}$$

to be a flow line for $\mathbf{F}$?

Compute $\nabla \cdot \mathbf{F}$ and $\nabla \times \mathbf{F}$ for the vector fields in Exercises 17 to 20.

17. $\mathbf{F} = 2x\mathbf{i} + 3y\mathbf{j} + 4z\mathbf{k}$

18. $\mathbf{F} = x^2\mathbf{i} + y^2\mathbf{j} + z^2\mathbf{k}$

19. $\mathbf{F} = (x + y)\mathbf{i} + (y + z)\mathbf{j} + (z + x)\mathbf{k}$

20. $\mathbf{F} = x\mathbf{i} + 3xy\mathbf{j} + z\mathbf{k}$

Compute the divergence and curl of the vector fields in Exercises 21 and 22 at the points indicated.

21. $\mathbf{F}(x, y, z) = y\mathbf{i} + z\mathbf{j} + x\mathbf{k}$, at the point $(1, 1, 1)$

22. $\mathbf{F}(x, y, z) = (x + y)^3\mathbf{i} + (\sin xy)\mathbf{j} + (\cos xyz)\mathbf{k}$, at the point $(2, 0, 1)$

Calculate the gradients of the functions in Exercises 23 to 26, and verify that $\nabla \times \nabla f = \mathbf{0}$.

23. $f(x, y) = e^{xy} + \cos(xy)$

24. $f(x, y) = \dfrac{x^2 - y^2}{x^2 + y^2}$

25. $f(x, y) = e^{x^2} - \cos(xy^2)$

26. $f(x, y) = \tan^{-1}(x^2 + y^2)$

27. (a) Let $f(x, y, z) = xyz^2$; compute ∇f.

(b) Let $\mathbf{F}(x, y, z) = xy\mathbf{i} + yz\mathbf{j} + zy\mathbf{k}$; compute $\nabla \times \mathbf{F}$.

(c) Compute $\nabla \times (f\mathbf{F})$ using identity 10 of the list of vector identities. Compare with a direct computation.

28. (a) Let $\mathbf{F} = 2xye^z\mathbf{i} + e^z x^2\mathbf{j} + (x^2ye^z + z^2)\mathbf{k}$. Compute $\nabla \cdot \mathbf{F}$ and $\nabla \times \mathbf{F}$.

(b) Find a function $f(x, y, z)$ such that $\mathbf{F} = \nabla f$.

29. Let $\mathbf{F}(x, y) = f(x^2 + y^2)[-y\mathbf{i} + x\mathbf{j}]$, as in Exercise 16. Calculate div $\mathbf{F}$ and curl $\mathbf{F}$ and discuss your answers in view of the results of Exercise 16.

30. Let a particle of mass m move along the elliptical helix $\mathbf{c}(\mathrm{t}) = (4\cos t, \sin t, t)$.

(a) Find the equation of the tangent line to the helix at $t = \pi/4$.

(b) Find the force acting on the particle at time $t = \pi/4$.

(c) Write an expression (in terms of an integral) for the arc length of the curve $\mathbf{c}(t)$ between $t = 0$ and $t = \pi/4$.

31. (a) Let $g(x, y, z) = x^3 + 5yz + z^2$ and let $h(u)$ be a function of one variable such that $h'(1) = 1/2$. Let $f = h \circ g$. Starting at (1, 0, 0), in what directions is f changing at 50% of its maximum rate?

(b) For $g(x, y, z) = x^3 + 5yz + z^2$, calculate $\mathbf{F} = \nabla g$, the gradient of g, and verify directly that $\nabla \times \mathbf{F} = \mathbf{0}$ at each point (x, y, z).

32. (a) Write in parametric form the curve that is the intersection of the surfaces $x^2 + y^2 + z^2 = 3$ and $y = 1$.

(b) Find the equation of the line tangent to this curve at (1, 1, 1).

(c) Write an integral expression for the arc length of this curve. What is the value of this integral?

33. In meteorology, the ***negative pressure gradient*** $\mathbf{G}$ is a vector quantity that points from regions of high pressure to regions of low pressure, normal to the lines of constant pressure (***isobars***).

(a) In an xy coordinate system,

$$\mathbf{G} = -\frac{\partial P}{\partial x}\mathbf{i} - \frac{\partial P}{\partial y}\mathbf{j}.$$

Write a formula for the magnitude of the negative pressure gradient.

(b) If the horizontal pressure gradient provided the only horizontal force acting on the air, the wind would blow directly across the isobars in the direction of $\mathbf{G}$, and for a given air mass, with acceleration proportional to the magnitude of $\mathbf{G}$. Explain, using Newton's second law.

(c) Because of the rotation of the earth, the wind does not blow in the direction that part (b) would suggest. Instead, it obeys ***Buys–Ballot's law***, which states: "If in the Northern Hemisphere, you stand with your back to the wind, the high pressure is on your right and the low pressure is on your left." Draw a figure and introduce xy coordinates so that $\mathbf{G}$ points in the proper direction.

(d) State and graphically illustrate Buys–Ballot's law for the Southern Hemisphere, in which the orientation of high and low pressure is reversed.

34. A sphere of mass m, radius a, and uniform density has potential u and gravitational force $\mathbf{F}$, at a distance r from the center (0, 0, 0), given by

$$u = \frac{3m}{2a} - \frac{mr^2}{2a^3}, \qquad \mathbf{F} = -\frac{m}{a^3}\mathbf{r} \qquad (r \leq a);$$

$$u = \frac{m}{r}, \qquad \mathbf{F} = -\frac{m}{r^3}\mathbf{r} \qquad (r > a).$$

Here, $r = \|\mathbf{r}\|$, $\mathbf{r} = x\mathbf{i} + y\mathbf{j} + z\mathbf{k}$.

(a) Verify that $\mathbf{F} = \nabla u$ on the inside and outside of the sphere.
(b) Check that u satisfies Poisson's equation: $\partial^2 u/\partial x^2 + \partial^2 u/\partial y^2 + \partial^2 u/\partial z^2 =$ constant inside the sphere.
(c) Show that u satisfies Laplace's equation: $\partial^2 u/\partial x^2 + \partial^2 u/\partial y^2 + \partial^2 u/\partial z^2 = 0$ outside the sphere.

35. A circular helix that lies on the cylinder $x^2 + y^2 = R^2$ with pitch ρ may be described parametrically by

$$x = R\cos\theta, \qquad y = R\sin\theta, \qquad z = \rho\theta, \qquad \theta \geq 0.$$

A particle slides under the action of gravity (which acts parallel to the z axis) without friction along the helix. If the particle starts out at the height $z_0 > 0$, then when it reaches the height z along the helix, its speed is given by

$$\frac{ds}{dt} = \sqrt{(z_0 - z)2g},$$

where s is arc length along the helix, g is the constant of gravity, t is time, and $0 \leq z \leq z_0$.

(a) Find the length of the part of the helix between the planes $z = z_0$ and $z = z_1$, $0 \leq z_1 < z_0$.
(b) Compute the time T_0 it takes the particle to reach the plane $z = 0$.

36. A sphere of radius 10 centimeters (cm) with center at (0, 0, 0) rotates about the z axis with angular velocity 4 in such a direction that the rotation looks counterclockwise from the positive z axis.

(a) Find the rotation vector $\boldsymbol{\omega}$ (see Example 9, in Section 4.4).
(b) Find the velocity $\mathbf{v} = \boldsymbol{\omega} \times \mathbf{r}$ when $\mathbf{r} = 5\sqrt{2}(\mathbf{i} - \mathbf{j})$ is on the "equator."
(c) Find the velocity of the point $(0, 5\sqrt{3}, 5)$ on the sphere.

37. Find the speed of the students in a classroom located at a latitude 49°N due to the rotation of the earth. (Ignore the motion of the earth about the sun, the sun in the galaxy, etc.; the radius of the earth is 3960 miles.)

5

Double and Triple Integrals

It is to Archimedes himself (c. 225 B.C.*) that we owe the nearest approach to actual integration to be found among the Greeks. His first noteworthy advance in this direction was concerned with his proof that the area of a parabolic segment is four thirds of the triangle with the same base and vertex, or two thirds of the circumscribed parallelogram.*

D. E. Smith, *History of Mathematics*

In this chapter and the next we study the integration of real-valued functions of several variables; this chapter treats integrals of functions of two and three variables, or *double* and *triple integrals*. The double integral has a basic geometric interpretation as volume, and can be defined rigorously as a limit of approximating sums. We shall present several techniques for evaluating double and triple integrals and consider some applications.

5.1 Introduction

This section discusses some geometric aspects of the double integral, deferring a more rigorous discussion in terms of Riemann sums until Section 5.2.

Double Integrals as Volumes

Consider a continuous function of two variables $f\colon R \subset \mathbb{R}^2 \to \mathbb{R}$ whose domain R is a rectangle with sides parallel to the coordinate axes. The rectangle R can be described in terms of the two closed intervals $[a, b]$ and $[c, d]$, representing the sides of R along the x and y axes, respectively, as in Figure 5.1.1. In this case, we say that R is the ***Cartesian product*** of $[a, b]$ and $[c, d]$ and write $R = [a, b] \times [c, d]$.

Assume that $f(x, y) \geq 0$ on R, so that the graph of $z = f(x, y)$ is a surface lying above the rectangle R. This surface, the rectangle R, and the four planes $x = a$, $x = b$, $y = c$, and $y = d$ form the boundary of a region V in space (see Figure 5.1.1).

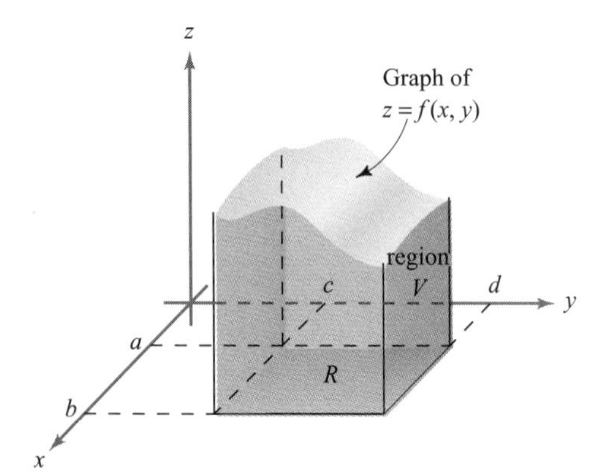

Figure 5.1.1 The region V in space is bounded by the graph of f, the rectangle R, and the four vertical sides indicated.

The problem of how to rigorously define the volume of V has to be faced, and we shall solve it in Section 5.2 by the classic method of exhaustion, or rather, in more modern terms, the method of Riemann sums. To gain an intuitive grasp of the double integral, we provisionally assume that the volume of a region has been defined.

> **Double Integrals** The volume of the region above R and under the graph of a nonnegative function f is called the ***(double) integral*** of f over R and is denoted by
>
> $$\iint_R f(x, y)\, dA, \qquad \text{or} \qquad \iint_R f(x, y)\, dx\, dy.$$

EXAMPLE 1 (a) If f is defined by $f(x, y) = k$, where k is a positive constant, then

$$\iint_R f(x, y)\, dA = k(b - a)(d - c),$$

because the integral is equal to the volume of a rectangular box with base R and height k.

(b) If $f(x, y) = 1 - x$ and $R = [0, 1] \times [0, 1]$, then

$$\iint_R f(x, y)\, dA = \frac{1}{2},$$

because the integral is equal to the volume of the triangular solid shown in Figure 5.1.2. ▲

EXAMPLE 2 Suppose $z = f(x, y) = x^2 + y^2$ and $R = [-1, 1] \times [0, 1]$. Then the integral $\iint_R (x^2 + y^2)\, dx\, dy$ is equal to the volume of the solid sketched in Figure 5.1.3. We shall compute this integral in Example 3. ▲

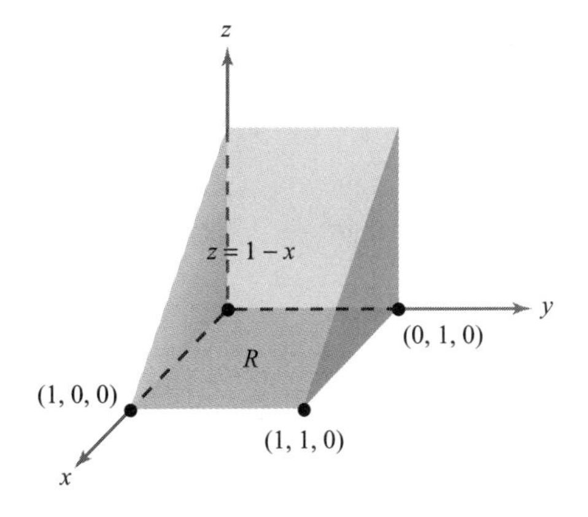

Figure 5.1.2 Volume under the graph $z = 1 - x$ and over $R = [0, 1] \times [0, 1]$.

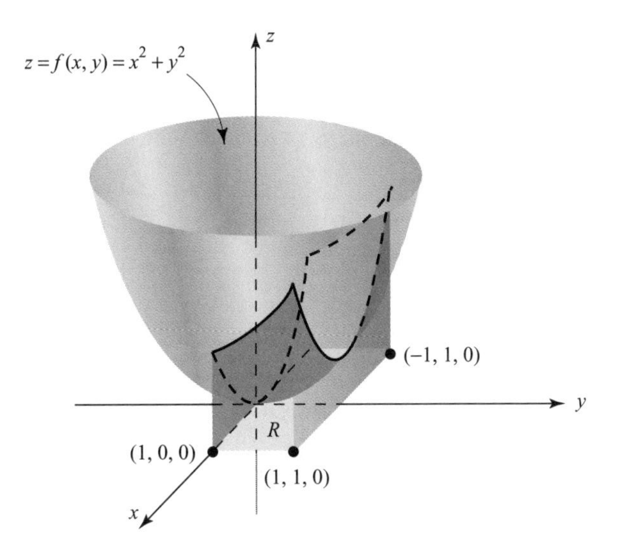

Figure 5.1.3 Volume under $z = x^2 + y^2$ and over $R = [-1, 1] \times [0, 1]$.

These ideas are similar to those for a single integral $\int_a^b f(x)\,dx$, which represents the area under the graph of f if $f \geq 0$; see Figure 5.1.4.[1]

Single integrals $\int_a^b f(x)\,dx$ can be rigorously defined, without recourse to the area concept, as a limit of Riemann sums. The idea is to approximate $\int_a^b f(x)\,dx$ by choosing a partition $a = x_0 < x_1 < \cdots < x_n = b$ of $[a, b]$, selecting points $c_i \in [x_i, x_{i+1}]$, and forming the Riemann sum

$$\sum_{i=0}^{n-1} f(c_i)(x_{i+1} - x_i) \approx \int_a^b f(x)\,dx$$

[1]Readers not already familiar with this idea should review the appropriate sections of their introductory calculus text.

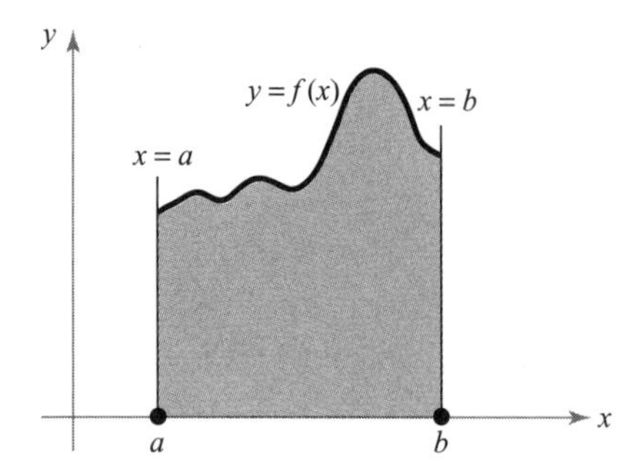

Figure 5.1.4 Area under the graph of a nonnegative continuous function f from $x = a$ to $x = b$ is $\int_a^b f(x)\,dx$.

(see Figure 5.1.5). We examine the analogous process for double integrals in the next section.

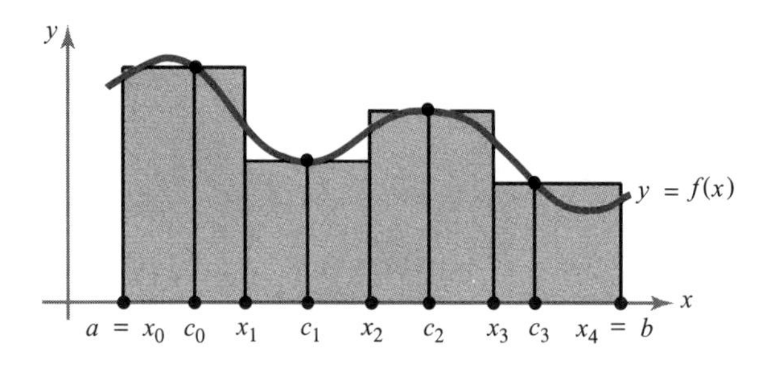

Figure 5.1.5 The sum of the areas of the shaded rectangles is a Riemann sum, which approximates the area under f from $x = a$ to $x = b$.

Cavalieri's Principle

There is a useful method for computing volumes, known as *Cavalieri's principle.* Suppose we have a solid body and we let $A(x)$ denote its cross-sectional area in a plane P_x measured at a distance x from a reference plane (Figure 5.1.6).

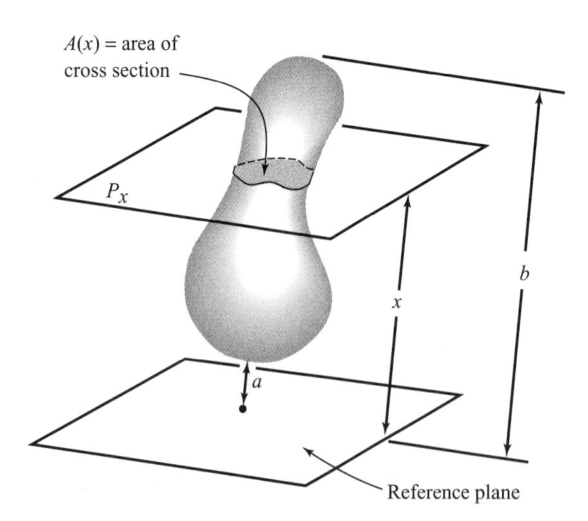

Figure 5.1.6 A solid body with cross-sectional area $A(x)$ at distance x from a reference plane.

According to Cavalieri's principle, the volume of the body is given by

$$\text{volume} = \int_a^b A(x)\,dx,$$

where a and b are the minimum and maximum distances from the reference plane. This can be made intuitively clear as follows. If we partition $[a, b]$ into $a = x_0 < x_1 < \cdots < x_n = b$, then an approximating Riemann sum for the preceding integral is

$$\sum_{i=0}^{n-1} A(c_i)(x_{i+1} - x_i).$$

But this sum also approximates the volume of the body, because $A(x)\,\Delta x$ is the volume of a slab with cross-sectional area $A(x)$ and thickness Δx (Figure 5.1.7). Therefore, it is reasonable to accept the preceding formula for the volume. A more careful justification of this method is given in the Internet supplement for Chapter 5.

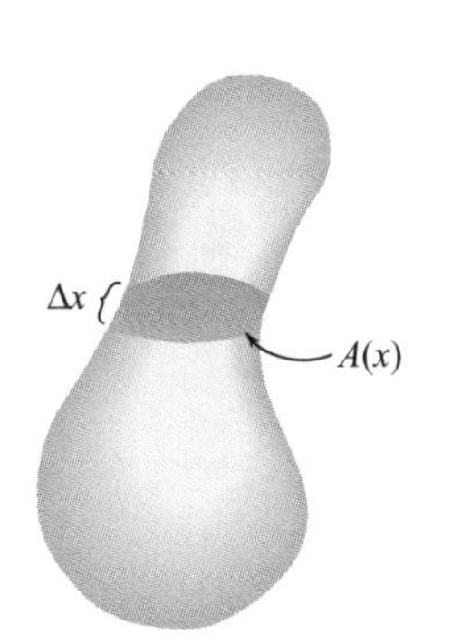

Figure 5.1.7 Volume of a slab with cross-sectional area $A(x)$ and thickness Δx equals $A(x)\Delta x$. The total volume of the body is $\int_a^b A(x)\,dx$.

The Slice Method—Cavalieri's Principle Let S be a solid and, for x satisfying $a \leq x \leq b$, let P_x be a family of parallel planes such that:

1. S lies between P_a and P_b;
2. The area of the slice of S cut by P_x is $A(x)$.

Then the volume of S is equal to

$$\int_a^b A(x)\,dx.$$

Historical Note

Bonaventura Cavalieri (1598–1647) was a pupil of Galileo and a professor in Bologna. His investigations into area and volume were important building blocks of the foundations of calculus. Although his methods were criticized by his contemporaries, similar ideas had been used by Archimedes in antiquity, and were later taken up by the "fathers" of calculus, Newton and Leibniz.

Reduction to Iterated Integrals

We now use Cavalieri's principle to evaluate double integrals. Consider the solid region under a graph $z = f(x, y)$ defined on the region $[a, b] \times [c, d]$, where f is continuous and greater than zero. There are two natural cross-sectional area functions: one obtained by using cutting planes perpendicular to the x axis, and the other obtained by using cutting planes perpendicular to the y axis. The cross section determined by a cutting plane $x = x_0$, of the first sort, is the plane region under the graph of $z = f(x_0, y)$ from $y = c$ to $y = d$ (Figure 5.1.8).

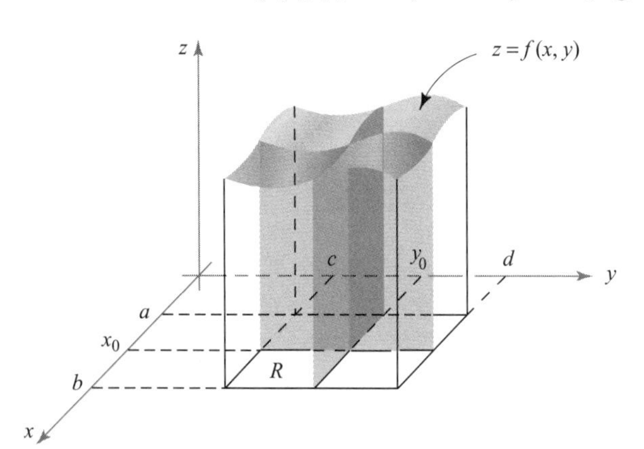

Figure 5.1.8 Two different cross sections sweeping out the volume under $z = f(x, y)$.

When we fix $x = x_0$, we obtain the function $y \mapsto f(x_0, y)$, which is continuous on $[c, d]$. The cross-sectional area $A(x_0)$ is, therefore, equal to the integral $\int_c^d f(x_0, y)\, dy$. Thus, the cross-sectional area function A has domain $[a, b]$, and is given by the rule $A: x \mapsto \int_c^d f(x, y)\, dy$. By Cavalieri's principle, the volume V of the region under $z = f(x, y)$ must be equal to

$$V = \int_a^b A(x)\, dx = \int_a^b \left[\int_c^d f(x, y)\, dy \right] dx.$$

The integral $\int_a^b \left[\int_c^d f(x, y)\, dy \right] dx$ is known as an ***iterated integral*** because it is obtained by integrating with respect to y and then integrating the result with respect to x. Because $\iint_R f(x, y)\, dA$ is equal to the volume V, we get the following result.

Double and Iterated Integrals

$$\iint_R f(x, y)\, dA = \int_a^b \left[\int_c^d f(x, y)\, dy\right] dx. \tag{1}$$

If we use cutting planes perpendicular to the y axis, we obtain

$$\iint_R f(x, y)\, dA = \int_c^d \left[\int_a^b f(x, y)\, dx\right] dy. \tag{2}$$

The expression on the right of formula (2) is the iterated integral obtained by integrating with respect to x and then integrating the result with respect to y.

Thus, if our intuition about volumes is correct, formulas (1) and (2) ought to be valid. This is in fact true when the concepts we are discussing are defined rigorously, and is known as *Fubini's theorem*. We give a proof of this theorem in the next section.

As the following examples illustrate, the notion of the iterated integral and equations (1) and (2) provide a powerful method for *computing* the double integral of a function of two variables.

EXAMPLE 3 Evaluate the integral

$$\iint_R (x^2 + y^2)\, dx\, dy,$$

where $R = [-1, 1] \times [0, 1]$.

SOLUTION By equation (2),

$$\iint_R (x^2 + y^2)\, dx\, dy = \int_0^1 \left[\int_{-1}^1 (x^2 + y^2)\, dx\right] dy.$$

To find $\int_{-1}^1 (x^2 + y^2)\, dx$, we treat y as a constant and integrate with respect to x. Because $x^3/3 + y^2x$ is an antiderivative of $x^2 + y^2$ with respect to x, we can integrate, using the fundamental theorem of calculus, to obtain

$$\int_{-1}^1 (x^2 + y^2)\, dx = \left[\frac{x^3}{3} + y^2x\right]_{x=-1}^1 = \frac{2}{3} + 2y^2.$$

Next, we integrate $\frac{2}{3} + 2y^2$ with respect to y from 0 to 1, to obtain

$$\int_0^1 \left(\frac{2}{3} + 2y^2\right) dy = \left[\frac{2}{3}y + \frac{2}{3}y^3\right]_{y=0}^1 = \frac{4}{3}.$$

Hence, the volume of the solid we saw in Figure 5.1.3 is $4/3$.

For completeness, let us evaluate $\iint_R (x^2 + y^2)\,dx\,dy$ using equation (1)—that is, integrating with respect to y first and then with respect to x. We have

$$\iint_R (x^2 + y^2)\,dx\,dy = \int_{-1}^{1} \left[\int_0^1 (x^2 + y^2)\,dy \right] dx.$$

Treating x as a constant in the y integration, we obtain

$$\int_0^1 (x^2 + y^2)\,dy = \left[x^2 y + \frac{y^3}{3} \right]_{y=0}^{1} = x^2 + \frac{1}{3}.$$

Next, we evaluate $\int_{-1}^{1} \left(x^2 + \frac{1}{3}\right) dx$ to obtain

$$\int_{-1}^{1} \left(x^2 + \frac{1}{3} \right) dx = \left[\frac{x^3}{3} + \frac{x}{3} \right]_{x=-1}^{1} = \frac{4}{3},$$

which agrees with our previous answer. ▲

EXAMPLE 4 Compute the double integral $\iint_S \cos x \sin y\,dx\,dy$, where S is the square $[0, \pi/2] \times [0, \pi/2]$ (see Figure 5.1.9).

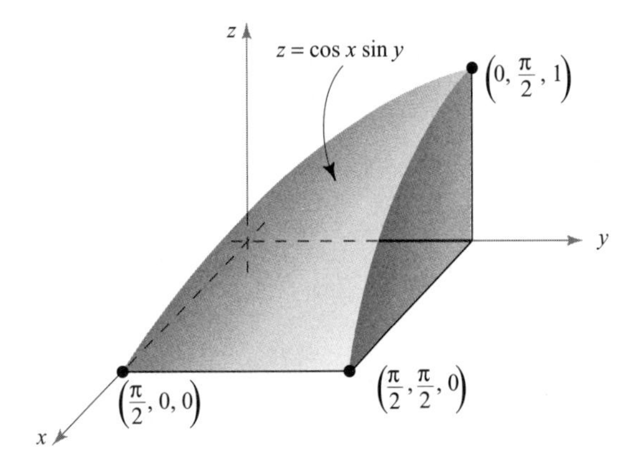

Figure 5.1.9 Volume under $z = \cos x \sin y$ and over the rectangle $[0, \pi/2] \times [0, \pi/2]$.

SOLUTION By equation (2),

$$\begin{aligned}\iint_S \cos x \sin y\,dx\,dy &= \int_0^{\pi/2} \left[\int_0^{\pi/2} \cos x \sin y\,dx \right] dy \\ &= \int_0^{\pi/2} \sin y \left[\int_0^{\pi/2} \cos x\,dx \right] dy = \int_0^{\pi/2} \sin y\,dy = 1. \quad ▲\end{aligned}$$

In the next section, we shall use Riemann sums to rigorously define the double integral for a large class of functions of two variables without recourse to the notion of volume. Although we shall drop the requirement that $f(x, y) \geq 0$, equations (1) and (2) will remain valid. Therefore, the iterated integral will again provide the key to computing the double integral. In Section 5.3, we treat double integrals over regions more general than rectangles.

Finally, we remark that it is common to delete the brackets in iterated integrals such as equations (1) and (2) and write

$$\int_a^b \int_c^d f(x, y)\, dy\, dx \qquad \text{in place of} \qquad \int_a^b \left[\int_c^d f(x, y)\, dy \right] dx$$

and

$$\int_c^d \int_a^b f(x, y)\, dx\, dy \qquad \text{in place of} \qquad \int_c^d \left[\int_a^b f(x, y)\, dx \right] dy.$$

EXERCISES

1. Evaluate the following iterated integrals:

(a) $\int_{-1}^{1} \int_0^1 (x^4 y + y^2)\, dy\, dx$

(b) $\int_0^{\pi/2} \int_0^1 (y \cos x + 2)\, dy\, dx$

(c) $\int_0^1 \int_0^1 (x y e^{x+y})\, dy\, dx$

(d) $\int_{-1}^{0} \int_1^2 (-x \log y)\, dy\, dx$

2. Evaluate the integrals in Exercise 1 by integrating with respect to x and then with respect to y. [The solution to part (b) only is in the Study Guide to this text.]

3. Use Cavalieri's principle to show that the volumes of two cylinders with the same base and height are equal (see Figure 5.1.10).

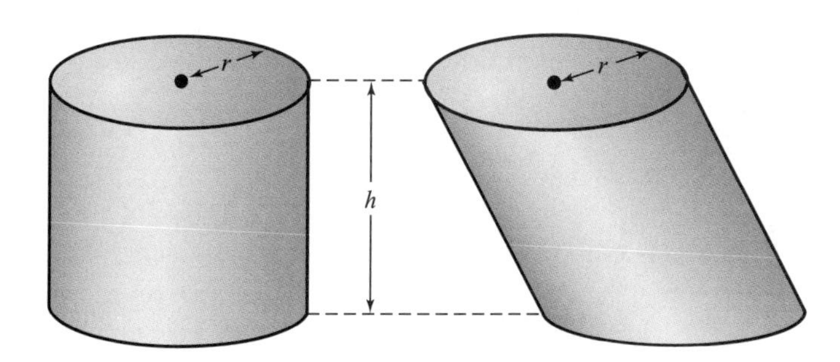

Figure 5.1.10 Two cylinders with the same base and height have the same volume.

4. Using Cavalieri's principle, compute the volume of the structure shown in Figure 5.1.11; each cross section is a rectangle of length 5 and width 3.

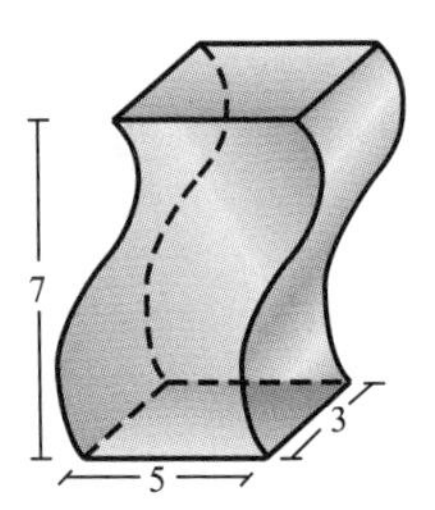

Figure 5.1.11 Compute this volume.

5. A lumberjack cuts out a wedge-shaped piece W of a cylindrical tree of radius r obtained by making two saw cuts to the tree's center, one horizontally and one at an angle θ. Compute the volume of the wedge W using Cavalieri's principle. (See Figure 5.1.12.)

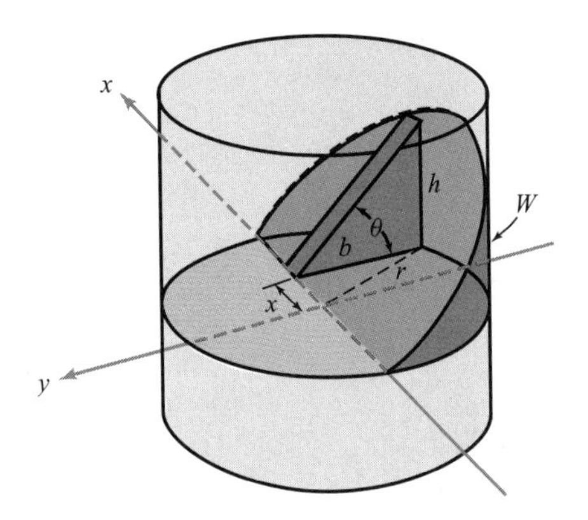

Figure 5.1.12 Find the volume of W.

6. (a) Show that the volume of the solid of revolution shown in Figure 5.1.13(a) is

$$\pi \int_a^b [f(x)]^2\, dx.$$

(b) Show that the volume of the region obtained by rotating the region under the graph of the parabola $y = -x^2 + 2x + 3$, $-1 \leq x \leq 3$, about the x axis is $512\pi/15$ [see Figure 5.1.13(b)].

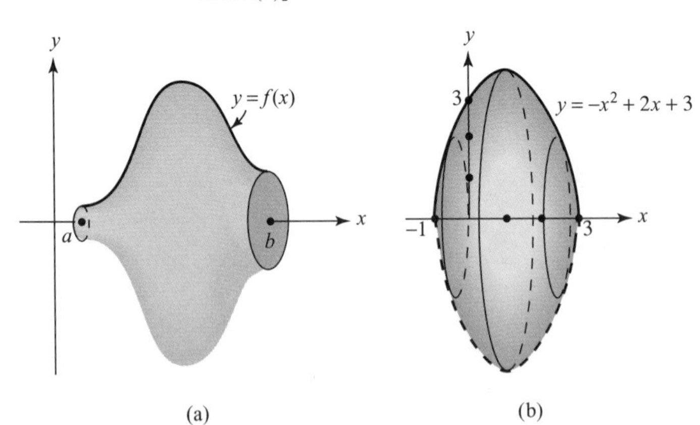

Figure 5.1.13 The solid of revolution (a) has volume $\pi \int_a^b [f(x)]^2\, dx$. Part (b) shows the region between the graph of $y = -x^2 + 2x + 3$ and the x axis rotated about the x axis.

Evaluate the double integrals in Exercises 7 to 9, where R is the rectangle $[0, 2] \times [-1, 0]$.

7. $\iint_R (x^2y^2 + x)\,dy\,dx$

8. $\iint_R \left(|y| \cos \frac{1}{4}\pi x\right) dy\,dx$

9. $\iint_R \left(-xe^x \sin \frac{1}{2}\pi y\right) dy\,dx$

10. Find the volume bounded by the graph of $f(x, y) = 1 + 2x + 3y$, the rectangle $[1, 2] \times [0, 1]$, and the four vertical sides of the rectangle R, as in Figure 5.1.1.

11. Repeat Exercise 10 for the function $f(x, y) = x^4 + y^2$ and the rectangle $[-1, 1] \times [-3, -2]$.

5.2 The Double Integral Over a Rectangle

We are ready to give a rigorous definition of the double integral as the limit of a sequence of sums. This will then be used to *define* the volume of the region under the graph of a function $f(x, y)$. We shall not require that $f(x, y) \geq 0$; but if $f(x, y)$ assumes negative values, we shall interpret the integral as a signed volume, just as for the area under the graph of a function of one variable. In addition, we shall discuss some of the fundamental algebraic properties of the double integral and prove Fubini's theorem, which states that the double integral can be calculated as an iterated integral. To begin, let us cstablish some notation for partitions and sums.

Definition of the Integral

Consider a closed rectangle $R \subset \mathbb{R}^2$; that is, R is a Cartesian product of two intervals: $R = [a, b] \times [c, d]$. By a ***regular partition*** of R of order n we mean the two ordered collections of $n + 1$ equally spaced points $\{x_j\}_{j=0}^n$ and $\{y_k\}_{k=0}^n$, that is, the points satisfying

$$a = x_0 < x_1 < \cdots < x_n = b, \qquad c = y_0 < y_1 < \cdots < y_n = d$$

and

$$x_{j+1} - x_j = \frac{b-a}{n}, \qquad y_{k+1} - y_k = \frac{d-c}{n}$$

(see Figure 5.2.1).

A function $f(x, y)$ is said to be ***bounded*** if there is a number $M > 0$ such that $-M \leq f(x, y) \leq M$ for all (x, y) in the domain of f. A continuous function on a *closed* rectangle is always bounded, but, for example, $f(x, y) = 1/x$ on $(0, 1] \times [0, 1]$ is continuous but is not bounded, because $1/x$ becomes arbitrarily large for x near 0. The rectangle $(0, 1] \times [0, 1]$ is not closed, because the endpoint 0 is missing in the first factor.

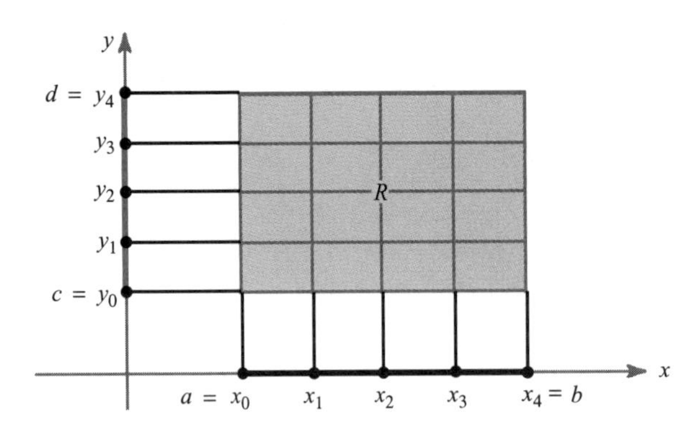

Figure 5.2.1 A regular partition of a rectangle R, with $n = 4$.

Let R_{jk} be the rectangle $[x_j, x_{j+1}] \times [y_k, y_{k+1}]$, and let $\mathbf{c}_{jk}$ be *any* point in R_{jk}. Suppose $f: R \to \mathbb{R}$ is a bounded real-valued function. Form the sum

$$S_n = \sum_{j,k=0}^{n-1} f(\mathbf{c}_{jk})\,\Delta x\,\Delta y = \sum_{j,k=0}^{n-1} f(\mathbf{c}_{jk})\,\Delta A, \tag{1}$$

where

$$\Delta x = x_{j+1} - x_j = \frac{b-a}{n}, \qquad \Delta y = y_{k+1} - y_k = \frac{d-c}{n},$$

and

$$\Delta A = \Delta x\,\Delta y.$$

This sum is taken over all j's and k's from 0 to $n-1$, and so there are n^2 terms. A sum of this type is called a ***Riemann sum*** for f.

DEFINITION: Double Integral If the sequence $\{S_n\}$ converges to a limit S as $n \to \infty$ and if the limit S is the same for any choice of points $\mathbf{c}_{jk}$ in the rectangles R_{jk}, then we say that f is ***integrable*** over R and we write

$$\iint_R f(x,y)\,dA, \qquad \iint_R f(x,y)\,dx\,dy, \qquad \text{or} \qquad \iint_R f\,dx\,dy$$

for the limit S.

Thus, we can rewrite integrability in the following way:

$$\operatorname*{limit}_{n\to\infty} \sum_{j,k=0}^{n-1} f(\mathbf{c}_{jk})\,\Delta x\,\Delta y = \iint_R f\,dx\,dy$$

for any choice of $\mathbf{c}_{jk} \in R_{jk}$.

Properties of the Integral

The proof of the following basic theorem is presented in the Internet supplement for Chapter 5.

THEOREM 1 Any continuous function defined on a closed rectangle R is integrable.

If $f(x, y) \geq 0$, the existence of $\text{limit}_{n\to\infty} S_n$ has a straightforward geometric meaning. Consider the graph of $z = f(x, y)$ as the top of a solid whose base is the rectangle R. If we take each $\mathbf{c}_{jk}$ to be a point where $f(x, y)$ has its minimum value[2] on R_{jk}, then $f(\mathbf{c}_{jk})\,\Delta x\,\Delta y$ represents the volume of a rectangular box with base R_{jk}. The sum $\sum_{j,k=0}^{n-1} f(\mathbf{c}_{jk})\,\Delta x\,\Delta y$ equals the volume of an inscribed solid, part of which is shown in Figure 5.2.2.

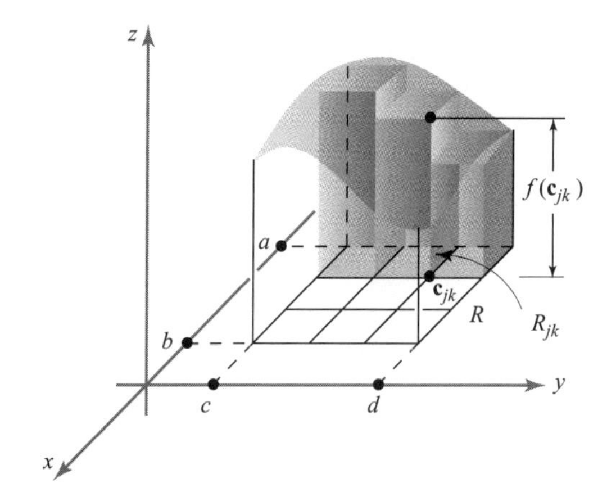

Figure 5.2.2 The sum of inscribed boxes approximates the volume under the graph of $z = f(x, y)$.

Similarly, if $\mathbf{c}_{jk}$ is a point where $f(x, y)$ has its maximum on R_{jk}, then the sum $\sum_{j,k=0}^{n-1} f(\mathbf{c}_{jk})\,\Delta x\,\Delta y$ is equal to the volume of a circumscribed solid (see Figure 5.2.3).

Therefore, if $\text{limit}_{n\to\infty} S_n$ exists and is independent of $\mathbf{c}_{jk} \in R_{jk}$, it follows that the volumes of the inscribed and circumscribed solids approach the same limit as $n \to \infty$. It is therefore reasonable to call this limit the exact volume of the solid under the graph of f. Thus, the method of Riemann sums supports the concepts introduced on an intuitive basis in Section 5.1.

There is a theorem guaranteeing the existence of the integral of certain discontinuous functions as well. We shall need this result in the next section in order to discuss the integrals of functions over regions more general than rectangles. We shall be specifically interested in functions whose discontinuities lie on curves in the xy plane. Figure 5.2.4 shows two functions defined on a rectangle R whose discontinuities

[2] Such $\mathbf{c}_{jk}$ exist by virtue of the continuity of f on R; see Theorem 7 in Section 3.3.

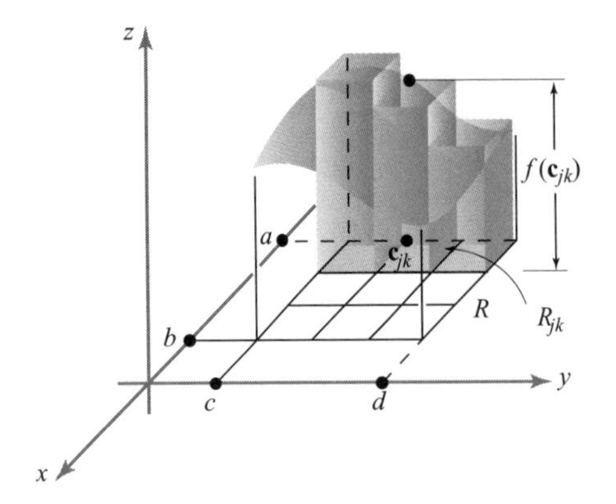

Figure 5.2.3 The volume of circumscribed boxes also approximates the volume under $z = f(x, y)$.

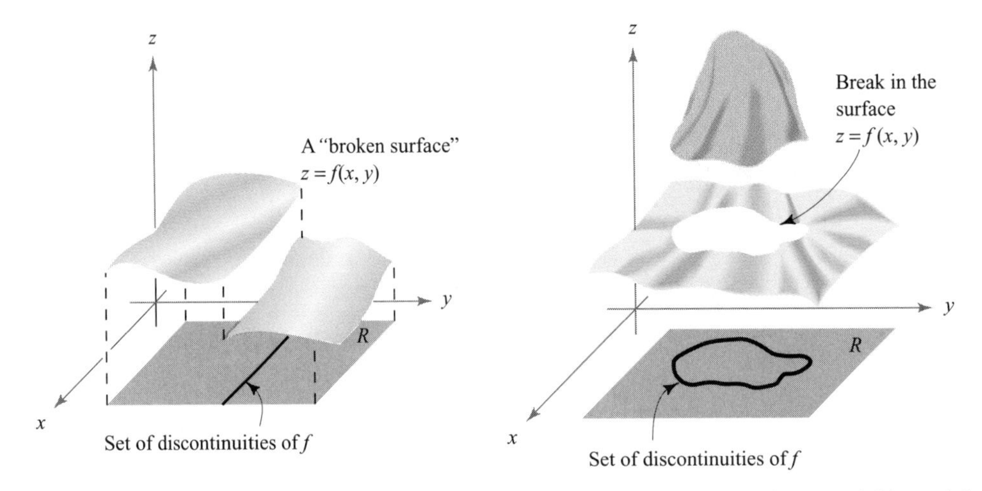

Figure 5.2.4 What the graphs of discontinuous functions of two variables might look like.

lie along curves. In other words, f is continuous at each point that is in R, but not necessarily on the curve.

Useful curves are graphs of functions such as $y = \phi(x)$, $a \leq x \leq b$, or $x = \psi(y)$, $c \leq y \leq d$, or finite unions of such graphs. Some examples are shown in Figure 5.2.5.

The next theorem provides an important criterion for determining whether a function is integrable. The proof is discussed in the Internet supplement.

THEOREM 2: Integrability of Bounded Functions Let $f: R \to \mathbb{R}$ be a bounded real-valued function on the rectangle R, and suppose that the set of points where f is discontinuous lies on a finite union of graphs of continuous functions. Then f is integrable over R.

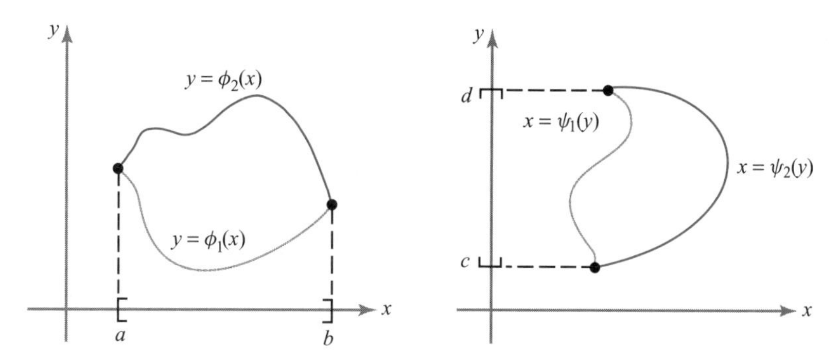

Figure 5.2.5 Curves in the plane represented as graphs.

Using Theorem 2 and the remarks preceding it, we see that the functions sketched in Figure 5.2.4 are integrable over R, because these functions are bounded and continuous except on graphs of continuous functions.

From the definition of the integral as a limit of sums and the limit theorems, we can deduce some fundamental properties of the integral $\iint_R f(x, y)\,dA$; these properties are essentially the same as for the integral of a real-valued function of a single variable.

Let f and g be integrable functions on the rectangle R, and let c be a constant. Then $f + g$ and cf are integrable, and

(i) Linearity

$$\iint_R [f(x, y) + g(x, y)]\,dA = \iint_R f(x, y)\,dA + \iint_R g(x, y)\,dA.$$

(ii) Homogeneity

$$\iint_R cf(x, y)\,dA = c\iint_R f(x, y)\,dA.$$

(iii) Monotonicity If $f(x, y) \geq g(x, y)$, then

$$\iint_R f(x, y)\,dA \geq \iint_R g(x, y)\,dA.$$

(iv) Additivity If R_i, $i = 1, \ldots, m$, are pairwise disjoint rectangles such that f is bounded and integrable over each R_i and if $Q = R_1 \cup R_2 \cup \cdots \cup R_m$ is a rectangle, then $f\colon Q \to \mathbb{R}$ is integrable over Q and

$$\iint_Q f(x, y)\,dA = \sum_{i=1}^{m} \iint_{R_i} f(x, y)\,dA.$$

Properties (i) and (ii) are a consequence of the definition of the integral as a limit of a sum and the following facts for convergent sequences $\{S_n\}$ and $\{T_n\}$, which are proved as with the limit theorems in Chapter 2:

$$\operatorname*{limit}_{n\to\infty} (T_n + S_n) = \operatorname*{limit}_{n\to\infty} T_n + \operatorname*{limit}_{n\to\infty} S_n$$

$$\operatorname*{limit}_{n\to\infty} (cS_n) = c \operatorname*{limit}_{n\to\infty} S_n.$$

To demonstrate monotonicity, we first observe that if $h(x, y) \geq 0$ and $\{S_n\}$ is a sequence of Riemann sums that converges to $\iint_R h(x, y)\, dA$, then $S_n \geq 0$ for all n, so that $\iint_R h(x, y)\, dA = \text{limit}_{n\to\infty} S_n \geq 0$. If $f(x, y) \geq g(x, y)$ for all $(x, y) \in R$, then $(f - g)(x, y) \geq 0$ for all (x, y), and, using properties (i) and (ii), we have

$$\iint_R f(x, y)\, dA - \iint_R g(x, y)\, dA = \iint_R [f(x, y) - g(x, y)]\, dA \geq 0.$$

This proves property (iii). The proof of property (iv) is more technical and a special case is proved in the Internet supplement. It should be intuitively obvious.

Another important result is the inequality

$$\left| \iint_R f\, dA \right| \leq \iint_R |f|\, dA. \tag{2}$$

To see why formula (2) is true, note that, by the definition of absolute value,

$$-|f| \leq f \leq |f|;$$

therefore, from the monotonicity and homogeneity of integration (with $c = -1$),

$$-\iint_R |f|\, dA \leq \iint_R f\, dA \leq \iint_R |f|\, dA,$$

which is equivalent to formula (2).

Fubini's Theorem

Although we have noted the integrability of a variety of functions, we have not yet established rigorously a general method of computing integrals. In the case of one variable, we avoid computing $\int_a^b f(x)\, dx$ from its definition as a limit of a sum by using the *fundamental theorem of integral calculus*. This important theorem tells us that *if f is continuous, then*

$$\int_a^b f(x)\, dx = F(b) - F(a),$$

where F is an antiderivative of f; that is, $F' = f$.

This technique will not work as stated for functions $f(x, y)$ of two variables. However, as we indicate in Section 5.1, we can often reduce a double integral over a rectangle to iterated single integrals; the fundamental theorem then applies to each of these single integrals. Fubini's theorem, which was mentioned in the last section, establishes this reduction to iterated integrals rigorously, by using Riemann sums. As we saw in Section 5.1, the reduction,

$$\iint_R f(x, y)\, dA = \int_a^b \left[\int_c^d f(x, y)\, dy\right] dx = \int_c^d \left[\int_a^b f(x, y)\, dx\right] dy,$$

is a consequence of Cavalieri's principle, at least if $f(x, y) \geq 0$. In terms of Riemann sums, it corresponds to the following equality:

$$\sum_{j,k=0}^{n-1} f(\mathbf{c}_{jk})\, \Delta x\, \Delta y = \sum_{j=0}^{n-1} \left(\sum_{k=0}^{n-1} f(\mathbf{c}_{jk})\, \Delta y\right) \Delta x = \sum_{k=0}^{n-1} \left(\sum_{j=0}^{n-1} f(\mathbf{c}_{jk})\, \Delta x\right) \Delta y,$$

which may be proved more generally as follows: *Let* $[a_{jk}]$ *be an* $n \times n$ *matrix, where* $0 \leq j \leq n-1$ *and* $0 \leq k \leq n-1$. *Let* $\sum_{j,k=0}^{n-1} a_{jk}$ *be the sum of the* n^2 *matrix entries. Then*

$$\sum_{j,k=0}^{n-1} a_{jk} = \sum_{j=0}^{n-1} \left(\sum_{k=0}^{n-1} a_{jk}\right) = \sum_{k=0}^{n-1} \left(\sum_{j=0}^{n-1} a_{jk}\right). \tag{3}$$

In the first equality, the right-hand side represents summing the matrix entries first by rows and then adding the results:

$$\left[\begin{array}{ccccccc} a_{00} & a_{01} & a_{02} & \cdots & \overrightarrow{a_{0k} \quad \cdots \quad a_{0(n-1)}} \\ \vdots & & & & \vdots \\ a_{j0} & a_{j1} & & \cdots & \overrightarrow{a_{jk} \quad \cdots \quad a_{j(n-1)}} \\ \vdots & & & & \vdots \\ a_{(n-1)0} & a_{(n-1)1} & & \cdots & \overrightarrow{a_{(n-1)k} \quad \cdots \quad a_{(n-1)(n-1)}} \end{array}\right] \quad \begin{array}{l} \sum_{k=0}^{n-1} a_{0k} \\ \vdots \\ \sum_{k=0}^{n-1} a_{jk} \\ \vdots \\ \sum_{k=0}^{n-1} a_{(n-1)k} \downarrow \\ \hline \sum_{j=0}^{n-1} \left(\sum_{k=0}^{n-1} a_{jk}\right). \end{array}$$

Clearly, this is equal to $\sum_{j,k=0}^{n-1} a_{jk}$, that is, the sum of all the a_{jk}. Similarly, the sum $\sum_{k=0}^{n-1} \left(\sum_{j=0}^{n-1} a_{jk}\right)$ represents a summing of the matrix entries by columns. This establishes equation (3) and makes the reduction to iterated integrals quite plausible if we remember that integrals can be approximated by the corresponding Riemann sums. The actual proof of Fubini's theorem exploits this idea.

THEOREM 3: Fubini's Theorem Let f be a continuous function with a rectangular domain $R = [a, b] \times [c, d]$. Then

$$\int_a^b \int_c^d f(x, y)\,dy\,dx = \int_c^d \int_a^b f(x, y)\,dx\,dy = \iint_R f(x, y)\,dA. \quad (4)$$

PROOF We shall first show that

$$\int_a^b \int_c^d f(x, y)\,dy\,dx = \iint_R f(x, y)\,dA.$$

Let $c = y_0 < y_1 < \cdots < y_n = d$ be a partition of $[c, d]$ into n equal parts. Define

$$F(x) = \int_c^d f(x, y)\,dy.$$

Then

$$F(x) = \sum_{k=0}^{n-1} \int_{y_k}^{y_{k+1}} f(x, y)\,dy.$$

Using the integral version of the mean-value theorem,[3] for each fixed x and for each k we have

$$\int_{y_k}^{y_{k+1}} f(x, y)\,dy = f(x, Y_k(x))(y_{k+1} - y_k)$$

(see Figure 5.2.6), where the point $Y_k(x)$ belongs to $[y_k, y_{k+1}]$ and may depend on x, k, and n.

We have thus shown that

$$F(x) = \sum_{k=0}^{n-1} f(x, Y_k(x))(y_{k+1} - y_k). \quad (5)$$

[3] This states that if $g(x)$ is continuous on $[a, b]$, then $\int_a^b g(x)\,dx = g(c)(b - a)$ for some point $c \in [a, b]$. The more general second mean-value theorem was proved in Section 3.2.

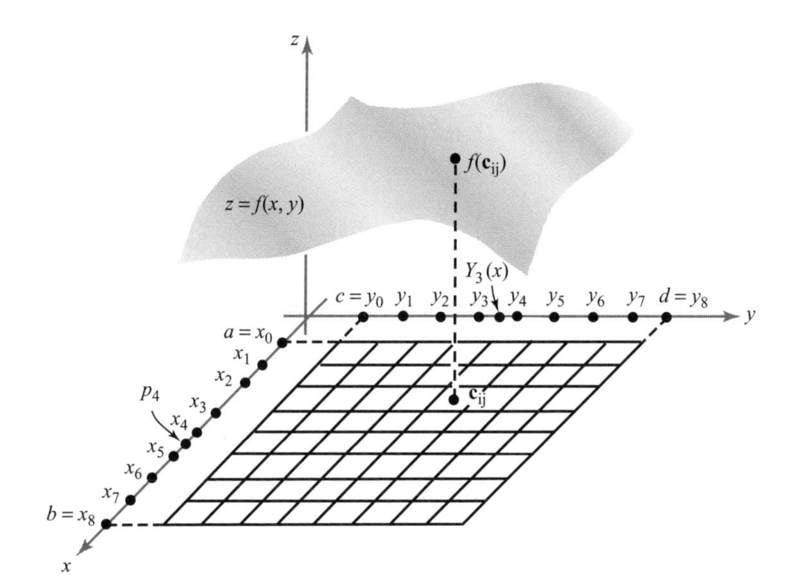

Figure 5.2.6 The notation needed in the proof of Fubini's theorem; $n = 8$.

By the definition of the integral in one variable as a limit of Riemann sums,

$$\int_a^b F(x)\,dx = \int_a^b \left[\int_c^d f(x, y)\,dy\right] dx = \lim_{n\to\infty} \sum_{j=0}^{n-1} F(p_j)(x_{j+1} - x_j),$$

where $a = x_0 < x_1 < \cdots < x_n = b$ is a partition of the interval $[a, b]$ into n equal parts and p_j is any point in $[x_j, x_{j+1}]$. Setting $\mathbf{c}_{jk} = (p_j, Y_k(p_j)) \in R_{jk}$, we have [substituting p_j for x in equation (5)]

$$F(p_j) = \sum_{k=0}^{n-1} f(\mathbf{c}_{jk})(y_{k+1} - y_k).$$

Therefore,

$$\begin{aligned}\int_a^b \int_c^d f(x, y)\,dy\,dx &= \int_a^b F(x)\,dx = \lim_{n\to\infty} \sum_{j=0}^{n-1} F(p_j)(x_{j+1} - x_j)\\ &= \lim_{n\to\infty} \sum_{j=0}^{n-1}\sum_{k=0}^{n-1} f(\mathbf{c}_{jk})(y_{k+1} - y_k)(x_{j+1} - x_j)\\ &= \iint_R f(x, y)\,dA.\end{aligned}$$

Thus, we have proved that

$$\int_a^b \int_c^d f(x, y)\, dy\, dx = \iint_R f(x, y)\, dA.$$

By the same reasoning we can show that

$$\int_c^d \int_a^b f(x, y)\, dx\, dy = \iint_R f(x, y)\, dA.$$

These two conclusions are exactly what we wanted to prove. ■

Fubini's theorem can be generalized to the case where f is not necessarily continuous. Although we shall not present a proof, we state here this more general version.

THEOREM 3′: Fubini's Theorem Let f be a bounded function with domain a rectangle $R = [a, b] \times [c, d]$, and suppose the discontinuities of f lie on a finite union of graphs of continuous functions. If the integral $\int_c^d f(x, y)\, dy$ exists for each $x \in [a, b]$, then

$$\int_a^b \left[\int_c^d f(x, y)\, dy \right] dx$$

exists and

$$\int_a^b \int_c^d f(x, y)\, dy\, dx = \iint_R f(x, y)\, dA.$$

Similarly, if $\int_a^b f(x, y)\, dx$ exists for each $y \in [c, d]$, then

$$\int_c^d \left[\int_a^b f(x, y)\, dx \right] dy$$

exists and

$$\int_c^d \int_a^b f(x, y)\, dx\, dy = \iint_R f(x, y)\, dA.$$

Thus, if all these conditions hold simultaneously,

$$\int_a^b \int_c^d f(x, y)\, dy\, dx = \int_c^d \int_a^b f(x, y)\, dx\, dy = \iint_R f(x, y)\, dA.$$

The assumptions made for this version of Fubini's theorem are more complicated than those we made in Theorem 3. They are necessary because if f is not continuous everywhere, for example, there is no guarantee that $\int_c^d f(x, y)\,dy$ will exist for each x.

EXAMPLE 1 Compute $\iint_R (x^2 + y)\,dA$, where R is the square $[0, 1] \times [0, 1]$.

SOLUTION By Fubini's theorem,

$$\iint_R (x^2 + y)\,dA = \int_0^1 \int_0^1 (x^2 + y)\,dx\,dy = \int_0^1 \left[\int_0^1 (x^2 + y)\,dx\right] dy.$$

By the fundamental theorem of calculus, the x integration may be performed:

$$\int_0^1 (x^2 + y)\,dx = \left[\frac{x^3}{3} + yx\right]_{x=0}^{1} = \frac{1}{3} + y.$$

Thus,

$$\iint_R (x^2 + y)\,dA = \int_0^1 \left[\frac{1}{3} + y\right] dy = \left[\frac{1}{3}y + \frac{y^2}{2}\right]_0^1 = \frac{5}{6}.$$

What we have done is hold y fixed, integrate with respect to x, and then evaluate the result between the given limits for the x variable. Next, we integrated the remaining function (of y alone) with respect to y to obtain the final answer. ▲

EXAMPLE 2 A consequence of Fubini's theorem is that interchanging the order of integration in the iterated integrals does not change the answer. Verify this for Example 1.

SOLUTION We carry out the integration in the other order:

$$\int_0^1 \int_0^1 (x^2 + y)\,dy\,dx = \int_0^1 \left[x^2 y + \frac{y^2}{2}\right]_{y=0}^{1} dx = \int_0^1 \left[x^2 + \frac{1}{2}\right] dx$$
$$= \left[\frac{x^3}{3} + \frac{x}{2}\right]_0^1 = \frac{5}{6}. \quad ▲$$

We have seen that when $f(x, y) \geq 0$ on $R = [a, b] \times [c, d]$, the integral $\iint_R f(x, y)\,dA$ can be interpreted as a volume. If the function also takes on negative values, then the double integral can be thought of as the sum of all volumes lying between the surface $z = f(x, y)$ and the plane $z = 0$, bounded by the planes $x = a, x = b, y = c$, and $y = d$; here the volumes above $z = 0$ are counted as positive and those below as negative. However, Fubini's theorem as stated remains valid in the case where $f(x, y)$ is negative or changes sign on R; that is, there is no restriction on the sign of f in the hypotheses of the theorem.

EXAMPLE 3 Let R be the rectangle $[-2, 1] \times [0, 1]$ and let f be defined by $f(x, y) = y(x^3 - 12x)$; $f(x, y)$ takes on both positive and negative values on R. Evaluate the integral $\iint_R f(x, y)\, dx\, dy = \iint_R y(x^3 - 12x)\, dx\, dy$.

SOLUTION By Fubini's theorem, we may write

$$\iint_R y(x^3 - 12x)\, dx\, dy = \int_0^1 \left[\int_{-2}^1 y(x^3 - 12x)\, dx \right] dy = \frac{57}{4} \int_0^1 y\, dy = \frac{57}{8}.$$

Alternatively, integrating first with respect to y, we find

$$\begin{aligned} \iint_R y(x^3 - 12x)\, dy\, dx &= \int_{-2}^1 \left[\int_0^1 (x^3 - 12x) y\, dy \right] dx \\ &= \frac{1}{2} \int_{-2}^1 (x^3 - 12x)\, dx = \frac{1}{2} \left[\frac{x^4}{4} - 6x^2 \right]_{-2}^1 = \frac{57}{8}. \end{aligned}$$ ▲

Historical Note

The Riemann Integral

The first time most mathematics students encounter the name of Bernhard Riemann is in their calculus courses, where they read about the Riemann integral. Leibniz had thought of the integral of a function of one variable as an infinite sum (the $\int$ standing for a sum) of infinitesimal areas $f(x)\, dx$, where dx is an "infinitesimal width" and $f(x)$ is the height of the corresponding "infinitesimally thin" rectangle. This intuitive approach sufficed for most purposes because the fundamental theorem

$$\int_a^b f(x)\, dx = F(b) - F(a)$$

showed how to evaluate this (nebulously defined) integral when one knows the antiderivative F of f.

However, Riemann was interested in applying integration to functions of one variable where the antiderivative was not known, and to functions in number theory or in general to those functions that "one need not find in nature."

Cauchy had already known that all continuous functions could be integrated and that the fundamental theorem was valid—that is, every continuous function had an antiderivative. However, his proofs were not entirely rigorous. For applications to number theory and to certain series (called *Fourier series*), Riemann needed a clear, precise definition of the integral, which he presented in a paper in 1854. In this paper he defines his integral and gives necessary and sufficient conditions for a bounded function f to be integrable over an interval $[a, b]$.

In 1876, the German mathematician Karl J. Thomae generalized Riemann's integral to apply to functions of several variables, as we do in this chapter. We further develop this approach in the Internet supplement.

In the first half of the nineteenth century, Cauchy had observed that for continuous function of two variables, Fubini's theorem was valid. But Cauchy also gave an example of an unbounded function of two variables for which the iterated integrals were not equal. In 1878, Thomae gave the first example of a bounded function of two variables where one iterated integral exists and the other does not. In these examples, the functions were not "Riemann integrable" in the sense described in this section. Cauchy and Thomae's examples demonstrated that one must apply caution and not necessarily assume that iterated integrals are always equal.

In 1902, the French mathematician Henri Lebesgue developed a truly sweeping generalization of the Riemann integral. Lebesgue's theory allowed integration of vastly more functions than did Riemann's approach. Perhaps, unforeseen by Lebesgue, his theory was to have a profound impact on the development of many areas of mathematics in the twentieth century—in particular the theory of partial differential equations. Mathematics students go into more depth about the Lebesgue integral in their first year of graduate study.

In 1907, the Italian mathematician Guido Fubini used the Lebesgue integral to state the most general form of the theorem on the equality of iterated integrals, the form that is studied today and used by working mathematicians and scientists in their research.

EXERCISES

1. Evaluate each of the following integrals if $R = [0, 1] \times [0, 1]$.

(a) $\iint_R (x^3 + y^2)\, dA$

(b) $\iint_R ye^{xy}\, dA$

(c) $\iint_R (xy)^2 \cos x^3\, dA$

(d) $\iint_R \ln [(x + 1)(y + 1)]\, dA$

2. Evaluate each of the following integrals if $R = [0, 1] \times [0, 1]$.

(a) $\iint_R (x^m y^n)\, dx\, dy$, where $m, n > 0$

(b) $\iint_R (ax + by + c)\, dx\, dy$

(c) $\iint_R \sin (x + y)\, dx\, dy$

(d) $\iint_R (x^2 + 2xy + y\sqrt{x})\, dx\, dy$

3. Compute the volume of the region over the rectangle $[0, 1] \times [0, 1]$ and under the graph of $z = xy$.

4. Compute the volume of the solid bounded by the xz plane, the yz plane, the xy plane, the planes $x = 1$ and $y = 1$, and the surface $z = x^2 + y^4$.

5. Let f be continuous on $[a, b]$ and g continuous on $[c, d]$. Show that

$$\iint_R [f(x)g(y)]\, dx\, dy = \left[\int_a^b f(x)\, dx\right]\left[\int_c^d g(y)\, dy\right],$$

where $R = [a, b] \times [c, d]$.

6. Compute the volume of the solid bounded by the surface $z = \sin y$, the planes $x = 1$, $x = 0$, $y = 0$, and $y = \pi/2$, and the xy plane.

7. Compute the volume of the solid bounded by the graph $z = x^2 + y$, the rectangle $R = [0, 1] \times [1, 2]$, and the "vertical sides" of R.

8. Let f be continuous on $R = [a, b] \times [c, d]$; for $a < x < b$, $c < y < d$, define

$$F(x, y) = \int_a^x \int_c^y f(u, v)\, dv\, du.$$

Show that $\partial^2 F/\partial x\, \partial y = \partial^2 F/\partial y\, \partial x = f(x, y)$. Use this example to discuss the relationship between Fubini's theorem and the equality of mixed partial derivatives.

9. Let $f\colon [0, 1] \times [0, 1] \to \mathbb{R}$ be defined by

$$f(x, y) = \begin{cases} 1 & x \text{ rational} \\ 2y & x \text{ irrational.} \end{cases}$$

Show that the iterated integral $\int_0^1 \left[\int_0^1 f(x, y)\, dy\right] dx$ exists but that f is not integrable.

10. Express $\iint_R \cosh xy\, dx\, dy$ as a convergent sequence, where $R = [0, 1] \times [0, 1]$.

11. Although Fubini's theorem holds for most functions met in practice, one must still exercise some caution. This exercise gives a function for which it fails. By using a substitution involving the tangent function, show that

$$\int_0^1 \int_0^1 \frac{x^2 - y^2}{(x^2 + y^2)^2}\, dy\, dx = \frac{\pi}{4}, \qquad \text{yet} \qquad \int_0^1 \int_0^1 \frac{x^2 - y^2}{(x^2 + y^2)^2}\, dx\, dy = -\frac{\pi}{4}.$$

Why does this not contradict Theorem 3 or 3′?

12. Let f be continuous, $f \geq 0$, on the rectangle R. If $\iint_R f\, dA = 0$, prove that $f = 0$ on R.

5.3 The Double Integral Over More General Regions

Our goal in this section is twofold: First, we wish to define the double integral of a function $f(x, y)$ over regions D more general than rectangles; second, we want to develop a technique for evaluating this type of integral. To accomplish this, we shall define three special types of subsets of the xy plane, and then extend the notion of the double integral to them.

Elementary Regions

Suppose we are given two continuous real-valued functions $\phi_1\colon [a, b] \to \mathbb{R}$ and $\phi_2\colon [a, b] \to \mathbb{R}$ that satisfy $\phi_1(x) \le \phi_2(x)$ for all $x \in [a, b]$. Let D be the set of all points (x, y) such that $x \in [a, b]$ and $\phi_1(x) \le y \le \phi_2(x)$. This region D is said to be ***y*-simple**. Figure 5.3.1 shows various examples of y-simple regions. The curves and straight-line segments that bound the region together constitute the ***boundary*** of D, denoted ∂D. We use the phrase y-simple because the region is described in a relatively simple way, using y as a function of x.

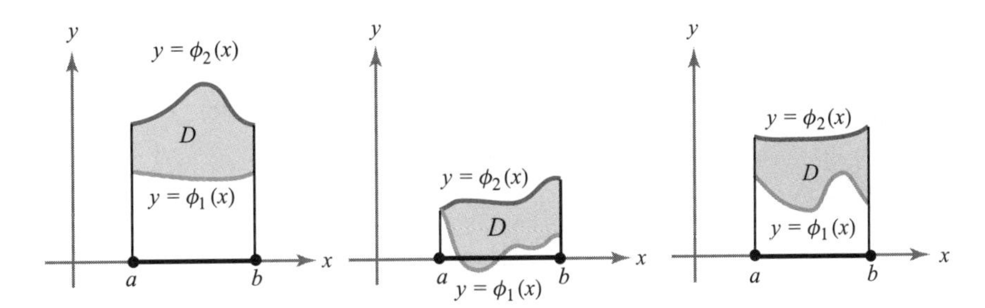

Figure 5.3.1 Some y-simple regions.

We say that a region D is ***x*-simple** if there are continuous functions ψ_1 and ψ_2 defined on $[c, d]$ such that D is the set of points (x, y) satisfying

$$y \in [c, d] \qquad \text{and} \qquad \psi_1(y) \le x \le \psi_2(y)$$

where $\psi_1(y) \le \psi_2(y)$ for all $y \in [c, d]$. Again, the curves that bound the region D constitute its boundary ∂D. Some examples of x-simple regions are shown in Figure 5.3.2. In this situation, x is the distinguished variable, given as a function of y. Thus, the phrase x-simple is appropriate.

Finally, a ***simple*** region is one that is both x- and y-simple; that is, a simple region can be described as both an x-simple region and a y-simple region. An example of a simple region is a unit disk (see Figure 5.3.3).

Sometimes we will refer to any of the regions as ***elementary regions***. Note that the boundary ∂D of an elementary region is the type of set of discontinuities of a function allowed in Theorem 2.

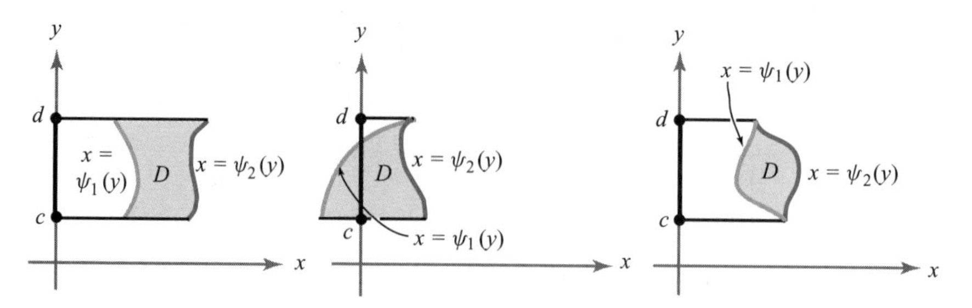

Figure 5.3.2 Some x-simple regions.

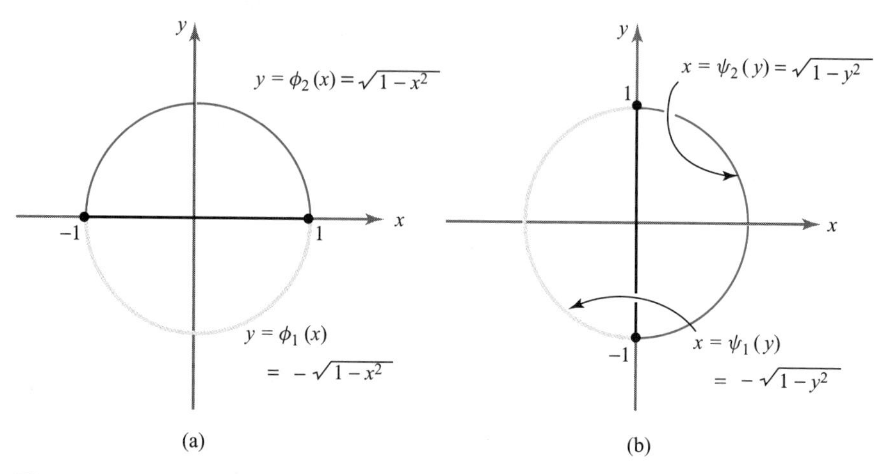

Figure 5.3.3 The unit disk, a simple region: (a) as a y-simple region, and (b) as an x-simple region.

The Integral over an Elementary Region

We can now use an interesting "trick" to extend the definition of the integral from rectangles to elementary regions.

DEFINITION: Integral over an Elementary Region If D is an elementary region in the plane, choose a rectangle R that contains D. Given $f: D \to \mathbb{R}$, where f is continuous (and hence bounded), define $\iint_D f(x, y)\, dA$, the ***integral of f over the set*** D as follows: Extend f to a function f^* defined on all of R by

$$f^*(x, y) = \begin{cases} f(x, y) & \text{if } (x, y) \in D \\ 0 & \text{if } (x, y) \notin D \text{ and } (x, y) \in R. \end{cases}$$

Note that f^* is bounded (because f is) and continuous except possibly on the boundary of D (see Figure 5.3.4). The boundary of D consists of graphs of continuous functions, and so f^* is integrable over R by Theorem 2, Section 5.2. Therefore, we can define

$$\iint_D f(x, y)\, dA = \iint_R f^*(x, y)\, dA.$$

When $f(x, y) \geq 0$ on D, we can interpret the integral $\iint_D f(x, y)\, dA$ as the volume of the three-dimensional region between the graph of f and D, as is evident from Figure 5.3.4.

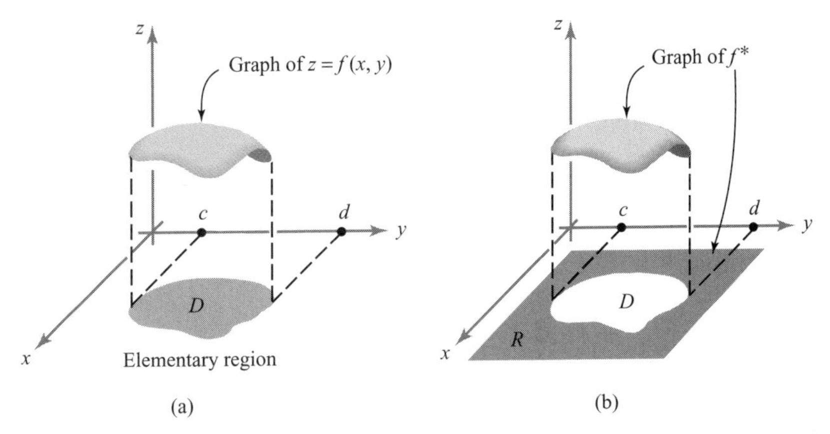

Figure 5.3.4 (a) Graph of $z = f(x, y)$ over an elementary region D. (b) Shaded region shows graph of $z = f^*(x, y)$ on some rectangle R containing D. From this picture we see that boundary points of D may be points of discontinuity of f^*, because the graph of $z = f^*(x, y)$ can be broken at these points.

We have defined $\iint_D f(x, y)\, dx\, dy$ by choosing a rectangle R that encloses D. It should be intuitively clear that the value of $\iint_D f(x, y)\, dx\, dy$ does not depend on the particular R we select; we shall demonstrate this fact at the end of this section.

Reduction to Iterated Integrals

If $R = [a, b] \times [c, d]$ is a rectangle containing D, we can use the results on iterated integrals in Section 5.2 to obtain

$$\iint_D f(x, y)\, dA = \iint_R f^*(x, y)\, dA = \int_a^b \int_c^d f^*(x, y)\, dy\, dx$$
$$= \int_c^d \int_a^b f^*(x, y)\, dx\, dy,$$

where f^* equals f in D and zero outside D, as before. Assume that D is a y-simple region determined by functions $\phi_1\colon [a, b] \to \mathbb{R}$ and $\phi_2\colon [a, b] \to \mathbb{R}$.

Consider the iterated integral

$$\int_a^b \int_c^d f^*(x, y)\, dy\, dx$$

and, in particular, the inner integral $\int_c^d f^*(x, y)\, dy$ for some fixed x (Figure 5.3.5). By definition, $f^*(x, y) = 0$ if $y < \phi_1(x)$ or $y > \phi_2(x)$, so we obtain

$$\int_c^d f^*(x, y)\, dy = \int_{\phi_1(x)}^{\phi_2(x)} f^*(x, y)\, dy = \int_{\phi_1(x)}^{\phi_2(x)} f(x, y)\, dy.$$

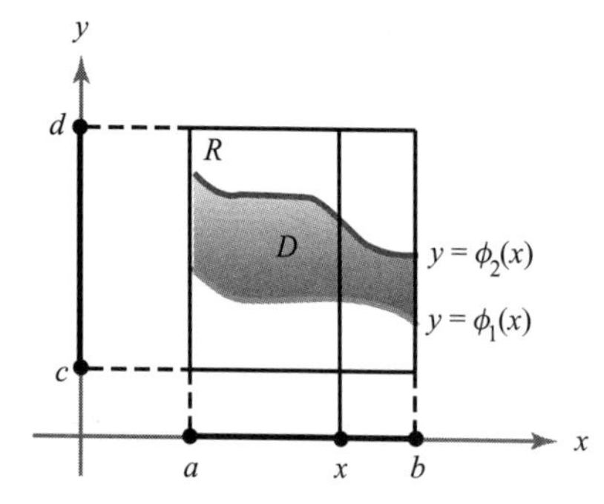

Figure 5.3.5 The region between two graphs—a y-simple region.

We summarize what we have obtained in the following.

THEOREM 4: Reduction to Iterated Integrals If D is a y-simple region, as shown in Figure 5.3.5, then

$$\iint_D f(x, y)\, dA = \int_a^b \int_{\phi_1(x)}^{\phi_2(x)} f(x, y)\, dy\, dx. \tag{1}$$

In the case $f(x, y) = 1$ for all $(x, y) \in D$, $\iint_D f(x, y)\, dA$ is the area of D. On the other hand, in this case, the right-hand side of formula (1) becomes:

$$\int_a^b \int_{\phi_1(x)}^{\phi_2(x)} f(x, y)\, dy\, dx = \int_a^b [\phi_2(x) - \phi_1(x)]\, dx = A(D),$$

which is the formula for the area of D learned in one-variable calculus. Thus, formula (1) checks in this case.

EXAMPLE 1 Find $\iint_T (x^3y + \cos x)\, dA$, where T is the triangle consisting of all points (x, y) such that $0 \le x \le \pi/2, 0 \le y \le x$.

SOLUTION Referring to Figure 5.3.6 and formula (1), we have

$$\begin{aligned}\iint_T (x^3y + \cos x)\, dA &= \int_0^{\pi/2} \int_0^x (x^3y + \cos x)\, dy\, dx \\ &= \int_0^{\pi/2} \left[\frac{x^3y^2}{2} + y \cos x\right]_{y=0}^{x} dx = \int_0^{\pi/2} \left(\frac{x^5}{2} + x \cos x\right) dx \\ &= \left[\frac{x^6}{12}\right]_0^{\pi/2} + \int_0^{\pi/2} (x \cos x)\, dx = \frac{\pi^6}{(12)(64)} + [x \sin x + \cos x]_0^{\pi/2} \\ &= \frac{\pi^6}{768} + \frac{\pi}{2} - 1. \quad \blacktriangle\end{aligned}$$

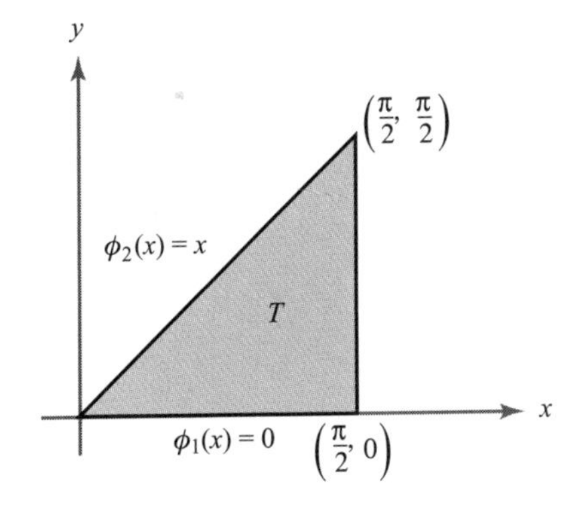

Figure 5.3.6 A triangle T represented as a y-simple region.

In the next example, we use formula (1) to find the volume of a solid whose base is a nonrectangular region D.

EXAMPLE 2 Find the volume of the tetrahedron bounded by the planes $y = 0$, $z = 0$, $x = 0$, and $y - x + z = 1$ (Figure 5.3.7).

SOLUTION We first note that the given tetrahedron has a triangular base D whose points (x, y) satisfy $-1 \le x \le 0$ and $0 \le y \le 1 + x$; hence, D is a y-simple region. In fact, D is a simple region; see Figure 5.3.8.

For any point (x, y) in D, the height of the surface z above (x, y) is $1 - y + x$. Thus, the volume we seek is given by the integral

$$\iint_D (1 - y + x)\, dA.$$

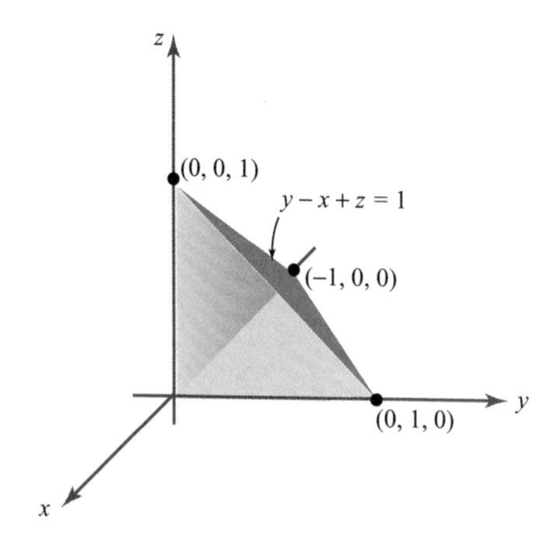

Figure 5.3.7 A tetrahedron bounded by the planes $y = 0, z = 0, x = 0$, and $y - x + z = 1$.

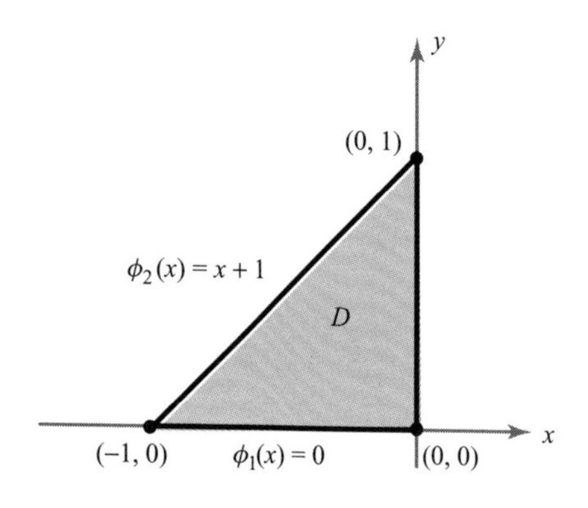

Figure 5.3.8 The base of the tetrahedron in Figure 5.3.7 represented as a y-simple region.

Using formula (1) with $\phi_1(x) = 0$ and $\phi_2(x) = x + 1$, we have

$$\iint_D (1 - y + x)\, dA = \int_{-1}^{0} \int_{0}^{1+x} (1 - y + x)\, dy\, dx = \int_{-1}^{0} \left[(1 + x)y - \frac{y^2}{2} \right]_{y=0}^{1+x} dx$$

$$= \int_{-1}^{0} \left[\frac{(1 + x)^2}{2} \right] dx = \left[\frac{(1 + x)^3}{6} \right]_{-1}^{0} = \frac{1}{6}. \quad \blacktriangle$$

EXAMPLE 3 Let D be a y-simple region. Describe its area $A(D)$ as a limit of Riemann sums.

SOLUTION If we recall the definition, $A(D) = \iint_D dx\, dy$ is the integral over a containing rectangle R of the function $f = 1$. A Riemann sum S_n for this integral is obtained by dividing R into subrectangles and forming the sum $S_n = \sum_{j,k=0}^{n-1} f^*(\mathbf{c}_{jk}) \Delta x\, \Delta y$, as in formula (1) of Section 5.2. Now $f^*(\mathbf{c}_{jk})$ is 1 or 0, depending on whether or not $\mathbf{c}_{jk}$ is in D. Consider those subrectangles R_{jk} that have nonvoid intersection with D, and choose $\mathbf{c}_{jk}$ in $D \cap R_{jk}$. Thus, S_n is the sum of the areas of the subrectangles that meet D and $A(D)$ is the limit of these as $n \to \infty$.

Thus, $A(D)$ is the limit of the areas of the rectangles "circumscribing" D. The reader should draw a figure to accompany this discussion. ▲

The methods for treating x-simple regions are entirely analogous. Specifically, we have the following.

THEOREM 4′: Iterated Integrals for x-Simple Regions Suppose that D is the set of points (x, y) such that $y \in [c, d]$ and $\psi_1(y) \leq x \leq \psi_2(y)$. If f is continuous on D, then

$$\iint_D f(x, y)\, dA = \int_c^d \left[\int_{\psi_1(y)}^{\psi_2(y)} f(x, y)\, dx \right] dy. \tag{2}$$

To find the area of D, we substitute $f = 1$ in formula (2); this yields

$$\iint_D dA = \int_c^d (\psi_2(y) - \psi_1(y))\, dy.$$

Again, this result for area agrees with the results of single-variable calculus for the area of a region between two curves.

Either the method for y-simple or the method for x-simple regions can be used for integrals over simple regions.

It follows from formulas (1) and (2) that $\iint_D f\, dA$ is independent of the choice of the rectangle R enclosing D used in the definition of $\iint_D f\, dA$, because, if we had picked another rectangle enclosing D, we would have arrived at the same formula (1).

EXERCISES

1. Evaluate the following iterated integrals and draw the regions D determined by the limits. State whether the regions are x-simple, y-simple, or simple.

(a) $\displaystyle\int_0^1 \int_0^{x^2} dy\, dx$

(b) $\displaystyle\int_1^2 \int_{2x}^{3x+1} dy\, dx$

(c) $\displaystyle\int_0^1 \int_1^{e^x} (x + y)\, dy\, dx$

(d) $\displaystyle\int_0^1 \int_{x^3}^{x^2} y\, dy\, dx$

2. Evaluate the following integrals and sketch the corresponding regions.

(a) $\displaystyle\int_{-3}^2 \int_0^{y^2} (x^2 + y)\, dx\, dy$

(b) $\displaystyle\int_{-1}^1 \int_{-2|x|}^{|x|} e^{x+y}\, dy\, dx$

(c) $\displaystyle\int_0^1 \int_0^{(1-x^2)^{1/2}} dy\, dx$

(d) $\displaystyle\int_0^{\pi/2} \int_0^{\cos x} y \sin x\, dy\, dx$

(e) $\displaystyle\int_0^1 \int_{y^2}^{y} (x^n + y^m)\, dx\, dy, \qquad m, n > 0$

(f) $\displaystyle\int_{-1}^0 \int_0^{2(1-x^2)^{1/2}} x\, dy\, dx$

3. Use double integrals to compute the area of a circle of radius r.

4. Using double integrals, determine the area of an ellipse with semiaxes of length a and b.

5. What is the volume of a barn that has a rectangular base 20 ft by 40 ft, vertical walls 30 ft high at the front (which we assume is on the 20-ft side of the barn), and 40 ft high at the rear? The barn has a flat roof. Use double integrals to compute the volume.

6. Let D be the region bounded by the positive x and y axes and the line $3x + 4y = 10$. Compute

$$\iint_D (x^2 + y^2)\, dA.$$

7. Let D be the region bounded by the y axis and the parabola $x = -4y^2 + 3$. Compute

$$\iint_D x^3 y\, dx\, dy.$$

8. Evaluate $\int_0^1 \int_0^{x^2} (x^2 + xy - y^2)\, dy\, dx$. Describe this iterated integral as an integral over a certain region D in the xy plane.

9. Let D be the region given as the set of (x, y) where $1 \leq x^2 + y^2 \leq 2$ and $y \geq 0$. Is D an elementary region? Evaluate $\iint_D f(x, y)\, dA$ where $f(x, y) = 1 + xy$.

10. Use the formula $A(D) = \iint_D dx\, dy$ to find the area enclosed by one period of the sine function $\sin x$, for $0 \leq x \leq 2\pi$, and the x axis.

11. Find the volume of the region inside the surface $z = x^2 + y^2$ and between $z = 0$ and $z = 10$.

12. Set up the integral required to calculate the volume of a cone of base radius r and height h.

13. Evaluate $\iint_D y\, dA$ where D is the set of points (x, y) such that $0 \leq 2x/\pi \leq y$, $y \leq \sin x$.

14. From Exercise 5, Section 5.2, $\int_a^b \int_c^d f(x) g(y)\, dy\, dx = \left(\int_a^b f(x)\, dx\right)\left(\int_c^d g(y)\, dy\right)$. Is it true that $\iint_D f(x) g(y)\, dx\, dy = \left(\int_a^b f(x)\, dx\right)\left(\int_{\phi_1(a)}^{\phi_2(b)} g(y)\, dy\right)$ for y-simple regions?

15. Let D be a region given as the set of (x, y) with $-\phi(x) \leq y \leq \phi(x)$ and $a \leq x \leq b$, where ϕ is a nonnegative continuous function on the interval $[a, b]$. Let $f(x, y)$ be a function on D such that $f(x, y) = -f(x, -y)$ for all $(x, y) \in D$. Argue that $\iint_D f(x, y)\, dA = 0$.

16. Use the methods of this section to show that the area of the parallelogram D determined by two planar vectors $\mathbf{a}$ and $\mathbf{b}$ is $|a_1 b_2 - a_2 b_1|$, where $\mathbf{a} = a_1\mathbf{i} + a_2\mathbf{j}$ and $\mathbf{b} = b_1\mathbf{i} + b_2\mathbf{j}$.

17. Describe the area $A(D)$ of a region as a limit of areas of inscribed rectangles, as in Example 3.

5.4 Changing the Order of Integration

Suppose that D is a simple region—that is, it is both x-simple and y-simple. Thus, it can be given as the set of points (x, y) such that

$$a \le x \le b, \qquad \phi_1(x) \le y \le \phi_2(x),$$

and also as the set of points (x, y) such that

$$c \le y \le d, \qquad \psi_1(y) \le x \le \psi_2(y).$$

Hence, we have the formulas

$$\iint_D f(x, y)\, dA = \int_a^b \int_{\phi_1(x)}^{\phi_2(x)} f(x, y)\, dy\, dx = \int_c^d \int_{\psi_1(y)}^{\psi_2(y)} f(x, y)\, dx\, dy.$$

If we are required to compute one of the preceding iterated integrals, we may do so by evaluating the other iterated integral; this technique is called *changing the order of integration*. It can be useful to make such a change when evaluating iterated integrals, because one of the iterated integrals may be more difficult to compute than the other.

EXAMPLE 1 By changing the order of integration, evaluate

$$\int_0^a \int_0^{(a^2-x^2)^{1/2}} (a^2 - y^2)^{1/2}\, dy\, dx.$$

SOLUTION Note that x varies between 0 and a, and for each such fixed x, we have $0 \le y \le (a^2 - x^2)^{1/2}$. Thus, the iterated integral is equivalent to the double integral

$$\iint_D (a^2 - y^2)^{1/2}\, dy\, dx,$$

where D is the set of points (x, y) such that $0 \le x \le a$ and $0 \le y \le (a^2 - x^2)^{1/2}$. But this is the representation of one quarter (the positive quadrant portion) of the disk of radius a; hence, D can also be described as the set of points (x, y) satisfying

$$0 \le y \le a, \qquad 0 \le x \le (a^2 - y^2)^{1/2}$$

(see Figure 5.4.1). Thus,

$$\begin{aligned}
\int_0^a \int_0^{(a^2-x^2)^{1/2}} (a^2 - y^2)^{1/2}\, dy\, dx &= \int_0^a \left[\int_0^{(a^2-y^2)^{1/2}} (a^2 - y^2)^{1/2}\, dx \right] dy \\
&= \int_0^a [x(a^2 - y^2)^{1/2}]_{x=0}^{(a^2-y^2)^{1/2}}\, dy \\
&= \int_0^a (a^2 - y^2)\, dy = \left[a^2 y - \frac{y^3}{3} \right]_0^a = \frac{2a^3}{3}. \quad \blacktriangle
\end{aligned}$$

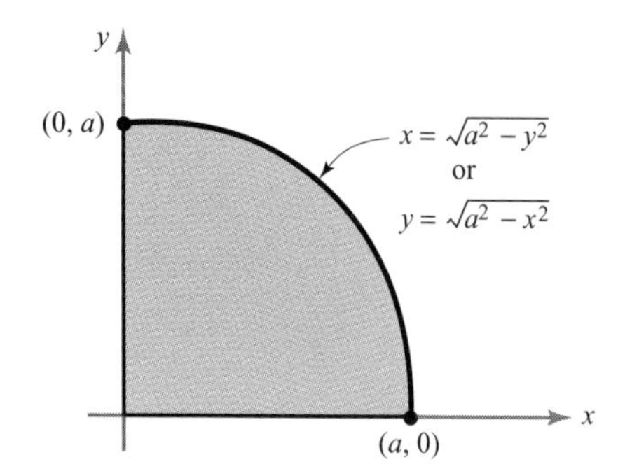

Figure 5.4.1 The positive-quadrant portion of a disk of radius a.

We could have evaluated the initial iterated integral directly, but, as the reader can easily verify, changing the order of integration makes the problem simpler. The next example shows that it may not be obvious how to evaluate an iterated integral, and yet it may be relatively simple to evaluate the iterated integral obtained by changing the order of integration.

EXAMPLE 2 Evaluate

$$\int_1^2 \int_0^{\log x} (x-1)\sqrt{1+e^{2y}}\, dy\, dx.$$

SOLUTION It will simplify matters if we first interchange the order of integration. First notice that the integral is equal to $\iint_D (x-1)\sqrt{1+e^{2y}}\, dA$, where D is the set of (x, y) such that

$$1 \le x \le 2 \qquad \text{and} \qquad 0 \le y \le \log x.$$

The region D is simple (see Figure 5.4.2) and can also be described by

$$0 \le y \le \log 2 \qquad \text{and} \qquad e^y \le x \le 2.$$

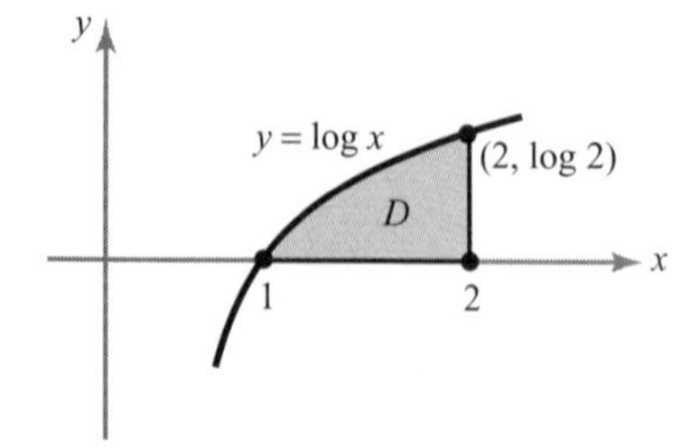

Figure 5.4.2 D is the region of integration for Example 2.

Thus, the given iterated integral is equal to

$$\int_0^{\log 2}\int_{e^y}^{2}(x-1)\sqrt{1+e^{2y}}\,dx\,dy = \int_0^{\log 2}\sqrt{1+e^{2y}}\left[\int_{e^y}^{2}(x-1)\,dx\right]dy$$

$$= \int_0^{\log 2}\sqrt{1+e^{2y}}\left[\frac{x^2}{2}-x\right]_{e^y}^{2}dy$$

$$= -\int_0^{\log 2}\left(\frac{e^{2y}}{2}-e^y\right)\sqrt{1+e^{2y}}\,dy$$

$$= -\frac{1}{2}\int_0^{\log 2}e^{2y}\sqrt{1+e^{2y}}\,dy + \int_0^{\log 2}e^y\sqrt{1+e^{2y}}\,dy. \tag{1}$$

In the first integral in expression (1), we substitute $u = e^{2y}$, and in the second, $v = e^y$. Hence, we obtain

$$-\frac{1}{4}\int_1^4\sqrt{1+u}\,du + \int_1^2\sqrt{1+v^2}\,dv. \tag{2}$$

Both integrals in expression (2) are easily found with techniques of one-variable calculus (or by consulting the table of integrals at the back of the book). For the first integral, we get

$$\frac{1}{4}\int_1^4\sqrt{1+u}\,du = \left[\frac{1}{6}(1+u)^{3/2}\right]_1^4 = \frac{1}{6}[(1+4)^{3/2}-2^{3/2}] = \frac{1}{6}[5^{3/2}-2^{3/2}]. \tag{3}$$

The second integral is

$$\int_1^2\sqrt{1+v^2}\,dv = \frac{1}{2}\left[v\sqrt{1+v^2}+\log\left(\sqrt{1+v^2}+v\right)\right]_1^2$$

$$= \frac{1}{2}\left[2\sqrt{5}+\log\left(\sqrt{5}+2\right)\right] - \frac{1}{2}\left[\sqrt{2}+\log\left(\sqrt{2}+1\right)\right] \tag{4}$$

(see formula 43 in the table of integrals at the back of the book). Finally, we subtract equation (3) from equation (4) to obtain the answer

$$\frac{1}{2}\left(2\sqrt{5}-\sqrt{2}+\log\frac{\sqrt{5}+2}{\sqrt{2}+1}\right) - \frac{1}{6}[5^{3/2}-2^{3/2}]. \quad \blacktriangle$$

Mean Value Inequality

We conclude with an inequality that helps us estimate integrals. Suppose there are numbers m and M such that for all $(x, y) \in D$, and $m \leq f(x, y) \leq M$, then integrating over D, we get

$$m \cdot A(D) \leq \iint_D f(x, y)\, dA \leq M \cdot A(D), \tag{5}$$

where $A(D)$ is the area of the region D. Even though this inequality is obvious, it can help us *estimate* integrals that we cannot easily evaluate *exactly*.

EXAMPLE 3 Consider the integral

$$\iint_D \frac{1}{\sqrt{1 + x^6 + y^8}}\, dx\, dy,$$

where D is the unit square $[0, 1] \times [0, 1]$. Because the integrand satisfies, for x and y between 0 and 1,

$$\frac{1}{\sqrt{3}} \leq \frac{1}{\sqrt{1 + x^6 + y^8}} \leq 1,$$

and because the square has area 1, we get:

$$\frac{1}{\sqrt{3}} \leq \iint_D \frac{1}{\sqrt{1 + x^6 + y^8}}\, dx\, dy \leq 1. \quad \blacktriangle$$

Mean Value Equality

The mean value inequality can be turned into an equality when f is continuous. Here is the formal statement.

THEOREM 5: Mean Value Theorem: Double Integrals Suppose $f: D \to \mathbb{R}$ is continuous and D is an elementary region. Then for some point (x_0, y_0) in D we have

$$\iint_D f(x, y)\, dA = f(x_0, y_0) A(D),$$

where $A(D)$ denotes the area of D.

PROOF We cannot prove this theorem with complete rigor, because it requires some concepts about continuous functions not proved in this course; but we can sketch the main ideas that underlie the proof.

Because f is continuous on D, it has a maximum value M and a minimum value m. Thus, $m \leq f(x, y) \leq M$ for all $(x, y) \in D$. Furthermore, $f(x_1, y_1) = m$ and $f(x_2, y_2) = M$ for some pairs (x_1, y_1) and (x_2, y_2) in D.

Dividing through inequality (5) by $A(D)$, we get

$$m \leq \frac{1}{A(D)} \iint_D f(x, y)\, dA \leq M. \tag{6}$$

Because a continuous function on D takes on every value between its maximum and minimum values (this is the two-variable *intermediate value theorem* proved in advanced calculus; see also Review Exercise 32), and because the number $[1/A(D)] \iint_D f(x, y)\, dA$ is, by inequality (6), between these values, there must be a point $(x_0, y_0) \in D$ with

$$f(x_0, y_0) = \frac{1}{A(D)} \iint_D f(x, y)\, dA,$$

which is precisely the conclusion of Theorem 5. ■

EXERCISES

1. In the following integrals, change the order of integration, sketch the corresponding regions, and evaluate the integral both ways.

(a) $\displaystyle\int_0^1 \int_x^1 xy\, dy\, dx$

(b) $\displaystyle\int_0^{\pi/2} \int_0^{\cos\theta} \cos\theta\, dr\, d\theta$

(c) $\displaystyle\int_0^1 \int_1^{2-y} (x+y)^2\, dx\, dy$

(d) $\displaystyle\int_a^b \int_a^y f(x, y)\, dx\, dy$ (express your answer in terms of antiderivatives).

2. Find

(a) $\displaystyle\int_{-1}^1 \int_{|y|}^1 (x+y)^2\, dx\, dy$

(b) $\displaystyle\int_{-3}^1 \int_{-\sqrt{(9-y^2)}}^{\sqrt{(9-y^2)}} x^2\, dx\, dy$

(c) $\displaystyle\int_0^4 \int_{y/2}^2 e^{x^2}\, dx\, dy$

(d) $\displaystyle\int_0^1 \int_{\tan^{-1} y}^{\pi/4} (\sec^5 x)\, dx\, dy$

3. If $f(x, y) = e^{\sin(x+y)}$ and $D = [-\pi, \pi] \times [-\pi, \pi]$, show that

$$\frac{1}{e} \leq \frac{1}{4\pi^2} \iint_D f(x, y)\, dA \leq e.$$

4. Show that

$$\frac{1}{2}(1 - \cos 1) \leq \iint_{[0,1]\times[0,1]} \frac{\sin x}{1 + (xy)^4}\, dx\, dy \leq 1.$$

5. If $D = [-1, 1] \times [-1, 2]$, show that

$$1 \leq \iint_D \frac{dx\, dy}{x^2 + y^2 + 1} \leq 6.$$

6. Using the mean value inequality, show that

$$\frac{1}{6} \le \iint_D \frac{dA}{y - x + 3} \le \frac{1}{4},$$

where D is the triangle with vertices $(0, 0)$, $(1, 1)$, and $(1, 0)$.

7. Compute the volume of an ellipsoid with semiaxes a, b, and c. (HINT: Use symmetry and first find the volume of one half of the ellipsoid.)

8. Compute $\iint_D f(x, y)\, dA$, where $f(x, y) = y^2\sqrt{x}$ and D is the set of (x, y) where $x > 0$, $y > x^2$, and $y < 10 - x^2$.

9. Find the volume of the region determined by $x^2 + y^2 + z^2 \le 10$, $z \ge 2$. Use the disk method from one-variable calculus and state how the method is related to Cavalieri's principle.

10. Evaluate $\iint_D e^{x-y}\, dx\, dy$, where D is the interior of the triangle with vertices $(0, 0)$, $(1, 3)$, and $(2, 2)$.

11. Evaluate $\iint_D y^3(x^2 + y^2)^{-3/2}\, dx\, dy$, where D is the region determined by the conditions $\frac{1}{2} \le y \le 1$ and $x^2 + y^2 \le 1$.

12. Given that the double integral $\iint_D f(x, y)\, dx\, dy$ of a positive continuous function f equals the iterated integral $\int_0^1 \left[\int_{x^2}^{x} f(x, y)\, dy\right] dx$, sketch the region D and interchange the order of integration.

13. Given that the double integral $\iint_D f(x, y)\, dx\, dy$ of a positive continuous function f equals the iterated integral $\int_0^1 \left[\int_y^{\sqrt{2-y^2}} f(x, y)\, dx\right] dy$, sketch the region D and interchange the order of integration.

14. Prove that $2\int_a^b \int_x^b f(x)f(y)\, dy\, dx = \left(\int_a^b f(x)\, dx\right)^2$. [HINT: Notice that $\left(\int_a^b f(x)\, dx\right)^2 = \iint_{[a,b]\times[a,b]} f(x)f(y)\, dx\, dy$.]

15. Show that (see Exercise 27, Section 2.5)

$$\frac{d}{dx}\int_a^x \int_c^d f(x, y, z)\, dz\, dy = \int_c^d f(x, y, z)\, dz + \int_a^x \int_c^d f_x(x, y, z)\, dz\, dy.$$

5.5 The Triple Integral

Triple integrals are needed for many physical problems. For example, if the temperature inside an oven is not uniform, determining the average temperature involves "summing" the values of the temperature function at all points in the solid region enclosed by the oven walls and then dividing the answer by the total volume of the oven. Such a sum is expressed mathematically as a triple integral.

Definition of the Triple Integral

Our objective now is to define the triple integral of a function $f(x, y, z)$ over a box (rectangular parallelepiped) $B = [a, b] \times [c, d] \times [p, q]$. Proceeding as in double integrals, we partition the three sides of B into n equal parts and form the sum

$$S_n = \sum_{i=0}^{n-1} \sum_{j=0}^{n-1} \sum_{k=0}^{n-1} f(\mathbf{c}_{ijk}) \Delta V,$$

where $\mathbf{c}_{ijk}$ is a point in B_{ijk}, the ijkth rectangular parallelepiped (or box) in the partition of B, and ΔV is the volume of B_{ijk} (see Figure 5.5.1).

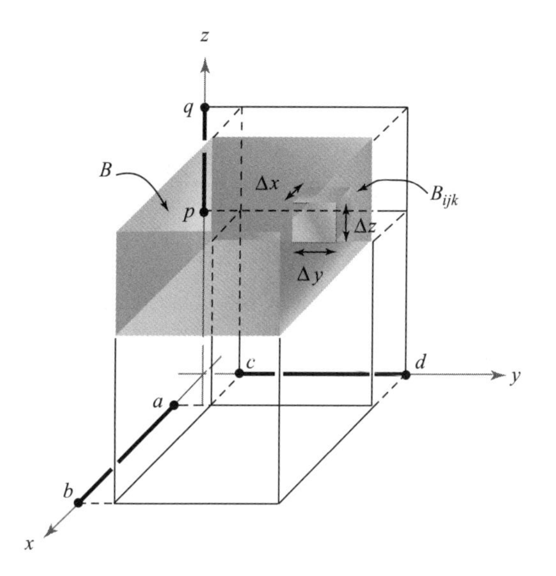

Figure 5.5.1 A partition of a box B into n^3 subboxes B_{ijk}.

DEFINITION: Triple Integrals Let f be a bounded function of three variables defined on B. If $\lim_{n\to\infty} S_n = S$ exists and is independent of any choice of $\mathbf{c}_{ijk}$, we call f ***integrable*** and call S the ***triple integral*** (or simply the integral) of f over B and denote it by

$$\iiint_B f\, dV, \qquad \iiint_B f(x, y, z)\, dV \qquad \text{or} \qquad \iiint_B f(x, y, z)\, dx\, dy\, dz.$$

Properties of Triple Integrals

As before, one can prove that continuous functions defined on B are integrable. Moreover, bounded functions whose discontinuities are confined to graphs of continuous functions [such as $x = \alpha(y, z)$, $y = \beta(x, z)$, or $z = \gamma(x, y)$] are integrable. The other basic properties (such as the fact that the integral of a sum is the sum of the integrals) for double integrals also hold for triple integrals. Especially important is the reduction to iterated integrals:

Reduction to Iterated Integrals Let $f(x, y, z)$ be integrable on the box $B = [a, b] \times [c, d] \times [p, q]$. Then any iterated integral that exists is equal to the triple integral; that is,

$$\iiint_B f(x, y, z)\, dx\, dy\, dz = \int_p^q \int_c^d \int_a^b f(x, y, z)\, dx\, dy\, dz$$

$$= \int_p^q \int_a^b \int_c^d f(x, y, z)\, dy\, dx\, dz$$

$$= \int_a^b \int_p^q \int_c^d f(x, y, z)\, dy\, dz\, dx,$$

and so on. (There are six possible orders altogether.)

EXAMPLE 1 (a) Let B be the box $[0, 1] \times \left[-\frac{1}{2}, 0\right] \times \left[0, \frac{1}{3}\right]$. Evaluate

$$\iiint_B (x + 2y + 3z)^2\, dx\, dy\, dz.$$

(b) Verify that we get the same answer if the integration is done in the order y first, then z, and then x.

SOLUTION (a) According to the principle of reduction to iterated integrals, this integral may be evaluated as

$$\int_0^{1/3} \int_{-1/2}^0 \int_0^1 (x + 2y + 3z)^2\, dx\, dy\, dz$$

$$= \int_0^{1/3} \int_{-1/2}^0 \left[\left. \frac{(x + 2y + 3z)^3}{3} \right|_{x=0}^{1} \right] dy\, dz$$

$$= \int_0^{1/3} \int_{-1/2}^0 \frac{1}{3} \left[(1 + 2y + 3z)^3 - (2y + 3z)^3 \right] dy\, dz$$

$$= \int_0^{1/3} \frac{1}{24} \left[(1 + 2y + 3z)^4 - (2y + 3z)^4 \right] \Bigg|_{y=-1/2}^{0} dz$$

$$= \int_0^{1/3} \frac{1}{24} \left[(3z + 1)^4 - 2(3z)^4 + (3z - 1)^4 \right] dz$$

$$= \frac{1}{24 \cdot 15} \left[(3z + 1)^5 - 2(3z)^5 + (3z - 1)^5 \right] \Bigg|_{z=0}^{1/3}$$

$$= \frac{1}{24 \cdot 15} (2^5 - 2) = \frac{1}{12}.$$

(b)

$$\begin{aligned}
&\iiint_B (x+2y+3z)^2\, dy\, dz\, dx \\
&= \int_0^1 \int_0^{1/3} \int_{-1/2}^0 (x+2y+3z)^2\, dy\, dz\, dx \\
&= \int_0^1 \int_0^{1/3} \left[\frac{(x+2y+3z)^3}{6} \Bigg|_{y=-1/2}^{0} \right] dz\, dx \\
&= \int_0^1 \int_0^{1/3} \frac{1}{6}\left[(x+3z)^3 - (x+3z-1)^3\right] dz\, dx \\
&= \int_0^1 \frac{1}{6} \left\{ \left[\frac{(x+3z)^4}{12} - \frac{(x+3z-1)^4}{12} \right] \Bigg|_{z=0}^{1/3} \right\} dx \\
&= \int_0^1 \frac{1}{72}\left[(x+1)^4 + (x-1)^4 - 2x^4\right] dx \\
&= \frac{1}{72}\frac{1}{5}\left[(x+1)^5 + (x-1)^5 - 2x^5\right]_{x=0}^{1} = \frac{1}{12}. \quad \blacktriangle
\end{aligned}$$

EXAMPLE 2 Integrate e^{x+y+z} over the box $[0, 1] \times [0, 1] \times [0, 1]$.

SOLUTION We perform the integrations in the standard order:

$$\begin{aligned}
\int_0^1 \int_0^1 \int_0^1 e^{x+y+z}\, dx\, dy\, dz &= \int_0^1 \int_0^1 (e^{x+y+z}|_{x=0}^{1})\, dy\, dz \\
&= \int_0^1 \int_0^1 (e^{1+y+z} - e^{y+z})\, dy\, dz = \int_0^1 \left[e^{1+y+z} - e^{y+z}\right]_{y=0}^{1} dz \\
&= \int_0^1 \left[e^{2+z} - 2e^{1+z} + e^z\right] dz = \left[e^{2+z} - 2e^{1+z} + e^z\right]_0^1 \\
&= e^3 - 3e^2 + 3e - 1 = (e-1)^3. \quad \blacktriangle
\end{aligned}$$

As in the two-variable case, we define the integral of a function f over a bounded region W by defining a new function f^*, equal to f on W and zero outside W, and then setting

$$\iiint_W f(x, y, z)\, dx\, dy\, dz = \iiint_B f^*(x, y, z)\, dx\, dy\, dz,$$

where B is any box containing the region W.

Elementary Regions

As before, we restrict our attention to particularly simple regions. An ***elementary region*** in three-dimensional space is one defined by restricting one of the variables to be between two functions of the remaining variables, the domains of these functions being an elementary (i.e., an x-simple or a y-simple) region in the plane. For example, if D is an elementary region in the xy plane and if $\gamma_1(x, y)$ and $\gamma_2(x, y)$ are two functions with $\gamma_2(x, y) \geq \gamma_1(x, y)$, an elementary region consists of all (x, y, z) such that (x, y) lies in D and $\gamma_1(x, y) \leq z \leq \gamma_2(x, y)$. Figure 5.5.2 shows two elementary regions.

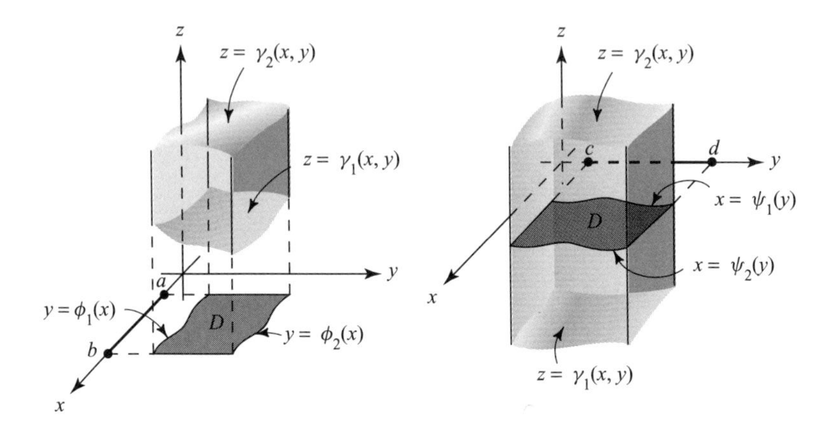

Figure 5.5.2 Two elementary regions in space. The domain D in the figure on the left is y-simple, while on the right it is x-simple.

EXAMPLE 3 Describe the unit ball $x^2 + y^2 + z^2 \leq 1$ as an elementary region.

SOLUTION This can be done in several ways. One, in which D is y-simple, is:

$$-1 \leq x \leq 1,$$
$$-\sqrt{1 - x^2} \leq y \leq \sqrt{1 - x^2},$$
$$-\sqrt{1 - x^2 - y^2} \leq z \leq \sqrt{1 - x^2 - y^2}.$$

In doing this, we first write the top and bottom hemispheres as $z = \sqrt{1 - x^2 - y^2}$ and $z = -\sqrt{1 - x^2 - y^2}$, respectively, where x and y vary over the unit disk (that is, $-\sqrt{1 - x^2} \leq y \leq \sqrt{1 - x^2}$ and x varies between -1 and 1). (See Figure 5.5.3.) We can describe the region in other ways by interchanging the roles of x, y, and z in the defining inequalities. ▲

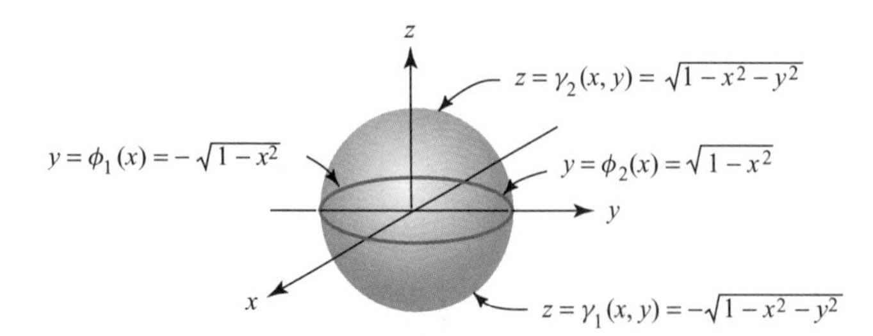

Figure 5.5.3 The unit ball as an elementary region in space.

Integrals over Elementary Regions

As with integrals in the plane, any function of three variables that is continuous over an elementary region is integrable on that region. An argument like that for double integrals shows that a triple integral over an elementary region can be rewritten as an iterated integral in which the limits of integration are functions. The formulas for such iterated integrals are given in the following box.

Triple Integrals by Iterated Integration Suppose that W is an elementary region described by bounding z between two functions of x and y. Then either

$$\iiint_W f(x, y, z)\, dx\, dy\, dz = \int_a^b \int_{\phi_1(x)}^{\phi_2(x)} \int_{\gamma_1(x,y)}^{\gamma_2(x,y)} f(x, y, z)\, dz\, dy\, dx$$

[see Figure 5.5.2 (left)] or

$$\iiint_W f(x, y, z)\, dx\, dy\, dz = \int_c^d \int_{\psi_1(y)}^{\psi_2(y)} \int_{\gamma_1(x,y)}^{\gamma_2(x,y)} f(x, y, z)\, dz\, dx\, dy$$

[see Figure 5.5.2 (right)].

If $f = 1$, we get the integral $\iiint_W dx\, dy\, dz$, which is the ***volume*** of the region W.

EXAMPLE 4 Verify the volume formula for the ball of radius 1:

$$\iiint_W dx\, dy\, dz = \frac{4}{3}\pi,$$

where W is the set of (x, y, z) with $x^2 + y^2 + z^2 \leq 1$.

SOLUTION We use the description of the unit ball from Example 3. From the first formula in the preceding box, the integral is

$$\int_{-1}^{1} \int_{-\sqrt{1-x^2}}^{\sqrt{1-x^2}} \int_{-\sqrt{1-x^2-y^2}}^{\sqrt{1-x^2-y^2}} dz\, dy\, dx.$$

Holding y and x fixed and integrating with respect to z yields

$$\int_{-1}^{1} \int_{-\sqrt{1-x^2}}^{\sqrt{1-x^2}} \left[z \Big|_{-\sqrt{1-x^2-y^2}}^{\sqrt{1-x^2-y^2}} \right] dy\, dx = 2 \int_{-1}^{1} \left[\int_{-\sqrt{1-x^2}}^{\sqrt{1-x^2}} (1 - x^2 - y^2)^{1/2}\, dy \right] dx.$$

Because x is fixed in the y-integral, it can be expressed as $\int_{-a}^{a} (a^2 - y^2)^{1/2}\, dy$, where $a = (1 - x^2)^{1/2}$. This integral is the area of a semicircular region of radius

a, so that

$$\int_{-a}^{a} (a^2 - y^2)^{1/2}\, dy = \frac{a^2}{2}\pi.$$

(We could also have used a trigonometric substitution or a table of integrals.) Thus,

$$\int_{-\sqrt{1-x^2}}^{\sqrt{1-x^2}} (1 - x^2 - y^2)^{1/2}\, dy = \frac{1 - x^2}{2}\pi,$$

and so

$$2\int_{-1}^{1}\int_{-\sqrt{1-x^2}}^{\sqrt{1-x^2}} (1 - x^2 - y^2)^{1/2}\, dy\, dx = 2\int_{-1}^{1} \pi \frac{1 - x^2}{2}\, dx$$

$$= \pi \int_{-1}^{1} (1 - x^2)\, dx = \pi\left(x - \frac{x^3}{3}\right)\Big|_{x=-1}^{1} = \frac{4}{3}\pi. \quad \blacktriangle$$

Other types of elementary regions are shown in Figure 5.5.4. For instance, in the second region, (y, z) lies in an elementary region in the yz plane and x lies between two graphs:

$$\rho_1(y, z) \le x \le \rho_2(y, z).$$

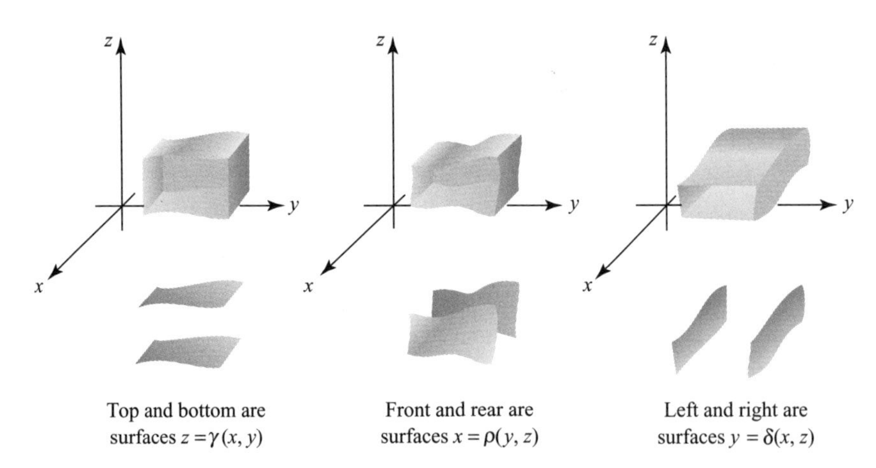

Figure 5.5.4 Types of elementary regions in space.

As shown in Figure 5.5.5, some elementary regions can be simultaneously described in all three ways. We shall call these regions ***symmetric elementary regions***.

Corresponding to each description of a region as an elementary region is an integration formula. For instance, if W is expressed as the set of all (x, y, z) such that

$$c \le y \le d, \qquad \psi_1(y) \le z \le \psi_2(y), \qquad \rho_1(y, z) \le x \le \rho_2(y, z),$$

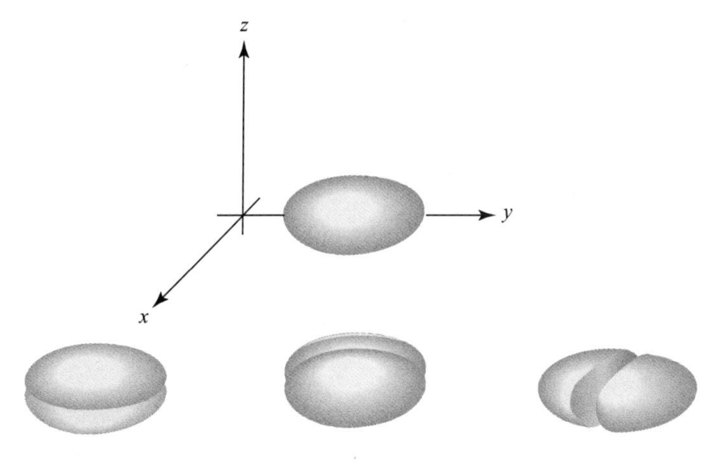

Figure 5.5.5 A symmetric elementary region can be described in three overall ways.

then

$$\iiint_W f(x, y, z)\, dx\, dy\, dz = \int_c^d \int_{\psi_1(y)}^{\psi_2(y)} \int_{\rho_1(y,z)}^{\rho_2(y,z)} f(x, y, z)\, dx\, dz\, dy.$$

EXAMPLE 5 Let W be the region bounded by the planes $x = 0$, $y = 0$, and $z = 2$, and the surface $z = x^2 + y^2$ and lying in the quadrant $x \geq 0$, $y \geq 0$. Compute $\iiint_W x\, dx\, dy\, dz$ and sketch the region.

SOLUTION *Method 1.* The region W is sketched in Figure 5.5.6. As indicated in the figure, we may describe this region by the inequalities

$$0 \leq x \leq \sqrt{2}, \qquad 0 \leq y \leq \sqrt{2 - x^2}, \qquad x^2 + y^2 \leq z \leq 2.$$

Therefore,

$$\begin{aligned}\iiint_W x\, dx\, dy\, dz &= \int_0^{\sqrt{2}} \left[\int_0^{\sqrt{2-x^2}} \left(\int_{x^2+y^2}^{2} x\, dz \right) dy \right] dx \\ &= \int_0^{\sqrt{2}} \int_0^{\sqrt{2-x^2}} x(2 - x^2 - y^2)\, dy\, dx \\ &= \int_0^{\sqrt{2}} x \left[(2 - x^2)^{3/2} - \frac{(2 - x^2)^{3/2}}{3} \right] dx \\ &= \int_0^{\sqrt{2}} \frac{2x}{3} (2 - x^2)^{3/2}\, dx = \left. \frac{-2(2 - x^2)^{5/2}}{15} \right|_0^{\sqrt{2}} \\ &= 2 \cdot \frac{2^{5/2}}{15} = \frac{8\sqrt{2}}{15}.\end{aligned}$$

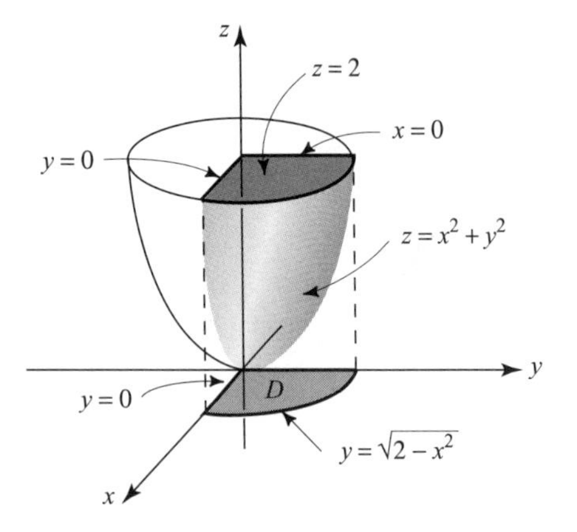

Figure 5.5.6 W is the region below the plane $z = 2$, above the paraboloid $z = x^2 + y^2$, and on the positive sides of the planes $x = 0$, $y = 0$.

Method 2. We can also place limits on x first and describe W by $0 \le x \le (z - y^2)^{1/2}$ and (y, z) in D, where D is the subset of the yz plane with $0 \le z \le 2$ and $0 \le y \le z^{1/2}$ (see Figure 5.5.7).

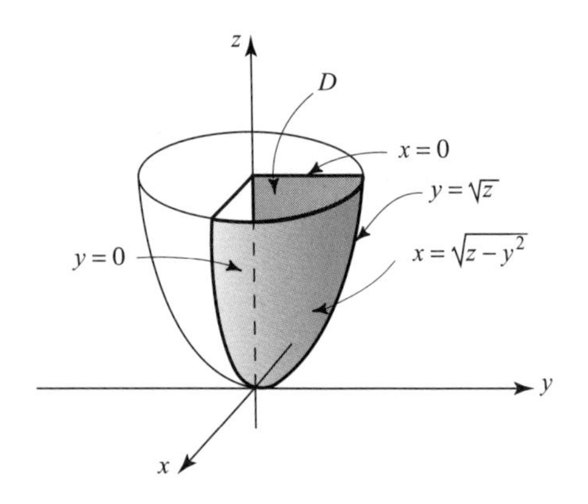

Figure 5.5.7 A different description of the region in Example 5.

Therefore,

$$
\begin{aligned}
\iiint_W x\,dx\,dy\,dz &= \iint_D \left(\int_0^{(z-y^2)^{1/2}} x\,dx \right) dy\,dz \\
&= \int_0^2 \left[\int_0^{z^{1/2}} \left(\int_0^{(z-y^2)^{1/2}} x\,dx \right) dy \right] dz \\
&= \int_0^2 \int_0^{z^{1/2}} \left(\frac{z - y^2}{2} \right) dy\,dz \\
&= \frac{1}{2} \int_0^2 \left(z^{3/2} - \frac{z^{3/2}}{3} \right) dz = \frac{1}{2} \int_0^2 \frac{2}{3} z^{3/2}\,dz \\
&= \left[\frac{2}{15} z^{5/2} \right]_0^2 = \frac{2}{15} 2^{5/2} = \frac{8\sqrt{2}}{15},
\end{aligned}
$$

which agrees with our previous answer. ▲

EXAMPLE 6 Evaluate

$$\int_0^1 \int_0^x \int_{x^2+y^2}^2 dz\,dy\,dx.$$

Sketch the region W of integration and interpret.

SOLUTION

$$\int_0^1 \int_0^x \int_{x^2+y^2}^2 dz\,dy\,dx = \int_0^1 \int_0^x (2 - x^2 - y^2)\,dy\,dx$$

$$= \int_0^1 \left(2x - x^3 - \frac{x^3}{3}\right) dx = 1 - \frac{1}{4} - \frac{1}{12} = \frac{2}{3}.$$

This integral is the volume of the region sketched in Figure 5.5.8. ▲

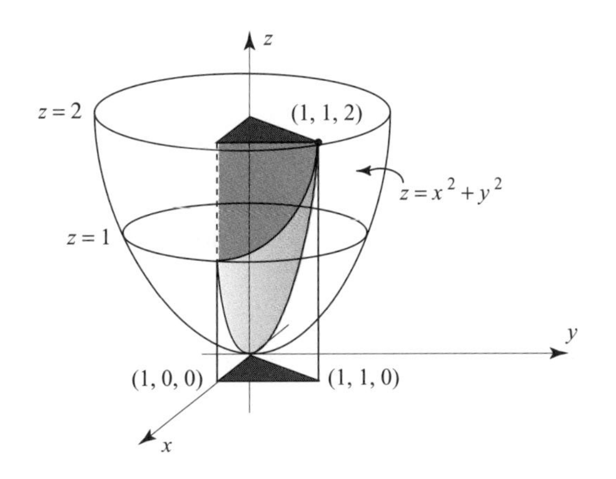

Figure 5.5.8 The region W lies between the paraboloid $z = x^2 + y^2$ and the plane $z = 2$, and above the region D.

EXERCISES

In Exercises 1 to 4, perform the indicated integration over the given box.

1. $\iiint_B x^2\,dx\,dy\,dz$, $B = [0, 1] \times [0, 1] \times [0, 1]$
2. $\iiint_B e^{-xy}\,y\,dx\,dy\,dz$, $B = [0, 1] \times [0, 1] \times [0, 1]$
3. $\iiint_B (2x + 3y + z)\,dx\,dy\,dz$, $B = [0, 2] \times [-1, 1] \times [0, 1]$
4. $\iiint_B ze^{x+y}\,dx\,dy\,dz$, $B = [0, 1] \times [0, 1] \times [0, 1]$

In Exercises 5 to 8, describe the given region as an elementary region.

5. The region between the cone $z = \sqrt{x^2 + y^2}$ and the paraboloid $z = x^2 + y^2$.

6. The region cut out of the ball $x^2 + y^2 + z^2 \leq 4$ by the elliptic cylinder $2x^2 + z^2 = 1$, that is, the region inside the cylinder and the ball.

7. The region inside the sphere $x^2 + y^2 + z^2 = 1$ and above the plane $z = 0$.

8. The region bounded by the planes $x = 0$, $y = 0$, $z = 0$, $x + y = 4$, and $x = z - y - 1$.

Find the volume of the region in Exercises 9 to 12.

9. The region bounded by $z = x^2 + y^2$ and $z = 10 - x^2 - 2y^2$.

10. The solid bounded by $x^2 + 2y^2 = 2$, $z = 0$, and $x + y + 2z = 2$.

11. The solid bounded by $x = y$, $z = 0$, $y = 0$, $x = 1$, and $x + y + z = 0$.

12. The region common to the intersecting cylinders $x^2 + y^2 \leq a^2$ and $x^2 + z^2 \leq a^2$.

Evaluate the integrals in Exercises 13 to 21.

13. $\displaystyle\int_0^1 \int_1^2 \int_2^3 \cos[\pi(x + y + z)]\, dx\, dy\, dz$

14. $\displaystyle\int_0^1 \int_0^x \int_0^y (y + xz)\, dz\, dy\, dx$

15. $\displaystyle\iiint_W (x^2 + y^2 + z^2)\, dx\, dy\, dz$; W is the region bounded by $x + y + z = a$ (where $a > 0$), $x = 0$, $y = 0$, and $z = 0$.

16. $\displaystyle\iiint_W z\, dx\, dy\, dz$; W is the region bounded by the planes $x = 0$, $y = 0$, $z = 0$, $z = 1$, and the cylinder $x^2 + y^2 = 1$, with $x \geq 0$, $y \geq 0$.

17. $\displaystyle\iiint_W x^2 \cos z\, dx\, dy\, dz$; W is the region bounded by $z = 0$, $z = \pi$, $y = 0$, $y = 1$, $x = 0$, and $x + y = 1$.

18. $\displaystyle\int_0^2 \int_0^x \int_0^{x+y} dz\, dy\, dx$

19. $\displaystyle\iiint_W (1 - z^2)\, dx\, dy\, dz$; W is the pyramid with top vertex at $(0, 0, 1)$ and base vertices at $(0, 0, 0)$, $(1, 0, 0)$, $(0, 1, 0)$, and $(1, 1, 0)$.

20. $\displaystyle\iiint_W (x^2 + y^2)\, dx\, dy\, dz$; W is the same pyramid as in Exercise 19.

21. $\displaystyle\int_0^1 \int_0^{2x} \int_{x^2+y^2}^{x+y} dz\, dy\, dx.$

22. (a) Sketch the region for the integral $\displaystyle\int_0^1 \int_0^x \int_0^y f(x, y, z)\, dz\, dy\, dx.$

(b) Write the integral with the integration order $dx\, dy\, dz$.

For the regions in Exercises 23 to 26, find the appropriate limits $\phi_1(x)$, $\phi_2(x)$, $\gamma_1(x, y)$, and $\gamma_2(x, y)$, and write the triple integral over the region W as an iterated integral in the form

$$\iiint_W f\,dV = \int_a^b \left\{ \int_{\phi_1(x)}^{\phi_2(x)} \left[\int_{\gamma_1(x,y)}^{\gamma_2(x,y)} f(x, y, z)\,dz \right] dy \right\} dx.$$

23. $W = \{(x, y, z) \mid \sqrt{x^2 + y^2} \leq z \leq 1\}$

24. $W = \{(x, y, z) \mid \frac{1}{2} \leq z \leq 1 \text{ and } x^2 + y^2 + z^2 \leq 1\}$

25. $W = \{(x, y, z) \mid x^2 + y^2 \leq 1, z \geq 0 \text{ and } x^2 + y^2 + z^2 \leq 4\}$

26. $W = \{(x, y, z) \mid |x| \leq 1, |y| \leq 1, z \geq 0 \text{ and } x^2 + y^2 + z^2 \leq 1\}$

27. Show that the formula using triple integrals for the volume under the graph of a positive function $f(x, y)$, on an elementary region D in the plane, reduces to the double integral of f over D.

28. Let W be the region bounded by the planes $x = 0$, $y = 0$, $z = 0$, $x + y = 1$, and $z = x + y$.

(a) Find the volume of W.
(b) Evalute $\iiint_W x\,dx\,dy\,dz$.
(c) Evalute $\iiint_W y\,dx\,dy\,dz$.

29. Let f be continuous and let B_ε be the ball of radius ε centered at the point (x_0, y_0, z_0). Let vol (B_ε) be the volume of B_ε. Prove that

$$\lim_{\varepsilon \to 0} \frac{1}{\operatorname{vol}(B_\varepsilon)} \iiint_{B_\varepsilon} f(x, y, z)\,dV = f(x_0, y_0, z_0).$$

REVIEW EXERCISES FOR CHAPTER 5

Evaluate the integrals in Exercises 1 to 4.

1. $\displaystyle\int_0^3 \int_{-x^2+1}^{x^2+1} xy\,dy\,dx$

2. $\displaystyle\int_0^1 \int_{\sqrt{x}}^1 (x + y)^2\,dy\,dx$

3. $\displaystyle\int_0^1 \int_{e^x}^{e^{2x}} x \ln y\,dy\,dx$

4. $\displaystyle\int_0^1 \int_1^2 \int_2^3 \cos\,[\pi(x + y + z)]\,dx\,dy\,dz.$

Reverse the order of integration of the integrals in Exercises 5 to 8 and evaluate.

5. The integral in Exercise 1.

6. The integral in Exercise 2.

7. The integral in Exercise 3.

8. The integral in Exercise 4.

9. Evaluate the integral $\int_0^1 \int_0^x \int_0^y (y + xz)\, dz\, dy\, dx$.

10. Evaluate $\int_0^1 \int_y^{y^2} e^{x/y}\, dx\, dy$.

11. Evaluate $\int_0^1 \int_0^{(\arcsin y)/y} y \cos xy\, dx\, dy$.

12. Change the order of integration and evaluate

$$\int_0^2 \int_{y/2}^1 (x + y)^2\, dx\, dy.$$

13. Show that evaluating $\iint_D dx\, dy$, where D is a y-simple region, reproduces the formula from one-variable calculus for the area between two curves.

14. Change the order of integration and evaluate

$$\int_0^1 \int_{y^{1/2}}^1 (x^2 + y^3 x)\, dx\, dy.$$

15. Let D be the region in the xy plane inside the unit circle $x^2 + y^2 = 1$. Evaluate $\iint_D f(x, y)\, dx\, dy$ in each of the following cases:

(a) $f(x, y) = xy$ (b) $f(x, y) = x^2 y^2$ (c) $f(x, y) = x^3 y^3$

16. Find $\iint_D y[1 - \cos(\pi x/4)]\, dx\, dy$, where D is the region in Figure 5.R.1.

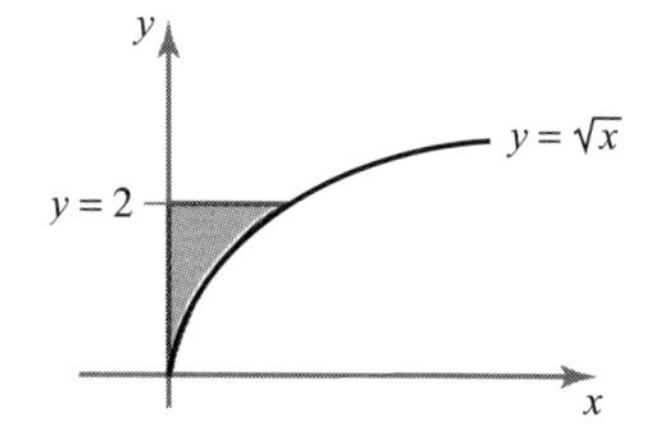

Figure 5.R.1 The region of integration for Exercise 16.

Evaluate the integrals in Exercises 17 to 24. Sketch and identify the type of the region (corresponding to the way the integral is written).

17. $\displaystyle\int_0^{\pi} \int_{\sin x}^{3 \sin x} x(1 + y)\, dy\, dx$

18. $\displaystyle\int_0^1 \int_{x-1}^{x \cos(\pi x/2)} (x^2 + xy + 1)\, dy\, dx$

19. $\displaystyle\int_{-1}^1 \int_{y^{2/3}}^{(2-y)^2} \left(\frac{3}{2}\sqrt{x} - 2y\right) dx\, dy$

20. $\displaystyle\int_0^2 \int_{-3(\sqrt{4-x^2})/2}^{3(\sqrt{4-x^2})/2} \left(\frac{5}{\sqrt{2 + x}} + y^3\right) dy\, dx$

21. $\displaystyle\int_0^1 \int_0^{x^2} (x^2 + xy - y^2)\, dy\, dx$

22. $\displaystyle\int_2^4 \int_{y^2-1}^{y^3} 3\, dx\, dy$

23. $\int_0^1 \int_{x^2}^{x} (x+y)^2 \, dy \, dx$

24. $\int_0^1 \int_0^{3y} e^{x+y} \, dx \, dy$

In Exercises 25 to 27, integrate the given function f over the given region D.

25. $f(x, y) = x - y$; D is the triangle with vertices (0, 0), (1, 0), and (2, 1).

26. $f(x, y) = x^3 y + \cos x$; D is the triangle defined by $0 \le x \le \pi/2, 0 \le y \le x$.

27. $f(x, y) = x^2 + 2xy^2 + 2$; D is the region bounded by the graph of $y = -x^2 + x$, the x axis, and the lines $x = 0$ and $x = 2$.

In Exercises 28 and 29, sketch the region of integration, interchange the order, and evaluate.

28. $\int_1^4 \int_1^{\sqrt{x}} (x^2 + y^2) \, dy \, dx$

29. $\int_0^1 \int_{1-y}^{1} (x + y^2) \, dx \, dy$

30. Show that

$$4e^5 \le \iint_{[1,3]\times[2,4]} e^{x^2+y^2} \, dA \le 4e^{25}.$$

31. Show that

$$4\pi \le \iint_D (x^2 + y^2 + 1) \, dx \, dy \le 20\pi,$$

where D is the disk of radius 2 centered at the origin.

32. Suppose W is a ***path-connected region***, that is, given any two points of W there is a continuous path joining them. If f is a continuous function on W, use the intermediate-value theorem to show that there is at least one point in W at which the value of f is equal to the average of f over W, that is, the integral of f over W divided by the volume of W. (Compare this with the mean-value theorem for double integrals.) What happens if W is not connected?

33. Prove: $\int_0^x [\int_0^t F(u) \, du] \, dt = \int_0^x (x - u) F(u) \, du$.

Evaluate the integrals in Exercises 34 to 36.

34. $\int_0^1 \int_0^z \int_0^y xy^2 z^3 \, dx \, dy \, dz$

35. $\int_0^1 \int_0^y \int_0^{x/\sqrt{3}} \frac{x}{x^2 + z^2} \, dz \, dx \, dy$

36. $\int_1^2 \int_1^z \int_{1/y}^{2} yz^2 \, dx \, dy \, dz$

37. Write the iterated integral $\int_0^1 \int_{1-x}^1 \int_x^1 f(x, y, z) \, dz \, dy \, dx$ as an integral over a region in $\mathbb{R}^3$ and then rewrite it in five other possible orders of integration.

6

The Change of Variables Formula and Applications of Integration

If you are stuck in a calculus problem and don't know what else to do, try integrating by parts or changing variables.

Jerry Kazdan

If that fails, go away, have a cup of coffee, and think!

Ute Müller

The change of variables formula is one of the most powerful integration methods in single-variable calculus; it enables us to evaluate integrals such as

$$\int_0^1 xe^{x^2}\,dx$$

by using the substitution, or *change of variables* $u = x^2$, which reduces the problem to the easy task of integrating e^u with respect to u. In this chapter, we develop the *multidimensional change of variables formula*, which is especially important and useful in evaluating multiple integrals in polar, cylindrical, and spherical coordinates.

One of the key ingredients in the change of variables formula is how to change variables in multidimensions. This involves the notion of mapping, which occurs in various interesting situations. For example, consider a deforming object, such as a swimming fish. As it changes its shape, one can imagine the instantaneous correspondence between points on the fish in its rest state and in its current shape. This type of correspondence is, in fact, the main idea behind a change of variables, in this case, of one three-dimensional region (the fish in its rest state) to another (the fish in its current shape).

The first section in this chapter describes the key concepts for mappings between regions of the plane. It goes on to develop the change of variables technique for double and then triple integrals. The chapter also includes some of the important physical applications of the integral.

6.1 The Geometry of Maps from $\mathbb{R}^2$ to $\mathbb{R}^2$

In this section, we shall be interested in maps from subsets of $\mathbb{R}^2$ to $\mathbb{R}^2$. The resulting geometric understanding will be useful in the next section, when we discuss the change of variables formula for multiple integrals.

Maps of One Region to Another

Let D^* be a subset of $\mathbb{R}^2$; suppose we consider a continuously differentiable map $T: D^* \to \mathbb{R}^2$, so T takes points in D^* to points in $\mathbb{R}^2$. We denote the set of image points by D or by $T(D^*)$; hence, $D = T(D^*)$ is the set of all points $(x, y) \in \mathbb{R}^2$ such that

$$(x, y) = T(x^*, y^*) \qquad \text{for some} \qquad (x^*, y^*) \in D^*.$$

One way to understand the geometry of a map T is to see how it *deforms* or changes D^*. For example, Figure 6.1.1 illustrates a map T that takes a slightly twisted region into a disk.

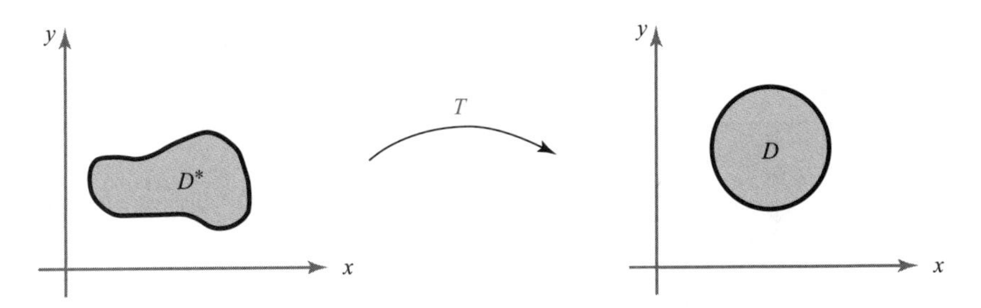

Figure 6.1.1 A function T from a region D^* to a disk D.

EXAMPLE 1 Let $D^* \subset \mathbb{R}^2$ be the rectangle $D^* = [0, 1] \times [0, 2\pi]$. Then all points in D^* are of the form (r, θ), where $0 \le r \le 1, 0 \le \theta \le 2\pi$. Let T be the polar coordinate "change of variables" defined by $T(r, \theta) = (r \cos\theta, r \sin\theta)$. Find the image set D.

SOLUTION Let $(x, y) = (r \cos\theta, r \sin\theta)$. Because of the identity $x^2 + y^2 = r^2 \cos^2\theta + r^2 \sin^2\theta = r^2 \le 1$, we see that the set of points $(x, y) \in \mathbb{R}^2$ such that $(x, y) \in D$ has the property that $x^2 + y^2 \le 1$, and so D is contained in the unit disk. In addition, any point (x, y) in the unit disk can be written as $(r \cos\theta, r \sin\theta)$ for some $0 \le r \le 1$ and $0 \le \theta \le 2\pi$. Thus, D is the unit disk (see Figure 6.1.2). ▲

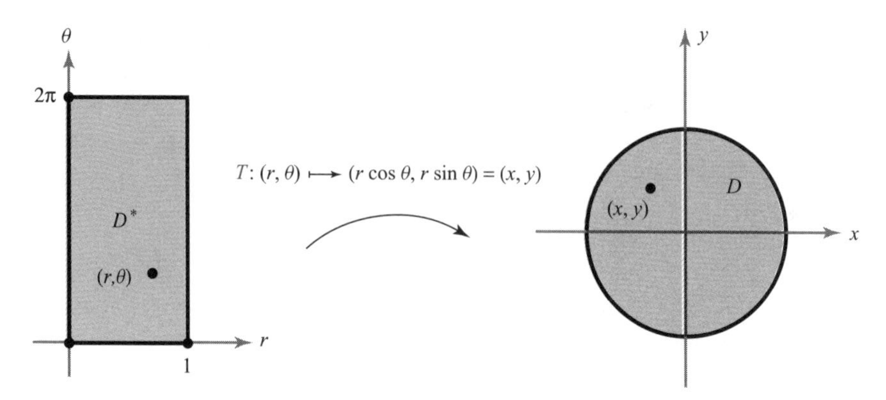

Figure 6.1.2 T gives a change of variables between Euclidean and polar coordinates. The unit circle is the image of a rectangle.

EXAMPLE 2 Let T be defined by $T(x, y) = ((x + y)/2, (x - y)/2)$ and let $D^* = [-1, 1] \times [-1, 1] \subset \mathbb{R}^2$ be a square with side of length 2 centered at the origin. Determine the image D obtained by applying T to D^*.

SOLUTION Let us first determine the effect of T on the line $\mathbf{c}_1(t) = (t, 1)$, where $-1 \leq t \leq 1$ (see Figure 6.1.3). We have $T(\mathbf{c}_1(t)) = ((t + 1)/2, (t - 1)/2)$. The map $t \mapsto T(\mathbf{c}_1(t))$ is a parametrization of the line $y = x - 1, 0 \leq x \leq 1$, because $(t - 1)/2 = (t + 1)/2 - 1$. This is the straight line segment joining $(1, 0)$ and $(0, -1)$.

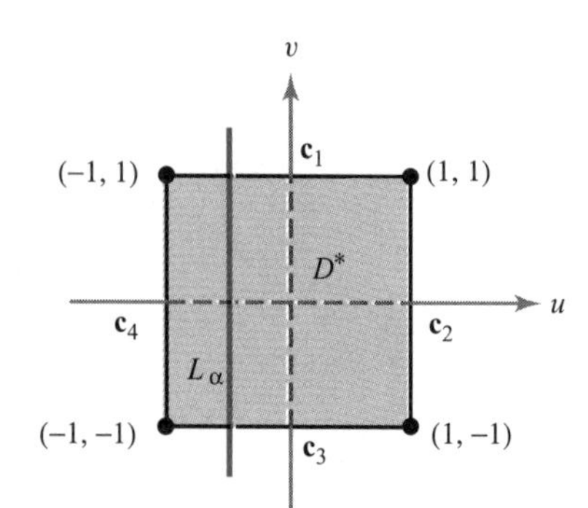

Figure 6.1.3 Domain for the transformation T of Example 2.

Let

$$\mathbf{c}_2(t) = (1, t), \qquad -1 \leq t \leq 1$$

$$\mathbf{c}_3(t) = (t, -1), \qquad -1 \leq t \leq 1$$

$$\mathbf{c}_4(t) = (-1, t), \qquad -1 \leq t \leq 1$$

be parametrizations of the other edges of the square D^*. Using the same argument as before, we see that $T \circ \mathbf{c}_2$ is a parametrization of the line $y = 1 - x, 0 \leq x \leq 1$

[the straight line segment joining (0, 1) and (1, 0)]; $T \circ \mathbf{c}_3$ is the line $y = x + 1$, $-1 \leq x \leq 0$ joining (0, 1) and $(-1, 0)$; and $T \circ \mathbf{c}_4$ is the line $y = -x - 1$, $-1 \leq x \leq 0$ joining $(-1, 0)$ and $(0, -1)$. By this time it seems reasonable to guess that T "flips" the square D^* over and takes it to the square D whose vertices are (1, 0), (0, 1), $(-1, 0)$, $(0, -1)$ (Figure 6.1.4).

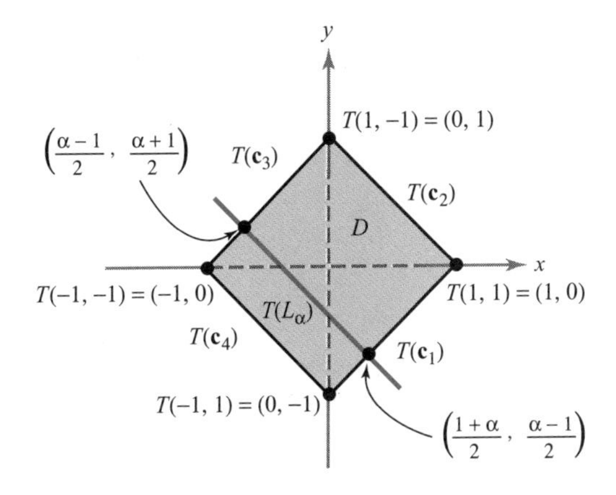

Figure 6.1.4 The effect of T on the region D^*.

To prove that this is indeed the case, let $-1 \leq \alpha \leq 1$ and let L_α (Figure 6.1.3) be a fixed line parametrized by $\mathbf{c}(t) = (\alpha, t)$, $-1 \leq t \leq 1$; then $T(\mathbf{c}(t)) = ((\alpha + t)/2, (\alpha - t)/2)$ is a parametrization of the line $y = -x + \alpha$, $(\alpha - 1)/2 \leq x \leq (\alpha + 1)/2$. This line begins, for $t = -1$, at the point $((\alpha - 1)/2, (1 + \alpha)/2)$ and ends at the point $((1 + \alpha)/2, (\alpha - 1)/2)$; as is easily checked, these points lie on the lines $T \circ \mathbf{c}_3$ and $T \circ \mathbf{c}_1$, respectively. Thus, as α varies between -1 and 1, L_α sweeps out the square D^* while $T(L_\alpha)$ sweeps out the square D determined by the vertices $(-1, 0)$, (0, 1), (1, 0), and $(0, -1)$. ▲

Images of Maps

The following theorem is a useful way to describe the image $T(D^*)$.

THEOREM 1 Let A be a 2×2 matrix with $\det A \neq 0$ and let T be the linear mapping of $\mathbb{R}^2$ to $\mathbb{R}^2$ given by $T(\mathbf{x}) = A\mathbf{x}$ (matrix multiplication). Then T transforms parallelograms into parallelograms and vertices into vertices. Moreover, if $T(D^*)$ is a parallelogram, D^* must be a parallelogram.

The proof of Theorem 1 is left as Exercise 10 at the end of this section. This theorem simplifies the result of Example 2, because we need only find the vertices of $T(D^*)$ and then connect them by straight lines.

One-to-One Maps

Although we cannot visualize the graph of a function $T: \mathbb{R}^2 \to \mathbb{R}^2$, it does help to consider how the function deforms subsets. However, simply looking at these deformations does not give us a complete picture of the behavior of T. We may characterize T further by using the notion of a one-to-one correspondence.

DEFINITION A mapping T is ***one-to-one*** on D^* if for (u, v) and $(u', v') \in D^*$, $T(u, v) = T(u', v')$ implies that $u = u'$ and $v = v'$.

This statement means that *two different points of D^* are not sent into the same point of D by T*. For example, the function $T(x, y) = (x^2 + y^2, y^4)$ is not one-to-one, because $T(1, -1) = (2, 1) = T(1, 1)$ and yet $(1, -1) \neq (1, 1)$.

EXAMPLE 3 Consider the polar coordinate mapping function $T: \mathbb{R}^2 \to \mathbb{R}^2$ described in Example 1, defined by $T(r, \theta) = (r \cos\theta, r \sin\theta)$. Show that T is not one-to-one if its domain is all of $\mathbb{R}^2$.

SOLUTION If $\theta_1 \neq \theta_2$, then $T(0, \theta_1) = T(0, \theta_2)$, and so T cannot be one-to-one. This observation implies that if L is the side of the rectangle $D^* = [0, 1] \times [0, 2\pi]$ where $0 \leq \theta \leq 2\pi$ and $r = 0$ (Figure 6.1.5), then T maps all of L into a single point, the center of the unit disk D. However, if we consider the set $S^* = (0, 1] \times [0, 2\pi)$, then $T: S^* \to S$ is one-to-one (see Exercise 1). Evidently, in determining whether a function is one-to-one, the domain chosen must be carefully considered. ▲

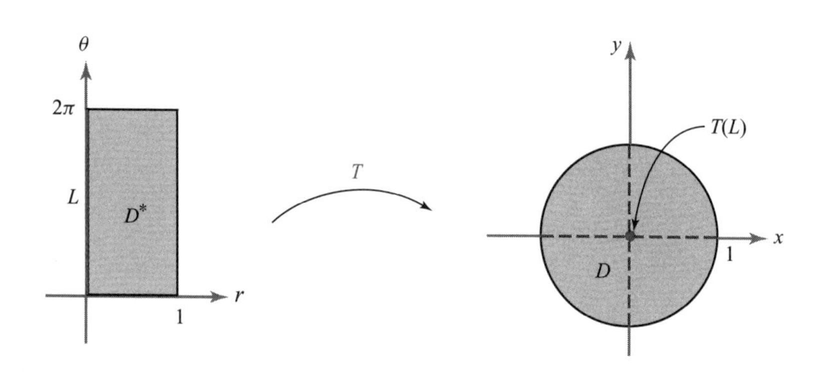

Figure 6.1.5 The polar-coordinate transformation T takes the line L to the point $(0, 0)$.

EXAMPLE 4 Show that the function $T: \mathbb{R}^2 \to \mathbb{R}^2$ of Example 2 is one-to-one.

SOLUTION Suppose $T(x, y) = T(x', y')$; then

$$\left(\frac{x+y}{2}, \frac{x-y}{2}\right) = \left(\frac{x'+y'}{2}, \frac{x'-y'}{2}\right)$$

and we have

$$x + y = x' + y',$$
$$x - y = x' - y'.$$

Adding, we have

$$2x = 2x'.$$

Thus, $x = x'$ and, similarly, subtracting gives $y = y'$, which shows that T is one-to-one (with domain all of $\mathbb{R}^2$). Actually, because T is linear and $T(\mathbf{x}) = A\mathbf{x}$, where A is a 2×2 matrix, it would also suffice to show that $\det A \neq 0$ (see Exercise 8). ▲

Onto Maps

In Examples 1 and 2, we have been determining the image $D = T(D^*)$ of a region D^* under a mapping T. What will be of interest to us in the next section is, in part, the inverse problem: Namely, given D and a one-to-one mapping T of $\mathbb{R}^2$ to $\mathbb{R}^2$, find D^* such that $T(D^*) = D$.

Before we examine this question in more detail, we introduce the notion of "onto."

DEFINITION The mapping T is ***onto*** D if for every point $(x, y) \in D$ there exists at least one point (u, v) in the domain of T such that $T(u, v) = (x, y)$.

Thus, if T is onto, we *can solve* the equation $T(u, v) = (x, y)$ for (u, v), given $(x, y) \in D$. If T is, in addition, one-to-one, this *solution is unique*.

For *linear* mappings T of $\mathbb{R}^2$ to $\mathbb{R}^2$ (or $\mathbb{R}^n$ to $\mathbb{R}^n$) it turns out that one-to-one and onto are equivalent notions (see Exercises 8 and 9).

If we are given a region D and a mapping T, the determination of a region D^* such that $T(D^*) = D$ will be possible only when for every $(x, y) \in D$ there is a (u, v) in the domain of T such that $T(u, v) = (x, y)$ (that is, T must be onto D). The next example shows that this cannot always be done.

EXAMPLE 5 Let $T: \mathbb{R}^2 \to \mathbb{R}^2$ be given by $T(u, v) = (u, 0)$. Let D be the square, $D = [0, 1] \times [0, 1]$. Because T takes all of $\mathbb{R}^2$ to one axis, it is impossible to find a D^* such that $T(D^*) = D$. ▲

Let us revisit Example 2 using these methods.

EXAMPLE 6 Let T be defined as in Example 2 and let D be the square whose vertices are $(1, 0), (0, 1), (-1, 0), (0, -1)$. Find a D^* with $T(D^*) = D$.

SOLUTION Because T is linear and $T(\mathbf{x}) = A\mathbf{x}$, where A is a 2×2 matrix satisfying $\det A \neq 0$, we know that $T: \mathbb{R}^2 \to \mathbb{R}^2$ is onto (see Exercises 8 and 9), and thus D^* can be found. By Theorem 1, D^* must be a parallelogram. In order to find D^*, it suffices to find the four points that are mapped onto vertices of D; then, by connecting these points, we will have found D^*. For the vertex $(1, 0)$ of D, we must solve $T(x, y) = (1, 0) = ((x + y)/2, (x - y)/2)$, so that $(x + y)/2 = 1$, $(x - y)/2 = 0$. Thus, $(x, y) = (1, 1)$ is a vertex of D^*. Solving for the other vertices, we find that $D^* = [-1, 1] \times [-1, 1]$. This is in agreement with what we found more laboriously in Example 2. ▲

EXAMPLE 7 Let D be the region in the first quadrant lying between the arcs of the circles $x^2 + y^2 = a^2$, $x^2 + y^2 = b^2$, $0 < a < b$ (see Figure 6.1.6). These circles have equations $r = a$ and $r = b$ in polar coordinates. Let T be the polar-coordinate transformation given by $T(r, \theta) = (r\cos\theta, r\sin\theta) = (x, y)$. Find D^* such that $T(D^*) = D$.

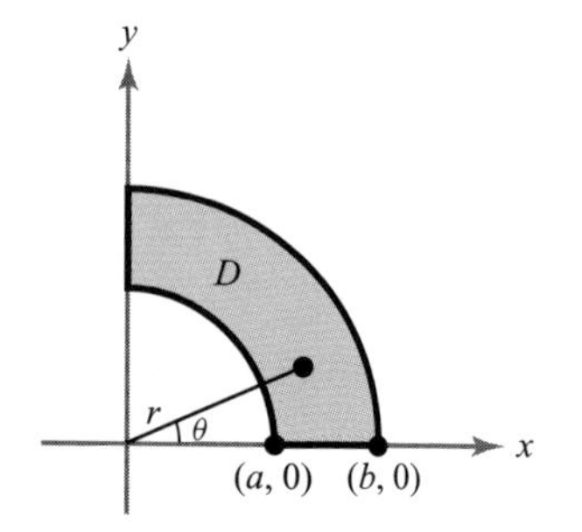

Figure 6.1.6 We seek a region D^* in the θr plane whose image under the polar-coordinate mapping is D.

SOLUTION In the region D, $a^2 \leq x^2 + y^2 \leq b^2$; and because $r^2 = x^2 + y^2$, we see that $a \leq r \leq b$. Clearly, for this region θ varies between $0 \leq \theta \leq \pi/2$. Thus, if $D^* = [a, b] \times [0, \pi/2]$, we have $T(D^*) = D$ and T is one-to-one. ▲

REMARK The inverse function theorem discussed in Section 3.5 is relevant to the material here. It states that if the determinant of $\mathbf{D}T(u_0, v_0)$ [which is the matrix of partial derivatives of T evaluated at (u_0, v_0)] is not zero, then for (u, v) near (u_0, v_0) and (x, y) near $(x_0, y_0) = T(u_0, v_0)$, the equation $T(u, v) = (x, y)$ can be *uniquely solved* for (u, v) as functions of (x, y). In particular, by uniqueness, T is one-to-one near (u_0, v_0); also, T is onto a neighborhood of (x_0, y_0), because $T(u, v) = (x, y)$ is solvable for (u, v) if (x, y) is near (x_0, y_0).

However, even if T is one-to-one near every point, and also onto, T need not be *globally* one-to-one. Thus, one must exercise caution (see Exercise 12).

Surprisingly, if D^* and D are elementary regions and $T: D^* \to D$ has the property that the determinant of $\mathbf{D}T(u, v)$ is not zero for any (u, v) in D^* and if T maps the boundary of D^* in a one-to-one and onto manner to the boundary of D, then T is one-to-one and onto from D^* to D. (This proof is beyond the scope of this text.)

In summary, we have:

One-to-One and Onto Mappings A mapping $T: D^* \to D$ is ***one-to-one*** when it maps distinct points to distinct points. It is ***onto*** when the image of D^* under T is all of D.

A *linear* transformation of $\mathbb{R}^n$ to $\mathbb{R}^n$ given by multiplication by a matrix A is one-to-one and onto when and only when $\det A \neq 0$.

EXERCISES

1. Let $S^* = (0, 1] \times [0, 2\pi)$ and define $T(r, \theta) = (r\cos\theta, r\sin\theta)$. Determine the image set S. Show that T is one-to-one on S^*.

2. Define

$$T(x^*, y^*) = \left(\frac{x^* - y^*}{\sqrt{2}}, \frac{x^* + y^*}{\sqrt{2}}\right).$$

Show that T rotates the unit square, $D^* = [0, 1] \times [0, 1]$.

3. Let $D^* = [0, 1] \times [0, 1]$ and define T on D^* by $T(u, v) = (-u^2 + 4u, v)$. Find the image D. Is T one-to-one?

4. Let D^* be the parallelogram bounded by the lines $y = 3x - 4$, $y = 3x$, $y = \frac{1}{2}x$, and $y = \frac{1}{2}(x + 4)$. Let $D = [0, 1] \times [0, 1]$. Find a T such that D is the image of D^* under T.

5. Let $D^* = [0, 1] \times [0, 1]$ and define T on D^* by $T(x^*, y^*) = (x^* y^*, x^*)$. Determine the image set D. Is T one-to-one? If not, can we eliminate some subset of D^* so that on the remainder T is one-to-one?

6. Let D^* be the parallelogram with vertices at $(-1, 3)$, $(0, 0)$, $(2, -1)$, and $(1, 2)$, and D be the rectangle $D = [0, 1] \times [0, 1]$. Find a T such that D is the image set of D^* under T.

7. Let $T: \mathbb{R}^3 \to \mathbb{R}^3$ be the spherical coordinate mapping defined by $(\rho, \phi, \theta) \mapsto (x, y, z)$, where

$$x = \rho \sin\phi \cos\theta, \qquad y = \rho \sin\phi \sin\theta, \qquad z = \rho \cos\phi.$$

Let D^* be the set of points (ρ, ϕ, θ) such that $\phi \in [0, \pi]$, $\theta \in [0, 2\pi]$, $\rho \in [0, 1]$. Find $D = T(D^*)$. Is T one-to-one? If not, can we eliminate some subset of D^* so that, on the remainder, T will be one-to-one?

In Exercises 8 and 9, let $T(\mathbf{x}) = A\mathbf{x}$, where A is a 2×2 matrix.

8. Show that T is one-to-one if and only if the determinant of A is not zero.

9. Show that $\det A \neq 0$ if and only if T is onto.

10. Suppose $T: \mathbb{R}^2 \to \mathbb{R}^2$ is linear and is given by $T(\mathbf{x}) = A\mathbf{x}$, where A is a 2×2 matrix. Show that if $\det A \neq 0$, then T takes parallelograms onto parallelograms. [HINT: The general parallelogram in $\mathbb{R}^2$ can be described by the set of points $\mathbf{q} = \mathbf{p} + \lambda\mathbf{v} + \mu\mathbf{w}$ for $\lambda, \mu \in (0, 1)$ where $\mathbf{p}$, $\mathbf{v}$, $\mathbf{w}$ are vectors in $\mathbb{R}^2$ with $\mathbf{v}$ not a scalar multiple of $\mathbf{w}$.]

11. Suppose $T: \mathbb{R}^2 \to \mathbb{R}^2$ is as in Exercise 10 and that $T(P^*) = P$ is a parallelogram. Show that P^* is a parallelogram.

12. Consider the map $T: D \to D$, where D is the unit disk in the plane, given by

$$T(r\cos\theta, r\sin\theta) = (r^2\cos 2\theta, r^2\sin 2\theta).$$

Using complex notation, $z = x + iy$, the map T can be written as $T(z) = z^2$. Show that the Jacobian determinant of T vanishes only at the origin. Thus, away from the origin, T is locally one-to-one. However, show that T is not globally one-to-one on $\mathbb{R}^2$ minus the origin.

6.2 The Change of Variables Theorem

Given two regions D and D^* in $\mathbb{R}^2$, a differentiable map T on D^* with image D, that is, $T(D^*) = D$, and any real-valued integrable function $f: D \to \mathbb{R}$, we would like to express $\iint_D f(x, y)\, dA$ as an integral over D^* of the composite function $f \circ T$. In this section we shall see how to do this.

Assume that D^* is a region in the uv plane and that D is a region in the xy plane. The map T is given by two coordinate functions:

$$T(u, v) = (x(u, v), y(u, v)) \qquad \text{for} \qquad (u, v) \in D^*.$$

At first, one might conjecture that

$$\iint_D f(x, y)\, dx\, dy \stackrel{?}{=} \iint_{D^*} f(x(u, v), y(u, v))\, du\, dv, \tag{1}$$

where $f \circ T(u, v) = f(x(u, v), y(u, v))$ is the composite function defined on D^*. However, if we consider the function $f: D \to \mathbb{R}^2$ where $f(x, y) = 1$, then equation (1) would imply

$$A(D) = \iint_D dx\, dy \stackrel{?}{=} \iint_{D^*} du\, dv = A(D^*). \tag{2}$$

But equation (2) will hold for only a few special cases and not for a general map T. For example, define T by $T(u, v) = (-u^2 + 4u, v)$. Restrict T to the unit square; that is, to the region $D^* = [0, 1] \times [0, 1]$ in the uv plane (see Figure 6.2.1). Then, as in Exercise 3, Section 6.1, T takes D^* onto $D = [0, 3] \times [0, 1]$. Clearly, $A(D) \neq A(D^*)$, and so formula (2) is *not valid.*

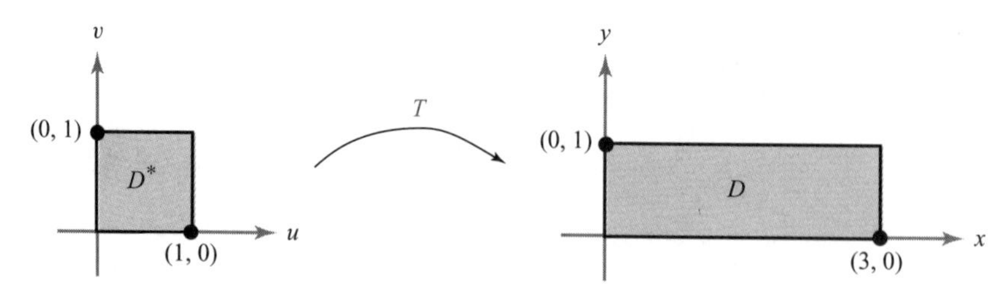

Figure 6.2.1 The map $T: (u, v) \mapsto (-u^2 + 4u, v)$ takes the square D^* onto the rectangle D.

Jacobian Determinants

To rectify the incorrect formula (1), we need a measure of how a transformation $T: \mathbb{R}^2 \to \mathbb{R}^2$ distorts the area of a region. This is given by the *Jacobian determinant*, which is defined as follows.

> DEFINITION: Jacobian Determinant Let $T: D^* \subset \mathbb{R}^2 \to \mathbb{R}^2$ be a C^1 transformation given by $x = x(u, v)$ and $y = y(u, v)$. The ***Jacobian determinant*** of T, written $\partial(x, y)/\partial(u, v)$, is the determinant of the derivative matrix $\mathbf{D}T(u, v)$ of T:
>
> $$\frac{\partial(x, y)}{\partial(u, v)} = \begin{vmatrix} \dfrac{\partial x}{\partial u} & \dfrac{\partial x}{\partial v} \\ \dfrac{\partial y}{\partial u} & \dfrac{\partial y}{\partial v} \end{vmatrix}.$$

EXAMPLE 1 The function from $\mathbb{R}^2$ to $\mathbb{R}^2$ that transforms polar coordinates into Cartesian coordinates is given by

$$x = r\cos\theta, \qquad y = r\sin\theta$$

and its Jacobian determinant is

$$\frac{\partial(x, y)}{\partial(r, \theta)} = \begin{vmatrix} \cos\theta & -r\sin\theta \\ \sin\theta & r\cos\theta \end{vmatrix} = r(\cos^2\theta + \sin^2\theta) = r. \quad \blacktriangle$$

Under suitable restrictions on the function T, we will argue below that the area of $D = T(D^*)$ is obtained by integrating the absolute value of the Jacobian $\partial(x, y)/\partial(u, v)$ over D^*; that is, we have the equation

$$A(D) = \iint_D dx\,dy = \iint_{D^*} \left|\frac{\partial(x, y)}{\partial(u, v)}\right| du\,dv. \tag{3}$$

To illustrate: From Example 1 in Section 6.1, take $T: D^* \to D$, where $D = T(D^*)$ is the set of (x, y) with $x^2 + y^2 \leq 1$ and $D^* = [0, 1] \times [0, 2\pi]$, and $T(r, \theta) = (r\cos\theta, r\sin\theta)$. By formula (3),

$$A(D) = \iint_{D^*} \left|\frac{\partial(x, y)}{\partial(r, \theta)}\right| dr\,d\theta = \iint_{D^*} r\,dr\,d\theta \tag{4}$$

(here r and θ play the role of u and v). From the preceding computation it follows that

$$\iint_{D^*} r\,dr\,d\theta = \int_0^{2\pi}\int_0^1 r\,dr\,d\theta = \int_0^{2\pi}\left[\frac{r^2}{2}\right]_0^1 d\theta = \frac{1}{2}\int_0^{2\pi} d\theta = \pi$$

is the area of the unit disk D, confirming formula (3) in this case. In fact, we may recall from first-year calculus that equation (4) is the correct formula for the area of a region in polar coordinates.

It is not so easy to rigorously prove assertion (3). However, looked at in the proper way, it becomes quite plausible. Recall that $A(D) = \iint_D dx\,dy$ was obtained by dividing up D into little rectangles, summing their areas, and then taking the limit of this sum as the size of the subrectangles tended to zero. The problem is that T may map rectangles into regions whose area is not easy to compute. The solution is to approximate these images by simpler regions whose area we can compute. A useful tool for doing this is the derivative of T, which we know (from Chapter 2) gives the best linear approximation to T.

Consider a small rectangle D^* in the uv plane as shown in Figure 6.2.2. Let T' denote the derivative of T evaluated at (u_0, v_0), so T' is a 2×2 matrix. From our work in Chapter 2, we know that a good approximation to $T(u, v)$ is given by

$$T(u_0, v_0) + T'\begin{pmatrix}\Delta u\\ \Delta v\end{pmatrix},$$

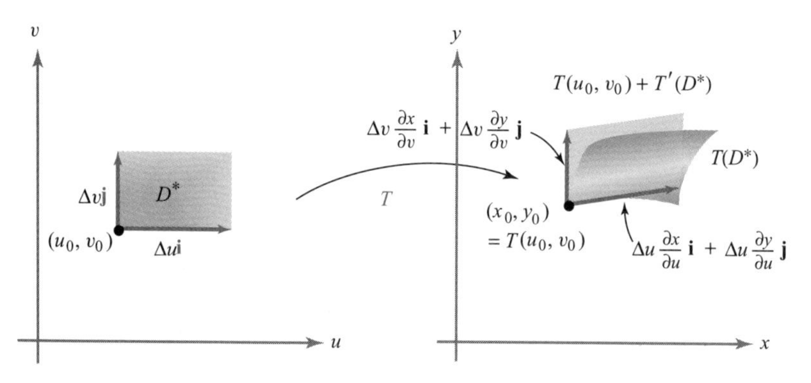

Figure 6.2.2 The effect of the transformation T on a small rectangle D^*.

where $\Delta u = u - u_0$ and $\Delta v = v - v_0$. This mapping T' takes D^* into a parallelogram with vertex at $T(u_0, v_0)$ and with adjacent sides given by the vectors

$$T'(\Delta u\mathbf{i}) = \begin{bmatrix}\dfrac{\partial x}{\partial u} & \dfrac{\partial x}{\partial v}\\[2mm] \dfrac{\partial y}{\partial u} & \dfrac{\partial y}{\partial v}\end{bmatrix}\begin{bmatrix}\Delta u\\ 0\end{bmatrix} = \Delta u\begin{bmatrix}\dfrac{\partial x}{\partial u}\\[2mm] \dfrac{\partial y}{\partial u}\end{bmatrix} = \Delta u\mathbf{T}_u$$

and

$$T'(\Delta v\mathbf{j}) = \begin{bmatrix} \dfrac{\partial x}{\partial u} & \dfrac{\partial x}{\partial v} \\ \dfrac{\partial y}{\partial u} & \dfrac{\partial y}{\partial v} \end{bmatrix} \begin{bmatrix} 0 \\ \Delta v \end{bmatrix} = \Delta v \begin{bmatrix} \dfrac{\partial x}{\partial v} \\ \dfrac{\partial y}{\partial v} \end{bmatrix} = \Delta v \mathbf{T}_v,$$

where

$$\mathbf{T}_u = \frac{\partial x}{\partial u}\mathbf{i} + \frac{\partial y}{\partial u}\mathbf{j} \qquad \text{and} \qquad \mathbf{T}_v = \frac{\partial x}{\partial v}\mathbf{i} + \frac{\partial y}{\partial v}\mathbf{j}$$

are evaluated at (u_0, v_0).

Recall from Section 1.3 that the area of the parallelogram with sides equal to the vectors $a\mathbf{i} + b\mathbf{j}$ and $c\mathbf{i} + d\mathbf{j}$ is equal to the absolute value of the determinant

$$\begin{vmatrix} a & b \\ c & d \end{vmatrix} = \begin{vmatrix} a & c \\ b & d \end{vmatrix}.$$

Thus, the area of $T(D^*)$ is approximately equal to the *absolute value* of

$$\begin{vmatrix} \dfrac{\partial x}{\partial u}\Delta u & \dfrac{\partial x}{\partial v}\Delta v \\ \dfrac{\partial y}{\partial u}\Delta u & \dfrac{\partial y}{\partial v}\Delta v \end{vmatrix} = \begin{vmatrix} \dfrac{\partial x}{\partial u} & \dfrac{\partial x}{\partial v} \\ \dfrac{\partial y}{\partial u} & \dfrac{\partial y}{\partial v} \end{vmatrix} \Delta u\, \Delta v = \frac{\partial(x, y)}{\partial(u, v)} \Delta u\, \Delta v$$

evaluated at (u_0, v_0).

This fact and a partitioning argument should make formula (3) plausible. Indeed, if we partition D^* into small rectangles with sides of length Δu and Δv, the images of these rectangles are approximated by parallelograms with sides $\mathbf{T}_u\, \Delta u$ and $\mathbf{T}_v\, \Delta v$, and hence with area $|\partial(x, y)/\partial(u, v)|\, \Delta u\, \Delta v$. Thus, the area of D^* is approximately $\sum \Delta u\, \Delta v$, where the sum is taken over all the rectangles R inside D^* (see Figure 6.2.3). Hence, the area of $T(D^*)$ is approximately the sum

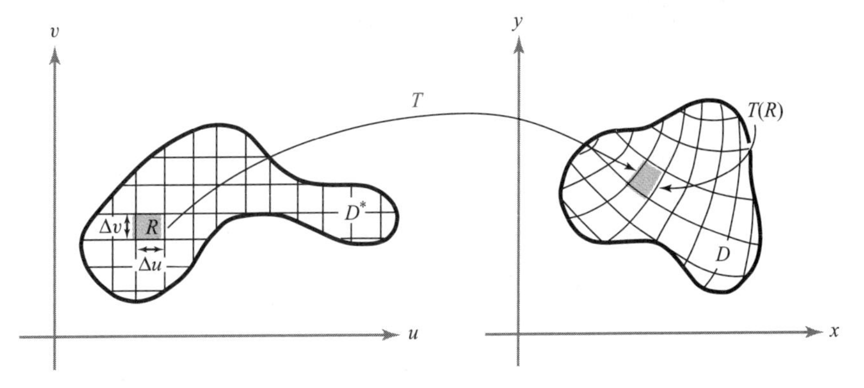

Figure 6.2.3 The area of the little rectangle R is $\Delta u\, \Delta v$. The area of $T(R)$ is approximately $|\partial(x, y)/\partial(u, v)|\Delta u \Delta v$.

$\sum |\partial(x, y)/\partial(u, v)| \Delta u\, \Delta v$. In the limit, this sum becomes

$$\iint_{D^*} \left| \frac{\partial(x, y)}{\partial(u, v)} \right| du\, dv.$$

Let us give another informal argument for the special case (4) of formula (3), that is, the case of polar coordinates. Consider a region D in the xy plane and a grid corresponding to a partition of the r and θ variables (Figure 6.2.4). The area of the shaded region shown is approximately $(\Delta r)(r_{jk}\, \Delta\theta)$, because the arc length of a segment of a circle of radius r subtending an angle ϕ is $r\phi$. The total area is then the limit of $\sum r_{jk}\, \Delta r\, \Delta\theta$; that is, $\iint_{D^*} r\, dr\, d\theta$. The key idea is thus that the jkth "polar rectangle" in the grid has area approximately equal to $r_{jk}\, \Delta r\, \Delta\theta$. (For n large, the jkth polar rectangle will look like a rectangle with sides of lengths $r_{jk}\, \Delta\theta$ and Δr). This should provide some insight into why we say *the "area element $dx\, dy$" is transformed into the "area element $r\, dr\, d\theta$."*

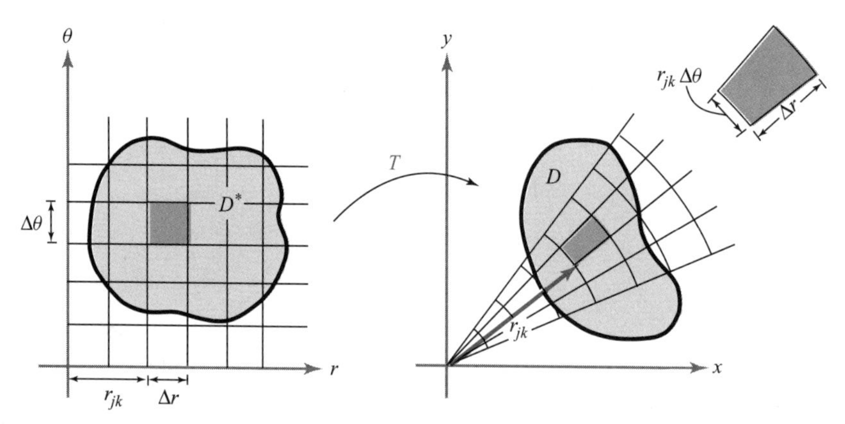

Figure 6.2.4 D^* is mapped to D under the polar-coordinate mapping T.

EXAMPLE 2 Let the elementary region D in the xy plane be bounded by the graph of a polar equation $r = f(\theta)$, where $\theta_0 \le \theta \le \theta_1$ and $f(\theta) \ge 0$ (see Figure 6.2.5). In the $r\theta$ plane we consider the r-simple region D^* where $\theta_0 \le \theta \le \theta_1$

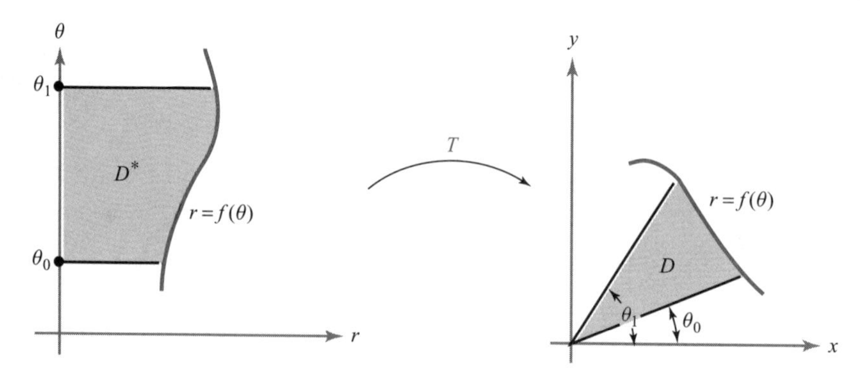

Figure 6.2.5 The effect on the region D^* of the polar-coordinate mapping.

and $0 \le r \le f(\theta)$. Under the transformation $x = r\cos\theta$, $y = r\sin\theta$, the region D^* is carried onto the region D. Use equation (4) to calculate the area of D.

SOLUTION

$$\begin{aligned} A(D) = \iint_D dx\,dy &= \iint_{D^*} \left|\frac{\partial(x,y)}{\partial(r,\theta)}\right| dr\,d\theta \\ &= \iint_{D^*} r\,dr\,d\theta = \int_{\theta_0}^{\theta_1}\left[\int_0^{f(\theta)} r\,dr\right] d\theta \\ &= \int_{\theta_0}^{\theta_1}\left[\frac{r^2}{2}\right]_0^{f(\theta)} d\theta = \int_{\theta_0}^{\theta_1}\frac{[f(\theta)]^2}{2} d\theta \end{aligned}$$

This formula for $A(D)$ should be familiar from one-variable calculus. ▲

Change of Variables Formula

Before stating the two-variable change of variables formula, which is the culmination of this discussion, let us recall the corresponding theorem from one-variable calculus that goes under the name the *method of substitution*:

$$\int_a^b f(x(u))\frac{dx}{du}\,du = \int_{x(a)}^{x(b)} f(x)\,dx, \tag{5}$$

where f is continuous and $u \mapsto x(u)$ is continuously differentiable on $[a, b]$.

PROOF Let F be an antiderivative of f; that is, $F' = f$, whose existence is guaranteed by the fundamental theorem of calculus. The right-hand side of equation (5) becomes

$$\int_{x(a)}^{x(b)} f(x)\,dx = F(x(b)) - F(x(a)).$$

To evaluate the left-hand side of equation (5), let $G(u) = F(x(u))$. By the chain rule, $G'(u) = F'(x(u))x'(u) = f(x(u))x'(u)$. Hence, again by the fundamental theorem,

$$\int_a^b f(x(u))x'(u)\,du = \int_a^b G'(u)\,du = G(b) - G(a) = F(x(b)) - F(x(a)),$$

as required. ■

Suppose now that we have a C^1 *function* $u \mapsto x(u)$ that is one-to-one on $[a, b]$. Thus, we must have either $dx/du \ge 0$ on $[a, b]$ or $dx/du \le 0$ on $[a, b]$.[1] Let I^* denote the interval $[a, b]$, and let I denote the closed interval with endpoints $x(a)$

[1] If dx/du is positive and then negative, the function $x = x(u)$ rises and then falls, and thus is not one-to-one; a similar statement applies if dx/du is negative and then positive.

and $x(b)$. (Thus, $I = [x(a), x(b)]$ if $u \mapsto x(u)$ is increasing and $I = [x(b), x(a)]$ if $u \mapsto x(u)$ is decreasing.) With these conventions we can rewrite formula (5) as

$$\int_{I^*} f(x(u)) \left| \frac{dx}{du} \right| du = \int_I f(x)\, dx.$$

This formula generalizes to double integrals, as was already given informally in formula (3): I^* becomes D^*, I becomes D, and $|dx/du|$ is replaced by $|\partial(x, y)/\partial(u, v)|$. Let us state the result formally (the technical proof is omitted).

THEOREM 2: Change of Variables: Double Integrals Let D and D^* be elementary regions in the plane and let $T\colon D^* \to D$ be of class C^1; suppose that T is one-to-one on D^*. Furthermore, suppose that $D = T(D^*)$. Then for any integrable function $f\colon D \to \mathbb{R}$, we have

$$\iint_D f(x, y)\, dx\, dy = \iint_{D^*} f(x(u, v), y(u, v)) \left| \frac{\partial(x, y)}{\partial(u, v)} \right| du\, dv. \tag{6}$$

One of the purposes of the change of variables theorem is to supply a method by which some double integrals can be simplified. One might encounter an integral $\iint_D f\, dA$ for which either the integrand f or the region D is complicated and for which direct evaluation is difficult. Therefore, a mapping T is chosen so that the integral is easier to evaluate with the new integrand $f \circ T$ and with the new region D^* [defined by $T(D^*) = D$]. Unfortunately, the problem may actually become more complicated if T is not selected carefully.

EXAMPLE 3 Let P be the parallelogram bounded by $y = 2x$, $y = 2x - 2$, $y = x$, and $y = x + 1$ (see Figure 6.2.6). Evaluate $\iint_P xy\, dx\, dy$ by making the

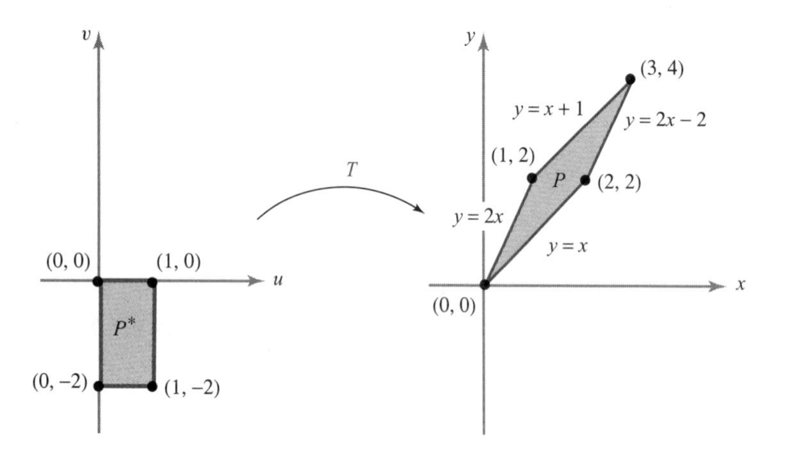

Figure 6.2.6 The effect of $T(u, v) = (u - v, 2u - v)$ on the rectangle P^*.

change of variables

$$x = u - v, \qquad y = 2u - v,$$

that is, $T(u, v) = (u - v, 2u - v)$.

SOLUTION The transformation T has nonzero determinant and so is one-to-one (see Exercise 8, Section 6.1). It is designed so that it takes the *rectangle* P^* bounded by $v = 0$, $v = -2$, $u = 0$, $u = 1$ onto P. The use of T simplifies the region of integration from P to P^*. Moreover,

$$\left|\frac{\partial(x, y)}{\partial(u, v)}\right| = \left|\det\begin{bmatrix} 1 & -1 \\ 2 & -1 \end{bmatrix}\right| = 1.$$

Therefore, by the change of variables formula,

$$\iint_P xy\,dx\,dy = \iint_{P^*} (u - v)(2u - v)\,du\,dv = \int_{-2}^{0}\int_{0}^{1} (2u^2 - 3vu + v^2)\,du\,dv$$

$$= \int_{-2}^{0}\left[\frac{2}{3}u^3 - \frac{3u^2v}{2} + v^2u\right]_0^1 dv = \int_{-2}^{0}\left[\frac{2}{3} - \frac{3}{2}v + v^2\right] dv$$

$$= \left[\frac{2}{3}v - \frac{3}{4}v^2 + \frac{v^3}{3}\right]_{-2}^{0} = -\left[\frac{2}{3}(-2) - 3 - \frac{8}{3}\right]$$

$$= -\left[-\frac{12}{3} - 3\right] = 7. \quad \blacktriangle$$

Integrals in Polar Coordinates

Suppose we consider the rectangle D^* defined by $0 \le \theta \le 2\pi$, $0 \le r \le a$ in the $r\theta$ plane. The transformation T given by $T(r, \theta) = (r\cos\theta, r\sin\theta)$ takes D^* onto the disk D with equation $x^2 + y^2 \le a^2$ in the xy plane. This transformation represents the change from Cartesian coordinates to polar coordinates. However, T does not satisfy the requirements of the change of variables theorem, because it is not one-to-one on D^*: In particular, T sends all points with $r = 0$ to $(0, 0)$ (see Figure 6.2.7 and Example 3 of Section 6.1). Nevertheless, the change of variables theorem is valid in this case. Basically, the reason for this is that the set of points where T is not one-to-one lies on an edge of D^*, which is the graph of a smooth curve and therefore, for the purpose of integration, can be neglected. In summary, the formula

Change of Variables---Polar Coordinates

$$\iint_D f(x, y)\,dx\,dy = \iint_{D^*} f(r\cos\theta, r\sin\theta)\,r\,dr\,d\theta \tag{7}$$

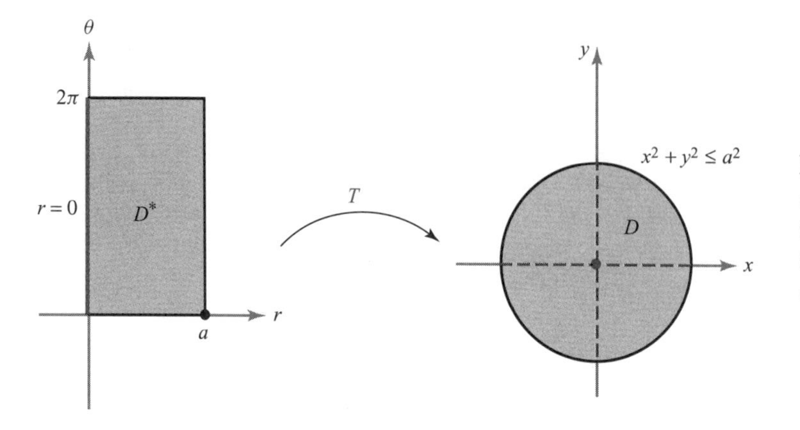

Figure 6.2.7 The image of the rectangle D^* under the polar-coordinate transformation is the disk D.

is valid when T sends D^* onto D in a one-to-one fashion except possibly for points on the boundary of D^*.

EXAMPLE 4 Evaluate $\iint_D \log(x^2 + y^2)\,dx\,dy$, where D is the region in the first quadrant lying between the arcs of the circles $x^2 + y^2 = a^2$ and $x^2 + y^2 = b^2$, where $0 < a < b$ (Figure 6.2.8).

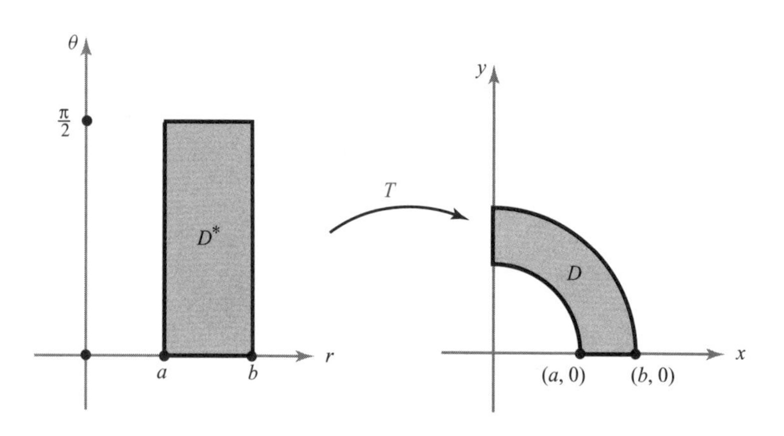

Figure 6.2.8 The polar-coordinate mapping takes a rectangle D^* onto part of an annulus D.

SOLUTION These circles have the simple equations $r = a$ and $r = b$ in polar coordinates. Moreover, $r^2 = x^2 + y^2$ appears in the integrand. Thus, a change to polar coordinates will simplify both the integrand and the region of integration. From Example 7, Section 6.1, the polar-coordinate transformation

$$x = r\cos\theta, \qquad y = r\sin\theta$$

sends the rectangle D^* given by $a \leq r \leq b, 0 \leq \theta \leq \pi/2$ onto the region D. This transformation is one-to-one on D^* and so, by formula (7), we have

$$\iint_D \log(x^2+y^2)\,dx\,dy = \int_a^b \int_0^{\pi/2} r \log r^2 \,d\theta\,dr$$
$$= \frac{\pi}{2}\int_a^b r \log r^2\,dr = \frac{\pi}{2}\int_a^b 2r \log r\,dr.$$

Applying integration by parts, or using the formula

$$\int x \log x\,dx = \frac{x^2}{2}\log x - \frac{x^2}{4}$$

from the table of integrals at the back of the book, we obtain the result

$$\frac{\pi}{2}\int_a^b 2r \log r\,dr = \frac{\pi}{2}\left[b^2 \log b - a^2 \log a - \frac{1}{2}(b^2 - a^2)\right]. \quad \blacktriangle$$

EXAMPLE 5 **The Gaussian Integral** One of the most beautiful applications of the change of variables formula, polar coordinates, and the reduction to iterated integrals is their application to the following formula, known as the ***Gaussian integral***:

$$\int_{-\infty}^{\infty} e^{-x^2}\,dx = \sqrt{\pi}.$$

Not only is this formula very attractive in its own right, but it is also useful in areas such as statistics. It also illustrates the unity of the transcendental numbers e and π nearly as well as does the classic formula $e^{i\pi} = -1$.

To carry out the integration of the Gaussian integral,[2] we first evaluate the double integral

$$\iint_{D_a} e^{-(x^2+y^2)}\,dx\,dy,$$

[2]The method that follows is admittedly not straightforward but requires a trick. The trick is to start with the desired formula and square both sides. You will then observe that the left-hand side resembles an iterated integral. There are several other ways to evaluate the Gaussian integral, but all of them require some nonobvious method. For the use of complex variables to evaluate it, see, for example, J. Marsden and M. Hoffman, *Basic Complex Analysis*, 3rd ed., W. H. Freeman, New York, 1998.

where D_a is the disk $x^2 + y^2 \leq a^2$. Because $r^2 = x^2 + y^2$, and $dx\,dy = r\,dr\,d\theta$, the change of variables formula gives

$$\iint_{D_a} e^{-(x^2+y^2)}\,dx\,dy = \int_0^{2\pi}\int_0^a e^{-r^2} r\,dr\,d\theta = \int_0^{2\pi} \left(-\frac{1}{2}e^{-r^2}\right)\Big|_0^a\,d\theta$$

$$= -\frac{1}{2}\int_0^{2\pi} (e^{-a^2} - 1)\,d\theta = \pi(1 - e^{-a^2}).$$

If we let $a \to \infty$ in this expression, we give meaning to the improper integral and get

$$\iint_{\mathbb{R}^2} e^{-(x^2+y^2)}\,dx\,dy = \pi.$$

Assuming (as shown in the Internet supplement) that we can also evaluate this improper integral as the limit of the integrals over the rectangles $R_a = [-a, a] \times [-a, a]$ as $a \to \infty$, we get

$$\lim_{a\to\infty} \iint_{R_a} e^{-(x^2+y^2)}\,dx\,dy = \pi.$$

By reduction to iterated integrals, we can write this as

$$\lim_{a\to\infty}\left[\int_{-a}^{a} e^{-x^2}\,dx \int_{-a}^{a} e^{-y^2}\,dy\right] = \left[\lim_{a\to\infty}\int_{-a}^{a} e^{-x^2}\,dx\right]^2 = \pi.$$

That is,

$$\left[\int_{-\infty}^{\infty} e^{-x^2}\,dx\right]^2 = \pi.$$

Thus, taking square roots, we arrive at the desired result.

Here is a variant of the Gaussian integral. Evaluate

$$\int_{-\infty}^{\infty} e^{-2x^2}\,dx.$$

To do this, use the change of variables formula $y = \sqrt{2}x$ to reduce the problem to the Gaussian integral just computed:

$$\int_{-\infty}^{\infty} e^{-2x^2}\,dx = \lim_{a\to\infty}\int_{-a}^{a} e^{-2x^2}\,dx = \lim_{a\to\infty}\int_{-\sqrt{2}a}^{\sqrt{2}a} e^{-y^2}\,\frac{dy}{\sqrt{2}}$$

$$= \frac{1}{\sqrt{2}}\int_{-\infty}^{\infty} e^{-y^2}\,dy = \frac{1}{\sqrt{2}}\sqrt{\pi} = \sqrt{\frac{\pi}{2}}. \quad \blacktriangle$$

Change of Variables Formula for Triple Integrals

To state this formula, we first define the Jacobian of a transformation from $\mathbb{R}^3$ to $\mathbb{R}^3$—it is a simple extension of the two-variable case.

DEFINITION Let $T\colon W \subset \mathbb{R}^3 \to \mathbb{R}^3$ be a C^1 function defined by the equations $x = x(u, v, w)$, $y = y(u, v, w)$, $z = z(u, v, w)$. Then the ***Jacobian*** of T, which is denoted $\partial(x, y, z)/\partial(u, v, w)$, is the determinant

$$\begin{vmatrix} \dfrac{\partial x}{\partial u} & \dfrac{\partial x}{\partial v} & \dfrac{\partial x}{\partial w} \\ \dfrac{\partial y}{\partial u} & \dfrac{\partial y}{\partial v} & \dfrac{\partial y}{\partial w} \\ \dfrac{\partial z}{\partial u} & \dfrac{\partial z}{\partial v} & \dfrac{\partial z}{\partial w} \end{vmatrix}.$$

The absolute value of this determinant is equal to the volume of the parallelepiped determined by the three vectors

$$\mathbf{T}_u = \frac{\partial x}{\partial u}\mathbf{i} + \frac{\partial y}{\partial u}\mathbf{j} + \frac{\partial z}{\partial u}\mathbf{k},$$

$$\mathbf{T}_v = \frac{\partial x}{\partial v}\mathbf{i} + \frac{\partial y}{\partial v}\mathbf{j} + \frac{\partial z}{\partial v}\mathbf{k},$$

$$\mathbf{T}_w = \frac{\partial x}{\partial w}\mathbf{i} + \frac{\partial y}{\partial w}\mathbf{j} + \frac{\partial z}{\partial w}\mathbf{k}.$$

Just as in the two-variable case, the Jacobian measures how the transformation T distorts the volume of its domain. Hence, for volume (triple) integrals, the change of variables formula takes the following form:

Change of Variables Formula: Triple Integrals

$$\begin{aligned} &\iiint_W f(x, y, z)\, dx\, dy\, dz \\ &= \iiint_{W^*} f(x(u, v, w), y(u, v, w), z(u, v, w)) \left| \frac{\partial(x, y, z)}{\partial(u, v, w)} \right| du\, dv\, dw, \end{aligned} \tag{8}$$

where W^* is an elementary region in uvw space corresponding to W in xyz space, under a mapping $T\colon (u, v, w) \mapsto (x(u, v, w), y(u, v, w), z(u, v, w))$, provided T is of class C^1 and is one-to-one, except possibly on a set that is the union of graphs of functions of two variables.

Cylindrical Coordinates

Let us apply formula (8) to cylindrical and then to spherical coordinates. First, we compute the Jacobian for the map defining the change to cylindrical coordinates. Because

$$x = r\cos\theta, \qquad y = r\sin\theta, \qquad z = z,$$

we have

$$\frac{\partial(x, y, z)}{\partial(r, \theta, z)} = \begin{vmatrix} \cos\theta & -r\sin\theta & 0 \\ \sin\theta & r\cos\theta & 0 \\ 0 & 0 & 1 \end{vmatrix} = r.$$

Thus, we obtain the formula

Change of Variables—Cylindrical Coordinates

$$\iiint_W f(x, y, z)\,dx\,dy\,dz = \iiint_{W^*} f(r\cos\theta, r\sin\theta, z)\,r\,dr\,d\theta\,dz. \tag{9}$$

Spherical Coordinates

Next we consider the spherical coordinate system. Recall that it is given by

$$x = \rho\sin\phi\cos\theta, \qquad y = \rho\sin\phi\sin\theta, \qquad z = \rho\cos\phi.$$

Therefore, we have

$$\frac{\partial(x, y, z)}{\partial(\rho, \theta, \phi)} = \begin{vmatrix} \sin\phi\cos\theta & -\rho\sin\phi\sin\theta & \rho\cos\phi\cos\theta \\ \sin\phi\sin\theta & \rho\sin\phi\cos\theta & \rho\cos\phi\sin\theta \\ \cos\phi & 0 & -\rho\sin\phi \end{vmatrix}.$$

Expanding along the last row, we get

$$\begin{aligned}\frac{\partial(x, y, z)}{\partial(\rho, \theta, \phi)} &= \cos\phi\begin{vmatrix} -\rho\sin\phi\sin\theta & \rho\cos\phi\cos\theta \\ \rho\sin\phi\cos\theta & \rho\cos\phi\sin\theta \end{vmatrix} \\ &\quad -\rho\sin\phi\begin{vmatrix} \sin\phi\cos\theta & -\rho\sin\phi\sin\theta \\ \sin\phi\sin\theta & \rho\sin\phi\cos\theta \end{vmatrix} \\ &= -\rho^2\cos^2\phi\sin\phi\sin^2\theta - \rho^2\cos^2\phi\sin\phi\cos^2\theta \\ &\quad -\rho^2\sin^3\phi\cos^2\theta - \rho^2\sin^3\phi\sin^2\theta \\ &= -\rho^2\cos^2\phi\sin\phi - \rho^2\sin^3\phi = -\rho^2\sin\phi.\end{aligned}$$

Thus, we arrive at the formula:

Change of Variables—Spherical Coordinates

$$\iiint_W f(x, y, z)\, dx\, dy\, dz = \iiint_{W^*} f(\rho \sin\phi \cos\theta, \rho \sin\phi \sin\theta, \rho \cos\phi)\rho^2 \sin\phi\, d\rho\, d\theta\, d\phi. \qquad (10)$$

To prove formula (10), one must show that the transformation S on the set W^* is one-to-one except on a set that is the union of finitely many graphs of continuous functions. We shall leave this verification as Exercise 34.

EXAMPLE 6 Evaluate

$$\iiint_W \exp(x^2 + y^2 + z^2)^{3/2}\, dV,$$

where W is the unit ball in $\mathbb{R}^3$.

SOLUTION First note that we cannot *easily* integrate this function using iterated integrals (try it!). Hence (employing the strategy in the quote that opened this chapter), let us try a change of variables. The transformation into spherical coordinates seems appropriate, because then the entire quantity $x^2 + y^2 + z^2$ can be replaced by one variable, namely, ρ^2. If W^* is the region such that

$$0 \le \rho \le 1, \qquad 0 \le \theta \le 2\pi, \qquad 0 \le \phi \le \pi,$$

we may apply formula (10) and write

$$\iiint_W \exp(x^2 + y^2 + z^2)^{3/2}\, dV = \iiint_{W^*} \rho^2 e^{\rho^3} \sin\phi\, d\rho\, d\theta\, d\phi.$$

This integral equals the iterated integral

$$\begin{aligned}\int_0^1 \int_0^\pi \int_0^{2\pi} e^{\rho^3} \rho^2 \sin\phi\, d\theta\, d\phi\, d\rho &= 2\pi \int_0^1 \int_0^\pi e^{\rho^3} \rho^2 \sin\phi\, d\phi\, d\rho \\ &= -2\pi \int_0^1 \rho^2 e^{\rho^3} [\cos\phi]_0^\pi\, d\rho \\ &= 4\pi \int_0^1 e^{\rho^3} \rho^2\, d\rho = \frac{4}{3}\pi \int_0^1 e^{\rho^3}(3\rho^2)\, d\rho \\ &= \left[\frac{4}{3}\pi e^{\rho^3}\right]_0^1 = \frac{4}{3}\pi(e - 1). \quad \blacktriangle\end{aligned}$$

EXAMPLE 7 Let W be the ball of radius R and center $(0, 0, 0)$ in $\mathbb{R}^3$. Find the volume of W.

SOLUTION The volume of W is $\iiint_W dx\,dy\,dz$. This integral may be evaluated by reducing it to iterated integrals or by regarding W as a volume of revolution, but let us evaluate it here by using spherical coordinates. We get

$$\iiint_W dx\,dy\,dz = \int_0^{\pi}\int_0^{2\pi}\int_0^R \rho^2 \sin\phi\, d\rho\, d\theta\, d\phi = \frac{R^3}{3}\int_0^{\pi}\int_0^{2\pi} \sin\phi\, d\theta\, d\phi$$
$$= \frac{2\pi R^3}{3}\int_0^{\pi} \sin\phi\, d\phi = \frac{2\pi R^3}{3}\{-[\cos(\pi) - \cos(0)]\} = \frac{4\pi R^3}{3},$$

which is the standard formula for the volume of a solid sphere. ▲

EXERCISES

1. Let D be the unit disk: $x^2 + y^2 \leq 1$. Evaluate

$$\iint_D \exp(x^2 + y^2)\, dx\, dy$$

by making a change of variables to polar coordinates.

2. Let D be the region $0 \leq y \leq x$ and $0 \leq x \leq 1$. Evaluate

$$\iint_D (x + y)\, dx\, dy$$

by making the change of variables $x = u + v$, $y = u - v$. Check your answer by evaluating the integral directly by using an iterated integral.

3. Let $T(u, v) = (x(u, v), y(u, v))$ be the mapping defined by $T(u, v) = (4u, 2u + 3v)$. Let D^* be the rectangle $[0, 1] \times [1, 2]$. Find $D = T(D^*)$ and evaluate

(a) $\displaystyle\iint_D xy\, dx\, dy$ (b) $\displaystyle\iint_D (x - y)\, dx\, dy$

by making a change of variables to evaluate them as integrals over D^*.

4. Repeat Exercise 3 for $T(u, v) = (u, v(1 + u))$.

5. Evaluate

$$\iint_D \frac{dx\, dy}{\sqrt{1 + x + 2y}},$$

where $D = [0, 1] \times [0, 1]$, by setting $T(u, v) = (u, v/2)$ and evaluating an integral over D^*, where $T(D^*) = D$.

6. Define $T(u, v) = (u^2 - v^2, 2uv)$. Let D^* be the set of (u, v) with $u^2 + v^2 \leq 1$, $u \geq 0$, $v \geq 0$. Find $T(D^*) = D$. Evaluate $\iint_D dx\, dy$.

7. Let $T(u, v)$ be as in Exercise 6. By making a change of variables, "formally" evaluate the "improper" integral

$$\iint_D \frac{dx\, dy}{\sqrt{x^2 + y^2}}.$$

[NOTE: This integral (and the one in the next exercise) is *improper*, because the integrand $1/\sqrt{x^2 + y^2}$ is neither continuous nor bounded on the domain of integration. (The theory of improper integrals is discussed in Section 6.4.)]

8. Calculate $\displaystyle\iint_R \frac{1}{x + y}\, dy\, dx$, where R is the region bounded by $x = 0$, $y = 0$, $x + y = 1$, $x + y = 4$, by using the mapping $T(u, v) = (u - uv, uv)$.

9. Evaluate $\displaystyle\iint_D (x^2 + y^2)^{3/2}\, dx\, dy$ where D is the disk $x^2 + y^2 \leq 4$.

10. Let D^* be a v-simple region in the uv plane bounded by $v = g(u)$ and $v = h(u) \leq g(u)$ for $a \leq u \leq b$. Let $T\colon \mathbb{R}^2 \to \mathbb{R}^2$ be the transformation given by $x = u$ and $y = \psi(u, v)$, where ψ is of class C^1 and $\partial\psi/\partial v$ is never zero. Assume that $T(D^*) = D$ is a y-simple region; show that if $f\colon D \to \mathbb{R}$ is continuous, then

$$\iint_D f(x, y)\, dx\, dy = \iint_{D^*} f(u, \psi(u, v)) \left|\frac{\partial\psi}{\partial v}\right| du\, dv.$$

11. Use double integrals to find the area inside the curve $r = 1 + \sin\theta$.

12. (a) Express $\int_0^1 \int_0^{x^2} xy\, dy\, dx$ as an integral over the triangle D^*, which is the set of (u, v) where $0 \leq u \leq 1$, $0 \leq v \leq u$. (HINT: Find a one-to-one mapping T of D^* onto the given region of integration.)

(b) Evaluate this integral directly and as an integral over D^*.

13. Integrate $ze^{x^2+y^2}$ over the cylinder $x^2 + y^2 \leq 4$, $2 \leq z \leq 3$.

14. Let D be the unit disk. Express $\displaystyle\iint_D (1 + x^2 + y^2)^{3/2}\, dx\, dy$ as an integral over $[0, 1] \times [0, 2\pi]$ and evaluate.

15. Using polar coordinates, find the area bounded by the *lemniscate* $(x^2 + y^2)^2 = 2a^2\,(x^2 - y^2)$.

16. Redo Exercise 11 of Section 5.3 using a change of variables and compare the effort involved in each method.

17. Calculate $\displaystyle\iint_R (x + y)^2 e^{x-y}\, dx\, dy$ where R is the region bounded by $x + y = 1$, $x + y = 4$, $x - y = -1$, and $x - y = 1$.

18. Let $T\colon \mathbb{R}^3 \to \mathbb{R}^3$ be defined by

$$T(u, v, w) = (u \cos v \cos w,\ u \sin v \cos w,\ u \sin w).$$

(a) Show that T is onto the unit sphere; that is, every (x, y, z) with $x^2 + y^2 + z^2 = 1$ can be written as $(x, y, z) = T(u, v, w)$ for some (u, v, w).

(b) Show that T is not one-to-one.

19. Integrate $x^2 + y^2 + z^2$ over the cylinder $x^2 + y^2 \le 2$, $-2 \le z \le 3$.

20. Evaluate $\int_0^\infty e^{-4x^2}\, dx$.

21. Let B be the unit ball. Evaluate

$$\iiint_B \frac{dx\, dy\, dz}{\sqrt{2 + x^2 + y^2 + z^2}}$$

by making the appropriate change of variables.

22. Evaluate $\iint_A [1/(x^2 + y^2)^2]\, dx\, dy$ where A is determined by the conditions $x^2 + y^2 \le 1$ and $x + y \ge 1$.

23. Evaluate $\iiint_W \frac{dx\, dy\, dz}{(x^2 + y^2 + z^2)^{3/2}}$, where W is the solid bounded by the two spheres $x^2 + y^2 + z^2 = a^2$ and $x^2 + y^2 + z^2 = b^2$, where $0 < b < a$.

24. Evaluate $\iint_D x^2\, dx\, dy$ where D is determined by the two conditions $0 \le x \le y$ and $x^2 + y^2 \le 1$.

25. Integrate $\sqrt{x^2 + y^2 + z^2}\, e^{-(x^2+y^2+z^2)}$ over the region described in Exercise 23.

26. Evaluate the following by using cylindrical coordinates.

(a) $\iiint_B z\, dx\, dy\, dz$ where B is the region within the cylinder $x^2 + y^2 = 1$ above the xy plane and below the cone $z = (x^2 + y^2)^{1/2}$.

(b) $\iiint_W (x^2 + y^2 + z^2)^{-1/2}\, dx\, dy\, dz$ where W is the region determined by the conditions $\frac{1}{2} \le z \le 1$ and $x^2 + y^2 + z^2 \le 1$.

27. Evaluate $\iint_B (x + y)\, dx\, dy$ where B is the rectangle in the xy plane with vertices at $(0, 1)$, $(1, 0)$, $(3, 4)$, and $(4, 3)$.

28. Evaluate $\iint_D (x + y)\, dx\, dy$ where D is the square with vertices at $(0, 0)$, $(1, 2)$, $(3, 1)$, and $(2, -1)$.

29. Let E be the ellipsoid $(x^2/a^2) + (y^2/b^2) + (z^2/c^2) \le 1$, where a, b, and c are positive.

(a) Find the volume of E.

(b) Evaluate $\iiint_E [(x^2/a^2) + (y^2/b^2) + (z^2/c^2)]\, dx\, dy\, dz$. (HINT: Change variables and then use spherical coordinates.)

30. Using spherical coordinates, compute the integral of $f(\rho, \phi, \theta) = 1/\rho$ over the region in the first octant of $\mathbb{R}^3$, which is bounded by the cones $\phi = \pi/4$, $\phi = \arctan 2$ and the sphere $\rho = \sqrt{6}$.

31. The mapping $T(u, v) = (u^2 - v^2, 2uv)$ transforms the rectangle $1 \le u \le 2, 1 \le v \le 3$ of the uv plane into a region R of the xy plane.

(a) Show that T is one-to-one.
(b) Find the area of R using the change of variables formula.

32. Let R denote the region inside $x^2 + y^2 = 1$, but outside $x^2 + y^2 = 2y$ with $x \ge 0, y \ge 0$.

(a) Sketch this region.
(b) Let $u = x^2 + y^2, v = x^2 + y^2 - 2y$. Sketch the region D in the uv plane, which corresponds to R under this change of coordinates.
(c) Compute $\iint_R xe^y\, dx\, dy$ using this change of coordinates.

33. Let D be the region bounded by $x^{3/2} + y^{3/2} = a^{3/2}$, for $x \ge 0, y \ge 0$, and the coordinate axes $x = 0, y = 0$. Express $\iint_D f(x, y)\, dx\, dy$ as an integral over the triangle D^*, which is the set of points $0 \le u \le a, 0 \le v \le a - u$. (Do not attempt to evaluate.)

34. Show that $S(\rho, \theta, \phi) = (\rho \sin\phi \cos\theta, \rho \sin\phi \sin\theta, \rho \cos\phi)$, the spherical change-of-coordinate mapping, is one-to-one except on a set that is a union of finitely many graphs of continuous functions.

6.3 Applications

In this section, we shall discuss average values, centers of mass, moments of inertia, and the gravitational potential as applications.

Averages

If $x_1, \ldots, x_n$ are n numbers, their ***average*** is defined by

$$[x_i]_{\mathrm{av}} = \frac{x_1 + \cdots + x_n}{n} = \frac{1}{n}\sum_{i=1}^{n} x_i.$$

Notice that if all the x_i happen to have a common value c, then their average, of course, also equals c.

This concept leads one to define the average values of functions as follows.

Average Values The ***average value*** of a function of one variable on the interval $[a, b]$ is defined by

$$[f]_{\mathrm{av}} = \frac{\int_a^b f(x)\, dx}{b - a}.$$

Likewise, for functions of two variables, the ratio of the integral to the area of D,

$$[f]_{\mathrm{av}} = \frac{\iint_D f(x, y)\, dx\, dy}{\iint_D dx\, dy}, \tag{1}$$

is called the ***average value*** of f over D. Similarly, the ***average value*** of a function f on a region W in three space is defined by

$$[f]_{\mathrm{av}} = \frac{\iiint_W f(x, y, z)\, dx\, dy\, dz}{\iiint_W dx\, dy\, dz}.$$

Again, notice that the denominator is chosen so that if f is a constant, say c, then $[f]_{\mathrm{av}} = c$.

EXAMPLE 1 Find the average value of $f(x, y) = x \sin^2(xy)$ on the region $D = [0, \pi] \times [0, \pi]$.

SOLUTION First, we compute

$$\begin{aligned}
\iint_D f(x, y)\, dx\, dy &= \int_0^{\pi} \int_0^{\pi} x \sin^2(xy)\, dx\, dy \\
&= \int_0^{\pi} \left[\int_0^{\pi} \frac{1 - \cos(2xy)}{2} x\, dy \right] dx \\
&= \int_0^{\pi} \left[\frac{y}{2} - \frac{\sin(2xy)}{4x} \right] x \Bigg|_{y=0}^{\pi} dx \\
&= \int_0^{\pi} \left[\frac{\pi x}{2} - \frac{\sin(2\pi x)}{4} \right] dx = \left[\frac{\pi x^2}{4} + \frac{\cos(2\pi x)}{8\pi} \right] \Bigg|_0^{\pi} \\
&= \frac{\pi^3}{4} + \frac{\cos(2\pi^2) - 1}{8\pi}.
\end{aligned}$$

Thus, the average value of f, by formula (1), is

$$\frac{\pi^3/4 + [\cos(2\pi^2) - 1]/8\pi}{\pi^2} = \frac{\pi}{4} + \frac{\cos(2\pi^2) - 1}{8\pi^3} \approx 0.7839. \quad \blacktriangle$$

EXAMPLE 2 The temperature at points in the cube $W = [-1, 1] \times [-1, 1] \times [-1, 1]$ is proportional to the square of the distance from the origin.

(a) What is the average temperature?

(b) At which points of the cube is the temperature equal to the average temperature?

SOLUTION (a) Let c be the constant of proportionality so $T = c(x^2 + y^2 + z^2)$ and the average temperature is $[T]_{\mathrm{av}} = \frac{1}{8} \iiint_W T\, dx\, dy\, dz$,because the volume of the cube is 8. Thus,

$$[T]_{\text{av}} = \frac{c}{8}\int_{-1}^{1}\int_{-1}^{1}\int_{-1}^{1}(x^2+y^2+z^2)\,dx\,dy\,dz.$$

The triple integral is the sum of the integrals of x^2, y^2, and z^2. Because x, y, and z enter symmetrically into the description of the cube, the three integrals will be equal, so that

$$[T]_{\text{av}} = \frac{3c}{8}\int_{-1}^{1}\int_{-1}^{1}\int_{-1}^{1} z^2\,dx\,dy\,dz = \frac{3c}{8}\int_{-1}^{1} z^2\left(\int_{-1}^{1}\int_{-1}^{1} dx\,dy\right)dz.$$

The inner integral is equal to the area of the square $[-1,1]\times[-1,1]$. The area of that square is 4, and so

$$[T]_{\text{av}} = \frac{3c}{8}\int_{-1}^{1} 4z^2\,dz = \frac{3c}{2}\left(\frac{z^3}{3}\right)\Bigg|_{-1}^{1} = c.$$

(b) The temperature is equal to the average temperature at all points satisfying $c(x^2+y^2+z^2)=c$, that is, which lie on the sphere $x^2+y^2+z^2=1$, which is inscribed in the cube W. ▲

Centers of Mass

If masses $m_1,\ldots,m_n$ are placed at points $x_1,\ldots,x_n$ on the x axis, their ***center of mass*** is defined to be

$$\bar{x} = \frac{\sum m_i x_i}{\sum m_i}. \tag{2}$$

This definition arises from the following observation: If one is balancing masses on a lever (Figure 6.3.1), the balance point $\bar{x}$ occurs where the total moment (mass times distance from the balance point) is zero, that is, where $\sum m_i(x_i-\bar{x})=0$. A physical principle, going back first to Archimedes and then in this generality to Newton, states that this condition means there is no tendency for the lever to rotate.

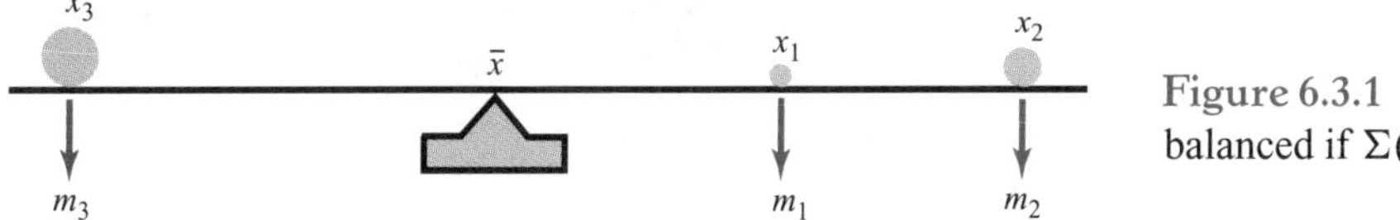

Figure 6.3.1 The lever is balanced if $\Sigma(x_i-\bar{x})m_i = 0$.

For a continuous mass density $\delta(x)$ along the lever (measured in, say, grams/cm), the analog of formula (2) is

$$\bar{x} = \frac{\int x\delta(x)\,dx}{\int \delta(x)\,dx}. \tag{3}$$

For two-dimensional plates, this generalizes to:

The Center of Mass of Two-Dimensional Plates

$$\bar{x} = \frac{\iint_D x\delta(x,y)\,dx\,dy}{\iint_D \delta(x,y)\,dx\,dy} \quad \text{and} \quad \bar{y} = \frac{\iint_D y\delta(x,y)\,dx\,dy}{\iint_D \delta(x,y)\,dx\,dy}, \tag{4}$$

where again $\delta(x, y)$ is the mass density (see Figure 6.3.2).

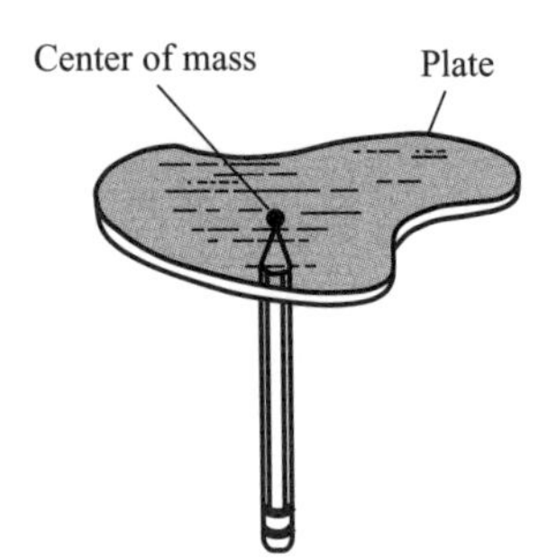

Figure 6.3.2 The plate balances when supported at its center of mass.

EXAMPLE 3 Find the center of mass of the rectangle $[0, 1] \times [0, 1]$ if the mass density is e^{x+y}.

SOLUTION First we compute the total mass:

$$\begin{aligned}\iint_D e^{x+y}\,dx\,dy &= \int_0^1\int_0^1 e^{x+y}\,dx\,dy = \int_0^1 \left(e^{x+y}|_{x=0}^1\right)dy = \int_0^1 (e^{1+y} - e^y)\,dy\\ &= (e^{1+y} - e^y)|_{y=0}^1 = e^2 - e - (e-1) = e^2 - 2e + 1.\end{aligned}$$

The numerator in formula (4) for $\bar{x}$ is

$$\begin{aligned}\int_0^1\int_0^1 xe^{x+y}\,dx\,dy &= \int_0^1 (xe^{x+y} - e^{x+y})|_{x=0}^1\,dy = \int_0^1 [e^{1+y} - e^{1+y} - (0e^y - e^y)]\,dy\\ &= \int_0^1 e^y\,dy = e^y|_{y=0}^1 = e - 1,\end{aligned}$$

so that

$$\bar{x} = \frac{e-1}{e^2 - 2e + 1} = \frac{e-1}{(e-1)^2} = \frac{1}{e-1} \approx 0.582.$$

The roles of x and y may be interchanged in all these calculations, so that $\bar{y} = 1/(e-1) \approx 0.582$ as well. ▲

For a region W in space with mass density $\delta(x, y, z)$, we know that

$$\text{volume} = \iiint_W dx\,dy\,dz, \tag{5}$$

$$\text{mass} = \iiint_W \delta(x, y, z)\,dx\,dy\,dz. \tag{6}$$

If one denotes the coordinates of the center of mass by $(\bar{x}, \bar{y}, \bar{z})$, then the generalization of the formulas in the preceding box are as follows.

Coordinates for the Center of Mass of Three-Dimensional Regions

$$\bar{x} = \frac{\iiint_W x\delta(x, y, z)\,dx\,dy\,dz}{\text{mass}},$$

$$\bar{y} = \frac{\iiint_W y\delta(x, y, z)\,dx\,dy\,dz}{\text{mass}}, \tag{7}$$

$$\bar{z} = \frac{\iiint_W z\delta(x, y, z)\,dx\,dy\,dz}{\text{mass}}.$$

EXAMPLE 4 The cube $[1, 2] \times [1, 2] \times [1, 2]$ has mass density given by $\delta(x, y, z) = (1 + x)e^z y$. Find the total mass of the box.

SOLUTION The mass of the box is, by formula (6),

$$\int_1^2 \int_1^2 \int_1^2 (1 + x)e^z y\,dx\,dy\,dz = \int_1^2 \int_1^2 \left[\left(x + \frac{x^2}{2}\right) e^z y\right]_{x=1}^{x=2} dy\,dz$$

$$= \int_1^2 \int_1^2 \frac{5}{2} e^z y\,dy\,dz = \int_1^2 \frac{15}{4} e^z\,dz = \left[\frac{15}{4} e^z\right]_{z=1}^{z=2} = \frac{15}{4}(e^2 - e). \quad \blacktriangle$$

If a region and its mass density are reflection-symmetric across a plane, then the center of mass lies on that plane. For example, in formula (7) for $\bar{x}$, if the region and mass density are symmetric in the yz plane, then the integrand is odd in x, and so $\bar{x} = 0$. This kind of use of symmetry is illustrated in the next example.

EXAMPLE 5 Find the center of mass of the hemispherical region W defined by the inequalities $x^2 + y^2 + z^2 \leq 1, z \geq 0$. (Assume that the density is unity.)

SOLUTION By symmetry, the center of mass must lie on the z axis, and so $\bar{x} = \bar{y} = 0$. To find $\bar{z}$, we must compute, by formula (7), the numerator $I = \iiint_W z\,dx\,dy\,dz$. The hemisphere is an elementary region, and thus the integral becomes

$$I = \int_0^1 \int_{-\sqrt{1-z^2}}^{\sqrt{1-z^2}} \int_{-\sqrt{1-y^2-z^2}}^{\sqrt{1-y^2-z^2}} z\,dx\,dy\,dz.$$

Because z is a constant for the x and y integrations, we can remove it from the first two integral signs, to obtain

$$I = \int_0^1 z \left(\int_{-\sqrt{1-z^2}}^{\sqrt{1-z^2}} \int_{-\sqrt{1-y^2-z^2}}^{\sqrt{1-y^2-z^2}} dx\,dy \right) dz.$$

Instead of calculating the inner two integrals explicitly, we observe that they equal the double integral $\iint_D dx\,dy$ over the disk $x^2 + y^2 \leq 1 - z^2$, considered as an x-simple region in the plane. The area of this disk is $\pi(1 - z^2)$, and so

$$I = \pi \int_0^1 z(1 - z^2)\,dz = \pi \int_0^1 (z - z^3)\,dz = \pi \left[\frac{z^2}{2} - \frac{z^4}{4} \right]_0^1 = \frac{\pi}{4}.$$

The volume of the hemisphere is $\frac{2}{3}\pi$, and so $\bar{z} = (\pi/4)/(\frac{2}{3}\pi) = \frac{3}{8}$. ▲

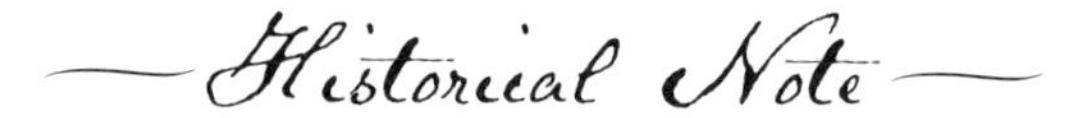

It is common knowledge that Archimedes observed the principle of the lever. Perhaps less known is that he was also responsible for discovering the concepts of center of mass and center of gravity. Only two of his works on mechanics have been handed down to us: *On Floating Bodies* and *On the Equilibrium and Centers of Mass of Plane Figures*. Both were translated into Latin by Niccolo Tartaglia, circa 1543.

In *Equilibrium*..., Archimedes began the field of applied mathematics, doing for mechanics what Euclid had accomplished for geometry. In this work he describes the principles behind all the machines of antiquity, including the lever, inclined plane, and pulley system.

Surprisingly, Archimedes never carefully defined the center of mass; the first proper definition was given by Pappus of Alexandria in 340 C.E. The concept of equilibrium was to have a profound effect on the development of mechanical engineering (through the introduction of gears), architecture, and in art, permitting the construction of complex machines, large-scale buildings, and sculptures. Figure 6.3.3 shows sketches by Leonardo DaVinci, illustrating equilibrium positions of the human body.

Figure 6.3.3 Equilibrium positions of the human body, to be observed by the painter. The center of mass should be supported to maintain equilibrium.

Moments of Inertia

Another important concept in mechanics, one that is needed in studying the dynamics of a rotating rigid body, is that of *moment of inertia*. If the solid W has uniform density δ, the ***moments of inertia*** I_x, I_y, and I_z about the x, y, and z axes, respectively, are defined by:

Moments of Inertia About the Coordinate Axes

$$I_x = \iiint_W (y^2 + z^2)\,\delta\,dx\,dy\,dz, \qquad I_y = \iiint_W (x^2 + z^2)\,\delta\,dx\,dy\,dz,$$
$$I_z = \iiint_W (x^2 + y^2)\,\delta\,dx\,dy\,dz. \tag{8}$$

The moment of inertia measures a body's response to efforts to rotate it; for example, as when one tries to rotate a merry-go-round. The moment of inertia is analogous to the mass of a body, which measures its response to efforts to translate it. In contrast to translational motion, however, the moments of inertia *depend on the shape and not just the total mass*. It is harder to spin up a large plate than a compact ball of the same mass.

For example, I_x measures the body's response to forces attempting to rotate it about the x axis. The factor $y^2 + z^2$, which is the square of the distance to the x axis, weights masses farther away from the rotation axis more heavily. This is in agreement with the intuition just explained.

EXAMPLE 6 Compute the moment of inertia I_z for the solid above the xy plane bounded by the paraboloid $z = x^2 + y^2$ and the cylinder $x^2 + y^2 = a^2$, assuming a and the mass density to be constants.

SOLUTION The paraboloid and cylinder intersect at the plane $z = a^2$. Using cylindrical coordinates, we find from equation (8),

$$I_z = \int_0^a \int_0^{2\pi} \int_0^{r^2} \delta r^2 \cdot r\,dz\,d\theta\,dr = \delta \int_0^a \int_0^{2\pi} \int_0^{r^2} r^3\,dz\,d\theta\,dr = \frac{\pi\delta a^6}{3}. \quad \blacktriangle$$

Gravitational Fields of Solid Objects

Another interesting physical application of triple integration is the determination of the gravitational fields of solid objects. Example 6, Section 2.6, showed that the gravitational force field $\mathbf{F}(x, y, z)$ of a particle is the negative of the gradient of a function $V(x, y, z)$ called the ***gravitational potential***. If there is a point mass M at (x, y, z), then the gravitational potential acting on a mass m at (x_1, y_1, z_1) due to this mass is $-GmM[(x - x_1)^2 + (y - y_1)^2 + (z - z_1)^2]^{-1/2}$, where G is the universal gravitational constant.

If our attracting object occupies a domain W with mass density $\delta(x, y, z)$, we may think of it as made of infinitesimal box-shaped regions with masses $dM = \delta(x, y, z)\,dx\,dy\,dz$ located at points (x, y, z). The total gravitational potential V for W is then obtained by "summing" the potentials from the infinitesimal masses. Thus, we arrive at the triple integral (see Figure 6.3.4):

$$V(x_1, y_1, z_1) = -Gm \iiint_W \frac{\delta(x, y, z)\,dx\,dy\,dz}{\sqrt{(x - x_1)^2 + (y - y_1)^2 + (z - z_1)^2}}. \tag{9}$$

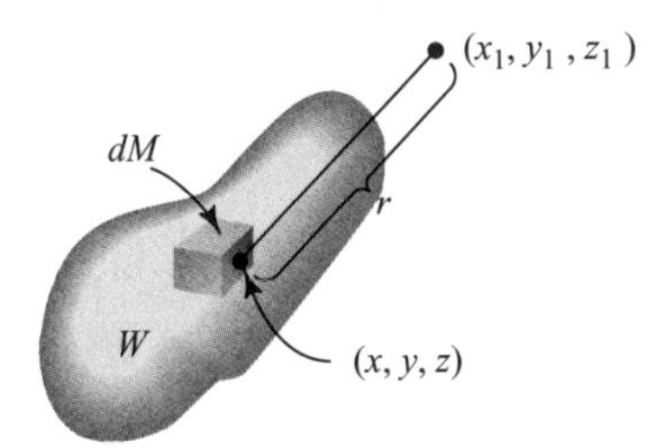

Figure 6.3.4 The gravitational potential that produces a force acting on a mass m at (x_1, y_1, z_1) arising from the mass $dM = \delta(x, y, z)\,dx\,dy\,dz$ at (x, y, z) is $-[Gm\delta(x, y, z)\,dx\,dy\,dz]/r$.

Historical Note

The theory of gravitational force fields and gravitational potentials was developed by Sir Isaac Newton (1642–1727). Newton withheld publication of his gravitational theories for quite some time. The result that a spherical planet has the same gravitational field as it would have if its mass were all concentrated at the planet's center first appeared in his famous *Philosophiae Naturalis Principia Mathematica,* the first edition of which appeared in 1687. Using multiple integrals and spherical coordinates, we shall solve Newton's problem here; remarkably, Newton's published solution used only Euclidean geometry.

EXAMPLE 7 Let W be a region of constant density and total mass M. Show that the gravitational potential is given by

$$V(x_1, y_1, z_1) = \left[\frac{1}{r}\right]_{\text{av}} GMm,$$

where $[1/r]_{\text{av}}$ is the average over W of

$$f(x, y, z) = \frac{1}{\sqrt{(x - x_1)^2 + (y - y_1)^2 + (z - z_1)^2}}.$$

SOLUTION According to formula (9),

$$
\begin{aligned}
-V(x_1, y_1, z_1) &= Gm \iiint_W \frac{\delta\, dx\, dy\, dz}{\sqrt{(x-x_1)^2 + (y-y_1)^2 + (z-z_1)^2}} \\
&= Gm\delta \iiint_W \frac{dx\, dy\, dz}{\sqrt{(x-x_1)^2 + (y-y_1)^2 + (z-z_1)^2}} \\
&= Gm[\delta \text{ volume } (W)] \frac{\displaystyle\iiint_W \frac{dx\, dy\, dz}{\sqrt{(x-x_1)^2 + (y-y_1)^2 + (z-z_1)^2}}}{\text{volume } (W)} \\
&= GmM \left[\frac{1}{r}\right]_{\text{av}}
\end{aligned}
$$

are required. ▲

Let us now use formula (9) and spherical coordinates to find the gravitational potential $V(x_1, y_1, z_1)$ for a region W with constant density between the concentric spheres $\rho = \rho_1$ and $\rho = \rho_2$, assuming the density is constant. Before evaluating the integral in formula (9), we make some observations that will simplify the computation. Because G, m, and the density are constants, we may ignore them at first. Because the attracting body, W, is symmetric with respect to rotations about the origin, the potential $V(x_1, y_1, z_1)$ should itself be symmetric—thus, $V(x_1, y_1, z_1)$ depends only on the distance $R = \sqrt{x_1^2 + y_1^2 + z_1^2}$ from the origin. Our computation will be simplest if we look at the point $(0, 0, R)$ on the z axis (see Figure 6.3.5). Thus, we need to evaluate the integral

$$
V(0, 0, R) = -\iiint_W \frac{dx\, dy\, dz}{\sqrt{x^2 + y^2 + (z-R)^2}}.
$$

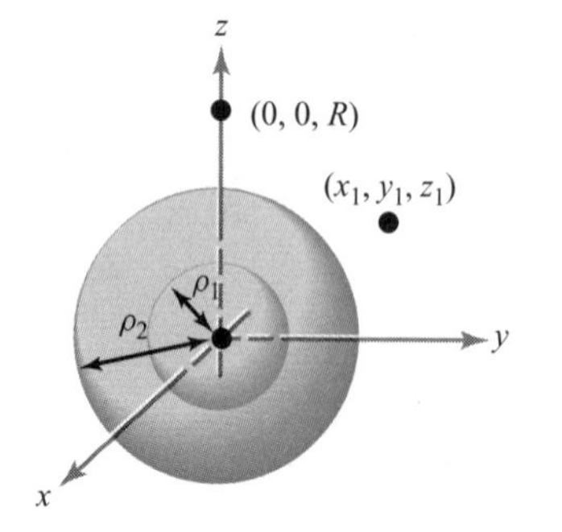

Figure 6.3.5 The gravitational potential at (x_1, y_1, z_1) is the same as at $(0, 0, R)$, where $R = \sqrt{x_1^2 + y_1^2 + z_1^2}$.

In spherical coordinates, W is described by the inequalities $\rho_1 \le \rho \le \rho_2$, $0 \le \theta \le 2\pi$, and $0 \le \phi \le \pi$, and so

$$-V(0,0,R) = \int_{\rho_1}^{\rho_2} \int_0^{\pi} \int_0^{2\pi} \frac{\rho^2 \sin\phi \, d\theta \, d\phi \, d\rho}{\sqrt{\rho^2 \sin^2\phi(\cos^2\theta + \sin^2\theta) + (\rho\cos\phi - R)^2}}.$$

Replacing $\cos^2\theta + \sin^2\theta$ by 1, so that the integrand no longer involves θ, we may integrate over θ to get

$$-V(0,0,R) = 2\pi \int_{\rho_1}^{\rho_2} \int_0^{\pi} \frac{\rho^2 \sin\phi \, d\phi \, d\rho}{\sqrt{\rho^2 \sin^2\phi + (\rho\cos\phi - R)^2}}$$

$$= 2\pi \int_{\rho_1}^{\rho_2} \rho^2 \left(\int_0^{\pi} \frac{\sin\phi \, d\phi}{\sqrt{\rho^2 - 2R\rho\cos\phi + R^2}} \right) d\rho.$$

The inner integral over ϕ may be evaluated using the substitution $u = -2R\rho\cos\phi$. We get

$$\frac{1}{2R\rho} \int_{-2R\rho}^{2R\rho} (\rho^2 + u + R^2)^{-1/2} \, du = \frac{2}{2R\rho}(\rho^2 + u + R^2)^{1/2} \Big|_{-2R\rho}^{2R\rho}$$

$$= \frac{1}{R\rho}[(\rho^2 + 2R\rho + R^2)^{1/2} - (\rho^2 - 2R\rho + R^2)^{1/2}$$

$$= \frac{1}{R\rho}\{[(\rho + R)^2]^{1/2} - [(\rho - R)^2]^{1/2}\}$$

$$= \frac{1}{R\rho}(\rho + R - |\rho - R|).$$

The expression $\rho + R$ is always positive, but $\rho - R$ may not be, so we must keep the absolute value sign. Substituting into the formula for V, we get

$$-V(0,0,R) = 2\pi \int_{\rho_1}^{\rho_2} \frac{\rho^2}{R\rho}(\rho + R - |\rho - R|) \, d\rho = \frac{2\pi}{R} \int_{\rho_1}^{\rho_2} \rho(\rho + R - |\rho - R|) \, d\rho.$$

We consider two possibilities for R, corresponding to the gravitational potential for objects *outside* and *inside* the hollow ball W.

Case 1. If $R \ge \rho_2$ [that is, if (x_1, y_1, z_1) is outside W], then $|\rho - R| = R - \rho$ for all ρ in the interval $[\rho_1, \rho_2]$, so that

$$-V(0,0,R) = \frac{2\pi}{R} \int_{\rho_1}^{\rho_2} \rho[\rho + R - (R - \rho)] \, d\rho = \frac{4\pi}{R} \int_{\rho_1}^{\rho_2} \rho^2 \, d\rho = \frac{1}{R}\frac{4\pi}{3}(\rho_2^3 - \rho_1^3).$$

The factor $(4\pi/3)(\rho_2^3 - \rho_1^3)$ equals the volume of W. Putting back the constants G, m, and the mass density, we find that *the gravitational potential* is $-GmM/R$, *where* M *is the mass of* W. *Thus,* V *is just as it would be if all the mass of* W *were concentrated at the central point.*

Case 2. If $R \le \rho_1$ [that is, if (x_1, y_1, z_1) is inside the hole], then $|\rho - R| = \rho - R$ for ρ in $[\rho_1, \rho_2]$, and so

$$\begin{aligned}-V(0,0,R) &= (Gm)\frac{2\pi}{R}\int_{\rho_1}^{\rho_2} \rho[\rho + R - (\rho - R)]\, d\rho = (Gm)4\pi\int_{\rho_1}^{\rho_2} \rho\, d\rho \\ &= (Gm)2\pi(\rho_2^2 - \rho_1^2).\end{aligned}$$

The result is independent of R, and so the potential V is *constant* inside the hole. Because the gravitational force is minus the gradient of V, we conclude that *there is no gravitational force inside a uniform hollow planet*!

We leave it to the reader to compute $V(0, 0, R)$ for the case $\rho_1 < R < \rho_2$.

A similar argument shows that the gravitational potential outside any *spherically symmetric* body of mass M (even if the density is variable) is $V = GMm/R$, where R is the distance to its center (which is its center of mass).

EXAMPLE 8 Find the gravitational potential acting on a unit mass of a spherical star with a mass $M = 3.02 \times 10^{30}$ kg at a distance of 2.25×10^{11} m from its center ($G = 6.67 \times 10^{-11}\ \mathrm{N \cdot m^2/kg^2}$).

SOLUTION The negative potential is

$$-V = \frac{GM}{R} = \frac{6.67 \times 10^{-11} \times 3.02 \times 10^{30}}{2.25 \times 10^{11}} = 8.95 \times 10^8\ \mathrm{m^2/s^2}. \quad \blacktriangle$$

EXERCISES

1. Find the average of $f(x, y) = y \sin xy$ over $D = [0, \pi] \times [0, \pi]$.

2. Find the average of $f(x, y) = e^{x+y}$ over the triangle with vertices $(0, 0)$, $(0, 1)$, and $(1, 0)$.

3. Find the center of mass of the region between $y = x^2$ and $y = x$ if the density is $x + y$.

4. Find the center of mass of the region between $y = 0$ and $y = x^2$, where $0 \le x \le \frac{1}{2}$.

5. A sculptured gold plate D is defined by $0 \le x \le 2\pi$ and $0 \le y \le \pi$ (centimeters) and has mass density $\delta(x, y) = y^2 \sin^2 4x + 2$ (grams per square centimeter). If gold sells for \$7 per gram, how much is the gold in the plate worth?

6. In Exercise 5, what is the average mass density in grams per square centimeter?

7. (a) Find the mass of the box $[0, \frac{1}{2}] \times [0, 1] \times [0, 2]$, assuming the density to be uniform.
(b) Same as part (a), but with a mass density $\delta(x, y, z) = x^2 + 3y^2 + z + 1$.

8. Find the mass of the solid bounded by the cylinder $x^2 + y^2 = 2x$ and the cone $z^2 = x^2 + y^2$ if the density is $\delta = \sqrt{x^2 + y^2}$.

9. Find the center of mass of the region bounded by $x + y + z = 2$, $x = 0$, $y = 0$, and $z = 0$, assuming the density to be uniform.

10. Find the center of mass of the cylinder $x^2 + y^2 \leq 1$, $1 \leq z \leq 2$ if the density is $\delta = (x^2 + y^2)z^2$.

11. Find the average value of $\sin^2 \pi z \cos^2 \pi x$ over the cube $[0, 2] \times [0, 4] \times [0, 6]$.

12. Find the average value of e^{-z} over the ball $x^2 + y^2 + z^2 \leq 1$.

13. A solid with constant density is bounded above by the plane $z = a$ and below by the cone described in spherical coordinates by $\phi = k$, where k is a constant $0 < k < \pi/2$. Set up an integral for its moment of inertia about the z axis.

14. Find the moment of inertia around the y axis for the ball $x^2 + y^2 + z^2 \leq R^2$ if the mass density is a constant δ.

15. Find the gravitational potential on a mass m of a spherical planet with mass $M = 3 \times 10^{26}$ kg, at a distance of 2×10^8 m from its center.

16. Find the gravitational force exerted on a 70-kg object at the position in Exercise 15.

17. A body W in xyz coordinates is called *symmetric with respect to a given plane* if for every particle on one side of the plane there is a particle of equal mass located at its mirror image through the plane.

(a) Discuss the planes of symmetry for an automobile shell.

(b) Let the plane of symmetry be the xy plane, and denote by W^+ and W^- the portions of W above and below the plane, respectively. By our assumption, the mass density $\delta(x, y, z)$ satisfies $\delta(x, y, -z) = \delta(x, y, z)$. Justify the following steps:

$$\begin{aligned}
\bar{z} \cdot \iiint_W \delta(x, y, z)\, dx\, dy\, dz &= \iiint_W z\delta(x, y, z)\, dx\, dy\, dz \\
&= \iiint_{W^+} z\delta(x, y, z)\, dx\, dy\, dz + \iiint_{W^-} z\delta(x, y, z)\, dx\, dy\, dz \\
&= \iiint_{W^+} z\delta(x, y, z)\, dx\, dy\, dz + \iiint_{W^+} -w\delta(u, v, -w)\, du\, dv\, dw \\
&= 0.
\end{aligned}$$

(c) Explain why part (b) proves that if a body is symmetrical with respect to a plane, then its center of mass lies in that plane.

(d) Derive this law of mechanics: *If a body is symmetric with respect to two planes, then its center of mass lies on their line of intersection.*

18. A uniform rectangular steel plate of sides a and b rotates about its center of mass with constant angular velocity ω.

(a) The kinetic energy equals $\frac{1}{2}$(mass)(velocity)^2. Argue that the kinetic energy of any element of mass $\delta\, dx\, dy$ (δ = constant) is given by $\delta(\omega^2/2)(x^2+y^2)\, dx\, dy$, provided the origin (0, 0) is placed at the center of mass of the plate.

(b) Justify the formula for kinetic energy:

$$\text{K.E.} = \iint_{\text{plate}} \delta \frac{\omega^2}{2}(x^2+y^2)\, dx\, dy.$$

(c) Evaluate the integral, assuming that the plate is described by the inequalities $-a/2 \le x \le a/2,\ -b/2 \le y \le b/2$.

19. As is well known, the density of a typical planet is not constant throughout the planet. Assume that planet C.M.W. has a radius of 5×10^8 cm and a mass density (in grams per cubic centimeter)

$$\rho(x, y, z) = \begin{cases} \dfrac{3 \times 10^4}{r}, & r \ge 10^4 \text{ cm}, \\ 3, & r \le 10^4 \text{ cm}, \end{cases}$$

where $r = \sqrt{x^2+y^2+z^2}$. Find a formula for the gravitational potential outside C.M.W.

6.4 Improper Integrals

In this section, we study improper integrals—that is, integrals in which the function may be unbounded or the region of integration is unbounded. We shall first recall the situation for functions of one variable.

One-Variable Improper Integrals

In the study of integrals of functions of one variable, one encounters various types of "improper" integrals; that is, integrals of unbounded functions defined on intervals or integrals of functions over unbounded intervals. For example,

$$\int_0^1 \frac{1}{\sqrt{x}}\, dx \quad \text{and} \quad \int_1^\infty \frac{dx}{x^2}$$

are improper integrals. They are evaluated using a limiting process; for instance,

$$\int_0^1 \frac{1}{\sqrt{x}}\, dx = \lim_{a\to 0} \int_a^1 \frac{1}{\sqrt{x}}\, dx = \lim_{a\to 0} \left(2\sqrt{x}\Big|_a^1\right) = \lim_{a\to 0} (2 - 2\sqrt{a}) = 2$$

and

$$\int_1^\infty \frac{dx}{x^2} = \lim_{b\to\infty} \int_1^b \frac{dx}{x^2} = \lim_{b\to\infty} \left(-\frac{1}{x}\bigg|_1^b \right) = \lim_{b\to\infty} \left(1 - \frac{1}{b}\right) = 1.$$

If, in such a limiting process, the limit does not exist (or is infinite), we say that the integral does not exist (or that the integral diverges).

Improper Integrals in the Plane

Next, we describe three types of improper integrals of two variables over a region D. The first two types are described in the text below, and the third type (integrals over unbounded regions) is left to the exercises. We will evaluate all integrals using a limiting process, as in the one-variable case.

For simplicity of exposition, we first restrict ourselves to nonnegative functions f—that is, $f(x, y) \geq 0$ for all points $(x, y) \in D$—and to y-simple regions described as the set of (x, y) such that

$$a \leq x \leq b, \qquad \phi_1(x) \leq y \leq \phi_2(x),$$

as in Figure 6.4.1.

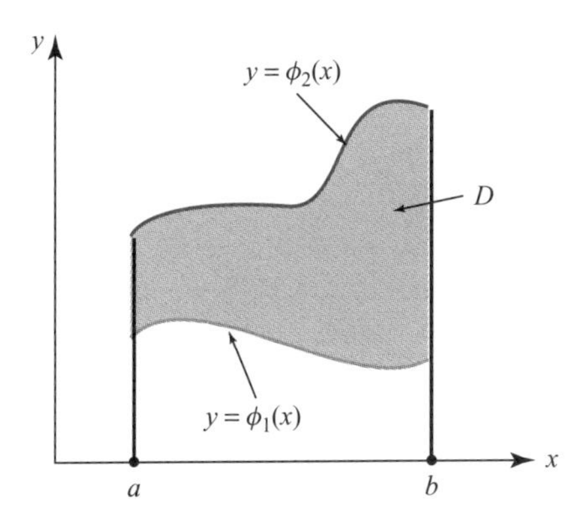

Figure 6.4.1 A y-simple domain.

In the first case we wish to consider, let's assume that $f: D \to \mathbb{R}$ is continuous except for points on the boundary of D. Consider, for example,

$$f(x, y) = \frac{1}{\sqrt{1 - x^2 - y^2}},$$

where D is the unit disk $D = \{(x, y) | x^2 + y^2 \leq 1\}$. Clearly, f is not defined on the boundary of D, where $x^2 + y^2 = 1$; yet it will be of practical interest to be able to evaluate $\iint_D f(x, y)\, dA$, because this integral represents the area of the upper hemisphere of the unit sphere in three space.

Exhausting Regions

Our basic idea will be to integrate such an f over a smaller region D', where we know the integral exists, and then let D' "tend" to D; that is, "exhaust" D and see if $\iint_D f\, dA$ tends to some limit. With this in mind, we pick a special kind of D', as follows.

Let $\eta > 0$ be small enough so that $a + \eta < b - \eta$. Let $\delta > 0$ be small enough so that $\phi_1(x) + \delta < \phi_2(x) - \delta$ for all x, $a \le x \le b$ (see Figure 6.4.2). If $\phi_2(x) = \phi_1(x)$ for some x, no such δ will exist, but we shall worry about this minor issue when it arises in our later examples. Then the region

$$D_{\eta,\delta} = \{(x, y) | a + \eta \le x \le b - \eta \quad \text{and} \quad \phi_2(x) + \delta \le y \le \phi_1(x) - \delta\}$$

is a subset of D, and as $(\eta, \delta) \to 0$, $D_{\eta,\delta}$ tends to D.

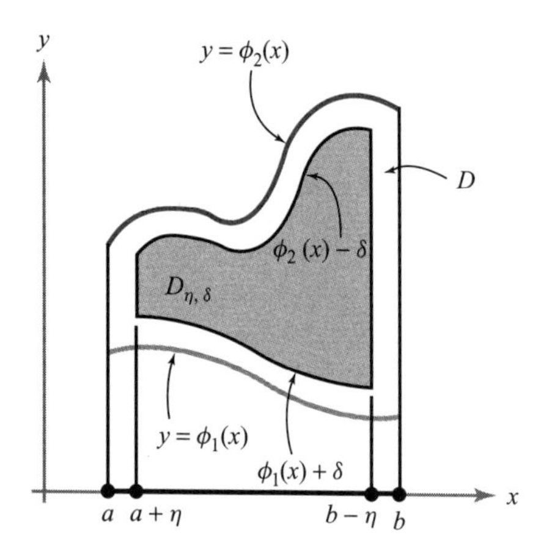

Figure 6.4.2 A shrunken domain $D_{\eta,\delta}$ for improper integrals.

Improper Integrals as Limits

Because f is continuous and bounded on $D_{\eta,\delta}$, the integral $\iint D_{\eta,\delta} f\, dA$ exists. We can now ask what happens as the region $D_{\eta,\delta}$ expands to fill the region, D—that is, as $(\eta, \delta) \to (0, 0)$. Provided that

$$\lim_{(\eta,\delta)\to(0,0)} \iint_{D_{\eta,\delta}} f\, dA$$

exists, we say that the integral of f over D is ***convergent*** or that f is ***integrable*** over D, and we define $\iint_D f\, dx\, dy$ to be equal to this limit.

EXAMPLE 1 Evaluate

$$\iint_D \frac{1}{\sqrt[3]{xy}}\, dA,$$

where D is the unit square $[0, 1] \times [0, 1]$.

SOLUTION D is clearly a y-simple region. Choose $\eta > 0$ and $\delta > 0$ so that $D_{\eta,\delta} \subset D$, as in Figure 6.4.3. Then, by Fubini's theorem:

$$\begin{aligned}\iint_{D_{\eta,\delta}} \frac{1}{\sqrt[3]{xy}}\, dA &= \int_{\eta}^{1-\eta} \int_{\delta}^{1-\delta} \frac{1}{\sqrt[3]{xy}}\, dy\, dx \\ &= \int_{\eta}^{1-\eta} \frac{1}{\sqrt[3]{x}}\, dx \int_{\delta}^{1-\delta} \frac{1}{\sqrt[3]{y}}\, dy \\ &= \frac{3}{2}\left((1-\eta)^{2/3} - \eta^{2/3}\right) \cdot \frac{3}{2}\left((1-\delta)^{2/3} - \delta^{2/3}\right).\end{aligned}$$

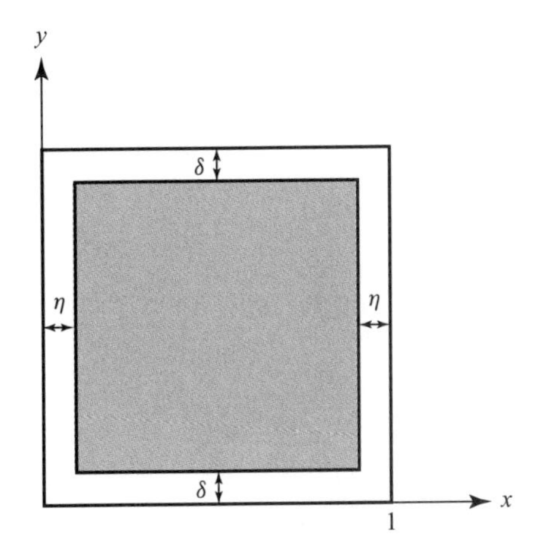

Figure 6.4.3 The slightly shrunken unit square.

Letting $(\eta, \delta) \to (0, 0)$, we see that

$$\lim_{(\eta,\delta)\to(0,0)} \iint_{D_{\eta,\delta}} \frac{1}{\sqrt[3]{xy}}\, dy\, dx = \frac{3}{2}\frac{3}{2} = \frac{9}{4}. \quad \blacktriangle$$

Unfortunately, it may not always be possible to evaluate such limits so directly and simply. This is often the case in the most interesting examples, as with the surface area of the hemisphere, mentioned earlier. It's as if the "real world" always presents the greatest challenges to the mathematician! So let us expand a bit on our theoretical discussion.

Improper Integrals as Limits of Iterated Integrals

Suppose f is integrable over $D_{\eta,\delta}$. We can then apply Fubini's theorem to obtain

$$\iint_{D_{\eta,\delta}} f\, dA = \int_{a+\eta}^{b-\eta} \int_{\phi_1(x)+\delta}^{\phi_2(x)-\delta} f(x, y)\, dy\, dx.$$

Hence, if f is integrable over D,

$$\iint_D f\,dA = \lim_{(\eta,\delta)\to(0,0)} \int_{a+\eta}^{b-\eta} \int_{\phi_1(x)+\delta}^{\phi_2(x)-\delta} f(x, y)\,dy\,dx. \tag{1}$$

Now $F(\eta, \delta) = \iint_{D_{\eta,\delta}} f\,dA$ is a function of two variables, η and δ, because as we change η and δ, we get another number. Now if f is integrable, then

$$\lim_{(\eta,\delta)\to 0} F(\eta, \delta) = L$$

exists. It follows that the iterated limits

$$\lim_{\eta\to 0}\lim_{\delta\to 0} F(\eta, \delta) \qquad \text{and} \qquad \lim_{\delta\to 0}\lim_{\eta\to 0} F(\eta, \delta)$$

also exist and are both equal to L, which in our case is $\iint_D f dA$. Thus, the iterated limit

$$\lim_{\eta\to 0}\lim_{\delta\to 0} \int_{a+\eta}^{b-\eta} \int_{\phi_1(x)+\delta}^{\phi_1(x)-\delta} f(x, y)\,dy\,dx$$

also exists. Conversely, if the iterated limits exist, it does not generally follow that the limit $\lim_{(\eta,\delta)\to(0,0)} F(\eta, \delta)$ exists.

For example, if it were to turn out in some way that $F(\eta, \delta) = \eta\delta/(\eta^2 + \delta^2)$, then $\lim_{\eta\to 0}\lim_{\delta\to 0} F(\eta, \delta) = \lim_{\delta\to 0}\lim_{\eta\to 0} F(\eta, \delta) = 0$; yet $\lim_{(\eta,\delta)\to 0} F(\eta, \delta)$ does not exist, because $F(\eta, \eta) = 1/2$ (see Section 2.2).

In view of this, consider expression (1) again. If f is integrable, then

$$\begin{aligned}\iint_D f(x, y)\,dA &= \lim_{(\eta,\delta)\to(0,0)} \int_{a+\eta}^{b-\eta} \int_{\phi_1(x)+\delta}^{\phi_2(x)-\delta} f(x, y)\,dy\,dx \\ &= \lim_{\eta\to(0)}\lim_{\delta\to 0} \int_{a+\eta}^{b-\eta} \int_{\phi_1(x)+\delta}^{\phi_2(x)-\delta} f(x, y)\,dy\,dx.\end{aligned}$$

Now suppose that for each x,

$$\lim_{\delta\to 0} \int_{\phi_1(x)+\delta}^{\phi_2(x)-\delta} f(x, y)\,dy$$

exists. Denote this by $\int_{\phi_1(x)}^{\phi_2(x)} f(x, y)\,dy$. Suppose further that

$$\lim_{\eta\to 0} \int_{a+\eta}^{b-\eta} \int_{\phi_1(x)}^{\phi_2(x)} f(x, y)\,dy$$

also exists. We denote this limit by $\int_a^b \int_{\phi_1(x)}^{\phi_2(x)} f(x, y)\,dy\,dx$. Then if all limits exist, *all limits must be equal*. Thus, f is integrable and the iterated improper integral exists, then necessarily

$$\iint_D f(x, y)\, dA = \int_a^b \int_{\phi_1(x)}^{\phi_2(x)} f(x, y)\, dy\, dx.$$

However, is it possible that the existence of *just* the iterated integrals *implies* the integrability of f? We turn to this important question next.

Fubini's Theorem for Improper Integrals

For *integrals*, something truly *remarkable* happens. Unlike the case for iterated limits (as in the counterexample considered earlier), the existence of the iterated limits *does imply* the integrability of f as long as $f \geq 0$. Thus, if $f \geq 0$ and if $\int_a^b \int_{\phi_1(x)}^{\phi_2(x)} f(x, y)\, dy\, dx$ exists as an iterated limit, then f is integrable and

$$\iint_D f(x, y)\, dA = \int_a^b \int_{\phi_1(x)}^{\phi_2(x)} f(x, y)\, dy\, dx.$$

If D is an x-simple region with the x coordinate lying between two functions ψ_1 and ψ_2, and if

$$\int_c^d \int_{\psi_1(y)}^{\psi_2(y)} f(x, y)\, dx\, dy$$

exists as an improper integral, it again follows that f is integrable and

$$\iint_D f(x, y)\, dA = \int_c^d \int_{\psi_1(y)}^{\psi_2(y)} f(x, y)\, dx\, dy.$$

All these results, which are the *improper* analogues of Theorems 4 and 4′ in Section 5.3, are known as *Fubini's theorem* for improper integrals, which we formally state.

THEOREM 3: Fubini's Theorem Let D be an elementary region in the plane and $f \geq 0$ a function continuous except for points possibly on the boundary of D. If either of the integrals

$$\iint_D f(x, y)\, dA,$$

$$\int_a^b \int_{\phi_1(x)}^{\phi_2(x)} f(x, y)\, dy\, dx, \qquad \text{for } y\text{-simple regions}$$

$$\int_c^d \int_{\psi_1(y)}^{\psi_2(y)} f(x, y)\, dx\, dy \qquad \text{for } x\text{-simple regions}$$

exist as improper integrals, f is integrable and they are all equal.

The proof of this involves advanced concepts of analysis, so we omit it here. This result can be quite useful in calculation, as the next example shows.

EXAMPLE 2 Let $f(x, y) = 1/\sqrt{1 - x^2 - y^2}$. Show that f is integrable and that $\iint_D f(x, y)\, dA = 2\pi$, half the surface area of the unit sphere.

SOLUTION For $-1 < x < 1$, we have

$$\begin{aligned}
\int_{-\sqrt{1-x^2}}^{\sqrt{1-x^2}} \frac{dy}{1 - x^2 - y^2} &= \lim_{\delta \to 0} \int_{-\sqrt{1-x^2}+\delta}^{\sqrt{1+x^2}+\delta} \frac{dy}{1 - x^2 - y^2} \\
&= \lim_{\delta \to 0} \sin^{-1}\left(\frac{y}{\sqrt{1 - x^2}}\right)\Bigg|_{-\sqrt{1-x^2}+\delta}^{\sqrt{1-x^2}-\delta} \\
&= \lim_{\delta \to 0} \left\{ \sin^{-1}\left(1 - \frac{\delta}{\sqrt{1 - x^2}}\right) - \sin^{-1}\left(-1 + \frac{\delta}{\sqrt{1 - x^2}}\right) \right\} \\
&= \sin^{-1}(1) - \sin^{-1}(-1) = \frac{\pi}{2} - \frac{(-\pi)}{2} = \pi.
\end{aligned}$$

Clearly,

$$\lim_{\eta \to 0} \int_{-1+\eta}^{1-\eta} \int_{-\sqrt{1-x^2}}^{\sqrt{1-x^2}} \frac{dy\, dx}{\sqrt{1 - x^2 - y^2}} = \lim_{\eta \to 0} \int_{-1+\eta}^{1-\eta} \pi\, dx = \lim_{\eta \to 0} \pi(2 - 2n) = 2\pi.$$

Thus, f is integrable. To see why this theorem is so useful, try to show directly from the definition that f is integrable. It is not easy to do so! ▲

EXAMPLE 3 Let $f(x, y) = 1/(x - y)$ and let D be the set of (x, y) satisfying $0 \le x \le 1$ and $0 \le y \le x$. Show that f is *not* integrable over D.

SOLUTION Because the denominator of f is zero on the line $y = x$, f is unbounded on part of the boundary of D. Let $0 < \eta < 1$ and $0 < \delta < \eta$, and let $D_{\eta,\delta}$ be the set of (x, y) with $\eta \le x \le 1 - \eta$ and $\delta \le y \le x - \delta$ (Figure 6.4.4).

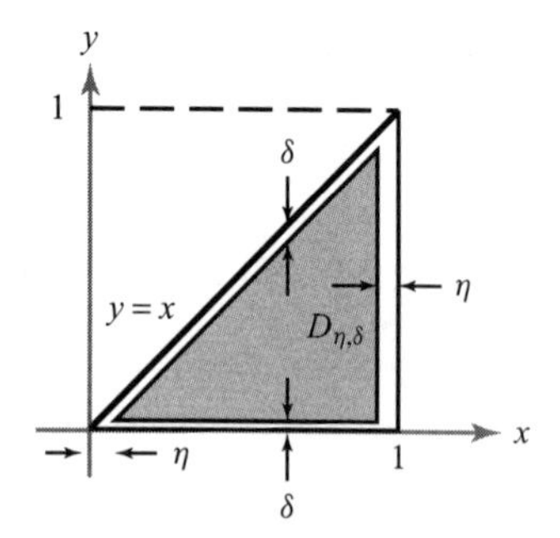

Figure 6.4.4 The shrunken domain $D_{\eta,\delta}$ for a triangular domain D.

Here the region D is y-simple with $\phi_1(x) = 0$, $\phi_2(x) = x$, and $\phi_1(0) = \phi_2(0)$. To ensure that $D_{\eta,\delta} \subset D$ and is depicted in the figure, we must choose δ a bit more carefully. A little geometry shows that we should choose $2\delta \leq \eta$. Consider

$$\begin{aligned}\iint_{D_{\eta,\delta}} f dA &= \int_{\eta}^{1-\eta}\int_{\delta}^{x-\delta} \frac{1}{x-y}\,dy\,dx\\ &= \int_{\eta}^{1-\eta} [-\log(x-y)]|_{y=\delta}^{x-\delta}\,dx\\ &= \int_{\eta}^{1-\eta} [-\log(\delta)+\log(x-\delta)]\,dx\\ &= [-\log\delta]\int_{\eta}^{1-\eta} dx + \int_{\eta}^{1-\eta}\log(x-\delta)\,dx\\ &= -(1-2\eta)\log\delta + [(x-\delta)\log(x-\delta)-(x-\delta)]|_{\eta}^{1-\eta}.\end{aligned}$$

In the last step, we used the fact that $\int \log u\, du = u \log u - u$. Continuing the preceding set of qualities, we have

$$\begin{aligned}\iint_{D_{\eta,\delta}} f\, dA = &-(1-2\eta)\log\delta + (1-\eta-\delta)\log(1-\eta-\delta)\\ &-1(1-\eta-\delta)-(\eta-\delta)\log(\eta-\delta)+(\eta-\delta).\end{aligned}$$

As $(\eta, \delta) \to (0, 0)$, the second term converges to $1 \log 1 = 0$, and the third and fifth terms converge to -1 and 0, respectively. Let $v = \eta - \delta$. Because $v \log v \to 0$ as $v \to 0$ (a limit established by using L'Hôpital's rule from calculus[3]), we see that the fourth term goes to zero as $(\eta, \delta) \to (0, 0)$. It is the first term that will give us trouble. Now:

$$-(1-2\eta)\log\delta = -\log\delta + 2\eta\log\delta, \tag{2}$$

and it is not hard to see that this does not converge as $(\eta, \delta) \to (0, 0)$. For example, let $\eta = 2\delta$; then expression (2) becomes $-\log\delta + 4\delta\log\delta$. As before, $4\delta\log\delta \to 0$ as $\delta \to 0$, but $-\log\delta \to +\infty$ as $\delta \to 0$, which shows that expression (2) does not converge. Hence, $\lim_{(\eta,\delta)\to(0,0)} \iint_{D_{\eta,\delta}} f\, dA$ does not exist and so f is not integrable. ▲

Functions Unbounded at Isolated Points

We now consider nonnegative functions f that become "infinite" or are undefined at isolated points in an x-simple or y-simple region D. For example, consider the function $f(x, y) = 1/\sqrt{x^2+y^2}$ on the unit disk $D = \{(x, y)|x^2 + y^2 \leq 1\}$. Again, $f \geq 0$, but f is unbounded and is not defined at the origin.

[3]L' Hôpital's rule was discovered by Bernoulli and was reported in L'Hôpital's textbook.

Let (x_0, y_0) be a point of a general region D where a nonnegative function f is undefined. Further, let $D_\delta = D_\delta(x_0, y_0)$ be the disk of radius δ centered at (x_0, y_0) and let $D \backslash D_\delta$ denote the region D with D_δ removed. Assume that f is continuous at every point of D except (x_0, y_0). Then $\iint_{D \backslash D_\delta} f\, dA$ is defined. We say that $\iint_D f\, dA$ is ***convergent***, or that f is ***integrable*** over D if

$$\lim_{\delta \to 0} \iint_{D \backslash D_\delta} f\, dA$$

exists.

EXAMPLE 4 Show that $f(x, y) = 1/\sqrt{x^2 + y^2}$ is integrable over the unit disk D and evaluate $\iint_D f dA$.

SOLUTION Let D_δ be the disk of radius δ centered at the origin. Then f is continuous everywhere on D except at (0, 0). Thus, $\iint_{D \backslash D_\delta} f\, dA$ exists. To evaluate this integral, we change variables to polar coordinates, $x = r\cos\theta$, $y = r\sin\theta$. Then $f(r\cos\theta, r\sin\theta) = 1/r$, and

$$\iint_{D \backslash D_\delta} f\, dA = \int_\delta^1 \int_0^{2\pi} \frac{1}{r} f\, d\theta\, dr = \int_\delta^1 \int_0^{2\pi} d\theta\, dr = 2\pi(1 - \delta).$$

Thus,

$$\iint_D f\, dA = \lim_{\delta \to 0} \iint_{D \backslash D_\delta} f\, dA = 2\pi. \quad \blacktriangle$$

More generally, one can, in an analogous manner, define the integral of nonnegative functions f that are continuous except at a finite number of points in D. One can also combine both types of improper integrals; that is, one may consider functions that are continuous except at a finite number of points on D or at points on the boundary of D, and define $\iint_D f\, dA$ appropriately.

If f takes both positive and negative values, one can use a more advanced integration theory, called the *Lebesgue integral*, to generalize the notion of convergent integral $\iint_D f dA$. Using this theory, it is possible to show that if $\iint_D f\, dA$ exists, it can then be evaluated as an iterated integral. This latter fact is also known as Fubini's theorem.

Unbounded Regions

As was mentioned previously, we will leave consideration of unbounded regions to the exercise section. However, we must point out that we have already addressed the main idea in Example 5 of Section 6.2 on the Gaussian integral. In that example, we integrated $\exp(-x^2 - y^2)$ over all of $\mathbb{R}^2$ by integrating first over a disk of radius a and then letting $a \to \infty$.

EXERCISES

In Exercises 1 to 4, evaluate the following integrals if they exist (discuss how you define the integral if it was not given in the text).

1. $\iint_D \frac{1}{\sqrt{xy}}\, dA$, where $D = [0, 1] \times [0, 1]$

2. $\iint_D \frac{1}{\sqrt{|x-y|}}\, dx\, dy$, where $D = \{(x, y) \mid 0 \le x \le 1, 0 \le y \le 1, y \le x\}$

3. $\iint_D (y/x)\, dx\, dy$, where D is bounded by $x = 1, x = y$, and $x = 2y$

4. $\int_0^1 \int_0^{e^y} \log x\, dx\, dy$

5. (a) Evaluate

$$\iint_D \frac{dA}{(x^2 + y^2)^{2/3}},$$

where D is the unit disk in $\mathbb{R}^2$

(b) Determine the real numbers λ for which the integral

$$\iint_D \frac{dA}{(x^2 + y^2)^{\lambda}}$$

is convergent, where again D is the unit disk.

6. (a) Discuss how you would define $\iint_D f\, dA$ if D is an unbounded region, for example, the set of (x, y) such that $a \le x < \infty$ and $\phi_1(x) \le y \le \phi_2(x)$, where $\phi_1 \le \phi_2$ are given (Figure 6.4.5).

(b) Evaluate $\iint_D xye^{-(x^2+y^2)}\, dx\, dy$ if $x \ge 0, 0 \le y \le 1$.

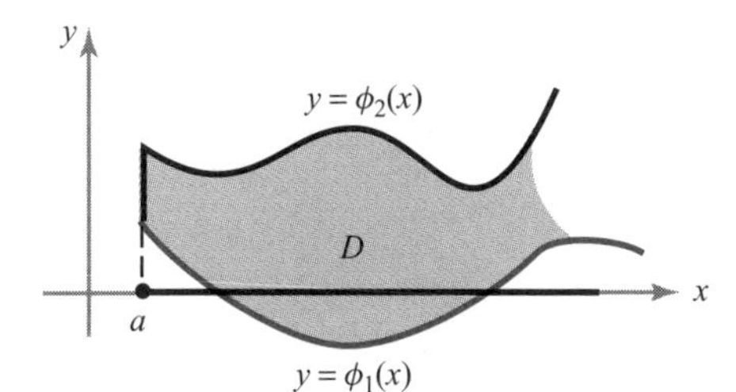

Figure 6.4.5 An unbounded region D.

7. Using Exercise 6, integrate e^{-xy} for $x \ge 0, 1 \le y \le 2$ in two ways. Assuming Fubini's theorem can be used, show that

$$\int_0^{\infty} \frac{e^{-x} - e^{-2x}}{x}\, dx = \log 2.$$

8. Show that the integral

$$\int_0^1 \int_0^a (x/\sqrt{a^2 - y^2})\, dy\, dx$$

exists, and compute its value.

9. Discuss whether the integral

$$\iint_D \frac{x+y}{x^2 + 2xy + y^2}\, dx\, dy$$

exists where $D = [0, 1] \times [0, 1]$. If it exists, compute its value.

10. One can also consider improper integrals of functions that fail to be continuous on entire curves lying in some region D. For example, by breaking $D = [0, 1] \times [0, 1]$ into two regions, define and then discuss the convergence of the integral

$$\iint_D \frac{1}{\sqrt{|x - y|}}\, dx\, dy.$$

11. Let W be the first octant of the ball $x^2 + y^2 + z^2 \leq a^2$, where $x \geq 0, y \geq 0, z \geq 0$. Evaluate the improper integral

$$\iiint_W \frac{(x^2 + y^2 + z^2)^{1/4}}{\sqrt{z + (x^2 + y^2 + z^2)^2}}\, dx\, dy\, dz$$

by changing variables.

12. Let f be a nonnegative function that may be unbounded and discontinuous on the boundary of an elementary region D. Let g be a similar function such that $f(x, y) \leq g(x, y)$ whenever both are defined. Suppose $\iint_D g(x, y)\, dA$ exists. Argue informally that this implies the existence of $\iint_D f(x, y)\, dA$.

13. Use Exercise 12 to show that

$$\iint_D \frac{\sin^2(x - y)}{\sqrt{1 - x^2 - y^2}}\, dy\, dx$$

exists where D is the unit disk $x^2 + y^2 \leq 1$.

14. Let f be as in Exercise 12 and let g be a function such that $0 \leq g(x, y) \leq f(x, y)$ whenever both are defined. Suppose that $\iint_D g(x, y)\, dA$ does not exist. Argue informally that $\iint_D f(x, y)\, dA$ cannot exist.

15. Use Exercise 14 to show that

$$\iint_D \frac{e^{x^2+y^2}}{x - y}\, dy\, dx$$

does not exist, where D is the set of (x, y) with $0 \leq x \leq 1$ and $0 \leq y \leq x$.

16. Let D be the unbounded region defined as the set of (x, y, z) with $x^2 + y^2 + z^2 \geq 1$. By making a change of variables, evaluate the improper integral

$$\iiint_D \frac{dx\, dy\, dz}{(x^2 + y^2 + z^2)^2}.$$

17. Evaluate

$$\int_0^1 \int_0^y \frac{x}{y}\, dx\, dy \qquad \text{and} \qquad \int_0^1 \int_x^1 \frac{x}{y}\, dy\, dx$$

Does Fubini's theorem apply?

18. In Exercise 11 of Section 5.2 we showed that

$$\int_0^1 \int_0^1 \frac{x^2 - y^2}{(x^2 + y^2)^2}\, dy\, dx \neq \int_0^1 \int_0^1 \frac{x^2 - y^2}{(x^2 + y^2)^2}\, dx\, dy.$$

Thus, Fubini's theorem does not hold here, even thought the iterated improper integrals both exist. What went wrong?

REVIEW EXERCISES FOR CHAPTER 6

1. (a) Find a linear transformation taking the square $S = [0, 1] \times [0, 1]$ to the parallelogram P with vertices $(0, 0)$, $(2, 0)$, $(1, 2)$, $(3, 2)$.

(b) Write down a change of variables formula appropriate to the transformation you found in part (a).

2. (a) Find the image of the square $[0, 1] \times [0, 1]$ under the transformation $T(x, y) = (2x, x + 3y)$.

(b) Write down a change of variables formula appropriate to the transformation and the region you found in part (a).

3. Let B be the region in the first quadrant bounded by the curves $xy = 1$, $xy = 3$, $x^2 - y^2 = 1$, and $x^2 - y^2 = 4$. Evaluate $\iint_B (x^2 + y^2)\, dx\, dy$ using the change of variables $u = x^2 - y^2$, $v = xy$.

4. In parts (a) to (d), make the indicated change of variables. (Do not evaluate.)

(a) $\displaystyle\int_0^1 \int_{-1}^1 \int_{-\sqrt{(1-y^2)}}^{\sqrt{(1-y^2)}} (x^2 + y^2)^{1/2}\, dx\, dy\, dz$, cylindrical coordinates

(b) $\displaystyle\int_{-1}^1 \int_{-\sqrt{(1-y^2)}}^{\sqrt{(1-y^2)}} \int_{-\sqrt{(4-x^2-y^2)}}^{\sqrt{(4-x^2-y^2)}} xyz\, dz\, dx\, dy$, cylindrical coordinates

(c) $\displaystyle\int_{-\sqrt{2}}^{\sqrt{2}} \int_{-\sqrt{(2-y^2)}}^{\sqrt{(2-y^2)}} \int_{\sqrt{(x^2+y^2)}}^{\sqrt{(4-x^2-y^2)}} z^2\, dz\, dx\, dy$, spherical coordinates

(d) $\displaystyle\int_0^1 \int_0^{\pi/4} \int_0^{2\pi} \rho^3 \sin 2\phi\, d\theta\, d\phi\, d\rho$, rectangular coordinates

5. Find the volume inside the surfaces $x^2 + y^2 = z$ and $x^2 + y^2 + z^2 = 2$.

6. Find the volume enclosed by the cone $x^2 + y^2 = z^2$ and the plane $2z - y - 2 = 0$.

7. A cylindrical hole of diameter 1 is bored through a sphere of radius 2. Assuming that the axis of the cylinder passes through the center of the sphere, find the volume of the solid that remains.

8. Let C_1 and C_2 be two cylinders of infinite extent, of diameter 2, and with axes on the x and y axes, respectively. Find the volume of their intersection, $C_1 \cap C_2$.

9. Find the volume bounded by $x/a + y/b + z/c = 1$ and the coordinate planes.

10. Find the volume determined by $z \leq 6 - x^2 - y^2$ and $z \geq \sqrt{x^2 + y^2}$.

11. The *tetrahedron* defined by $x \geq 0, y \geq 0, z \geq 0, x + y + z \leq 1$ is to be sliced into n segments of equal volume by planes parallel to the plane $x + y + z = 1$. Where should the slices be made?

12. Let E be the solid ellipsoid $E = \{(x, y, z) \mid (x^2/a^2) + (y^2/b^2) + (z^2/c^2) \leq 1\}$ where $a > 0, b > 0$, and $c > 0$. Evaluate

$$\iiint xyz \, dx \, dy \, dz$$

(a) over the whole ellipsoid; and
(b) over that part of it in the first quadrant:

$$x \geq 0, \qquad y \geq 0, \qquad \text{and} \qquad z \geq 0, \qquad \frac{x^2}{a^2} + \frac{y^2}{b^2} + \frac{z^2}{c^2} \leq 1.$$

13. Find the volume of the "ice cream cone" defined by the inequalities $x^2 + y^2 \leq \frac{1}{5}z^2$, and $0 \leq z \leq 5 + \sqrt{5 - x^2 - y^2}$.

14. Let ρ, θ, ϕ be spherical coordinates in $\mathbb{R}^3$ and suppose that a surface surrounding the origin is described by a continuous positive function $\rho = f(\theta, \phi)$. Show that the volume enclosed by the surface is

$$V = \frac{1}{3} \int_0^{2\pi} \int_0^{\pi} [f(\theta, \phi)]^3 \sin\phi \, d\phi \, d\theta.$$

15. Using an appropriate change of variables, evaluate

$$\iint_B \exp\,[(y - x)/(y + x)] \, dx \, dy$$

where B is the interior of the triangle with vertices at $(0, 0)$, $(0, 1)$, and $(1, 0)$.

16. Suppose the density of a solid of radius R is given by $(1 + d^3)^{-1}$ where d is the distance to the center of the sphere. Find the total mass of the sphere.

17. The density of the material of a spherical shell whose inner radius is 1 m and whose outer radius is 2 m is $0.4d^2$ g/cm^3, where d is the distance to the center of the sphere in meters. Find the total mass of the shell.

18. If the shell in Exercise 17 were dropped into a large tank of pure water, would it float? What if the shell leaked? (Assume that the density of water is exactly 1 g/cm^3.)

19. The temperature at points in the cube $C = \{(x, y, z) \mid -1 \leq x \leq 1, -1 \leq y \leq 1$, and $-1 \leq z \leq 1\}$ is $32d^2$, where d is the distance to the origin.

(a) What is the average temperature?
(b) At what points of the cube is the temperature equal to the average temperature?

20. Use cylindrical coordinates to find the center of mass of the region defined by

$$y^2 + z^2 \leq \frac{1}{4}, \qquad (x-1)^2 + y^2 + z^2 \leq 1, \qquad x \geq 1.$$

21. Find the center of mass of the solid hemisphere

$$V = \{(x, y, z) \mid x^2 + y^2 + z^2 \leq a^2 \text{ and } z \geq 0\}$$

if the density is constant.

22. Evaluate $\iint_B e^{-x^2-y^2}\, dx\, dy$ where B consists of those (x, y) satisfying $x^2 + y^2 \leq 1$ and $y \leq 0$.

23. Evaluate

$$\iiint_S \frac{dx\, dy\, dz}{(x^2 + y^2 + z^2)^{3/2}},$$

where S is the solid bounded by the spheres $x^2 + y^2 + z^2 = a^2$ and $x^2 + y^2 + z^2 = b^2$, where $a > b > 0$.

24. Evaluate $\displaystyle\iiint_D (x^2 + y^2 + z^2) xyz\, dx\, dy\, dz$ over each of the following regions.

(a) The sphere $D = \{(x, y, z) \mid x^2 + y^2 + z^2 \leq R^2\}$
(b) The hemisphere $D = \{(x, y, z) \mid x^2 + y^2 + z^2 \leq R^2 \text{ and } z \geq 0\}$
(c) The octant $D = \{(x, y, z) \mid x \geq 0, y \geq 0, z \geq 0, \text{ and } z^2 + y^2 + z^2 \leq R^2\}$

25. Let C be the cone-shaped region $\{(x, y, z) \mid \sqrt{x^2 + y^2} \leq z \leq 1\}$ in $\mathbb{R}^3$ and evaluate the integral $\displaystyle\iiint_C (1 + \sqrt{x^2 + y^2})\, dx\, dy\, dz$.

26. Find $\displaystyle\iiint_{\mathbb{R}^3} f(x, y, z)\, dx\, dy\, dz$ where $f(x, y, z) = \exp\,[-(x^2 + y^2 + z^2)^{3/2}]$.

27. The *flexural rigidity* EI of a uniform beam is the product of its Young's modulus of elasticity E and the moment of inertia I of the cross section of the beam with respect to a horizontal line l passing through the center of gravity of this cross section. Here

$$I = \iint_R [d(x, y)]^2\, dx\, dy,$$

where $d(x, y) =$ the distance from (x, y) to l and $R =$ the cross section of the beam being considered.

(a) Assume that the cross section R is the rectangle $-1 \leq x \leq 1, -1 \leq y \leq 2$, and l is the line $y = 1/2$. Find I.

(b) Assume the cross section R is a circle of radius 4 and l is the x axis. Find I, using polar coordinates.

28. Find, $\iiint_{\mathbb{R}^3} f(x, y, z)\, dx\, dy\, dz$ where

$$f(x, y, z) = \frac{1}{[1 + (x^2 + y^2 + z^2)^{3/2}]^{3/2}}.$$

29. Suppose D is the unbounded region of $\mathbb{R}^2$ given by the set of (x, y) with $0 \leq x < \infty$, $0 \leq y \leq x$. Let $f(x, y) = x^{-3/2} e^{y-x}$. Does the improper integral $\iint_D f(x, y)\, dx\, dy$ exist?

30. If the world were two-dimensional, the laws of physics would predict that the gravitational potential of a mass point is proportional to the logarithm of the distance from the point. Using polar coordinates, write an integral giving the gravitational potential of a disk of constant density.

31. (a) Evaluate the improper integral

$$\int_0^\infty \int_0^y x e^{-y^3}\, dx\, dy.$$

(b) Evaluate

$$\iint_B (x^4 + 2x^2 y^2 + y^4)\, dx\, dy,$$

where B is the portion of the disk of radius 2 [centered at (0, 0) in the first quadrant].

32. Let f be a nonnegative function on an x-simple or a y-simple region $D \subset \mathbb{R}^2$, and that is continuous except for points on the boundary of D and at most finitely many points interior to D. Give a suitable definition of $\iint_D f\, dA$.

33. Evaluate $\iint_{\mathbb{R}^2} f(x, y)\, dx\, dy$ where $f(x, y) = 1/(1 + x^2 + y^2)^{3/2}$. (Hint: You may assume that changing variables and Fubini's theorem are valid for improper integrals.)

7

Integrals Over Paths and Surfaces

I hold in fact: (1) That small portions of space are of a nature analogous to little hills on a surface which is on the average flat. (2) That this property of being curved or distorted is continually passed on from one portion of space to another after the manner of a wave. (3) That this variation of curvature of space is really what happens in that phenomenon which we call the motion of matter whether ponderable or ethereal. (4) That in this physical world nothing else takes place but this variation, subject, possibly, to the law of continuity.

W. K. Clifford (1870)

In Chapter 5, we studied integration over regions in $\mathbb{R}^2$ and $\mathbb{R}^3$. In this chapter, we study integration over paths and surfaces. This is basic to an understanding of Chapter 8, in which we discuss the basic relation between vector differential calculus (Chapter 4) and vector integral calculus (this chapter), a relation that generalizes the fundamental theorem of calculus to several variables. This generalization is summarized in the theorems of Green, Gauss, and Stokes.

7.1 The Path Integral

This section introduces the concept of a path integral; this is one of the several ways in which integrals of functions of one variable can be generalized to functions of several variables. Besides those in Chapter 5, there are other generalizations, to be discussed in later sections.

Suppose we are given a scalar function $f: \mathbb{R}^3 \to \mathbb{R}$, so that f sends points in $\mathbb{R}^3$ to real numbers. It will be useful to define the integral of such a function f along a path

$\mathbf{c}: I = [a, b] \to \mathbb{R}^3$, where $\mathbf{c}(t) = (x(t), y(t), z(t))$. To relate this notion to something tangible, suppose that the image of $\mathbf{c}$ represents a wire. We can let $f(x, y, z)$ denote the mass density at (x, y, z) and the integral of f will be the total mass of the wire. By letting $f(x, y, z)$ indicate temperature, we can also use the integral to determine the average temperature along the wire. We first give the formal definition of the path integral and then, after the following example, further motivate it.

DEFINITION: Path Integrals The ***path integral***, or the ***integral of*** $f(x, y, z)$ ***along the path*** $\mathbf{c}$, is defined when $\mathbf{c}: I = [a, b] \to \mathbb{R}^3$ is of class C^1 and when the composite function $t \mapsto f(x(t), y(t), z(t))$ is continuous on I. We define this integral by the equation

$$\int_{\mathbf{c}} f\, ds = \int_a^b f(x(t), y(t), z(t)) \|\mathbf{c}'(t)\|\, dt.$$

Sometimes $\int_{\mathbf{c}} f\, ds$ is denoted

$$\int_{\mathbf{c}} f(x, y, z)\, ds$$

or

$$\int_a^b f(\mathbf{c}(t)) \|\mathbf{c}'(t)\|\, dt.$$

If $\mathbf{c}(t)$ is only piecewise C^1 or $f(\mathbf{c}(t))$ is piecewise continuous, we define $\int_{\mathbf{c}} f\, ds$ by breaking $[a, b]$ into pieces over which $f(\mathbf{c}(t))\|\mathbf{c}'(t)\|$ is continuous, and summing the integrals over the pieces.

When $f = 1$, we recover the definition of the arc length of $\mathbf{c}$. Also note that f need only be defined on the image curve C of $\mathbf{c}$ and not necessarily on the whole space in order for the preceding definition to make sense.

EXAMPLE 1 Let $\mathbf{c}$ be the helix $\mathbf{c}: [0, 2\pi] \to \mathbb{R}^3, t \mapsto (\cos t, \sin t, t)$ (see Figure 2.4.9), and let $f(x, y, z) = x^2 + y^2 + z^2$. Evaluate the integral $\int_{\mathbf{c}} f(x, y, z)\, ds$.

SOLUTION First we compute $\|\mathbf{c}'(t)\|$:

$$\|\mathbf{c}'(t)\| = \sqrt{\left[\frac{d(\cos t)}{dt}\right]^2 + \left[\frac{d(\sin t)}{dt}\right]^2 + \left[\frac{dt}{dt}\right]^2} = \sqrt{\sin^2 t + \cos^2 t + 1} = \sqrt{2}.$$

Next, we substitute for x, y, and z in terms of t to obtain

$$f(x, y, z) = x^2 + y^2 + z^2 = \cos^2 t + \sin^2 t + t^2 = 1 + t^2$$

along $\mathbf{c}$. Inserting this information into the definition of the path integral yields

$$\int_{\mathbf{c}} f(x, y, z)\, ds = \int_0^{2\pi} (1 + t^2)\sqrt{2}\, dt = \sqrt{2}\left[t + \frac{t^3}{3}\right]_0^{2\pi} = \frac{2\sqrt{2}\pi}{3}(3 + 4\pi^2). \quad \blacktriangle$$

To motivate the definition of the path integral, we shall consider "Riemann-like" sums S_N in the same general way we did to define arc length in Section 4.2. For simplicity, let $\mathbf{c}$ be of class C^1 on I. Subdivide the interval $I = [a, b]$ by means of a partition

$$a = t_0 < t_1 < \cdots < t_N = b.$$

This leads to a decomposition of $\mathbf{c}$ into paths $\mathbf{c}_i$ (Figure 7.1.1) defined on $[t_i, t_{i+1}]$ for $0 \le i \le N - 1$. Denote the arc length of $\mathbf{c}_i$ by Δs_i; thus,

$$\Delta s_i = \int_{t_i}^{t_{i+1}} \|\mathbf{c}'(t)\|\, dt.$$

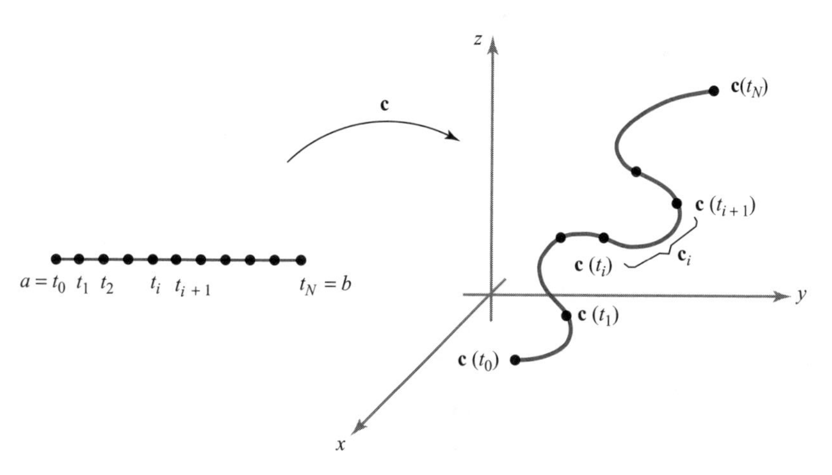

Figure 7.1.1 Breaking $\mathbf{c}$ into smaller $\mathbf{c}_i$.

When N is large, the arc length Δs_i is small and $f(x, y, z)$ is approximately constant for points on $\mathbf{c}_i$. We consider the sums

$$S_N = \sum_{i=0}^{N-1} f(x_i, y_i, z_i)\, \Delta s_i,$$

where $(x_i, y_i, z_i) = \mathbf{c}(t)$ for some $t \in [t_i, t_{i+1}]$. By the mean-value theorem we know that $\Delta s_i = \|\mathbf{c}'(t_i^*)\| \Delta t_i$, where $t_i \le t_i^* \le t_{i+1}$ and $\Delta t_i = t_{i+1} - t_i$. From the theory of

Riemann sums, it can be shown that

$$\lim_{N\to\infty} S_N = \lim_{N\to\infty} \sum_{i=0}^{N-1} f(x_i, y_i, z_i)\|\mathbf{c}'(t_i^*)\|\Delta t_i = \int_I f(x(t), y(t), z(t))\|\mathbf{c}'(t)\|\, dt$$

$$= \int_{\mathbf{c}} f(x, y, z)\, ds.$$

The Path Integral for Planar Curves

An important special case of the path integral occurs when the path $\mathbf{c}$ describes a plane curve. Suppose that all points $\mathbf{c}(t)$ lie in the xy plane and f is a real-valued function of two variables. The path integral of f along $\mathbf{c}$ is

$$\int_{\mathbf{c}} f(x, y)\, ds = \int_a^b f(x(t), y(t))\sqrt{x'(t)^2 + y'(t)^2}\, dt.$$

When $f(x, y) \geq 0$, this integral has a geometric interpretation as the "area of a fence." We can construct a "fence" with base the image of $\mathbf{c}$ and with height $f(x, y)$ at (x, y) (Figure 7.1.2). If $\mathbf{c}$ moves only once along the image of $\mathbf{c}$, the integral $\int_{\mathbf{c}} f(x, y)\, ds$ represents the area of a side of this fence. Readers should try to justify this interpretation for themselves, using an argument like the one used to justify the arc-length formula.

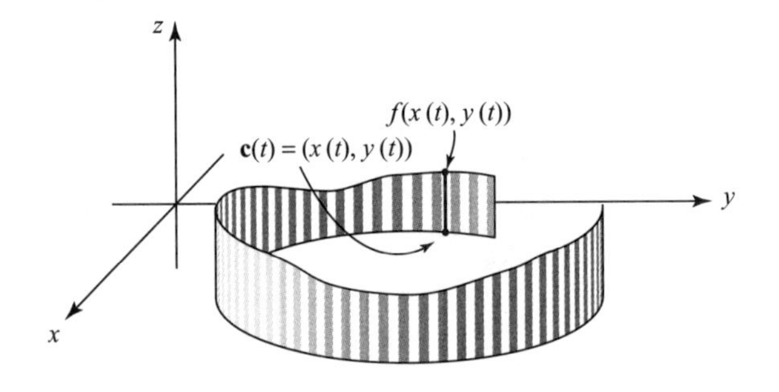

Figure 7.1.2 The path integral as the area of a fence.

EXAMPLE 2 Tom Sawyer's aunt has asked him to whitewash both sides of the old fence shown in Figure 7.1.3. Tom estimates that for each 25 ft² of whitewashing he lets someone do for him, the willing victim will pay 5 cents. How much can Tom hope to earn, assuming his aunt will provide whitewash free of charge?

SOLUTION From Figure 7.1.3, the base of the fence in the first quadrant is the path $\mathbf{c}$: $[0, \pi/2] \to \mathbb{R}^2, t \mapsto (30\cos^3 t, 30\sin^3 t)$, and the height of the fence at (x, y) is $f(x, y) = 1 + y/3$. The area of one side of the half of the fence is equal to the *integral* $\int_{\mathbf{c}} f(x, y)\, ds = \int_{\mathbf{c}}(1 + y/3)\, ds$. Because $\mathbf{c}'(t) = (-90\cos^2 t\sin t, 90\sin^2 t\cos t)$, we

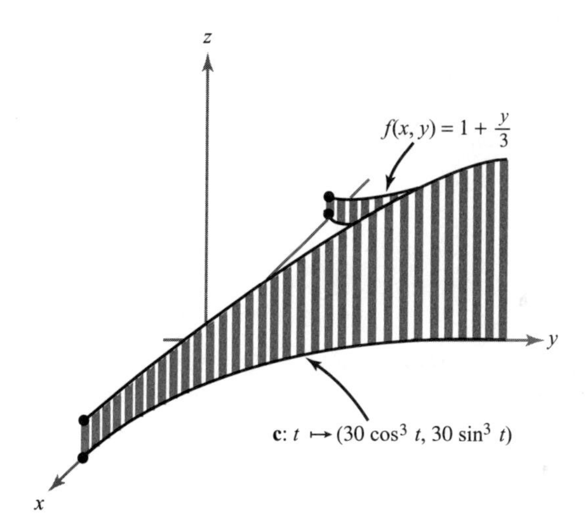

Figure 7.1.3 Tom Sawyer's fence.

have $\|\mathbf{c}'(t)\| = 90 \sin t \cos t$. Thus, the integral is

$$\begin{aligned}\int_{\mathbf{c}} \left(1 + \frac{y}{3}\right) ds &= \int_0^{\pi/2} \left(1 + \frac{30 \sin^3 t}{3}\right) 90 \sin t \cos t \, dt \\ &= 90 \int_0^{\pi/2} (\sin t + 10 \sin^4 t) \cos t \, dt \\ &= 90 \left[\frac{\sin^2 t}{2} + 2 \sin^5 t\right]_0^{\pi/2} = 90\left(\frac{1}{2} + 2\right) = 225,\end{aligned}$$

which is the area in the first quadrant. Hence, the area of one side of the fence is 450 ft^2. Because both sides are to be whitewashed, we must multiply by 2 to find the total area, which is 900 ft^2. Dividing by 25 and then multiplying by 5, we find that Tom could realize as much as \$1.80 for the job. ▲

This concludes our study of integration of *scalar* functions over paths. In the next section we shall turn our attention to the integration of *vector fields* over paths, and we shall see many further applications of the path integral in Chapter 8, when we study vector analysis.

Supplement to Section 7.1: The Total Curvature of a Curve

Exercises 12 to 17 of Section 4.2 described the notions of curvature κ and torsion τ of a smooth curve C in space. If $\mathbf{c}$: $[a, b] \rightarrow C \subset \mathbb{R}^3$ is a unit-speed parametrization of C, so that $\|\mathbf{c}'(t)\| = 1$, then the ***curvature*** $\kappa(p)$ at $p \in C$ is defined by $\kappa(p) = \|\mathbf{c}''(t)\|$, where $p = \mathbf{c}(t)$. A result of differential geometry is that two unit-speed curves with the same curvature and torsion can be obtained from one another by a rigid rotation, translation, or reflection.

The curvature $\kappa\colon C \to \mathbb{R}$ is a real-valued function on the set C, so we define the ***total curvature*** as its path integral over C: $\int_C \kappa\, ds$. There are some surprising facts that mathematicians have been able to prove about the total curvature. For one thing, if C is a closed [that is, $\mathbf{c}(a) = \mathbf{c}(b)$] planar curve, then

$$\int_C \kappa\, ds \geq 2\pi,$$

and equals 2π only when C is a circle. If C is a closed space curve with

$$\int_C \kappa\, ds \leq 4\pi,$$

then C is "unknotted"; that is, C can be continuously deformed (without ever intersecting itself) into a planar circle. Therefore, for knotted curves,

$$\int_C \kappa\, ds > 4\pi.$$

See Figure 7.1.4.

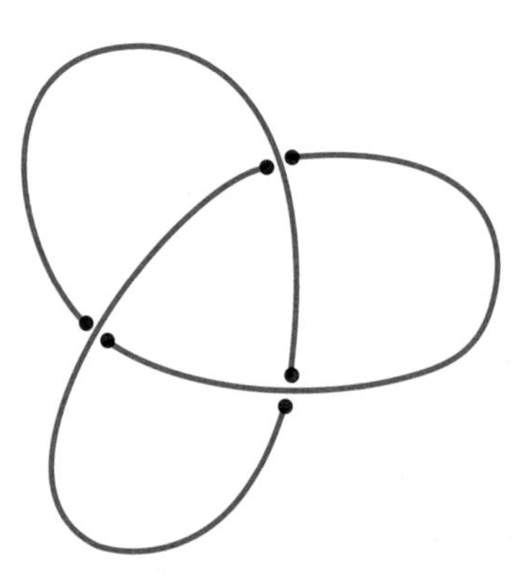

Figure 7.1.4 A knotted curve in $\mathbb{R}^3$.

The formal statement of this fact is known as the *Fary–Milnor* theorem. Legend has it that John Milnor, a contemporary of John Nash's[1] at Princeton University, was asleep in a math class as the professor wrote three *unsolved* knot theory problems on the blackboard. At the end of the class, Milnor (still an undergraduate) woke up and, thinking the blackboard problems were assigned as homework, quickly wrote

[1]John Nash is the subject of Sylvia Nasar's best-selling biography, *A Beautiful Mind*, a fictionalized version of which was made into a movie in 2001.

them down. The following week he turned in the solution to all three problems—one of which was a proof of the Fary–Milnor theorem! Some years later, he was appointed a professor at Princeton, and in 1962 he was awarded (albeit for other work) a Fields medal, mathematics' highest honor, generally regarded as the mathematical Nobel Prize.

EXERCISES

1. Let $f(x, y, z) = y$ and $\mathbf{c}(t) = (0, 0, t)$, $0 \le t \le 1$. Prove that $\int_{\mathbf{c}} f\, ds = 0$.

2. Evaluate the following path integrals $\int_{\mathbf{c}} f(x, y, z)\, ds$, where

(a) $f(x, y, z) = x + y + z$ and $\mathbf{c}\colon t \mapsto (\sin t, \cos t, t)$, $t \in [0, 2\pi]$
(b) $f(x, y, z) = \cos z$, $\mathbf{c}$ as in part (a)

3. Evaluate the following path integrals $\int_{\mathbf{c}} f(x, y, z)\, ds$, where

(a) $f(x, y, z) = \exp \sqrt{z}$, and $\mathbf{c}\colon t \mapsto (1, 2, t^2)$, $t \in [0, 1]$
(b) $f(x, y, z) = yz$, and $\mathbf{c}\colon t \mapsto (t, 3t, 2t)$, $t \in [1, 3]$

4. Evaluate the integral of $f(x, y, z)$ along the path $\mathbf{c}$, where

(a) $f(x, y, z) = x \cos z$, $\mathbf{c}\colon t \mapsto t\mathbf{i} + t^2\mathbf{j}$, $t \in [0, 1]$
(b) $f(x, y, z) = (x + y)/(y + z)$, and $\mathbf{c}\colon t \mapsto \left(t, \frac{2}{3}t^{3/2}, t\right)$, $t \in [1, 2]$

5. Let $f\colon \mathbb{R}^3 \backslash \{xz \text{ plane}\} \to \mathbb{R}$ be defined by $f(x, y, z) = 1/y^3$. Evaluate $\int_{\mathbf{c}} f(x, y, z)\, ds$, where $\mathbf{c}\colon [1, e] \to \mathbb{R}^3$ is given by $\mathbf{c}(t) = (\log t)\mathbf{i} + t\mathbf{j} + 2\mathbf{k}$.

6. (a) Show that the path integral of $f(x, y)$ along a path given in polar coordinates by $r = r(\theta)$, $\theta_1 \le \theta \le \theta_2$, is

$$\int_{\theta_1}^{\theta_2} f(r \cos \theta, r \sin \theta) \sqrt{r^2 + \left(\frac{dr}{d\theta}\right)^2}\, d\theta.$$

(b) Compute the arc length of the path $r = 1 + \cos \theta$, $0 \le \theta \le 2\pi$.

7. Let $f(x, y) = 2x - y$, and consider the path $x = t^4$, $y = t^4$, $-1 \le t \le 1$.

(a) Compute the integral of f along this path and interpret the answer geometrically.
(b) Evaluate the arc-length function $s(t)$ and redo part (a) in terms of s (you may wish to consult Exercise 2, Section 4.2).

Exercises 8 to 11 are concerned with the application of the path integral to the problem of defining the average value of a scalar function along a path. Define the number

$$\frac{\int_{\mathbf{c}} f(x, y, z)\, ds}{l(\mathbf{c})}$$

to be the ***average value*** *of f along* **c**. *Here $l(\mathbf{c})$ is the length of the path:*

$$l(\mathbf{c}) = \int_{\mathbf{c}} \|\mathbf{c}'(t)\|\, dt.$$

(This is analogous to the average of a function over a region defined in Section 6.3.)

8. (a) Justify the formula $[\int_{\mathbf{c}} f(x, y, z)\, ds]/l(\mathbf{c})$ for the average value of f along **c** using Riemann sums.

(b) Show that the average value of f along **c** in Example 1 is $(1 + \frac{4}{3}\pi^2)$.

(c) In Exercise 2(a) and (b) above, find the average value of f over the given curves.

9. Find the average y coordinate of the points on the semicircle parametrized by $\mathbf{c}\colon [0, \pi] \to \mathbb{R}^3, \theta \mapsto (0, a \sin\theta, a \cos\theta); a > 0$.

10. Suppose the semicircle in Exercise 9 is made of a wire with a uniform density of 2 grams per unit length.

(a) What is the total mass of the wire?

(b) Where is the center of mass of this configuration of wire? (Consult Section 6.3.)

11. Let **c** be the path given by $\mathbf{c}(t) = (t^2, t, 3)$ for $t \in [0, 1]$.

(a) Find $l(\mathbf{c})$, the length of the path.

(b) Find the average y coordinate along the path **c**.

12. If $f\colon [a, b] \to \mathbb{R}$ is piecewise continuously differentiable, let the *length of the graph* of f on $[a, b]$ be defined as the length of the path $t \mapsto (t, f(t))$ for $t \in [a, b]$.

(a) Show that the length of the graph of f on $[a, b]$ is

$$\int_a^b \sqrt{1 + [f'(x)]^2}\, dx.$$

(b) Find the length of the graph of $y = \log x$ from $x = 1$ to $x = 2$.

13. Find the mass of a wire formed by the intersection of the sphere $x^2 + y^2 + z^2 = 1$ and the plane $x + y + z = 0$ if the density at (x, y, z) is given by $\rho(x, y, z) = x^2$ grams per unit length of wire.

14. Evaluate $\int_{\mathbf{c}} f\, ds$ where $f(x, y, z) = z$ and $\mathbf{c}(t) = (t\cos t, t\sin t, t)$ for $0 \le t \le t_0$.

15. Write the following limit as a path integral of $f(x, y, z) = xy$ over some path **c** on $[0, 1]$ and evaluate:

$$\operatorname*{limit}_{N\to\infty} \sum_{i=1}^{N-1} t_i^2 \left(t_{i+1}^2 - t_i^2\right),$$

where $t_1, \ldots, t_N$ is a partition of $[0, 1]$.

16. Consider paths that connect the points $A = (0, 1)$ and $B = (1, 0)$ in the xy plane, as in Figure 7.1.5.[2]

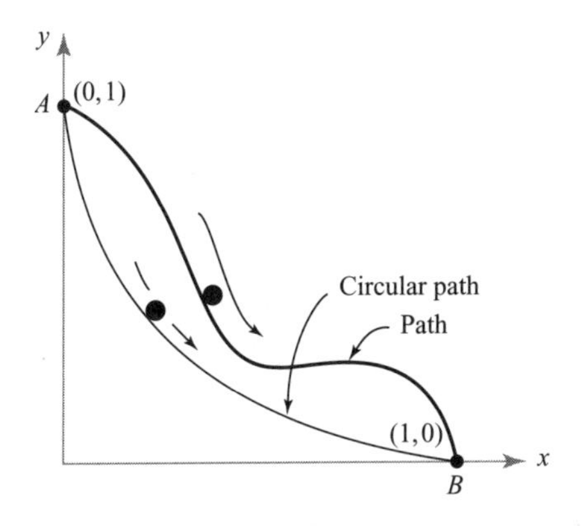

Figure 7.1.5 A curve joining the points A and B.

Galileo contemplated the following question: does a bead falling under the influence of gravity from a point A to a point B along a curve do so in *the least possible time* if that curve is a circular arc? For any given path, the time of transit T is a path integral

$$T = \int \frac{dt}{v},$$

where the bead's velocity is $v = \sqrt{2gy}$, where g is the gravitational constant. In 1697, Johann Bernoulli challenged the mathematical world to find the path in which the bead would roll from A to B in the least time. This solution would determine whether Galileo's considerations had been correct.

(a) Calculate T for the straight-line path $y = 1 - x$.
(b) Write a formula for T for Galileo's circular path, given by $(x - 1)^2 + (y - 1)^2 = 1$.

Incidentally, Newton was the first to send his solution [which turned out to be a cycloid—the same curve (inverted) that we studied in Example 2.4.4], but he did so anonymously. Bernoulli was not fooled, however. When he received the solution, he immediately knew its author, exclaiming, "I know the Lion from his paw." While the solution of this problem is a cycloid, it is known in the literature as the *brachistrochrone*. This was the beginning of the important field called the *calculus of variations*.

7.2 Line Integrals

We now consider the problem of integrating a *vector field* along a path. We will begin by considering the notion of *work* to motivate the general definition.

Work Done by Force Fields

If **F** is a force field in space, then a test particle (for example, a small unit charge in an electric force field or a unit mass in a gravitational field) will experience the force **F**.

---

[2] We thank Tanya Leise for suggesting this exercise.

Suppose the particle moves along the image of a path $\mathbf{c}$ while being acted upon by $\mathbf{F}$. A fundamental concept is the *work done* by $\mathbf{F}$ on the particle as it traces out the path $\mathbf{c}$. If $\mathbf{c}$ is a straight-line displacement given by the vector $\mathbf{d}$ and if $\mathbf{F}$ is a constant force, then the work done by $\mathbf{F}$ in moving the particle along the path is the dot product $\mathbf{F} \cdot \mathbf{d}$:

$$\mathbf{F} \cdot \mathbf{d} = (\text{magnitude of force}) \times (\text{displacement in direction of force}).$$

If the path is curved, we can imagine that it is made up of a succession of infinitesimal straight-line displacements or that it is *approximated* by a finite number of straight-line displacements. Then (as in our derivation of the formulas for the path integral in the preceding section) we are led to the following formula for the work done by the force field $\mathbf{F}$ on a particle moving along a path $\mathbf{c}\colon [a, b] \to \mathbb{R}^3$:

$$\text{work done by } \mathbf{F} = \int_a^b \mathbf{F}(\mathbf{c}(t)) \cdot \mathbf{c}'(t)\, dt.$$

We can further justify this derivation as follows. As t ranges over a small interval t to $t + \Delta t$, the particle moves from $\mathbf{c}(t)$ to $\mathbf{c}(t + \Delta t)$, a vector displacement of $\Delta \mathbf{s} = \mathbf{c}(t + \Delta t) - \mathbf{c}(t)$ (see Figure 7.2.1).

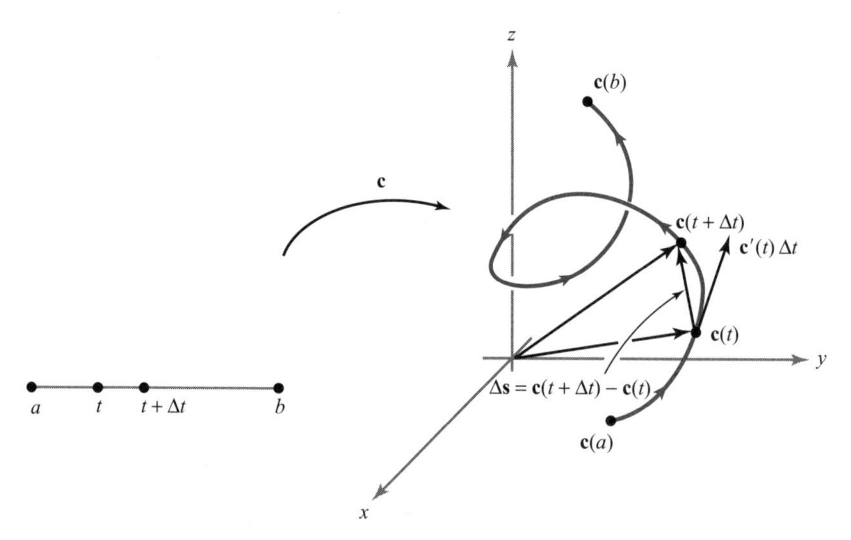

Figure 7.2.1 For small Δt, $\Delta \mathbf{s} = \mathbf{c}(t + \Delta t) - \mathbf{c}(t) \approx \mathbf{c}'(t)\Delta t$.

From the definition of the derivative, we get the approximation $\Delta \mathbf{s} \approx \mathbf{c}'(t)\Delta t$. The work done in going from $\mathbf{c}(t)$ to $\mathbf{c}(t + \Delta t)$ is therefore approximately

$$\mathbf{F}(\mathbf{c}(t)) \cdot \Delta \mathbf{s} \approx \mathbf{F}(\mathbf{c}(t)) \cdot \mathbf{c}'(t)\Delta t.$$

If we subdivide the interval $[a, b]$ into n equal parts $a = t_0 < t_1 < \cdots < t_n = b$, with $\Delta t = t_{i+1} - t_i$, then the work done by $\mathbf{F}$ is approximately

$$\sum_{i=0}^{n-1} \mathbf{F}(\mathbf{c}(t_i)) \cdot \Delta \mathbf{s} \approx \sum_{i=0}^{n-1} \mathbf{F}(\mathbf{c}(t_i)) \cdot \mathbf{c}'(t_i) \Delta t.$$

As $n \to \infty$, this approximation becomes better and better, and so it is reasonable to take as our definition of work to be the limit of the sum just given as $n \to \infty$. This limit is given by the integral

$$\int_a^b \mathbf{F}(\mathbf{c}(t)) \cdot \mathbf{c}'(t)\, dt.$$

Definition of the Line Integral

The previous discussion of work motivates the following definition.

DEFINITION: Line Integrals Let $\mathbf{F}$ be a vector field on $\mathbb{R}^3$ that is continuous on the C^1 path $\mathbf{c}$: $[a, b] \to \mathbb{R}^3$. We define $\int_{\mathbf{c}} \mathbf{F} \cdot d\mathbf{s}$, the ***line integral*** of $\mathbf{F}$ along $\mathbf{c}$, by the formula

$$\int_{\mathbf{c}} \mathbf{F} \cdot d\mathbf{s} = \int_a^b \mathbf{F}(\mathbf{c}(t)) \cdot \mathbf{c}'(t)\, dt;$$

that is, we integrate the dot product of $\mathbf{F}$ with $\mathbf{c}'$ over the interval $[a, b]$.

As is the case with scalar functions, we can also define $\int_{\mathbf{c}} \mathbf{F} \cdot d\mathbf{s}$ if $\mathbf{F}(\mathbf{c}(t)) \cdot \mathbf{c}'(t)$ is only piecewise continuous.

For paths $\mathbf{c}$ that satisfy $\mathbf{c}'(t) \neq \mathbf{0}$, there is another useful formula for the line integral: Namely, if $\mathbf{T}(t) = \mathbf{c}'(t)/\|\mathbf{c}'(t)\|$ denotes the unit tangent vector, we have

$$\begin{aligned} \int \mathbf{F} \cdot d\mathbf{s} &= \int_a^b \mathbf{F}(\mathbf{c}(t)) \cdot \mathbf{c}'(t)\, dt && \text{(by definition)} \\ &= \int_a^b \left[\mathbf{F}(\mathbf{c}(t)) \cdot \frac{\mathbf{c}'(t)}{\|\mathbf{c}'(t)\|} \right] \|\mathbf{c}'(t)\|\, dt && \text{(canceling } \|\mathbf{c}'(t)\|) \qquad (1) \\ &= \int_a^b [\mathbf{F}(\mathbf{c}(t)) \cdot \mathbf{T}(t)] \|\mathbf{c}'(t)\|\, dt. \end{aligned}$$

This formula says that $\int_{\mathbf{c}} \mathbf{F} \cdot d\mathbf{s}$ is equal to something that looks like the path integral of the tangential component $\mathbf{F}(\mathbf{c}(t)) \cdot \mathbf{T}(t)$ of $\mathbf{F}$ along $\mathbf{c}$. In fact, the last part of formula (1) is analogous to the path integral of a scalar function f along $\mathbf{c}$.[3]

To compute a line integral in any particular case, one can either use the original definition or integrate the tangential component of $\mathbf{F}$ along $\mathbf{c}$, as prescribed by formula (1), whichever is easier or more appropriate.

EXAMPLE 1 Let $\mathbf{c}(t) = (\sin t, \cos t, t)$ with $0 \leq t \leq 2\pi$. Let the vector field $\mathbf{F}$ be defined by $\mathbf{F}(x, y, z) = x\mathbf{i} + y\mathbf{j} + z\mathbf{k}$. Compute $\int_{\mathbf{c}} \mathbf{F} \cdot d\mathbf{s}$.

SOLUTION Here, $\mathbf{F}(\mathbf{c}(t)) = \mathbf{F}(\sin t, \cos t, t) = (\sin t)\mathbf{i} + (\cos t)\mathbf{j} + t\mathbf{k}$, and $\mathbf{c}'(t) = (\cos t)\mathbf{i} - (\sin t)\mathbf{j} + \mathbf{k}$. Therefore,

$$\mathbf{F}(\mathbf{c}(t)) \cdot \mathbf{c}'(t) = \sin t \cos t - \cos t \sin t + t = t,$$

and so

$$\int_{\mathbf{c}} \mathbf{F} \cdot d\mathbf{s} = \int_0^{2\pi} t\, dt = 2\pi^2. \quad \blacktriangle$$

Another common way of writing line integrals is

$$\int_{\mathbf{c}} \mathbf{F} \cdot d\mathbf{s} = \int_{\mathbf{c}} F_1\, dx + F_2\, dy + F_3\, dz,$$

where F_1, F_2, and F_3 are the components of the vector field $\mathbf{F}$. We call the expression $F_1\, dx + F_2\, dy + F_3\, dz$ a ***differential form***.[4] By *definition*, the integral of a differential form along a path $\mathbf{c}$, where $\mathbf{c}(t) = (x(t), y(t), z(t))$, is

$$\int_{\mathbf{c}} F_1\, dx + F_2\, dy + F_3\, dz = \int_a^b \left(F_1 \frac{dx}{dt} + F_2 \frac{dy}{dt} + F_3 \frac{dz}{dt} \right) dt = \int_{\mathbf{c}} \mathbf{F} \cdot d\mathbf{s}.$$

Note that we may think of $d\mathbf{s}$ as the differential form $d\mathbf{s} = dx\mathbf{i} + dy\mathbf{j} + dz\mathbf{k}$. Thus, the differential form $F_1\, dx + F_2\, dy + F_3\, dz$ may be written as the dot product $\mathbf{F} \cdot d\mathbf{s}$.

EXAMPLE 2 Evaluate the line integral

$$\int_{\mathbf{c}} x^2\, dx + xy\, dy + dz,$$

where $\mathbf{c}\colon [0, 1] \to \mathbb{R}^3$ is given by $\mathbf{c}(t) = (t, t^2, 1) = (x(t), y(t), z(t))$.

---

[3] If $\mathbf{c}$ does not intersect itself (that is, if $\mathbf{c}(t_1) = \mathbf{c}(t_2)$ implies $t_1 = t_2$), then each point P on C (the image curve of $\mathbf{c}$) can be written uniquely as $\mathbf{c}(t)$ for some t. If we define $f(\mathrm{P}) = f(\mathbf{c}(t)) = \mathbf{F}(\mathbf{c}) \cdot \mathbf{T}(t)$, f is a function on C; by definition, its path integral along $\mathbf{c}$ is given by formula (1) and there is no difficulty in literally interpreting $\int_{\mathbf{c}} \mathbf{F} \cdot d\mathbf{s}$ as a path integral. If $\mathbf{c}$ intersects itself, we cannot define f as a function on C as before (why?); however, in this case it is still useful to think of the right side of formula (1) as a path integral.

[4] See Section 8.6 for a brief discussion of the general theory of differential forms.

SOLUTION We compute $dx/dt = 1$, $dy/dt = 2t$, $dz/dt = 0$; therefore,

$$\int_{\mathbf{c}} x^2\,dx + xy\,dy + dz = \int_0^1 \left([x(t)]^2 \frac{dx}{dt} + [x(t)y(t)]\frac{dy}{dt}\right) dt$$

$$= \int_0^1 (t^2 + 2t^4)\,dt = \left[\frac{1}{3}t^3 + \frac{2}{5}t^5\right]_0^1 = \frac{11}{15}. \quad \blacktriangle$$

EXAMPLE 3 Evaluate the line integral

$$\int_{\mathbf{c}} \cos z\,dx + e^x\,dy + e^y\,dz,$$

where the path $\mathbf{c}$ is defined by $\mathbf{c}(t) = (1, t, e^t)$ and $0 \le t \le 2$.

SOLUTION We compute $dx/dt = 0$, $dy/dt = 1$, $dz/dt = e^t$, and so

$$\int_{\mathbf{c}} \cos z\,dx + e^x\,dy + e^y\,dz = \int_0^2 (0 + e + e^{2t})\,dt$$

$$= \left[et + \frac{1}{2}e^{2t}\right]_0^2 = 2e + \frac{1}{2}e^4 - \frac{1}{2}. \quad \blacktriangle$$

EXAMPLE 4 Let $\mathbf{c}$ be the path

$$x = \cos^3\theta, \qquad y = \sin^3\theta, \qquad z = \theta, \qquad 0 \le \theta \le \frac{7\pi}{2}$$

(see Figure 7.2.2). Evaluate the integral $\int_{\mathbf{c}}(\sin z\,dx + \cos z\,dy - (xy)^{1/3}\,dz)$.

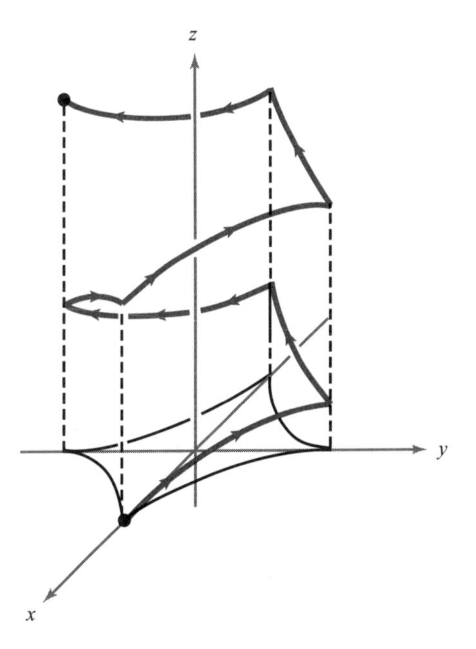

Figure 7.2.2 The image of the path $x = \cos^3\theta, y = \sin^3\theta, z = \theta; 0 \le \theta \le 7\pi/2$.

SOLUTION In this case, we have

$$\frac{dx}{d\theta} = -3\cos^2\theta\sin\theta, \qquad \frac{dy}{d\theta} = 3\sin^2\theta\cos\theta, \qquad \frac{dz}{d\theta} = 1,$$

so the integral is

$$\int_{\mathbf{c}} \sin z\,dx + \cos z\,dy - (xy)^{1/3}\,dz$$
$$= \int_0^{7\pi/2} (-3\cos^2\theta\sin^2\theta + 3\sin^2\theta\cos^2\theta - \cos\theta\sin\theta)\,d\theta.$$

The first two terms cancel, and so we get

$$-\int_0^{7\pi/2} \cos\theta\sin\theta\,d\theta = -\left[\frac{1}{2}\sin^2\theta\right]_0^{7\pi/2} = -\frac{1}{2}. \quad \blacktriangle$$

EXAMPLE 5 Suppose $\mathbf{F}$ is the force vector field $\mathbf{F}(x, y, z) = x^3\mathbf{i} + y\mathbf{j} + z\mathbf{k}$. Parametrize the circle of radius a in the yz plane by letting $\mathbf{c}(\theta)$ have components

$$x = 0, \qquad y = a\cos\theta, \qquad z = a\sin\theta, \qquad 0 \le \theta \le 2\pi.$$

Because $\mathbf{F}(\mathbf{c}(\theta)) \cdot \mathbf{c}'(\theta) = 0$, the force field $\mathbf{F}$ is normal to the circle at every point on the circle, so $\mathbf{F}$ will not do any work on a particle moving along the circle (Figure 7.2.3).

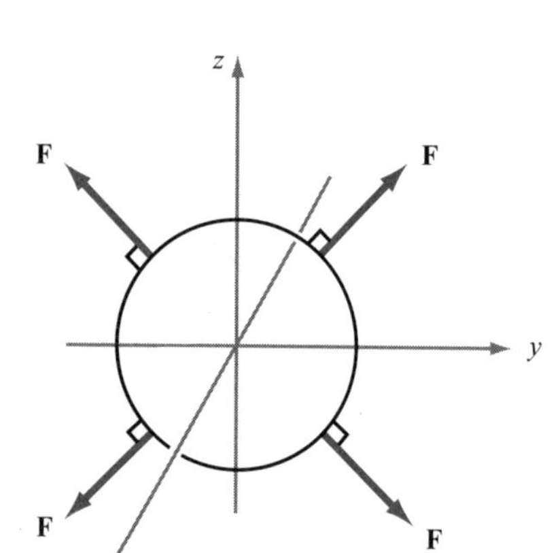

Figure 7.2.3 A vector field $\mathbf{F}$ normal to a circle in the yz plane.

We can verify by direct computation that the work done by $\mathbf{F}$ is zero:

$$W = \int_{\mathbf{c}} \mathbf{F} \cdot d\mathbf{s} = \int_{\mathbf{c}} x^3\,dx + y\,dy + z\,dz$$
$$= \int_0^{2\pi} (0 - a^2\cos\theta\sin\theta + a^2\cos\theta\,\sin\theta)\,d\theta = 0. \quad \blacktriangle$$

EXAMPLE 6 If we consider the field and curve of Example 4, we see that the work done by the field is $-\frac{1}{2}$, a negative quantity. This means that the field impedes movement along the path. ▲

Reparametrizations

The line integral $\int_{\mathbf{c}} \mathbf{F} \cdot d\mathbf{s}$ depends not only on the field $\mathbf{F}$ but also on the path $\mathbf{c}\colon [a, b] \to \mathbb{R}^3$. In general, if $\mathbf{c}_1$ and $\mathbf{c}_2$ are two different paths in $\mathbb{R}^3$, $\int_{\mathbf{c}_1} \mathbf{F} \cdot d\mathbf{s} \neq \int_{\mathbf{c}_2} \mathbf{F} \cdot d\mathbf{s}$. On the other hand, we shall see that it is true that $\int_{\mathbf{c}_1} \mathbf{F} \cdot d\mathbf{s} = \pm \int_{\mathbf{c}_2} \mathbf{F} \cdot d\mathbf{s}$ for every vector field $\mathbf{F}$ if $\mathbf{c}_1$ is what we call a ***reparametrization*** of $\mathbf{c}_2$; roughly speaking, this means that $\mathbf{c}_1$ and $\mathbf{c}_2$ are different descriptions of the same geometric curve.

DEFINITION Let $h\colon I \to I_1$ be a C^1 real-valued function that is a one-to-one map of an interval $I = [a, b]$ onto another interval $I_1 = [a_1, b_1]$. Let $\mathbf{c}\colon I_1 \to \mathbb{R}^3$ be a piecewise C^1 path. Then we call the composition

$$\mathbf{p} = \mathbf{c} \circ h\colon I \to \mathbb{R}^3$$

a ***reparametrization*** of $\mathbf{c}$.

This means that $\mathbf{p}(t) = \mathbf{c}(h(t))$, and so h changes the variable; alternatively, one can think of h as changing the speed at which a point moves along the path. Indeed, observe that $\mathbf{p}'(t) = \mathbf{c}'(h(t))h'(t)$, so that the velocity vector for $\mathbf{p}$ equals that for $\mathbf{c}$ but is multiplied by the scalar factor $h'(t)$.

It is implicit in the definition that h must carry endpoints to endpoints; that is, either $h(a) = a_1$ and $h(b) = b_1$, or $h(a) = b_1$ and $h(b) = a_1$. We thus distinguish two types of reparametrizations. If $\mathbf{c} \circ h$ is a reparametrization of $\mathbf{c}$, then either

$$(\mathbf{c} \circ h)(a) = \mathbf{c}(a_1) \quad \text{and} \quad (\mathbf{c} \circ h)(b) = \mathbf{c}(b_1)$$

or

$$(\mathbf{c} \circ h)(a) = \mathbf{c}(b_1) \quad \text{and} \quad (\mathbf{c} \circ h)(b) = \mathbf{c}(a_1).$$

In the first case, the reparametrization is said to be ***orientation-preserving***, and a particle tracing the path $\mathbf{c} \circ h$ moves in the *same direction* as a particle tracing $\mathbf{c}$. In the second case, the reparametrization is described as ***orientation-reversing***, and a particle tracing the path $\mathbf{c} \circ h$ moves in the *opposite direction* to that of a particle tracing $\mathbf{c}$ (Figure 7.2.4).

For example, if C is the image of a path $\mathbf{c}$, as shown in Figure 7.2.5, that is, $C = \mathbf{c}([a_1, b_1])$, and if h is orientation-preserving, then $\mathbf{c} \circ h(t)$ will go from $\mathbf{c}(a_1)$ to $\mathbf{c}(b_1)$ as t goes from a to b; and if h is orientation-reversing, $\mathbf{c} \circ h(t)$ will go from $\mathbf{c}(b_1)$ to $\mathbf{c}(a_1)$ as t goes from a to b.

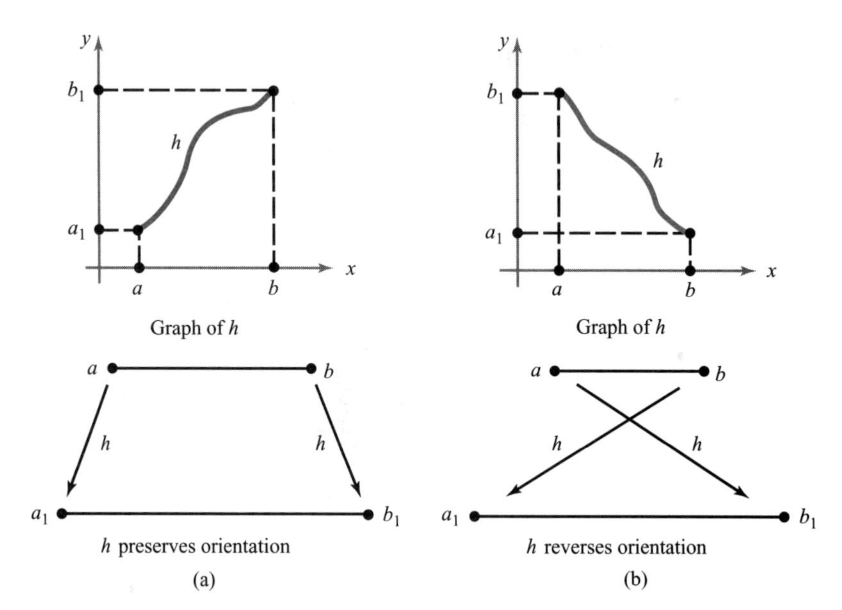

Figure 7.2.4 Illustrating (a) an orientation-preserving reparametrization, and (b) an orientation-reversing reparametrization.

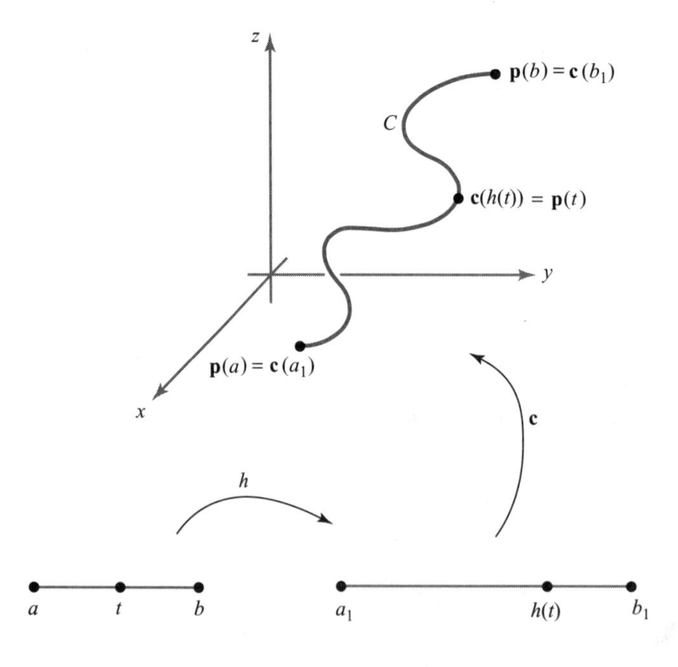

Figure 7.2.5 The path $\mathbf{p} = \mathbf{c} \circ h$ is a reparametrization of $\mathbf{c}$.

EXAMPLE 7 Let $\mathbf{c}: [a, b] \to \mathbb{R}^3$ be a piecewise C^1 path. Then:

(a) The path $\mathbf{c}_{\text{op}}: [a, b] \to \mathbb{R}^3, t \mapsto \mathbf{c}(a + b - t)$, is reparametrization of $\mathbf{c}$ corresponding to the map $h: [a, b] \to [a, b], t \mapsto a + b - t$; we call $\mathbf{c}_{\text{op}}$ the ***opposite path*** to $\mathbf{c}$. This reparametrization is orientation-reversing.

(b) The path $\mathbf{p}: [0, 1] \to \mathbb{R}^3, t \mapsto \mathbf{c}(a + (b - a)t)$, is an orientation-preserving reparametrization of $\mathbf{c}$ corresponding to a change of coordinates $h: [0, 1] \to [a, b], t \mapsto a + (b - a)t$. ▲

THEOREM 1: Change of Parametrization for Line Integrals
Let $\mathbf{F}$ be a vector field continuous on the C^1 path $\mathbf{c}: [a_1, b_1] \to \mathbb{R}^3$, and let $\mathbf{p}: [a, b] \to \mathbb{R}^3$ be a reparametrization of $\mathbf{c}$. If $\mathbf{p}$ is orientation-preserving, then

$$\int_{\mathbf{p}} \mathbf{F} \cdot d\mathbf{s} = \int_{\mathbf{c}} \mathbf{F} \cdot d\mathbf{s},$$

and if $\mathbf{p}$ is orientation-reversing, then

$$\int_{\mathbf{p}} \mathbf{F} \cdot d\mathbf{s} = -\int_{\mathbf{c}} \mathbf{F} \cdot d\mathbf{s}.$$

PROOF By hypothesis, we have a map h such that $\mathbf{p} = \mathbf{c} \circ h$. By the chain rule,

$$\mathbf{p}'(t) = \mathbf{c}'(h(t))h'(t),$$

and so

$$\int_{\mathbf{p}} \mathbf{F} \cdot d\mathbf{s} = \int_a^b [\mathbf{F}(\mathbf{c}(h(t))) \cdot \mathbf{c}'(h(t))]h'(t)\,dt.$$

Changing variables with $s = h(t)$, this becomes

$$\int_{h(a)}^{h(b)} \mathbf{F}(\mathbf{c}(s)) \cdot \mathbf{c}'(s)\,ds$$

$$= \begin{cases} \displaystyle\int_{a_1}^{b_1} \mathbf{F}(\mathbf{c}(s)) \cdot \mathbf{c}'(s)\,ds = \int_{\mathbf{c}} \mathbf{F} \cdot d\mathbf{s} & \text{if } \mathbf{p} \text{ is orientation-preserving} \\ \displaystyle\int_{b_1}^{a_1} \mathbf{F}(\mathbf{c}(s)) \cdot \mathbf{c}'(s)\,ds = -\int_{\mathbf{c}} \mathbf{F} \cdot d\mathbf{s} & \text{if } \mathbf{p} \text{ is orientation-reversing.} \end{cases}$$ ■

Theorem 1 also holds for piecewise C^1 paths, as may be seen by breaking up the intervals into segments on which the paths are of class C^1 and summing the integrals over the separate intervals.

Thus, if it is convenient to reparametrize a path when evaluating an integral, Theorem 1 assures us that the value of the integral will not be affected, except possibly for the sign, depending on the orientation.

EXAMPLE 8 Let $\mathbf{F}(x, y, z) = yz\mathbf{i} + xz\mathbf{j} + xy\mathbf{k}$ and $\mathbf{c}$: $[-5, 10] \to \mathbb{R}^3$ be defined by $t \mapsto (t, t^2, t^3)$. Evaluate $\int_{\mathbf{c}} \mathbf{F} \cdot d\mathbf{s}$ and $\int_{\mathbf{c}_{\text{op}}} \mathbf{F} \cdot d\mathbf{s}$.

SOLUTION For the path $\mathbf{c}$, we have $dx/dt = 1$, $dy/dt = 2t$, $dz/dt = 3t^2$, and $\mathbf{F}(\mathbf{c}(t)) = t^5\mathbf{i} + t^4\mathbf{j} + t^3\mathbf{k}$. Therefore,

$$\int_{\mathbf{c}} \mathbf{F} \cdot d\mathbf{s} = \int_{-5}^{10} \left(F_1 \frac{dx}{dt} + F_2 \frac{dy}{dt} + F_3 \frac{dz}{dt} \right) dt = \int_{-5}^{10} (t^5 + 2t^5 + 3t^5)\, dt = [t^6]_{-5}^{10} = 984{,}375.$$

On the other hand, for

$$\mathbf{c}_{\text{op}}\colon [-5, 10] \to \mathbb{R}^3, t \mapsto \mathbf{c}(5 - t) = (5 - t, (5 - t)^2, (5 - t)^3),$$

we have $dx/dt = -1$, $dy/dt = -10 + 2t = -2(5 - t)$, $dz/dt = -75 + 30t - 3t^2 = -3(5 - t)^2$, and $\mathbf{F}(\mathbf{c}_{\text{op}}(t)) = (5 - t)^5\mathbf{i} + (5 - t)^4\mathbf{j} + (5 - t)^3\mathbf{k}$. Therefore,

$$\int_{\mathbf{c}_{\text{op}}} \mathbf{F} \cdot d\mathbf{s} = \int_{-5}^{10} [-(5 - t)^5 - 2(5 - t)^5 - 3(5 - t)^5]\, dt = [(5 - t)^6]_{-5}^{10} = -984{,}375. \quad \blacktriangle$$

We are interested in reparametrizations, because if the image of a particular $\mathbf{c}$ can be represented in many ways, we want to be sure that path and line integrals depend only on the image curve and not on the particular parametrization. For example, for some problems the unit circle may be conveniently represented by the map $\mathbf{p}$ given by

$$x(t) = \cos 2t, \qquad y(t) = \sin 2t, \qquad 0 \le t \le \pi.$$

Theorem 1 guarantees that any integral computed for this representation will be the same as when we represent the circle by the map $\mathbf{c}$ given by

$$x(t) = \cos t, \qquad y(t) = \sin t, \qquad 0 \le t \le 2\pi,$$

because $\mathbf{p} = \mathbf{c} \circ h$, where $h(t) = 2t$, and thus $\mathbf{p}$ is an orientation-preserving reparametrization of $\mathbf{c}$. However, notice that the map γ given by

$$\gamma(t) = (\cos t, \sin t), \qquad 0 \leq t \leq 4\pi$$

is *not* a reparametrization of $\mathbf{c}$. Although it traces out the same image (the circle), it does so twice. (Why does this imply that γ is not a reparametrization of $\mathbf{c}$?)

The line integral is an *oriented integral*, in that a change of sign occurs (as we have seen in Theorem 1) if the orientation of the curve is reversed. The *path integral* does not have this property. This follows from the fact that changing t to $-t$ (reversing orientation) just changes the sign of $\mathbf{c}'(t)$, not its length. This is one of the differences between line and path integrals. The following theorem, which is proved by the same method as Theorem 1, shows that path integrals are unchanged under reparametrizations—even orientation-reversing ones.

THEOREM 2: Change of Parametrization for Path Integrals Let $\mathbf{c}$ be piecewise C^1, let f be a continuous (real-valued) function on the image of $\mathbf{c}$, and let $\mathbf{p}$ be any reparametrization of $\mathbf{c}$. Then

$$\int_{\mathbf{c}} f(x, y, z)\, ds = \int_{\mathbf{p}} f(x, y, z)\, ds. \tag{2}$$

Line Integrals of Gradient Fields

We next consider a useful technique for evaluating certain types of line integrals. Recall that a vector field $\mathbf{F}$ is a *gradient vector field* if $\mathbf{F} = \nabla f$ for some real-valued function f. Thus,

$$\mathbf{F} = \frac{\partial f}{\partial x}\mathbf{i} + \frac{\partial f}{\partial y}\mathbf{j} + \frac{\partial f}{\partial z}\mathbf{k}.$$

Suppose g and G are real-valued continuous functions defined on a closed interval $[a, b]$, that G is differentiable on (a, b), and that $G' = g$. Then by the fundamental theorem of calculus

$$\int_a^b g(x)\, dx = G(b) - G(a).$$

Thus, the value of the integral of g depends only on the value of G at the endpoints of the interval $[a, b]$. Because ∇f represents the derivative of f, one can ask whether $\int_{\mathbf{c}} \nabla f \cdot d\mathbf{s}$ is completely determined by the value of f at the endpoints $\mathbf{c}(a)$ and $\mathbf{c}(b)$. The answer is contained in the following *generalization of the fundamental theorem of calculus*.

THEOREM 3: Line Integrals of Gradient Vector Fields Suppose that $f\colon \mathbb{R}^3 \to \mathbb{R}$ is of class C^1 and that $\mathbf{c}\colon [a, b] \to \mathbb{R}^3$ is a piecewise C^1 path. Then

$$\int_{\mathbf{c}} \nabla f \cdot d\mathbf{s} = f(\mathbf{c}(b)) - f(\mathbf{c}(a)).$$

PROOF Apply the chain rule to the composite function

$$F\colon t \mapsto f(\mathbf{c}(t))$$

to obtain

$$F'(t) = (f \circ \mathbf{c})'(t) = \nabla f(\mathbf{c}(t)) \cdot \mathbf{c}'(t).$$

The function F is a real-valued function of the variable t, and so, by the fundamental theorem of single-variable calculus,

$$\int_a^b F'(t)\, dt = F(b) - F(a) = f(\mathbf{c}(b)) - f(\mathbf{c}(a)).$$

Therefore,

$$\int_{\mathbf{c}} \nabla f \cdot d\mathbf{s} = \int_a^b \nabla f(\mathbf{c}(t)) \cdot \mathbf{c}'(t)\, dt = \int_a^b F'(t)\, dt = F(b) - F(a)$$
$$= f(\mathbf{c}(b)) - f(\mathbf{c}(a)). \quad \blacksquare$$

EXAMPLE 9 Let $\mathbf{c}$ be the path $\mathbf{c}(t) = (t^4/4, \sin^3(t\pi/2), 0)$, $t \in [0, 1]$. Evaluate

$$\int_{\mathbf{c}} y\, dx + x\, dy$$

(which means $\int_{\mathbf{c}} y\, dx + x\, dy + 0\, dz$).

SOLUTION We recognize $y\, dx + x\, dy$, or equivalently, the vector field $y\mathbf{i} + x\mathbf{j} + 0\mathbf{k}$, as the gradient of the function $f(x, y, z) = xy$. Thus,

$$\int_{\mathbf{c}} y\, dx + x\, dy = f(\mathbf{c}(1)) - f(\mathbf{c}(0)) = \frac{1}{4} \cdot 1 - 0 = \frac{1}{4}. \quad \blacktriangle$$

Obviously, if one can recognize the integrand as a gradient, then evaluation of the integral becomes much easier. For example, the reader should try to work out the integral in Example 9 directly. In one-variable calculus, every integral is, in principle, obtainable by finding an antiderivative. For vector fields, however, this is not always true, because a given vector field need not be a gradient. This point will be examined in detail in Section 8.3, where we derive a test to determine when a vector field $\mathbf{F}$ is a gradient; that is, when $\mathbf{F} = \nabla f$ for some f.

Line Integrals Over Geometric Curves

We have seen how to define path integrals (integrals of scalar functions) and line integrals (integrals of vector functions) over parametrized curves. We have also seen that our work is simplified if we make a judicious choice of parametrization. Because these integrals are independent of the parametrization (except possibly for the sign), it seems natural to express the theory in a way that is independent of the parametrization, and that is thereby more "geometric." We do this briefly and somewhat informally in the following discussion.

DEFINITION We define a ***simple curve*** C to be the image of a piecewise C^1 map $\mathbf{c}\colon I \to \mathbb{R}^3$ that is one-to-one on an interval I; $\mathbf{c}$ is called a ***parametrization*** of C. Thus, a simple curve is one that does not intersect itself (Figure 7.2.6). If $I = [a, b]$, we call $\mathbf{c}(a)$ and $\mathbf{c}(b)$ ***endpoints*** of the curve.

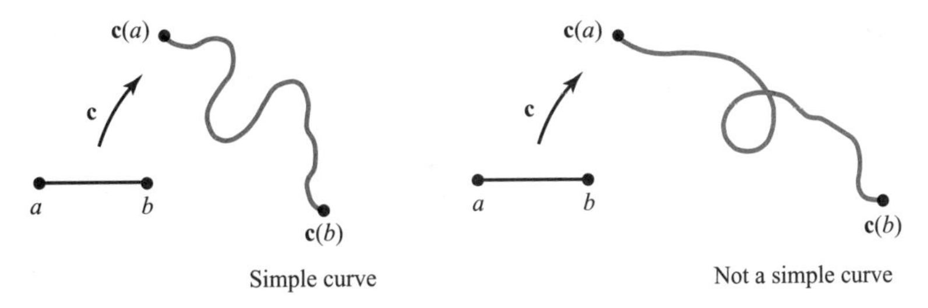

Figure 7.2.6 A simple curve that has no self-intersections is shown on the left. On the right is a curve with a self-intersection, so it is not simple.

Each simple curve C has two orientations or directions associated with it. If P and Q are the endpoints of the curve, then we can consider C as directed either from P to Q or from Q to P. The simple curve C together with a sense of direction is called an ***oriented simple curve*** or ***directed simple curve*** (Figure 7.2.7).

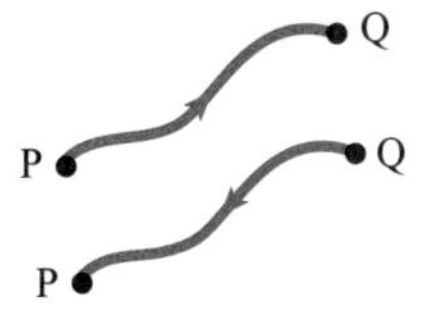

Figure 7.2.7 There are two possible senses of direction on a curve joining P and Q.

> DEFINITION: Simple Closed Curves By a ***simple closed curve*** we mean the image of a piecewise C^1 map $\mathbf{c}\colon [a, b] \to \mathbb{R}^3$ that is one-to-one on $[a, b)$ and satisfies $\mathbf{c}(a) = \mathbf{c}(b)$ (Figure 7.2.8). If $\mathbf{c}$ satisfies the condition $\mathbf{c}(a) = \mathbf{c}(b)$, but is not necessarily one-to-one on $[a, b)$, we call its image a ***closed curve***. Simple closed curves have two orientations, corresponding to the two possible directions of motion along the curve (Figure 7.2.9).

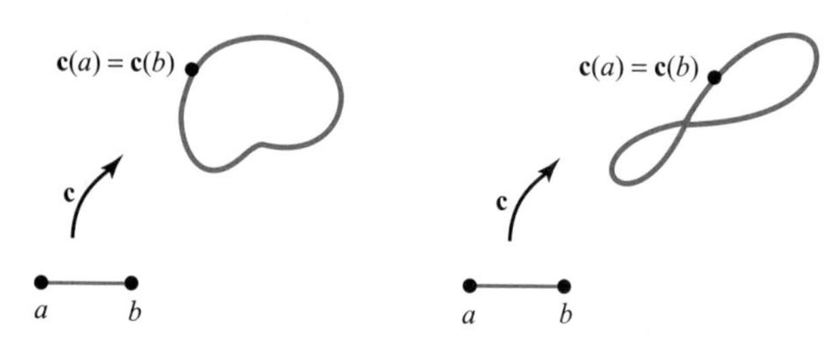

Figure 7.2.8 A simple closed curve (left) and a closed curve that is not simple (right).

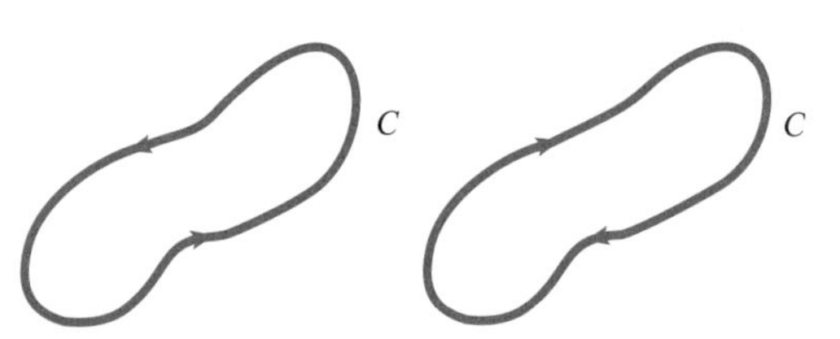

Figure 7.2.9 Two possible orientations for a simple closed curve C.

If C is an oriented simple curve or an oriented simple closed curve, we may unambiguously define line integrals along them.

> Line Integrals and Path Integrals Over Oriented Simple Curves and Simple Closed Curves C:
>
> $$\int_C \mathbf{F}\cdot d\mathbf{s} = \int_{\mathbf{c}} \mathbf{F}\cdot d\mathbf{s} \qquad \text{and} \qquad \int_C f\,ds = \int_{\mathbf{c}} f\,ds, \tag{3}$$
>
> where $\mathbf{c}$ is any *orientation-preserving* parametrization of C.

These integrals do not depend on the choice of $\mathbf{c}$ as long as $\mathbf{c}$ is one-to-one (except possibly at the endpoints) by virtue of Theorems 1 and 2.[5] The point we want to make here is that *although a curve must be parametrized to make integration along it tractable, it is not necessary to include the parametrization in our notation for the integral.*

[5] We have not proved that any two one-to-one paths $\mathbf{c}$ and $\mathbf{p}$ with the same image must be reparametrizations of each other, but this technical point will be omitted.

EXAMPLE 10 If $I = [a, b]$ is a closed interval on the x axis, then I, as a curve, has two orientations: one corresponding to motion from a to b (left to right) and the other corresponding to motion from b to a (right to left). If f is a real-valued function continuous on I, then denoting I with the first orientation by I^+ and I with the second orientation by I^-, we have

$$\int_{I^+} f(x)\,dx = \int_a^b f(x)\,dx = -\int_b^a f(x)\,dx = -\int_{I^-} f(x)\,dx. \quad \blacktriangle$$

A given simple closed curve can be parametrized in many different ways. Figure 7.2.10 shows C represented as the image of a map $\mathbf{p}$, with $\mathbf{p}(t)$ progressing in a prescribed direction around an oriented curve C as t ranges from a to b. Note that $\mathbf{p}'(t)$ points in this direction also. The speed with which we traverse C may vary from parametrization to parametrization, but as long as the orientation is preserved, the integral will not, according to Theorems 1 and 2.

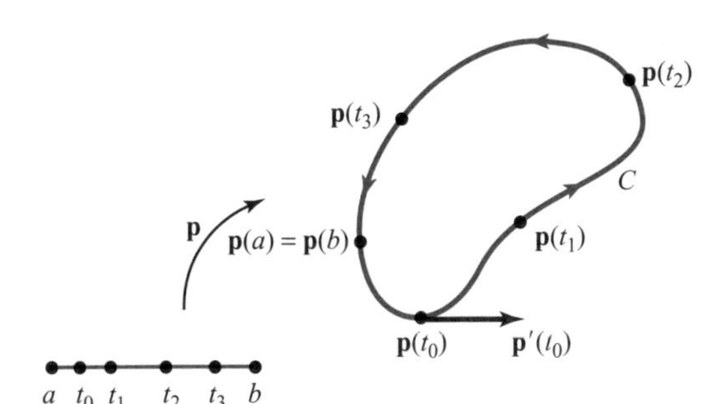

Figure 7.2.10 As t goes from a to b, $\mathbf{p}(t)$ moves around the curve C in some fixed direction.

The following precaution should be noted in regard to these remarks. It is possible to have two mappings $\mathbf{c}$ and $\mathbf{p}$ with the same image, and inducing the same orientation on the image, such that

$$\int_{\mathbf{c}} \mathbf{F}\cdot d\mathbf{s} \neq \int_{\mathbf{p}} \mathbf{F}\cdot d\mathbf{s}.$$

For an example, let $\mathbf{c}(t) = (\cos t, \sin t, 0)$ and $\mathbf{p}(t) = (\cos 2t, \sin 2t, 0)$, $0 \le t \le 2\pi$, with $\mathbf{F}(x, y, z) = (y, 0, 0)$. Then $F_1(x, y, z) = y$, $F_2(x, y, z) = 0$, and $F_3(x, y, z) = 0$, so

$$\int_{\mathbf{c}} \mathbf{F}\cdot d\mathbf{s} = \int_0^{2\pi} F_1(\mathbf{c}(t))\,\frac{dx}{dt}\,dt = -\int_0^{2\pi} \sin^2 t\,dt = -\pi.$$

But $\int_{\mathbf{p}} \mathbf{F}\cdot d\mathbf{s} = -2\int_0^{2\pi} \sin^2 2t\,dt = -2\pi$. Clearly, $\mathbf{c}$ and $\mathbf{p}$ have the same image, namely, the unit circle in the xy plane. Moreover, they traverse the unit circle in the same direction; yet $\int_{\mathbf{c}} \mathbf{F}\cdot d\mathbf{s} \neq \int_{\mathbf{p}} \mathbf{F}\cdot d\mathbf{s}$. The reason for this is that $\mathbf{c}$ is one-to-one, but $\mathbf{p}$ is not ($\mathbf{p}$ traverses the unit circle *twice* in a counterclockwise direction); therefore, $\mathbf{p}$ is not a parametrization of the unit circle as a simple closed curve.

As a consequence of Theorem 1 and generalizing the notation in Example 10, we introduce the following convention:

Line Integrals Over Curves with Opposite Orientations Let C^- be the same curve as C, but with the opposite orientation. Then

$$\int_C \mathbf{F} \cdot d\mathbf{s} = -\int_{C^-} \mathbf{F} \cdot d\mathbf{s}.$$

We also have:

Line Integrals Over Curves Consisting of Several Components Let C be an oriented curve that is made up of several oriented component curves C_i, $i = 1, \ldots, k$, as in Figure 7.2.11. Then we shall write $C = C_1 + C_2 + \cdots + C_k$. Because we can parametrize C by parametrizing the pieces $C_1, \ldots, C_k$ separately, one can prove that

$$\int_C \mathbf{F} \cdot d\mathbf{s} = \int_{C_1} \mathbf{F} \cdot d\mathbf{s} + \int_{C_2} \mathbf{F} \cdot d\mathbf{s} + \cdots + \int_{C_k} \mathbf{F} \cdot d\mathbf{s}. \tag{4}$$

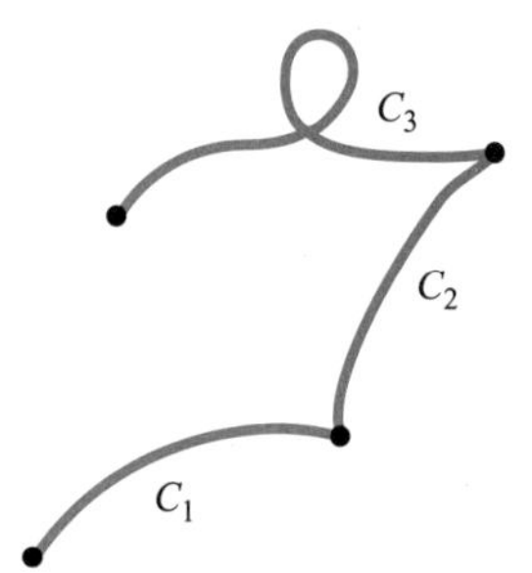

Figure 7.2.11 A curve can be made up of several components.

One reason for writing a curve as a sum of components is that it may be easier to parametrize the components C_i individually than it is to parametrize C as a whole. If that is the case, formula (4) provides a convenient way of evaluating $\int_C \mathbf{F} \cdot d\mathbf{s}$.

The $d\mathbf{r}$ Notation for Line Integrals

Sometimes one writes, as we occasionally do later, the line integral using the notation

$$\int_C \mathbf{F} \cdot d\mathbf{r}.$$

The reason is that we think of describing a C^1 path $\mathbf{c}$ in terms of a moving *position vector* based at the origin and ending at the point $\mathbf{c}(t)$ at time t. Position vectors are often denoted by $\mathbf{r} = x\mathbf{i} + y\mathbf{j} + z\mathbf{k}$, and so the curve is described using the notation $\mathbf{r}(t) = x(t)\mathbf{i} + y(t)\mathbf{j} + z(t)\mathbf{k}$ in place of $\mathbf{c}(t)$. By definition, the line integral is given by

$$\int_a^b \mathbf{F}(\mathbf{r}(t)) \cdot \frac{d\mathbf{r}}{dt}\, dt.$$

Formally canceling the dt's, and using the parametrization independence to replace the limits of integration with the geometric curve C, we arrive at the notation $\int_C \mathbf{F} \cdot d\mathbf{r}$.

EXAMPLE 11 Consider C, the perimeter of the unit square in $\mathbb{R}^2$, oriented in the counterclockwise sense (see Figure 7.2.12). Evaluate the line integral

$$\int_C x^2\, dx + xy\, dy.$$

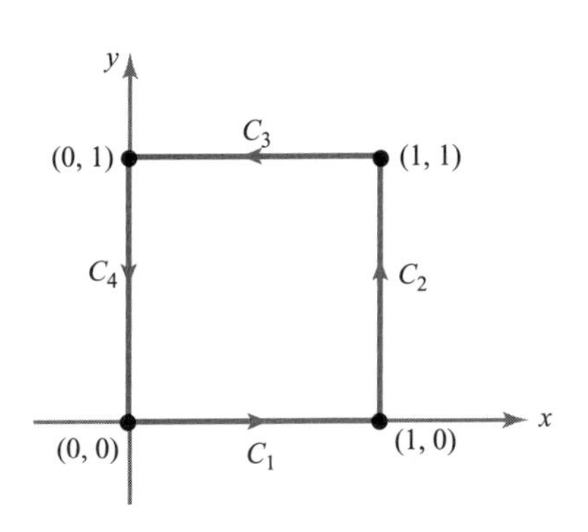

Figure 7.2.12 The perimeter of the unit square, parametrized in four pieces.

SOLUTION We evaluate the integral using a convenient parametrization of C that induces the given orientation. For example:

$$\mathbf{c}\colon [0,4] \to \mathbb{R}^2, \qquad t \mapsto \begin{cases} (t, 0) & 0 \le t \le 1 \\ (1, t-1) & 1 \le t \le 2 \\ (3-t, 1) & 2 \le t \le 3 \\ (0, 4-t) & 3 \le t \le 4. \end{cases}$$

Then

$$\begin{aligned} \int_C x^2\, dx + xy\, dy &= \int_0^1 (t^2 + 0)\, dt + \int_1^2 [0 + (t-1)]\, dt \\ &\quad + \int_2^3 [-(3-t)^2 + 0]\, dt + \int_3^4 (0+0)\, dt \\ &= \frac{1}{3} + \frac{1}{2} + \left(-\frac{1}{3}\right) + 0 = \frac{1}{2}. \end{aligned}$$

Now let us reevaluate this line integral, using formula (4) and parametrizing the C_i separately. Notice that $C = C_1 + C_2 + C_3 + C_4$, where C_i are the oriented curves pictured in Figure 7.2.12. These can be parametrized as follows:

$$C_1\colon\ \mathbf{c}_1(t) = (t, 0),\ 0 \le t \le 1$$
$$C_2\colon\ \mathbf{c}_2(t) = (1, t),\ 0 \le t \le 1$$
$$C_3\colon\ \mathbf{c}_3(t) = (1 - t, 1),\ 0 \le t \le 1$$
$$C_4\colon\ \mathbf{c}_4(t) = (0, 1 - t),\ 0 \le t \le 1,$$

and so

$$\int_{C_1} x^2\,dx + xy\,dy = \int_0^1 t^2\,dt = \frac{1}{3}$$
$$\int_{C_2} x^2\,dx + xy\,dy = \int_0^1 t\,dt = \frac{1}{2}$$
$$\int_{C_3} x^2\,dx + xy\,dy = \int_0^1 -(1-t)^2\,dt = -\frac{1}{3}$$
$$\int_{C_4} x^2\,dx + xy\,dy = \int_0^1 0\,dt = 0.$$

Thus, again,

$$\int_C x^2\,dx + xy\,dy = \frac{1}{3} + \frac{1}{2} - \frac{1}{3} + 0 = \frac{1}{2}. \quad \blacktriangle$$

EXAMPLE 12 An interesting application of the line integral is the mathematical formulation of Ampère's law, which relates electric currents to their magnetic effects.[6] Suppose $\mathbf{H}$ denotes a magnetic field in $\mathbb{R}^3$, and let C be a closed oriented curve in $\mathbb{R}^3$. In appropriate physical units, Ampère's law states that

$$\int_C \mathbf{H} \cdot d\mathbf{s} = I,$$

where I is the net current that passes through any surface bounded by C (see Figure 7.2.13). ▲

Finally, let us mention that the line integral has another important physical meaning, specifically, the interpretation of $\int_C \mathbf{V} \cdot d\mathbf{s}$ as *circulation*, where $\mathbf{V}$ is the velocity

[6]The discovery that electric currents produce magnetic effects was made by Haas Christian Oersted circa 1820. See any elementary physics text for discussions of the physical basis of these ideas.

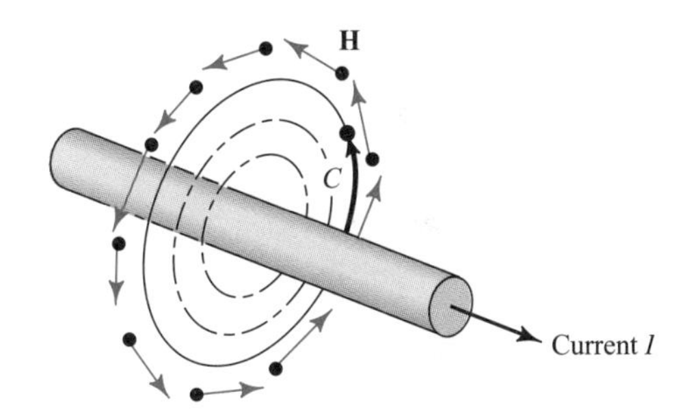

Figure 7.2.13 The magnetic field $\mathbf{H}$ surrounding a wire carrying a current I satisfies Ampère's law: $\int_C \mathbf{H} \cdot d\mathbf{s} = I$.

field of a fluid, as we shall discuss in Section 8.2. Thus, a wide variety of physical concepts, from the notion of work to electromagnetic fields and the motions of fluids, can be analyzed with the help of line integrals.

EXERCISES

1. Let $\mathbf{F}(x, y, z) = x\mathbf{i} + y\mathbf{j} + z\mathbf{k}$. Evaluate the integral of $\mathbf{F}$ along each of the following paths:

(a) $\mathbf{c}(t) = (t, t, t), \quad 0 \le t \le 1$

(b) $\mathbf{c}(t) = (\cos t, \sin t, 0), \quad 0 \le t \le 2\pi$

(c) $\mathbf{c}(t) = (\sin t, 0, \cos t), \quad 0 \le t \le 2\pi$

(d) $\mathbf{c}(t) = (t^2, 3t, 2t^3), \quad -1 \le t \le 2$

2. Evaluate each of the following line integrals:

(a) $\int_{\mathbf{c}} x\, dy - y\, dx, \quad \mathbf{c}(t) = (\cos t, \sin t), \quad 0 \le t \le 2\pi$

(b) $\int_{\mathbf{c}} x\, dx + y\, dy, \quad \mathbf{c}(t) = (\cos \pi t, \sin \pi t), \quad 0 \le t \le 2$

(c) $\int_{\mathbf{c}} yz\, dx + xz\, dy + xy\, dz$, where $\mathbf{c}$ consists of straight-line segments joining $(1, 0, 0)$ to $(0, 1, 0)$ to $(0, 0, 1)$

(d) $\int_{\mathbf{c}} x^2\, dx - xy\, dy + dz$, where $\mathbf{c}$ is the parabola $z = x^2, y = 0$ from $(-1, 0, 1)$ to $(1, 0, 1)$.

3. Consider the force field $\mathbf{F}(x, y, z) = x\mathbf{i} + y\mathbf{j} + z\mathbf{k}$. Compute the work done in moving a particle along the parabola $y = x^2, z = 0$, from $x = -1$ to $x = 2$.

4. Let $\mathbf{c}$ be a smooth path.

(a) Suppose $\mathbf{F}$ is perpendicular to $\mathbf{c}'(t)$ at the point $\mathbf{c}(t)$. Show that

$$\int_{\mathbf{c}} \mathbf{F} \cdot d\mathbf{s} = 0.$$

(b) If $\mathbf{F}$ is parallel to $\mathbf{c}'(t)$ at $\mathbf{c}(t)$, show that

$$\int_{\mathbf{c}} \mathbf{F} \cdot d\mathbf{s} = \int_{\mathbf{c}} \|\mathbf{F}\|\, ds.$$

[By parallel to $\mathbf{c}'(t)$ we mean that $\mathbf{F}(\mathbf{c}(t)) = \lambda(t)\mathbf{c}'(t)$, where $\lambda(t) > 0$.]

5. Suppose the path $\mathbf{c}$ has length l, and $\|\mathbf{F}\| \le M$. Prove that

$$\left| \int_{\mathbf{c}} \mathbf{F} \cdot d\mathbf{s} \right| \le Ml.$$

6. Evaluate $\int_{\mathbf{c}} \mathbf{F} \cdot d\mathbf{s}$ where $\mathbf{F}(x, y, z) = y\mathbf{i} + 2x\mathbf{j} + y\mathbf{k}$ and the path $\mathbf{c}$ is defined by $\mathbf{c}(t) = t\mathbf{i} + t^2\mathbf{j} + t^3\mathbf{k}$, $0 \le t \le 1$.

7. Evaluate

$$\int_{\mathbf{c}} y\,dx + (3y^3 - x)\,dy + z\,dz$$

for each of the paths $\mathbf{c}(t) = (t, t^n, 0)$, $0 \le t \le 1$, where $n = 1, 2, 3, \ldots$.

8. This exercise refers to Example 12. Let L be a very long wire, a planar section of which (with the plane perpendicular to the wire) is shown in Figure 7.2.14. Suppose this plane is the xy plane. Experiments show that $\mathbf{H}$ is tangent to every circle in the xy plane whose center is the axis of L, and that the magnitude of $\mathbf{H}$ is constant on every such circle C. Thus, $\mathbf{H} = H\mathbf{T}$, where $\mathbf{T}$ is a unit tangent vector to C and H is some scalar. Using this information, show that $H = I/2\pi r$, where r is the radius of circle C and I is the current flowing in the wire.

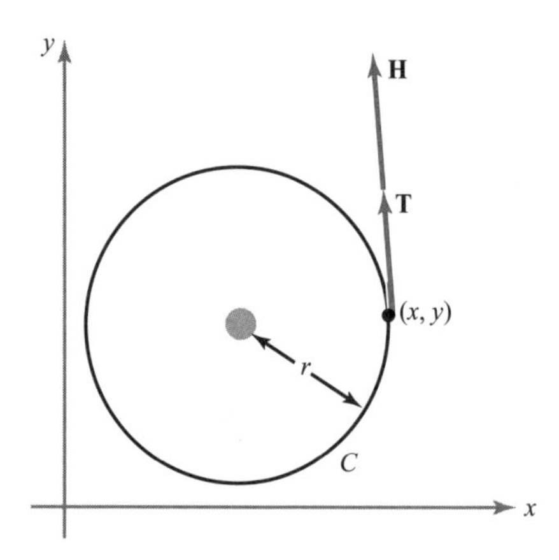

Figure 7.2.14 A planar section of a long wire and a curve C about the wire.

9. The image of the path $t \mapsto (\cos^3 t, \sin^3 t)$, $0 \le t \le 2\pi$ in the plane is shown in Figure 7.2.15. Evaluate the integral of the vector field $\mathbf{F}(x, y) = x\mathbf{i} + y\mathbf{j}$ around this curve.

10. Suppose $\mathbf{c}_1$ and $\mathbf{c}_2$ are two paths with the same endpoints and $\mathbf{F}$ is a vector field. Show that $\int_{\mathbf{c}_1} \mathbf{F} \cdot d\mathbf{s} = \int_{\mathbf{c}_2} \mathbf{F} \cdot d\mathbf{s}$ is equivalent to $\int_C \mathbf{F} \cdot d\mathbf{s} = 0$, where C is the closed curve obtained by first moving along $\mathbf{c}_1$ and then moving along $\mathbf{c}_2$ in the opposite direction.

11. Let $\mathbf{c}(t)$ be a path and $\mathbf{T}$ the unit tangent vector. What is $\int_{\mathbf{c}} \mathbf{T} \cdot d\mathbf{s}$?

12. Let $\mathbf{F} = (z^3 + 2xy)\mathbf{i} + x^2\mathbf{j} + 3xz^2\mathbf{k}$. Show that the integral of $\mathbf{F}$ around the circumference of the unit square with vertices $(\pm 1, \pm 1)$ is zero.

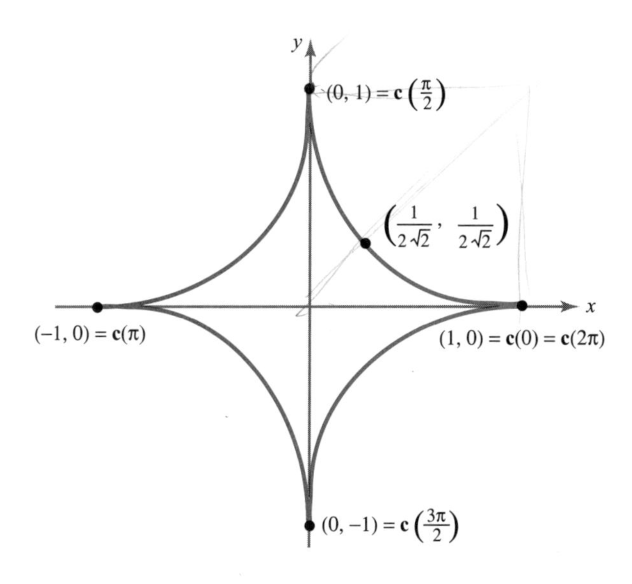

Figure 7.2.15 The hypocycloid $\mathbf{c}(t) = (\cos^3 t, \sin^3 t)$ (Exercise 9).

13. Using the path in Exercise 9, observe that a C^1 map $\mathbf{c}$: $[a, b] \to \mathbb{R}^3$ can have an image that does not "look smooth." Do you think this could happen if $\mathbf{c}'(t)$ were always nonzero?

14. What is the value of the integral of a gradient field around a closed curve C?

15. Evaluate the line integral

$$\int_C 2xyz\, dx + x^2 z\, dy + x^2 y\, dz,$$

where C is an oriented simple curve connecting $(1, 1, 1)$ to $(1, 2, 4)$.

16. Suppose $\nabla f(x, y, z) = 2xyze^{x^2}\mathbf{i} + ze^{x^2}\mathbf{j} + ye^{x^2}\mathbf{k}$. If $f(0, 0, 0) = 5$, find $f(1, 1, 2)$.

17. Consider the gravitational force field (with $G = m = M = 1$) defined [for $(x, y, z) \neq (0, 0, 0)$] by

$$\mathbf{F}(x, y, z) = -\frac{1}{(x^2 + y^2 + z^2)^{3/2}}(x\mathbf{i} + y\mathbf{j} + z\mathbf{k}).$$

Show that the work done by the gravitational force as a particle moves from (x_1, y_1, z_1) to (x_2, y_2, z_2) along any path depends only on the radii $R_1 = \sqrt{x_1^2 + y_1^2 + z_1^2}$ and $R_2 = \sqrt{x_2^2 + y_2^2 + z_2^2}$.

18. A cyclist rides up a mountain along the path shown in Figure 7.2.16. She makes one complete revolution around the mountain in reaching the top, while her vertical rate of climb is constant. Throughout the trip she exerts a force described by the vector field

$$\mathbf{F}(x, y, z) = y\mathbf{i} + x\mathbf{j} + \mathbf{k}.$$

What is the work done by the cyclist in traveling from A to B? What is unrealistic about this model of a cyclist?

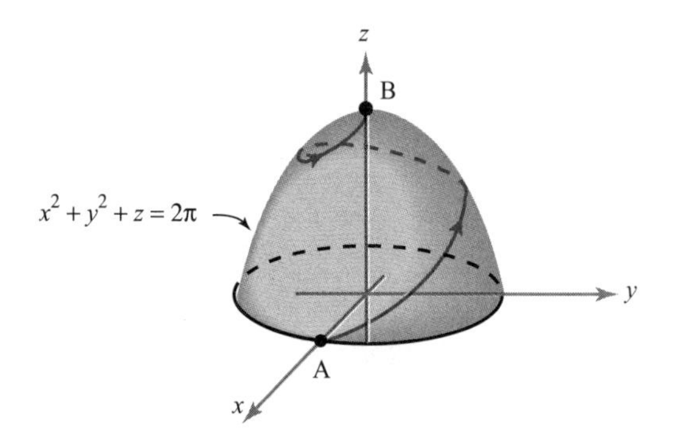

Figure 7.2.16 How much work is done in cycling up this mountain?

19. Let $\mathbf{c}$: $[a, b] \to \mathbb{R}^3$ be a path such that $\mathbf{c}'(t) \neq \mathbf{0}$. Recall from Section 4.1 that when this condition holds, $\mathbf{c}$ is said to be ***regular***. Let the function f be defined by the formula $f(x) = \int_a^x \|\mathbf{c}'(t)\|\, dt$.

(a) What is df/dx?

(b) Using the answer to part (a), prove that f: $[a, b] \to [0, L]$, where L is the length of $\mathbf{c}$, has a differentiable inverse g: $[0, L] \to [a, b]$ satisfying $f \circ g(s) = s$, $g \circ f(x) = x$. (You may use the one-variable inverse function theorem stated at the beginning of Section 3.5.)

(c) Compute dg/ds.

(d) Recall that a path $s \mapsto \mathbf{b}(s)$ is said to be of unit speed, or parametrized by arc length, if $\|\mathbf{b}'(s)\| = 1$. Show that the reparametrization of $\mathbf{c}$ given by $\mathbf{b}(s) = \mathbf{c} \circ g(s)$ is of unit speed. Conclude that any regular path can be reparametrized by arc length. (Thus, for example, the Frenet formulas in Exercise 17 of Section 4.2 can be applied to the reparametrization $\mathbf{b}$.)

20. Along a "thermodynamic path" C in (V, T, P) space,

(i) The heat gained is $\int_C \Lambda_V\, dV + K_V\, dT$, where Λ_V, K_V are functions of (V, T, P), depending on the particular physical system.

(ii) The work done is $\int_C P\, dV$.

For a van der Waals gas, we have

$$P(V, T) = \frac{RT}{V - b} - \frac{a}{V^2}, \qquad J\Lambda_V = \frac{RT}{V - b}, \qquad \text{and} \qquad K_V = \text{constant},$$

where R, b, a, and J are known constants. Initially the gas is at a temperature T_0 and volume V_0.

(a) An ***adiabatic*** process is a thermodynamic motion $(V(t), T(t), P(t))$ for which

$$\frac{dT}{dV} = \frac{dT/dt}{dV/dt} = -\frac{\Lambda_V}{K_V}.$$

If the van der Waals gas undergoes an adiabatic process in which the volume doubles to $2V_0$, compute

(1) the heat gained;
(2) the work done; and
(3) the final volume, temperature, and pressure.

(b) After the process indicated in part (a), the gas is cooled (or heated) at constant volume until the original temperature T_0 is achieved. Compute

(1) the heat gained;
(2) the work done; and
(3) the final volume, temperature, and pressure.

(c) After the process indicated in part (b), the gas is compressed until the gas returns to its original volume V_0. The temperature is held constant throughout the process. Compute

(1) the heat gained;
(2) the work done; and
(3) the final volume, temperature, and pressure.

(d) For the cyclic process described in parts (a), (b), (c), compute

(1) the total heat gained; and
(2) the total work done.

7.3 Parametrized Surfaces

In Sections 7.1 and 7.2, we studied integrals of scalar and vector functions along curves. Now we turn to integrals over surfaces and begin by studying the geometry of surfaces themselves.

Graphs Are Too Restrictive

We are already used to one kind of surface, namely, the graph of a function $f(x, y)$. Graphs were extensively studied in Chapter 2, and we know how to compute their tangent planes. However, it would be unduly limiting to restrict ourselves to this case. For example, many surfaces arise as level surfaces of functions. Suppose our surface S is the set of points (x, y, z) where $x - z + z^3 = 0$. Here S is a sheet that (relative to the xy plane) doubles back on itself (see Figure 7.3.1). Obviously, we want to

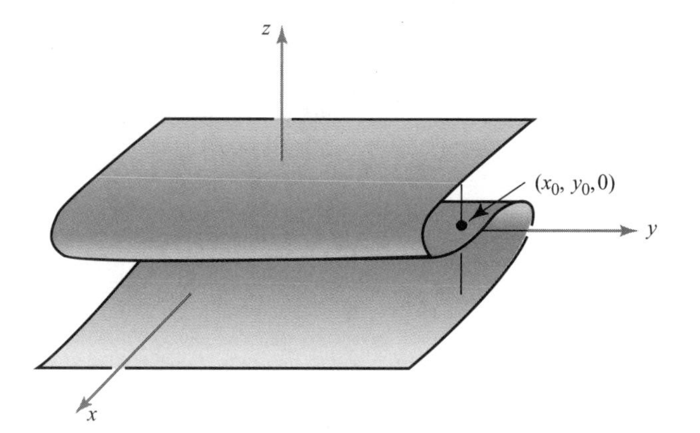

Figure 7.3.1 A surface that is not the graph of a function $z = f(x, y)$.

call S a surface, because it is just a plane with a wrinkle. However, S is *not* the graph of some function $z = f(x, y)$, because this means that for each $(x_0, y_0) \in \mathbb{R}^2$ there must be *one* z_0 with $(x_0, y_0, z_0) \in S$. As Figure 7.3.1 illustrates, this condition is violated.

Another example is the torus, or surface of a doughnut, which is depicted in Figure 7.3.2. Anyone would call a torus a surface; yet, by the same reasoning as before, a torus cannot be the graph of a differentiable function of two variables. These observations encourage us to extend our definition of a surface.

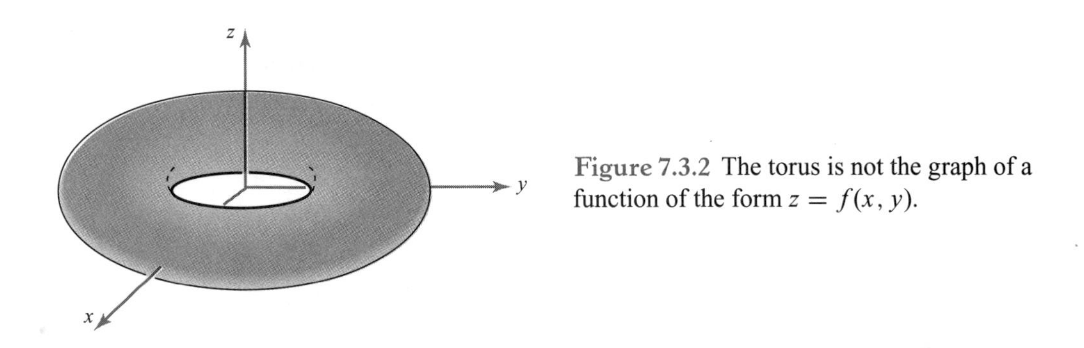

Figure 7.3.2 The torus is not the graph of a function of the form $z = f(x, y)$.

The motivation for the extended definition that follows is partly that a surface can be thought of as being obtained from the plane by "rolling," "bending," and "pushing." For example, to get a torus, we take a portion of the plane and roll it (see Figure 7.3.3), then take the two "ends" and bring them together until they meet (Figure 7.3.4).

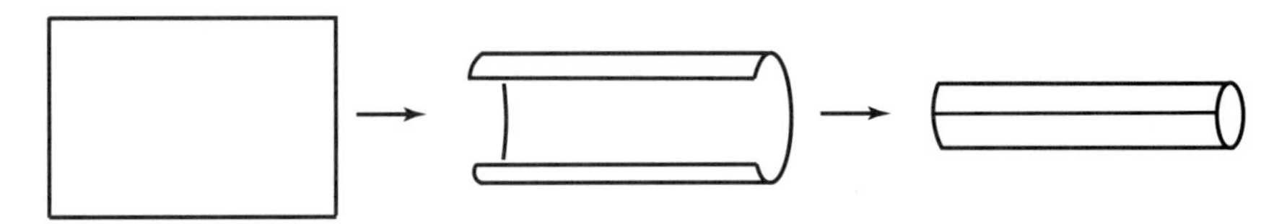

Figure 7.3.3 The first step in obtaining a torus from a rectangle is to make a cylinder.

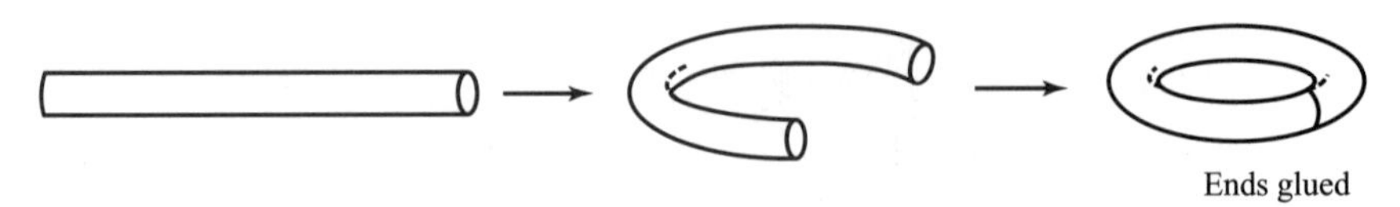

Figure 7.3.4 Bend the cylinder and glue the ends to get a torus.

Parametrized Surfaces as Mappings

In our study of differential calculus we dealt with mappings $f: A \subset \mathbb{R}^n \to \mathbb{R}^m$. Taking $n = 2$ and $m = 3$ corresponds to the case of a two-dimensional surface in 3-space. With surfaces, just as with curves, we want to distinguish a map (a parametrization) from its image (a geometric object). This leads us to the following definition.

> DEFINITION: Parametrized Surfaces A ***parametrization of a surface*** is a function $\mathbf{\Phi}: D \subset \mathbb{R}^2 \to \mathbb{R}^3$, where D is some domain in $\mathbb{R}^2$. The ***surface*** S corresponding to the function $\mathbf{\Phi}$ is its image: $S = \mathbf{\Phi}(D)$. We can write
>
> $$\mathbf{\Phi}(u, v) = (x(u, v), y(u, v), z(u, v)).$$
>
> If $\mathbf{\Phi}$ is differentiable or is of class C^1 [which is the same as saying that $x(u, v)$, $y(u, v)$, and $z(u, v)$ are differentiable or C^1 functions of (u, v)], we call S a ***differentiable or a*** C^1 ***surface***.

We can think of $\mathbf{\Phi}$ as twisting or bending the region D in the plane to yield the surface S (see Figure 7.3.5). Thus, each point (u, v) in D becomes a label for a point $(x(u, v), y(u, v), z(u, v))$ on S.

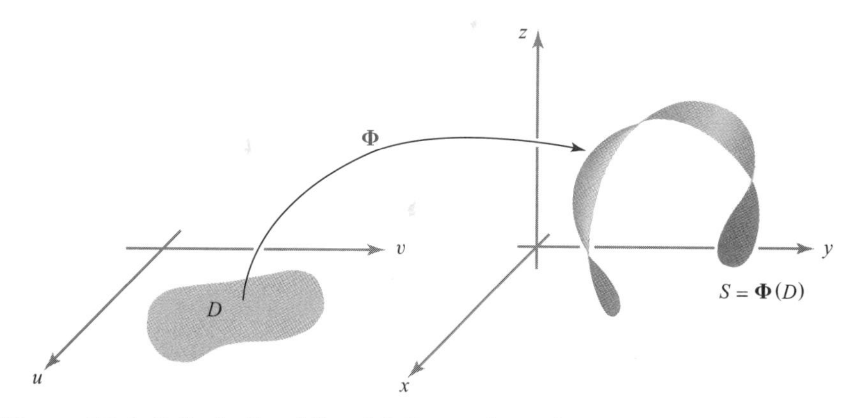

Figure 7.3.5 $\mathbf{\Phi}$ "twists" and "bends" D onto the surface $S = \mathbf{\Phi}(D)$.

Of course, surfaces need not bend or twist at all. In fact, planes are flat, as shown in our first, and simplest, example.

EXAMPLE 1 In Section 1.3 we studied the equation of a plane P. We did so in terms of graphs and level sets. Now we examine the same notion using a parametrization.

Let P be a plane that is parallel to two vectors α and β and that passes through the tip of another vector γ, as in Figure 7.3.6.

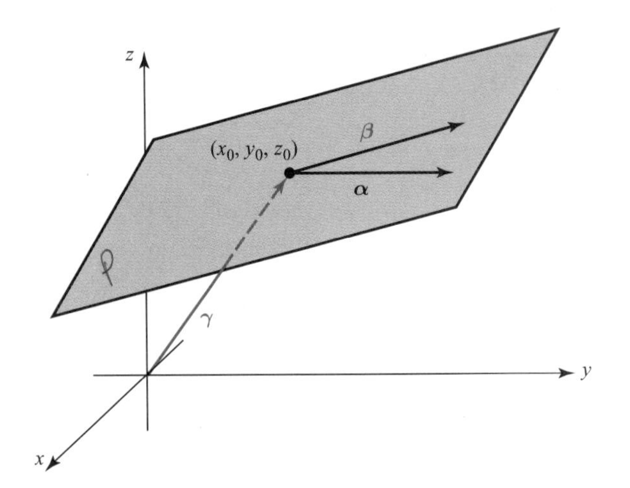

Figure 7.3.6 Describing a plane parametrically.

Our goal in this example is to find a parametrization of this plane. Notice that the vector $\alpha \times \beta = \mathbf{N}$, which we also write as $A\mathbf{i} + B\mathbf{j} + C\mathbf{k}$, is normal to P. If the tip of γ is the point (x_0, y_0, z_0), then the equation of P as a level set (as discussed in Section 1.3) is given by:

$$A(x - x_0) + B(y - y_0) + C(z - z_0) = 0.$$

However, the set of all points on the plane P can also be described by the set of all vectors that are γ plus a linear combination of α and β. Using our preferred choice of real parameters u and v, we arrive at the ***parametric equation of the plane*** P:

$$\Phi(u, v) = \alpha u + \beta v + \gamma. \quad \blacktriangle$$

Tangent Vectors to Parametrized Surfaces

Suppose that Φ is a parametrized surface that is differentiable at $(u_0, v_0) \in \mathbb{R}^2$. Fixing u at u_0, we get a map $\mathbb{R} \to \mathbb{R}^3$ given by $t \mapsto \Phi(u_0, t)$, whose image is a curve on the surface (Figure 7.3.7). From Chapters 2 and 4 we know that the vector tangent to this curve at the point $\Phi(u_0, v_0)$, which we denote by $\mathbf{T}_v$, is given by

$$\mathbf{T}_v = \frac{\partial \Phi}{\partial v} = \frac{\partial x}{\partial v}(u_0, v_0)\mathbf{i} + \frac{\partial y}{\partial v}(u_0, v_0)\mathbf{j} + \frac{\partial z}{\partial v}(u_0, v_0)\mathbf{k}.$$

Similarly, if we fix v and consider the curve $t \mapsto \Phi(t, v_0)$, we obtain the tangent vector to this curve at $\Phi(u_0, v_0)$, given by

$$\mathbf{T}_u = \frac{\partial \Phi}{\partial u} = \frac{\partial x}{\partial u}(u_0, v_0)\mathbf{i} + \frac{\partial y}{\partial u}(u_0, v_0)\mathbf{j} + \frac{\partial z}{\partial u}(u_0, v_0)\mathbf{k}.$$

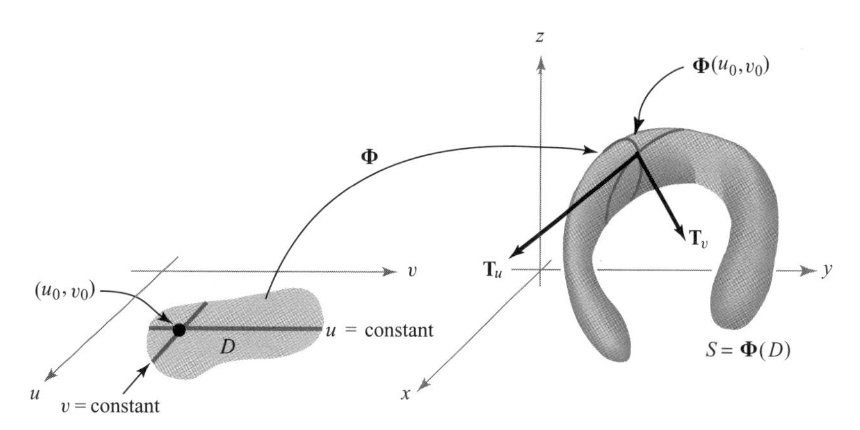

Figure 7.3.7 The tangent vectors $\mathbf{T}_u$ and $\mathbf{T}_v$ that are tangent to the curve on a surface S, and hence tangent to S.

Regular Surfaces

Because the vectors $\mathbf{T}_u$ and $\mathbf{T}_v$ are tangent to two curves on the surface at a given point, the vector $\mathbf{T}_u \times \mathbf{T}_v$ ought to be normal to the surface at the same point.

We say that the surface S is ***regular*** or ***smooth***[7] at $\Phi(u_0, v_0)$, provided that $\mathbf{T}_u \times \mathbf{T}_v \neq \mathbf{0}$ at (u_0, v_0). The surface is called ***regular*** if it is regular at all points $\Phi(u_0, v_0) \in S$. The nonzero vector $\mathbf{T}_u \times \mathbf{T}_v$ is *normal* to S (recall that the vector product of $\mathbf{T}_u$ and $\mathbf{T}_v$ is perpendicular to the plane spanned by $\mathbf{T}_u$ and $\mathbf{T}_v$); the fact that it is nonzero ensures that there will be a tangent plane. Intuitively, a smooth surface has no "corners."[8]

EXAMPLE 2 Consider the surface given by the equations

$$x = u \cos v, \qquad y = u \sin v, \qquad z = u, \qquad u \geq 0.$$

Is this surface differentiable? Is it regular?

SOLUTION These equations describe the surface $z = \sqrt{x^2 + y^2}$ (square the equations for x, y, and z to check this), which is shown in Figure 7.3.8. This surface is a cone with a "point" at $(0, 0, 0)$; it is a differentiable surface because each component function is differentiable as a function of u and v. However, *the surface*

---

[7] Strictly speaking, regularity depends on the parametrization Φ and not just on its image S. Therefore, this terminology is somewhat imprecise; however, it is descriptive and should not cause confusion. (See Exercise 15.)

[8] In Section 3.5, we showed that level surfaces $f(x, y, z) = 0$ were in fact graphs of functions of two variables in some neighborhood of a point (x_0, y_0, z_0) satisfying $\nabla f(x_0, y_0, z_0) \neq 0$. This united two concepts of a surface—graphs and level sets. Again, using the implicit function theorem, it is likewise possible to show that the image of a parametrized surface Φ in the neighborhood of a point (u_0, v_0) where $\mathbf{T}_u \times \mathbf{T}_v \neq \mathbf{0}$ is also the graph of a function of two variables. Thus, all definitions of a surface are consistent. (See Exercise 16.)

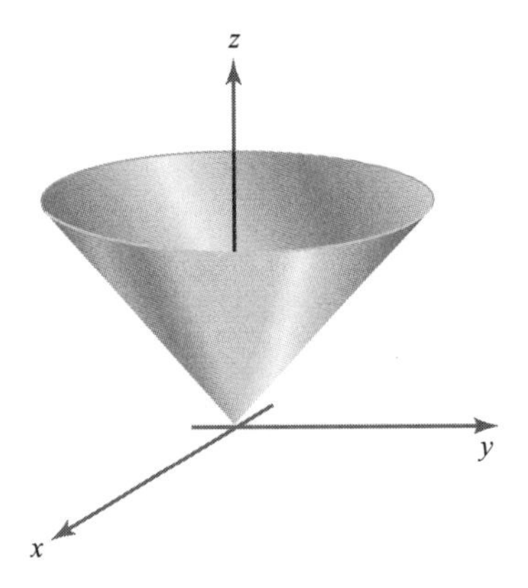

Figure 7.3.8 The surface $z = \sqrt{x^2 + y^2}$ is a cone. It is not regular at its tip.

is not regular at $(0, 0, 0)$. To see this, compute $\mathbf{T}_u$ and $\mathbf{T}_v$ at $(0, 0) \in \mathbb{R}^2$:

$$\mathbf{T}_u = \frac{\partial \mathbf{\Phi}}{\partial u} = \frac{\partial x}{\partial u}(0, 0)\mathbf{i} + \frac{\partial y}{\partial u}(0, 0)\mathbf{j} + \frac{\partial z}{\partial u}(0, 0)\mathbf{k} = (\cos 0)\mathbf{i} + (\sin 0)\mathbf{j} + \mathbf{k} = \mathbf{i} + \mathbf{k},$$

and similarly,

$$\mathbf{T}_v = \frac{\partial \mathbf{\Phi}}{\partial v} = 0(-\sin 0)\mathbf{i} + 0(\cos 0)\mathbf{j} + 0\mathbf{k} = \mathbf{0}.$$

Thus, $\mathbf{T}_u \times \mathbf{T}_v = \mathbf{0}$, and so, by definition, the surface is not regular at $(0, 0, 0)$. ▲

Tangent Plane to a Parametrized Surface

We can use the fact that $\mathbf{n} = \mathbf{T}_u \times \mathbf{T}_v$ is normal to the surface to both formally define the tangent plane and to compute it.

DEFINITION: The Tangent Plane to a Surface If a parametrized surface $\mathbf{\Phi}: D \subset \mathbb{R}^2 \to \mathbb{R}^3$ is *regular* at $\mathbf{\Phi}(u_0, v_0)$, that is, if $\mathbf{T}_u \times \mathbf{T}_v \neq \mathbf{0}$ at (u_0, v_0), we define the ***tangent plane*** of the surface at $\mathbf{\Phi}(u_0, v_0)$ to be the plane determined by the vectors $\mathbf{T}_u$ and $\mathbf{T}_v$. Thus, $\mathbf{n} = \mathbf{T}_u \times \mathbf{T}_v$ is a normal vector, and an equation of the tangent plane at (x_0, y_0, z_0) on the surface is given by

$$(x - x_0, y - y_0, z - z_0) \cdot \mathbf{n} = 0, \tag{1}$$

where $\mathbf{n}$ is evaluated at (u_0, v_0); that is, the tangent plane is the set of (x, y, z) satisfying (1). If $\mathbf{n} = (n_1, n_2, n_3) = n_1\mathbf{i} + n_2\mathbf{j} + n_3\mathbf{k}$, then formula (1) becomes

$$n_1(x - x_0) + n_2(y - y_0) + n_3(z - z_0) = 0. \tag{1'}$$

EXAMPLE 3 Let $\mathbf{\Phi}: \mathbb{R}^2 \to \mathbb{R}^3$ be given by

$$x = u\cos v, \qquad y = u\sin v, \qquad z = u^2 + v^2.$$

Where does a tangent plane exist? Find the tangent plane at $\mathbf{\Phi}(1, 0)$.

SOLUTION We compute

$$\mathbf{T}_u = (\cos v)\mathbf{i} + (\sin v)\mathbf{j} + 2u\mathbf{k} \qquad \text{and} \qquad \mathbf{T}_v = -u(\sin v)\mathbf{i} + u(\cos v)\mathbf{j} + 2v\mathbf{k},$$

so the tangent plane at the point $\mathbf{\Phi}(u_0, v_0)$ is the set of vectors through $\mathbf{\Phi}(u_0, v_0)$ perpendicular to

$$(\mathbf{T}_u \times \mathbf{T}_v)(u_0, v_0) = (-2u_0^2\cos v_0 + 2v_0\sin v_0, -2u_0^2\sin v_0 - 2v_0\cos v_0, u_0)$$

if this vector is nonzero. Because $\mathbf{T}_u \times \mathbf{T}_v$ is equal to $\mathbf{0}$ at $(u_0, v_0) = (0, 0)$, we cannot find a tangent plane at $\mathbf{\Phi}(0, 0) = (0, 0, 0)$. However, we can find an equation of the tangent plane at all the other points, where $\mathbf{T}_u \times \mathbf{T}_v \neq \mathbf{0}$. At the point $\mathbf{\Phi}(1, 0) = (1, 0, 1)$,

$$\mathbf{n} = (\mathbf{T}_u \times \mathbf{T}_v)(1, 0) = (-2, 0, 1) = -2\mathbf{i} + \mathbf{k}.$$

Because we have the vector $\mathbf{n}$ normal to the surface and a point $(1, 0, 1)$ on the surface, we can use formula (1′) to obtain an equation of the tangent plane:

$$-2(x-1) + (z-1) = 0; \text{ that is, } z = 2x - 1. \quad ▲$$

EXAMPLE 4 Suppose a surface S is the graph of a differentiable function $g: \mathbb{R}^2 \to \mathbb{R}$. Write S in parametric form and show that the surface is smooth at all points $(u_0, v_0, g(u_0, v_0)) \in \mathbb{R}^3$.

SOLUTION Write S in parametric form as follows:

$$x = u, \qquad y = v, \qquad z = g(u, v),$$

which is the same as $z = g(x, y)$. Then at the point (u_0, v_0),

$$\mathbf{T}_u = \mathbf{i} + \frac{\partial g}{\partial u}(u_0, v_0)\mathbf{k} \qquad \text{and} \qquad \mathbf{T}_v = \mathbf{j} + \frac{\partial g}{\partial v}(u_0, v_0)\mathbf{k},$$

and for $(u_0, v_0) \in \mathbb{R}^2$,

$$\mathbf{n} = \mathbf{T}_u \times \mathbf{T}_v = -\frac{\partial g}{\partial u}(u_0, v_0)\mathbf{i} - \frac{\partial g}{\partial v}(u_0, v_0)\mathbf{j} + \mathbf{k} \neq \mathbf{0}. \tag{2}$$

This is nonzero because the coefficient of **k** is 1; consequently, the parametrization $(u, v) \mapsto (u, v, g(u, v))$ is regular at all points. Moreover, the tangent plane at the point $(x_0, y_0, z_0) = (u_0, v_0, g(u_0, v_0))$ is given, by formula (1), as

$$(x - x_0, y - y_0, z - z_0) \cdot \left(-\frac{\partial g}{\partial u}, -\frac{\partial g}{\partial v}, 1\right) = 0,$$

where the partial derivatives are evaluated at (u_0, v_0). Remembering that $x = u$ and $y = v$, we can write this as

$$z - z_0 = \left(\frac{\partial g}{\partial x}\right)(x - x_0) + \left(\frac{\partial g}{\partial y}\right)(y - y_0), \tag{3}$$

where $\partial g/\partial x$ and $\partial g/\partial y$ are evaluated at (x_0, y_0). ▲

This example also shows that the definition of the tangent plane for parametrized surfaces agrees with the one for surfaces obtained as graphs, because equation (3) is the same formula we derived (in Chapter 2) for the plane tangent to S at the point $(x_0, y_0, z_0) \in S$.

It is also useful to consider piecewise smooth surfaces, that is, surfaces composed of a certain number of images of smooth parametrized surfaces. For example, the surface of a cube in $\mathbb{R}^3$ is such a surface. These surfaces are considered in Section 7.4.

EXAMPLE 5 Find a parametrization for the hyperboloid of one sheet:

$$x^2 + y^2 - z^2 = 1.$$

SOLUTION Because x and y appear in the combination $x^2 + y^2$, the surface is invariant under rotation about the z axis, and so it is natural to write

$$x = r\cos\theta, \qquad y = r\sin\theta.$$

Then $x^2 + y^2 - z^2 = 1$ becomes $r^2 - z^2 = 1$. This we can conveniently parametrize by[9]

$$r = \cosh u, \qquad z = \sinh u.$$

Thus, a parametrization is

$$x = (\cosh u)(\cos\theta), \qquad y = (\cosh u)(\sin\theta), \qquad z = \sinh u,$$

where $0 \le \theta < 2\pi$, $-\infty < u < \infty$. ▲

[9] Recall from one-variable calculus that $\cosh u = (e^u + e^{-u})/2$ and $\sinh u = (e^u - e^{-u})/2$. One easily verifies from these definitions that $\cosh^2 u - \sinh^2 u = 1$.

EXERCISES

In Exercises 1 to 3, find an equation for the plane tangent to the given surface at the specified point.

1. $x = 2u, \quad y = u^2 + v, \quad z = v^2, \quad at\ (0, 1, 1)$

2. $x = u^2 - v^2, \quad y = u + v, \quad z = u^2 + 4v, \quad at\ (-\frac{1}{4}, \frac{1}{2}, 2)$

3. $x = u^2, \quad y = u \sin e^v, \quad z = \frac{1}{3} u \cos e^v, \quad at\ (13, -2, 1)$

4. At what points are the surfaces in Exercises 1 and 2 regular?

5. Find an expression for a unit vector normal to the surface

$$x = \cos v \sin u, \quad y = \sin v \sin u, \quad z = \cos u$$

at the image of a point (u, v) for u in $[0, \pi]$ and v in $[0, 2\pi]$. Identify this surface.

6. Repeat Exercise 5 for the surface

$$x = 3\cos\theta \sin\phi, \quad y = 2\sin\theta\sin\phi, \quad z = \cos\phi$$

for θ in $[0, 2\pi]$ and ϕ in $[0, \pi]$.

7. Repeat Exercise 5 for the surface

$$x = \sin v, \quad y = u, \quad z = \cos v$$

for $0 \le v \le 2\pi$ and $-1 \le u \le 3$.

8. Repeat Exercise 5 for the surface

$$x = (2 - \cos v)\cos u, \quad y = (2 - \cos v)\sin u, \quad z = \sin v$$

for $-\pi \le u \le \pi, -\pi \le v \le \pi$. Is this surface regular?

9. (a) Develop a formula for the plane tangent to the surface $x = h(y, z)$.
(b) Obtain a similar formula for $y = k(x, z)$.

10. Find the equation of the plane tangent to the surface $x = u^2, y = v^2, z = u^2 + v^2$ at the point $u = 1, v = 1$.

11. Find a parametrization of the surface $z = 3x^2 + 8xy$ and use it to find the tangent plane at $x = 1, y = 0, z = 3$. Compare your answer with that using graphs.

12. Find a parametrization of the surface $x^3 + 3xy + z^2 = 2, z > 0$, and use it to find the tangent plane at the point $x = 1, y = 1/3, z = 0$. Compare your answer with that using level sets.

13. Consider the surface in $\mathbb{R}^3$ parametrized by

$$\mathbf{\Phi}(r, \theta) = (r\cos\theta, r\sin\theta, \theta), \quad 0 \le r \le 1 \quad \text{and} \quad 0 \le \theta \le 4\pi.$$

(a) Sketch and describe the surface.
(b) Find an expression for a unit normal to the surface.
(c) Find an equation for the plane tangent to the surface at the point (x_0, y_0, z_0).
(d) If (x_0, y_0, z_0) is a point on the surface, show that the horizontal line segment of unit length from the z axis through (x_0, y_0, z_0) is contained in the surface and in the plane tangent to the surface at (x_0, y_0, z_0).

14. Given a sphere of radius 2 centered at the origin, find the equation for the plane that is tangent to it at the point $(1, 1, \sqrt{2})$ by considering the sphere as:

(a) a surface parametrized by $\mathbf{\Phi}(\theta, \phi) = (2\cos\theta \sin\phi, 2\sin\theta \sin\phi, 2\cos\phi)$;
(b) a level surface of $f(x, y, z) = x^2 + y^2 + z^2$; and
(c) the graph of $g(x, y) = \sqrt{4 - x^2 - y^2}$.

15. (a) Find a parametrization for the hyperboloid $x^2 + y^2 - z^2 = 25$.
(b) Find an expression for a unit normal to this surface.
(c) Find an equation for the plane tangent to the surface at $(x_0, y_0, 0)$, where $x_0^2 + y_0^2 = 25$.
(d) Show that the lines $(x_0, y_0, 0) + t(-y_0, x_0, 5)$ and $(x_0, y_0, 0) + t(y_0, -x_0, 5)$ lie in the surface *and* in the tangent plane found in part (c).

16. A parametrized surface is described by a differentiable function $\mathbf{\Phi}\colon \mathbb{R}^2 \to \mathbb{R}^3$. According to Chapter 2, the derivative should give a linear approximation that yields a representation of the tangent plane. This exercise demonstrates that this is indeed the case.

(a) Assuming $\mathbf{T}_u \times \mathbf{T}_v \neq \mathbf{0}$, show that the range of the linear transformation $\mathbf{D\Phi}(u_0, v_0)$ is the plane spanned by $\mathbf{T}_u$ and $\mathbf{T}_v$. [Here $\mathbf{T}_u$ and $\mathbf{T}_v$ are evaluated at (u_0, v_0).]
(b) Show that $\mathbf{w} \perp (\mathbf{T}_u \times \mathbf{T}_v)$ if and only if $\mathbf{w}$ is in the range of $\mathbf{D\Phi}(u_0, v_0)$.
(c) Show that the tangent plane as defined in this section is the same as the "parametrized plane"

$$(u, v) \mapsto \mathbf{\Phi}(u_0, v_0) + \mathbf{D\Phi}(u_0, v_0)\begin{bmatrix} u - u_0 \\ v - v_0 \end{bmatrix}.$$

17. Consider the surfaces $\mathbf{\Phi}_1(u, v) = (u, v, 0)$ and $\mathbf{\Phi}_2(u, v) = (u^3, v^3, 0)$.

(a) Show that the image of $\mathbf{\Phi}_1$ and of $\mathbf{\Phi}_2$ is the xy plane.
(b) Show that $\mathbf{\Phi}_1$ describes a regular surface, yet $\mathbf{\Phi}_2$ does not. Conclude that the notion of regularity of a surface S depends on the existence of at least one regular parametrization for S.
(c) Prove that the tangent plane of S is well defined independently of the regular (one-to-one) parametrization (you will need to use the inverse function theorem from Section 3.5).
(d) After these remarks, do you think you can find a regular parametrization of the cone of Figure 7.3.7?

18. Let $\mathbf{\Phi}$ be a regular surface at (u_0, v_0); that is, $\mathbf{\Phi}$ is of class C^1 and $\mathbf{T}_u \times \mathbf{T}_v \neq \mathbf{0}$ at (u_0, v_0).

(a) Use the implicit function theorem (Section 3.5) to show that the image of $\mathbf{\Phi}$ *near* (u_0, v_0) is the graph of a C^1 function of two variables. If the z component of $\mathbf{T}_u \times \mathbf{T}_v$ is nonzero, we can write it as $z = f(x, y)$.
(b) Show that the tangent plane at $\mathbf{\Phi}(u_0, v_0)$ defined by the plane spanned by $\mathbf{T}_u$ and $\mathbf{T}_v$ coincides with the tangent plane of the graph of $z = f(x, y)$ at this point.

7.4 Area of a Surface

Before proceeding to general surface integrals, let us first consider the problem of computing the area of a surface, just as we considered the problem of finding the arc length of a curve before discussing path integrals.

Definition of Surface Area

In Section 7.3, we defined a parametrized surface S to be the *image* of a function $\mathbf{\Phi}\colon D \subset \mathbb{R}^2 \to \mathbb{R}^3$, written as $\mathbf{\Phi}(u, v) = (x(u, v), y(u, v), z(u, v))$. The map $\mathbf{\Phi}$ was called the parametrization of S and S was said to be regular at $\mathbf{\Phi}(u, v) \in S$ provided that $\mathbf{T}_u \times \mathbf{T}_v \neq \mathbf{0}$, where

$$\mathbf{T}_u = \frac{\partial x}{\partial u}(u, v)\mathbf{i} + \frac{\partial y}{\partial u}(u, v)\mathbf{j} + \frac{\partial z}{\partial u}(u, v)\mathbf{k}$$

and

$$\mathbf{T}_v = \frac{\partial x}{\partial v}(u, v)\mathbf{i} + \frac{\partial y}{\partial v}(u, v)\mathbf{j} + \frac{\partial z}{\partial v}(u, v)\mathbf{k}.$$

Recall that a regular surface (loosely speaking) is one that has no corners or breaks.

In the rest of this chapter and in the next one, we shall consider only piecewise regular surfaces that are unions of images of parametrized surfaces $\mathbf{\Phi}_i\colon D_i \to \mathbb{R}^3$ for which:

(i) D_i is an elementary region in the plane;

(ii) $\mathbf{\Phi}_i$ is of class C^1 and one-to-one, except possibly on the boundary of D_i; and

(iii) S_i, the image of $\mathbf{\Phi}_i$ is regular, except possibly at a finite number of points.

DEFINITION: Area of a Parametrized Surface We define the ***surface area***[10] $A(S)$ of a parametrized surface by

$$A(S) = \iint_D \|\mathbf{T}_u \times \mathbf{T}_v\|\, du\, dv \tag{1}$$

where $\|\mathbf{T}_u \times \mathbf{T}_v\|$ is the norm of $\mathbf{T}_u \times \mathbf{T}_v$. If S is a union of surfaces S_i, its area is the sum of the areas of the S_i.

[10] As we have not yet discussed the independence of parametrization, it may seem that $A(S)$ depends on the parametrization $\mathbf{\Phi}$. We shall discuss independence of parametrization in Section 7.6; the use of this notation here should not cause confusion.

As the reader can easily verify, we have

$$\|\mathbf{T}_u \times \mathbf{T}_v\| = \sqrt{\left[\frac{\partial(x,y)}{\partial(u,v)}\right]^2 + \left[\frac{\partial(y,z)}{\partial(u,v)}\right]^2 + \left[\frac{\partial(x,z)}{\partial(u,v)}\right]^2}, \tag{2}$$

where

$$\frac{\partial(x,y)}{\partial(u,v)} = \begin{vmatrix} \dfrac{\partial x}{\partial u} & \dfrac{\partial x}{\partial v} \\ \dfrac{\partial y}{\partial u} & \dfrac{\partial y}{\partial v} \end{vmatrix},$$

and so on. Thus, formula (1) becomes

$$A(S) = \iint_D \sqrt{\left[\frac{\partial(x,y)}{\partial(u,v)}\right]^2 + \left[\frac{\partial(y,z)}{\partial(u,v)}\right]^2 + \left[\frac{\partial(x,z)}{\partial(u,v)}\right]^2}\, du\, dv. \tag{3}$$

Justification of the Area Formula

We can justify the definition of surface area by analyzing the integral $\iint_D \|\mathbf{T}_u \times \mathbf{T}_v\|\, du\, dv$ in terms of Riemann sums. For simplicity, suppose D is a rectangle; consider the nth regular partition of D, and let R_{ij} be the ijth rectangle in the partition, with vertices $(u_i, v_j), (u_{i+1}, v_j), (u_i, v_{j+1})$, and $(u_{i+1}, v_{j+1}), 0 \le i \le n-1, 0 \le j \le n-1$. Denote the values of $\mathbf{T}_u$ and $\mathbf{T}_v$ at (u_i, v_j) by $\mathbf{T}_{u_i}$ and $\mathbf{T}_{v_j}$. We can think of the vectors $\Delta u \mathbf{T}_{u_i}$ and $\Delta v \mathbf{T}_{v_j}$ as tangent to the surface at $\mathbf{\Phi}(u_i, v_j) = (x_{ij}, y_{ij}, z_{ij})$, where $\Delta u = u_{i+1} - u_i$, $\Delta v = v_{j+1} - v_j$. Then these vectors form a parallelogram P_{ij} that lies in the plane tangent to the surface at (x_{ij}, y_{ij}, z_{ij}) (see Figure 7.4.1).

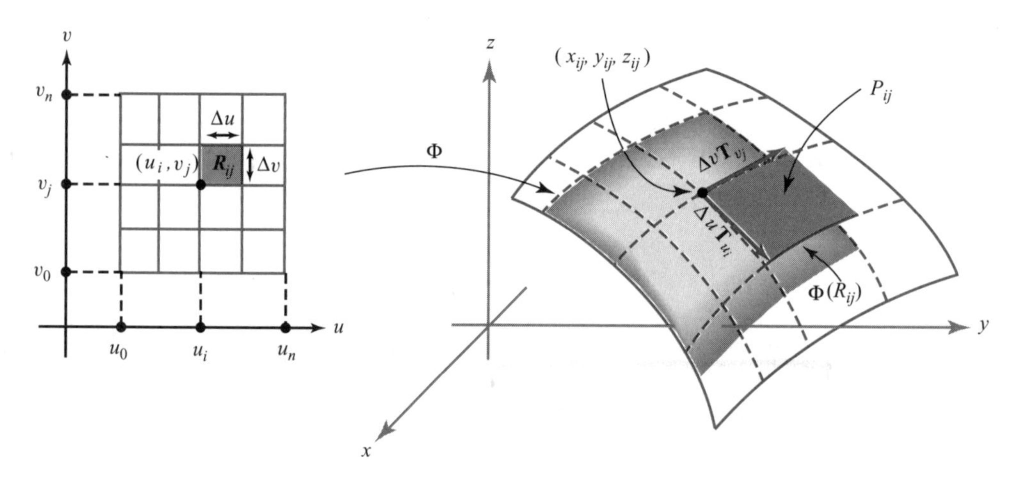

Figure 7.4.1 $\|\mathbf{T}_{u_i} \times \mathbf{T}_{v_j}\|\, \Delta u\, \Delta v$ is equal to the area of a parallelogram that approximates the area of a patch on a surface $S = \mathbf{\Phi}(D)$.

We thus have a "patchwork cover" of the surface by the P_{ij}. For n large, the area of P_{ij} is a good approximation to the area of $\mathbf{\Phi}(R_{ij})$. Because the area of the parallelogram spanned by two vectors $\mathbf{v}_1$ and $\mathbf{v}_2$ is $\|\mathbf{v}_1 \times \mathbf{v}_2\|$ (see Chapter 1), we see that

$$A(P_{ij}) = \|\Delta u \mathbf{T}_{u_i} \times \Delta v \mathbf{T}_{v_j}\| = \|\mathbf{T}_{u_i} \times \mathbf{T}_{v_j}\| \Delta u\, \Delta v.$$

Therefore, the area of the patchwork cover is

$$A_n = \sum_{i=0}^{n-1} \sum_{j=0}^{n-1} A(P_{ij}) = \sum_{i=0}^{n-1} \sum_{j=0}^{n-1} \|\mathbf{T}_{u_i} \times \mathbf{T}_{v_j}\|\, \Delta u\, \Delta v.$$

As $n \to \infty$, the sums A_n converge to the integral

$$\iint_D \|\mathbf{T}_u \times \mathbf{T}_v\|\, du\, dv.$$

Because A_n should approximate the surface area better and better as $n \to \infty$, we are led to formula (1) as a reasonable definition of $A(S)$.

EXAMPLE 1 Let D be the region determined by $0 \le \theta \le 2\pi$, $0 \le r \le 1$ and let the function $\mathbf{\Phi}\colon D \to \mathbb{R}^3$, defined by

$$x = r\cos\theta, \qquad y = r\sin\theta, \qquad z = r$$

be a parametrization of a cone S (see Figure 7.3.8). Find its surface area.

SOLUTION In formula (3),

$$\frac{\partial(x,y)}{\partial(r,\theta)} = \begin{vmatrix} \cos\theta & -r\sin\theta \\ \sin\theta & r\cos\theta \end{vmatrix} = r,$$

$$\frac{\partial(y,z)}{\partial(r,\theta)} = \begin{vmatrix} \sin\theta & r\cos\theta \\ 1 & 0 \end{vmatrix} = -r\cos\theta,$$

and

$$\frac{\partial(x,z)}{\partial(r,\theta)} = \begin{vmatrix} \cos\theta & -r\sin\theta \\ 1 & 0 \end{vmatrix} = r\sin\theta,$$

so the area integrand is

$$\|\mathbf{T}_r \times \mathbf{T}_\theta\| = \sqrt{r^2 + r^2\cos^2\theta + r^2\sin^2\theta} = r\sqrt{2}.$$

Clearly, $\|\mathbf{T}_r \times \mathbf{T}_\theta\|$ vanishes for $r = 0$, but $\mathbf{\Phi}(0, \theta) = (0, 0, 0)$ for any θ. Thus, $(0, 0, 0)$ is the only point where the surface is not regular. We have

$$\iint_D \|\mathbf{T}_r \times \mathbf{T}_\theta\| \, dr \, d\theta = \int_0^{2\pi} \int_0^1 \sqrt{2} r \, dr \, d\theta = \int_0^{2\pi} \frac{1}{2}\sqrt{2} \, d\theta = \sqrt{2}\pi.$$

To confirm that this is the area of $\mathbf{\Phi}(D)$, we must verify that $\mathbf{\Phi}$ is one-to-one (for points not on the boundary of D). Let D^0 be the set of (r, θ) with $0 < r < 1$ and $0 < \theta < 2\pi$. Hence, D^0 is D without its boundary. To see that $\mathbf{\Phi}\colon D^0 \to \mathbb{R}^3$ is one-to-one, assume that $\mathbf{\Phi}(r, \theta) = \mathbf{\Phi}(r', \theta')$ for (r, θ) and $(r', \theta') \in D^0$. Then

$$r \cos\theta = r' \cos\theta', \qquad r \sin\theta = r' \sin\theta', \qquad r = r'.$$

From these equations it follows that $\cos\theta = \cos\theta'$ and $\sin\theta = \sin\theta'$. Thus, either $\theta = \theta'$ or $\theta = \theta' + 2\pi n$. But the second case is impossible for n a nonzero integer, because both θ and θ' belong to the open interval $(0, 2\pi)$, and thus cannot be more than 2π radians apart. This proves that off the boundary, $\mathbf{\Phi}$ is one-to-one. (Is $\mathbf{\Phi}\colon D \to \mathbb{R}^3$ one-to-one?) In future examples, we shall not usually verify that the parametrization is one-to-one when it is intuitively clear. ▲

EXAMPLE 2 A ***helicoid*** is defined by $\mathbf{\Phi}\colon D \to \mathbb{R}^3$, where

$$x = r\cos\theta, \qquad y = r\sin\theta, \qquad z = \theta$$

and D is the region where $0 \le \theta \le 2\pi$ and $0 \le r \le 1$ (Figure 7.4.2). Find its area.

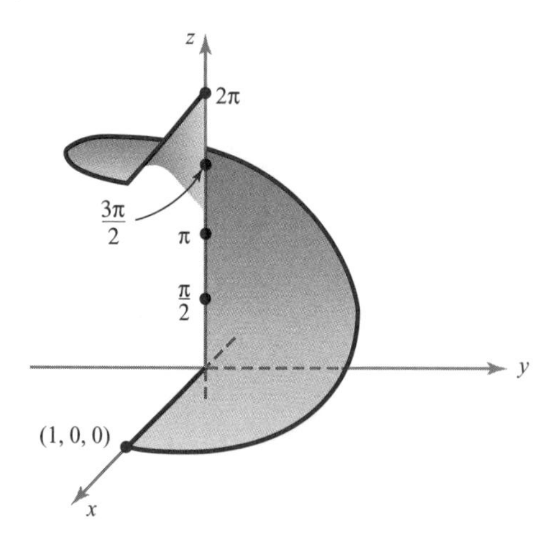

Figure 7.4.2 The helicoid $x = r\cos\theta$, $y = r\sin\theta$, $z = \theta$.

SOLUTION We compute $\partial(x, y)/\partial(r, \theta) = r$ as in Example 1, and

$$\frac{\partial(y, z)}{\partial(r, \theta)} = \begin{vmatrix} \sin\theta & r\cos\theta \\ 0 & 1 \end{vmatrix} = \sin\theta,$$

$$\frac{\partial(x, z)}{\partial(r, \theta)} = \begin{vmatrix} \cos\theta & -r\sin\theta \\ 0 & 1 \end{vmatrix} = \cos\theta.$$

The area integrand is therefore $\sqrt{r^2 + 1}$, which never vanishes, so the surface is regular. The area of the helicoid is

$$\iint_D \|\mathbf{T}_r \times \mathbf{T}_\theta\| \, dr\, d\theta = \int_0^{2\pi} \int_0^1 \sqrt{r^2 + 1}\, dr d\theta = 2\pi \int_0^1 \sqrt{r^2 + 1}\, dr.$$

After a little computation (using the table of integrals), we find that this integral is equal to

$$\pi[\sqrt{2} + \log(1 + \sqrt{2})]. \quad \blacktriangle$$

Surface Area of a Graph

A surface S given in the form $z = g(x, y)$, where $(x, y) \in D$, admits the parametrization

$$x = u, \qquad y = v, \qquad z = g(u, v)$$

for $(u, v) \in D$. When g is of class C^1, this parametrization is smooth, and the formula for surface area reduces to

$$A(S) = \iint_D \left(\sqrt{\left(\frac{\partial g}{\partial x}\right)^2 + \left(\frac{\partial g}{\partial y}\right)^2 + 1} \right) dA, \tag{4}$$

after applying the formulas

$$\mathbf{T}_u = \mathbf{i} + \frac{\partial g}{\partial u}\mathbf{k}, \qquad \mathbf{T}_v = \mathbf{j} + \frac{\partial g}{\partial v}\mathbf{k},$$

and

$$\mathbf{T}_u \times \mathbf{T}_v = -\frac{\partial g}{\partial u}\mathbf{i} - \frac{\partial g}{\partial v}\mathbf{j} + \mathbf{k} = -\frac{\partial g}{\partial x}\mathbf{i} - \frac{\partial g}{\partial y}\mathbf{j} + \mathbf{k},$$

as noted in Example 4 of Section 7.3.

Surfaces of Revolution

In most books on one-variable calculus, it is shown that the lateral surface area generated by revolving the graph of a function $y = f(x)$ about the x axis is given by

$$A = 2\pi \int_a^b (|f(x)|\sqrt{1 + [f'(x)]^2})\, dx. \tag{5}$$

If the graph is revolved about the y axis, the surface area is

$$A = 2\pi \int_a^b (|x|\sqrt{1 + [f'(x)]^2})\, dx. \tag{6}$$

We shall derive formula (5) by using the methods just developed; one can obtain formula (6) in a similar fashion (Exercise 10).

To derive formula (5) from formula (3), we must give a parametrization of S. Define the parametrization by

$$x = u, \qquad y = f(u)\cos v, \qquad z = f(u)\sin v$$

over the region D given by

$$a \le u \le b, \qquad 0 \le v \le 2\pi.$$

This is indeed a parametrization of S, because for fixed u, the point

$$(u, f(u)\cos v, f(u)\sin v)$$

traces out a circle of radius $|f(u)|$ with the center $(u, 0, 0)$ (Figure 7.4.3).

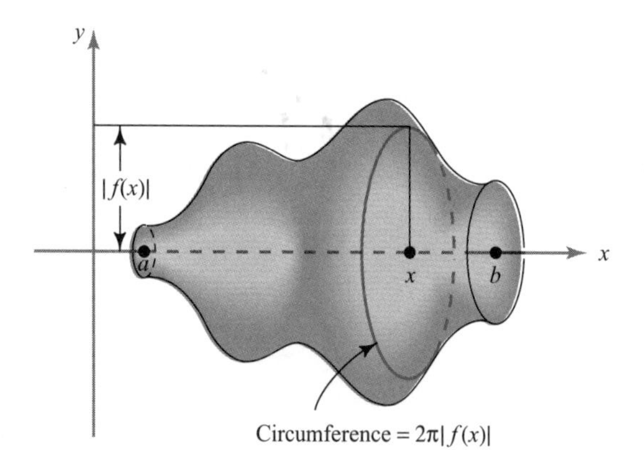

Figure 7.4.3 The curve $y = f(x)$ rotated about the x axis.

We calculate

$$\frac{\partial(x, y)}{\partial(u, v)} = -f(u)\sin v, \qquad \frac{\partial(y, z)}{\partial(u, v)} = f(u)f'(u), \qquad \frac{\partial(x, z)}{\partial(u, v)} = f(u)\cos v,$$

and so by formula (3)

$$\begin{aligned} A(S) &= \iint_D \sqrt{\left[\frac{\partial(x, y)}{\partial(u, v)}\right]^2 + \left[\frac{\partial(y, z)}{\partial(u, v)}\right]^2 + \left[\frac{\partial(x, z)}{\partial(u, v)}\right]^2}\, du\, dv \\ &= \iint_D \sqrt{[f(u)]^2 \sin^2 v + [f(u)]^2[f'(u)]^2 + [f(u)]^2 \cos^2 v}\, du\, dv \\ &= \iint_D |f(u)|\sqrt{1 + [f'(u)]^2}\, du\, dv \\ &= \int_a^b \int_0^{2\pi} |f(u)|\sqrt{1 + [f'(u)]^2}\, dv\, du \\ &= 2\pi \int_a^b |f(u)|\sqrt{1 + [f'(u)]^2}\, du, \end{aligned}$$

which is formula (5).

If S is the surface of revolution, then $2\pi\,|f(x)|$ is the circumference of the vertical cross section to S at the point x (Figure 7.4.3). Observe that we can write

$$2\pi \int_a^b |f(x)|\sqrt{1 + [f'(x)]^2}\, dx = \int_{\mathbf{c}} 2\pi\,|f(x)|\, ds,$$

where the expression on the right is the path integral of $2\pi\,|f(x)|$ along the path given by $\mathbf{c}\colon [a, b] \to \mathbb{R}^2$, $t \mapsto (t, f(t))$. Therefore, *the lateral surface of a solid of revolution is obtained by integrating the cross-sectional circumference along the path that is the graph of the given function.*

The most famous mathematician in ancient times was Archimedes. In addition to being an extraordinarily gifted mathematician, he was also an engineering genius on a scale never before seen and was greatly admired by his contemporaries and by later writers for his insights into mechanics. It was these talents that helped the people of the city of Syracuse in 214 B.C. to defend their city against the onslaught of the Roman legions under their commander Marcellus.

When the Romans besieged the city, they encountered an enemy whom Archimedes had supplied—totally unexpectedly—with powerful weapons, including artillery and burning mirrors, which, as legend has it, incinerated the Roman fleet.

The siege of Syracuse lasted two years, and the city finally fell as a result of acts of treason. In the aftermath of the assault, the old scientist was slain by a Roman soldier, even though the commander had asked his men to spare Archimedes' life. As the story goes, Archimedes was sitting in front of his house studying some geometric figures he had drawn in the sand. When a Roman soldier approached, Archimedes shouted out, "Don't disturb my figures!" The ruffian, feeling insulted, slew Archimedes.

To honor this great man, Marcellus erected a tomb for Archimedes on which, according to Archimedes's own wishes, were depicted a cone, a sphere, and a cylinder (Figure 7.4.4).

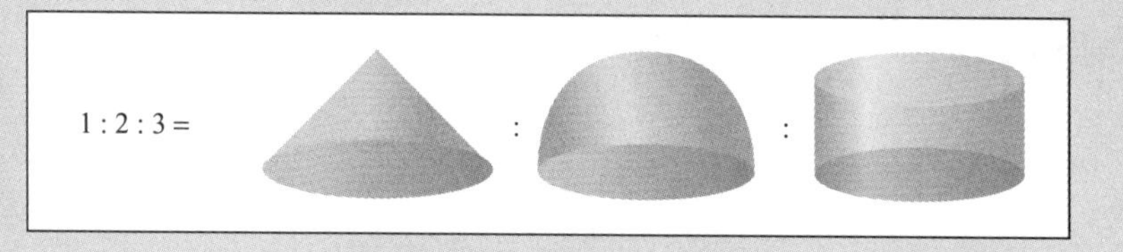

Figure 7.4.4 Archimedes' theorem: The ratios of the volumes of a cone, a half ball, and a cylinder, all of the same height and radius, are 1:2:3.

Archimedes was incredibly proud of his calculation of the volume and surface area of the sphere, which justifiably were seen as truly outstanding accomplishments for their time. As in his works on centers of gravity, for which he provided no clear definition, Archimedes was able to compute the surface area of the sphere without having a clear definition of precisely what it was. However, as with many mathematical works, one knows the answer long before a proof or even the correct definition can be found.

The problem of properly defining surface areas is a difficult one. To Archimedes' credit, a careful theory of surface areas was not achieved until the twentieth century, after a long development that began in the seventeenth century with the discovery of calculus.

Christiaan Huygens (1629–1695) was the first person since Archimedes to give results on the areas of special surfaces beyond the sphere, and he obtained the areas of portions of surfaces of revolution, such as the paraboloid and hyperboloid.

The brilliant and prolific mathematician Leonhard Euler (1707–1783) presented the first fundamental work on the theory of surfaces in 1760 with *Recherches sur la courbure des surfaces*. However, it was in 1728, in a paper on shortest paths on surfaces, that Euler defined a surface as a graph $z = f(x,y)$. Euler was interested in studying the curvature of surfaces, and in 1771 he introduced the notion of the parametric surfaces that are described in this section.

After the rapid development of calculus in the early eighteenth century, formulas for the lengths of curves and areas of surfaces were developed. Although we do not know when all the area formulas presented in this section first appeared, they were certainly common by the end of the eighteenth century. The underlying concepts of the length of a curve and the area of a surface were understood intuitively before this time, and the use of formulas from calculus to compute areas was considered a great achievement.

Augustin-Louis Cauchy (1789–1857) was the first to take the step of defining the quantities of length and surface areas by integrals as we have presented in this book. The question of defining surface area independent of integrals was taken up somewhat later, but this posed many difficult problems that were not properly resolved until this century.

We end this section by describing the fascinating classic area problem of Plateau, which has enjoyed a long history in mathematics. The Belgian physicist Joseph Plateau (1801–1883) carried out many experiments from 1830 to 1869 on surface tension and capillary phenomena, experiments that had enormous impact at the time and were repeated by notable nineteenth-century physicists, such as Michael Faraday (1791–1867). The corresponding collection of mathematical problems relating to soap films was named in 1904 after Plateau by the great French mathematician Henri Lebesgue (1875–1941).

If a wire is dipped into a soap or glycerine solution, then one usually withdraws a soap film spanning the wire. Some examples are given in Figure 7.4.5, although readers might like to perform the experiment for themselves. Plateau raised the mathematical question: For a given boundary (wire), how does one prove the existence of such a surface (soap film) and how many surfaces can there be? The underlying physical principle is that nature tends to minimize area; that is, the surface that forms should be a surface of least area among all possible surfaces that have the given curve as their boundary. This again is another example of the action principle of Maupertuis and Leibniz (c.f. Section 3.3).

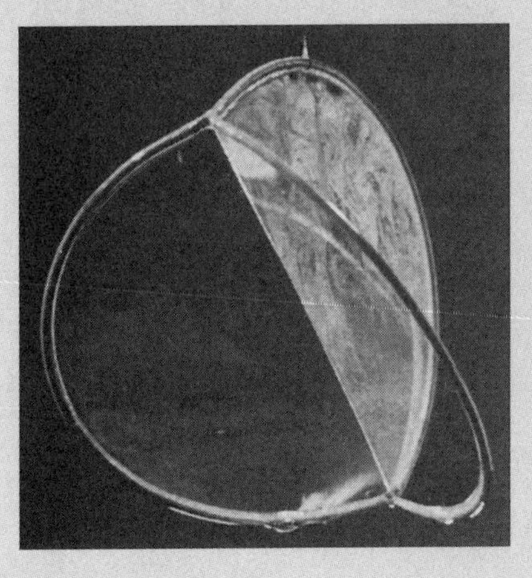

Figure 7.4.5 Two soap films spanning wires.

For soap film surfaces that are disklike, the problem can be formulated in the following way. Let $D \subset \mathbb{R}^2$ be the unit disk defined to be the set $\{(x, y) \mid x^2 + y^2 \leq 1\}$ and let ∂D be its boundary. Furthermore, suppose that the image Γ of $\mathbf{c}: [0, 2\pi] \to \mathbb{R}^3$ is a simple closed curve, Γ representing a wire in $\mathbb{R}^3$.

Let $\mathcal{S}$ be the set of all maps $\mathbf{\Phi}: D \to \mathbb{R}^3$ such that $\mathbf{\Phi}(\partial D) = \Gamma$, $\mathbf{\Phi}$ is of class C^1, and $\mathbf{\Phi}$ is one-to-one on ∂D. Each $\mathbf{\Phi} \in \mathcal{S}$ represents a parametric C^1 "disklike" surface "spanning" the wire Γ.

The soap films in Figure 7.4.5 are not disklike, but represent a system of multiple disklike surfaces. Figure 7.4.6 shows a contour that bounds two disklike surfaces and one nondisklike surface.

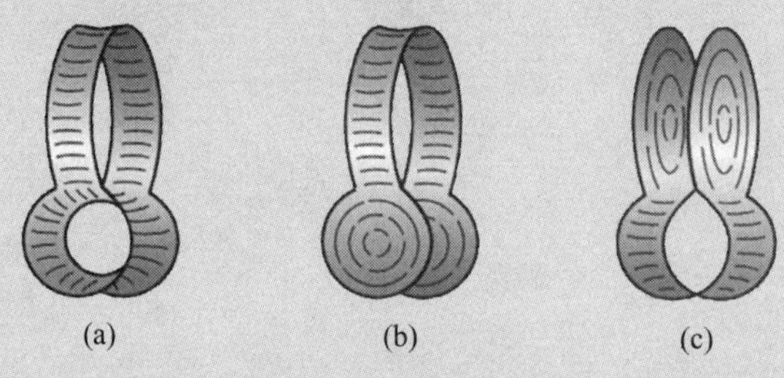

Figure 7.4.6 Soap film surfaces; (b) and (c) are disklike surfaces, but (a) is not.

For each $\mathbf{\Phi} \in \mathcal{S}$, consider the area of the image surface, namely, $A(\mathbf{\Phi}) = \iint_D \|\mathbf{T}_u \times \mathbf{T}_v\| \, du \, dv$. This area is a function that assigns to each parametric surface its area. Plateau asked whether A has a minimum on $\mathcal{S}$; that is, does there exist a $\mathbf{\Phi}_0$ such that $A(\mathbf{\Phi}_0) \leq A(\mathbf{\Phi})$ for all $\mathbf{\Phi} \in \mathcal{S}$? Unfortunately, the methods of this book are not adequate to solve this problem. We can tackle questions of finding minima of real-valued functions of several variables, but in no way can the set $\mathcal{S}$ be thought of as a region in $\mathbb{R}^n$ for *any n*!

In his own study of surfaces of least area, Weierstrass showed that if a minimum

$$\mathbf{\Phi}_0(u, v) = (x(u, v), y(u, v), z(u, v))$$

existed at all, it would have to satisfy (after suitable normalizations) the partial differential equations

(i) $\nabla^2 \mathbf{\Phi}_0 = 0$

(ii) $\dfrac{\partial \mathbf{\Phi}_0}{\partial u} \cdot \dfrac{\partial \mathbf{\Phi}_0}{\partial v} = 0$

(iii) $\left\| \dfrac{\partial \mathbf{\Phi}_0}{\partial u} \right\| = \left\| \dfrac{\partial \mathbf{\Phi}_0}{\partial v} \right\|$

where $\|\mathbf{w}\|$ denotes the "norm"or length of the vector $\mathbf{w}$. This example illustrates the intimate connections between problems of maxima and minima (the calculus of variations) and the subject of partial differential equations.

For well over 70 years, mathematicians such as Riemann, Weierstrass, H. A. Schwarz, Darboux, and Lebesgue puzzled over the challenge posed by Plateau. In 1931 the question was finally settled when Jesse Douglas showed that such a $\mathbf{\Phi}_0$ existed. However, many questions about soap films remain unsolved, and this area of research is still active today.[11]

EXERCISES

1. Find the surface area of the unit sphere S represented parametrically by $\mathbf{\Phi}\colon D \to S \subset \mathbb{R}^3$, where D is the rectangle $0 \le \theta \le 2\pi$, $0 \le \phi \le \pi$ and $\mathbf{\Phi}$ is given by the equations

$$x = \cos\theta \sin\phi, \qquad y = \sin\theta \sin\phi, \qquad z = \cos\phi.$$

Note that we can represent the entire sphere parametrically, but we cannot represent it in the form $z = f(x, y)$.

2. In Exercise 1, what happens if we allow ϕ to vary from $-\pi/2$ to $\pi/2$? From 0 to 2π? Why do we obtain different answers?

3. Find the area of the helicoid in Example 2 if the domain D is $0 \le r \le 1$ and $0 \le \theta \le 3\pi$.

4. The torus T can be represented parametrically by the function $\mathbf{\Phi}\colon D \to \mathbb{R}^3$, where $\mathbf{\Phi}$ is given by the coordinate functions $x = (R + \cos\phi)\cos\theta$, $y = (R + \cos\phi)\sin\theta$, $z = \sin\phi$; D is the rectangle $[0, 2\pi] \times [0, 2\pi]$, that is, $0 \le \theta \le 2\pi$, $0 \le \phi \le 2\pi$; and $R > 1$ is fixed (see Figure 7.4.7). Show that $A(T) = (2\pi)^2 R$, first by using formula (3) and then by using formula (6).

5. Let $\mathbf{\Phi}(u, v) = (u - v, u + v, uv)$ and let D be the unit disk in the uv plane. Find the area of $\mathbf{\Phi}(D)$.

6. Find the area of the portion of the unit sphere that is cut out by the cone $z \ge \sqrt{x^2 + y^2}$ (see Exercise 1).

7. Show that the surface $x = 1/\sqrt{y^2 + z^2}$, where $1 \le x < \infty$, can be filled but not painted!

8. Find a parametrization of the surface $x^2 - y^2 = 1$, where $x > 0$, $-1 \le y \le 1$ and $0 \le z \le 1$. Use your answer to express the area of the surface as an integral.

[11] For more information on this fascinating subject, the reader may consult *The Parsimonious Universe: Shape and Form in the Natural World*, by S. Hildebrandt and A. Tromba, Springer-Verlag, New York/Heidelberg, 1995.

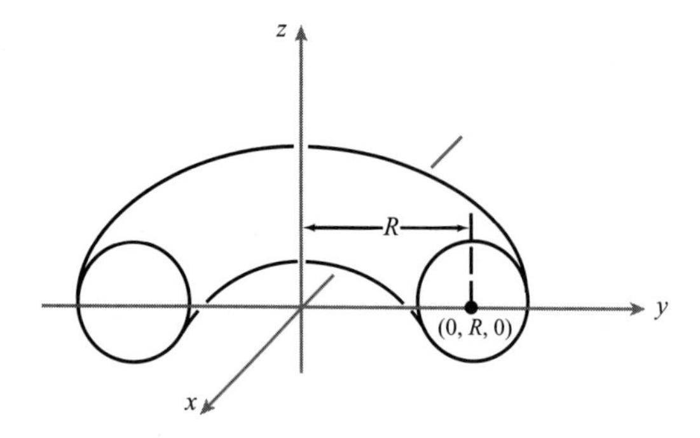

Figure 7.4.7 A cross section of a torus.

9. Represent the ellipsoid E:

$$\frac{x^2}{a^2} + \frac{y^2}{b^2} + \frac{z^2}{c^2} = 1$$

parametrically and write out the integral for its surface area $A(E)$. (Do not evaluate the integral.)

10. Let the curve $y = f(x)$, $a \leq x \leq b$, be rotated about the y axis. Show that the area of the surface swept out is given by equation (6); that is,

$$A = 2\pi \int_a^b |x|\sqrt{1 + [f'(x)]^2}\, dx.$$

Interpret the formula geometrically using arc length and slant height.

11. Find the area of the surface obtained by rotating the curve $y = x^2$, $0 \leq x \leq 1$, about the y axis.

12. Use formula (4) to compute the surface area of the cone in Example 1.

13. Find the area of the surface defined by $x + y + z = 1$, $x^2 + 2y^2 \leq 1$.

14. Show that for the vectors $\mathbf{T}_u$ and $\mathbf{T}_v$, we have the formula

$$\|\mathbf{T}_u \times \mathbf{T}_v\| = \sqrt{\left[\frac{\partial(x, y)}{\partial(u, v)}\right]^2 + \left[\frac{\partial(y, z)}{\partial(u, v)}\right]^2 + \left[\frac{\partial(x, z)}{\partial(u, v)}\right]^2}.$$

15. Compute the area of the surface given by

$$x = r\cos\theta, \qquad y = 2r\cos\theta, \qquad z = \theta, \qquad 0 \leq r \leq 1, \qquad 0 \leq \theta \leq 2\pi.$$

Sketch.

16. Prove *Pappus' theorem*: Let $\mathbf{c}$: $[a, b] \to \mathbb{R}^2$ be a C^1 path whose image lies in the right half plane and is a simple closed curve. The area of the lateral surface generated by rotating the image of $\mathbf{c}$ about the y axis is equal to $2\pi\bar{x}l(\mathbf{c})$, where $\bar{x}$ is the average value of the x coordinates of points on $\mathbf{c}$ and $l(\mathbf{c})$ is the length of $\mathbf{c}$. (See Exercises 8 to 11, Section 7.1, for a discussion of average values.)

17. The cylinder $x^2 + y^2 = x$ divides the unit sphere S into two regions S_1 and S_2, where S_1 is inside the cylinder and S_2 outside. Find the ratio of areas $A(S_2)/A(S_1)$.

18. Suppose a surface S that is the graph of a function $z = f(x, y)$, where $(x, y) \in D \subset \mathbb{R}^2$ can also be described as the set of $(x, y, z) \in \mathbb{R}^3$ with $F(x, y, z) = 0$ (a level surface). Derive a formula for $A(S)$ that involves only F.

19. Calculate the area of the frustum shown in Figure 7.4.8 using (a) geometry alone and, second, (b) a surface area formula.

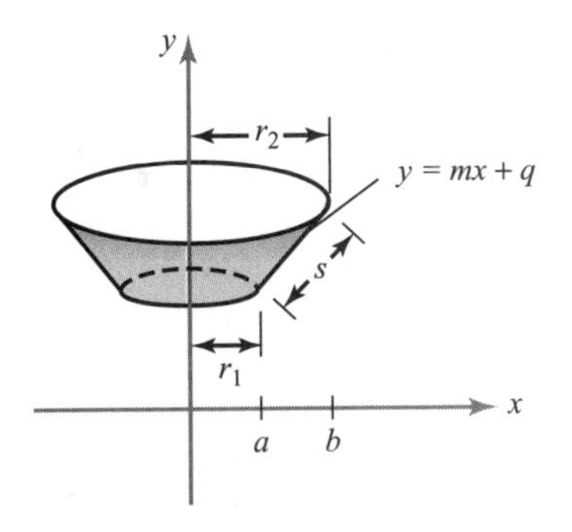

Figure 7.4.8 A line segment revolved around the y axis becomes a frustum of a cone.

20. A cylindrical hole of radius 1 is bored through a solid ball of radius 2 to form a ring coupler, as shown in Figure 7.4.9. Find the volume and outer surface area of this coupler.

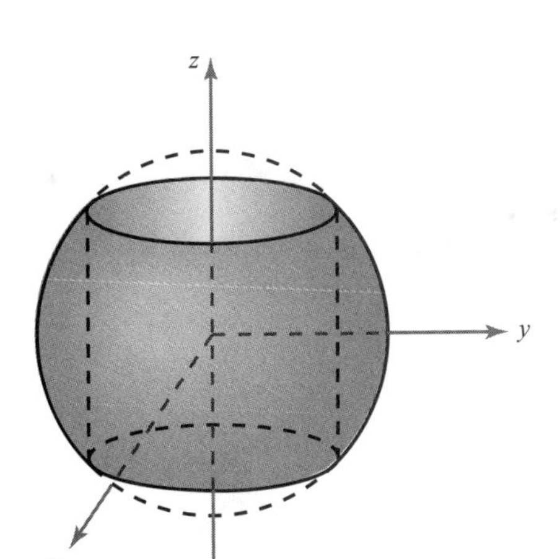

Figure 7.4.9 Find the outer surface area and volume of the shaded region.

21. Find the area of the graph of the function $f(x, y) = \frac{2}{3}(x^{3/2} + y^{3/2})$ that lies over the domain $[0, 1] \times [0, 1]$.

22. Express the surface area of the following graphs over the indicated region D as a double integral. Do not evaluate.

(a) $(x + 2y)^2$; $D = [-1, 2] \times [0, 2]$
(b) $xy + x/(y + 1)$; $D = [1, 4] \times [1, 2]$
(c) $xy^3e^{x^2y^2}$; $D =$ unit circle centered at the origin
(d) $y^3 \cos^2 x$; $D =$ triangle with vertices $(-1, 1)$, $(0, 2)$, and $(1, 1)$

23. Show that the surface area of the upper hemisphere of radius R, $z = \sqrt{R^2 - x^2 - y^2}$, can be computed by formula (4), evaluated as an improper integral.

7.5 Integrals of Scalar Functions Over Surfaces

Now we are ready to define the integral of a *scalar* function f *over a surface* S. This concept is a natural generalization of the area of a surface, which corresponds to the integral over S of the scalar function $f(x, y, z) = 1$. This is quite analogous to considering the path integral as a generalization of arc length. In the next section we shall deal with the integral of a *vector* function $\mathbf{F}$ over a surface. These concepts will play a crucial role in the vector analysis treated in the final chapter.

Let us start with a surface S parametrized by a mapping $\mathbf{\Phi}: D \to S \subset \mathbb{R}^3$, where D is an elementary region, which we write as $\mathbf{\Phi}(u, v) = (x(u, v), y(u, v), z(u, v))$.

DEFINITION: The Integral of a Scalar Function Over a Surface If $f(x, y, z)$ is a real-valued continuous function defined on a parametrized surface S, we define the ***integral of*** f ***over*** S to be

$$\iint_S f(x, y, z)\, dS = \iint_S f\, dS = \iint_D f(\mathbf{\Phi}(u, v)) \|\mathbf{T}_u \times \mathbf{T}_v\|\, du\, dv. \quad (1)$$

Written out, equation (1) becomes

$$\iint_S f\, dS = \iint_D f(x(u, v), y(u, v), z(u, v)) \sqrt{\left[\frac{\partial(x, y)}{\partial(u, v)}\right]^2 + \left[\frac{\partial(y, z)}{\partial(u, v)}\right]^2 + \left[\frac{\partial(x, z)}{\partial(u, v)}\right]^2}\, du\, dv. \quad (2)$$

Thus, if f is identically 1, we recover the area formula (3) of Section 7.4. Like surface area, the surface integral is independent of the particular parametrization used. This will be discussed in Section 7.6.

We can gain some intuitive knowledge about this integral by considering it as a limit of sums. Let D be a rectangle partitioned into n^2 rectangles R_{ij} with areas $\Delta u\, \Delta v$. Let $S_{ij} = \mathbf{\Phi}(R_{ij})$ *be the portion of the surface* $\mathbf{\Phi}(D)$ corresponding to R_{ij} (see Figure 7.5.1), and let $A(S_{ij})$ be the area of this portion of the surface. For large n, f will be approximately constant on S_{ij}, and we form the sum

$$S_n = \sum_{i=0}^{n-1} \sum_{j=0}^{n-1} f(\mathbf{\Phi}(u_i, v_j)) A(S_{ij}), \tag{3}$$

where $(u_i, v_j) \in R_{ij}$. From Section 7.4 we have a formula for $A(S_{ij})$:

$$A(S_{ij}) = \iint_{R_{ij}} \|\mathbf{T}_u \times \mathbf{T}_v\|\, du\, dv,$$

which, by the mean-value theorem for integrals, equals $\|\mathbf{T}_{u_i^*} \times \mathbf{T}_{v_j^*}\|\, \Delta u\, \Delta v$ for some point (u_i^*, v_j^*) in R_{ij}. Hence, our sum becomes

$$S_n = \sum_{i=0}^{n-1} \sum_{j=0}^{n-1} f(\mathbf{\Phi}(u_i, v_j)) \|\mathbf{T}_{u_i^*} \times \mathbf{T}_{v_j^*}\| \Delta u\, \Delta v,$$

which is an approximating sum for the last integral in formula (1). Therefore,

$$\operatorname*{limit}_{n \to \infty} S_n = \iint_S f\, dS.$$

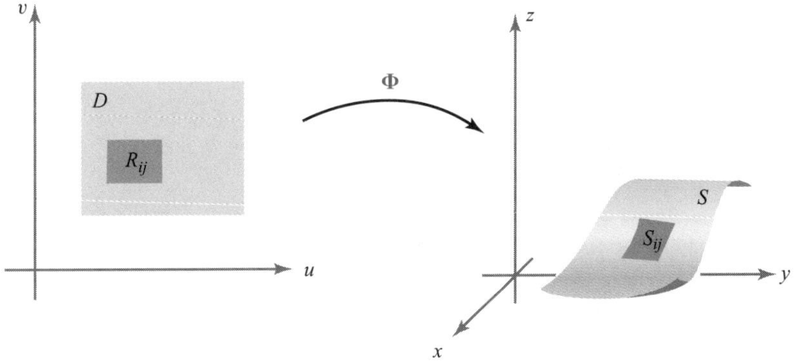

Figure 7.5.1 $\mathbf{\Phi}$ takes a portion R_{ij} of D to a portion of S.

Note that each term in the sum in formula (3) is the value of f at some point $\mathbf{\Phi}(u_i, v_j)$ times the area of S_{ij}. Compare this with the Riemann-sum interpretation of the path integral in Section 7.1.

If S is a union of parametrized surfaces S_i, $i = 1, \ldots, N$, that do not intersect except possibly along curves defining their boundaries, then the integral of f over S is defined by

$$\iint_S f\,dS = \sum_{i=1}^{N} \iint_{S_i} f\,dS,$$

as we should expect. For example, the integral over the surface of a cube may be expressed as the sum of the integrals over the six sides.

EXAMPLE 1 Suppose a helicoid is described as in Example 2, Section 7.4, and let f be given by $f(x, y, z) = \sqrt{x^2 + y^2 + 1}$. Find $\iint_S f\,dS$.

SOLUTION As in Examples 1 and 2 of Section 7.4,

$$\frac{\partial(x, y)}{\partial(r, \theta)} = r, \qquad \frac{\partial(y, z)}{\partial(r, \theta)} = \sin\theta, \qquad \frac{\partial(x, z)}{\partial(r, \theta)} = \cos\theta.$$

Also, $f(r\cos\theta, r\sin\theta, \theta) = \sqrt{r^2 + 1}$. Therefore,

$$\begin{aligned}\iint_S f(x, y, z)\,dS &= \iint_D f(\mathbf{\Phi}(r, \theta))\|\mathbf{T}_r \times \mathbf{T}_\theta\|\,dr\,d\theta \\ &= \int_0^{2\pi}\int_0^1 \sqrt{r^2 + 1}\sqrt{r^2 + 1}\,dr\,d\theta = \int_0^{2\pi} \frac{4}{3}\,d\theta = \frac{8}{3}\pi. \quad \blacktriangle\end{aligned}$$

Surface Integrals Over Graphs

Suppose S is the graph of a C^1 function $z = g(x, y)$. Recall from Section 7.4 that we can parametrize S by

$$x = u, \qquad y = v, \qquad z = g(u, v),$$

and that in this case

$$\|\mathbf{T}_u \times \mathbf{T}_v\| = \sqrt{1 + \left(\frac{\partial g}{\partial u}\right)^2 + \left(\frac{\partial g}{\partial v}\right)^2},$$

so

$$\iint_S f(x, y, z)\, dS = \iint_D f(x, y, g(x, y)) \sqrt{1 + \left(\frac{\partial g}{\partial x}\right)^2 + \left(\frac{\partial g}{\partial y}\right)^2}\, dx\, dy. \quad (4)$$

EXAMPLE 2 Let S be the surface defined by $z = x^2 + y$, where D is the region $0 \le x \le 1$, $-1 \le y \le 1$. Evaluate $\iint_S x\, dS$.

SOLUTION If we let $z = g(x, y) = x^2 + y$, formula (4) gives

$$\begin{aligned}\iint_S x\, dS &= \iint_D x \sqrt{1 + \left(\frac{\partial g}{\partial x}\right)^2 + \left(\frac{\partial g}{\partial y}\right)^2}\, dx\, dy = \int_{-1}^{1}\int_0^1 x\sqrt{1 + 4x^2 + 1}\, dx\, dy \\ &= \frac{1}{8}\int_{-1}^{1}\left[\int_0^1 (2 + 4x^2)^{1/2}(8x\, dx)\right] dy = \frac{2}{3}\cdot\frac{1}{8}\int_{-1}^{1} [(2 + 4x^2)^{3/2}]|_0^1\, dy \\ &= \frac{1}{12}\int_{-1}^{1}(6^{3/2} - 2^{3/2})\, dy = \frac{1}{6}(6^{3/2} - 2^{3/2}) = \sqrt{6} - \frac{\sqrt{2}}{3} \\ &= \sqrt{2}\left(\sqrt{3} - \frac{1}{3}\right). \quad \blacktriangle\end{aligned}$$

EXAMPLE 3 Evaluate $\iint_S z^2\, dS$, where S is the unit sphere $x^2 + y^2 + z^2 = 1$.

SOLUTION For this problem, it is convenient to use spherical coordinates and to represent the sphere parametrically by the equations $x = \cos\theta \sin\phi$, $y = \sin\theta\sin\phi$, $z = \cos\phi$, over the region D in the $\theta\phi$ plane given by the inequalities $0 \le \phi \le \pi$, $0 \le \theta \le 2\pi$. From equation (1) we get

$$\iint_S z^2\, dS = \iint_D (\cos\phi)^2 \|\mathbf{T}_\theta \times \mathbf{T}_\phi\|\, d\theta\, d\phi.$$

A little computation [use formula (2) of Section 7.4; see Exercise 6] shows that

$$\|\mathbf{T}_\theta \times \mathbf{T}_\phi\| = \sin\phi.$$

(Note that for $0 \le \phi \le \pi$, we have $\sin\phi \ge 0$). Thus,

$$\begin{aligned}\iint_S z^2\, dS &= \int_0^{2\pi}\int_0^{\pi} \cos^2\phi \sin\phi\, d\phi\, d\theta \\ &= \frac{1}{3}\int_0^{2\pi} [-\cos^3\phi]_0^{\pi}\, d\theta = \frac{2}{3}\int_0^{2\pi} d\theta = \frac{4\pi}{3}. \quad \blacktriangle\end{aligned}$$

This example also shows that on a sphere of radius R,

$$\iint_S f\,ds = \int_0^{2\pi}\int_0^{\pi} f(\phi,\theta)R^2 \sin\phi\, d\phi\, d\theta,$$

or, for short, the *area element on the sphere* is given by

$$dS = R^2 \sin\phi\, d\phi\, d\theta.$$

Integrals Over Graphs

We now develop another formula for surface integrals when the surface can be represented as a graph. To do so, we let S be the graph of $z = g(x, y)$ and consider formula (4). We claim that

$$\iint_S f(x, y, z)\, dS = \iint_D \frac{f(x, y, g(x, y))}{\cos\theta}\, dx\, dy, \tag{5}$$

where θ is the angle the normal to the surface makes with the unit vector $\mathbf{k}$ at the point $(x, y, g(x, y))$ (see Figure 7.5.2). Describing the surface by the equation $\phi(x, y, z) = z - g(x, y) = 0$, a normal vector $\mathbf{N}$ is $\nabla\phi$; that is,

$$\mathbf{N} = -\frac{\partial g}{\partial x}\mathbf{i} - \frac{\partial g}{\partial y}\mathbf{j} + \mathbf{k} \tag{6}$$

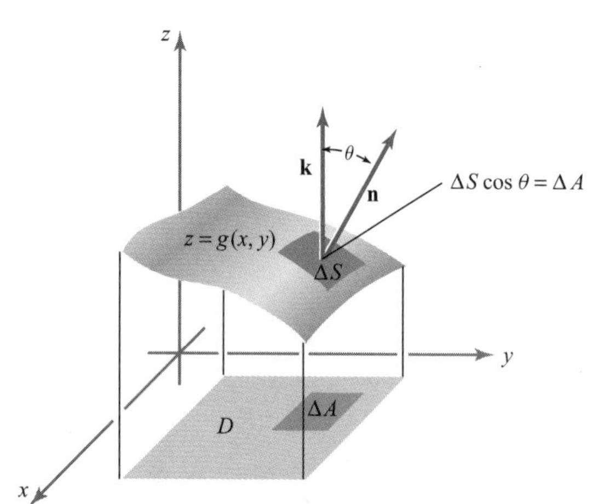

Figure 7.5.2 The area of a patch of area ΔS over a patch ΔA is $\Delta S = \Delta A/\cos\theta$, where θ is the angle the unit normal $\mathbf{n}$ makes with $\mathbf{k}$.

[see Example 4 of Section 7.3, or recall that the normal to a surface with equation $g(x, y, z) =$ constant is given by ∇g]. Thus,

$$\cos\theta = \frac{\mathbf{N}\cdot\mathbf{k}}{\|\mathbf{N}\|} = \frac{1}{\sqrt{(\partial g/\partial x)^2 + (\partial g/\partial y)^2 + 1}}.$$

Substitution of this formula into equation (4) gives equation (5). Note that $\cos\theta = \mathbf{n}\cdot\mathbf{k}$, where $\mathbf{n} = \mathbf{N}/\|\mathbf{N}\|$ is the unit normal. Thus, we can write

$$d\mathbf{S} = \frac{dx\,dy}{\mathbf{n}\cdot\mathbf{k}}$$

The result is, in fact, obvious geometrically, for if a small rectangle in the xy plane has area ΔA, then the area of the portion above it on the surface is $\Delta S = \Delta A/\cos\theta$ (Figure 7.5.2). This intuitive approach can help us to remember formula (5) and to apply it in problems.

EXAMPLE 4 Compute $\iint_S x\,dS$, where S is the triangle with vertices (1, 0, 0), (0, 1, 0), (0, 0, 1) (see Figure 7.5.3).

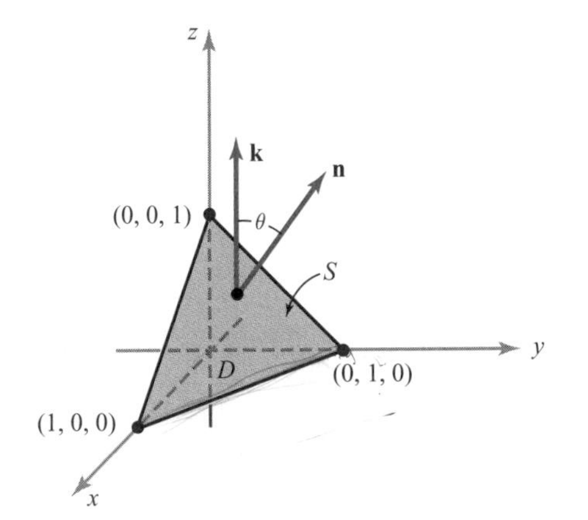

Figure 7.5.3 In computing a specific surface integral, one finds a formula for the unit normal **n** and computes the angle θ in preparation for formula (5).

SOLUTION This surface is the plane described by the equation $x + y + z = 1$. Because the surface is a plane, the angle θ is constant and a unit normal vector is $\mathbf{n} = (1/\sqrt{3}, 1/\sqrt{3}, 1/\sqrt{3})$. Thus, $\cos\theta = \mathbf{n}\cdot\mathbf{k} = 1/\sqrt{3}$, and by equation (5),

$$\iint_S x\,dS = \sqrt{3}\iint_D x\,dx\,dy,$$

where D is the domain in the xy plane. But

$$\sqrt{3}\iint_D x\,dx\,dy = \sqrt{3}\int_0^1\int_0^{1-x} x\,dy\,dx = \sqrt{3}\int_0^1 x(1-x)\,dx = \frac{\sqrt{3}}{6}. \quad \blacktriangle$$

Integrals of functions over surfaces are useful for computing the mass of a surface when the mass density function m is known. The total mass of a surface with mass density (per unit area) m is given by

$$M(S) = \iint_S m(x, y, z)\, dS. \tag{7}$$

EXAMPLE 5 Let $\mathbf{\Phi}\colon D \to \mathbb{R}^3$ be the parametrization of the helicoid $S = \mathbf{\Phi}(D)$ of Example 2 of Section 7.4. Recall that $\mathbf{\Phi}(r, \theta) = (r \cos\theta, r \sin\theta, \theta)$, where $0 \leq \theta \leq 2\pi$, and $0 \leq r \leq 1$. Suppose S has a mass density at $(x, y, z) \in S$ equal to twice the distance of (x, y, z) from the central axis (see Figure 7.4.2), that is, $m(x, y, z) = 2\sqrt{x^2 + y^2} = 2r$, in the cylindrical coordinate system. Find the total mass of the surface.

SOLUTION Applying formula (7),

$$M(S) = \iint_S 2\sqrt{x^2 + y^2}\, dS = \iint_D 2r\, dS = \iint_D 2r \|\mathbf{T}_r \times \mathbf{T}_\theta\|\, dr\, d\theta.$$

From Example 2 of Section 7.4, we see that $\|\mathbf{T}_r \times \mathbf{T}_\theta\| = \sqrt{1 + r^2}$. Thus,

$$\begin{aligned} M(S) &= \iint_D 2r\sqrt{1 + r^2}\, dr\, d\theta = \int_0^{2\pi} \int_0^1 2r\sqrt{1 + r^2}\, dr\, d\theta \\ &= \int_0^{2\pi} \left[\frac{2}{3}(1 + r^2)^{3/2}\right]_0^1 d\theta = \int_0^{2\pi} \frac{2}{3}(2^{3/2} - 1)\, d\theta = \frac{4\pi}{3}(2^{3/2} - 1). \quad \blacktriangle \end{aligned}$$

EXERCISES

1. Compute $\iint_S xy\, dS$, where S is the surface of the tetrahedron with sides $z = 0$, $y = 0$, $x + z = 1$, and $x = y$.

2. Evaluate $\iint_S xyz\, dS$, where S is the triangle with vertices $(1, 0, 0)$, $(0, 2, 0)$, and $(0, 1, 1)$.

3. Evaluate $\iint_S z\, dS$, where S is the upper hemisphere of radius a, that is, the set of (x, y, z) with $z = \sqrt{a^2 - x^2 - y^2}$.

4. Evaluate $\iint_S (x + y + z)\, dS$, where S is the boundary of the unit ball B; that is, S is the set of (x, y, z) with $x^2 + y^2 + z^2 = 1$. (HINT: Use the symmetry of the problem.)

5. (a) Compute the area of the portion of the cone $x^2 + y^2 = z^2$ with $z \geq 0$ that is inside the sphere $x^2 + y^2 + z^2 = 2Rz$, where R is a positive constant.

(b) What is the area of that portion of the sphere that is inside the cone?

6. Verify that in spherical coordinates, on a sphere of radius R,

$$\|\mathbf{T}_\phi \times \mathbf{T}_\theta\|\, d\phi\, d\theta = R^2 \sin\phi\, d\phi\, d\theta.$$

7. Evaluate $\iint_S z\, dS$, where S is the surface $z = x^2 + y^2$, $x^2 + y^2 \leq 1$.

8. Evaluate the surface integral $\iint_S z^2\, dS$, where S is the boundary of the cube $C = [-1, 1] \times [-1, 1] \times [-1, 1]$. (HINT: Do each face separately and add the results.)

9. Find the mass of a spherical surface S of radius R such that at each point $(x, y, z) \in S$ the mass density is equal to the distance of (x, y, z) to some fixed point $(x_0, y_0, z_0) \in S$.

10. A metallic surface S is in the shape of a hemisphere $z = \sqrt{R^2 - x^2 - y^2}$, where (x, y) satisfies $0 \le x^2 + y^2 \le R^2$. The mass density at $(x, y, z) \in S$ is given by $m(x, y, z) = x^2 + y^2$. Find the total mass of S.

11. Let S be the sphere of radius R.

(a) Argue by symmetry that

$$\iint_S x^2\, dS = \iint_S y^2\, dS = \iint_S z^2\, dS.$$

(b) Use this fact and some clever thinking to evaluate, with very little computation, the integral

$$\iint_S x^2\, dS.$$

(c) Does this help in Exercise 10?

12. (a) Use Riemann sums to justify the formula

$$\frac{1}{A(S)} \iint_S f(x, y, z)\, dS$$

for the *average value* of f over the surface S.

(b) In Example 3 of this section, show that the average of $f(x, y, z) = z^2$ over the sphere is 1/3.

(c) Define the ***center of gravity*** $(\bar{x}, \bar{y}, \bar{z})$ of a surface S to be such that $\bar{x}$, $\bar{y}$, and $\bar{z}$ are the average values of the x, y, and z coordinates on S. Show that the center of gravity of the triangle in Example 4 of this section is $(\frac{1}{3}, \frac{1}{3}, \frac{1}{3})$.

13. Find the x, y, and z coordinates of the center of gravity of the octant of the solid sphere of radius R and centered at the origin determined by $x \ge 0$, $y \ge 0$, $z \ge 0$. (HINT: Write this octant as a parametrized surface—see Example 3 of this section and Exercise 12.)

14. Find the z coordinate of the center of gravity (the average z coordinate) of the surface of a hemisphere ($z \le 0$) with radius r (see Exercise 12). Argue by symmetry that the average x and y coordinates are both zero.

15. Let $\mathbf{\Phi}\colon D \subset \mathbb{R}^2 \to \mathbb{R}^3$ be a parametrization of a surface S defined by

$$x = x(u, v), \qquad y = y(u, v), \qquad z = z(u, v).$$

(a) Let

$$\frac{\partial \mathbf{\Phi}}{\partial u} = \left(\frac{\partial x}{\partial u}, \frac{\partial y}{\partial u}, \frac{\partial z}{\partial u}\right) \quad \text{and} \quad \frac{\partial \mathbf{\Phi}}{\partial v} = \left(\frac{\partial x}{\partial v}, \frac{\partial y}{\partial v}, \frac{\partial z}{\partial v}\right),$$

that is, $\partial \mathbf{\Phi}/\partial u = \mathbf{T}_u$ and $\partial \mathbf{\Phi}/\partial v = \mathbf{T}_v$, and set

$$E = \left\|\frac{\partial \mathbf{\Phi}}{\partial u}\right\|^2, \qquad F = \frac{\partial \mathbf{\Phi}}{\partial u} \cdot \frac{\partial \mathbf{\Phi}}{\partial v}, \qquad G = \left\|\frac{\partial \mathbf{\Phi}}{\partial v}\right\|^2.$$

Show that

$$\sqrt{EG - F^2} = \|\mathbf{T}_u \times \mathbf{T}_v\|,$$

and that the surface area of S is

$$A(S) = \iint_D \sqrt{EG - F^2}\, du\, dv.$$

In this notation, how can we express $\iint_S f\, dS$ for a general function of f?

(b) What does the formula for $A(S)$ become if the vectors $\partial \mathbf{\Phi}/\partial u$ and $\partial \mathbf{\Phi}/\partial v$ are orthogonal?

(c) Use parts (a) and (b) to compute the surface area of a sphere of radius a.

16. ***Dirichlet's functional*** for a parametrized surface $\mathbf{\Phi}\colon D \to \mathbb{R}^3$ is defined by[12]

$$J(\mathbf{\Phi}) = \frac{1}{2} \iint_D \left(\left\|\frac{\partial \mathbf{\Phi}}{\partial u}\right\|^2 + \left\|\frac{\partial \mathbf{\Phi}}{\partial v}\right\|^2\right) du\, dv.$$

Use Exercise 15 to argue that the area $A(\mathbf{\Phi}) \leq J(\mathbf{\Phi})$ and equality holds if

$$\text{(a)} \quad \left\|\frac{\partial \mathbf{\Phi}}{\partial u}\right\|^2 = \left\|\frac{\partial \mathbf{\Phi}}{\partial v}\right\|^2 \qquad \text{and} \qquad \text{(b)} \quad \frac{\partial \mathbf{\Phi}}{\partial u} \cdot \frac{\partial \mathbf{\Phi}}{\partial v} = 0.$$

Compare these equations with Exercise 15 and the remarks at the end of Section 7.4. A parametrization $\mathbf{\Phi}$ that satisfies conditions (a) and (b) is said to be ***conformal***.

17. Let $D \subset \mathbb{R}^2$ and $\mathbf{\Phi}\colon D \to \mathbb{R}^2$ be a smooth function $\mathbf{\Phi}(u, v) = (x(u, v), y(u, v))$ satisfying conditions (a) and (b) of Exercise 16 and assume that

$$\det \begin{bmatrix} \dfrac{\partial x}{\partial u} & \dfrac{\partial x}{\partial v} \\[2ex] \dfrac{\partial y}{\partial u} & \dfrac{\partial y}{\partial v} \end{bmatrix} > 0.$$

[12]Dirichlet's functional played a major role in the mathematics of the nineteenth century. The mathematician Georg Friedrich Bernhard Riemann (1826–1866) used it to develop his complex function theory and to give a proof of the famous Riemann mapping theorem. Today it is still used extensively as a tool in the study of partial differential equations.

Show that x and y satisfy the ***Cauchy–Riemann equations*** $\partial x/\partial u = \partial y/\partial v$, $\partial x/\partial v = -\partial y/\partial u$. Conclude that $\nabla^2 \mathbf{\Phi} = 0$ (i.e., each component of $\mathbf{\Phi}$ is harmonic).

18. Let S be a sphere of radius r and $\mathbf{p}$ be a point inside or outside the sphere (but not on it). Show that

$$\iint_S \frac{1}{\|\mathbf{x} - \mathbf{p}\|}\, dS = \begin{cases} 4\pi r & \text{if } \mathbf{p} \text{ is inside } S \\ 4\pi r^2/d & \text{if } \mathbf{p} \text{ is outside } S, \end{cases}$$

where d is the distance from $\mathbf{p}$ to the center of the sphere and the integration is over the sphere.

19. Find the surface area of that part of the cylinder $x^2 + z^2 = a^2$ that is inside the cylinder $x^2 + y^2 = 2ay$ and also in the positive octant ($x \geq 0, y \geq 0, z \geq 0$). Assume $a > 0$.

20. Let a surface S be defined implicitly by $F(x, y, z) = 0$ for (x, y) in a domain D of $\mathbb{R}^2$. Show that

$$\iint_S \left| \frac{\partial F}{\partial z} \right| dS = \iint_D \sqrt{\left(\frac{\partial F}{\partial x}\right)^2 + \left(\frac{\partial F}{\partial y}\right)^2 + \left(\frac{\partial F}{\partial z}\right)^2}\, dx\, dy.$$

Compare with Exercise 18 of Section 7.4.

7.6 Surface Integrals of Vector Fields

The goal of this section is to develop the notion of the integral of a vector field over a surface. Recall that the definition of the line integral of a vector field was motivated by the fundamental physical notion of *work*. Similarly, there is a basic physical notion of *flux* that motivates the definition of the surface integral of a vector field.

For example, if the vector field is the velocity field of a fluid (perhaps the velocity field of a flowing river), and one puts an imagined mathematical surface into the fluid, one can ask: "What is the rate at which fluid is crossing the given surface (measured in, say, cubic meters per second)?" The answer is given by the surface integral of the fluid velocity vector field over the surface.

We shall come back to the physical interpretation shortly and reconcile it with the formal definition that we give first.

Definition of the Surface Integral

We now define the integral of a vector field, denoted $\mathbf{F}$ over a surface S. We first give the definition and later in this section give its physical interpretation. This can also be used as a *motivation* for the definition if the reader so desires. Also, we shall start with a parametrized surface $\mathbf{\Phi}$ and later study the question of independence of parametrization.

DEFINITION: The Surface Integral of Vector Fields Let $\mathbf{F}$ be a vector field defined on S, the image of a parametrized surface $\mathbf{\Phi}$. The ***surface integral*** of $\mathbf{F}$ over $\mathbf{\Phi}$, denoted by

$$\iint_{\mathbf{\Phi}} \mathbf{F} \cdot d\mathbf{S},$$

is defined by (see Figure 7.6.1))

$$\iint_{\mathbf{\Phi}} \mathbf{F} \cdot d\mathbf{S} = \iint_{D} \mathbf{F} \cdot (\mathbf{T}_u \times \mathbf{T}_v)\, du\, dv.$$

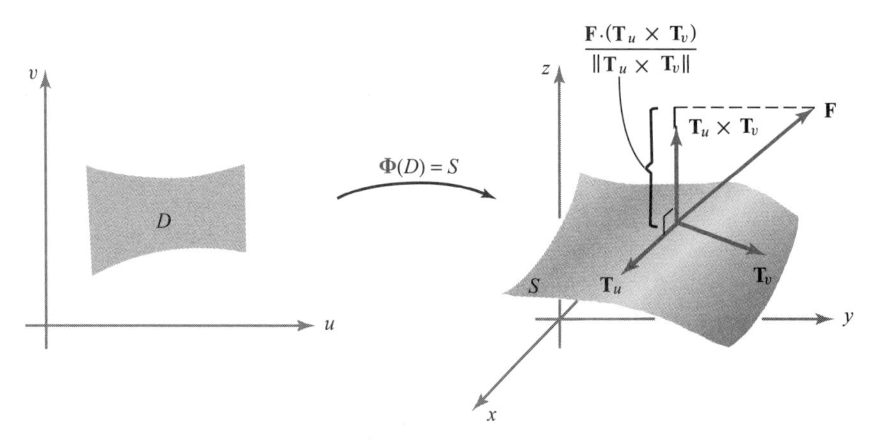

Figure 7.6.1 The geometric significance of $\mathbf{F} \cdot (\mathbf{T}_u \times \mathbf{T}_v)$.

EXAMPLE 1 Let D be the rectangle in the $\theta\phi$ plane defined by

$$0 \le \theta \le 2\pi, \qquad 0 \le \phi \le \pi,$$

and let the surface S be defined by the parametrization $\mathbf{\Phi}$: $D \to \mathbb{R}^3$ given by

$$x = \cos\theta \sin\phi, \qquad y = \sin\theta \sin\phi, \qquad z = \cos\phi.$$

(Thus, θ and ϕ are the angles of spherical coordinates, and S is the unit sphere parametrized by $\mathbf{\Phi}$.) Let $\mathbf{r}$ be the position vector $\mathbf{r}(x, y, z) = x\mathbf{i} + y\mathbf{j} + z\mathbf{k}$. Compute $\iint_{\mathbf{\Phi}} \mathbf{r} \cdot d\mathbf{S}$.

SOLUTION First we find

$$\mathbf{T}_\theta = (-\sin\phi \sin\theta)\mathbf{i} + (\sin\phi \cos\theta)\mathbf{j}$$

$$\mathbf{T}_\phi = (\cos\theta \cos\phi)\mathbf{i} + (\sin\theta \cos\phi)\mathbf{j} - (\sin\phi)\mathbf{k},$$

and hence

$$\mathbf{T}_\theta \times \mathbf{T}_\phi = (-\sin^2\phi \cos\theta)\mathbf{i} - (\sin^2\phi \sin\theta)\mathbf{j} - (\sin\phi \cos\phi)\mathbf{k}.$$

Then we evaluate

$$\begin{aligned}\mathbf{r}\cdot(\mathbf{T}_\theta \times \mathbf{T}_\phi) &= (x\mathbf{i}+y\mathbf{j}+z\mathbf{k})\cdot(\mathbf{T}_\theta \times \mathbf{T}_\phi)\\ &= [(\cos\theta\sin\phi)\mathbf{i} + (\sin\theta\sin\phi)\mathbf{j} + (\cos\phi)\mathbf{k}]\\ &\quad \cdot(-\sin\phi)[(\sin\phi\cos\theta)\mathbf{i} + (\sin\phi\sin\theta)\mathbf{j} + (\cos\phi)\mathbf{k}]\\ &= (-\sin\phi)(\sin^2\phi\cos^2\theta + \sin^2\phi\sin^2\theta + \cos^2\phi) = -\sin\phi.\end{aligned}$$

Thus,

$$\iint_\Phi \mathbf{r}\cdot d\mathbf{S} = \iint_D -\sin\phi\, d\phi\, d\theta = \int_0^{2\pi}(-2)\,d\theta = -4\pi. \quad \blacktriangle$$

Orientation

An analogy can be drawn between the surface integral $\iint_\Phi \mathbf{F}\cdot d\mathbf{S}$ and the line integral $\int_\mathbf{c} \mathbf{F}\cdot d\mathbf{s}$. Recall that the line integral is an oriented integral. We needed the notion of orientation of a curve to extend the definition of $\int_\mathbf{c} \mathbf{F}\cdot d\mathbf{s}$ to line integrals $\int_C \mathbf{F}\cdot d\mathbf{s}$ over oriented curves. We extend the definition of $\iint_\Phi \mathbf{F}\cdot d\mathbf{S}$ to *oriented surfaces* in a similar fashion; that is, given a surface S parametrized by a mapping $\mathbf{\Phi}$, we want to define $\iint_S \mathbf{F}\cdot d\mathbf{S} = \iint_\Phi \mathbf{F}\cdot d\mathbf{S}$ and show that it is independent of the parametrization, except possibly for the sign. To accomplish this, we need the notion of orientation of a surface.

DEFINITION: Oriented Surfaces An ***oriented surface*** is a two-sided surface with one side specified as the ***outside*** or ***positive side***; we call the other side the ***inside*** or ***negative side***.[13] At each point $(x, y, z) \in S$ there are two unit normal vectors $\mathbf{n}_1$ and $\mathbf{n}_2$, where $\mathbf{n}_1 = -\mathbf{n}_2$ (see Figure 7.6.2). Each of these two normals can be associated with one side of the surface. Thus, to specify a side of a surface S, at each point we choose a unit normal vector $\mathbf{n}$ that points away from the positive side of S at that point.

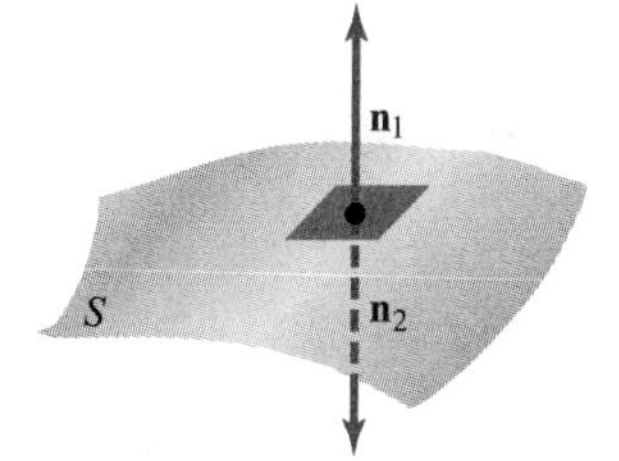

Figure 7.6.2 The two possible unit normals to a surface at a point.

[13] We use the term "side" in an intuitive sense. This concept can be developed rigorously, but this will not be done here. Also, the choice of the side to be named the "outside" is often dictated by the surface itself, as, for example, is the case with a sphere. In other cases, the naming is somewhat arbitrary (see the piece of surface depicted in Figure 7.6.2, for instance).

This definition assumes that our surface does have two sides. In fact, this is necessary, because there are examples of surfaces with only one side! The first known example of such a surface was the Möbius strip (named after the German mathematician and astronomer A. F. Möbius, who, along with the mathematician J. B. Listing, discovered it in 1858). Pictures of such a surface are given in Figures 7.6.3 and 7.6.4. At each point of M there are two unit normals, $\mathbf{n}_1$ and $\mathbf{n}_2$. However, $\mathbf{n}_1$ does not determine a unique side of M, and neither does $\mathbf{n}_2$. To see this intuitively, we can slide $\mathbf{n}_2$ around the closed curve C (Figure 7.6.3). When $\mathbf{n}_2$ returns to a fixed point p on C it will coincide with $\mathbf{n}_1$, showing that both $\mathbf{n}_1$ and $\mathbf{n}_2$ point away from the same side of M and, consequently, that M has only one side.

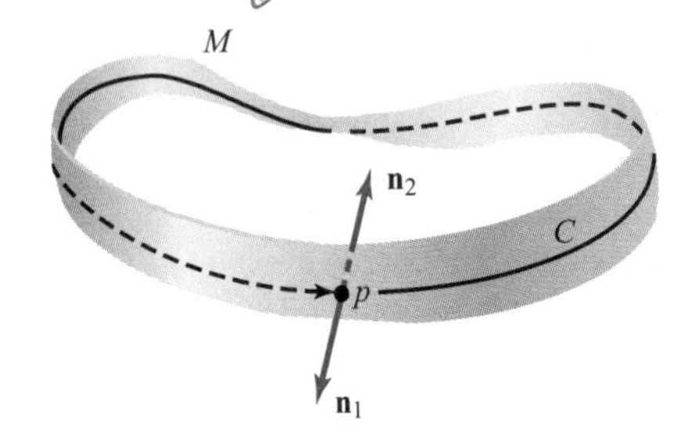

Figure 7.6.3 The Möbius strip: Slide $\mathbf{n}_2$ around C once; when $\mathbf{n}_2$ returns to its initial point, it will coincide with $\mathbf{n}_1 = -\mathbf{n}_2$.

Figure 7.6.4 is a Möbius strip as drawn by the well-known twentieth-century mathematician and artist M. C. Escher. It depicts ants crawling along the Möbius band. After one trip around the band (without crossing an edge) they end up on the "opposite side" of the surface.

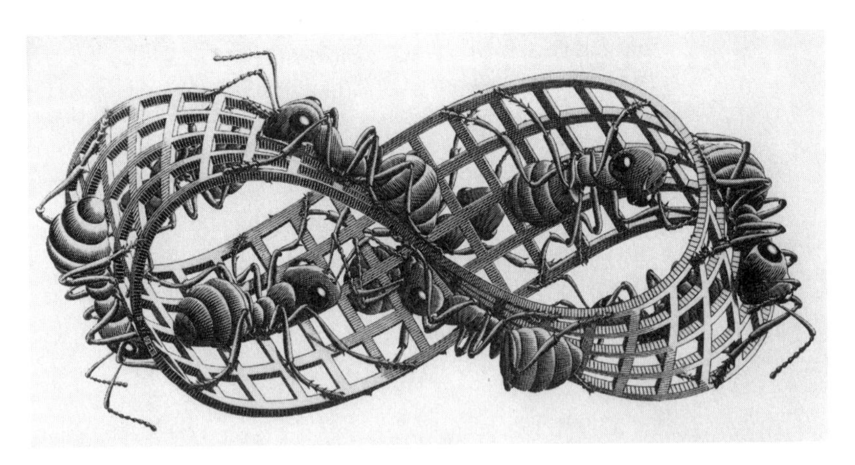

Figure 7.6.4 Ants walking on a Möbius strip.

Let $\mathbf{\Phi}: D \to \mathbb{R}^3$ be a parametrization of an oriented surface S and suppose S is regular at $\mathbf{\Phi}(u_0, v_0)$, $(u_0, v_0) \in D$; thus, the vector $(\mathbf{T}_{u_0} \times \mathbf{T}_{v_0})/\|\mathbf{T}_{u_0} \times \mathbf{T}_{v_0}\|$ is defined. If $\mathbf{n}(\mathbf{\Phi}(u_0, v_0))$ denotes the unit normal to S at $\mathbf{\Phi}(u_0, v_0)$, it follows that

$$(\mathbf{T}_{u_0} \times \mathbf{T}_{v_0})/\|\mathbf{T}_{u_0} \times \mathbf{T}_{v_0}\| = \pm\mathbf{n}(\mathbf{\Phi}(u_0, v_0)).$$

The parametrization $\mathbf{\Phi}$ is said to be ***orientation-preserving*** if we have the $+$ sign; that is, if $(\mathbf{T}_u \times \mathbf{T}_v)/\|\mathbf{T}_u \times \mathbf{T}_v\| = \mathbf{n}(\mathbf{\Phi}(u, v))$ at all $(u, v) \in D$ for which S is smooth at $\mathbf{\Phi}(u, v)$. In other words, $\mathbf{\Phi}$ is orientation-preserving if the vector $\mathbf{T}_u \times \mathbf{T}_v$ points to the outside of the surface. If $\mathbf{T}_u \times \mathbf{T}_v$ points to the inside of the surface at all points $(u, v) \in D$ for which S is regular at $\mathbf{\Phi}(u, v)$, then $\mathbf{\Phi}$ is said to be ***orientation-reversing***. Using the preceding notation, this condition corresponds to the choice $(\mathbf{T}_u \times \mathbf{T}_v)/\|\mathbf{T}_u \times \mathbf{T}_v\| = -\mathbf{n}(\mathbf{\Phi}(u, v))$.

It follows from this discussion that the Möbius band M cannot be parametrized by a single parametrization for which $\mathbf{n} = \mathbf{T}_u \times \mathbf{T}_v \neq \mathbf{0}$ and $\mathbf{n}$ is continuous over the whole surface[14] (if there were such a parameterization, then M would indeed have two sides, one determined by $\mathbf{n}$ and one determined by $-\mathbf{n}$). The sphere in Example 1 can be parametrized by a single parametrization, but not by one that is everywhere one-to-one—see the discussion at the beginning of Section 7.4.

Thus, any one-to-one parametrized surface for which $\mathbf{T}_u \times \mathbf{T}_v$ never vanishes can be considered as an oriented surface with a positive side determined by the direction of $\mathbf{T}_u \times \mathbf{T}_v$.

EXAMPLE 2 We can give the unit sphere $x^2 + y^2 + z^2 = 1$ in $\mathbb{R}^3$ (Figure 7.6.5) an orientation by selecting the unit vector $\mathbf{n}(x, y, z) = \mathbf{r}$, where $\mathbf{r} = x\mathbf{i} + y\mathbf{j} + z\mathbf{k}$, which points to the outside of the surface. This choice corresponds to our intuitive notion of outside for the sphere.

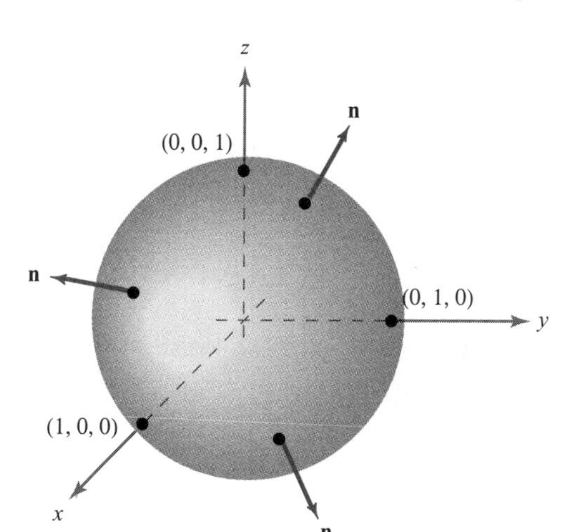

Figure 7.6.5 The unit sphere oriented by its outward normal **n**.

Now that the sphere S is an oriented surface, consider the parametrization $\mathbf{\Phi}$ of S given in Example 1. The cross product of the tangent vectors $\mathbf{T}_\theta$ and $\mathbf{T}_\phi$—that is,

[14] There is a single parametrization obtained by cutting a strip of paper, twisting it, and gluing the ends, but it produces a discontinuous **n** on the surface.

a normal to S is given by

$$(-\sin\phi)[(\cos\theta\sin\phi)\mathbf{i} + (\sin\theta\sin\phi)\mathbf{j} + (\cos\phi)\mathbf{k}] = -\mathbf{r}\sin\phi.$$

Because $-\sin\phi \leq 0$ for $0 \leq \phi \leq \pi$, this normal vector points inward from the sphere. Thus, the given parametrization $\mathbf{\Phi}$ is *orientation-reversing*. By swapping the order of θ and ϕ, we would get an orientation-preserving parametrization. ▲

Orientation and the Vector Surface Element of a Sphere

Consider the sphere of radius R, namely, $x^2 + y^2 + z^2 = R^2$. It is standard practice to orient the sphere with the *outward unit normal*. In terms of the position vector $\mathbf{r} = x\mathbf{i} + y\mathbf{j} + z\mathbf{k}$, the outward unit normal is given by

positive orientation

$$\mathbf{n} = \frac{\mathbf{r}}{R}.$$

The order of spherical coordinates that goes along with this orientation, as is evident from Example 2, is given by the order (ϕ, θ). The computation in Example 2 shows that the surface-area element is then given by

$$d\mathbf{S} = \mathbf{n}\cdot(\mathbf{T}_\phi \times \mathbf{T}_\theta)\, d\phi\, d\theta = \mathbf{r}R\sin\phi\, d\phi\, d\theta = \mathbf{n}R^2\sin\phi\, d\phi\, d\theta.$$

The Orientation of Graphs

The next example discusses the orientation conventions for graphs. We shall compute the area element on graphs later in this section.

EXAMPLE 3 Let S be a surface described by $z = g(x, y)$. As in equation (6), Section 7.5, there are two unit normal vectors to S at $(x_0, y_0, g(x_0, y_0))$, namely, $\pm\mathbf{n}$, where

$$\mathbf{n} = \frac{-\dfrac{\partial g}{\partial x}(x_0, y_0)\mathbf{i} - \dfrac{\partial g}{\partial y}(x_0, y_0)\mathbf{j} + \mathbf{k}}{\sqrt{\left[\dfrac{\partial g}{\partial x}(x_0, y_0)\right]^2 + \left[\dfrac{\partial g}{\partial y}(x_0, y_0)\right]^2 + 1}}.$$

We can orient all such surfaces by taking the positive side of S to be the side away from which $\mathbf{n}$ points (Figure 7.6.6). Thus, the positive side of such a surface is

of a graph

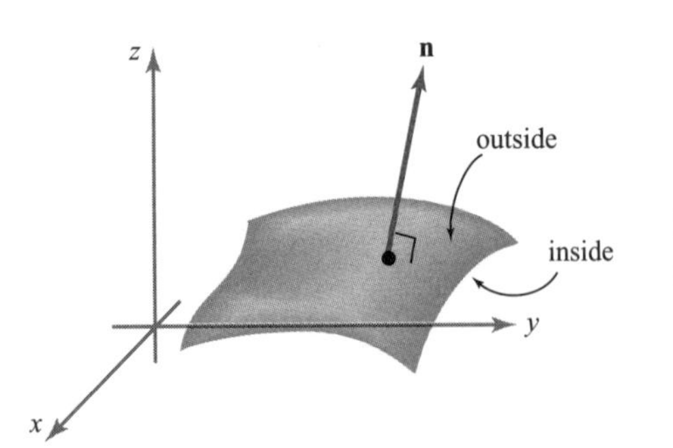

Figure 7.6.6 $\mathbf{n}$ points away from the outside of the surface.

determined by the unit normal $\mathbf{n}$ with positive $\mathbf{k}$ component—that is, it is *upward-pointing*. If we parametrize this surface by $\mathbf{\Phi}(u, v) = (u, v, g(u, v))$, then $\mathbf{\Phi}$ will be orientation-preserving. ▲

Independence of Parametrization

We now state without proof a theorem showing that the integral over an oriented surface is independent of the parametrization. The proof of this theorem is analogous to that of Theorem 1 (Section 7.2); the heart of the proof is again the change of variables formula—this time applied to double integrals.

THEOREM 4: Independence of Surface Integrals on Parametrizations Let S be an oriented surface and let $\mathbf{\Phi}_1$ and $\mathbf{\Phi}_2$ be two regular orientation-preserving parametrizations, with $\mathbf{F}$ a continuous vector field defined on S. Then

$$\iint_{\mathbf{\Phi}_1} \mathbf{F} \cdot d\mathbf{S} = \iint_{\mathbf{\Phi}_2} \mathbf{F} \cdot d\mathbf{S}.$$

If $\mathbf{\Phi}_1$ is orientation-preserving and $\mathbf{\Phi}_2$ orientation-reversing, then

$$\iint_{\mathbf{\Phi}_1} \mathbf{F} \cdot d\mathbf{S} = -\iint_{\mathbf{\Phi}_2} \mathbf{F} \cdot d\mathbf{S}.$$

If f is a real-valued continuous function defined on S, and if $\mathbf{\Phi}_1$ and $\mathbf{\Phi}_2$ are parametrizations of S, then

$$\iint_{\mathbf{\Phi}_1} f\, dS = \iint_{\mathbf{\Phi}_2} f\, dS.$$

Note that if $f = 1$, we obtain

$$A(S) = \iint_{\mathbf{\Phi}_1} dS = \iint_{\mathbf{\Phi}_2} dS,$$

thus showing that area is independent of parametrization.

We can therefore unambiguously use the notation

$$\iint_S \mathbf{F} \cdot d\mathbf{S} = \iint_{\mathbf{\Phi}} \mathbf{F} \cdot d\mathbf{S}$$

(or a sum of such integrals, if S is a union of parametrized surfaces that intersect only along their boundary curves) where $\mathbf{\Phi}$ is an orientation-preserving parametrization.

Theorem 4 guarantees that the value of the integral does not depend on the selection of $\mathbf{\Phi}$.

Relation with Scalar Integrals

Recall from formula (1) of Section 7.2 that a line integral $\int_{\mathbf{c}} \mathbf{F} \cdot d\mathbf{s}$ can be thought of as the path integral of the tangential component of $\mathbf{F}$ along $\mathbf{c}$ (although for the case in which $\mathbf{c}$ intersects itself, the integral obtained is technically not a path integral). A similar situation holds for surface integrals, because we are assuming that the mappings $\mathbf{\Phi}$ defining the surface S are one-to-one except perhaps on the boundary of D, which can be ignored for the purposes of integration. Thus, in defining integrals over surfaces, we assume in this book that the surfaces are nonintersecting.

For an oriented smooth surface S and an orientation-preserving parametrization $\mathbf{\Phi}$ of S, we can express $\iint_S \mathbf{F} \cdot d\mathbf{S}$ as an integral of a real-valued function f over the surface. Let $\mathbf{n} = (\mathbf{T}_u \times \mathbf{T}_v)/\|\mathbf{T}_u \times \mathbf{T}_v\|$ be the unit normal pointing to the outside of S. Then

$$\begin{aligned}\iint_S \mathbf{F} \cdot d\mathbf{S} &= \iint_{\mathbf{\Phi}} \mathbf{F} \cdot d\mathbf{S} = \iint_D \mathbf{F} \cdot (\mathbf{T}_u \times \mathbf{T}_v)\, du\, dv \\ &= \iint_D \mathbf{F} \cdot \left(\frac{\mathbf{T}_u \times \mathbf{T}_v}{\|\mathbf{T}_u \times \mathbf{T}_v\|} \right) \|\mathbf{T}_u \times \mathbf{T}_v\|\, du\, dv \\ &= \iint_D (\mathbf{F} \cdot \mathbf{n}) \|\mathbf{T}_u \times \mathbf{T}_v\|\, du\, dv = \iint_S (\mathbf{F} \cdot \mathbf{n})\, dS = \iint_S f\, dS,\end{aligned}$$

where $f = \mathbf{F} \cdot \mathbf{n}$. We have thus proved the following theorem.

THEOREM 5 $\iint_S \mathbf{F} \cdot d\mathbf{S}$, the surface integral of $\mathbf{F}$ over S, is equal to the integral of the normal component of $\mathbf{F}$ over the surface. In short,

$$\iint_S \mathbf{F} \cdot d\mathbf{S} = \iint_S \mathbf{F} \cdot \mathbf{n}\, dS.$$

The observation in Theorem 5 can often save computational effort, as Example 4 demonstrates.

The Physical Interpretation of Surface Integrals

The geometric and physical significance of the surface integral can be understood by expressing it as a limit of Riemann sums. For simplicity, we assume D is a rectangle. Fix a parametrization $\mathbf{\Phi}$ of S that preserves orientation and partition the region D into n^2 pieces D_{ij}, $0 \le i \le n-1$, $0 \le j \le n-1$. We let Δu denote the length of the horizontal side of D_{ij} and Δv denote the length of the vertical side of D_{ij}. Let (u, v) be a point in D_{ij}, and $(x, y, z) = \mathbf{\Phi}(u, v)$ the corresponding point on the surface. We consider the parallelogram with sides $\Delta u\, \mathbf{T}_u$ and $\Delta v\, \mathbf{T}_v$ lying in the plane tangent to S at (x, y, z) and the parallelepiped formed by $\mathbf{F}$, $\Delta u\, \mathbf{T}_u$, and $\Delta v\, \mathbf{T}_v$. The volume of the parallelepiped is the absolute value of the triple product

$$\mathbf{F}\cdot(\Delta u\,\mathbf{T}_u \times \Delta v\,\mathbf{T}_v) = \mathbf{F}\cdot(\mathbf{T}_u \times \mathbf{T}_v)\,\Delta u\,\Delta v.$$

The vector $\mathbf{T}_u \times \mathbf{T}_v$ is normal to the surface at (x, y, z) and points away from the outside of the surface. Thus, the number $\mathbf{F}\cdot(\mathbf{T}_u \times \mathbf{T}_v)$ is positive when the parallelepiped lies on the outside of the surface (Figure 7.6.7).

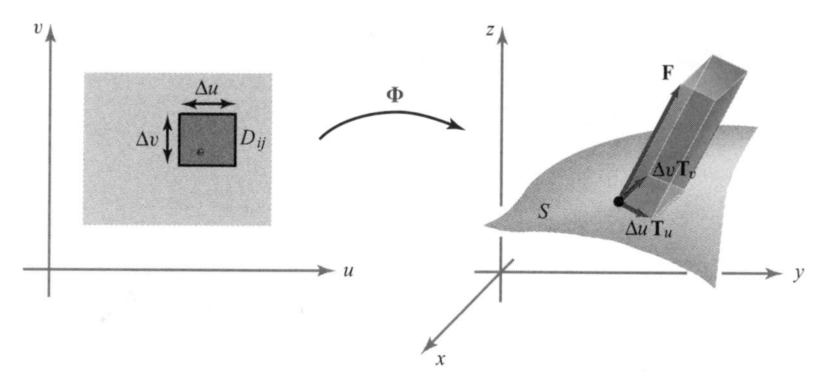

Figure 7.6.7 $\mathbf{F}\cdot(\mathbf{T}_u \times \mathbf{T}_v) > 0$ when the parallelpiped formed by $\Delta v\,\mathbf{T}_v$, $\Delta u\,\mathbf{T}_u$, and $\mathbf{F}$ lies to the "outside" of the surface S.

In general, the parallelepiped lies on that side of the surface away from which $\mathbf{F}$ is pointing. If we think of $\mathbf{F}$ as the velocity field of a fluid, $\mathbf{F}(x, y, z)$ is pointing in the direction in which fluid is moving across the surface near (x, y, z). Moreover, the number

$$|\mathbf{F}\cdot(\mathbf{T}_u\,\Delta u \times \mathbf{T}_v\,\Delta v)|$$

measures the amount of fluid that passes through the tangent parallelogram per unit time. Because the sign of $\mathbf{F}\cdot(\Delta u\,\mathbf{T}_u \times \Delta v\,\mathbf{T}_v)$ is positive if the vector $\mathbf{F}$ is pointing outward at (x, y, z) and negative if $\mathbf{F}$ is pointing inward, $\sum_{i,j}\mathbf{F}\cdot(\mathbf{T}_u \times \mathbf{T}_v)\,\Delta u\,\Delta v$ is an approximate measure of the net quantity of fluid to flow outward across the surface per unit time. (Remember that "outward" or "inward" depends on our choice of parametrization. Figure 7.6.8 illustrates $\mathbf{F}$ directed outward and inward, given $\mathbf{T}_u$ and $\mathbf{T}_v$.) Hence, *the integral* $\iint_S \mathbf{F}\cdot d\mathbf{S}$ *is the net quantity of fluid to flow across the surface per unit time, that is, the rate of fluid flow.* This integral is also called the ***flux*** of $\mathbf{F}$ across the surface.

In the case where $\mathbf{F}$ represents an electric or a magnetic field, $\iint_S \mathbf{F}\cdot d\mathbf{S}$ is also commonly known as the flux. The reader may be familiar with physical laws (such as *Faraday's law*) that relate flux of a vector field to a circulation (or current) in a bounding loop. This is the historical and physical basis of Stokes's theorem, which we will discuss in Section 8.2. The corresponding principle in fluid mechanics is called *Kelvin's circulation theorem.*

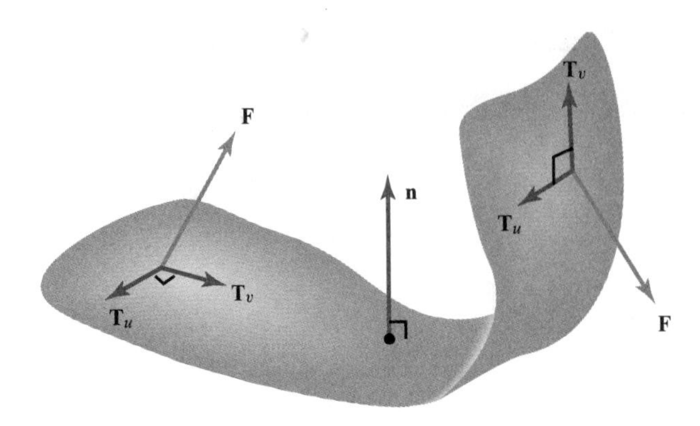

Figure 7.6.8 When $\mathbf{F} \cdot (\mathbf{T}_u \times \mathbf{T}_v) > 0$ (left), $\mathbf{F}$ points outward; when $\mathbf{F} \cdot (\mathbf{T}_u \times \mathbf{T}_v) < 0$ (right), $\mathbf{F}$ points inward.

Surface integrals also apply to the study of heat flow. Let $T(x, y, z)$ be the temperature at a point $(x, y, z) \in W \subset \mathbb{R}^3$, where W is some region and T is a C^1 function. Then

$$\nabla T = \frac{\partial T}{\partial x}\mathbf{i} + \frac{\partial T}{\partial y}\mathbf{j} + \frac{\partial T}{\partial z}\mathbf{k}$$

represents the temperature gradient, and heat "flows" with the vector field $-k\,\nabla T = \mathbf{F}$, where k is a positive constant (see Section 8.5). Therefore, $\iint_S \mathbf{F} \cdot d\mathbf{S}$ is the total rate of heat flow or flux across the surface S.

EXAMPLE 4 Suppose a temperature function is given in $\mathbb{R}^3$ by the formula $T(x, y, z) = x^2 + y^2 + z^2$, and let S be the unit sphere $x^2 + y^2 + z^2 = 1$ oriented with the outward normal (see Example 2). Find the heat flux across the surface S if $k = 1$.

SOLUTION We have

$$\mathbf{F} = -\nabla T(x, y, z) = -2x\mathbf{i} - 2y\mathbf{j} - 2z\mathbf{k}.$$

On S, the vector $\mathbf{n}(x, y, z) = x\mathbf{i} + y\mathbf{j} + z\mathbf{k}$ is the unit "outward" normal to S at (x, y, z), and $f(x, y, z) = \mathbf{F} \cdot \mathbf{n} = -2x^2 - 2y^2 - 2z^2 = -2$ is the normal component of $\mathbf{F}$. From Theorem 5 we can see that the surface integral of $\mathbf{F}$ is equal to the integral of its normal component $f = \mathbf{F} \cdot \mathbf{n}$ over S. Thus,

$$\iint_S \mathbf{F} \cdot d\mathbf{S} = \iint_S f\, dS = -2 \iint_S dS = -2A(S) = -2(4\pi) = -8\pi.$$

The flux of heat is directed toward the center of the sphere (why toward?). Clearly, our observation that $\iint_S \mathbf{F} \cdot d\mathbf{S} = \iint_S f\, dS$ has saved us considerable computational time.

In this example, $\mathbf{F}(x, y, z) = -2x\mathbf{i} - 2y\mathbf{j} - 2z\mathbf{k}$ could also represent an electric field, in which case $\iint_S \mathbf{F} \cdot d\mathbf{S} = -8\pi$ would be the electric flux across S. ▲

EXAMPLE 5 **Gauss' Law** There is an important physical law, due to the great mathematician and physicist K. F. Gauss, that relates the flux of an electric field $\mathbf{E}$ over a "closed" surface S (for example, a sphere or an ellipsoid) to the net charge Q enclosed by the surface, namely (in suitable units),

$$\iint_S \mathbf{E} \cdot d\mathbf{S} = Q \tag{1}$$

(see Figure 7.6.9). Gauss' law will be discussed in detail in Chapter 8. This law is analogous to Ampère's law (see Example 12, Section 7.2).

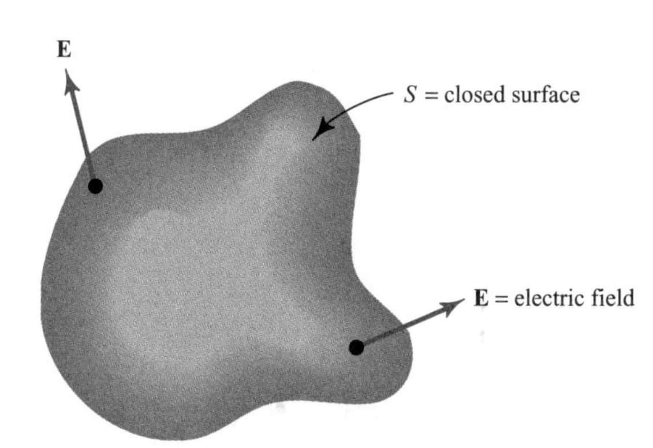

Figure 7.6.9 Gauss' law: $\iint_S \mathbf{E} \cdot d\mathbf{S} = Q$, where Q is the net charge inside S.

Suppose that $\mathbf{E} = E\mathbf{n}$; that is, $\mathbf{E}$ is a constant scalar multiple of the unit normal to S. Then Gauss' law, equation (1) in Example 5, becomes

$$\iint_S \mathbf{E} \cdot d\mathbf{S} = \iint_S E\, dS = E \iint_S dS = Q$$

because $E = \mathbf{E} \cdot \mathbf{n}$. Thus,

$$E = \frac{Q}{A(S)}. \tag{2}$$

In the case where S is the sphere of radius R, equation (2) becomes

$$E = \frac{Q}{4\pi R^2} \tag{3}$$

(see Figure 7.6.10).

Now suppose that $\mathbf{E}$ arises from an isolated point charge, Q. From symmetry it is reasonable that $\mathbf{E} = E\mathbf{n}$, where $\mathbf{n}$ is the unit normal to any sphere centered at Q. Hence, equation (3) holds. Consider a second point charge, Q_0, located at a distance

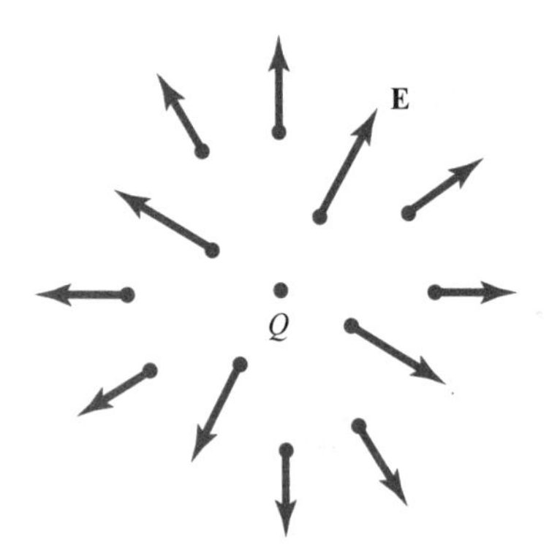

Figure 7.6.10 The field $\mathbf{E}$ due to a point charge Q is $\mathbf{E} = Q\mathbf{n}/4\pi R^2$.

R from Q. The force $\mathbf{F}$ that acts on this second charge, Q_0, is given by

$$\mathbf{F} = \mathbf{E}Q_0 = EQ_0\mathbf{n} = \frac{QQ_0}{4\pi R^2}\mathbf{n}.$$

If F is the magnitude of $\mathbf{F}$, we have

$$F = \frac{QQ_0}{4\pi R^2},$$

which is ***Coulomb's law*** for the force between two point charges.[15] ▲

Surface Integrals Over Graphs

Finally, let us derive the surface-integral formulas for vector fields $\mathbf{F}$ over surfaces S that are graphs of functions. Consider the surface S described by $z = g(x, y)$, where $(x, y) \in D$, where S is oriented with the upward pointing unit normal:

$$\mathbf{n} = \frac{-\dfrac{\partial g}{\partial x}\mathbf{i} - \dfrac{\partial g}{\partial y}\mathbf{j} + \mathbf{k}}{\sqrt{\left(\dfrac{\partial g}{\partial x}\right)^2 + \left(\dfrac{\partial g}{\partial y}\right)^2 + 1}}.$$

We have seen that we can parametrize S by $\mathbf{\Phi}\colon D \to \mathbb{R}^3$ given by $\mathbf{\Phi}(x, y) = (x, y, g(x, y))$. In this case, $\iint_S \mathbf{F}\cdot d\mathbf{S}$ can be written in a particularly simple form. We have

$$\mathbf{T}_x = \mathbf{i} + \frac{\partial g}{\partial x}\mathbf{k}, \qquad \mathbf{T}_y = \mathbf{j} + \frac{\partial g}{\partial y}\mathbf{k}.$$

[15] Sometimes one sees the formula $F = (1/4\pi\varepsilon_0)QQ_0/R^2$. The extra constant ε_0 appears when MKS units are used for measuring charge. We are using CGS, or Gaussian, units.

Thus, $\mathbf{T}_x \times \mathbf{T}_y = -(\partial g/\partial x)\mathbf{i} - (\partial g/\partial y)\mathbf{j} + \mathbf{k}$. If $\mathbf{F} = F_1\mathbf{i} + F_2\mathbf{j} + F_3\mathbf{k}$ is a continuous vector field, then we get

The Surface Integral of a Vector Field Over a Graph S

$$\begin{aligned}\iint_S \mathbf{F}\cdot d\mathbf{S} &= \iint_D \mathbf{F}\cdot(\mathbf{T}_x \times \mathbf{T}_y)\,dx\,dy \\ &= \iint_D \left[F_1\left(-\frac{\partial g}{\partial x}\right) + F_2\left(-\frac{\partial g}{\partial y}\right) + F_3\right]dx\,dy.\end{aligned} \tag{4}$$

EXAMPLE 6 The equations

$$z = 12, \qquad x^2 + y^2 \leq 25$$

describe a disk of radius 5 lying in the plane $z = 12$. Suppose $\mathbf{r}$ is the vector field

$$\mathbf{r}(x, y, z) = x\mathbf{i} + y\mathbf{j} + z\mathbf{k}.$$

Compute $\iint_S \mathbf{r}\cdot d\mathbf{S}$.

SOLUTION We shall do this in three ways. First, we have $\partial z/\partial x = \partial z/\partial y = 0$, because $z = 12$ is constant on the disk, so

$$\mathbf{r}(x, y, z)\cdot(\mathbf{T}_x \times \mathbf{T}_y) = \mathbf{r}(x, y, z)\cdot(\mathbf{i}\times\mathbf{j}) = \mathbf{r}(x, y, z)\cdot\mathbf{k} = z.$$

Using the original definition at the beginning of this section, the integral becomes

$$\iint_S \mathbf{r}\cdot d\mathbf{S} = \iint_D z\,dx\,dy = \iint_D 12\,dx\,dy = 12(\text{area of } D) = 300\pi.$$

A second solution: Because the disk is parallel to the xy plane, the outward unit normal is $\mathbf{k}$. Hence, $\mathbf{n}(x, y, z) = \mathbf{k}$ and $\mathbf{r}\cdot\mathbf{n} = z$. However, $\|\mathbf{T}_x \times \mathbf{T}_y\| = \|\mathbf{k}\| = 1$, and so we know from the discussion preceding Theorem 5 that

$$\iint_S \mathbf{r}\cdot d\mathbf{S} = \iint_S \mathbf{r}\cdot\mathbf{n}\,dS = \iint_S z\,dS = \iint_D 12\,dx\,dy = 300\pi.$$

Third, we may solve this problem by using formula (4) directly, with $g(x, y) = 12$ and D the disk $x^2 + y^2 \leq 25$:

$$\iint_S \mathbf{r} \cdot d\mathbf{S} = \iint_D (x \cdot 0 + y \cdot 0 + 12)\, dx\, dy = 12(\text{area of } D) = 300\pi. \quad \blacktriangle$$

Summary: Formulas for Surface Integrals

1. **Parametrized Surface: $\mathbf{\Phi}(u, v)$**

(a) Integral of a scalar function f:

$$\iint_S f\, dS = \iint_D f(\mathbf{\Phi}(u, v)) \|\mathbf{T}_u \times \mathbf{T}_v\|\, du\, dv$$

(b) Scalar surface element:

$$dS = \|\mathbf{T}_u \times \mathbf{T}_v\|\, du\, dv$$

(c) Integral of a vector field $\mathbf{F}$:

$$\iint_S \mathbf{F} \cdot d\mathbf{S} = \iint_D \mathbf{F} \cdot (\mathbf{T}_u \times \mathbf{T}_v)\, du\, dv$$

(d) Vector surface element:

$$d\mathbf{S} = (\mathbf{T}_u \times \mathbf{T}_v)\, du\, dv = \mathbf{n}\, dS$$

2. **Graph**: $z = g(x, y)$

(a) Integral of a scalar function f:

$$\iint_S f\, dS = \iint_D \frac{f(x, y, g(x, y))}{\cos\theta}\, dx\, dy$$

(b) Scalar surface element:

$$dS = \frac{dx\, dy}{\cos\theta} = \sqrt{\left(\frac{\partial g}{\partial x}\right)^2 + \left(\frac{\partial g}{\partial y}\right)^2 + 1}\, dx\, dy,$$

where $\cos\theta = \mathbf{n} \cdot \mathbf{k}$, and $\mathbf{n}$ is a unit normal vector to the surface.

(c) Integral of a vector field $\mathbf{F}$:

$$\iint_S \mathbf{F} \cdot d\mathbf{S} = \iint_D \left(-F_1 \frac{\partial g}{\partial x} - F_2 \frac{\partial g}{\partial y} + F_3 \right) dx\, dy$$

(d) Vector surface element:

$$d\mathbf{S} = \mathbf{n} \cdot dS = \left(-\frac{\partial g}{\partial x}\mathbf{i} - \frac{\partial g}{\partial y}\mathbf{j} + \mathbf{k} \right) dx\, dy$$

3. **Sphere:** $x^2 + y^2 + z^2 = R^2$

(a) Scalar surface element:

$$dS = R^2 \sin\phi\, d\phi\, d\theta$$

(b) Vector surface element:

$$d\mathbf{S} = (x\mathbf{i} + y\mathbf{j} + z\mathbf{k}) R \sin\phi\, d\phi\, d\theta = \mathbf{r} R \sin\phi\, d\phi\, d\theta = \mathbf{n} R^2 \sin\phi\, d\phi\, d\theta$$

EXERCISES

1. Let the temperature of a point in $\mathbb{R}^3$ be given by $T(x, y, z) = 3x^2 + 3z^2$. Compute the heat flux across the surface $x^2 + z^2 = 2$, $0 \le y \le 2$, if $k = 1$.

2. Compute the heat flux across the unit sphere S if $T(x, y, z) = x$. Can you interpret your answer physically?

3. Let S be the closed surface that consists of the hemisphere $x^2 + y^2 + z^2 = 1$, $z \ge 0$, and its base $x^2 + y^2 \le 1$, $z = 0$. Let $\mathbf{E}$ be the electric field defined by $\mathbf{E}(x, y, z) = 2x\mathbf{i} + 2y\mathbf{j} + 2z\mathbf{k}$. Find the electric flux across S. (HINT: Break S into two pieces S_1 and S_2 and evaluate $\iint_{S_1} \mathbf{E} \cdot d\mathbf{S}$ and $\iint_{S_2} \mathbf{E} \cdot d\mathbf{S}$ separately.)

4. Let the velocity field of a fluid be described by $\mathbf{F} = \sqrt{y}\mathbf{i}$ (measured in meters per second). Compute how many cubic meters of fluid per second are crossing the surface $x^2 + z^2 = 1$, $0 \le y \le 1$, $0 \le x \le 1$.

5. Evaluate $\iint_S (\nabla \times \mathbf{F}) \cdot d\mathbf{S}$, where S is the surface $x^2 + y^2 + 3z^2 = 1$, $z \le 0$, and $\mathbf{F}$ is the vector field $\mathbf{F} = y\mathbf{i} - x\mathbf{j} + zx^3y^2\mathbf{k}$. (Let $\mathbf{n}$, the unit normal, be upward pointing.)

6. Evaluate $\iint_S (\nabla \times \mathbf{F}) \cdot d\mathbf{S}$ where $\mathbf{F} = (x^2 + y - 4)\mathbf{i} + 3xy\mathbf{j} + (2xz + z^2)\mathbf{k}$ and S is the surface $x^2 + y^2 + z^2 = 16$, $z \ge 0$. (Let $\mathbf{n}$, the unit normal, be upward pointing.)

7. Calculate the integral $\iint_S \mathbf{F} \cdot d\mathbf{S}$, where S is the entire surface of the solid half ball $x^2 + y^2 + z^2 \leq 1, z \geq 0$, and $\mathbf{F} = (x + 3y^5)\mathbf{i} + (y + 10xz)\mathbf{j} + (z - xy)\mathbf{k}$. (Let S be oriented by the outward pointing normal.)

8.* A restaurant is being built on the side of a mountain. The architect's plans are shown in Figure 7.6.11.

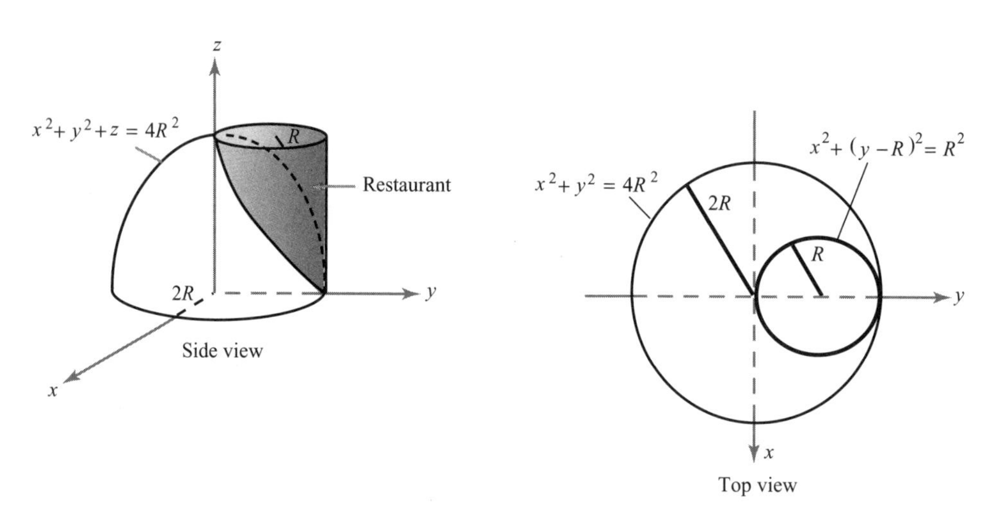

Figure 7.6.11 Restaurant plans.

(a) The vertical curved wall of the restaurant is to be built of glass. What will be the surface area of this wall?

(b) To be large enough to be profitable, the consulting engineer informs the developer that the volume of the interior must exceed $\pi R^4/2$. For what R does the proposed structure satisfy this requirement?

(c) During a typical summer day, the environs of the restaurant are subject to a temperature field given by

$$T(x, y, z) = 3x^2 + (y - R)^2 + 16z^2.$$

A heat flux density $\mathbf{V} = -k\,\nabla T$ (k is a constant depending on the grade of insulation to be used) through all sides of the restaurant (including the top and the contact with the hill) produces a heat flux. What is this total heat flux? (Your answer will depend on R and k.)

9. Find the flux of the vector field $\mathbf{V}(x, y, z) = 3xy^2\mathbf{i} + 3x^2y\mathbf{j} + z^3\mathbf{k}$ out of the unit sphere.

10. Evaluate the surface integral $\iint_S \mathbf{F} \cdot \mathbf{n}\, dA$, where $\mathbf{F}(x, y, z) = \mathbf{i} + \mathbf{j} + z(x^2 + y^2)^2\mathbf{k}$ and S is the surface of the cylinder $x^2 + y^2 \leq 1, 0 \leq z \leq 1$.

---

*The solution to this problem may be somewhat time-consuming.

11. Let S be the surface of the unit sphere. Let $\mathbf{F}$ be a vector field and F_r its radial component. Prove that

$$\iint_S \mathbf{F} \cdot d\mathbf{S} = \int_{\theta=0}^{2\pi} \int_{\phi=0}^{\pi} F_r \sin\phi \, d\phi \, d\theta.$$

What is the corresponding formula for real-valued functions f?

12. Prove the following mean-value theorem for surface integrals: If $\mathbf{F}$ is a continuous vector field, then

$$\iint_S \mathbf{F} \cdot \mathbf{n} \, dS = [\mathbf{F}(\mathrm{Q}) \cdot \mathbf{n}(\mathrm{Q})] A(S)$$

for some point $\mathrm{Q} \in S$, where $A(S)$ is the area of S. [HINT: Prove it for real functions first, by reducing the problem to one of a double integral: Show that if $g \geq 0$, then

$$\iint_D fg \, dA = f(\mathrm{Q}) \iint_D g \, dA$$

for some $\mathrm{Q} \in D$ (do it by considering $(\iint_D fg \, dA)/(\iint_D g \, dA)$ and using the intermediate value theorem).]

13. Work out a formula like that in Exercise 11 for integration over the surface of a cylinder.

14. Let S be a surface in $\mathbb{R}^3$ that is actually a subset D of the xy plane. Show that the integral of a scalar function $f(x, y, z)$ over S reduces to the double integral of $f(x, y, z)$ over D. What does the surface integral of a vector field over S become? (Make sure your answer is compatible with Example 6.)

15. Let the velocity field of a fluid be described by $\mathbf{F} = \mathbf{i} + x\mathbf{j} + z\mathbf{k}$ (measured in meters per second). Compute how many cubic meters of fluid per second are crossing the surface described by $x^2 + y^2 + z^2 = 1, z \geq 0$.

16. (a) A uniform fluid that flows vertically downward (heavy rain) is described by the vector field $\mathbf{F}(x, y, z) = (0, 0, -1)$. Find the total flux through the cone $z = (x^2 + y^2)^{1/2}$, $x^2 + y^2 \leq 1$.

(b) The rain is driven sideways by a strong wind so that it falls at a 45° angle, and it is described by $\mathbf{F}(x, y, z) = -(\sqrt{2}/2, 0, \sqrt{2}/2)$. Now what is the flux through the cone?

17. For $a > 0, b > 0, c > 0$, let S be the upper half ellipsoid

$$S = \left\{ (x, y, z) \,\middle|\, \frac{x^2}{a^2} + \frac{y^2}{b^2} + \frac{z^2}{c^2} = 1, z \geq 0 \right\},$$

with orientation determined by the upward normal. Compute $\iint_S \mathbf{F} \cdot d\mathbf{S}$ where $\mathbf{F}(x, y, z) = (x^3, 0, 0)$.

18. If S is the upper hemisphere $\{(x, y, z) \mid x^2 + y^2 + z^2 = 1, z \geq 0\}$ oriented by the normal pointing out of the sphere, compute $\iint_S \mathbf{F} \cdot d\mathbf{S}$ for parts (a) and (b).

(a) $\mathbf{F}(x, y, z) = x\mathbf{i} + y\mathbf{j}$
(b) $\mathbf{F}(x, y, z) = y\mathbf{i} + x\mathbf{j}$
(c) For each of these vector fields, compute $\iint_S (\nabla \times \mathbf{F}) \cdot d\mathbf{S}$ and $\int_C \mathbf{F} \cdot d\mathbf{s}$ where C is the unit circle in the xy plane traversed in the counterclockwise direction (as viewed from the positive z axis). (Notice that C is the boundary of S. The phenomenon illustrated here will be studied more thoroughly in the next chapter, using Stokes' theorem.)

7.7 Applications to Differential Geometry, Physics, and Forms of Life*

In the first half of the nineteenth century, the great German mathematician Karl Friedrich Gauss developed a theory of curved surfaces in $\mathbb{R}^3$. More than a century earlier, Isaac Newton had defined a measure of the curvature of a space curve, and Gauss was able to find extensions of this idea of curvature that would apply to surfaces. In so doing, Gauss made several remarkable discoveries.

Curvature of Surfaces

For paths $\mathbf{c}\colon [a, b] \to \mathbb{R}^3$ that have unit speed—that is, $\|\mathbf{c}'(t)\| = 1$—the curvature κ of the image curve $\kappa(\mathbf{c}(t))$ at the point $\mathbf{c}(t)$ is defined to be the length of the acceleration vector. That is, $\|\mathbf{c}''(t)\| = \kappa(\mathbf{c}(t))$. For paths $\mathbf{c}$ in space, the curvature is a true measure of the curvature of the geometric image curve C. As we saw at the end of Section 7.1, the "total curvature" $\int \kappa\, ds$ over C has "topological" implications. The same, and even more, will hold for Gauss' definition of the total curvature of a surface. We begin with some definitions.

Let $\mathbf{\Phi}\colon D \to \mathbb{R}^3$ be a smooth parametrized surface. Then, as we know,

$$\mathbf{T}_u = \frac{\partial \mathbf{\Phi}}{\partial u} \qquad \text{and} \qquad \mathbf{T}_v = \frac{\partial \Phi}{\partial v}$$

are tangent vectors to the image surface $S = \mathbf{\Phi}(D)$ at the point $\mathbf{\Phi}(u, v)$. We will also assume that there is a well-defined normal vector; that is, we assume the surface is regular: $\mathbf{T}_u \times \mathbf{T}_v \neq \mathbf{0}$.

Let

$$E = \left\| \frac{\partial \mathbf{\Phi}}{\partial u} \right\|^2, \qquad F = \frac{\partial \mathbf{\Phi}}{\partial u} \cdot \frac{\partial \mathbf{\Phi}}{\partial v}, \qquad G = \left\| \frac{\partial \mathbf{\Phi}}{\partial v} \right\|^2.$$

In Exercise 15 of Section 7.5, we saw that

$$\|\mathbf{T}_u \times \mathbf{T}_v\|^2 = EG - F^2.$$

---

*This section can be skipped on a first reading without loss of continuity.

For notational reasons, we denote $EG - F^2$ by W. Furthermore, we let

$$\mathbf{N} = \frac{\mathbf{T}_u \times \mathbf{T}_v}{\|\mathbf{T}_u \times \mathbf{T}_v\|} = \frac{\mathbf{T}_u \times \mathbf{T}_v}{\sqrt{W}}$$

denote the *unit* normal vector to the image surface at $p = \mathbf{\Phi}(u, v)$. Next we will define two new measures of the curvature of a surface at p—the "Gauss curvature," $K(p)$, and the "mean curvature," $H(p)$. Both of these curvatures have deep connections to the curvature of space curves, which illuminate the meaning of their definitions, but we do not explore these here.

To define these two curvatures, we first define three new functions ℓ, m, n on S as follows:

$$\begin{aligned} \ell(p) &= \mathbf{N}(u, v) \cdot \frac{\partial^2 \mathbf{\Phi}}{\partial u^2} = \mathbf{N}(u, v) \cdot \mathbf{\Phi}_{uu} \\ m(p) &= \mathbf{N}(u, v) \cdot \frac{\partial^2 \mathbf{\Phi}}{\partial u \partial v} = \mathbf{N}(u, v) \cdot \mathbf{\Phi}_{uv} \\ n(p) &= \mathbf{N}(u, v) \cdot \frac{\partial^2 \mathbf{\Phi}}{\partial v^2} = \mathbf{N}(u, v) \cdot \mathbf{\Phi}_{vv}. \end{aligned} \tag{1}$$

The ***Gauss curvature*** $K(p)$ of S at p is given by

$$K(p) = \frac{\ell n - m^2}{W}, \tag{2}$$

and the ***mean curvature*** $H(p)$ of S at p is defined by[16]

$$H(p) = \frac{G\ell + En - 2Fm}{2W}, \tag{3}$$

where the right-hand sides of both expressions are calculated at the point $p = \mathbf{\Phi}(u, v)$.

EXAMPLE 1 **Planes Have Zero Curvature** Let $\mathbf{\Phi}(u, v) = \boldsymbol{\alpha} u + \boldsymbol{\beta} v + \boldsymbol{\gamma}$, $(u, v) \in \mathbb{R}^2$, where $\boldsymbol{\alpha}, \boldsymbol{\beta}, \boldsymbol{\gamma}$ are vectors in $\mathbb{R}^3$. According to Example 1 of Section 7.3, this determines a parametrized plane in $\mathbb{R}^3$. Show that at every point, both the Gauss and mean curvatures are zero, and hence K and H vanish identically.

SOLUTION Because $\mathbf{\Phi}_{uu} = \mathbf{\Phi}_{uv} = \mathbf{\Phi}_{vv} \equiv 0$, the functions ℓ, m, n vanish everywhere, and so do H and K. Thus, a plane has "zero" curvature. Hence, at least in this example, we ought to be convinced that H and K actually do measure the *flatness* of the plane. Conversely, one can show that if H and K vanish identically, then S is part of a plane (see Exercise 10). ▲

---

[16]Technically speaking, $K(p)$ and $H(p)$ could, in principle, depend on the parametrization $\mathbf{\Phi}$ of S, but one can show that they are, in fact, independent of $\mathbf{\Phi}$.

EXAMPLE 2 **Curvature of a Hemisphere** Let

$$\mathbf{\Phi}(u, v) = (u, v, g(u, v)),$$

where $g(u, v) = \sqrt{R^2 - u^2 - v^2}$ is a parametrization of the "upper hemisphere" of radius R. Show that the Gauss curvature at every point is $1/R^2$ and the mean curvature is $1/R$.

SOLUTION We must first calculate the following quantities:

$$\mathbf{T}_u, \mathbf{T}_v, \mathbf{T}_u \times \mathbf{T}_v, \mathbf{\Phi}_{uu}, \mathbf{\Phi}_{vv}, \mathbf{\Phi}_{uv}, E, G, F, \ell, m, n.$$

First of all, we have

$$\mathbf{\Phi}_u = \mathbf{T}_u = \mathbf{i} - \frac{u}{\sqrt{R^2 - u^2 - v^2}}\mathbf{k}$$

$$\mathbf{\Phi}_v = \mathbf{T}_v = \mathbf{j} - \frac{v}{\sqrt{R^2 - v^2 - v^2}}\mathbf{k}.$$

From formula (2) in Section 7.3, we have

$$\begin{aligned} \mathbf{T}_u \times \mathbf{T}_v &= -\frac{\partial g}{\partial u}\mathbf{i} - \frac{\partial g}{\partial v}\mathbf{j} + \mathbf{k} \\ &= \frac{u}{\sqrt{R^2 - u^2 - v^2}}\mathbf{i} + \frac{v}{\sqrt{R^2 - u^2 - v^2}}\mathbf{j} + \mathbf{k}. \end{aligned}$$

Therefore,

$$E = \|\mathbf{\Phi}_u\|^2 = 1 + \frac{u^2}{R^2 - u^2 - v^2} = \frac{R^2 - v^2}{R^2 - u^2 - v^2}$$

$$G = \|\mathbf{\Phi}_v\|^2 = \frac{R^2 - u^2}{R^2 - u^2 - v^2}$$

$$F = \mathbf{\Phi}_u \cdot \mathbf{\Phi}_v = \frac{uv}{R^2 - u^2 - v^2}.$$

From Exercise 15 of Section 7.5, we know that

$$\begin{aligned} \|\mathbf{T}_u \times \mathbf{T}_v\|^2 = EG - F^2 &= \frac{(R^2 - v^2)(R^2 - u^2) - u^2v^2}{(R^2 - u^2 - v^2)^2} \\ &= \frac{R^4 - R^2u^2 - R^2v^2}{(R^2 - u^2 - v^2)^2} = \frac{R^2}{(R^2 - u^2 - v^2)} = W. \end{aligned}$$

Now a direct calculation shows that

$$\mathbf{\Phi}_{uu} = \frac{R^2 - v^2}{(R^2 - u^2 - v^2)^{3/2}}\mathbf{k}$$

$$\mathbf{\Phi}_{vv} = \frac{R^2 - u^2}{(R^2 - u^2 - v^2)^{3/2}}\mathbf{k}$$

$$\mathbf{\Phi}_{uv} = \frac{uv}{(R^2 - u^2 - v^2)^{3/2}}\mathbf{k}.$$

Furthermore,

$$\begin{aligned}\mathbf{N} &= \frac{\mathbf{T}_u \times \mathbf{T}_v}{\|\mathbf{T}_u \times \mathbf{T}_v\|} = \frac{\mathbf{T}_u \times \mathbf{T}_v}{\sqrt{W}} \\ &= \frac{\sqrt{R^2 - u^2 - v^2}}{R} \cdot \left(\frac{u}{\sqrt{R^2 - u^2 - v^2}}\mathbf{i} + \frac{v}{\sqrt{R^2 - u^2 - v^2}}\mathbf{j} + \mathbf{k} \right) \\ &= \frac{1}{R}\left(u\mathbf{i} + v\mathbf{j} + \sqrt{R^2 - u^2 - v^2}\,\mathbf{k} \right).\end{aligned}$$

Thus,

$$\ell = \mathbf{N} \cdot \mathbf{\Phi}_{uv} = \frac{1}{R}\left(\frac{R^2 - v^2}{R^2 - u^2 - v^2} \right)$$

$$n = \mathbf{N} \cdot \mathbf{\Phi}_{vv} = \frac{1}{R}\left(\frac{R^2 - u^2}{R^2 - u^2 - v^2} \right)$$

$$m = \mathbf{N} \cdot \mathbf{\Phi}_{uv} = \frac{1}{R}\left(\frac{uv}{R^2 - u^2 - v^2} \right).$$

Therefore,

$$\begin{aligned}\ell n - m^2 &= \frac{1}{R^2}\left(\frac{(R^2 - v^2)(R^2 - u^2) - u^2 v^2}{(R^2 - u^2 - v^2)^2} \right) \\ &= \frac{1}{R^2 - u^2 - v^2}.\end{aligned}$$

Dividing this by W yields $K = 1/R^2$. Thus, the Gauss curvature does not change from point to point on the hemisphere; that is, it is constant. This conforms to our intuition that the sphere is perfectly symmetrical and that its curvature is everywhere equal. Hence, the mean curvature should also be constant. This is verified by the following calculation:

$$
\begin{aligned}
H &= \frac{G\ell + En - 2Fm}{2W} \\
&= \frac{1}{2W}\left\{\left(\frac{R^2 - u^2}{R^2 - u^2 - v^2}\right)\frac{1}{R}\left(\frac{R^2 - v^2}{R^2 - u^2 - v^2}\right)\right. \\
&\quad \left. + \left(\frac{R^2 - v^2}{R^2 - u^2 - v^2}\right)\frac{1}{R}\left(\frac{R^2 - u^2}{R^2 - u^2 - v^2}\right) - 2\frac{u^2v^2}{(R^2 - u^2 - v^2)^2}\right\} \\
&= \frac{1}{W}\left\{\frac{R}{R^2 - u^2 - v^2}\right\} = \frac{1}{R}. \quad \blacktriangle
\end{aligned}
$$

Surfaces of Constant Curvature

Surfaces of constant Gauss and mean curvature are of great interest to mathematicians. It was known in the nineteenth century that the only closed and bounded smooth surfaces with "no boundary" and with constant Gauss curvature were spheres. In the twentieth century, the Russian mathematician Alexandrov showed that the only closed and bounded smooth surfaces without a boundary that do not intersect themselves and that have constant mean curvature must also be spheres. Mathematicians believed that Alexandrov's result held even if the surface *was* allowed to intersect itself, but no one could find a proof. In 1984, Professor Henry Wente (Toledo, Ohio) startled the world by finding a self-intersecting torus of constant mean curvature.

Surfaces of constant mean curvature are physically relevant and occur throughout nature. Soap bubble formations have constant nonzero mean curvature (see Figure 7.7.1), and soap film formations (containing no air) have constant mean curvature zero (see Figures 7.7.2 and 7.7.3).

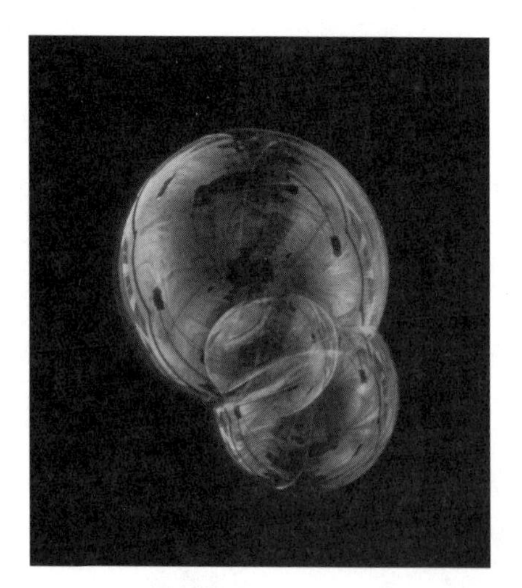

Figure 7.7.1 Soap bubble formation; $H =$ constant.

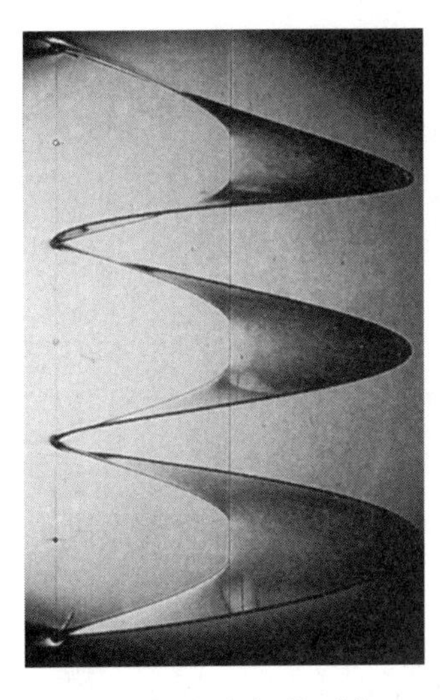

Figure 7.7.2 A helicoid, $H = 0$.

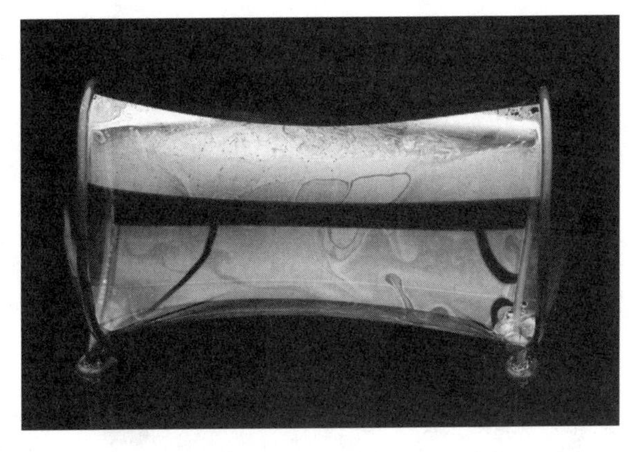

Figure 7.7.3 A soap film, $H = 0$, spanning two circular wires; this one is the catenoid.

In the early nineteenth century, the French mathematician Delaunay discovered all surfaces of revolution that have constant mean curvature. They are the cylinder, sphere, catenoid, unduloid, and nodoid. The catenoid exists as a soap film surface spanning two circular contours.

Optimal Shapes in Nature

Throughout the ages, people have speculated on why things are shaped the way they are. Why are the earth and the stars "round" and not cubical? Why are life forms shaped the way they are?

In 1917, the British natural philosopher D'Arcy Thompson published a provocative work entitled *On Growth and Form*, in which he investigated the forces behind the creation of living forms in nature. He wrote:

> In an organism, great or small, it is not merely the nature of the motions of the living substance which we must interpret in terms of force (according to kinetics), but also the conformation of the organism itself, whose permanence or equilibrium is explained by the interaction or balance of forces, as described in statics.

Surprisingly, Thompson discovered *all* of Delaunay's surfaces in the form of unicellular organisms (see Figure 7.7.4). The constant mean curvature of these organisms can be explained by minimum principles similar to those described in the Historical Note in Section 3.3. In 1952, Watson and Crick determined that the structure of DNA is that of a double helix, a discovery that set the stage for the genetic revolution. We know from soap films, as in Figure 7.2.2, that nature likes helicoid forms, and nature tends to repeat patterns. A better understanding of the scientific principles underlying life may ultimately help mathematics play a more prominent role in this area of theoretical biology.

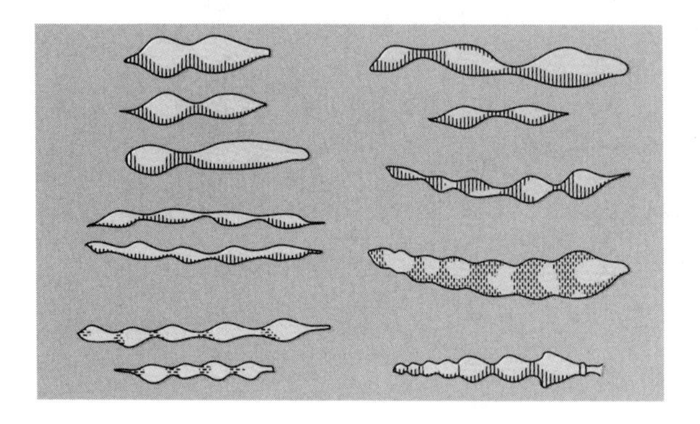

Figure 7.7.4 Surfaces of revolution of constant mean curvature as unicellulars.

Curvature and Physics

The theory of curved surfaces, initiated by Gauss, has had a profound effect on physics. Gauss realized that the Gauss curvature K of a surface depended only on the measure of distance *on the surface itself*; that is, curvature was *intrinsic to the surface.* This is not true of the mean curvature H. Thus, beings "living" on the surface would be able to tell that the surface was curving, without any reference to an "external" world. Gauss himself found this mathematical result to be so striking that he named it *theorema egregium*, or "remarkable theorem." Gauss' theory was generalized by his student Bernhard Riemann to n-dimensional surfaces for which one could describe a notion of curvature.

Recall that Newton created the idea of a gravitational force acting over vast galactic distances, pulling galaxies together as well as pushing them apart (see Figure 7.7.5). In the early 1900s, Albert Einstein used Riemann's ideas to develop the *general theory of relativity*, a theory of gravitation that eliminated the need to consider forces (as Newton did) acting over great distances. Einstein's theory explained the bending of light by the sun, black holes, the expansion of the universe, the formation of galaxies, and the Big Bang itself. For most applications, including the dynamics of our solar system, Newton's theory suffices and is commonly used today by NASA to plan space missions, as we saw in Section 4.1. But for

Figure 7.7.5 The Andromeda Galaxy. It will collide with the Milky Way in roughly 2 billion years.

cosmological applications on the grand scale, Einstein's theory replaced that of Isaac Newton, published in his *Principia* in 1687.

As a testament to his genius, and despite the astounding success of this theory, Newton was nevertheless disturbed by questions about *how* this gravitational force acted. He could give no other explanation than to say, "I have not been able to deduce from phenomena the reason for these properties of gravitation, and I do not invent hypotheses; for anything which cannot be deduced from phenomena should be called an hypothesis." Moreover, in a letter to his friend, Richard Bentley, Newton wrote:

> That gravity should be innate, inherent and essential to matter, so that one body may act upon another at a distance, through which their action may be conveyed from one to another, is to me so great an absurdity that I believe no man, who has in philosophical matters a competent faculty of thinking, can ever fall into it.

Newton coined the term *action at a distance* (which means "force acting at a distance") to describe the mysterious effect of gravitation over large distances. This effect is as difficult to understand today as it was in Newton's time.

Johann Bernoulli found it difficult to believe in the concept of a force that acts through a vacuum of space over distances of even hundreds of millions of miles. He viewed this force as a concept revolting to minds unaccustomed to accepting

Figure 7.7.6 Albert Einstein (1879–1955) at his desk in the Patent Office, Bern, 1905.

any principle in physics, save those that are incontestable and evident. Additionally, Leibniz considered gravitation to be an incorporeal and inexplicable power, philosophically void.

It was perhaps Albert Einstein's greatest inspiration (see Figure 7.7.6) to replace Newton's model of gravitation with a model that would have thrilled the early Greeks—*a geometric model of gravitation*. In Einstein's theory, the concept of a force acting through great distances has been replaced by the *curvature* of a space–time[17] world. As the quote at the beginning of the chapter illustrates, W. K. Clifford had a premonition of events to come! In order to elucidate Einstein's scheme, we shall present an oversimplified model that conveys some of his basic ideas.

We represent space by a surface that we imagine as an originally flat trampoline (the vacuum state), which is at some point strongly deformed by the weight of a gigantic steel ball (the sun). A tiny steel ball rolling on the trampoline is our planet Earth (see Figure 7.7.7)

If we roll the small steel ball across the flat trampoline, it will travel in a straight-line path. However, if we now place the gigantic steel ball in the center, it will cause the trampoline to bend, or "curve," even "far away" from the large ball. If we then give our little ball a push, it will no longer travel in a straight line but in a curved

[17]Space–time is locally like $\mathbb{R}^4$ with three space coordinates and one time coordinate.

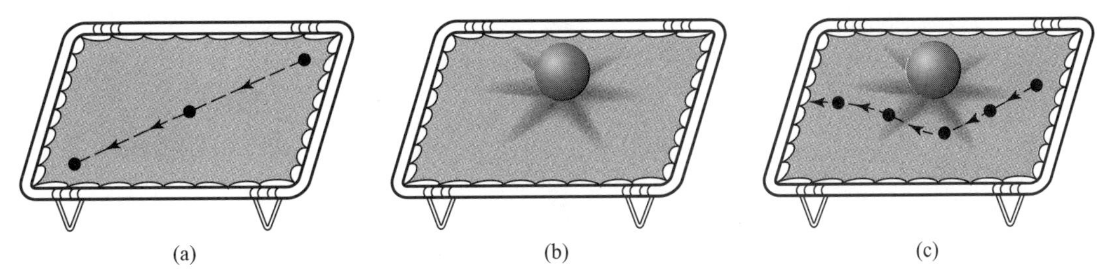

Figure 7.7.7 (a) A particle on a taut trampoline moves in a straight line. (b) A heavy steel ball distorts the trampoline. (c) A particle on the distorted trampoline follows a curved path.

path. The big ball affects the trajectory of the little ball by curving the space around it. With just the right push, the little ball might even orbit the big one for a while. This trampoline model explains how a large body could, by curving space, influence a small one over great distance.

Einstein stated that space–time is curved by matter and energy. In this curved space–time, even light rays are *bent* as they pass near massive objects like our sun. Thanks to Gauss and Riemann, the curvature of space–time requires no external "universe" in which it curves.

The equations that tell one how much space and time are curved by matter and energy are known as *Einstein's field equations.* A description of them is beyond the scope of this book, but the mathematical kernel from which these equations arise is not; this kernel is based on another remarkable result of the research of Gauss and Bonnet.

Gauss–Bonnet Theorem

In Example 2, we computed the Gauss curvature K of the sphere $x^2 + y^2 + z^2 = R^2$ of radius R and found it to be the constant $1/R^2$. The Gauss curvature K is a scalar-valued function over the surface, and as such we can integrate it over the surface. We wish to consider a constant times this surface integral, namely,

$$\frac{1}{2\pi} \iint_S K \, dA.$$

For the sphere of radius R, this quantity becomes

$$\frac{1}{2\pi R^2} \iint_S dA = \frac{4\pi R^2}{2\pi R^2} = 2.$$

What Gauss and Bonnet discovered was that if S is *any* "sphere-like" closed surface (closed and bounded, but with no boundary, as in Figure 7.7.8), then

$$\frac{1}{2\pi}\iint_S K\, dA = 2$$

still holds.[18]

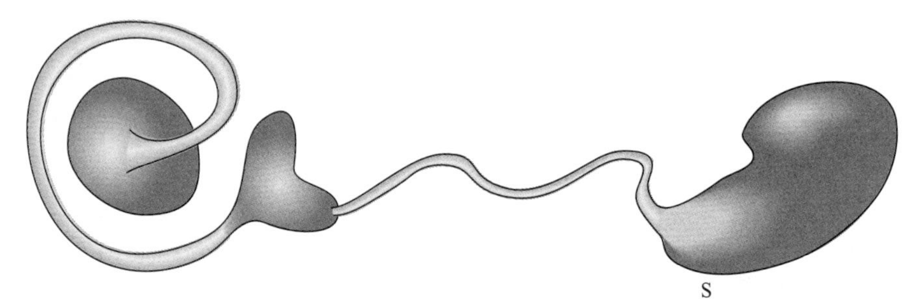

Figure 7.7.8 A deformed sphere. $1/2\pi \iint_S K\, dA = 2$.

Thus, the integral

$$\frac{1}{2\pi}\iint_S K\, dA$$

always equals the integer 2, and is therefore a *topological invariant* of the surface. That the integral of curvature should be an interesting quantity should be already clear from the discussion at the end of Section 7.1.

Now consider a torus, or doughnut. The torus can be considered as coming from the sphere by cutting out two disks and gluing in a handle (see Figure 7.7.9).

Moreover, we can continue this process adding $1, 2, 3, \ldots, g$ handles to the sphere. If g handles are attached, we call the resulting surface a surface of genus g, as in Figure 7.7.10. Notice that the torus has genus 1.

If two surfaces have a different genus, they are topologically distinct, and thus cannot be obtained from one another by bending or stretching. Interestingly, even two surfaces with the same genus can sit in space in quite different and complex ways, as in Figure 7.7.11. Astonishingly, even though the integral (or total curvature) given by $(1/2\pi)\iint_S K\, dA$ depends on the genus, it does not depend on how the surface sits in space (and thus not on K).

[18]Roughly speaking, this means that S can be obtained from the sphere by bending and stretching (like with a balloon) but not tearing (the balloon bursts!)

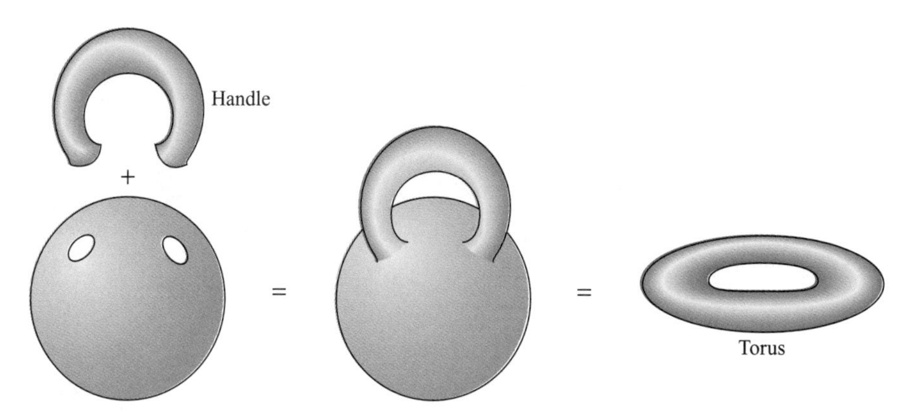

Figure 7.7.9 Gluing a handle to a sphere to obtain a torus.

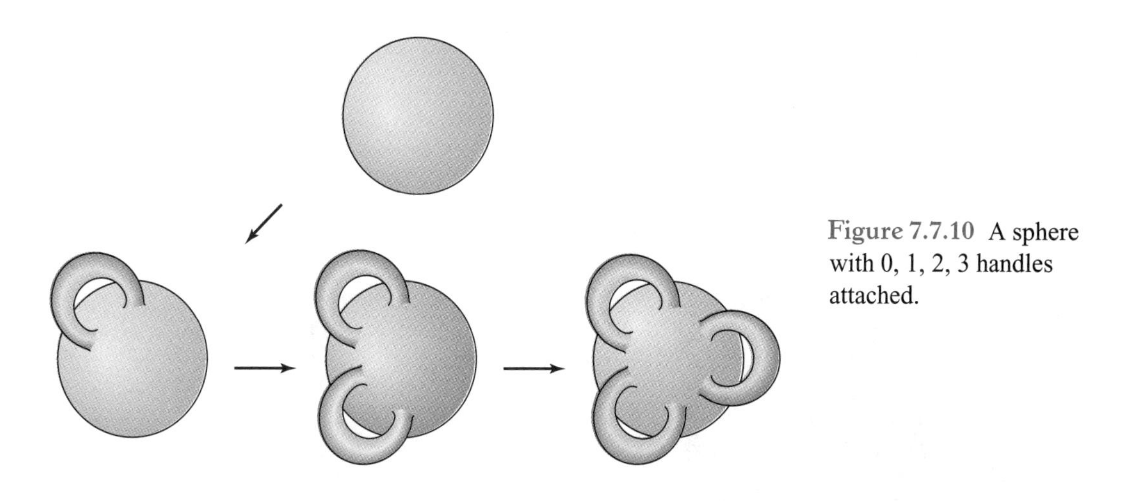

Figure 7.7.10 A sphere with 0, 1, 2, 3 handles attached.

Gauss and Bonnet proved that

$$\frac{1}{2\pi}\iint_S K\, dA = 2 - 2g.$$

Thus, for the sphere ($g = 0$), it is always 2 (already verified); for the torus, it is always 0 (see Exercise 8).

There is something even more remarkable connected to the theorem of Gauss–Bonnet, observed by the great German mathematician David Hilbert (Figure 7.7.12).

Simple double doughnut

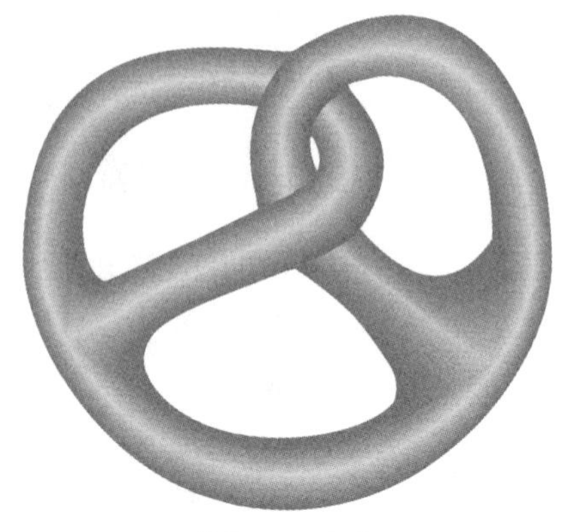

Baker's pretzel

Figure 7.7.11 Two manifestations of a surface S in $\mathbb{R}^3$ of genus 2.

Figure 7.7.12 David Hilbert (1862–1943) was a leading mathematician of his time.

Hilbert observed that the Gauss–Bonnet theorem is, in effect, a two-dimensional version of Einstein's field equations. In the physics literature, this fact is known as *Hilbert's action principle* in general relativity.[19] Not surprisingly, similar geometric ideas are being employed by contemporary researchers in an effort to unify gravity and quantum mechanics—to "quantize" gravity, so to speak.

[19]See C. Misner, K. Thorne, and A. Wheeler, *Gravitation*, Freeman, New York, 1972.

EXERCISES

1. The helicoid can be described by

$$\mathbf{\Phi}(u, v) = (u \cos v, u \sin v, bv), \text{ where } b \neq 0.$$

Show that $H = 0$ and that $K = -b^2/(b^2 + u^2)^2$. In Figures 7.7.1 and 7.7.5, we see that the helicoid is actually a soap film surface. Surfaces in which $H = 0$ are called ***minimal surfaces.***

2. Consider the saddle surface $z = xy$. Show that

$$K = \frac{-1}{(1 + x^2 + y^2)^2},$$

and that

$$H = \frac{-xy}{(1 + x^2 + y^2)^{3/2}}.$$

3. Show that $\mathbf{\Phi}(u, v) = (u, v, \log \cos v - \log \cos u)$ has mean curvature zero (and is thus a minimal surface; see Exercise 1).

4. Find the Gauss curvature of the elliptic paraboloid

$$z = \frac{x^2}{a^2} + \frac{y^2}{b^2}.$$

5. Find the Gauss curvature of the hyperbolic paraboloid

$$z = \frac{x^2}{a^2} - \frac{y^2}{b^2}.$$

6. Compute the Gauss curvature of the ellipsoid

$$\frac{x^2}{a^2} + \frac{y^2}{b^2} + \frac{z^2}{c^2} = 1.$$

7. Show that Enneper's surface

$$\mathbf{\Phi}(u, v) = \left(u - \frac{u^3}{3} + uv^2, v - \frac{v^3}{3} + u^2 v, u^2 - v^2\right)$$

is a minimal surface ($H = 0$).

8. Consider the torus T given in Exercise 4, Section 7.4. Compute its Gauss curvature and verify the theorem of Gauss–Bonnet. [HINT: Show that $\|\mathbf{T}_\theta \times \mathbf{T}_\phi\|^2 = (R + \cos\phi)^2$ and $K = \cos\phi/(R + \cos\phi)$.]

9. Let $\mathbf{\Phi}(u, v) = (u, h(u) \cos v, h(u) \sin v)$, $h > 0$, be a surface of revolution. Show that $K = -h''/h\{1 + (h')^2\}^2$.

10. A parametrization Φ of a surface S is said to be ***conformal*** (see Section 7.4), provided that $E = G$, $F = 0$. Assume that Φ conformally parametrizes S.[20] Show that if H and K vanish identically, then S must be part of a plane in $\mathbb{R}^3$.

REVIEW EXERCISES FOR CHAPTER 7

1. Integrate $f(x, y, z) = xyz$ along the following paths:

(a) $\mathbf{c}(t) = (e^t \cos t, e^t \sin t, 3), 0 \le t \le 2\pi$
(b) $\mathbf{c}(t) = (\cos t, \sin t, t), 0 \le t \le 2\pi$
(c) $\mathbf{c}(t) = \frac{3}{2}t^2\mathbf{i} + 2t^2\mathbf{j} + t\mathbf{k}, 0 \le t \le 1$
(d) $\mathbf{c}(t) = t\mathbf{i} + (1/\sqrt{2})t^2\mathbf{j} + \frac{1}{3}t^3\mathbf{k}, 0 \le t \le 1$

2. Compute the integral of f along the path $\mathbf{c}$ in each of the following cases:

(a) $f(x, y, z) = x + y + yz; \mathbf{c}(t) = (\sin t, \cos t, t), 0 \le t \le 2\pi$
(b) $f(x, y, z) = x + \cos^2 z; \mathbf{c}(t) = (\sin t, \cos t, t), 0 \le t \le 2\pi$
(c) $f(x, y, z) = x + y + z; \mathbf{c}(t) = (t, t^2, \frac{2}{3}t^3), 0 \le t \le 1$

3. Compute each of the following line integrals:

(a) $\int_C (\sin \pi x)\, dy - (\cos \pi y)\, dz$, where C is the triangle whose vertices are $(1, 0, 0)$, $(0, 1, 0)$, and $(0, 0, 1)$, in that order

(b) $\int_C (\sin z)\, dx + (\cos z)\, dy - (xy)^{1/3}\, dz$, where C is the path $\mathbf{c}(\theta) = (\cos^3 \theta, \sin^3 \theta, \theta)$, $0 \le \theta \le 7\pi/2$

4. If $\mathbf{F}(\mathbf{x})$ is orthogonal to $\mathbf{c}'(t)$ at each point on the curve $\mathbf{x} = \mathbf{c}(t)$, what can you say about $\int_\mathbf{c} \mathbf{F} \cdot d\mathbf{s}$?

5. Find the work done by the force $\mathbf{F}(x, y) = (x^2 - y^2)\mathbf{i} + 2xy\mathbf{j}$ in moving a particle counterclockwise around the square with corners $(0, 0)$, $(a, 0)$, (a, a), $(0, a)$, $a > 0$.

6. A ring in the shape of the curve $x^2 + y^2 = a^2$ is formed of thin wire weighing $|x| + |y|$ grams per unit length at (x, y). Find the mass of the ring.

7. Find a parametrization for each of the following surfaces;

(a) $x^2 + y^2 + z^2 - 4x - 6y = 12$
(b) $2x^2 + y^2 + z^2 - 8x = 1$
(c) $4x^2 + 9y^2 - 2z^2 = 8$

8. Find the area of the surface defined by $\Phi: (u, v) \mapsto (x, y, z)$, where

$$x = h(u, v) = u + v, \qquad y = g(u, v) = u, \qquad z = f(u, v) = v;$$

$0 \le u \le 1, 0 \le v \le 1$. Sketch.

---

[20]Gauss proved that conformal parametrization of a surface always exists. The result of this exercise remains valid even if Φ is not conformal, but the proof is more difficult.

9. Write a formula for the surface area of $\mathbf{\Phi}$: $(r, \theta) \mapsto (x, y, z)$, where

$$x = r\cos\theta, \qquad y = 2r\sin\theta, \qquad z = r;$$

$0 \le r \le 1, 0 \le \theta \le 2\pi$. Describe the surface.

10. Suppose $z = f(x, y)$ and $(\partial f/\partial x)^2 + (\partial f/\partial y)^2 = c, c > 0$. Show that the area of the graph of f lying over a region D in the xy plane is $\sqrt{1+c}$ times the area of D.

11. Compute the integral of $f(x, y, z) = x^2 + y^2 + z^2$ over the surface in Review Exercise 8.

12. Find $\iint_S f\, dS$ in each of the following cases:

(a) $f(x, y, z) = x$; S is the part of the plane $x + y + z = 1$ in the positive octant defined by $x \ge 0, y \ge 0, z \ge 0$

(b) $f(x, y, z) = x^2$; S is the part of the plane $x = z$ inside the cylinder $x^2 + y^2 = 1$

(c) $f(x, y, z) = x$; S is the part of the cylinder $x^2 + y^2 = 2x$ with $0 \le z \le \sqrt{x^2 + y^2}$

13. Compute the integral of $f(x, y, z) = xyz$ over the rectangle with vertices $(1, 0, 1)$, $(2, 0, 0)$, $(1, 1, 1)$, and $(2, 1, 0)$.

14. Compute the integral of $x + y$ over the surface of the unit sphere.

15. Compute the surface integral of x over the triangle with vertices $(1, 1, 1)$, $(2, 1, 1)$, and $(2, 0, 3)$.

16. A paraboloid of revolution S is parametrized by $\mathbf{\Phi}(u, v) = (u\cos v, u\sin v, u^2)$, $0 \le u \le 2, 0 \le v \le 2\pi$.

(a) Find an equation in x, y, and z describing the surface.

(b) What are the geometric meanings of the parameters u and v?

(c) Find a unit vector orthogonal to the surface at $\mathbf{\Phi}(u, v)$.

(d) Find the equation for the tangent plane at $\mathbf{\Phi}(u_0, v_0) = (1, 1, 2)$ and express your answer in the following two ways:

(i) parametrized by u and v; and

(ii) in terms of x, y, and z.

(e) Find the area of S.

17. Let $f(x, y, z) = xe^y \cos\pi z$.

(a) Compute $\mathbf{F} = \nabla f$.

(b) Evaluate $\int_C \mathbf{F} \cdot d\mathbf{s}$ where $\mathbf{c}(t) = (3\cos^4 t, 5\sin^7 t, 0), 0 \le t \le \pi$.

18. Let $\mathbf{F}(x, y, z) = x\mathbf{i} + y\mathbf{j} + z\mathbf{k}$. Evaluate $\iint_S \mathbf{F} \cdot d\mathbf{S}$ where S is the upper hemisphere of the unit sphere $x^2 + y^2 + z^2 = 1$.

19. Let $\mathbf{F}(x, y, z) = x\mathbf{i} + y\mathbf{j} + z\mathbf{k}$. Evaluate $\int_{\mathbf{c}} \mathbf{F} \cdot d\mathbf{s}$ where $\mathbf{c}(t) = (e^t, t, t^2), 0 \le t \le 1$.

20. Let $\mathbf{F} = \nabla f$ for a given scalar function. Let $\mathbf{c}(t)$ be a closed curve, that is, $\mathbf{c}(b) = \mathbf{c}(a)$. Show that $\int_{\mathbf{c}} \mathbf{F} \cdot d\mathbf{s} = 0$.

21. Consider the surface $\mathbf{\Phi}(u, v) = (u^2 \cos v, u^2 \sin v, u)$. Compute the unit normal at $u = 1, v = 0$. Compute the equation of the tangent plane at this point.

22. Let S be the part of the cone $z^2 = x^2 + y^2$ with z between 1 and 2 oriented by the normal pointing out of the cone. Compute $\iint_S \mathbf{F} \cdot d\mathbf{S}$ where $\mathbf{F}(x, y, z) = (x^2, y^2, z^2)$.

23. Let $\mathbf{F} = x\mathbf{i} + x^2\mathbf{j} + yz\mathbf{k}$ represent the velocity field of a fluid (velocity measured in meters per second). Compute how many cubic meters of fluid per second are crossing the xy plane through the square $0 \le x \le 1, 0 \le y \le 1$.

24. Show that the surface area of the part of the sphere $x^2 + y^2 + z^2 = 1$ lying above the rectangle $[-a, a] \times [-a, a]$, where $2a^2 < 1$, in the xy plane is

$$A = 2 \int_{-a}^{a} \sin^{-1}\left(\frac{a}{\sqrt{1 - x^2}}\right) dx.$$

25. Let S be a surface and C a closed curve bounding S. Verify the equality

$$\iint_S (\nabla \times \mathbf{F}) \cdot d\mathbf{S} = \int_C \mathbf{F} \cdot d\mathbf{s}$$

if $\mathbf{F}$ is a gradient field (use Review Exercise 20).

26. Calculate $\iint_S \mathbf{F} \cdot d\mathbf{S}$ where $\mathbf{F}(x, y, z) = (x, y, -y)$ and S is the cylindrical surface defined by $x^2 + y^2 = 1, 0 \le z \le 1$, with normal pointing out of the cylinder.

27. Let S be the portion of the cylinder $x^2 + y^2 = 4$ between the planes $z = 0$ and $z = x + 3$. Compute the following:

(a) $\iint_S x^2\, dS$ (b) $\iint_S y^2\, dS$ (c) $\iint_S z^2\, dS$

28. Let Γ be the curve of intersection of the plane $z = ax + by$ with the cylinder $x^2 + y^2 = 1$. Find all values of the real numbers a and b such that $a^2 + b^2 = 1$ and

$$\int_\Gamma y\, dx + (z - x)\, dy - y\, dz = 0.$$

29. A circular helix that lies on the cylinder $x^2 + y^2 = R^2$ with pitch p may be described parametrically by

$$x = R\cos\theta, \qquad y = R\sin\theta, \qquad z = p\theta, \qquad \theta \ge 0.$$

A particle slides under the action of gravity (which acts parallel to the z axis) without friction along the helix. If the particle starts out at the height $z_0 > 0$, then when it reaches the height $z, 0 \leq z < z_0$, along the helix, its speed is given by

$$\frac{ds}{dt} = \sqrt{(z_0 - z)2g},$$

where s is arc length along the helix, g is the constant of gravity, and t is time.

(a) Find the length of the part of the helix between the planes $z = z_0$ and $z = z_1$, $0 \leq z_1 < z_0$.

(b) Compute the time T_0 it takes the particle to reach the plane $z = 0$.

8

The Integral Theorems of Vector Analysis

All the theory of the motion of fluids has just been reduced to the solution of analytic formulas.

Leonhard Euler

Fluids are a lot easier to drink than they are to understand.

Alan Newell

We are now prepared to tie together vector differential calculus and vector integral calculus. This will be done by means of the important theorems of Green, Gauss, and Stokes. We shall also point out some of the physical applications of these theorems to the study of electricity and magnetism, hydrodynamics, heat conduction, and differential equations.

The basic integral theorems in vector analysis had their origins in applications. For example, Green's theorem, discovered about 1828, arose in connection with potential theory (this includes gravitational and electrical potentials). Gauss' theorem—the divergence theorem—arose in connection with the study of capillarity (this theorem should be jointly credited to the Russian mathematician Ostrogradsky, who discovered the theorem around the same time as Gauss). Stokes' theorem was first suggested in a letter to Stokes from the physicist Lord Kelvin in 1850 and was used by Stokes on the examination for the Smith Prize in 1854.

8.1 Green's Theorem

Green's theorem relates a line integral along a closed curve C in the plane $\mathbb{R}^2$ to a double integral over the region enclosed by C. This important result will be generalized in the following sections to curves and surfaces in $\mathbb{R}^3$. We shall be referring to line integrals around curves that are the boundaries of elementary regions (see Section 5.3). To understand the ideas in this section, you may also need to refer to Section 7.2.

Simple and Elementary Regions and their Boundaries

A simple closed curve C that is the boundary of an elementary region has two orientations—counterclockwise (positive) and clockwise (negative). We denote C with the counterclockwise orientation as C^+, and with the clockwise orientation as C^- (Figure 8.1.1).

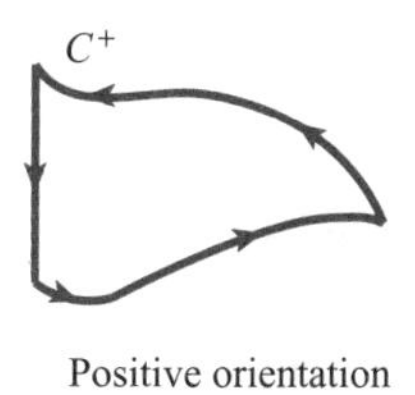

Positive orientation

(a)

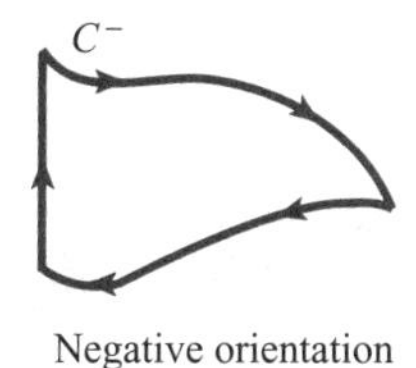

Negative orientation

(b)

Figure 8.1.1 (a) Positive orientation of C, and (b) negative orientation of C.

The boundary C of a y-simple region can be decomposed into bottom and top portions, C_1 and C_2, and (if applicable) left and right vertical portions, B_1 and B_2. Following Figure 8.1.2, we write,

$$C^+ = C_1^+ + B_2^+ + C_2^- + B_1^-,$$

where the pluses denote the curves oriented in the direction of left to right or bottom to top, and the minuses denote the curves oriented from right to left or from top to bottom.

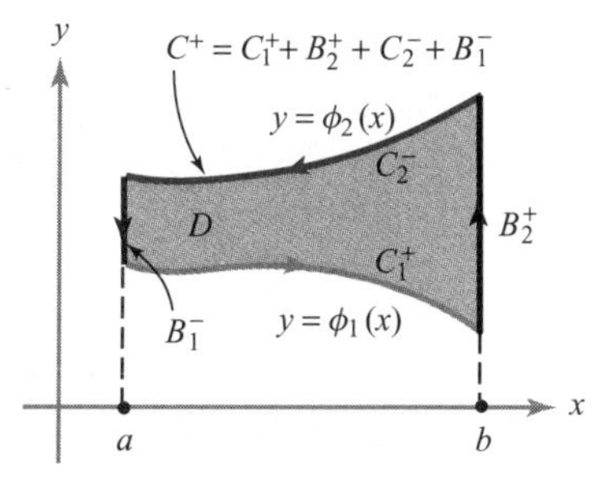

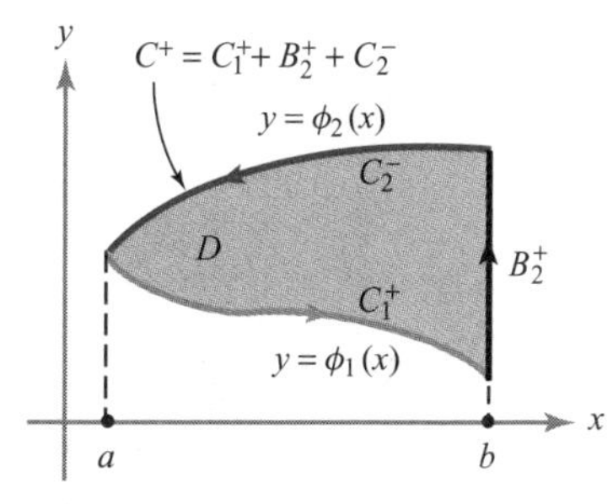

Figure 8.1.2 Two examples showing how to break the positively oriented boundary of a y-simple region D into oriented components.

We may make a similar decomposition of the boundary of an x-simple region into left and right portions, and upper and lower horizontal portions (if applicable) (Figure 8.1.3).

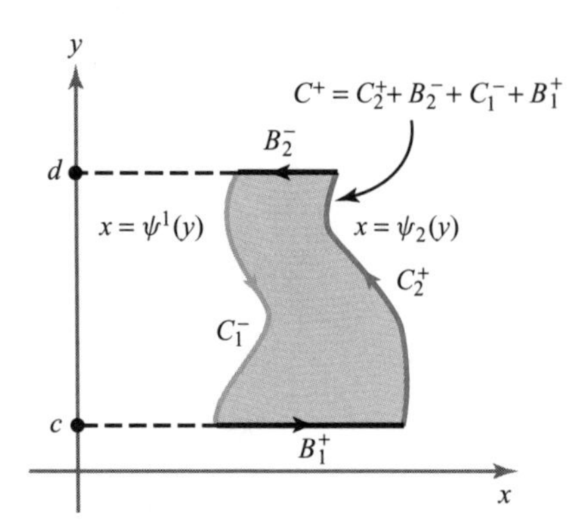

Figure 8.1.3 An example showing how to break the positively oriented boundary of an x-simple region D into oriented components.

Similarly, the boundary of a simple region has *two* decompositions—one into upper and lower halves, the other into left and right halves.

Green's Theorem

We shall now prove two lemmas in preparation for Green's theorem.

LEMMA 1 Let D be a y-simple region and let C be its boundary. Suppose $P: D \to \mathbb{R}$ is of class C^1. Then

$$\int_{C^+} P\,dx = -\iint_D \frac{\partial P}{\partial y}\,dx\,dy.$$

(The left-hand side denotes the line integral $\int_{C^+} P\,dx + Q\,dy$ where $Q = 0$.)

PROOF Suppose D is described by

$$a \le x \le b \qquad \phi_1(x) \le y \le \phi_2(x).$$

We decompose C^+ by writing $C^+ = C_1^+ + B_2^+ + C_2^- + B_1^-$ (see Figure 8.1.2). By Fubini's theorem, we may evaluate the double integral as an iterated integral and then use the fundamental theorem of calculus:

$$\begin{aligned}\iint_D \frac{\partial P}{\partial y}(x, y)\,dx\,dy &= \int_a^b \int_{\phi_1(x)}^{\phi_2(x)} \frac{\partial P}{\partial y}(x, y)\,dy\,dx \\ &= \int_a^b [P(x, \phi_2(x)) - P(x, \phi_1(x))]\,dx.\end{aligned}$$

However, because C_1^+ can be parametrized by $x \mapsto (x, \phi_1(x))$, $a \leq x \leq b$, and C_2^+ can be parametrized by $x \mapsto (x, \phi_2(x))$, $a \leq x \leq b$, we have

$$\int_a^b P(x, \phi_1(x))\, dx = \int_{C_1^+} P(x, y)\, dx$$

and

$$\int_a^b P(x, \phi_2(x))\, dx = \int_{C_2^+} P(x, y)\, dx.$$

Thus, by reversing orientations,

$$-\int_a^b P(x, \phi_2(x))\, dx = \int_{C_2^-} P(x, y)\, dx.$$

Hence,

$$\iint_D \frac{\partial P}{\partial y}\, dx\, dy = -\int_{C_1^+} P\, dx - \int_{C_2^-} P\, dx.$$

Because x is constant on B_2^+ and B_1^-, we have

$$\int_{B_2^+} P\, dx = 0 = \int_{B_1^-} P\, dx,$$

so

$$\int_{C^+} P\, dx = \int_{C_1^+} P\, dx + \int_{B_2^+} P\, dx + \int_{C_2^-} P\, dx + \int_{B_1^-} P\, dx = \int_{C_1^+} P\, dx + \int_{C_2^-} P\, dx.$$

Thus,

$$\iint_D \frac{\partial P}{\partial y}\, dx\, dy = -\int_{C_1^+} P\, dx - \int_{C_2^-} P\, dx = -\int_{C^+} P\, dx. \quad \blacksquare$$

We now prove the analogous lemma with the roles of x and y interchanged.

LEMMA 2 Let D be an x-simple region with boundary C. Then if $Q\colon D \to \mathbb{R}$ is C^1,

$$\int_{C^+} Q\, dy = \iint_D \frac{\partial Q}{\partial x}\, dx\, dy.$$

The negative sign does not occur here, because reversing the role of x and y corresponds to a change of orientation for the plane.

PROOF Suppose D is given by

$$\psi_1(y) \le x \le \psi_2(y), \qquad c \le y \le d.$$

Using the notation of Figure 8.1.3, and noting that y is constant on B_1^+ and B_2^-, we have

$$\int_{C^+} Q\,dy = \int_{C_1^- + B_1^+ + C_2^+ + B_2^-} Q\,dy = \int_{C_2^+} Q\,dy + \int_{C_1^-} Q\,dy,$$

where C_2^+ is the curve parametrized by $y \mapsto (\psi_2(y), y)$, $c \le y \le d$, and C_1^+ is the curve $y \mapsto (\psi_1(y), y)$, $c \le y \le d$. Applying Fubini's theorem and the fundamental theorem of calculus, we obtain

$$\begin{aligned}\iint_D \frac{\partial Q}{\partial x}\,dx\,dy &= \int_c^d \int_{\psi_1(y)}^{\psi_2(y)} \frac{\partial Q}{\partial x}\,dx\,dy = \int_c^d [Q(\psi_2(y), y) - Q(\psi_1(y), y)]\,dy \\ &= \int_{C_2^+} Q\,dy - \int_{C_1^+} Q\,dy = \int_{C_2^+} Q\,dy + \int_{C_1^-} Q\,dy = \int_{C^+} Q\,dy. \quad \blacksquare\end{aligned}$$

Adding the results of Lemmas 1 and 2 proves the following important theorem.

THEOREM 1: Green's Theorem Let D be a simple region and let C be its boundary. Suppose $P\colon D \to \mathbb{R}$ and $Q\colon D \to \mathbb{R}$ are of class C^1. Then

$$\int_{C^+} P\,dx + Q\,dy = \iint_D \left(\frac{\partial Q}{\partial x} - \frac{\partial P}{\partial y}\right) dx\,dy.$$

The correct (positive) orientation for the boundary curves of region D can be remembered by the following device: *If you walk along the curve C with the correct orientation, the region D will be on your left* (see Figure 8.1.4).

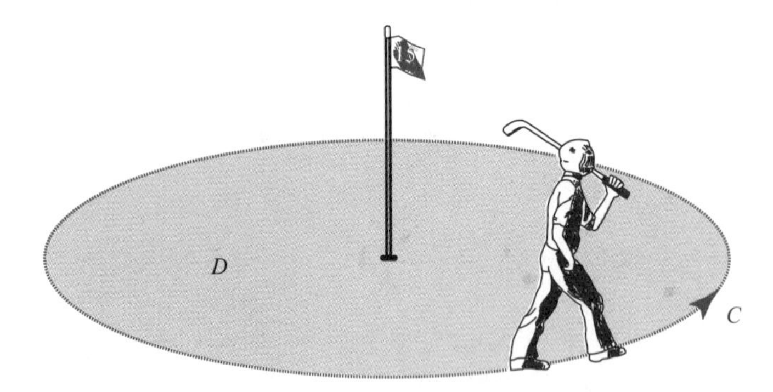

Figure 8.1.4 The correct orientation for the boundary of a region D.

Generalizing Green's Theorem

Green's theorem actually applies to any "decent" region in $\mathbb{R}^2$. For instance, Green's theorem applies to regions that are not simple, but that can be broken up into pieces, each of which is simple. An example is shown in Figure 8.1.5. The region D is an annulus; its boundary consists of two curves $C = C_1 + C_2$ with the indicated orientations. (Note that for the inner region the correct orientation to ensure the validity of Green's theorem is *clockwise*; *the device in Figure 8.1.4 still works for remembering the orientation*!) If Theorem 1 is applied to each of the regions D_1, D_2, D_3, and D_4 and the results are summed, the equality of Green's theorem will be obtained for D and its boundary curve C. This works because the integrals along the interior lines in opposite directions cancel. This trick, in fact, shows that Green's theorem holds for virtually all regions with reasonable boundaries that one is likely to encounter (see Exercise 8).

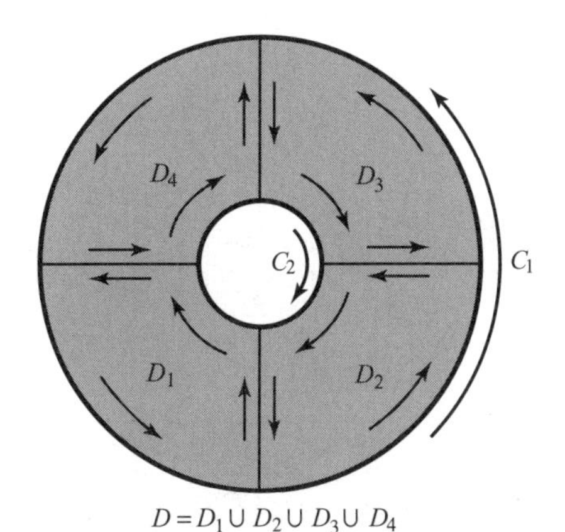

Figure 8.1.5 Green's theorem applies to $D = D_1 \cup D_2 \cup D_3 \cup D_4$.

Let us use the notation ∂D for the oriented curve C^+, that is, the boundary curve of D oriented in the sense as described by the device in Figure 8.1.4. Then we may write Green's theorem as

$$\int_{\partial D} P\,dx + Q\,dy = \iint_D \left(\frac{\partial Q}{\partial x} - \frac{\partial P}{\partial y} \right) dx\,dy.$$

Green's theorem is very useful because it relates a line integral around the boundary of a region to an area integral over the interior of the region, and in many cases it is easier to evaluate the line integral than the area integral or vice versa. For example, if we know that P vanishes on the boundary, we can immediately conclude that $\iint_D (\partial P/\partial y)\,dx\,dy = 0$ even though $\partial P/\partial y$ need not vanish on the interior. (Can you construct such a P on the unit square?)

EXAMPLE 1 Verify Green's theorem for $P(x, y) = x$ and $Q(x, y) = xy$ where D is the unit disk $x^2 + y^2 \leq 1$.

SOLUTION We do this by evaluating both sides in Green's theorem directly. The boundary of D is the unit circle parametrized by $x = \cos t$, $y = \sin t$, $0 \leq t \leq 2\pi$, and so

$$\int_{\partial D} P\,dx + Q\,dy = \int_0^{2\pi} [(\cos t)(-\sin t) + \cos t \sin t \cos t]\,dt$$
$$= \left[\frac{\cos^2 t}{2}\right]_0^{2\pi} + \left[-\frac{\cos^3 t}{3}\right]_0^{2\pi} = 0.$$

On the other hand,

$$\iint_D \left(\frac{\partial Q}{\partial x} - \frac{\partial P}{\partial y}\right) dx\,dy = \iint_D y\,dx\,dy,$$

which is also zero by symmetry. Thus, Green's theorem is verified in this case. ▲

Areas

We can use Green's theorem to obtain a formula for the area of a region bounded by a simple closed curve.

THEOREM 2: Area of a Region If C is a simple closed curve that bounds a region to which Green's theorem applies, then the area of the region D bounded by $C = \partial D$ is

$$A = \frac{1}{2}\int_{\partial D} x\,dy - y\,dx.$$

PROOF Let $P(x, y) = -y$, $Q(x, y) = x$; then by Green's theorem we have

$$\frac{1}{2}\int_{\partial D} x\,dy - y\,dx = \frac{1}{2}\iint_D \left[\frac{\partial x}{\partial x} - \frac{\partial(-y)}{\partial y}\right] dx\,dy$$
$$= \frac{1}{2}\iint_D [1 + 1]\,dx\,dy = \iint_D dx\,dy = A. \quad ■$$

EXAMPLE 2 Let $a > 0$. Compute the area (see Figure 8.1.6) of the region enclosed by the hypocycloid defined by $x^{2/3} + y^{2/3} = a^{2/3}$ using the parametrization

$$x = a\cos^3\theta, \qquad y = a\sin^3\theta, \qquad 0 \le \theta \le 2\pi.$$

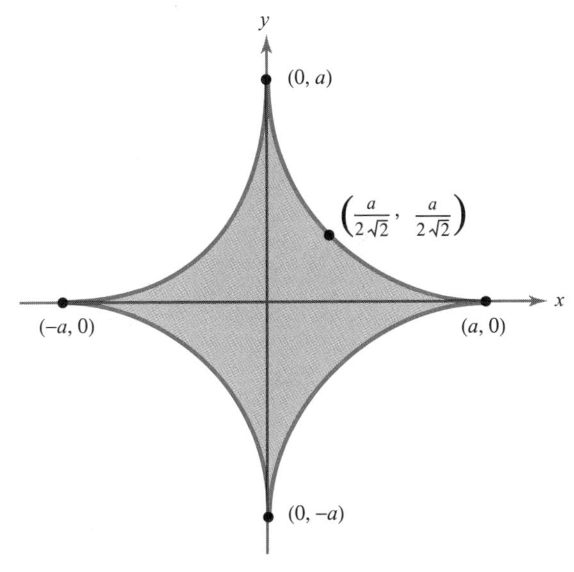

Figure 8.1.6 The hypocycloid $x = a\cos^3\theta, y = a\sin^3\theta, 0 \le \theta \le 2\pi$.

SOLUTION From the preceding box, and using the trigonometric identities $\cos^2\theta + \sin^2\theta = 1$, $\sin 2\theta = 2\sin\theta\cos\theta$, and $\sin^2\phi = (1 - \cos 2\phi)/2$, we get

$$\begin{aligned}
A &= \frac{1}{2}\int_{\partial D} x\,dy - y\,dx \\
&= \frac{1}{2}\int_0^{2\pi} [(a\cos^3\theta)(3a\sin^2\theta\cos\theta) - (a\sin^3\theta)(-3a\cos^2\theta\sin\theta)]\,d\theta \\
&= \frac{3}{2}a^2\int_0^{2\pi} (\sin^2\theta\cos^4\theta + \cos^2\theta\sin^4\theta)\,d\theta = \frac{3}{2}a^2\int_0^{2\pi} \sin^2\theta\cos^2\theta\,d\theta \\
&= \frac{3}{8}a^2\int_0^{2\pi} \sin^2 2\theta\,d\theta = \frac{3}{8}a^2\int_0^{2\pi}\left(\frac{1-\cos 4\theta}{2}\right)d\theta \\
&= \frac{3}{16}a^2\int_0^{2\pi} d\theta - \frac{3}{16}a^2\int_0^{2\pi}\cos 4\theta\,d\theta = \frac{3}{8}\pi a^2. \quad \blacktriangle
\end{aligned}$$

Vector Form using the Curl

The statement of Green's theorem can be neatly rewritten in the language of vector fields. As we will see, this points the way to one possible generalization to $\mathbb{R}^3$.

THEOREM 3: Vector Form of Green's Theorem Let $D \subset \mathbb{R}^2$ be a region to which Green's theorem applies, let ∂D be its (positively oriented) boundary, and let $\mathbf{F} = P\mathbf{i} + Q\mathbf{j}$ be a C^1 vector field on D. Then

$$\int_{\partial D} \mathbf{F}\cdot d\mathbf{s} = \iint_D (\operatorname{curl}\mathbf{F})\cdot\mathbf{k}\,dA = \iint_D (\nabla\times\mathbf{F})\cdot\mathbf{k}\,dA$$

(see Figure 8.1.7).

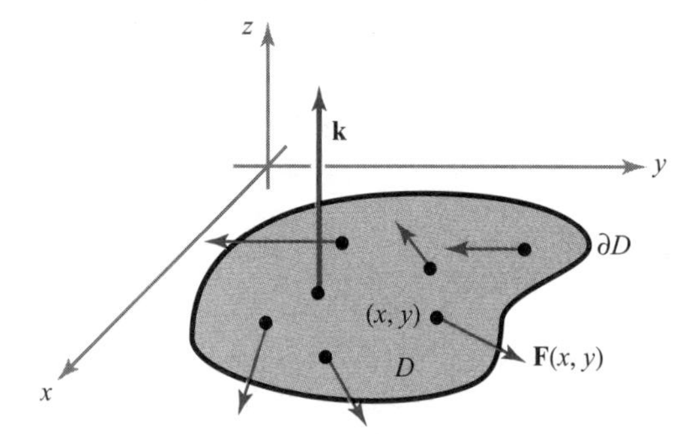

Figure 8.1.7 The vector form of Green's theorem.

This result follows from Theorem 1 and the fact that $(\nabla \times \mathbf{F}) \cdot \mathbf{k} = \partial Q/\partial x - \partial P/\partial y$. We ask the reader to supply the details in Exercise 14.

EXAMPLE 3 Let $\mathbf{F} = (xy^2, y + x)$. Integrate $(\nabla \times \mathbf{F}) \cdot \mathbf{k}$ over the region in the first quadrant bounded by the curves $y = x^2$ and $y = x$.

SOLUTION *Method 1.* We first compute the curl

$$\nabla \times \mathbf{F} = \left(0, 0, \frac{\partial F_2}{\partial x} - \frac{\partial F_1}{\partial y}\right) = (1 - 2xy)\mathbf{k}.$$

Thus, $(\nabla \times \mathbf{F}) \cdot \mathbf{k} = 1 - 2xy$. This can be integrated over the given region D (see Figure 8.1.8) using an iterated integral as follows:

$$\begin{aligned}\iint_D (\nabla \times \mathbf{F}) \cdot \mathbf{k}\, dx\, dy &= \int_0^1 \int_{x^2}^{x} (1 - 2xy)\, dy\, dx = \int_0^1 [y - xy^2]|_{x^2}^{x}\, dx \\ &= \int_0^1 [x - x^3 - x^2 + x^5]\, dx = \frac{1}{2} - \frac{1}{4} - \frac{1}{3} + \frac{1}{6} = \frac{1}{12}.\end{aligned}$$

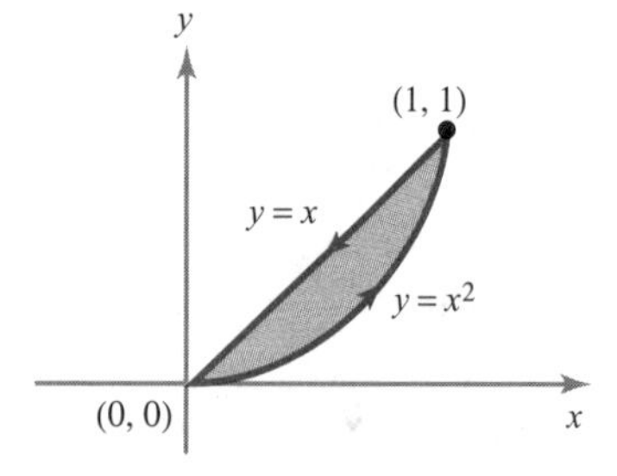

Figure 8.1.8 The region bounded by the curves $y = x^2$ and $y = x$.

Method 2. Here we use Theorem 3 to obtain

$$\iint_D (\nabla \times \mathbf{F}) \cdot \mathbf{k}\, dx\, dy = \int_{\partial D} \mathbf{F} \cdot d\mathbf{s}.$$

The line integral of $\mathbf{F}$ along the curve $y = x$ from left to right is

$$\int_0^1 F_1\,dx + F_2\,dy = \int_0^1 (x^3 + 2x)\,dx = \frac{1}{4} + 1 = \frac{5}{4}.$$

Along the curve $y = x^2$ we get

$$\int_0^1 F_1\,dx + F_2\,dy = \int_0^1 x^5\,dx + (x + x^2)(2x\,dx) = \frac{1}{6} + \frac{2}{3} + \frac{1}{2} = \frac{4}{3}.$$

Thus, remembering that the integral along $y = x$ is to be taken from right to left, as in Figure 8.1.8,

$$\int_{\partial D} \mathbf{F} \cdot d\mathbf{s} = \frac{4}{3} - \frac{5}{4} = \frac{1}{12}. \quad \blacktriangle$$

Vector Form Using the Divergence

There is another form of Green's theorem that can be generalized to $\mathbb{R}^3$.

THEOREM 4: Divergence Theorem in the Plane Let $D \subset \mathbb{R}^2$ be a region to which Green's theorem applies and let ∂D be its boundary. Let $\mathbf{n}$ denote the outward unit normal to ∂D. If $\mathbf{c}$: $[a, b] \to \mathbb{R}^2$, $t \mapsto \mathbf{c}(t) = (x(t), y(t))$ is a positively oriented parametrization of ∂D, $\mathbf{n}$ is given by

$$\mathbf{n} = \frac{(y'(t), -x'(t))}{\sqrt{[x'(t)]^2 + [y'(t)]^2}}$$

(see Figure 8.1.9). Let $\mathbf{F} = P\mathbf{i} + Q\mathbf{j}$ be a C^1 vector field on D. Then

$$\int_{\partial D} \mathbf{F} \cdot \mathbf{n}\,ds = \iint_D \operatorname{div} \mathbf{F}\,dA.$$

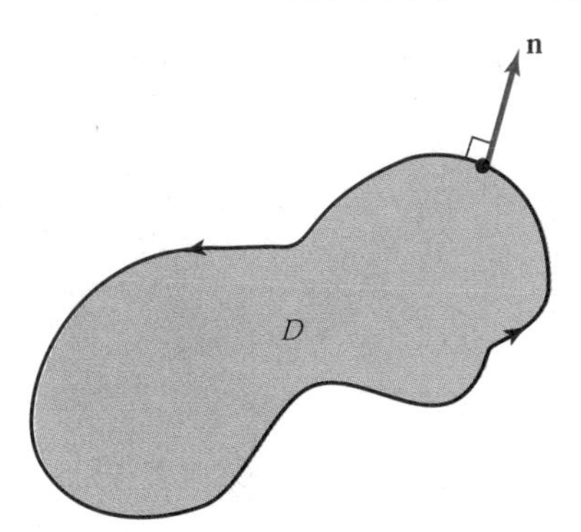

Figure 8.1.9 $\mathbf{n}$ is the outward unit normal to ∂D.

PROOF Recall that $\mathbf{c}'(t) = (x'(t), y'(t))$ is tangent to ∂D, and note that $\mathbf{n} \cdot \mathbf{c}' = 0$. Thus, $\mathbf{n}$ is normal to the boundary. The sign of $\mathbf{n}$ is chosen to make it correspond to the *outward* (rather than the inward) direction. By the definition of the line integral (see Section 7.2),

$$\begin{aligned}\int_{\partial D} \mathbf{F} \cdot \mathbf{n}\, ds &= \int_a^b \frac{P(x(t), y(t))y'(t) - Q(x(t), y(t))x'(t)}{\sqrt{[x'(t)]^2 + [y'(t)]^2}} \sqrt{[x'(t)]^2 + [y'(t)]^2}\, dt \\ &= \int_a^b [P(x(t), y(t))y'(t) - Q(x(t), y(t))x'(t)]\, dt \\ &= \int_{\partial D} P\, dy - Q\, dx.\end{aligned}$$

By Green's theorem, this equals

$$\iint_D \left(\frac{\partial P}{\partial x} + \frac{\partial Q}{\partial y} \right) dx\, dy = \iint_D \operatorname{div} \mathbf{F}\, dA. \quad ■$$

EXAMPLE 4 Let $\mathbf{F} = y^3\mathbf{i} + x^5\mathbf{j}$. Compute the integral of the normal component of $\mathbf{F}$ around the unit square.

SOLUTION This can be done using the divergence theorem. Indeed,

$$\int_{\partial D} \mathbf{F} \cdot \mathbf{n}\, ds = \iint_D \operatorname{div} \mathbf{F}\, dA.$$

But $\operatorname{div} \mathbf{F} = 0$, and so the integral is zero. ▲

EXERCISES

1. Evaluate $\int_C y\, dx - x\, dy$ where C is the boundary of the square $[-1, 1] \times [-1, 1]$ oriented in the counterclockwise direction, using Green's theorem.

2. Find the area of the disk D of radius R using Green's theorem.

3. Verify Green's theorem for the disk D with center $(0, 0)$ and radius R and the functions:

(a) $P(x, y) = xy^2,\ Q(x, y) = -yx^2$
(b) $P(x, y) = x + y,\ Q(x, y) = y$
(c) $P(x, y) = xy = Q(x, y)$
(d) $P(x, y) = 2y,\ Q(x, y) = x$

4. Using the divergence theorem, show that $\int_{\partial D} \mathbf{F} \cdot \mathbf{n}\, ds = 0$ where $\mathbf{F}(x, y) = y\mathbf{i} - x\mathbf{j}$ and D is the unit disk. Verify this directly.

5. Find the area bounded by one arc of the cycloid $x = a(\theta - \sin\theta),\ y = a(1 - \cos\theta)$, where $a > 0$, and $0 \le \theta \le 2\pi$, and the x axis (use Green's theorem).

6. Under the conditions of Green's theorem, prove that

(a) $$\int_{\partial D} PQ\,dx + PQ\,dy = \iint_D \left[Q\left(\frac{\partial P}{\partial x} - \frac{\partial P}{\partial y}\right) + P\left(\frac{\partial Q}{\partial x} - \frac{\partial Q}{\partial y}\right)\right] dx\,dy$$

(b) $$\int_{\partial D} \left(Q\frac{\partial P}{\partial x} - P\frac{\partial Q}{\partial x}\right) dx + \left(P\frac{\partial Q}{\partial y} - Q\frac{\partial P}{\partial y}\right) dy$$
$$= 2\iint_D \left(P\frac{\partial^2 Q}{\partial x\,\partial y} - Q\frac{\partial^2 P}{\partial x\,\partial y}\right) dx\,dy$$

7. Evaluate the line integral

$$\int_C (2x^3 - y^3)\,dx + (x^3 + y^3)\,dy,$$

where C is the unit circle, and verify Green's theorem for this case.

8. Prove the following generalization of Green's theorem: Let D be a region in the xy plane with boundary a finite number of oriented simple closed curves. Suppose that by means of a finite number of line segments parallel to the coordinate axes, D can be decomposed into a finite number of simple regions D_i with the boundary of each D_i oriented counterclockwise (see Figure 8.1.5). Then if P and Q are of class C^1 on D,

$$\iint_D \left(\frac{\partial Q}{\partial x} - \frac{\partial P}{\partial y}\right) dx\,dy = \int_{\partial D} P\,dx + Q\,dy,$$

where ∂D is the oriented boundary of D. (HINT: Apply Green's theorem to each D_i.)

9. Verify Green's theorem for the integrand of Exercise 7 (that is, with $P = 2x^3 - y^3$ and $Q = x^3 + y^3$) and the annular region D described by $a \le x^2 + y^2 \le b$, with boundaries oriented as in Figure 8.1.5.

10. Let D be a region for which Green's theorem holds. Suppose f is harmonic; that is,

$$\frac{\partial^2 f}{\partial x^2} + \frac{\partial^2 f}{\partial y^2} = 0$$

on D. Prove that

$$\int_{\partial D} \frac{\partial f}{\partial y}\,dx - \frac{\partial f}{\partial x}\,dy = 0.$$

11. (a) Verify the divergence theorem for $\mathbf{F} = x\mathbf{i} + y\mathbf{j}$ and D the unit disk $x^2 + y^2 \le 1$.

(b) Evaluate the integral of the normal component of $2xy\mathbf{i} - y^2\mathbf{j}$ around the ellipse defined by $x^2/a^2 + y^2/b^2 = 1$.

12. Let $P(x, y) = -y/(x^2 + y^2)$ and $Q(x, y) = x/(x^2 + y^2)$. Assuming D is the unit disk, investigate why Green's theorem fails for this P and Q.

13. Use Green's theorem to evaluate $\int_{C^+} (y^2 + x^3)\,dx + x^4\,dy$, where C^+ is the perimeter of the square $[0, 1] \times [0, 1]$ in the counterclockwise direction.

14. Verify Theorem 3 by showing that $(\nabla \times \mathbf{F}) \cdot \mathbf{k} = \partial Q/\partial x - \partial P/\partial y$.

15. Use Theorem 2 to compute the area inside the ellipse $x^2/a^2 + y^2/b^2 = 1$.

16. Use Theorem 2 to recover the formula $A = \frac{1}{2}\int_a^b r^2\, d\theta$ for a region in polar coordinates.

17. Sketch the proof of Green's theorem for the region shown in Figure 8.1.10.

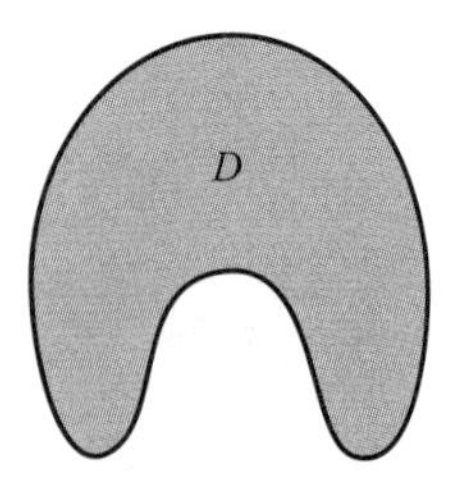

Figure 8.1.10 Prove Green's theorem for this region.

18. Prove the identity

$$\int_{\partial D} \phi \nabla \phi \cdot \mathbf{n}\, ds = \iint_D (\phi \nabla^2 \phi + \nabla \phi \cdot \nabla \phi)\, dA.$$

19. Use Green's theorem to find the area of one loop of the four-leafed rose $r = 3\sin 2\theta$. (HINT: $x\,dy - y\,dx = r^2 d\theta$).

20. Show that if C is a simple closed curve that bounds a region to which Green's theorem applies, then the area of the region D bounded by C is

$$A = \int_{\partial D} x\, dy = -\int_{\partial D} y\, dx.$$

Show how this implies Theorem 2.

Exercises 21 to 29 illustrate the application of Green's theorem to partial differential equations. (Further applications are given in the Internet supplement.) They are particularly concerned with solutions to Laplace's equation, that is, with harmonic functions. For these exercises, let D be an open region in $\mathbb{R}^2$ with boundary ∂D. Let $u\colon D \cup \partial D \to \mathbb{R}$ be a continuous function that is of class C^2 on D. Suppose $\mathbf{p} \in D$ and the closed disks $B_\rho = B_\rho(\mathbf{p})$ of radius ρ centered at $\mathbf{p}$ are contained in D for $0 < \rho \le R$. Define $I(\rho)$ by

$$I(\rho) = \frac{1}{\rho}\int_{\partial B_\rho} u\, ds.$$

21. Show that $\text{limit}_{\rho \to 0}\, I(\rho) = 2\pi u(\mathbf{p})$.

22. Let $\mathbf{n}$ denote the outward unit normal to ∂B_ρ and $\partial u/\partial n = \nabla u \cdot \mathbf{n}$. Show that

$$\int_{\partial B_\rho} \frac{\partial u}{\partial n}\, ds = \iint_{B_\rho} \nabla^2 u\, dA.$$

23. Using Exercise 22, show that $I'(\rho) = (1/\rho) \iint_{B_\rho} \nabla^2 u \, dA$.

24. Suppose u satisfies Laplace's equation: $\nabla^2 u = 0$ on D. Use the preceding exercises to show that

$$u(\mathbf{p}) = \frac{1}{2\pi R} \int_{\partial B_R} u \, ds.$$

(This expresses the fact that the value of a harmonic function at a point is the average of its values on the circumference of any disk centered about it.)

25. Use Exercise 24 to show that if u is harmonic (i.e., if $\nabla^2 u = 0$), then $u(\mathbf{p})$ can be expressed as an area integral

$$u(\mathbf{p}) = \frac{1}{\pi R^2} \iint_{B_R} u \, dA.$$

26. Suppose u is a harmonic function defined on D (i.e., $\nabla^2 u = 0$ on D) and that u has a local maximum (or minimum) at a point $\mathbf{p}$ in D.

(a) Show that u must be constant on some disk centered at $\mathbf{p}$. (HINT: Use the results of Exercise 25.)

(b) Suppose that D is path-connected [i.e., for any points $\mathbf{p}$ and $\mathbf{q}$ in D, there is a continuous path $\mathbf{c}\colon [0, 1] \to D$ such that $\mathbf{c}(0) = \mathbf{p}$ and $\mathbf{c}(1) = \mathbf{q}$] and that for some $\mathbf{p}$ the maximum or minimum at $\mathbf{p}$ is absolute; thus, $u(\mathbf{q}) \le u(\mathbf{p})$ or $u(\mathbf{q}) \ge u(\mathbf{p})$ for every $\mathbf{q}$ in D. Show that u must be constant on D.

(The result in this Exercise is called a ***strong maximum*** or ***minimum principle*** for harmonic functions. Compare this with Exercises 36 to 40 in Section 3.3.)

27. A function is said to be ***subharmonic*** on D if $\nabla^2 u \ge 0$ everywhere in D. It is said to be ***superharmonic*** if $\nabla^2 u \le 0$.

(a) Derive a strong maximum principle for subharmonic functions.

(b) Derive a strong minimum principle for superharmonic functions.

28. Suppose D is the disk $\{(x, y) \mid x^2 + y^2 < 1\}$ and C is the circle $\{(x, y) \mid x^2 + y^2 = 1\}$. In the Internet supplement, we shall show that if f is a continuous real-valued function on C, then there is a continuous function u on $D \cup C$ that agrees with f on C and is harmonic on D. That is, f has a harmonic extension to the disk. Assuming this, show the following:

(a) If q is a nonconstant continuous function on $D \cup C$ that is subharmonic (but not harmonic) on D, then there is a continuous function u on $D \cup C$ that is harmonic on D such that u agrees with q on C and $q < u$ everywhere on D.

(b) The same assertion holds if "subharmonic" is replaced by "superharmonic" and "$q < u$" by "$q > u$."

29. Let D be as in Exercise 28. Let $f\colon D \to \mathbb{R}$ be continuous. Show that a solution to the equation $\nabla^2 u = 0$ satisfying $u(\mathbf{x}) = f(\mathbf{x})$ for all $\mathbf{x} \in \partial D$ is unique.

30. Use Green's theorem to prove the change of variables formula in the following special case:

$$\iint_D dx\,dy = \iint_{D^*} \left|\frac{\partial(x,y)}{\partial(u,v)}\right| du\,dv$$

for a transformation $(u, v) \mapsto (x(u, v), y(u, v))$.

8.2 Stokes' Theorem

Stokes' theorem relates the line integral of a vector field around a simple closed curve C in $\mathbb{R}^3$ to an integral over a surface S for which C is the boundary. In this regard it is very much like Green's theorem.

Stokes' Theorem for Graphs

Let us begin by recalling a few facts from Chapter 7. Consider a surface S that is the graph of a function $f(x, y)$, so that S is parametrized by

$$\begin{cases} x = u \\ y = v \\ z = f(u, v) = f(x, y) \end{cases}$$

for (u, v) in some domain D in the plane. The integral of a vector function $\mathbf{F}$ over S was developed in Section 7.6 as

$$\iint_S \mathbf{F}\cdot d\mathbf{S} = \iint_D \left[F_1\left(-\frac{\partial z}{\partial x}\right) + F_2\left(-\frac{\partial z}{\partial y}\right) + F_3\right] dx\,dy, \qquad (1)$$

where $\mathbf{F} = F_1\mathbf{i} + F_2\mathbf{j} + F_3\mathbf{k}$.

In Section 8.1, we first assumed that the regions D under consideration were simple; while this was used in our proof of Green's theorem, we noted there that the theorem is valid for a wider class of regions. In this section we assume that D is a region whose boundary is a simple closed curve and to which Green's theorem applies. Green's theorem involves choosing an orientation on the boundary of D, as was explained in Section 8.1. The choice of orientation that validates Green's theorem will be called ***positive***. Recall that if D is simple, then the positive orientation is the counterclockwise one.

Suppose that $\mathbf{c}\colon [a, b] \to \mathbb{R}^2$, $\mathbf{c}(t) = (x(t), y(t))$ is a parametrization of ∂D in the positive direction. Then we define the ***boundary curve*** ∂S to be the oriented simple closed curve that is the image of the mapping $\mathbf{p}\colon t \mapsto (x(t), y(t), f(x(t), y(t)))$ with the orientation induced by $\mathbf{p}$ (Figure 8.2.1).

To remember this orientation (that is, the positive direction) on ∂S, imagine that you are an "observer" walking along the boundary of the surface with the normal as your upright direction; you are moving in the positive direction if the surface is on your left. This orientation on ∂S is often called the ***orientation induced by an upward normal*** $\mathbf{n}$.

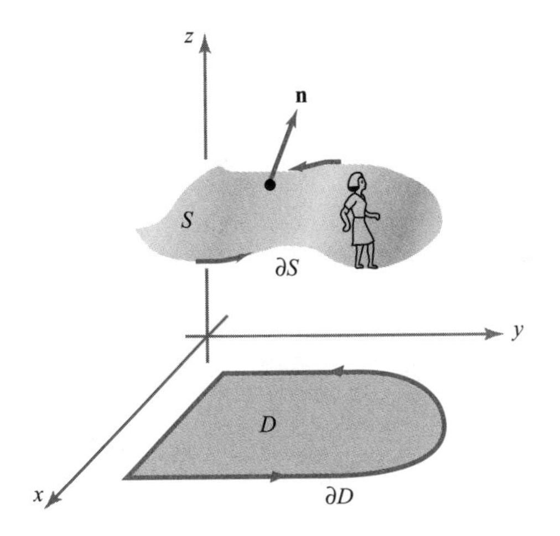

Figure 8.2.1 The induced orientation on ∂S: As you walk around the boundary, the surface should be on your left.

THEOREM 5: Stokes' Theorem for Graphs Let S be the oriented surface defined by a C^2 function $z = f(x, y)$, where $(x, y) \in D$, a region to which Green's theorem applies, and let $\mathbf{F}$ be a C^1 vector field on S. Then if ∂S denotes the oriented boundary curve of S as just defined, we have

$$\iint_S \text{curl } \mathbf{F} \cdot d\mathbf{S} = \iint_S (\nabla \times \mathbf{F}) \cdot d\mathbf{S} = \int_{\partial S} \mathbf{F} \cdot d\mathbf{s}.$$

Remember that $\int_{\partial S} \mathbf{F} \cdot d\mathbf{s}$ is the integral around ∂S of the tangential component of $\mathbf{F}$, while $\iint_S \mathbf{G} \cdot d\mathbf{S}$ is the integral over S of $\mathbf{G} \cdot \mathbf{n}$, the normal component of $\mathbf{G}$ (see Sections 7.2 and 7.6). Thus, Stokes' theorem says that the integral of *the normal component of the curl of a vector field* $\mathbf{F}$ *over a surface* S *is equal to the integral of the tangential component of* $\mathbf{F}$ *around the boundary* ∂S.

PROOF If $\mathbf{F} = F_1\mathbf{i} + F_2\mathbf{j} + F_3\mathbf{k}$, then

$$\text{curl } \mathbf{F} = \left(\frac{\partial F_3}{\partial y} - \frac{\partial F_2}{\partial z}\right)\mathbf{i} + \left(\frac{\partial F_1}{\partial z} - \frac{\partial F_3}{\partial x}\right)\mathbf{j} + \left(\frac{\partial F_2}{\partial x} - \frac{\partial F_1}{\partial y}\right)\mathbf{k}.$$

Therefore, we use formula (1) to write

$$\iint_S \text{curl } \mathbf{F} \cdot d\mathbf{S} = \iint_D \left[\left(\frac{\partial F_3}{\partial y} - \frac{\partial F_2}{\partial z}\right)\left(-\frac{\partial z}{\partial x}\right) + \left(\frac{\partial F_1}{\partial z} - \frac{\partial F_3}{\partial x}\right)\left(-\frac{\partial z}{\partial y}\right) + \left(\frac{\partial F_2}{\partial x} - \frac{\partial F_1}{\partial y}\right)\right] dA. \tag{2}$$

On the other hand,

$$\int_{\partial S} \mathbf{F} \cdot d\mathbf{s} = \int_{\mathbf{p}} \mathbf{F} \cdot d\mathbf{s} = \int_{\mathbf{p}} F_1\,dx + F_2\,dy + F_3\,dz,$$

where $\mathbf{p}\colon [a, b] \to \mathbb{R}^3$, $\mathbf{p}(t) = (x(t), y(t), f(x(t), y(t)))$ is the orientation-preserving parametrization of the oriented simple closed curve ∂S discussed earlier. Thus,

$$\int_{\partial S} \mathbf{F} \cdot d\mathbf{s} = \int_a^b \left(F_1 \frac{dx}{dt} + F_2 \frac{dy}{dt} + F_3 \frac{dz}{dt} \right) dt. \tag{3}$$

By the chain rule

$$\frac{dz}{dt} = \frac{\partial z}{\partial x}\frac{dx}{dt} + \frac{\partial z}{\partial y}\frac{dy}{dt}.$$

Substituting this expression into equation (3), we obtain

$$\begin{aligned}
\int_{\partial S} \mathbf{F} \cdot d\mathbf{s} &= \int_a^b \left[\left(F_1 + F_3 \frac{\partial z}{\partial x} \right) \frac{dx}{dt} + \left(F_2 + F_3 \frac{\partial z}{\partial y} \right) \frac{dy}{dt} \right] dt \\
&= \int_{\mathbf{c}} \left(F_1 + F_3 \frac{\partial z}{\partial x} \right) dx + \left(F_2 + F_3 \frac{\partial z}{\partial y} \right) dy \qquad (4) \\
&= \int_{\partial D} \left(F_1 + F_3 \frac{\partial z}{\partial x} \right) dx + \left(F_2 + F_3 \frac{\partial z}{\partial y} \right) dy.
\end{aligned}$$

Applying Green's theorem to equation (4) yields (we are assuming that Green's theorem applies to D)

$$\iint_D \left[\frac{\partial(F_2 + F_3\,\partial z/\partial y)}{\partial x} - \frac{\partial(F_1 + F_3\,\partial z/\partial x)}{\partial y} \right] dA.$$

Now we use the chain rule, remembering that F_1, F_2, and F_3 are functions of x, y, and z and that z is a function of x and y, to obtain

$$\begin{aligned}
\iint_D \Bigg[&\left(\frac{\partial F_2}{\partial x} + \frac{\partial F_2}{\partial z}\frac{\partial z}{\partial x} + \frac{\partial F_3}{\partial x}\frac{\partial z}{\partial y} + \frac{\partial F_3}{\partial z}\frac{\partial z}{\partial x}\frac{\partial z}{\partial y} + F_3 \frac{\partial^2 z}{\partial x \partial y} \right) \\
&- \left(\frac{\partial F_1}{\partial y} + \frac{\partial F_1}{\partial z}\frac{\partial z}{\partial y} + \frac{\partial F_3}{\partial y}\frac{\partial z}{\partial x} + \frac{\partial F_3}{\partial z}\frac{\partial z}{\partial y}\frac{\partial z}{\partial x} + F_3 \frac{\partial^2 z}{\partial y \partial x} \right) \Bigg] dA.
\end{aligned}$$

Because mixed partials are equal, the last two terms in each parenthesis cancel each other, and we can rearrange terms to obtain the integral of equation (2), which completes the proof. ■

EXAMPLE 1 Let $\mathbf{F} = ye^z\mathbf{i} + xe^z\mathbf{j} + xye^z\mathbf{k}$. Show that the integral of $\mathbf{F}$ around an oriented simple closed curve C that is the boundary of a surface S is 0. (Assume S is the graph of a function, as in Theorem 5.)

SOLUTION Indeed, $\int_C \mathbf{F} \cdot d\mathbf{s} = \iint_S (\nabla \times \mathbf{F}) \cdot d\mathbf{S}$, by Stokes' theorem. But we compute

$$\nabla \times \mathbf{F} = \begin{vmatrix} \mathbf{i} & \mathbf{j} & \mathbf{k} \\ \dfrac{\partial}{\partial x} & \dfrac{\partial}{\partial y} & \dfrac{\partial}{\partial z} \\ ye^z & xe^z & xye^z \end{vmatrix} = \mathbf{0},$$

and so $\int_C \mathbf{F} \cdot d\mathbf{s} = 0$. Alternatively, we can observe that $\mathbf{F} = \nabla(xye^z)$, so its integral around a closed curve is zero. ▲

EXAMPLE 2 Use Stokes' theorem to evaluate the line integral

$$\int_C -y^3\, dx + x^3\, dy - z^3\, dz,$$

where C is the intersection of the cylinder $x^2 + y^2 = 1$ and the plane $x + y + z = 1$, and the orientation on C corresponds to counterclockwise motion in the xy plane.

SOLUTION The curve C bounds the surface S defined by the equation $z = 1 - x - y = f(x, y)$ for (x, y) in the set $D = \{(x, y) \mid x^2 + y^2 \leq 1\}$ (Figure 8.2.2). We set $\mathbf{F} = -y^3\mathbf{i} + x^3\mathbf{j} - z^3\mathbf{k}$, which has curl $\nabla \times \mathbf{F} = (3x^2 + 3y^2)\mathbf{k}$. Then, by

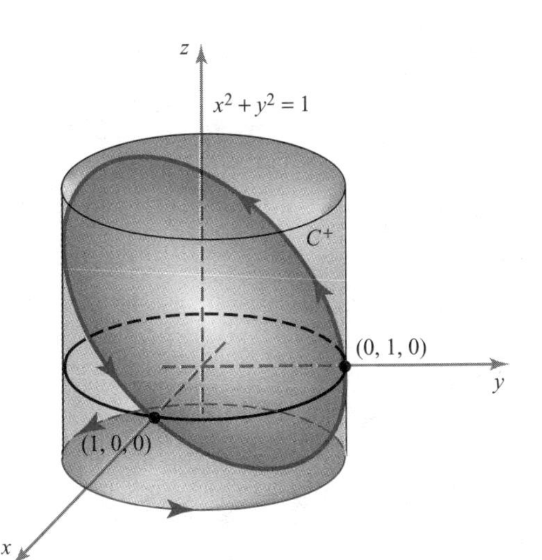

Figure 8.2.2 The curve C is the intersection of the cylinder $x^2 + y^2 = 1$ and the plane $x + y + z = 1$.

Stokes' theorem, the line integral is equal to the surface integral

$$\iint_S (\nabla \times \mathbf{F}) \cdot d\mathbf{S}.$$

But $\nabla \times \mathbf{F}$ has only a $\mathbf{k}$ component. Thus, by formula (1) we have

$$\iint_S (\nabla \times \mathbf{F}) \cdot d\mathbf{S} = \iint_D (3x^2 + 3y^2)\, dx\, dy.$$

This integral can be evaluated by changing to polar coordinates. Doing this, we get

$$3 \iint_D (x^2 + y^2)\, dx\, dy = 3 \int_0^1 \int_0^{2\pi} r^2 \cdot r\, d\theta\, dr = 6\pi \int_0^1 r^3\, dr = \frac{6\pi}{4} = \frac{3\pi}{2}.$$

Let us verify this result by *directly* evaluating the line integral

$$\int_C -y^3\, dx + x^3\, dy - z^3\, dz.$$

We can parametrize the curve ∂D by the equations

$$x = \cos t, \qquad y = \sin t, \qquad z = 0, \qquad 0 \le t \le 2\pi.$$

The curve C is therefore parametrized by the equations

$$x = \cos t, \qquad y = \sin t, \qquad z = 1 - \sin t - \cos t, \qquad 0 \le t \le 2\pi.$$

Thus,

$$\begin{aligned} &\int_C -y^3\, dx + x^3\, dy - z^3\, dz \\ &\quad = \int_0^{2\pi} [(-\sin^3 t)(-\sin t) + (\cos^3 t)(\cos t) \\ &\qquad\qquad - (1 - \sin t - \cos t)^3(-\cos t + \sin t)]\, dt \\ &\quad = \int_0^{2\pi} (\cos^4 t + \sin^4 t)\, dt - \int_0^{2\pi} (1 - \sin t - \cos t)^3(-\cos t + \sin t)\, dt. \end{aligned}$$

The second integrand is of the form $u^3\, du$, where $u = 1 - \sin t - \cos t$, and thus the integral is equal to

$$\frac{1}{4}[(1 - \sin t - \cos t)^4]_0^{2\pi} = 0.$$

Hence, we are left with

$$\int_0^{2\pi} (\cos^4 t + \sin^4 t)\, dt.$$

This integral can be evaluated using formulas (18) and (19) of the table of integrals. We can also proceed as follows. Using the trigonometric identities

$$\sin^2 t = \frac{1 - \cos 2t}{2}, \qquad \cos^2 t = \frac{1 + \cos 2t}{2},$$

substituting and squaring these expressions, we reduce the preceding integral to

$$\frac{1}{2}\int_0^{2\pi} (1 + \cos^2 2t)\, dt = \pi + \frac{1}{2}\int_0^{2\pi} \cos^2 2t\, dt.$$

Again using the identity $\cos^2 2t = (1 + \cos 4t)/2$, we find

$$\begin{aligned}\pi + \frac{1}{4}\int_0^{2\pi} (1 + \cos 4t)\, dt &= \pi + \frac{1}{4}\int_0^{2\pi} dt + \frac{1}{4}\int_0^{2\pi} \cos 4t\, dt \\ &= \pi + \frac{\pi}{2} + 0 = \frac{3\pi}{2}. \quad \blacktriangle\end{aligned}$$

Stokes' Theorem for Parametrized Surfaces

To simplify the proof of Stokes' theorem given earlier, we assumed that the surface S could be described as the graph of a function $z = f(x, y), (x, y) \in D$, where D is some region to which Green's theorem applies. However, without too much more effort we can obtain a *more general theorem* for oriented parametrized surfaces S. The main complication is in the definition of ∂S.

Suppose $\mathbf{\Phi}\colon D \to \mathbb{R}^3$ is a parametrization of a surface S and $\mathbf{c}(t) = (u(t), v(t))$ is a parametrization of ∂D. We might be tempted to define ∂S as the curve parametrized by $t \mapsto \mathbf{p}(t) = \mathbf{\Phi}(u(t), v(t))$. However, with this definition, ∂S might not be the boundary of S in any reasonable geometric sense.

For example, we would conclude that the boundary of the unit sphere S parametrized by spherical coordinates in $\mathbb{R}^3$ is half of the great circle on S lying in the xz plane, but clearly in a geometric sense S is a smooth surface (no points or cusps) with no boundary or edge at all (see Figure 8.2.3 and Exercise 20). Thus, this great circle is in some sense the "mistaken" boundary of S.

We can get around this difficulty by assuming that $\mathbf{\Phi}$ is one-to-one on all of D. Then the image of ∂D under $\mathbf{\Phi}$, namely, $\mathbf{\Phi}(\partial D)$, will be the geometric boundary of $S = \mathbf{\Phi}(D)$. If $\mathbf{c}(t) = (u(t), v(t))$ is a parametrization of ∂D in the positive direction,

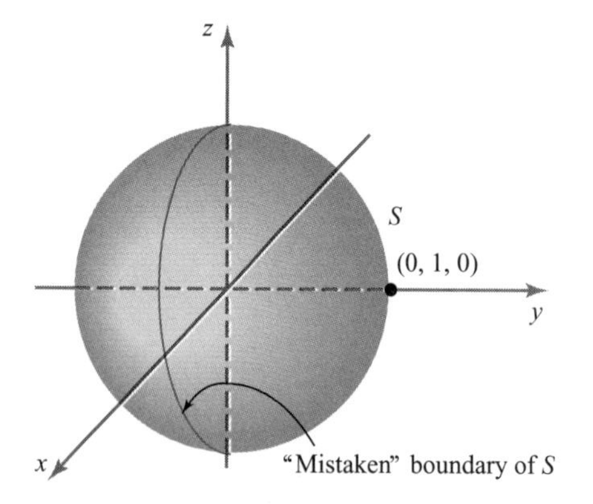

Figure 8.2.3 The surface S is a portion of a sphere.

we define ∂S to be the oriented simple closed curve that is the image of the mapping $\mathbf{p}: t \mapsto \mathbf{\Phi}(u(t), v(t))$, with the orientation of ∂S induced by $\mathbf{p}$ (see Figure 8.2.1).

THEOREM 6: Stokes' Theorem: Parametrized Surfaces Let S be an oriented surface defined by a one-to-one parametrization $\mathbf{\Phi}: D \subset \mathbb{R}^2 \to S$, where D is a region to which Green's theorem applies. Let ∂S denote the oriented boundary of S and let $\mathbf{F}$ be a C^1 vector field on S. Then

$$\iint_S (\nabla \times \mathbf{F}) \cdot d\mathbf{S} = \int_{\partial S} \mathbf{F} \cdot d\mathbf{s}.$$

If S has no boundary, and this includes surfaces such as the sphere, then the integral on the left is zero (see Exercise 17).

This is proved in the same way as Theorem 5.

EXAMPLE 3 Let S be the surface shown in Figure 8.2.4, with the indicated orientation. Let $\mathbf{F} = y\mathbf{i} - x\mathbf{j} + e^{xz}\mathbf{k}$. Evaluate $\iint_S (\nabla \times \mathbf{F}) \cdot d\mathbf{S}$.

SOLUTION This surface could be parametrized using spherical coordinates based at the center of the sphere. However, we need not explicitly find $\mathbf{\Phi}$ in order to solve this problem. By Theorem 6, $\iint_S (\nabla \times \mathbf{F}) \cdot d\mathbf{S} = \int_{\partial S} \mathbf{F} \cdot d\mathbf{s}$, and so if we parametrize ∂S by $x(t) = \cos t$, $y(t) = \sin t$, $0 \le t \le 2\pi$, we determine

$$\int_{\partial S} \mathbf{F} \cdot d\mathbf{s} = \int_0^{2\pi} \left(y \frac{dx}{dt} - x \frac{dy}{dt} \right) dt = \int_0^{2\pi} (-\sin^2 t - \cos^2 t)\, dt = -\int_0^{2\pi} dt = -2\pi$$

and therefore $\iint_S (\nabla \times \mathbf{F}) \cdot d\mathbf{S} = -2\pi$. ▲

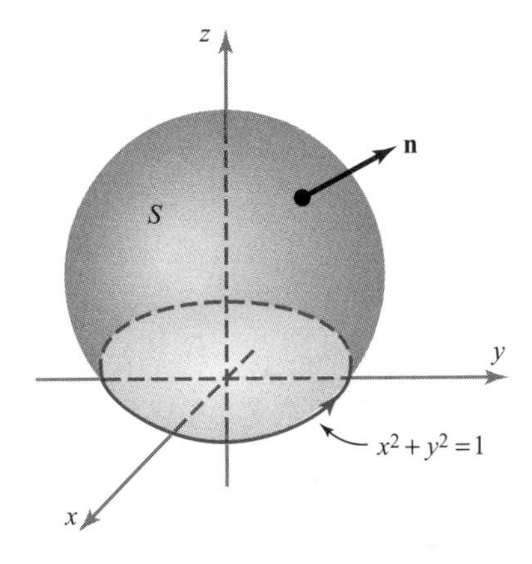

Figure 8.2.4 This surface S is a portion of a sphere sitting on top of the circle $x^2 + y^2 = 1$. It does *not* include the disk $x^2 + y^2 < 1$ in the xy plane.

The Curl as Circulation per Unit Area

Let us now use Stokes' theorem to justify the physical interpretation of $\nabla \times \mathbf{F}$ in terms of paddle wheels that was proposed in Chapter 4. Paraphrasing Theorem 6, we have

$$\iint_S (\text{curl } \mathbf{F}) \cdot \mathbf{n}\, dS = \iint_S (\text{curl } \mathbf{F}) \cdot d\mathbf{S} = \int_{\partial S} \mathbf{F} \cdot d\mathbf{s} = \int_{\partial S} F_T\, ds,$$

where F_T is the tangential component of $\mathbf{F}$. This says that the integral of the normal component of the curl of a vector field over an oriented surface S is equal to the line integral of $\mathbf{F}$ along ∂S, which in turn is equal to the path integral of the tangential component of $\mathbf{F}$ over ∂S.

Suppose $\mathbf{V}$ represents the velocity vector field of a fluid. Consider a point P and a unit vector $\mathbf{n}$. Let S_ρ denote the disk of radius ρ and center P, which is perpendicular to $\mathbf{n}$. By Stokes' theorem,

$$\iint_{S_\rho} \text{curl } \mathbf{V} \cdot d\mathbf{S} = \iint_{S_\rho} \text{curl } \mathbf{V} \cdot \mathbf{n}\, dS = \int_{\partial S_\rho} \mathbf{V} \cdot d\mathbf{s},$$

where ∂S_ρ has the orientation induced by $\mathbf{n}$ (see Figure 8.2.5).

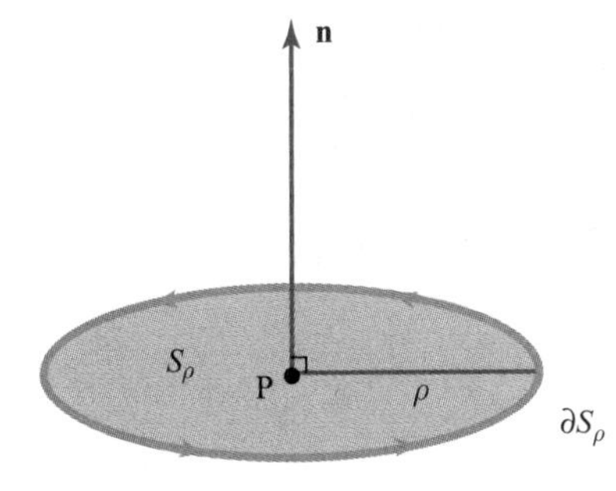

Figure 8.2.5 A normal $\mathbf{n}$ induces an orientation on the boundary ∂S_ρ of the disk S_ρ.

By the mean-value theorem for integrals (Exercise 12, Section 7.6), there is a point Q in S_ρ such that

$$\iint_{S_\rho} \text{curl}\, \mathbf{V} \cdot \mathbf{n}\, dS = [\text{curl}\, \mathbf{V}(\text{Q}) \cdot \mathbf{n}] A(S_\rho)$$

where $A(S_\rho) = \pi\rho^2$ is the area of S_ρ, curl $\mathbf{V}(\text{Q})$ is the value of curl $\mathbf{V}$ at Q. Thus,

$$\begin{aligned}\underset{\rho\to 0}{\text{limit}}\, \frac{1}{A(S_\rho)} \int_{\partial S_\rho} \mathbf{V} \cdot d\mathbf{s} &= \underset{\rho\to 0}{\text{limit}}\, \frac{1}{A(S_\rho)} \iint_{S_\rho} (\text{curl}\, \mathbf{V}) \cdot d\mathbf{S} \\ &= \underset{\rho\to 0}{\text{limit}}\, \text{curl}\, \mathbf{V}(\text{Q}) \cdot \mathbf{n} = \text{curl}\, \mathbf{V}(\text{P}) \cdot \mathbf{n}.\end{aligned}$$

Thus,[1]

$$\text{curl}\, \mathbf{V}(\text{P}) \cdot \mathbf{n} = \underset{\rho\to 0}{\text{limit}}\, \frac{1}{A(S_\rho)} \int_{\partial S_\rho} \mathbf{V} \cdot d\mathbf{s}. \tag{5}$$

Let us pause to consider the physical meaning of $\int_C \mathbf{V} \cdot d\mathbf{s}$ when $\mathbf{V}$ is the velocity field of a fluid. Suppose, for example, that $\mathbf{V}$ points in the direction tangent to the oriented curve C (Figure 8.2.6). Then clearly $\int_C \mathbf{V} \cdot d\mathbf{s} > 0$, and particles on C tend to rotate counterclockwise. If $\mathbf{V}$ is pointing in the opposite direction, then $\int_C \mathbf{V} \cdot d\mathbf{s} < 0$ and particles tend to rotate clockwise. If $\mathbf{V}$ is perpendicular to C, then particles don't rotate on C at all and $\int_C \mathbf{V} \cdot d\mathbf{s} = 0$. In general, $\int_C \mathbf{V} \cdot d\mathbf{s}$, being the integral of the tangential component of $\mathbf{V}$, represents the net amount of turning of the fluid in a counterclockwise direction around C. One therefore refers to $\int_C \mathbf{V} \cdot d\mathbf{s}$ as the ***circulation*** of $\mathbf{V}$ around C (see Figure 8.2.7).

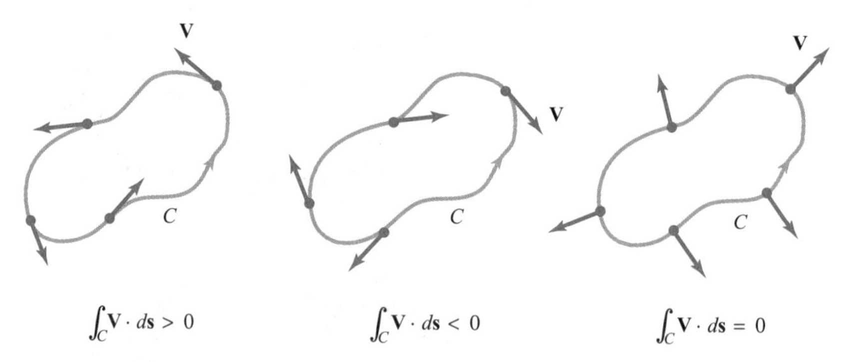

Figure 8.2.6 The intuitive meaning of the possible signs of $\int_C \mathbf{V} \cdot d\mathbf{s}$.

[1] Some informal texts adopt equation (5) as the *definition* of the curl, and use it to "prove" Stokes' theorem. However, this raises the danger of circular reasoning, for to show that equation (5) really defines a vector "curl $\mathbf{V}(\text{P})$" requires Stokes' theorem, or some similar argument.

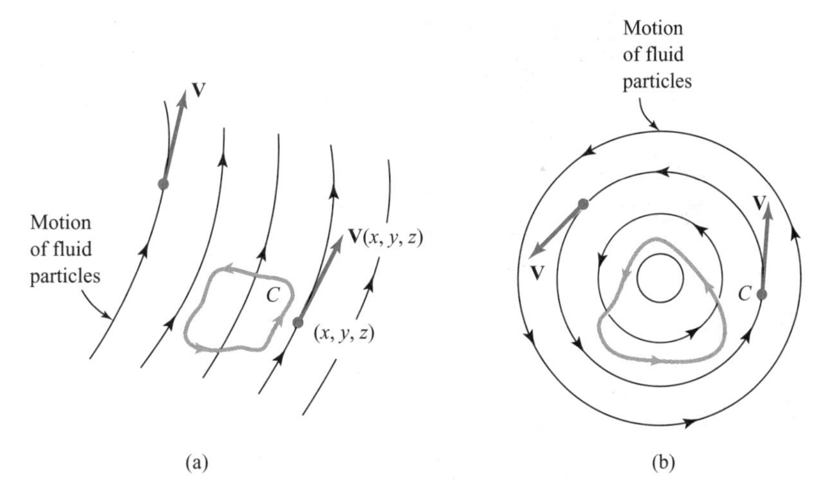

Figure 8.2.7 Circulation of a vector field (velocity field of a fluid): (a) Circulation about C is zero; (b) nonzero circulation about C ("whirlpool").

These results allow us to see just what curl $\mathbf{V}$ means for the motion of a fluid. The circulation $\int_{\partial S_\rho} \mathbf{V} \cdot d\mathbf{s}$ is the net velocity of the fluid around ∂S_ρ, so that $(\text{curl } \mathbf{V}) \cdot \mathbf{n}$ represents the turning or rotating effect of the fluid around the axis $\mathbf{n}$.

Circulation and Curl The dot product of curl $\mathbf{V}(\mathrm{P})$ with a unit vector $\mathbf{n}$, namely, curl $\mathbf{V}(\mathrm{P}) \cdot \mathbf{n}$, equals the circulation of $\mathbf{V}$ per unit area at P on a surface perpendicular to $\mathbf{n}$.

Observe that the magnitude of curl $\mathbf{V}(\mathrm{P}) \cdot \mathbf{n}$ is maximized when $\mathbf{n} = \text{curl } \mathbf{V}/\|\text{curl } \mathbf{V}\|$ (evaluated at P). Therefore, the rotating effect at P is greatest about the axis that is parallel to curl $\mathbf{V}/\|\text{curl } \mathbf{V}\|$. Thus, curl $\mathbf{V}$ is aptly called the ***vorticity vector***.

We can use these ideas to compute the curl in cylindrical coordinates.

EXAMPLE 4 Let the unit vectors $\mathbf{e}_r$, $\mathbf{e}_\theta$, $\mathbf{e}_z$ associated to cylindrical coordinates be as shown in Figure 8.2.8. Let $\mathbf{F} = F_r \mathbf{e}_r + F_\theta \mathbf{e}_\theta + F_z \mathbf{e}_z$. (The subscripts here denote components of $\mathbf{F}$, not partial derivatives.) Find a formula for the $\mathbf{e}_r$ component of $\nabla \times \mathbf{F}$ in cylindrical coordinates.

SOLUTION Let S be the surface shown in Figure 8.2.9.

The area of S is $r\,d\theta\,dz$ and the unit normal is $\mathbf{e}_r$. The integral of $\mathbf{F}$ around the edges of S is approximately

$$[F_\theta(r, \theta, z) - F_\theta(r, \theta, z + dz)]r\,d\theta + [F_z(r, \theta + d\theta, z) - F_z(r, \theta, z)]\,dz$$
$$\approx -\frac{\partial F_\theta}{\partial z}\,dz\,r\,d\theta + \frac{\partial F_z}{\partial \theta}\,d\theta\,dz.$$

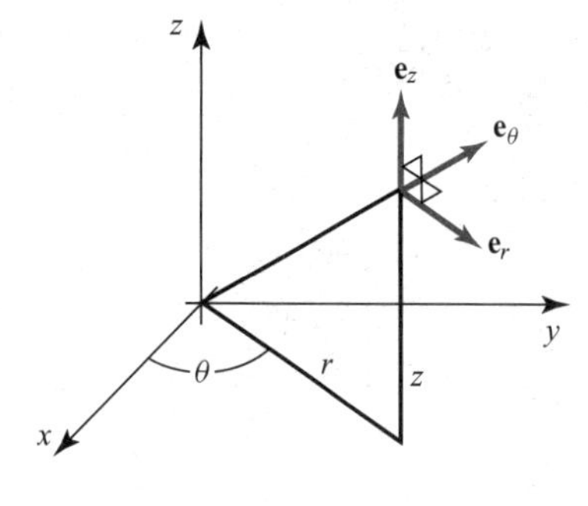

Figure 8.2.8 Orthonormal vectors $\mathbf{e}_r$, $\mathbf{e}_\theta$, and $\mathbf{e}_z$ associated with cylindrical coordinates. The vector $\mathbf{e}_r$ is parallel to the line labeled r.

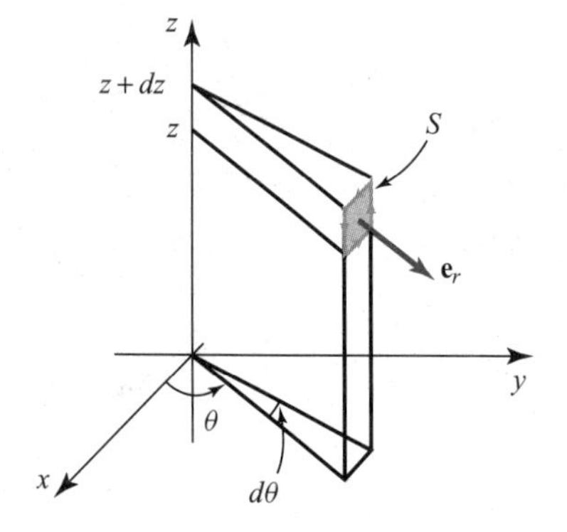

Figure 8.2.9 A surface element in cylindrical coordinates.

Thus, the circulation per unit area is this expression divided by $r\,d\theta dz$, namely,

$$\frac{1}{r}\frac{\partial F_z}{\partial \theta} - \frac{\partial F_\theta}{\partial z}.$$

According to the previous box, this must be the $\mathbf{e}_r$ component of the curl. ▲

Gradient, Divergence, and Curl in Cylindrical and Spherical Coordinates

By similar arguments to Example 4, one finds that the curl in cylindrical coordinates is given by

$$\nabla \times \mathbf{F} = \frac{1}{r}\begin{vmatrix} \mathbf{e}_r & r\mathbf{e}_\theta & \mathbf{e}_z \\ \dfrac{\partial}{\partial r} & \dfrac{\partial}{\partial \theta} & \dfrac{\partial}{\partial z} \\ F_r & rF_\theta & F_z \end{vmatrix}.$$

We can find other important vector quantities expressed in different coordinate systems. For example, the chain rule shows that the gradient in cylindrical coordinates is

$$\nabla f = \frac{\partial f}{\partial r}\,\mathbf{e}_r + \frac{1}{r}\frac{\partial f}{\partial \theta}\,\mathbf{e}_\theta + \frac{\partial f}{\partial z}\,\mathbf{e}_z,$$

and in Section 8.4 we will establish related techniques that give the following formula for the divergence in cylindrical coordinates:

$$\nabla \cdot \mathbf{F} = \frac{1}{r}\left[\frac{\partial}{\partial r}(rF_r) + \frac{\partial F_\theta}{\partial \theta} + \frac{\partial}{\partial z}(rF_z)\right].$$

Corresponding formulas for gradient, divergence, and curl in spherical coordinates are

$$\nabla f = \frac{\partial f}{\partial \rho}\mathbf{e}_\rho + \frac{1}{\rho}\frac{\partial f}{\partial \phi}\mathbf{e}_\phi + \frac{1}{\rho \sin\phi}\frac{\partial f}{\partial \theta}\mathbf{e}_\theta$$

$$\nabla \cdot \mathbf{F} = \frac{1}{\rho^2}\frac{\partial}{\partial \rho}(\rho^2 F_\rho) + \frac{1}{\rho \sin\phi}\frac{\partial}{\partial \phi}(\sin\phi F_\phi) + \frac{1}{\rho \sin\phi}\frac{\partial F_\theta}{\partial \theta}$$

and

$$\nabla \times \mathbf{F} = \left[\frac{1}{\rho \sin\phi}\frac{\partial}{\partial \phi}(\sin\phi F_\theta) - \frac{1}{\rho \sin\phi}\frac{\partial F_\phi}{\partial \theta}\right]\mathbf{e}_\rho$$
$$+ \left[\frac{1}{\rho \sin\phi}\frac{\partial F_\rho}{\partial \theta} - \frac{1}{\rho}\frac{\partial}{\partial \rho}(\rho F_\theta)\right]\mathbf{e}_\phi + \left[\frac{1}{\rho}\frac{\partial}{\partial \rho}(\rho F_\phi) - \frac{1}{\rho}\frac{\partial F_\rho}{\partial \phi}\right]\mathbf{e}_\theta$$

where $\mathbf{e}_\rho$, $\mathbf{e}_\phi$, $\mathbf{e}_\theta$ are as shown in Figure 8.2.10 and where $\mathbf{F} = F_\rho \mathbf{e}_\rho + F_\phi \mathbf{e}_\phi + F_\theta \mathbf{e}_\theta$.

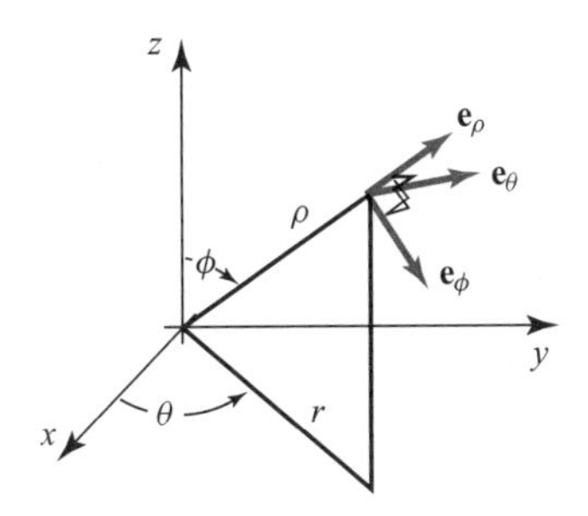

Figure 8.2.10 Orthonormal vectors $\mathbf{e}_\rho$, $\mathbf{e}_\phi$, and $\mathbf{e}_\theta$ associated with spherical coordinates.

Faraday's Law

Vector calculus plays an essential role in the theory of electromagnetism. The next example shows how Stokes' theorem applies.

EXAMPLE 5 Let **E** and **H** be time-dependent electric and magnetic fields, respectively, in space. Let S be a surface with boundary C. We define

$$\int_C \mathbf{E} \cdot d\mathbf{s} = \text{voltage around } C,$$

$$\iint_S \mathbf{H} \cdot d\mathbf{S} = \text{magnetic flux across } S.$$

Faraday's law (see Figure 8.2.11) states that the *voltage around C equals the negative rate of change of magnetic flux through S.* Show that Faraday's law follows from the following differential equation (one of the Maxwell equations):

$$\nabla \times \mathbf{E} = -\frac{\partial \mathbf{H}}{\partial t}.$$

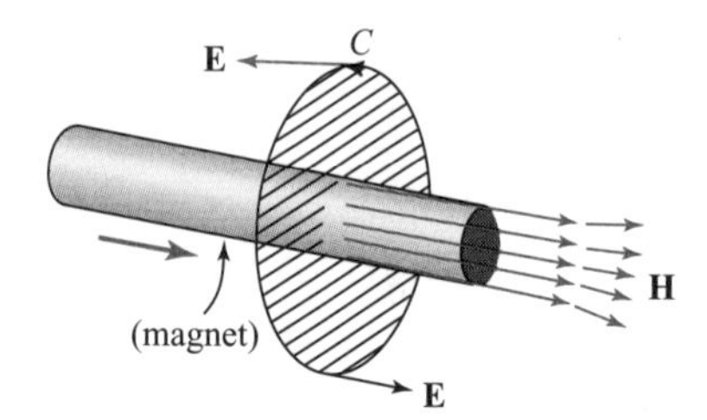

Figure 8.2.11 Faraday's law.

SOLUTION Assume that $-\partial \mathbf{H}/\partial t = \nabla \times \mathbf{E}$ holds. By Stokes' theorem,

$$\int_C \mathbf{E} \cdot d\mathbf{s} = \iint_S (\nabla \times \mathbf{E}) \cdot d\mathbf{S}.$$

Assuming that we can move $\partial/\partial t$ under the integral sign, we get

$$-\frac{\partial}{\partial t} \iint_S \mathbf{H} \cdot d\mathbf{S} = \iint_S -\frac{\partial \mathbf{H}}{\partial t} \cdot d\mathbf{S} = \iint_S (\nabla \times \mathbf{E}) \cdot d\mathbf{S} = \int_C \mathbf{E} \cdot d\mathbf{s}$$

and so

$$\int_C \mathbf{E} \cdot d\mathbf{s} = -\frac{\partial}{\partial t} \iint_S \mathbf{H} \cdot d\mathbf{S},$$

which is Faraday's law. ▲

Supplement to Section 8.2: Stokes' Theorem, Astronauts, and Falling Cats

Falling Cats

Have you ever wondered how a falling cat can right itself? Released from a resting position with its feet above its head, the cat is able to execute a 180° reorientation and land safely on its feet. This well-known phenomenon has fascinated people for many years—especially in cities like New York, where cats have been known to survive falls of 8 to 30 stories!

There have been many incorrect explanations as to how cats are able to right themselves, including the idea that it has to do with how the cat twirls its tail. This cannot be right because Manx cats, which have no tails, can also perform this feat!

Figure 8.2.12 The falling cat rights itself by wriggling its body parts.

One observes, as in Figure 8.2.12, that the cat achieves this net change in orientation by wriggling, to create *changes in its internal shape or configuration.* On the surface, this provides a seeming contradiction; because the cat is dropped from a resting position, it has zero angular momentum at the beginning of the fall and hence, according to a basic law of physics called *conservation of angular momentum,* the cat has zero angular momentum throughout the duration of its fall.[2] Amazingly, the cat has effectively changed its angular position while maintaining zero angular momentum!

The exact process by which this occurs is subtle; intuitive reasoning can lead one astray and, as we have indicated, many false explanations have been offered throughout the history of trying to solve this mystery.[3] Recently, new and interesting insights have been discovered using geometric methods that, in fact, are related to curvature (see Section 7.7).[4]

The way that curvature and geometry are related to the falling cat phenomenon is not easy to explain in full detail, but we can explain a similar phenomenon that is easy to understand. One of the points to emphasize is that *Stokes' theorem is the key to proving all of the relevant theorems.*

Reorienting Astronauts

Another example to help visualize this effect is to consider astronauts who wish to reorient themselves in a free-space environment. As with the falling cat, this motion can again be achieved using internal gyrations, or *shape changes.* For instance, consider astronauts moving their arms much like the motion of arms stirring liquid in a large kettle. The arms are held out forward, to lie in a horizontal plane that goes through the shoulders, parallel to the floor; the hands are clasped together and remain in this horizontal plane during the circular stirring motion. At the point of maximum extension of the arms, the inertia of the body about a vertical axis is also at a maximum. Conservation of angular momentum requires that the body rotate in an opposite and proportional manner to the motion of the arms. As the arms rotate around and are brought in, however, the inertia of the body is reduced. The motion of the body in reaction is therefore also reduced. Thus, in one complete cycle of arm movement, the body undergoes a net rotation opposite the direction of arm motion. When the desired

---

[2] We saw an instance of the law of conservation of angular momentum in Section 4.1, Exercise 20.

[3] Another favorite fallacious argument, showing that a cat *cannot* turn itself over(!), is this: "Accept from physics that angular momentum is the moment of inertia times angular velocity [moments of inertia are discussed in Section 6.3]. But the angular momentum of the cat is zero, so the angular velocity must also be zero. Because angular velocity is the rate of change of the angular position, the angular position is constant. Thus, the cat cannot turn itself over." What is wrong? This argument ignores the fact that the cat changes its *shape,* and hence its moment of inertia, during the fall.

[4] See T. R. Kane and M. Scher, "A Dynamical Explanation of the Falling Cat Phenomenon," *Int. J. Solids Struct.,* 5 (1969): 663–670. See also R. Montgomery, "Isoholonomic Problems and Some Applications," *Commun. Math. Phys.,* 128 (1990): 565–592; R. Montgomery, "How Much Does a Rigid Body Rotate? A Berry's Phase from the 18th Century," *Am. J. Pys.,* 59 (1991b): 394–398. See also J. E. Marsden and J. Ostrowski, "Symmetries in Motion: Geometric Foundations of Motion Control," *Nonlinear Science Today* (1998), http://link.springer-ny.com; R. Batterman, "Falling Cats, Parallel Parking, and Polarized Light," *Philos. Soc. Arch.* (2002); http://philsci-archive.pitt.edu/documents/disk0/00/00/05/83, http://www.its.caltech.edu/~mleok/falling_cats.htm, and references therein.

orientation is achieved, the astronaut needs merely to stop the arm motion in order to come to rest. One often refers to the extra motion that is achieved as ***geometric phase***.

Link with Non-Euclidean Geometry

The theory of geometric phases also shows up in an interesting way in non-Euclidean geometry—as in the geometry of triangles drawn on a sphere. A simple way to explain this link is as follows. Hold your hand at arm's length, but allow rotation in your shoulder joints. Move your hand along three great circles, forming a triangle on the sphere; during the motion along each arc, always keep your thumb *parallel*; that is, it should move in such a way that it forms a *fixed* angle with the direction of motion along each arc and does not rotate when switching arcs. After completing the circuit around the triangle, your thumb will return rotated through an angle relative to its starting position (see Figure 8.2.13). Can you see in Figure 8.2.13 that the angle of rotation is 90° (or $\pi/2$ radians) and that this is what happens when you do the thumb experiment yourself?

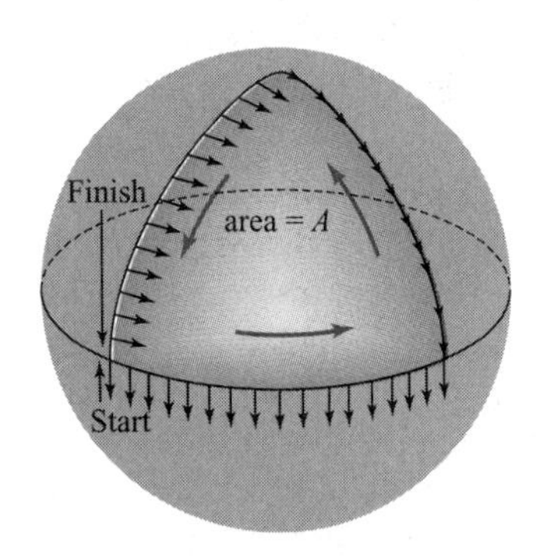

Figure 8.2.13 A parallel movement of your thumb around a spherical triangle produces a phase shift.

For general spherical triangles, this angle (in radians) is given by $\Theta = \Delta - \pi$, where Δ is the sum of the angles of the triangle. The fact that Θ is strictly positive (!) is one of the basic truths of non-Euclidean geometry—the sum of the angles of a right triangle on a sphere is greater than π! This angle is also related to the *area* A enclosed by the triangle through the relation $\Theta = A/r^2$, where r is the radius of the sphere. The rotational shift of the thumb during the course of its cyclic journey around the spherical triangle is directly related to the curvature of the sphere and to the area enclosed by the path that is traced out. Notice first that for a spherical triangle that is 1/8 of the sphere, $A = 4\pi r^2/8 = \pi r^2/2$. Thus, $A/r^2 = \pi/2$. Notice also that when $r \to \infty$, the sphere becomes flatter and thus approaches a Euclidean plane, in which case $\Theta = 0$.

The cyclic journey of the thumb around the closed path is analogous to the cyclic internal motions made by the cat during its fall; the 90° shift in the direction of the thumb after one trip around is analogous to the 180° reorientation of the cat. A deeper look at the underlying mathematics shows that, in fact, they are both instances of the same phenomenon (called *holonomy*)—and Stokes' theorem is the key to understanding it.

EXERCISES

1. Redo Exercise 5 of Section 7.6 using Stokes' theorem.

2. Redo Exercise 6 of Section 7.6 using Stokes' theorem.

3. Verify Stokes' theorem for the upper hemisphere $z = \sqrt{1 - x^2 - y^2}$, $z \geq 0$, and the radial vector field $\mathbf{F}(x, y, z) = x\mathbf{i} + y\mathbf{j} + z\mathbf{k}$.

4. Let S be a surface with boundary ∂S, and suppose $\mathbf{E}$ is an electric field that is perpendicular to ∂S. Show that the induced magnetic flux across S is constant in time. (HINT: Use Faraday's law.)

5. Let S be the capped cylindrical surface shown in Figure 8.2.14. S is the union of two surfaces, S_1 and S_2, where S_1 is the set of (x, y, z) with $x^2 + y^2 = 1$, $0 \leq z \leq 1$, and S_2 is the set of (x, y, z) with $x^2 + y^2 + (z - 1)^2 = 1$, $z \geq 1$. Set $\mathbf{F}(x, y, z) = (zx + z^2y + x)\mathbf{i} + (z^3yx + y)\mathbf{j} + z^4x^2\mathbf{k}$. Compute $\iint_S (\nabla \times \mathbf{F}) \cdot d\mathbf{S}$. (HINT: Stokes' theorem holds for this surface.)

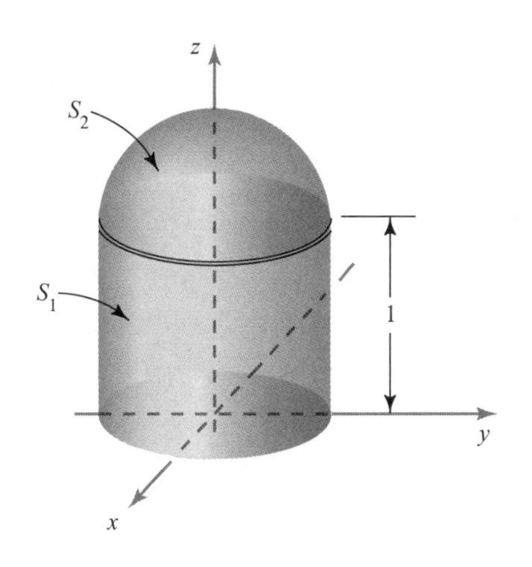

Figure 8.2.14 The capped cylinder is the union of S_1 and S_2.

6. Let $\mathbf{c}$ consist of straight lines joining $(1, 0, 0)$, $(0, 1, 0)$, and $(0, 0, 1)$ and let S be the triangle with these vertices. Verify Stokes' theorem directly with $\mathbf{F} = yz\mathbf{i} + xz\mathbf{j} + xy\mathbf{k}$.

7. Evaluate the integral $\iint_S (\nabla \times \mathbf{F}) \cdot d\mathbf{S}$, where S is the portion of the surface of a sphere defined by $x^2 + y^2 + z^2 = 1$ and $x + y + z \geq 1$, and where $\mathbf{F} = \mathbf{r} \times (\mathbf{i} + \mathbf{j} + \mathbf{k})$, $\mathbf{r} = x\mathbf{i} + y\mathbf{j} + z\mathbf{k}$.

8. Show that the calculation in Exercise 7 can be simplified by observing that $\int_{\partial S} \mathbf{F} \cdot d\mathbf{r} = \int_{\partial \Sigma} \mathbf{F} \cdot d\mathbf{r}$ for any other surface Σ. By picking Σ appropriately, $\iint_\Sigma (\nabla \times \mathbf{F}) \cdot d\mathbf{S}$ may be easy to compute. Show that this is the case if Σ is taken to be the portion of the plane $x + y + z = 1$ inside the circle ∂S.

9. Calculate the surface integral $\iint_S (\nabla \times \mathbf{F}) \cdot d\mathbf{S}$ where S is the hemisphere $x^2+y^2+z^2 = 1$, $x \geq 0$ and $\mathbf{F} = x^3\mathbf{i} - y^3\mathbf{j}$.

10. Find $\iint_S (\nabla \times \mathbf{F}) \cdot d\mathbf{S}$, where S is the ellipsoid $x^2 + y^2 + 2z^2 = 10$ and $\mathbf{F}$ is the vector field $\mathbf{F} = (\sin xy)\mathbf{i} + e^x\mathbf{j} - yz\mathbf{k}$.

11. Let $\mathbf{F} = y\mathbf{i} - x\mathbf{j} + zx^3y^2\mathbf{k}$. Evaluate $\iint_S (\nabla \times \mathbf{F}) \cdot \mathbf{n}\, dA$, where S is the surface defined by $x^2 + y^2 + z^2 = 1, z \leq 0$.

12. A hot-air balloon has the truncated spherical shape shown in Figure 8.2.15. The hot gases escape through the porous envelope with a velocity vector field

$$\mathbf{V}(x, y, z) = \nabla \times \mathbf{\Phi}(x, y, z) \qquad \text{where} \qquad \mathbf{\Phi}(x, y, z) = -y\mathbf{i} + x\mathbf{j}.$$

If $R = 5$, compute the volume flow rate of the gases through the surface.

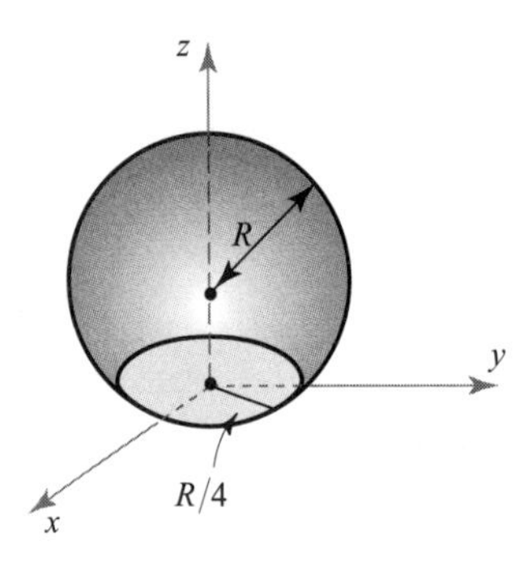

Figure 8.2.15 A hot-air balloon.

13. Prove that Faraday's law implies $\nabla \times \mathbf{E} = -\partial \mathbf{H}/\partial t$.

14. Let S be a surface and let $\mathbf{F}$ be perpendicular to the tangent to the boundary of S. Show that

$$\iint_S (\nabla \times \mathbf{F}) \cdot d\mathbf{S} = 0.$$

What does this mean physically if $\mathbf{F}$ is an electric field?

15. Consider two surfaces S_1, S_2 with the same boundary ∂S. Describe with sketches how S_1 and S_2 must be oriented to ensure that

$$\iint_{S_1} (\nabla \times \mathbf{F}) \cdot d\mathbf{S} = \iint_{S_2} (\nabla \times \mathbf{F}) \cdot d\mathbf{S}.$$

16. For a surface S and a fixed vector $\mathbf{v}$, prove that

$$2 \iint_S \mathbf{v} \cdot \mathbf{n}\, dS = \int_{\partial S} (\mathbf{v} \times \mathbf{r}) \cdot d\mathbf{s},$$

where $\mathbf{r}(x, y, z) = (x, y, z)$.

17. Argue informally that if S is a closed surface, then

$$\iint_S (\nabla \times \mathbf{F}) \cdot d\mathbf{S} = 0$$

(see Exercise 15). (A *closed surface* is one that forms the boundary of a region in space; thus, for example, a sphere is a closed surface.)

18. If C is a closed curve that is the boundary of a surface S, and f and g are C^2 functions, show that

(a) $\displaystyle\int_C f\nabla g \cdot d\mathbf{s} = \iint_S (\nabla f \times \nabla g) \cdot d\mathbf{S}$

(b) $\displaystyle\int_C (f\nabla g + g\nabla f) \cdot d\mathbf{s} = 0$

19. (a) If C is a closed curve that is the boundary of a surface S, and $\mathbf{v}$ is a constant vector, show that

$$\int_C \mathbf{v} \cdot d\mathbf{s} = 0.$$

(b) Show that this is true even if C is not the boundary of a surface S.

20. Show that $\boldsymbol{\Phi}$: $D \to \mathbb{R}^3$, $D = [0, \pi] \times [0, 2\pi]$, $\boldsymbol{\Phi}(\phi, \theta) = (\cos\theta \sin\phi, \sin\theta \sin\phi, \cos\phi)$, which parametrizes the unit sphere, takes the boundary of D to half of a great circle on S.

21. Verify Theorem 6 for the helicoid $\boldsymbol{\Phi}(r, \theta) = (r\cos\theta, r\sin\theta, \theta)$, $(r, \theta) \in [0, 1] \times [0, \pi/2]$, and the vector field $\mathbf{F}(x, y, z) = (z, x, y)$.

22. Prove Theorem 6.

23. Let $\mathbf{F} = x^2\mathbf{i} + (2xy + x)\mathbf{j} + z\mathbf{k}$. Let C be the circle $x^2 + y^2 = 1$ and S the disk $x^2 + y^2 \leq 1$ within the plane $z = 0$.

(a) Determine the flux of $\mathbf{F}$ out of S.
(b) Determine the circulation of $\mathbf{F}$ around C.
(c) Find the flux of $\nabla \times \mathbf{F}$. Verify Stokes' theorem directly in this case.

24. Let S be a surface with boundary ∂S, and suppose that $\mathbf{E}$ is an electric field that is perpendicular to ∂S. Use Faraday's law to show that the induced magnetic flux across S is constant in time.

25. Integrate $\nabla \times \mathbf{F}$, $\mathbf{F} = (3y, -xz, -yz^2)$ over the portion of the surface $2z = x^2 + y^2$ below the plane $z = 2$, both directly and by using Stokes' theorem.

26. ***Ampère's law*** states that if the electric current density is described by a vector field $\mathbf{J}$ and the induced magnetic field is $\mathbf{H}$, then the circulation of $\mathbf{H}$ around the boundary C of a surface S equals the integral of $\mathbf{J}$ over S (i.e., the total current crossing S). See Figure 8.2.16. Show that this is implied by the steady-state *Maxwell equation* $\nabla \times \mathbf{H} = \mathbf{J}$.

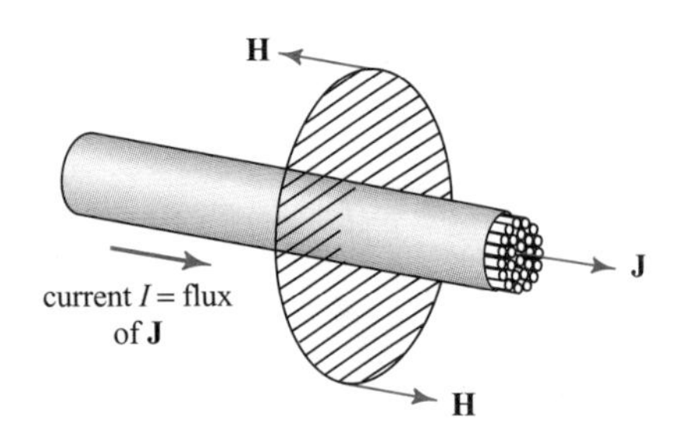

Figure 8.2.16 Ampère's law.

27. Faraday's law relates the line integral of the electric field around a loop C to the surface integral of the rate of change of the magnetic field over a surface S with boundary C. Regarding the equation $\nabla \times \mathbf{E} = -\partial \mathbf{H}/\partial t$ as the basic equation, Faraday's law is a consequence of Stokes' theorem, as we have seen in Example 4.

Suppose we are given electric and magnetic fields in space that satisfy $\nabla \times \mathbf{E} = -\partial \mathbf{H}/\partial t$. Suppose C is the boundary of the Möbius band shown in Figures 7.6.3 and 7.6.4. Because the Möbius band cannot be oriented, Stokes' theorem does not apply. What becomes of Faraday's law? What do you guess $\int_C \mathbf{E} \cdot d\mathbf{s}$ equals?

28. (a) If in spherical coordinates, we write

$$\mathbf{e}_r = \alpha \mathbf{i} + \beta \mathbf{j} + \gamma \mathbf{k}, \text{ find } \alpha, \beta, \text{ and } \gamma.$$

(b) Find similar formulas for $\mathbf{e}_\phi$ and $\mathbf{e}_\theta$.

8.3 Conservative Fields

We saw in Section 7.2 that for a gradient force field $\mathbf{F} = \nabla f$, line integrals of $\mathbf{F}$ were evaluated as follows:

$$\int_{\mathbf{c}} \mathbf{F} \cdot d\mathbf{s} = f(\mathbf{c}(b)) - f(\mathbf{c}(a)).$$

The value of the integral depends only on the endpoints $\mathbf{c}(b)$ and $\mathbf{c}(a)$ of the path. In other words, if we used another path with the same endpoints, we would still get the same answer. This leads us to say that the integral is *path-independent.*

Gradient fields are important in many physical problems. For example, if $V = -f$ represents a potential energy (gravitational, electrical, and so on), then $\mathbf{F}$ represents a force.[5] Consider the example of a particle of mass m in the field of the earth; in this case, one takes f to be GmM/r or $V = -GmM/r$, where G is the gravitational constant, M is the mass of the earth, and r is the distance from the center of the

[5] If the minus sign is used, then V is *decreasing* in the direction $\mathbf{F}$.

earth. The corresponding force is $\mathbf{F} = -(GmM/r^3)\mathbf{r} = -(GmM/r^2)\mathbf{n}$, where $\mathbf{n}$ is the unit radial vector. Note that $\mathbf{F}$ fails to be defined at the point $r = 0$.

When are Vector Fields Gradients?

We wish to characterize those vector fields that can be written as a gradient. Our task is simplified considerably by Stokes' theorem.

THEOREM 7: Conservative Fields Let $\mathbf{F}$ be a C^1 vector field defined on $\mathbb{R}^3$ except possibly for a finite number of points. The following conditions on $\mathbf{F}$ are all equivalent:

(i) For any oriented simple closed curve C, $\int_C \mathbf{F} \cdot d\mathbf{s} = 0$.

(ii) For any two oriented simple curves C_1 and C_2 that have the same endpoints,

$$\int_{C_1} \mathbf{F} \cdot d\mathbf{s} = \int_{C_2} \mathbf{F} \cdot d\mathbf{s}.$$

(iii) $\mathbf{F}$ is the gradient of some function f; that is, $\mathbf{F} = \nabla f$ (and if $\mathbf{F}$ has one or more exceptional points where it fails to be defined, f is also undefined there).

(iv) $\nabla \times \mathbf{F} = \mathbf{0}$.

A vector field satisfying one (and, hence, all) of the conditions (i)–(iv) is called a ***conservative vector field***.[6]

PROOF We shall establish the following chain of implications, which will prove the theorem:

$$\text{(i)} \Rightarrow \text{(ii)} \Rightarrow \text{(iii)} \Rightarrow \text{(iv)} \Rightarrow \text{(i)}.$$

First we show that condition (i) implies condition (ii). Suppose $\mathbf{c}_1$ and $\mathbf{c}_2$ are parametrizations representing C_1 and C_2, with the same endpoints. Construct the closed curve $\mathbf{c}$ obtained by first traversing $\mathbf{c}_1$ and then $-\mathbf{c}_2$ (Figure 8.3.1), or, symbolically, the curve $\mathbf{c} = \mathbf{c}_1 - \mathbf{c}_2$. Assuming $\mathbf{c}$ is simple, condition (i) gives

$$\int_{\mathbf{c}} \mathbf{F} \cdot d\mathbf{s} = \int_{\mathbf{c}_1} \mathbf{F} \cdot d\mathbf{s} - \int_{\mathbf{c}_2} \mathbf{F} \cdot d\mathbf{s} = 0,$$

and so condition (ii) holds. (If $\mathbf{c}$ is not simple, an additional argument, omitted here, is needed.)

---

[6] In the plane $\mathbb{R}^2$, exceptional points are not allowed (see Exercise 12). Theorem 7 can be proved in the same way if $\mathbf{F}$ is defined and is of class C^1 only on an open convex set in $\mathbb{R}^2$ or $\mathbb{R}^3$. (A set D is convex if $P, Q \in D$ implies the line joining P and Q also belongs to D.)

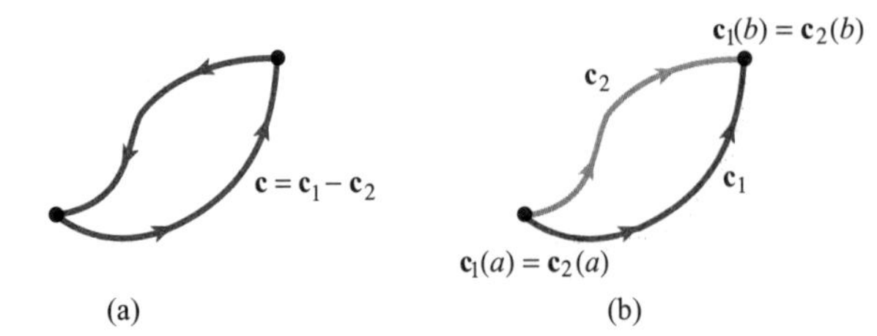

Figure 8.3.1 Constructing (a) an oriented simple closed curve $\mathbf{c}_1 - \mathbf{c}_2$ from (b) two oriented simple curves.

Next we prove that condition (ii) implies condition (iii). Let C be any oriented simple curve joining a point such as $(0, 0, 0)$ to (x, y, z), and suppose C is represented by the parametrization $\mathbf{c}$ [if $(0, 0, 0)$ is the exceptional point of $\mathbf{F}$, we can choose a different starting point for $\mathbf{c}$ without affecting the argument]. Define $f(x, y, z)$ to be $\int_{\mathbf{c}} \mathbf{F} \cdot d\mathbf{s}$. By hypothesis (ii), $f(x, y, z)$ is independent of C. We shall show that $\mathbf{F} = \operatorname{grad} f$. Indeed, choose $\mathbf{c}$ to be the path shown in Figure 8.3.2, so that

$$f(x, y, z) = \int_0^x F_1(t, 0, 0)\, dt + \int_0^y F_2(x, t, 0)\, dt + \int_0^z F_3(x, y, t)\, dt,$$

where $\mathbf{F} = (F_1, F_2, F_3)$.

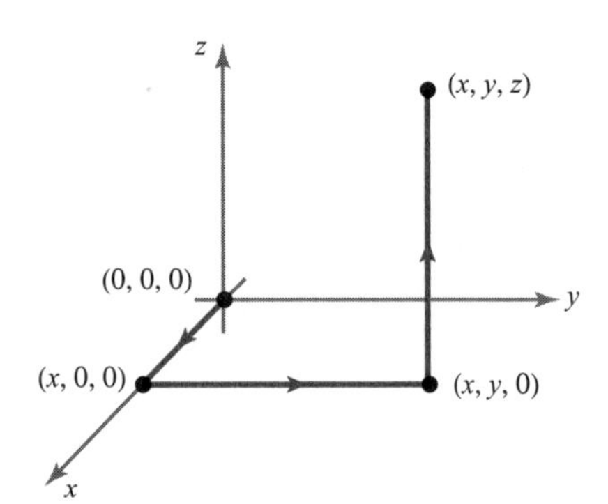

Figure 8.3.2 A path joining $(0, 0, 0)$ to (x, y, z).

It follows from the fundamental theorem of calculus that $\partial f/\partial z = F_3$. We can repeat this processing using two other paths from $(0, 0, 0)$ to (x, y, z) [for example, by drawing the lines from $(0, 0, 0)$ to $(0, y, 0)$ to $(x, y, 0)$ to (x, y, z)], and we can similarly show that $\partial f/\partial x = F_1$ and $\partial f/\partial y = F_2$ (see Exercise 22). Thus, $\nabla f = \mathbf{F}$.

Third, condition (iii) implies condition (iv), because, as proved in Section 4.4,

$$\nabla \times \nabla f = \mathbf{0}.$$

Finally, let $\mathbf{c}$ represent a closed curve C and let S be any surface whose boundary is $\mathbf{c}$ (if $\mathbf{F}$ has exceptional points, choose S to avoid them). Figure 8.3.3 indicates that we can probably always find such a surface; however, a formal proof of this would require the development of more sophisticated mathematical ideas than we can present here.

By Stokes' theorem,

$$\int_C \mathbf{F} \cdot d\mathbf{s} = \int_{\mathbf{c}} \mathbf{F} \cdot d\mathbf{s} = \iint_S (\nabla \times \mathbf{F}) \cdot \mathbf{n}\, dS = \iint_S (\text{curl}\ \mathbf{F}) \cdot \mathbf{n}\, dS.$$

Because $\nabla \times \mathbf{F} = \mathbf{0}$, this integral vanishes, so that condition (iv) $\Rightarrow$ condition (i). ■

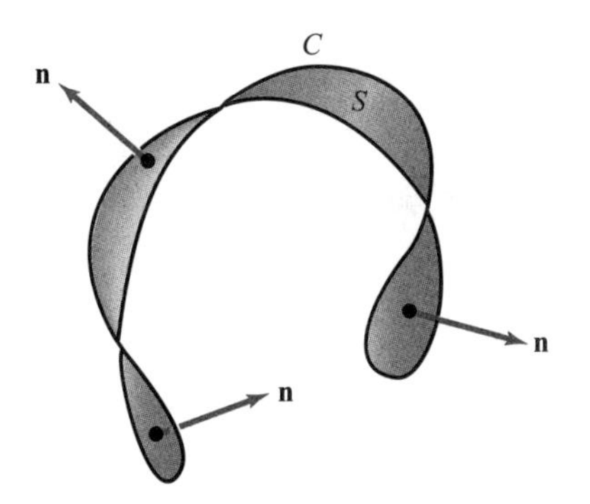

Figure 8.3.3 A surface S spanning a curve C.

Physical Interpretations of $\int_C \mathbf{F} \cdot d\mathbf{s}$

We have already seen that one interpretation of the line integral is as the work done by $\mathbf{F}$ in moving a particle along C. A second interpretation is the notion of circulation, which we encountered at the end of the last section. Recall that in this case, we think of $\mathbf{F}$ as the velocity field of a fluid; that is, to each point P in space, $\mathbf{F}$ assigns the velocity vector of the fluid at P (in the last section, $\mathbf{F}$ was designated $\mathbf{V}$). Take C to be a closed curve, and let $\Delta\mathbf{s}$ be a small directed chord of C. Then $\mathbf{F} \cdot \Delta\mathbf{s}$ is approximately the tangential component of $\mathbf{F}$ times $\|\Delta\mathbf{s}\|$. The circulation $\int_C \mathbf{F} \cdot d\mathbf{s}$ is the net tangential component around C. A small paddle wheel placed in the fluid would rotate if it is centered at a point where $\mathbf{F}$ vanishes and if the circulation of the fluid is nonzero, or $\int_C \mathbf{F} \cdot d\mathbf{s} \neq 0$ for small loops C (see Figure 8.3.4).

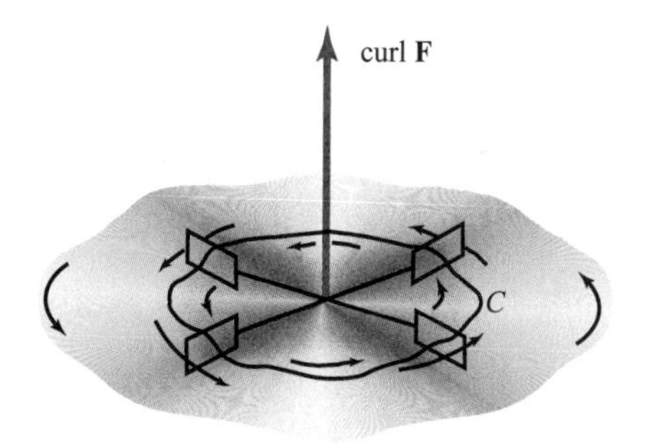

Figure 8.3.4 $\int_C \mathbf{F} \cdot d\mathbf{s} \neq 0$ implies that a paddle wheel in a fluid with velocity field $\mathbf{F}$ will rotate around its axis.

There is a similar interpretation in electromagnetic theory: If $\mathbf{F}$ represents an electric field, then a current will flow around a loop C if $\int_C \mathbf{F} \cdot d\mathbf{s} \neq 0$.

By Theorem 7, a field $\mathbf{F}$ has no circulation if and only if curl $\mathbf{F} = \nabla \times \mathbf{F} = \mathbf{0}$. Hence, a vector field $\mathbf{F}$ with curl $\mathbf{F} = \mathbf{0}$ is called ***irrotational***. We have therefore proved that *a vector field in* $\mathbb{R}^3$ *is irrotational if and only if it is a gradient field for some function, that is, if and only if* $\mathbf{F} = \nabla f$. The function f is called a ***potential*** for $\mathbf{F}$.

EXAMPLE 1 Consider the vector field $\mathbf{F}$ on $\mathbb{R}^3$ defined by

$$\mathbf{F}(x, y, z) = y\mathbf{i} + (z\cos yz + x)\mathbf{j} + (y\cos yz)\mathbf{k}.$$

Show that $\mathbf{F}$ is irrotational and find a scalar potential for $\mathbf{F}$.

SOLUTION We compute $\nabla \times \mathbf{F}$:

$$\begin{aligned}\nabla \times \mathbf{F} &= \begin{vmatrix} \mathbf{i} & \mathbf{j} & \mathbf{k} \\ \dfrac{\partial}{\partial x} & \dfrac{\partial}{\partial y} & \dfrac{\partial}{\partial z} \\ y & x + z\cos yz & y\cos yz \end{vmatrix} \\ &= (\cos yz - yz\sin yz - \cos yz + yz\sin yz)\mathbf{i} + (0 - 0)\mathbf{j} + (1 - 1)\mathbf{k} \\ &= 0\mathbf{i} + 0\mathbf{j} + 0\mathbf{k} = \mathbf{0},\end{aligned}$$

so $\mathbf{F}$ is irrotational. Thus, a potential exists by Theorem 7. We can find it in several ways.

Method 1. By the technique used to prove that condition (ii) implies condition (iii) in Theorem 7, we can set

$$\begin{aligned}f(x, y, z) &= \int_0^x F_1(t, 0, 0)\, dt + \int_0^y F_2(x, t, 0)\, dt + \int_0^z F_3(x, y, t)\, dt \\ &= \int_0^x 0\, dt + \int_0^y x\, dt + \int_0^z y\cos yt\, dt \\ &= 0 + xy + \sin yz = xy + \sin yz.\end{aligned}$$

One easily verifies that $\nabla f = \mathbf{F}$, as required:

$$\nabla f = \frac{\partial f}{\partial x}\mathbf{i} + \frac{\partial f}{\partial y}\mathbf{j} + \frac{\partial f}{\partial z}\mathbf{k} = y\mathbf{i} + (x + z\cos yz)\mathbf{j} + (y\cos yz)\mathbf{k}.$$

Method 2. Because we know that f exists, we know that we can solve the system of equations

$$\frac{\partial f}{\partial x} = y, \qquad \frac{\partial f}{\partial y} = x + z\cos yz, \qquad \frac{\partial f}{\partial z} = y\cos yz,$$

for $f(x, y, z)$. These are equivalent to the simultaneous equations

(a) $f(x, y, z) = xy + h_1(y, z)$

(b) $f(x, y, z) = \sin yz + xy + h_2(x, z)$

(c) $f(x, y, z) = \sin yz + h_3(x, y)$

for functions h_1, h_2, h_3 independent of x, y, and z (respectively). When $h_1(y, z) = \sin yz$, $h_2(x, z) = 0$, and $h_3(x, y) = xy$, the three equations agree and so yield a potential for **F**. However, we have only guessed at the values of h_1, h_2, and h_3. To derive the formula for f more systematically, we note that because $f(x, y, z) = xy + h_1(y, z)$ and $\partial f/\partial z = y \cos yz$, we find that

$$\frac{\partial h_1(y, z)}{\partial z} = y \cos yz$$

or

$$h_1(y, z) = \int y \cos yz \, dz + g(y) = \sin yz + g(y).$$

Therefore, substituting this back into equation (a), we get

$$f(x, y, z) = xy + \sin yz + g(y);$$

but by equation (b),

$$g(y) = h_2(x, z).$$

Because the right side of this equation is a function of x and z and the left side is a function of y alone, we conclude that they must equal some constant C. Thus,

$$f(x, y, z) = xy + \sin yz + C$$

and we have determined f up to a constant. ▲

EXAMPLE 2 A mass M at the origin in $\mathbb{R}^3$ exerts a force on a mass m located at $\mathbf{r} = (x, y, z)$ with magnitude GmM/r^2 and directed toward the origin. Here, G is the gravitational constant, which depends on the units of measurement, and $r = \|\mathbf{r}\| = \sqrt{x^2 + y^2 + z^2}$. If we remember that $-\mathbf{r}/r$ is a unit vector directed toward the origin, then we can write the force field as

$$\mathbf{F}(x, y, z) = -\frac{GmM\mathbf{r}}{r^3}.$$

Show that $\mathbf{F}$ is irrotational and find a scalar potential for $\mathbf{F}$. (Notice that $\mathbf{F}$ is not defined at the origin, but Theorem 7 still applies, because it allows an exceptional point.)

SOLUTION First let us verify that $\nabla \times \mathbf{F} = \mathbf{0}$. Referring to formula 10 in the table of vector identities in Section 4.4, we get

$$\nabla \times \mathbf{F} = -GmM\left[\nabla\left(\frac{1}{r^3}\right) \times \mathbf{r} + \frac{1}{r^3}\nabla \times \mathbf{r}\right].$$

But $\nabla(1/r^3) = -3\mathbf{r}/r^5$ (see Exercise 30, Section 4.4), and so the first term vanishes, because $\mathbf{r} \times \mathbf{r} = \mathbf{0}$. The second term vanishes, because

$$\nabla \times \mathbf{r} = \begin{vmatrix} \mathbf{i} & \mathbf{j} & \mathbf{k} \\ \dfrac{\partial}{\partial x} & \dfrac{\partial}{\partial y} & \dfrac{\partial}{\partial z} \\ x & y & z \end{vmatrix} = \left(\frac{\partial z}{\partial y} - \frac{\partial y}{\partial z}\right)\mathbf{i} + \left(\frac{\partial x}{\partial z} - \frac{\partial z}{\partial x}\right)\mathbf{j} + \left(\frac{\partial y}{\partial x} - \frac{\partial x}{\partial y}\right)\mathbf{k} = \mathbf{0}.$$

Hence, $\nabla \times \mathbf{F} = 0$ (for $\mathbf{r} \neq 0$).

If we recall the formula $\nabla(r^n) = nr^{n-2}\mathbf{r}$ (again, see Exercise 30, Section 4.4), then we can read off a scalar potential for $\mathbf{F}$ by inspection. We have $\mathbf{F} = -\nabla V$, where $V(x, y, z) = -GmM/r$ is called the ***gravitational potential energy***.

[We observe in passing that by Theorem 3 of Section 7.2, the work done by $\mathbf{F}$ in moving a particle of mass m from a point P_1 to a point P_2 is given by

$$V(\mathrm{P}_1) - V(\mathrm{P}_2) = GmM\left(\frac{1}{r_2} - \frac{1}{r_1}\right),$$

where r_1 is the radial distance of P_1 from the origin, with r_2 similarly defined.] ▲

The Planar Case

By the same proof, Theorem 7 is also true for C^1 vector fields $\mathbf{F}$ on $\mathbb{R}^2$. In this case, we require that $\mathbf{F}$ has *no* exceptional points; that is, $\mathbf{F}$ is smooth everywhere (see Exercise 12). Notice, however, that the conclusion *might* still hold even if there are exceptional points, an example being $(x\mathbf{i} + y\mathbf{j})/(x^2 + y^2)^{3/2}$. An example where the conclusion does *not* hold is $(-y\mathbf{i} + x\mathbf{j})/(x^2 + y^2)$, as shown in Exercise 12.

If $\mathbf{F} = P\mathbf{i} + Q\mathbf{j}$, then

$$\nabla \times \mathbf{F} = \left(\frac{\partial Q}{\partial x} - \frac{\partial P}{\partial y}\right)\mathbf{k}.$$

Sometimes $\partial Q/\partial x - \partial P/\partial y$ is called the ***scalar curl*** of $\mathbf{F}$. Therefore, the condition $\nabla \times \mathbf{F} = 0$ reduces to

$$\frac{\partial P}{\partial y} = \frac{\partial Q}{\partial x}.$$

Thus, we have:

COROLLARY 1 If $\mathbf{F}$ is a C^1 vector field on $\mathbb{R}^2$ of the form $P\mathbf{i} + Q\mathbf{j}$ that satisfies $\partial P/\partial y = \partial Q/\partial x$, then $\mathbf{F} = \nabla f$ for some f on $\mathbb{R}^2$.

We emphasize again that this corollary can be false if $\mathbf{F}$ fails to be of class C^1 at even a single point (an example is given in Exercise 12). In $\mathbb{R}^3$, however, as already noted, exceptions at single points are allowed (see Theorem 7).

EXAMPLE 3 (a) Determine whether the vector field

$$\mathbf{F} = e^{xy}\mathbf{i} + e^{x+y}\mathbf{j}$$

is a gradient field.

(b) Repeat part (a) for

$$\mathbf{F} = (2x \cos y)\mathbf{i} - (x^2 \sin y)\mathbf{j}.$$

SOLUTION (a) Here $P(x, y) = e^{xy}$ and $Q(x, y) = e^{x+y}$, and so we compute

$$\frac{\partial P}{\partial y} = xe^{xy}, \qquad \frac{\partial Q}{\partial x} = e^{x+y}.$$

These are not equal, and so $\mathbf{F}$ cannot have a potential function.

(b) In this case, we find

$$\frac{\partial P}{\partial y} = -2x \sin y = \frac{\partial Q}{\partial x},$$

and so $\mathbf{F}$ has a potential function f. To compute f we solve the equations

$$\frac{\partial f}{\partial x} = 2x \cos y, \qquad \frac{\partial f}{\partial y} = -x^2 \sin y.$$

Thus, $f(x, y) = x^2 \cos y + h_1(y)$ and $f(x, y) = x^2 \cos y + h_2(x)$. If h_1 and h_2 are the same constant, then both equations are satisfied, and so $f(x, y) = x^2 \cos y$ is a potential for $\mathbf{F}$. ▲

EXAMPLE 4 Let $\mathbf{c}$: $[1, 2] \to \mathbb{R}^2$ be given by $x = e^{t-1},\ \ y = \sin(\pi/t)$. Compute the integral

$$\int_{\mathbf{c}} \mathbf{F} \cdot d\mathbf{s} = \int_{\mathbf{c}} 2x \cos y \, dx - x^2 \sin y \, dy,$$

where $\mathbf{F} = (2x \cos y)\mathbf{i} - (x^2 \sin y)\mathbf{j}$.

SOLUTION The endpoints are $\mathbf{c}(1) = (1, 0)$ and $\mathbf{c}(2) = (e, 1)$. Because $\partial(2x \cos y)/\partial y = \partial(-x^2 \sin y)/\partial x$, $\mathbf{F}$ is irrotational and hence a gradient vector field (as we saw in Example 3). Thus, by Theorem 7, we can replace $\mathbf{c}$ by any piecewise C^1 curve having the same endpoints, in particular, by the polygonal path from $(1, 0)$ to $(e, 0)$ to $(e, 1)$. Thus, the line integral must be equal to

$$\begin{aligned}\int_{\mathbf{c}} \mathbf{F} \cdot d\mathbf{s} &= \int_1^e 2t \cos 0 \, dt + \int_0^1 -e^2 \sin t \, dt = (e^2 - 1) + e^2(\cos 1 - 1) \\ &= e^2 \cos 1 - 1.\end{aligned}$$

Alternatively, using Theorem 3 of Section 7.2, we have

$$\int_{\mathbf{c}} 2x \cos y \, dx - x^2 \sin y \, dy = \int_{\mathbf{c}} \nabla f \cdot d\mathbf{s} = f(\mathbf{c}(2)) - f(\mathbf{c}(1)) = e^2 \cos 1 - 1,$$

because $f(x, y) = x^2 \cos y$ is a potential function for $\mathbf{F}$. Evidently, this technique is simpler than computing the integral directly. ▲

We conclude this section with a theorem that is quite similar in spirit to Theorem 7. Theorem 7 was motivated partly as a converse to the result that curl $\nabla f = \mathbf{0}$ for any C^1 function $f\colon \mathbb{R}^3 \to \mathbb{R}$—or, if curl $\mathbf{F} = 0$, then $\mathbf{F} = \nabla f$. We also know [formula (9) in the table of vector identities in Section 4.4] that div (curl $\mathbf{G}$) $= 0$ for any C^2 vector field $\mathbf{G}$. We may ask about the converse statement: If div $\mathbf{F} = 0$, is $\mathbf{F}$ the curl of a vector field $\mathbf{G}$? The following theorem answers this in the affirmative.

THEOREM 8 If $\mathbf{F}$ is a C^1 vector field on all of $\mathbb{R}^3$ with div $\mathbf{F} = 0$, then there exists a C^1 vector field $\mathbf{G}$ with $\mathbf{F} =$ curl $\mathbf{G}$.

The proof is outlined in Exercise 16. We should warn the reader at this point that, unlike the $\mathbf{F}$ in Theorem 7, the vector field $\mathbf{F}$ in Theorem 8 is not allowed to have an exceptional point. For example, the gravitational force field $\mathbf{F} = -(GmM\mathbf{r}/r^3)$ has the property that div $\mathbf{F} = 0$, and yet there is no $\mathbf{G}$ for which $\mathbf{F} =$ curl $\mathbf{G}$ (see Exercise 25). Theorem 8 does not apply, because the gravitational force field $\mathbf{F}$ is not defined at $\mathbf{0} \in \mathbb{R}^3$.

EXERCISES

1. Show that any two potential functions for a vector field on $\mathbb{R}^3$ differ at most by a constant.

2. (a) Let $\mathbf{F}(x, y) = (xy, y^2)$ and let $\mathbf{c}$ be the path $y = 2x^2$ joining $(0, 0)$ to $(1, 2)$ in $\mathbb{R}^2$. Evaluate $\int_{\mathbf{c}} \mathbf{F} \cdot d\mathbf{s}$.

(b) Does the integral in part (a) depend on the path joining $(0, 0)$ to $(1, 2)$?

3. Let $\mathbf{F}(x, y, z) = (2xyz + \sin x)\mathbf{i} + x^2 z\mathbf{j} + x^2 y\mathbf{k}$. Find a function f such that $\mathbf{F} = \nabla f$.

4. Evaluate $\int_{\mathbf{c}} \mathbf{F} \cdot d\mathbf{s}$, where $\mathbf{c}(t) = (\cos^5 t, \sin^3 t, t^4)$, $0 \le t \le \pi$, and $\mathbf{F}$ is as in Exercise 3.

5. If $f(x)$ is a smooth function of one variable, must $\mathbf{F}(x, y) = f(x)\mathbf{i} + f(y)\mathbf{j}$ be a gradient?

6. (a) Show that $\mathbf{F} = -\mathbf{r}/\|\mathbf{r}\|^3$ is the gradient of $f(x, y, z) = 1/r$.

(b) What is the work done by the force $\mathbf{F} = -\mathbf{r}/\|\mathbf{r}\|^3$ in moving a particle from a point $\mathbf{r}_0 \in \mathbb{R}^3$ "to ∞," where $\mathbf{r}(x, y, z) = (x, y, z)$?

7. Let $\mathbf{F}(x, y, z) = xy\mathbf{i} + y\mathbf{j} + z\mathbf{k}$. Can there exist a function f such that $\mathbf{F} = \nabla f$?

8. Let $\mathbf{F} = F_1\mathbf{i} + F_2\mathbf{j} + F_3\mathbf{k}$ and suppose each F_k satisfies the homogeneity condition

$$F_k(tx, ty, tz) = tF_k(x, y, z), \qquad k = 1, 2, 3.$$

Suppose also $\nabla \times \mathbf{F} = \mathbf{0}$. Prove that $\mathbf{F} = \nabla f$, where

$$2f(x, y, z) = xF_1(x, y, z) + yF_2(x, y, z) + zF_3(x, y, z).$$

(HINT: Use Review Exercise 23, Chapter 2.)

9. Let $\mathbf{F}(x, y, z) = (e^x \sin y)\mathbf{i} + (e^x \cos y)\mathbf{j} + z^2\mathbf{k}$. Evaluate the integral $\int_{\mathbf{c}} \mathbf{F} \cdot d\mathbf{s}$, where $\mathbf{c}(t) = (\sqrt{t}, t^3, \exp\sqrt{t}),\ 0 \le t \le 1$.

10. Let a fluid have the velocity field $\mathbf{F}(x, y, z) = xy\mathbf{i} + yz\mathbf{j} + xz\mathbf{k}$. What is the circulation around the unit circle in the xy plane? Interpret your answer.

11. The mass of the earth is approximately 6×10^{27} g and that of the sun is 330,000 times as much. The gravitational constant is 6.7×10^{-8} cm^3/s$^2 \cdot$ g. The distance of the earth from the sun is about 1.5×10^{12} cm. Compute, approximately, the work necessary to increase the distance of the earth from the sun by 1 cm.

12. (a) Show that $\int_C (x\,dy - y\,dx)/(x^2 + y^2) = 2\pi$, where C is the unit circle.

(b) Conclude that the associated vector field $[-y/(x^2 + y^2)]\mathbf{i} + [x/(x^2 + y^2)]\mathbf{j}$ is not a conservative field.

(c) Show, however, that $\partial P/\partial y = \partial Q/\partial x$. Does this contradict the corollary to Theorem 7? If not, why not?

13. Determine which of the following vector fields $\mathbf{F}$ in the plane is the gradient of a scalar function f. If such an f exists, find it.

(a) $\mathbf{F}(x, y) = x\mathbf{i} + y\mathbf{j}$
(b) $\mathbf{F}(x, y) = xy\mathbf{i} + xy\mathbf{j}$
(c) $\mathbf{F}(x, y) = (x^2 + y^2)\mathbf{i} + 2xy\mathbf{j}$

14. Repeat Exercise 13 for the following vector fields:

(a) $\mathbf{F}(x, y) = (\cos xy - xy \sin xy)\mathbf{i} - (x^2 \sin xy)\mathbf{j}$
(b) $\mathbf{F}(x, y) = (x\sqrt{x^2y^2 + 1})\mathbf{i} + (y\sqrt{x^2y^2 + 1})\mathbf{j}$
(c) $\mathbf{F}(x, y) = (2x \cos y + \cos y)\mathbf{i} - (x^2 \sin y + x \sin y)\mathbf{j}$

15. Show that the following vector fields are conservative. Calculate $\int_C \mathbf{F} \cdot d\mathbf{s}$ for the given curve.

(a) $\mathbf{F} = (xy^2 + 3x^2y)\mathbf{i} + (x + y)x^2\mathbf{j}$; C is the curve consisting of line segments from $(1, 1)$ to $(0, 2)$ to $(3, 0)$.

(b) $\mathbf{F} = \dfrac{2x}{y^2+1}\mathbf{i} - \dfrac{2y(x^2+1)}{(y^2+1)^2}\mathbf{j}$; C is parametrized by $x = t^3 - 1$, $y = t^6 - t$, $0 \le t \le 1$.

(c) $\mathbf{F} = [\cos(xy^2) - xy^2 \sin(xy^2)]\mathbf{i} - 2x^2y \sin(xy^2)\mathbf{j}$; C is the curve (e^t, e^{t+1}), $-1 \le t \le 0$.

16. Prove Theorem 8. [HINT: Define $\mathbf{G} = G_1\mathbf{i} + G_2\mathbf{j} + G_3\mathbf{k}$ by

$$G_1(x, y, z) = \int_0^z F_2(x, y, t)\,dt - \int_0^y F_3(x, t, 0)\,dt$$

$$G_2(x, y, z) = -\int_0^z F_1(x, y, t)\,dt$$

and $G_3(x, y, z) = 0$.]

17. Is each of the following vector fields the curl of some other vector field? If so, find the vector field.

(a) $\mathbf{F} = x\mathbf{i} + y\mathbf{j} + z\mathbf{k}$

(b) $\mathbf{F} = (x^2 + 1)\mathbf{i} + (z - 2xy)\mathbf{j} + y\mathbf{k}$

18. Let $\mathbf{F} = xz\mathbf{i} - yz\mathbf{j} + y\mathbf{k}$. Verify that $\nabla \cdot \mathbf{F} = 0$. Find a $\mathbf{G}$ such that $\mathbf{F} = \nabla \times \mathbf{G}$.

19. Repeat Exercise 18 for $\mathbf{F} = y^2\mathbf{i} + z^2\mathbf{j} + x^2\mathbf{k}$.

20. Let $\mathbf{F} = xe^y\mathbf{i} - (x \cos z)\mathbf{j} - ze^y\mathbf{k}$. Find a $\mathbf{G}$ such that $\mathbf{F} = \nabla \times \mathbf{G}$.

21. Let $\mathbf{F} = (x \cos y)\mathbf{i} - (\sin y)\mathbf{j} + (\sin x)\mathbf{k}$. Find a $\mathbf{G}$ such that $\mathbf{F} = \nabla \times \mathbf{G}$.

22. By using different paths from $(0, 0, 0)$ to (x, y, z), show that the function f defined in the proof of Theorem 7 for "condition (ii) implies condition (iii)" satisfies $\partial f/\partial x = F_1$ and $\partial f/\partial y = F_2$.

23. Let $\mathbf{F}$ be the vector field on $\mathbb{R}^3$ given by $\mathbf{F} = -y\mathbf{i} + x\mathbf{j}$.

(a) Show that $\mathbf{F}$ is rotational, that is, $\mathbf{F}$ is not irrotational.

(b) Suppose $\mathbf{F}$ represents the velocity vector field of a fluid. Show that if we place a cork in this fluid it will revolve in a plane parallel to the xy plane, in a circular trajectory about the z axis.

(c) In what direction does the cork revolve?

24. Let $\mathbf{G}$ be the vector field on $\mathbb{R}^3\backslash\{z \text{ axis}\}$ defined by

$$\mathbf{G} = \frac{-y}{x^2+y^2}\mathbf{i} + \frac{x}{x^2+y^2}\mathbf{j}.$$

(a) Show that **G** is irrotational.
(b) Show that the result of Exercise 23(b) holds for **G** also.
(c) How can we resolve the fact that the trajectories of **F** and **G** are both the same (circular about the z axis) yet **F** is rotational and **G** is not? (HINT: The property of being rotational is a local condition, that is, a property of the fluid in the neighborhood of a point.)

25. Let $\mathbf{F} = -(GmM\mathbf{r}/r^3)$ be the gravitational force field defined on $\mathbb{R}^3\backslash\{\mathbf{0}\}$.

(a) Show that div $\mathbf{F} = 0$.
(b) Show that $\mathbf{F} \neq$ curl $\mathbf{G}$ for any C^1 vector field $\mathbf{G}$ on $\mathbb{R}^3\backslash\{\mathbf{0}\}$.

8.4 Gauss' Theorem

Gauss' theorem states that the flux of a vector field out of a closed surface equals the integral of the divergence of that vector field over the volume enclosed by the surface. The result parallels Stokes' theorem and Green's theorem in that it relates an integral over a closed geometrical object (curve or surface) to an integral over a contained region (surface or volume).

Elementary Regions and their Boundaries

We shall begin by asking the reader to review the various elementary regions in space that were introduced when we considered the volume integral; these regions are illustrated in Figures 5.5.2 and 5.5.4. As these figures indicate, the boundary of an elementary region in $\mathbb{R}^3$ is a surface made up of a finite number (at most six, at least two) of surfaces that can be described as graphs of functions from $\mathbb{R}^2$ to $\mathbb{R}$. This kind of surface is called a *closed surface.* The surfaces $S_1, S_2, \ldots, S_N$ composing such a closed surface are called its *faces*.

EXAMPLE 1 The cube in Figure 8.4.1(a) is an elementary region, and in fact a symmetric elementary region, with six rectangles composing its boundary. The sphere in Figure 8.4.1(b) is the boundary of a solid ball, which is also a symmetric elementary region. ▲

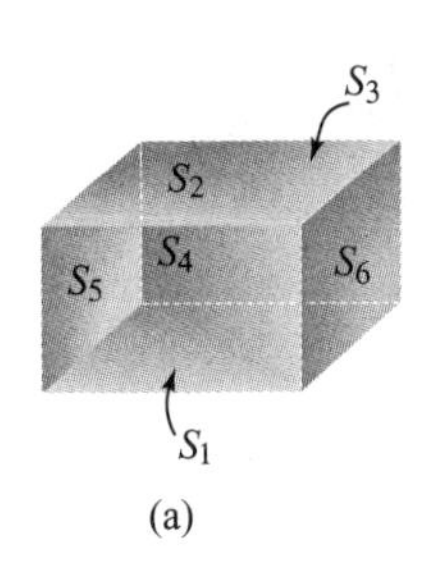

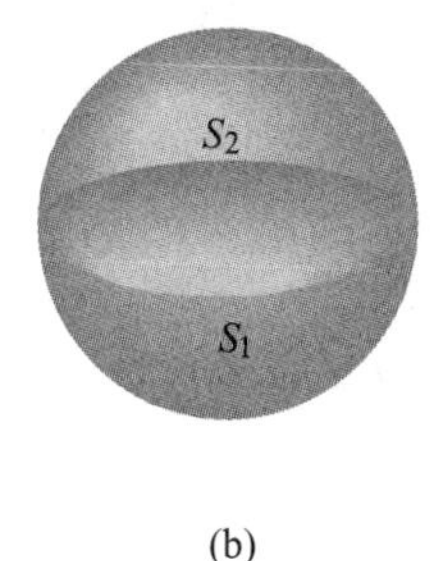

Figure 8.4.1 (a) Symmetric elementary regions and (b) the surface S_i composing their boundaries.

Closed surfaces can be oriented in two ways. The outward orientation makes the normal point outward into space, and the inward orientation makes the normal point into the bounded region (Figure 8.4.2).

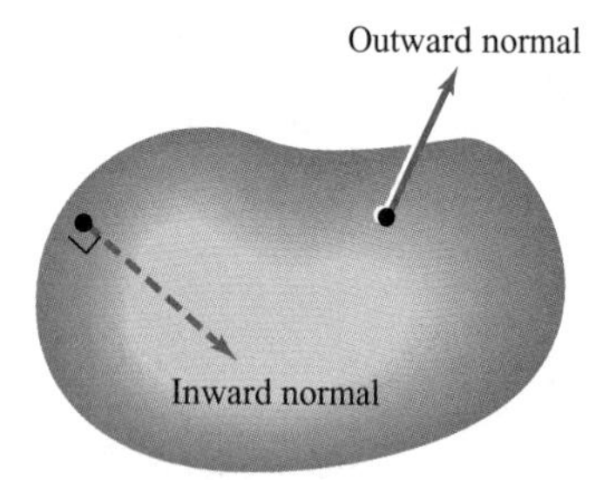

Figure 8.4.2 Two possible orientations for a closed surface.

Suppose S is a closed surface oriented in one of these two ways and $\mathbf{F}$ is a vector field on S. Then, as we defined it in Section 7.6,

$$\iint_S \mathbf{F} \cdot d\mathbf{S} = \sum_i \iint_{S_i} \mathbf{F} \cdot d\mathbf{S}.$$

If S is given the outward orientation, the integral $\iint_S \mathbf{F} \cdot d\mathbf{S}$ measures the total flux of $\mathbf{F}$ outward across S. That is, if we think of $\mathbf{F}$ as the velocity field of a fluid, $\iint_S \mathbf{F} \cdot d\mathbf{S}$ indicates the amount of fluid leaving the region bounded by S per unit time. If S is given the inward orientation, the integral $\iint_S \mathbf{F} \cdot d\mathbf{S}$ measures the total flux of $\mathbf{F}$ inward across S.

We recall another common way of writing these surface integrals, a way that explicitly specifies the orientation of S. Let the orientation of S be given by a unit normal vector $\mathbf{n}(x, y, z)$ at each point of S. Then we have the oriented integral

$$\iint_S \mathbf{F} \cdot d\mathbf{S} = \iint_S (\mathbf{F} \cdot \mathbf{n})\, dS,$$

that is, the integral of the normal component of $\mathbf{F}$ over S. In the remainder of this section, if S is a closed surface enclosing a region W, we adopt the default convention that $S = \partial W$ is given the outward orientation, with outward unit normal $\mathbf{n}(x, y, z)$ at each point $(x, y, z) \in S$. Furthermore, we denote the surface with the opposite (inward) orientation by ∂W_{op}. Then the associated unit normal direction for this orientation is $-\mathbf{n}$. Thus,

$$\iint_{\partial W} \mathbf{F} \cdot d\mathbf{S} = \iint_S (\mathbf{F} \cdot \mathbf{n})\, dS = -\iint_S [\mathbf{F} \cdot (-\mathbf{n})]\, dS = -\iint_{\partial W_{\mathrm{op}}} \mathbf{F} \cdot d\mathbf{S}.$$

EXAMPLE 2 The unit cube W given by

$$0 \le x \le 1, \qquad 0 \le y \le 1, \qquad 0 \le z \le 1$$

is a symmetric elementary region in space (see Figures 8.4.3 and 5.5.5).

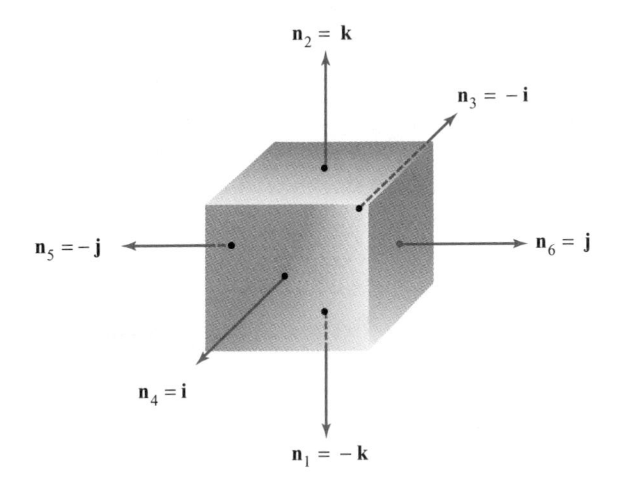

Figure 8.4.3 The outward orientation on the cube.

We write the faces as

$$
\begin{aligned}
&S_1\colon z = 0, && 0 \le x \le 1, && 0 \le y \le 1\\
&S_2\colon z = 1, && 0 \le x \le 1, && 0 \le y \le 1\\
&S_3\colon x = 0, && 0 \le y \le 1, && 0 \le z \le 1\\
&S_4\colon x = 1, && 0 \le y \le 1, && 0 \le z \le 1\\
&S_5\colon y = 0, && 0 \le x \le 1, && 0 \le z \le 1\\
&S_6\colon y = 1, && 0 \le x \le 1, && 0 \le z \le 1.
\end{aligned}
$$

From Figure 8.4.3 we see that

$$
\begin{aligned}
\mathbf{n}_2 &= \mathbf{k} = -\mathbf{n}_1,\\
\mathbf{n}_4 &= \mathbf{i} = -\mathbf{n}_3,\\
\mathbf{n}_6 &= \mathbf{j} = -\mathbf{n}_5,
\end{aligned}
$$

and so for a continuous vector field $\mathbf{F} = F_1\mathbf{i} + F_2\mathbf{j} + F_3\mathbf{k}$,

$$
\begin{aligned}
\iint_{\partial W} \mathbf{F}\cdot d\mathbf{S} = \iint_S \mathbf{F}\cdot\mathbf{n}\,dS = -\iint_{S_1} F_3\,dS + \iint_{S_2} F_3\,dS - \iint_{S_3} F_1\,dS\\
+ \iint_{S_4} F_1\,dS - \iint_{S_5} F_2\,dS + \iint_{S_6} F_2\,dS. \quad \blacktriangle
\end{aligned}
$$

Gauss' Theorem

We have now come to the last of the three central theorems of this chapter. This theorem relates surface integrals to volume integrals; in other words, the theorem states that if W is a region in $\mathbb{R}^3$, then the flux of a vector field $\mathbf{F}$ outward across the closed surface ∂W is equal to the integral of div $\mathbf{F}$ over W. We begin by assuming that W is a symmetric elementary region (Figure 5.5.5).

THEOREM 9: Gauss' Divergence Theorem Let W be a symmetric elementary region in space. Denote by ∂W the oriented closed surface that bounds W. Let $\mathbf{F}$ be a smooth vector field defined on W. Then

$$\iiint_W (\nabla \cdot \mathbf{F})\, dV = \iint_{\partial W} \mathbf{F} \cdot d\mathbf{S}$$

or, alternatively,

$$\iiint_W (\text{div } \mathbf{F})\, dV = \iint_{\partial W} (\mathbf{F} \cdot \mathbf{n})\, dS.$$

PROOF If $\mathbf{F} = P\mathbf{i} + Q\mathbf{j} + R\mathbf{k}$, then by definition, the divergence of $\mathbf{F}$ is given by div $\mathbf{F} = \partial P/\partial x + \partial Q/\partial y + \partial R/\partial z$, so we can write (using additivity of the volume integral)

$$\iiint_W \text{div } \mathbf{F}\, dV = \iiint_W \frac{\partial P}{\partial x}\, dV + \iiint_W \frac{\partial Q}{\partial y}\, dV + \iiint_W \frac{\partial R}{\partial z}\, dV.$$

On the other hand, the surface integral in question is

$$\begin{aligned}\iint_{\partial W} \mathbf{F} \cdot \mathbf{n}\, dS &= \iint_{\partial W} (P\mathbf{i} + Q\mathbf{j} + R\mathbf{k}) \cdot \mathbf{n}\, dS \\ &= \iint_{\partial W} P\mathbf{i} \cdot \mathbf{n}\, dS + \iint_{\partial W} Q\mathbf{j} \cdot \mathbf{n}\, dS + \iint_{\partial W} R\mathbf{k} \cdot \mathbf{n}\, dS.\end{aligned}$$

The theorem will follow if we establish the three equalities

$$\iint_{\partial W} P\mathbf{i} \cdot \mathbf{n}\, dS = \iiint_W \frac{\partial P}{\partial x}\, dV, \tag{1}$$

$$\iint_{\partial W} Q\mathbf{j} \cdot \mathbf{n}\, dS = \iiint_W \frac{\partial Q}{\partial y}\, dV, \tag{2}$$

and

$$\iint_{\partial W} R\mathbf{k}\cdot\mathbf{n}\,dS = \iiint_W \frac{\partial R}{\partial z}\,dV. \tag{3}$$

We shall prove equation (3); the other two equalities can be proved in an analogous fashion.

Because W is a symmetric elementary region, there is a pair of functions

$$z = g_1(x, y), \qquad z = g_2(x, y),$$

with common domain an elementary region D in the xy plane, such that W is the set of all points (x, y, z) satisfying

$$g_1(x, y) \le z \le g_2(x, y), \qquad (x, y) \in D.$$

By reduction to iterated integrals, we have

$$\iiint_W \frac{\partial R}{\partial z}\,dV = \iint_D \left(\int_{z=g_1(x,y)}^{z=g_2(x,y)} \frac{\partial R}{\partial z}\,dz\right) dx\,dy,$$

and so, by the fundamental theorem of calculus,

$$\iiint_W \frac{\partial R}{\partial z}\,dV = \iint_D [R(x, y, g_2(x, y)) - R(x, y, g_1(x, y))]\,dx\,dy. \tag{4}$$

The boundary of W is a closed surface whose top S_2 is the graph of $z = g_2(x, y)$, where $(x, y) \in D$, and whose bottom S_1 is the graph of $z = g_1(x, y)$, $(x, y) \in D$. The four other sides of ∂W consist of surfaces S_3, S_4, S_5, and S_6, whose normals are always perpendicular to the z axis. (See Figure 8.4.4. Note that some of the other four sides

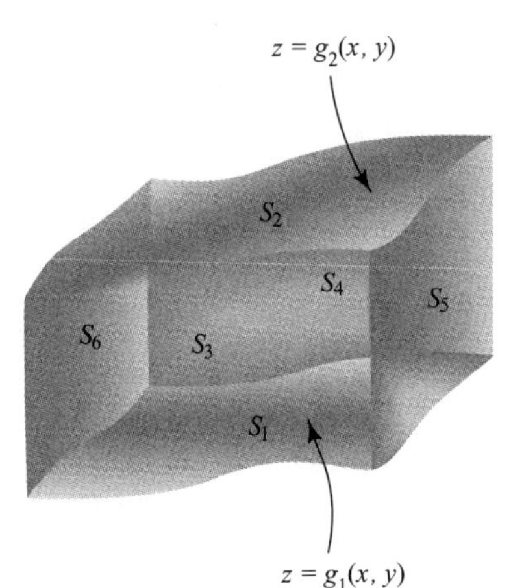

Figure 8.4.4 A symmetric elementary region W for which $\iint_{\partial W} R\mathbf{k}\cdot d\mathbf{S} = \iiint_W (\partial R/\partial z)\,dV$. The four sides of ∂W, which are S_3, S_4, S_5, S_6 have normals perpendicular to the z axis.

might be absent; for instance, if W is a solid ball and ∂W is a sphere.) By definition,

$$\iint_{\partial W} R\mathbf{k}\cdot\mathbf{n}\, dS = \iint_{S_1} R\mathbf{k}\cdot\mathbf{n}_1\, dS + \iint_{S_2} R\mathbf{k}\cdot\mathbf{n}_2\, dS + \sum_{i=3}^{6}\iint_{S_i} R\mathbf{k}\cdot\mathbf{n}_i\, dS.$$

Because the normal $\mathbf{n}_i$ is perpendicular to $\mathbf{k}$ on each of S_3, S_4, S_5, S_6, we have $\mathbf{k}\cdot\mathbf{n} = 0$ along these faces, and so the integral reduces to

$$\iint_{\partial W} R\mathbf{k}\cdot\mathbf{n}\, ds = \iint_{S_1} R\mathbf{k}\cdot d\mathbf{S}_1 + \iint_{S_2} R\mathbf{k}\cdot d\mathbf{S}_2. \tag{5}$$

The surface S_1 is defined by $z = g_1(x, y)$, and

$$d\mathbf{S}_1 = \left(\frac{\partial g_1}{\partial x}\mathbf{i} + \frac{\partial g_1}{\partial y}\mathbf{j} - \mathbf{k}\right) dx\, dy$$

(the negative of the general formula for $d\mathbf{S}$ for graphs from Section 7.6, because the normal is downward pointing). Therefore,

$$\iint_{S_1} R\mathbf{k}\cdot d\mathbf{S}_1 = -\iint_D R(x, y, g_1(x, y))\, dx\, dy. \tag{6}$$

Similarly, for the top face S_2,

$$d\mathbf{S}_2 = \left(-\frac{\partial g_2}{\partial x}\mathbf{i} - \frac{\partial g_2}{\partial y}\mathbf{j} + \mathbf{k}\right) dx\, dy.$$

Therefore,

$$\iint_{S_2} R\mathbf{k}\cdot d\mathbf{S}_2 = \iint_D R(x, y, g_2(x, y))\, dx\, dy. \tag{7}$$

Substituting equations (6) and (7) into equation (5) and then comparing with equation (4), we obtain

$$\iiint_W \frac{\partial R}{\partial z}\, dV = \iint_{\partial W} R(\mathbf{k}\cdot\mathbf{n})\, dS.$$

The remaining equalities, (1) and (2), can be established in the same way to complete the proof. ■

Generalizing Gauss' Theorem

The reader should note that the proof of Gauss' theorem is similar to that of Green's theorem. By the procedure used in Exercise 8 of Section 8.1, we can extend Gauss' theorem to any region that can be broken up into symmetric elementary regions. This includes all regions of interest to us. An example of a region to which Gauss' theorem applies is the region between two closed surfaces, one inside the other. The surface of this region consists of two pieces oriented as shown in Figure 8.4.5. We shall apply the divergence theorem to such a region when we prove Gauss' law in Theorem 10.

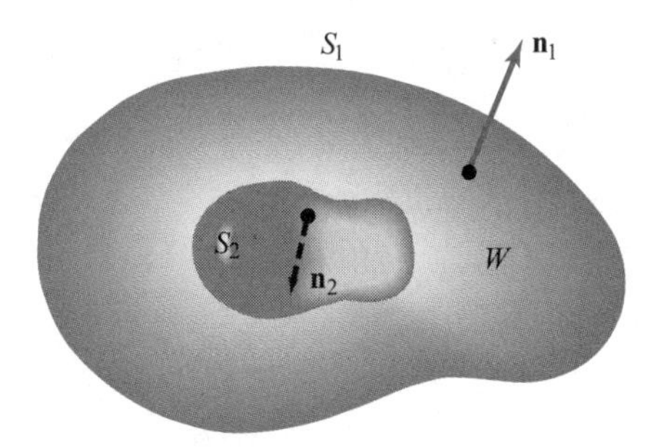

Figure 8.4.5 A more general region to which Gauss' theorem applies.

EXAMPLE 3 Consider $\mathbf{F} = 2x\mathbf{i} + y^2\mathbf{j} + z^2\mathbf{k}$. Let S be the unit sphere defined by $x^2 + y^2 + z^2 = 1$. Evaluate $\iint_S \mathbf{F} \cdot \mathbf{n}\, dS$.

SOLUTION By Gauss' theorem,

$$\iint_S \mathbf{F} \cdot \mathbf{n}\, dS = \iiint_W (\text{div}\, \mathbf{F})\, dV,$$

where W is the ball bounded by the sphere. The integral on the right is

$$2\iiint_W (1 + y + z)\, dV = 2\iiint_W dV + 2\iiint_W y\, dV + 2\iiint_W z\, dV.$$

By symmetry, we can argue that $\iiint_W y\, dV = \iiint_W z\, dV = 0$ (see Exercise 17, Section 6.3). Thus, because a sphere of radius R has volume $4\pi R^3/3$,

$$2\iiint_W (1 + y + z)\, dV = 2\iiint_W dV = \frac{8\pi}{3}.$$

Readers can convince themselves that direct computation of $\iint_S \mathbf{F} \cdot \mathbf{n}\, dS$ proves unwieldy. ▲

EXAMPLE 4 Use the divergence theorem to evaluate

$$\iint_{\partial W} (x^2 + y + z)\, dS,$$

where W is the solid ball $x^2 + y^2 + z^2 \leq 1$.

SOLUTION To apply Gauss' divergence theorem, we find a vector field $\mathbf{F} = F_1\mathbf{i} + F_2\mathbf{j} + F_3\mathbf{k}$ on W with $\mathbf{F} \cdot \mathbf{n} = x^2 + y + z$. At any point $(x, y, z) \in \partial W$, the outward unit normal $\mathbf{n}$ to ∂W is

$$\mathbf{n} = x\mathbf{i} + y\mathbf{j} + z\mathbf{k},$$

because on ∂W, $x^2 + y^2 + z^2 = 1$ and the radius vector $\mathbf{r} = x\mathbf{i} + y\mathbf{j} + z\mathbf{k}$ is normal to the sphere ∂W (Figure 8.4.6).

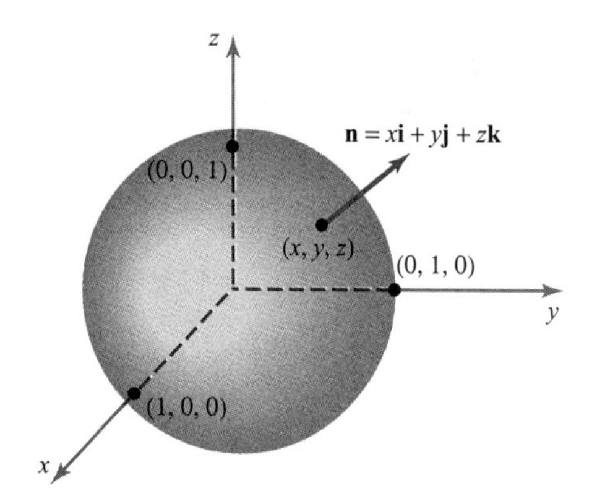

Figure 8.4.6 $\mathbf{n}$ is the unit normal to ∂W, the boundary of the ball W.

Therefore, if $\mathbf{F}$ is the desired vector field, then

$$\mathbf{F} \cdot \mathbf{n} = F_1 x + F_2 y + F_2 z.$$

We set

$$F_1 x = x^2, \qquad F_2 y = y, \qquad F_3 z = z$$

and solve for F_1, F_2, and F_3 to find that $\mathbf{F} = x\mathbf{i} + \mathbf{j} + \mathbf{k}$. Computing div $\mathbf{F}$, we get

$$\text{div}\, \mathbf{F} = 1 + 0 + 0 = 1.$$

Thus, by Gauss' divergence theorem,

$$\iint_{\partial W} (x^2 + y + z)\, dS = \iiint_W dV = \text{volume}\,(W) = \frac{4}{3}\pi. \quad \blacktriangle$$

The Divergence as the Flux per Unit Volume

The physical meaning of divergence is that at a point P, div $\mathbf{F}$(P) is the rate of net outward flux at P per unit volume. This follows from Gauss' theorem and the mean-value theorem for integrals: If W_ρ is a ball in $\mathbb{R}^3$ of radius ρ centered at P, then there is a point $Q \in W_\rho$ such that

$$\iint_{\partial W_\rho} \mathbf{F} \cdot \mathbf{n}\, dS = \iiint_{W_\rho} \operatorname{div} \mathbf{F}\, dV = \operatorname{div} \mathbf{F}(Q) \cdot \text{volume}\,(W_\rho)$$

and so

$$\operatorname{div} \mathbf{F}(P) = \underset{\rho \to 0}{\text{limit}} \operatorname{div} \mathbf{F}(Q) = \underset{\rho \to 0}{\text{limit}} \frac{1}{V(W_\rho)} \iint_{\partial W_\rho} \mathbf{F} \cdot \mathbf{n}\, dS.$$

This is analogous to the limit formulation of the curl given at the end of Section 8.2. Thus, if div $\mathbf{F}(P) > 0$, we consider P to be a ***source***, for there is a net outward flow near P. If div $\mathbf{F}(P) < 0$, P is called a ***sink*** for $\mathbf{F}$.

A C^1 vector field $\mathbf{F}$ defined on $\mathbb{R}^3$ is said to be ***divergence free*** if div $\mathbf{F} = 0$. If $\mathbf{F}$ is divergence-free, we have $\iint_S \mathbf{F} \cdot d\mathbf{S} = 0$ for all closed surfaces S. The converse can also be demonstrated readily using Gauss' theorem: If $\iint_S \mathbf{F} \cdot d\mathbf{S} = 0$ for all closed surfaces S, then $\mathbf{F}$ is divergence-free. If $\mathbf{F}$ is divergence-free, we thus see that the flux of $\mathbf{F}$ across any closed surface S is 0, so that if $\mathbf{F}$ is the velocity field of a fluid, the net amount of fluid that flows out of any region will be 0. Thus, exactly as much fluid must flow into the region as flows out (in unit time). A fluid with this property is therefore described as ***incompressible***.

EXAMPLE 5 Evaluate $\iint_S \mathbf{F} \cdot d\mathbf{S}$, where $\mathbf{F}(x, y, z) = xy^2\mathbf{i} + x^2y\mathbf{j} + y\mathbf{k}$ and S is the surface of the cylinder $x^2 + y^2 = 1$, bounded by the planes $z = 1$ and $z = -1$, and including the portions $x^2 + y^2 \le 1$ when $z = \pm 1$.

SOLUTION One can compute this integral directly, but it is easier to use the divergence theorem.

Now S is the boundary of the region W given by $x^2 + y^2 \le 1, -1 \le z \le 1$. Thus, $\iint_S \mathbf{F} \cdot d\mathbf{S} = \iiint_W (\operatorname{div} \mathbf{F})\, dV$. Moreover,

$$\iiint_W (\operatorname{div} \mathbf{F})\, dV = \iiint_W (x^2 + y^2)\, dx\, dy\, dz = \int_{-1}^{1} \left(\iint_{x^2+y^2 \le 1} (x^2 + y^2)\, dx\, dy \right) dz$$

$$= 2 \iint_{x^2+y^2 \le 1} (x^2 + y^2)\, dx\, dy.$$

Before evaluating the double integral, we note that the surface integral satisfies

$$\iint_{\partial W} \mathbf{F} \cdot \mathbf{n}\, dS = 2 \iint_{x^2+y^2 \le 1} (x^2 + y^2)\, dx\, dy > 0.$$

This means that $\iint_{\partial W} \mathbf{F} \cdot d\mathbf{S}$, the net flux of $\mathbf{F}$ out of the cylinder, is positive.

We change variables to polar coordinates to evaluate the double integral:

$$x = r\cos\theta, \qquad y = r\sin\theta, \qquad 0 \le r \le 1, \qquad 0 \le \theta \le 2\pi.$$

Hence, we have $\partial(x, y)/\partial(r, \theta) = r$ and $x^2 + y^2 = r^2$. Thus,

$$\iint_{x^2+y^2 \le 1} (x^2 + y^2)\, dx\, dy = \int_0^{2\pi} \left(\int_0^1 r^3\, dr \right) d\theta = \frac{1}{2}\pi.$$

Therefore, $\iiint_W \operatorname{div} \mathbf{F}\, dV = \pi$. ▲

Gauss' Law

As we remarked earlier, Gauss' divergence theorem can be applied to regions in space more general than symmetric elementary regions. To conclude this section, we shall use this observation *in the proof* of the following important results.

THEOREM 10: Gauss' Law Let M be a symmetric elementary region in $\mathbb{R}^3$. Then if $(0, 0, 0) \notin \partial M$, we have

$$\iint_{\partial M} \frac{\mathbf{r} \cdot \mathbf{n}}{r^3}\, dS = \begin{cases} 4\pi & \text{if } (0,0,0) \in M \\ 0 & \text{if } (0,0,0) \notin M, \end{cases}$$

where

$$\mathbf{r}(x, y, z) = x\mathbf{i} + y\mathbf{j} + z\mathbf{k}$$

and

$$r(x, y, z) = \|\mathbf{r}(x, y, z)\| = \sqrt{x^2 + y^2 + z^2}.$$

PROOF OF GAUSS' LAW First suppose $(0, 0, 0) \notin M$. Then $\mathbf{r}/r^3$ is a C^1 vector field on M and ∂M, and so by the divergence theorem,

$$\iint_{\partial M} \frac{\mathbf{r} \cdot \mathbf{n}}{r^3}\, dS = \iiint_M \nabla \cdot \left(\frac{\mathbf{r}}{r^3} \right) dV.$$

But $\nabla \cdot (\mathbf{r}/r^3) = 0$ for $r \ne 0$, as the reader can easily verify (see Exercise 30, Section 4.4). Thus,

$$\iint_{\partial M} \frac{\mathbf{r} \cdot \mathbf{n}}{r^3}\, dS = 0.$$

Now let us suppose $(0, 0, 0) \in M$. We can no longer use the preceding method, because $\mathbf{r}/r^3$ is not smooth on M, on account of the zero denominator at $\mathbf{r} = (0, 0, 0)$. Because $(0, 0, 0) \in M$ and $(0, 0, 0) \notin \partial M$, there is an $\varepsilon > 0$ such that the ball N of radius ε centered at $(0, 0, 0)$ is contained completely inside M. Let W be the region between M and N. Then W has boundary $\partial N \cup \partial M = S$. But the orientation on ∂N induced by the *outward* normal on W is the opposite of that obtained from N (see Figure 8.4.7).

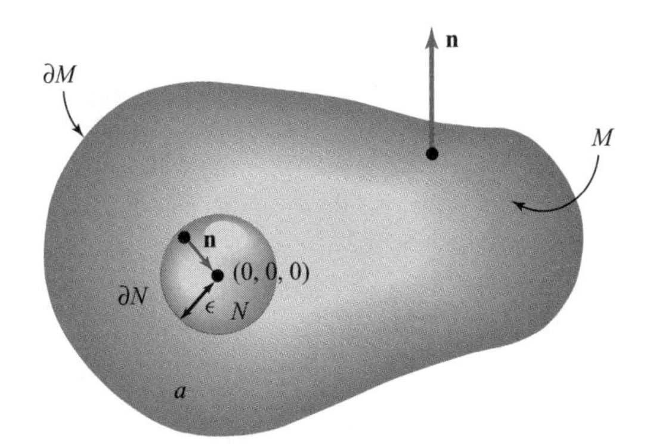

Figure 8.4.7 Induced outward orientation on S; W is M minus the ball N.

Now $\nabla \cdot (\mathbf{r}/r^3) = 0$ on W, and so, by the divergence theorem applied to the (nonelementary) region W,

$$\iint_S \frac{\mathbf{r} \cdot \mathbf{n}}{r^3}\, dS = \iiint_W \nabla \cdot \left(\frac{\mathbf{r}}{r^3}\right) dV = 0.$$

Because

$$\iint_S \frac{\mathbf{r} \cdot \mathbf{n}}{r^3}\, dS = \iint_{\partial M} \frac{\mathbf{r} \cdot \mathbf{n}}{r^3}\, dS + \iint_{\partial N} \frac{\mathbf{r} \cdot \mathbf{n}}{r^3}\, dS,$$

where $\mathbf{n}$ is the outward normal to S, we have

$$\iint_{\partial M} \frac{\mathbf{r} \cdot \mathbf{n}}{r^3}\, dS = -\iint_{\partial N} \frac{\mathbf{r} \cdot \mathbf{n}}{r^3}\, dS.$$

However, on ∂N, $\mathbf{n} = -\mathbf{r}/r$ and $r = \varepsilon$, because ∂N is a sphere of radius ε, so that

$$-\iint_{\partial N} \frac{\mathbf{r} \cdot \mathbf{n}}{r^3}\, dS = \iint_{\partial N} \frac{\varepsilon^2}{\varepsilon^4}\, dS = \frac{1}{\varepsilon^2} \iint_{\partial N} dS.$$

But $\iint_{\partial N} dS = 4\pi\varepsilon^2$, the surface area of the sphere of radius ε. This proves the result. ■

EXAMPLE 6 Gauss' law has the following physical interpretation. The potential due to a point charge Q at $(0, 0, 0)$ is given by

$$\phi(x, y, z) = \frac{Q}{4\pi r} = \frac{Q}{4\pi\sqrt{x^2 + y^2 + z^2}},$$

and the corresponding electric field is

$$\mathbf{E} = -\nabla\phi = \frac{Q}{4\pi}\left(\frac{\mathbf{r}}{r^3}\right).$$

Thus, Theorem 10 states that the total electric flux $\iint_{\partial M} \mathbf{E} \cdot d\mathbf{S}$ (that is, the flux of $\mathbf{E}$ out of a closed surface ∂M) equals Q if the charge lies inside M and zero otherwise. Note that even if $(0, 0, 0) \notin M$, $\mathbf{E}$ will still be *nonzero* on M.

For a continuous charge distribution described by a charge density ρ in a region W, the field $\mathbf{E}$ is related to the density ρ by

$$\text{div } \mathbf{E} = \nabla \cdot \mathbf{E} = \rho.$$

Thus, by Gauss' theorem,

$$\iint_{\partial W} \mathbf{E} \cdot d\mathbf{S} = \iiint_W \rho\, dV = Q;$$

that is, the flux out of a surface is equal to the total charge inside. ▲

Divergence in Spherical Coordinates

We next use Gauss' theorem to derive the formula

$$\text{div } \mathbf{F} = \frac{1}{\rho^2}\frac{\partial}{\partial\rho}(\rho^2 F_\rho) + \frac{1}{\rho\sin\phi}\frac{\partial}{\partial\phi}(\sin\phi\, F_\phi) + \frac{1}{\rho\,\sin\phi}\frac{\partial F_\theta}{\partial\theta} \tag{8}$$

for the divergence of a vector field $\mathbf{F}$ in spherical coordinates, which was stated in Section 8.2. (Again, the subscripts here denote *components*, *not* partial derivatives.) The method is to use the formula

$$\text{div } \mathbf{F}(\mathrm{P}) = \underset{W \to P}{\text{limit}} \frac{1}{V(W)} \iint_{\partial W} \mathbf{F} \cdot \mathbf{n}\, dS, \tag{9}$$

where W is a region with volume $V(W)$, which shrinks down to a point $\mathbf{P}$ (in the main text we took a ball, but one can use regions of any shape). Let W be the shaded region in Figure 8.4.8.

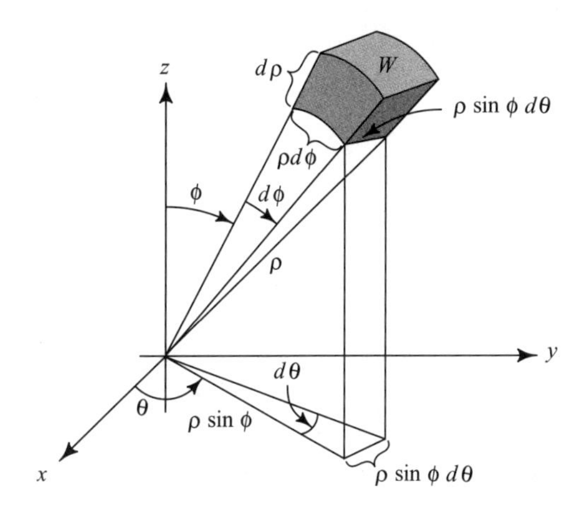

Figure 8.4.8 Infinitesimal volume determined by $d\rho, d\theta, d\phi$ at (ρ, θ, ϕ).

For the two faces orthogonal to the radial direction, the surface integral in equation (9) is, approximately,

$$\begin{aligned} &F_\rho(\rho + d\rho, \phi, \theta) \times \text{ (area of outer face)} - F_\rho(\rho, \phi, \theta) \times \text{ (area of inner face)} \\ &\quad \approx F_\rho(\rho + d\rho, \phi, \theta)(\rho + d\rho)^2 \sin\phi \, d\phi \, d\theta - F_\rho(\rho, \phi, \theta)\rho^2 \sin\phi \, d\phi \, d\theta \\ &\quad \approx \frac{\partial}{\partial \rho}(F_\rho \rho^2 \sin\phi) \, d\rho \, d\phi \, d\theta \end{aligned} \tag{10}$$

by the one-variable mean-value theorem. Dividing by the volume of the region W, namely, $\rho^2 \sin\phi \, d\rho \, d\phi \, d\theta$, we see that the contribution to the right-hand side of equation (9) is

$$\frac{1}{\rho^2}\frac{\partial}{\partial \rho}(\rho^2 F_\rho) \tag{11}$$

for these faces. Likewise, the contribution from the faces orthogonal to the ϕ direction is

$$\frac{1}{\rho \sin\phi}\frac{\partial}{\partial \phi}(\sin\phi F_\phi), \qquad \text{and for the } \theta \text{ direction,} \qquad \frac{1}{\rho \sin\phi}\frac{\partial F_\theta}{\partial \theta}.$$

Substituting (11) and these expressions in equation (9) and taking the limit gives equation (8).

EXERCISES

1. Use the divergence theorem to calculate the flux of $\mathbf{F} = (x - y)\mathbf{i} + (y - z)\mathbf{j} + (z - x)\mathbf{k}$ out of the unit sphere.

2. Let $\mathbf{F} = x^3\mathbf{i} + y^3\mathbf{j} + z^3\mathbf{k}$. Evaluate the surface integral of $\mathbf{F}$ over the unit sphere.

3. Evaluate $\iint_{\partial W} \mathbf{F} \cdot d\mathbf{S}$, where $\mathbf{F} = x\mathbf{i} + y\mathbf{j} + z\mathbf{k}$ and W is the unit cube (in the first octant). Perform the calculation directly and check by using the divergence theorem.

4. Repeat Exercise 3 for

(a) $\mathbf{F} = \mathbf{i} + \mathbf{j} + \mathbf{k}$
(b) $\mathbf{F} = x^2\mathbf{i} + x^2\mathbf{j} + z^2\mathbf{k}$

5. Let $\mathbf{F} = y\mathbf{i} + z\mathbf{j} + xz\mathbf{k}$. Evaluate $\iint_{\partial W} \mathbf{F} \cdot d\mathbf{S}$ for each of the following regions W:

(a) $x^2 + y^2 \leq z \leq 1$
(b) $x^2 + y^2 \leq z \leq 1$ and $x \geq 0$
(c) $x^2 + y^2 \leq z \leq 1$ and $x \leq 0$

6. Repeat Exercise 5 for $\mathbf{F} = (x - y)\mathbf{i} + (y - z)\mathbf{j} + (z - x)\mathbf{k}$. [The solution to part (b) only is in the Study Guide to this text.]

7. Find the flux of the vector field $\mathbf{F} = (x - y^2)\mathbf{i} + y\mathbf{j} + x^3\mathbf{k}$ out of the rectangular solid $[0, 1] \times [1, 2] \times [1, 4]$.

8. Evaluate $\iint_S \mathbf{F} \cdot d\mathbf{S}$, where $\mathbf{F} = 3xy^2\mathbf{i} + 3x^2y\mathbf{j} + z^3\mathbf{k}$ and S is the surface of the unit sphere.

9. Evaluate $\iint_{\partial W} \mathbf{F} \cdot \mathbf{n}\, dA$, where $\mathbf{F}(x, y, z) = x\mathbf{i} + y\mathbf{j} - z\mathbf{k}$ and W is the unit cube in the first octant. Perform the calculation directly and check by using the divergence theorem.

10. Evaluate the surface integral $\iint_{\partial S} \mathbf{F} \cdot \mathbf{n}\, dA$, where $\mathbf{F}(x, y, z) = \mathbf{i} + \mathbf{j} + z(x^2 + y^2)^2\mathbf{k}$ and ∂S is the surface of the cylinder $x^2 + y^2 \leq 1, 0 \leq z \leq 1$.

11. Prove that

$$\iiint_W (\nabla f) \cdot \mathbf{F}\, dx\, dy\, dz = \iint_{\partial W} f\mathbf{F} \cdot \mathbf{n}\, dS - \iiint_W f\nabla \cdot \mathbf{F}\, dx\, dy\, dz.$$

12. Prove the identity

$$\nabla \cdot (\mathbf{F} \times \mathbf{G}) = \mathbf{G} \cdot (\nabla \times \mathbf{F}) - \mathbf{F} \cdot (\nabla \times \mathbf{G}).$$

13. Show that $\iiint_W (1/r^2)\, dx\, dy\, dz = \iint_{\partial W} (\mathbf{r} \cdot \mathbf{n}/r^2)\, dS$, where $\mathbf{r} = x\mathbf{i} + y\mathbf{j} + z\mathbf{k}$.

14. Fix vectors $\mathbf{v}_1, \ldots, \mathbf{v}_k \in \mathbb{R}^3$ and numbers ("charges") $q_1, \ldots, q_k$. Define the function ϕ by $\phi(x, y, z) = \sum_{i=1}^{k} q_i/(4\pi\, \|\mathbf{r} - \mathbf{v}_i\|)$, where $\mathbf{r} = (x, y, z)$. Show that for a closed surface S and $\mathbf{E} = -\nabla\phi$,

$$\iint_S \mathbf{E} \cdot d\mathbf{S} = Q,$$

where Q is the total charge inside S. (Assume that Gauss' law from Theorem 10 applies and that none of the charges are on S.)

15. Prove ***Green's identities***

$$\iint_{\partial W} f\nabla g \cdot \mathbf{n}\, dS = \iiint_W (f\nabla^2 g + \nabla f \cdot \nabla g)\, dV$$

and

$$\iint_{\partial W} (f\nabla g - g\nabla f)\cdot \mathbf{n}\, dS = \iiint_W (f\nabla^2 g - g\nabla^2 f)\, dV.$$

16. Suppose $\mathbf{F}$ satisfies div $\mathbf{F} = 0$ and curl $\mathbf{F} = \mathbf{0}$ on all of $\mathbb{R}^3$. Show that we can write $\mathbf{F} = \nabla f$, where $\nabla^2 f = 0$.

17. Let ρ be a continuous function on $\mathbb{R}^3$ such that $\rho(\mathbf{q}) = 0$ except for $\mathbf{q}$ in some region W. Let $\mathbf{q} \in W$ be denoted by $\mathbf{q} = (x, y, z)$. The ***potential*** of ρ is defined to be the function

$$\phi(\mathbf{p}) = \iiint_W \frac{\rho(\mathbf{q})}{4\pi\,\|\mathbf{p} - \mathbf{q}\|}\, dV(\mathbf{q}),$$

where $\|\mathbf{p} - \mathbf{q}\|$ is the distance between $\mathbf{p}$ and $\mathbf{q}$.

(a) Using the method of Theorem 10, show that

$$\iint_{\partial W} \nabla\phi \cdot \mathbf{n}\, dS = -\iiint_W \rho\, dV$$

for those regions W that can be partitioned into a finite union of symmetric elementary regions.

(b) Show that ϕ satisfies ***Poisson's equation***

$$\nabla^2\phi = -\rho.$$

[HINT: Use part (a).] (Notice that if ρ is a charge density, then the integral defining ϕ may be thought of as the sum of the potential at $\mathbf{p}$ caused by point charges distributed over W according to the density ρ.)

18. Suppose $\mathbf{F}$ is tangent to the closed surface $S = \partial W$ of a region W. Prove that

$$\iiint_W (\text{div } \mathbf{F})\, dV = 0.$$

19. Use Gauss' law and symmetry to prove that the electric field due to a charge Q evenly spread over the surface of a sphere is the same outside the surface as the field from a point charge Q located at the center of the sphere. What is the field inside the sphere?

20. Reformulate Exercise 19 in terms of gravitational fields.

21. Show how Gauss' law can be used to solve part (b) of Exercise 25 in Section 8.3.

22. Let S be a closed surface. Use Gauss' theorem to show that if $\mathbf{F}$ is a C^2 vector field, then we have $\iint_S (\nabla \times \mathbf{F}) \cdot d\mathbf{S} = 0$.

23. Let S be the surface of region W. Show that

$$\iint_S \mathbf{r}\cdot\mathbf{n}\,dS = 3\text{ volume }(W).$$

Explain this geometrically.

8.5 Some Differential Equations of Mechanics and Technology

Isaac Newton reputedly said, "All in nature reduces to differential equations." This point of view was paraphrased by Max Planck (see the Historical Note in Section 3.3): "... Present day physics, as far as it is theoretically organized, is completely governed by a system of space–time differential equations."

In this section, we apply the central theorems of vector analysis to the derivation of the differential equations governing heat transfer, electromagnetism, and the motion of some fluids.

Keep in mind the importance of these problems in modern technology. For example, a good understanding of fluids and the ability to do computations to solve their governing equations is at the heart of how one builds a modern airplane or designs a submarine. For instance, the flow of air (the fluid in this case) over the wings of an aircraft is very subtle, even though the governing equations are relatively simple. We shall derive a slightly idealized form of these equations in this section. Likewise, the equations of electromagnetism, as we will discuss in the following paragraphs, is central to the communications industry; wireless, television, and much of the operation of modern electronic devices, including computers, depends on these and related fundamental equations.

Conservation Laws

As preparation for deriving the equations of a fluid, let us first discuss an important equation that is referred to as a ***conservation*** equation. For fluids, it expresses the conservation of mass; for electromagnetic theory, it expresses the conservation of charge. We shall apply these ideas to the equation for heat conduction and to electromagnetism.

Let $\mathbf{V}(t, x, y, z)$ be a C^1 vector field on $\mathbb{R}^3$ for each t and let $\rho(t, x, y, z)$ be a C^1 real-valued function. By the ***law of conservation of mass*** for $\mathbf{V}$ and ρ, we mean that the condition

$$\frac{d}{dt}\iiint_W \rho\,dV = -\iint_{\partial W} \mathbf{J}\cdot\mathbf{n}\,dS$$

holds for all regions W in $\mathbb{R}^3$, where $\mathbf{J} = \rho\mathbf{V}$ (see Figure 8.5.1).

If we think of ρ as a mass density (ρ could also be a charge density)—that is, the mass per unit volume—and of $\mathbf{V}$ as the velocity field of a fluid, the condition simply says that the rate of change of total mass in W equals the rate at which mass flows *into* W. Recall that $\iint_{\partial W} \mathbf{J}\cdot\mathbf{n}\,dS$ is called the *flux* of $\mathbf{J}$. We need the following result.

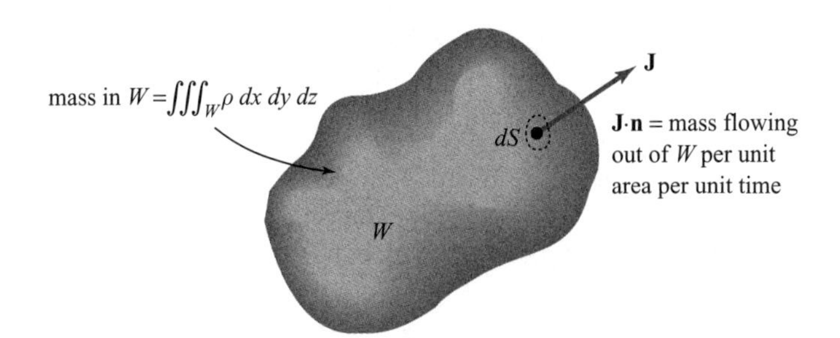

Figure 8.5.1 The rate of change of mass in W equals the rate at which mass crosses ∂W.

THEOREM 11 For $\mathbf{V}$ and ρ (a smooth vector field and a scalar field on $\mathbb{R}^3$), the law of conservation of mass for $\mathbf{V}$ and ρ is equivalent to the condition

$$\text{div}\, \mathbf{J} + \frac{\partial \rho}{\partial t} = 0. \tag{1}$$

That is,

$$\rho \,\text{div}\, \mathbf{V} + \mathbf{V} \cdot \nabla \rho + \frac{\partial \rho}{\partial t} = 0. \tag{1$'$}$$

Here, div $\mathbf{J}$ means that we compute div $\mathbf{J}$ for t held fixed, and $\partial \rho / \partial t$ means we differentiate ρ with respect to t for x, y, z fixed.

PROOF First, observe that by differentiating under the integral, we get

$$\frac{d}{dt} \iiint_W \rho \, dx\, dy\, dz = \iiint_W \frac{\partial \rho}{\partial t} \, dx\, dy\, dz$$

and also

$$\iint_{\partial W} \mathbf{J} \cdot \mathbf{n} \, dS = \iiint_W \text{div}\, \mathbf{J} \, d\mathbf{V}$$

by the divergence theorem. Thus, conservation of mass is equivalent to the condition

$$\iiint_W \left(\text{div}\, \mathbf{J} + \frac{\partial \rho}{\partial t} \right) dx\, dy\, dz = 0.$$

Because this is to hold for all regions W, it is equivalent to div $\mathbf{J} + \partial \rho / \partial t = 0$. ■

The equation div $\mathbf{J} + \partial \rho / \partial t = 0$ is called the ***equation of continuity***. An interesting remark is that using the change of variables formula, the law of conservation of mass may be shown to be equivalent to the condition

$$\frac{d}{dt}\iiint_{W_t} \rho \, dV = 0,$$

where W_t is the image of W obtained by moving each point in W along flow lines of $\mathbf{V}$ for time t. This result is a special case of the *transport theorem* that we discuss next.

The Transport Theorem

The transport theorem is an interesting application of the divergence theorem that will be needed in our derivation of the equations of a fluid.

THEOREM 12 Let $\mathbf{F}$ be a vector field on $\mathbb{R}^3$ and denote the flow line of $\mathbf{F}$ starting at $\mathbf{x}$ after time t by $\phi(\mathbf{x}, t)$. (See the Internet supplement to Section 4.4 for more information.) Let $J(\mathbf{x}, t)$ be the Jacobian of the map $\phi_t \colon \mathbf{x} \mapsto \phi(\mathbf{x}, t)$ for t fixed. Then

$$\frac{\partial}{\partial t} J(\mathbf{x}, t) = [\operatorname{div} \mathbf{F}(\phi(\mathbf{x}, t))] J(\mathbf{x}, t).$$

For a given function $f(x, y, z, t)$ and a region $W \subset \mathbb{R}^3$, the ***transport equation*** holds:

$$\frac{d}{dt}\iiint_{W_t} f(x, y, z, t)\, dx\, dy\, dz = \iiint_{W_t} \left(\frac{Df}{Dt} + f \operatorname{div} \mathbf{F}\right) dx\, dy\, dz,$$

where $W_t = \phi_t(W)$, which is the region moving with the flow, and where

$$\frac{Df}{Dt} = \partial f/\partial t + \nabla f \cdot \mathbf{F}$$

is the ***material derivative***.

Taking $f = 1$, Theorem 12 implies that the following assertions are equivalent (which justifies the use of the term *incompressible*):

1. $\operatorname{div} \mathbf{F} = 0$
2. $\text{volume}\,(W_t) = \text{volume}\,(W)$
3. $J(\mathbf{x}, t) = 1$

Let ϕ, J, $\mathbf{F}$, f be as just defined. There is also a vector form of the transport theorem, namely,

$$\frac{d}{dt}\iiint_{W_t} (f\mathbf{F})\, dx\, dy\, dz$$

$$= \iiint_{W_t} \left[\frac{\partial}{\partial t}(f\mathbf{F}) + \mathbf{F} \cdot \nabla(f\mathbf{F}) + (f\mathbf{F}) \operatorname{div} \mathbf{F}\right] dx\, dy\, dz,$$

where $\mathbf{F} \cdot \nabla(f\mathbf{F})$ denotes the 3×3 derivative matrix $\mathbf{D}(f\mathbf{F})$ operating on the column vector $\mathbf{F}$; in Cartesian coordinates, $\mathbf{F} \cdot \nabla \mathbf{G}$ is the vector whose ith component is

$$\sum_{j=1}^{3} F_j \frac{\partial G^i}{\partial x_j} = F_1 \frac{\partial G^i}{\partial x} + F_2 \frac{\partial G^i}{\partial y} + F_3 \frac{\partial G^i}{\partial z}.$$

We shall leave the proofs of these results, which are extensions of the arguments used to prove Theorem 11, to the reader (see the exercises).

Derivation of Euler's Equation of a Perfect Fluid

The continuity equation is not sufficient to completely determine the motion of a fluid—we need other conditions.

The fluids that the continuity equation governs can be compressible. If $\operatorname{div} \mathbf{V} = 0$ (incompressible case) and ρ is constant, equation (1′) follows automatically. But in general, even for incompressible fluids, the equation is not automatic, because ρ can depend on (x, y, z) and t. Thus, even if the equation $\operatorname{div} \mathbf{V} = 0$ holds, $\operatorname{div}(\rho \mathbf{V}) \neq 0$ may still be true.

Here we discuss Euler's equation for a perfect fluid. Consider a nonviscous fluid moving in space with a velocity field $\mathbf{V}$. When we say that the fluid is *perfect*, we mean that if W is any portion of the fluid, forces of pressure act on the boundary of W along its normal. We assume that the force per unit area acting on ∂W is $-p\mathbf{n}$, where $p(x, y, z, t)$ is some function called the ***pressure*** (see Figure 8.5.2). Thus, the total pressure force acting on W is

$$\mathbf{F}_{\partial W} = \text{force} = -\iint_{\partial W} p\mathbf{n}\, dS.$$

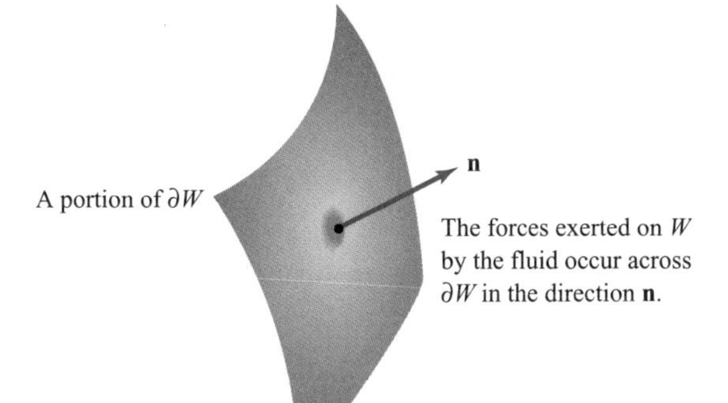

Figure 8.5.2 The force acting on ∂W per unit area is $-p\mathbf{n}$.

This is a *vector* quantity; the ith component of $\mathbf{F}_{\partial W}$ is the integral of the ith component of $p\mathbf{n}$ over the surface ∂W (this is therefore the surface integral of a real-valued function). If $\mathbf{e}$ is any fixed vector in space, we have

$$\mathbf{F}_{\partial W} \cdot \mathbf{e} = -\iint_{\partial W} p\mathbf{e} \cdot \mathbf{n}\, dS,$$

which is the integral of a scalar over ∂W. By the divergence theorem and identity (7) in the table of vector identities (Section 4.4), we get

$$\mathbf{E} \cdot \mathbf{F}_{\partial W} = -\iiint_W \operatorname{div}(p\mathbf{E})\, dx\, dy\, dz = -\iiint_W (\operatorname{grad} p) \cdot \mathbf{E}\, dx\, dy\, dz,$$

so that

$$\mathbf{F}_{\partial W} = -\iiint_W \nabla p\, dx\, dy\, dz.$$

Now we apply *Newton's second law* to a moving region W_t. As in the transport theorem, $W_t = \phi_t(W)$, where $\phi_t(\mathbf{x}) = \phi(\mathbf{x}, t)$ denotes the flow of $\mathbf{V}$. The rate of change of momentum of the fluid in W_t equals the force acting on it:

$$\frac{d}{dt}\iiint_{W_t} p\mathbf{V}\, dx\, dy\, dz = \mathbf{F}_{\partial W_t} = \iiint_{W_t} \nabla p\, dx\, dy\, dz.$$

We apply the vector form of the transport theorem to the left-hand side to get

$$\iiint_{W_t} \left[\frac{\partial}{\partial t}(\rho\mathbf{V}) + \mathbf{V} \cdot \nabla(\rho\mathbf{V}) + p\mathbf{V} \operatorname{div} \mathbf{V}\right] dx\, dy\, dz = -\iiint_{W_t} \nabla p\, dx\, dy\, dz.$$

Because W_t is arbitrary, this is equivalent to

$$\frac{\partial}{\partial t}(\rho\mathbf{V}) + \mathbf{V} \cdot \nabla(\rho\mathbf{V}) + \rho\mathbf{V} \operatorname{div} \mathbf{V} = -\nabla p.$$

Simplification using the equation of continuity, namely, formula (1′), gives

$$\rho\left(\frac{\partial \mathbf{V}}{\partial t} + \mathbf{V} \cdot \nabla\mathbf{V}\right) = -\nabla p. \tag{2}$$

This is ***Euler's equation for a perfect fluid***. For compressible fluids, p is a given function of ρ (for instance, for many gases, $p = A\rho^\gamma$ for constants A and γ). On the other hand, if the fluid is incompressible, ρ is to be determined from the condition $\operatorname{div} \mathbf{V} = 0$. Equations (1) and (2) then govern the motion of the fluid.

Historical Note

The equations describing the motion of a fluid were first derived by Leonhard Euler in 1755, in a paper entitled "General Principles of the Motion of Fluids." Euler did basic work in mechanics as well as voluminous work in pure mathematics, a small part of which has already been discussed in this book; he essentially began the subject of analytical mechanics (as opposed to the Euclidean geometric methods used by Newton). He is responsible for the equations of a rigid body (equations that apply, for example, to a tumbling satellite) and the formulation of many basic equations of mechanics in terms of variational principles; that is, by the methods of maxima and minima of real-valued functions. Euler wrote the first comprehensive textbook on calculus and contributed to virtually all branches of mathematics. He wrote several books and hundreds of research papers even after he became totally blind, and he was working on a new treatise on fluid mechanics at the time of his death in 1783. Euler's equations for a fluid were eventually modified by Navier and Stokes to include viscous effects; the resulting Navier–Stokes equations are described in virtually every textbook on fluid mechanics.[7] Stokes is, of course, also responsible for developing Stokes' theorem, one of the main results discussed in this text!

Conservation of Energy and the Derivation of the Heat Equation

If $T(t, x, y, z)$ (a C^2 function) denotes the temperature in a body at time t, then ∇T represents the temperature gradient: Heat "flows" with the vector field $-\nabla T = \mathbf{F}$. Note that ∇T points in the direction of *increasing* T. Because heat flows from hot to cold, we have inserted a minus sign to reflect this physically observable fact. The energy density, that is, the energy per unit volume, is $c\rho_0 T$, where c is a constant (called the specific heat) and ρ_0 is the mass density, assumed constant. (We accept these assertions from elementary physics.) The ***energy flux vector*** is defined to be $\mathbf{J} = k\mathbf{F}$, where k is a constant called the ***conductivity***.

One now makes the hypothesis that energy is conserved. This means that $\mathbf{J}$ and $\rho = c\rho_0 T$ should obey the law of conservation of mass, with ρ playing the role of "mass" (note that it is *energy density*, not mass); that is,

$$\frac{d}{dt}\iiint_W \rho\, dV = -\iint_{\partial W} \mathbf{J}\cdot\mathbf{n}\, dS.$$

By Theorem 11, this assertion is equivalent to

$$\operatorname{div}\mathbf{J} + \frac{\partial \rho}{\partial t} = 0.$$

[7]The Clay Foundation has offered a prize of $1 million to anyone who shows that for the incompressible Navier–Stokes equations, smooth data at $t = 0$ lead to smooth solutions for all $t > 0$.

But

$$\operatorname{div} \mathbf{J} = \operatorname{div}(-k\nabla T) = -k\nabla^2 T.$$

(Recall that $\nabla^2 T = \partial^2 T/\partial x^2 + \partial^2 T/\partial y^2 + \partial^2 T/\partial z^2$ and ∇^2 is the Laplace operator.) Continuing, we have

$$\frac{\partial \rho}{\partial t} = \frac{\partial (c\rho_0 T)}{\partial t} = c\rho_0 \frac{\partial T}{\partial t}.$$

Thus, the equation $\operatorname{div} \mathbf{J} + \partial\rho/\partial t = 0$ becomes

$$\frac{\partial T}{\partial t} = \frac{k}{c\rho_0}\nabla^2 T = k\nabla^2 T, \tag{3}$$

where $\kappa = k/c\rho_0$ is called the ***diffusivity***. Equation (3) is the important ***heat equation***.

Just as equations (1) and (2) govern the flow of an ideal fluid, equation (3) governs the conduction of heat in the following sense. If $T(0, x, y, z)$ is a given initial temperature distribution, then a unique $T(t, x, y, z)$ is determined that satisfies equation (3). In other words, the initial condition at $t = 0$ gives the result for $t > 0$. Notice that if T does not change with time (the steady-state case), then we must have $\nabla^2 T = 0$ (Laplace's equation).

Maxwell's Equations and the Prediction of Radio Waves: The Communication Revolution Begins

We now return to *Maxwell's equations*, which govern the propagation of electromagnetic fields. The form of these equations depends on the physical units one is employing, and changing units introduces factors like 4π and the velocity of light. We shall choose the system in which Maxwell's equations are simplest.

Let $\mathbf{E}$ and $\mathbf{H}$ be C^1 functions of (t, x, y, z) that are vector fields for each t. They satisfy (by definition) ***Maxwell's equation with charge density*** $\rho(t, x, y, z)$ and ***current density*** $\mathbf{J}(t, x, y, z)$ when the following conditions hold:

$$\nabla \cdot \mathbf{E} = \rho \text{ (Gauss' law)}, \tag{4}$$

$$\nabla \cdot \mathbf{H} = 0 \text{ (no negative sources)}, \tag{5}$$

$$\nabla \times \mathbf{E} + \frac{\partial \mathbf{H}}{\partial t} = \mathbf{0} \text{ (Faraday's law)}, \tag{6}$$

and

$$\nabla \times \mathbf{H} - \frac{\partial \mathbf{E}}{\partial t} = J \text{ (Ampère's law)}, \tag{7}$$

Of these laws, equations (4) and (6) were described in integral form in Sections 8.2 and 8.4; historically, they arose in these forms as physically observed laws. Ampère's law was mentioned for a special case in Section 7.2, Example 12.

Physically, one interprets $\mathbf{E}$ as the ***electric field*** and $\mathbf{H}$ as the ***magnetic field***. According to the preceding equations, as time t progresses, these fields interact with each other, and with any charges and currents that are present. For example, the propagation of electromagnetic waves (TV signals, radio waves, light from the sun, etc.) in a vacuum is governed by these equations with $\mathbf{J} = 0$ and $\rho = 0$.

Because $\nabla \cdot \mathbf{H} = 0$, we can apply Theorem 8 (from Section 8.3) to conclude that $\mathbf{H} = \nabla \times \mathbf{A}$ for some vector field $\mathbf{A}$. (We are assuming that $\mathbf{H}$ is defined on all of $\mathbb{R}^3$ for each time t.) The vector field $\mathbf{A}$ is not unique, and we can use $\mathbf{A}' = \mathbf{A} + \nabla f$ equally well for any function $f(t, x, y, z)$, because $\nabla \times \nabla f = 0$. (This freedom in the choice of $\mathbf{A}$ is called *gauge freedom*.) For any such choice of $\mathbf{A}$, we have, by equation (6),

$$\begin{aligned} \mathbf{0} = \nabla \times \mathbf{E} + \frac{\partial \mathbf{H}}{\partial t} &= \nabla \times \mathbf{E} + \frac{\partial}{\partial t} \nabla \times \mathbf{A} \\ &= \nabla \times \mathbf{E} \times \nabla \times \frac{\partial \mathbf{A}}{\partial t} \\ &= \nabla \times \left(\mathbf{E} + \frac{\partial \mathbf{A}}{\partial t} \right). \end{aligned}$$

Applying Theorem 7 (from Section 8.3), there is a real-valued function ϕ on $\mathbb{R}^3$ such that

$$\mathbf{E} + \frac{\partial \mathbf{A}}{\partial t} = -\nabla \phi.$$

Substituting this equation and $\mathbf{H} = \nabla \times \mathbf{A}$ into equation (7), and using the vector identity (whose proof we leave as an exercise)

$$\nabla \times (\nabla \times \mathbf{A}) = \nabla(\nabla \cdot \mathbf{A}) - \nabla^2 \mathbf{A},$$

we get

$$\begin{aligned} \mathbf{J} = \nabla \times \mathbf{H} - \frac{\partial \mathbf{E}}{\partial t} &= \nabla \times (\nabla \times \mathbf{A}) - \frac{\partial}{\partial t} \left(-\frac{\partial \mathbf{A}}{\partial t} - \nabla \phi \right) \\ &= \nabla(\nabla \cdot \mathbf{A}) - \nabla^2 \mathbf{A} + \frac{\partial^2 \mathbf{A}}{\partial t^2} + \frac{\partial}{\partial t}(\nabla \phi). \end{aligned}$$

Thus,

$$\nabla^2 \mathbf{A} - \frac{\partial^2 \mathbf{A}}{\partial t^2} = -\mathbf{J} + \nabla(\nabla \cdot \mathbf{A}) + \frac{\partial}{\partial t}(\nabla \phi).$$

That is,

$$\nabla^2 \mathbf{A} - \frac{\partial^2 \mathbf{A}}{\partial t^2} = -\mathbf{J} + \nabla \left(\nabla \cdot \mathbf{A} + \frac{\partial \phi}{\partial t} \right). \tag{8}$$

Again using the equation $\mathbf{E} + \partial \mathbf{A}/\partial t = -\nabla \phi$ and the equation $\nabla \cdot \mathbf{E} = \rho$, we obtain

$$\rho = \nabla \cdot \mathbf{E} = \nabla \cdot \left(-\nabla \phi - \frac{\partial \mathbf{A}}{\partial t} \right) = -\nabla^2 \phi - \frac{\partial (\nabla \cdot \mathbf{A})}{\partial t}.$$

That is,

$$\nabla^2 \phi = -\rho - \frac{\partial (\nabla \cdot \mathbf{A})}{\partial t}. \tag{9}$$

Now let us exploit the freedom in our choice of $\mathbf{A}$. We impose the "condition"

$$\nabla \cdot \mathbf{A} + \frac{\partial \phi}{\partial t} = 0. \tag{10}$$

We must be sure we can do this. Supposing we have a given $\mathbf{A}_0$ and a corresponding ϕ_0, can we choose a new $\mathbf{A} = \mathbf{A}_0 + \nabla f$ and then a new ϕ such that $\nabla \cdot \mathbf{A} + \partial \phi/\partial t = 0$? With this new $\mathbf{A}$, the new ϕ is $\phi_0 - \partial f/\partial t$; we leave verification as an exercise for the reader. Condition (10) on f then becomes

$$0 = \nabla \cdot (\mathbf{A}_0 + \nabla f) = \frac{\partial (\phi_0 - \partial f/\partial t)}{\partial t} = \nabla \cdot \mathbf{A}_0 + \nabla^2 f + \frac{\partial \phi_0}{\partial t} - \frac{\partial^2 f}{\partial t^2}$$

or

$$\nabla^2 f - \frac{\partial^2 f}{\partial t^2} = -\left(\nabla \cdot \mathbf{A}_0 + \frac{\partial \phi_0}{\partial t} \right). \tag{11}$$

Thus, to be able to choose $\mathbf{A}$ and ϕ satisfying $\nabla \cdot \mathbf{A} + \partial \phi/\partial t = 0$, we must be able to solve equation (11) for f. One can indeed do this under general conditions, although we do not prove it here. Equation (11) is called the ***inhomogeneous wave equation***.

If we accept that $\mathbf{A}$ and ϕ can be chosen to satisfy $\nabla \cdot \mathbf{A} + \partial \phi/\partial t = 0$, then equations (8) and (9) for $\mathbf{A}$ and ϕ become

$$\nabla^2 \mathbf{A} - \frac{\partial^2 \mathbf{A}}{\partial t^2} = -\mathbf{J}; \tag{8$'$}$$

$$\nabla^2 \phi - \frac{\partial^2 \phi}{\partial t^2} = -\rho. \tag{9$'$}$$

Conversely, if $\mathbf{A}$ and ϕ satisfy the equations $\nabla \cdot \mathbf{A} + \partial\phi/\partial t = 0$, $\nabla^2\phi - \partial^2\phi/\partial t^2 = -\rho$, and $\nabla^2\mathbf{A} - \partial^2\mathbf{A}/\partial t^2 = -\mathbf{J}$, then $\mathbf{E} = -\nabla\phi - \partial\mathbf{A}/\partial t$ and $\mathbf{H} = \nabla \times \mathbf{A}$ satisfy Maxwell's equations. *This procedure then "reduces" Maxwell's equations to a study of the wave equation.*[8]

Since the eighteenth century, solutions to the wave equation have been well studied (one learns these in most courses on differential equations). To indicate the wavelike nature of the solutions, for example, observe that for any function f,

$$\phi(t, x, y, z) = f(x - t)$$

solves the wave equation $\nabla^2\phi - (\partial^2\phi/\partial t^2) = 0$. This solution just propagates the graph of f like a wave; thus, one might conjecture that solutions of Maxwell's equations are wavelike in nature. Historically, all of this was Maxwell's great achievement, and it soon led to Hertz's discovery of radio waves. To quote from the Feynman *Lectures on Physics* (Vol. II):

> From a long view of the history of mankind—seen from, say, ten thousand years from now—there can be little doubt that the most significant event of the nineteenth century will be judged as Maxwell's discovery of the laws of electrodynamics. The American Civil War will pale into provincial insignificance in comparison with this important scientific event of the same decade.

Mathematics again shows its uncanny ability not only to *describe* but to *predict* natural phenomena.

There are other techniques (called Green's function methods) for dealing with the basic equations of mechanics and mathematical physics that also rely on vector calculus. Some of these methods are discussed in the Internet supplement for this book.

EXERCISES

1. Use a direct argument (or the proof of Theorem 1 in the Internet supplement to Section 4.4) to show that

$$\frac{\partial}{\partial t} J(\mathbf{x}, t) = [\text{div } \mathbf{F}(\phi(\mathbf{x}, t))] J(\mathbf{x}, t).$$

---

[8]There are variations on this procedure. For further details, see, for example, *Differential Equations of Applied Mathematics*, by G. F. D. Duff and D. Naylor, Wiley, New York, 1966, or books on electromagnetic theory, such as *Classical Electrodynamics*, by J. D. Jackson, Wiley, New York, 1962.

2. Using the change of variables theorem and Exercise 1, show that if $f(x, y, z, t)$ is a given function and $W \subset \mathbb{R}^3$ is any region, then the ***transport equation*** holds:

$$\frac{d}{dt}\iiint_{W_t} f(x, y, z, t)\, dx\, dy\, dz = \iiint_{W_t}\left(\frac{Df}{Dt} + f \operatorname{div} \mathbf{F}\right) dx\, dy\, dz$$

where $W_t = \phi_t(W)$, which is the region moving with the flow, and where $Df/Dt = \partial f/\partial t + \nabla f \cdot \mathbf{F}$ is the material derivative.

3. Use the transport equation to show that

$$\frac{d}{dt}\iiint_{W_t} \rho\, dx\, dy\, dz = 0$$

is equivalent to the law of conservation of mass.

4. Using Exercise 3 and the change of variables theorem, show that $\rho(\mathbf{x}, t)$ can be expressed in terms of the Jacobian $J(\mathbf{x}, t)$ of the flow map $\phi(\mathbf{x}, t)$ and $\rho(\mathbf{x}, 0)$ by the equation

$$\rho(\mathbf{x}, t) J(\mathbf{x}, t) = \rho(\mathbf{x}, 0).$$

What can you conclude from this for incompressible flow?

5. Prove the vector form of the transport theorem, namely,

$$\frac{d}{dt}\iiint_{W_t} (f\mathbf{F})\, dx\, dy\, dz = \iiint_{W_t}\left[\frac{\partial}{\partial t}(f\mathbf{F}) + \mathbf{F} \cdot \nabla(f\mathbf{F}) + (f\mathbf{F}) \operatorname{div} \mathbf{F}\right] dx\, dy\, dz,$$

where $\mathbf{F} \cdot \nabla(f\mathbf{F})$ denotes the 3×3 derivative matrix $\mathbf{D}(f\mathbf{F})$ operating on the column vector $\mathbf{F}$; in Cartesian coordinates, $\mathbf{F} \cdot \nabla\mathbf{G}$ is the vector whose ith component is

$$\sum_{j=1}^{3} F_j \frac{\partial G^i}{\partial x_j} = F_1 \frac{\partial G^i}{\partial x} + F_2 \frac{\partial G^i}{\partial y} + F_3 \frac{\partial G^i}{\partial z}.$$

6. Let $\mathbf{V}$ be a vector field with flow $\phi(\mathbf{x}, t)$ and let $\mathbf{V}$ and ρ satisfy the law of conservation of mass. Let W_t be the region transported with the flow. Prove the following version of the transport theorem:

$$\frac{d}{dt}\iiint_{W_t} \rho f\, dx\, dy\, dz = \iiint_{W_t} \rho \frac{Df}{Dt}\, dx\, dy\, dz.$$

7. (*Bernoulli's law*) (a) Let $\mathbf{V}$, ρ satisfy the law of conservation of mass and equation (2) (Euler's equation for a perfect fluid). Suppose $\mathbf{V}$ is irrotational and hence that $\mathbf{V} = \nabla\phi$ for a function ϕ. Show that if C is a path connecting two points P_1 and P_2, then

$$\left(\frac{\partial \phi}{\partial t} + \frac{1}{2}\|\mathbf{V}\|^2\right)\Bigg|_{\mathrm{P}_1}^{\mathrm{P}_2} + \int_C \frac{dp}{\rho} = 0.$$

[HINT: You may need the vector identity, $(\mathbf{V} \cdot \nabla)\mathbf{V} = \frac{1}{2}\nabla(\|\mathbf{V}\|^2) + (\nabla \times \mathbf{V}) \times \mathbf{V}$.]

(b) If in part (a), $\mathbf{V}$ is stationary—that is, $\partial\mathbf{V}/\partial t = 0$—and ρ is constant, show that

$$\frac{1}{2}\|\mathbf{V}\|^2 + \frac{p}{\rho}$$

is constant in space. Deduce that, in this situation, *higher pressure is associated with lower fluid speed.*

8. Using Exercise 7, show that if ϕ satisfies Laplace's equation $\nabla^2\phi = 0$, then $\mathbf{V} = \nabla\phi$ is a stationary solution to Euler's equation for a perfect *incompressible* fluid with constant density.

9. Verify that Maxwell's equations imply the equation of continuity for $\mathbf{J}$ and ρ.

10. For a steady-state charge distribution and divergence-free current distribution, the electric and magnetic fields $\mathbf{E}(x, y, z)$ and $\mathbf{H}(x, y, z)$ satisfy

$$\nabla \times \mathbf{E} = \mathbf{0}, \qquad \nabla \cdot \mathbf{H} = 0, \qquad \nabla \cdot \mathbf{J} = 0, \qquad \nabla \cdot \mathbf{E} = \rho, \qquad \text{and} \qquad \nabla \times \mathbf{H} = \mathbf{J}.$$

Here $\rho = \rho(x, y, z)$ and $\mathbf{J}(x, y, z)$ are assumed to be known. The radiation that the fields produce through a surface S is determined by a radiation flux density vector field, called the ***Poynting*** vector field,

$$\mathbf{P} = \mathbf{E} \times \mathbf{H}.$$

(a) If S is a *closed* surface, show that the radiation flux—that is, the flux of $\mathbf{P}$ through S—is given by

$$\iint_S \mathbf{P} \cdot d\mathbf{S} = -\iiint_V \mathbf{E} \cdot \mathbf{J}\, dV,$$

where V is the region enclosed by S.

(b) Examples of such fields are

$$\mathbf{E}(x, y, z) = z\mathbf{j} + y\mathbf{k} \qquad \text{and} \qquad \mathbf{H}(x, y, z) = -xy\mathbf{i} + x\mathbf{j} + yz\mathbf{k}.$$

In this case, find the flux of the Poynting vector through the hemispherical shell shown in Figure 8.5.3. (Notice that it is an *open* surface.)

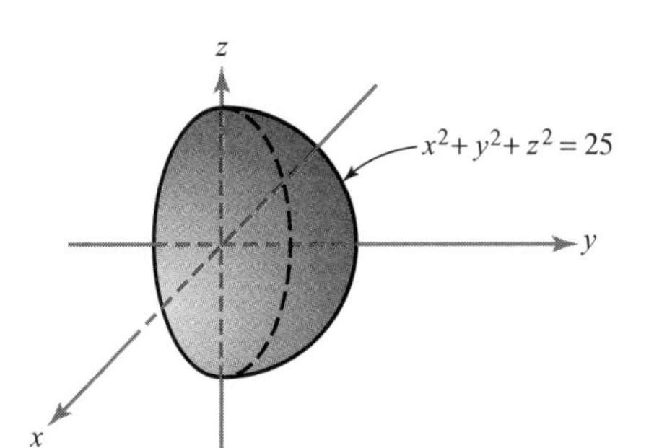

Figure 8.5.3 The surface for Exercise 10.

(c) The fields of part (b) produce a Poynting vector field passing through the toroidal surface shown in Figure 8.5.4. What is the flux through this torus?

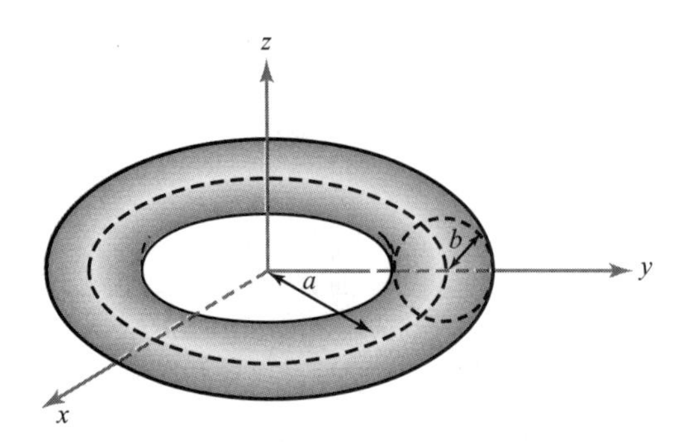

Figure 8.5.4 The surface for Exercise 10(c).

8.6 Differential Forms

The theory of differential forms provides an elegant way of formulating Green's, Stokes', and Gauss' theorems as one statement, the ***fundamental theorem of calculus***. The birth of the concept of a differential form is another dramatic example of how mathematics speaks to mathematicians and drives its own development. These three theorems are, in reality, generalizations of the fundamental theorem of calculus of Newton and Leibniz for functions of one variable,

$$\int_a^b f'(x)\,dx = f(b) - f(a)$$

to two and three dimensions.

Recall that Bernhard Riemann created the concept of n-dimensional spaces. If the fundamental theorem of calculus was truly *fundamental,* then it should generalize to arbitrary dimensions. But wait! The cross product, and therefore the curl, does not generalize to higher dimensions, as we remarked in footnote 3, in Section 1.3. Thus, some new idea is needed.

Recall that Hamilton searched for almost 15 years for his quaternions, which ultimately led to the discovery of the cross product. What is the nonexistence of a cross product in higher dimensions telling us? If the fundamental theorem of calculus is the core concept, this suggests the existence of a mathematical language in which it can be formulated in n-dimensions. In order to achieve this, mathematicians realized that they were forced to move away from vectors and on to the discovery of *dual* vectors and an entirely new mathematical object, a *differential form.* In this new language, all of the theorems of Green, Stokes, and Gauss have the same elegant and extraordinarily simple form.

Simply and very briefly stated, an expression of the type $P\,dx + Q\,dy$ is a 1-form, or a *differential one-form* on a region in the xy plane, and $F\,dx\,dy$ is a 2-form. Analogously, one can define the notion of an n-form. There is an operation d, which takes n-forms to $n+1$-forms. It is like a generalized curl and has the property that for $\omega = P\,dx + Q\,dy$, we have

$$d\omega = \left(\frac{\partial Q}{\partial x} - \frac{\partial P}{\partial y}\right) dx\,dy$$

and so in this notation, Green's theorem becomes

$$\int_{\partial D} \omega = \int_D d\omega,$$

which, interestingly, just switches the boundary operator ∂ with the d operator. However, differential forms are more than just notation. They create a beautiful theory that generalizes to n-dimensions.

In general, if M is an oriented surface of dimension n with an $(n-1)$-dimensional boundary ∂M and if ω is an $(n-1)$-form on M, then the fundamental theorem of calculus (also called the *generalized Stokes' theorem*) reads

$$\boxed{\int_{\partial M} \omega = \int_M d\omega.}$$

A useful thing for the reader to contemplate at this stage is the sense in which the fundamental theorem of calculus becomes a special instance of this result.

In this section, we shall give a very elementary exposition of the theory of forms. Because our primary goal is to show that the theorems of Green, Stokes, and Gauss can be unified under a single theorem, we shall be satisfied with less than the strongest possible version of these theorems. Moreover, we shall introduce forms in a purely axiomatic and nonconstructive manner, thereby avoiding the tremendous number of formal algebraic preliminaries that are usually required for their construction. To the purist our approach will be far from complete, but to the student it may be comprehensible. We hope that this will motivate some students to delve further into the theory of differential forms.

We shall begin by introducing the notion of a 0-form.

0-Forms

Let K be an open set in $\mathbb{R}^3$. A 0-***form*** on K is a real-valued function $f\colon K \to \mathbb{R}$. When we differentiate f once, it is assumed to be of class C^1, and C^2 when we differentiate twice.

Given two 0-forms f_1 and f_2 on K, we can add them in the usual way to get a new 0-form $f_1 + f_2$ or multiply them to get a 0-form $f_1 f_2$.

EXAMPLE 1 $f_1(x, y, z) = xy + yz$ and $f_2(x, y, z) = y \sin xz$ are 0-forms on $\mathbb{R}^3$:

$$(f_1 + f_2)(x, y, z) = xy + yz + y \sin xz$$

and

$$(f_1 f_2)(x, y, z) = y^2 x \sin xz + y^2 z \sin xz. \quad \blacktriangle$$

1-Forms

The ***basic* 1-*forms*** are the expressions dx, dy, and dz. At present we consider these to be only formal symbols. A 1-***form*** ω on an open set K is a formal linear combination

$$\omega = P(x, y, z)\, dx + Q(x, y, z)\, dy + R(x, y, z)\, dz,$$

or simply

$$\omega = P\, dx + Q\, dy + R\, dz,$$

where P, Q, and R are real-valued functions on K. By the expression $P\, dx$ we mean the 1-form $P\, dx + 0 \cdot dy + 0 \cdot dz$, and similarly for $Q\, dy$ and $R\, dz$. Also, the order of $P\, dx$, $Q\, dy$, and $R\, dz$ is immaterial, and so

$$P\, dx + Q\, dy + R\, dz = R\, dz + P\, dx + Q\, dy, \text{ etc.}$$

Given two 1-forms $\omega_1 = P_1\, dx + Q_1\, dy + R_1\, dz$ and $\omega_2 = P_2\, dx + Q_2\, dy + R_2\, dz$, we can add them to get a new 1-form $\omega_1 + \omega_2$ defined by

$$\omega_1 + \omega_2 = (P_1 + P_2)\, dx + (Q_1 + Q_2)\, dy + (R_1 + R_2)\, dz,$$

and given a 0-form f, we can form the 1-form $f\omega_1$ defined by

$$f\omega_1 = (fP_1)\, dx + (fQ_1)\, dy + (fR_1)\, dz.$$

EXAMPLE 2 Let $\omega_1 = (x + y^2)\, dx + (zy)\, dy + (e^{xyz})\, dz$ and $\omega_2 = \sin y\, dx + \sin x\, dy$ be 1-forms. Then

$$\omega_1 + \omega_2 = (x + y^2 + \sin y)\, dx + (zy + \sin x)\, dy + (e^{xyz})\, dz.$$

If $f(x, y, z) = x$, then

$$f\omega_2 = x \sin y\, dx + x \sin x\, dy. \quad \blacktriangle$$

2-Forms

The ***basic* 2-*forms*** are the formal expressions $dx\, dy$, $dy\, dz$, and $dz\, dx$. These expressions should be thought of as products of dx and dy, dy and dz, and dz and dx.

A 2-***form*** η on K is a formal expression

$$\eta = F\,dx\,dy + G\,dy\,dz + H\,dz\,dx,$$

where F, G, and H are real-valued functions on K. The order of $F\,dx\,dy$, $G\,dy\,dz$, and $H\,dz\,dx$ is immaterial; for example,

$$F\,dx\,dy + G\,dy\,dz + H\,dz\,dx = H\,dz\,dx + F\,dx\,dy + G\,dy\,dz, \text{ etc.}$$

At this point it is useful to note that in a 2-form the basic 1-forms dx, dy, and dz always appear in cyclic pairs (see Figure 8.6.1), that is, $dx\,dy$, $dy\,dz$, and $dz\,dx$.

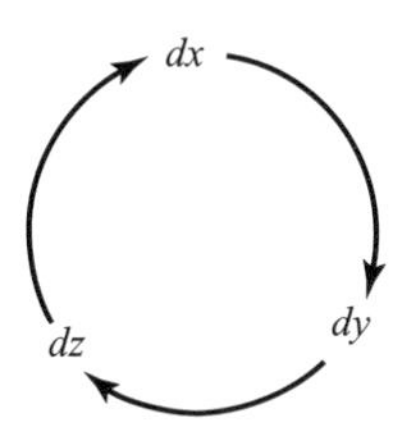

Figure 8.6.1 The cyclic order of *dx, dy,* and *dz.*

By analogy with 0-forms and 1-forms, we can add two 2-forms

$$\eta_i = F_i\,dx\,dy + G_i\,dy\,dz + H_i\,dz\,dx,$$

$i = 1$ and 2, to obtain a new 2-form,

$$\eta_1 + \eta_2 = (F_1 + F_2)\,dx\,dy + (G_1 + G_2)\,dy\,dz + (H_1 + H_2)\,dz\,dx.$$

Similarly, if f is a 0-form and η is a 2-form, we can take the product

$$f\eta = (fF)\,dx\,dy + (fG)\,dy\,dz + (fH)\,dz\,dx.$$

Finally, by the expression $F\,dx\,dy$ we mean the 2-form $F\,dx\,dy + 0 \cdot dy\,dz + 0 \cdot dz\,dx$.

EXAMPLE 3 The expressions

$$\eta_1 = x^2\,dx\,dy + y^3x\,dy\,dz + \sin zy\,dz\,dx$$

and

$$\eta_2 = y\,dy\,dz$$

are 2-forms. Their sum is

$$\eta_1 + \eta_2 = x^2\, dx\, dy + (y^3 x + y)\, dy\, dz + \sin zy\, dz\, dx.$$

If $f(x, y, z) = xy$, then

$$f\eta_2 = xy^2\, dy\, dz. \quad \blacktriangle$$

3-Forms

A ***basic*** *3-**form*** is a formal expression $dx\, dy\, dz$ (in this specific cyclic order, as in Figure 8.6.1). A *3-**form*** ν on an open set $K \subset \mathbb{R}^3$ is an expression of the form $\nu = f(x, y, z)\, dx\, dy\, dz$, where f is a real-valued function on K.

We can add two 3-forms and we can multiply them by 0-forms in the obvious way. There seems to be little difference between a 0-form and a 3-form, because both involve a single real-valued function. But we distinguish them for a purpose that will become clear when we multiply and differentiate forms.

EXAMPLE 4 Let $\nu_1 = y\, dx\, dy\, dz$, $\nu_2 = e^{x^2}\, dx\, dy\, dz$, and $f(x, y, z) = xyz$. Then $\nu_1 + \nu_2 = (y + e^{x^2})\, dx\, dy\, dz$ and $f\nu_1 = y^2 xz\, dx\, dy\, dz$. $\blacktriangle$

Although we can add two 0-forms, two 1-forms, two 2-forms, or two 3-forms, we do not need to add a k-form and a j-form if $k \neq j$. For example, we shall not need to write

$$f(x, y, z)\, dx\, dy + g(x, y, z)\, dz.$$

Now that we have defined these formal objects (forms), one can legitimately ask what they are good for, how they are used, and, perhaps most important, what they mean. The answer to the first question will become clear as we proceed, but we can immediately describe how to use and interpret them.

A real-valued function on a domain K in $\mathbb{R}^3$ is a rule that assigns a real number to each point in K. Differential forms are, in some sense, generalizations of the real-valued functions we have studied in calculus. In fact, 0-forms on an open set K are just functions on K. Thus, a 0-form f takes points in K to real numbers.

We should like to interpret differential k-forms (for $k \geq 1$) not as functions on points in K, but as functions on geometric objects such as curves and surfaces. Many of the early Greek geometers viewed lines and curves as being made up of infinitely many points, and planes and surfaces as being made up of infinitely many curves. Consequently, there is at least some historical justification for applying this geometric hierarchy to the interpretation of differential forms.

Given an open subset $K \subset \mathbb{R}^3$, we shall distinguish four types of subsets of K (see Figure 8.6.2):

(i) points in K,

(ii) oriented simple curves and oriented simple closed curves, C, in K,

(iii) oriented surfaces, $S \subset K$,

(iv) elementary subregions, $R \subset K$.

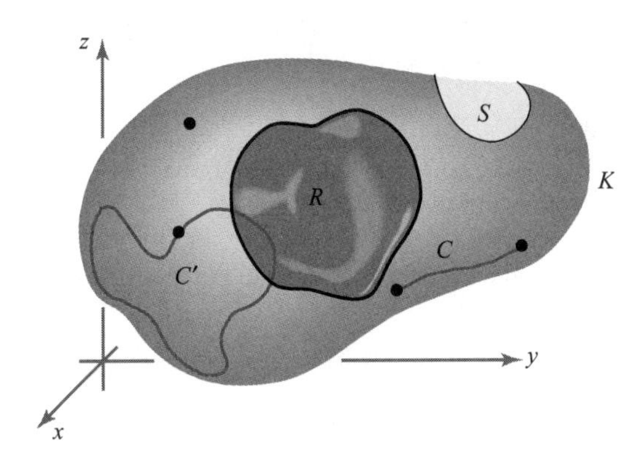

Figure 8.6.2 The four geometric types of subsets of an open set $K \subset \mathbb{R}^3$ to which the theory of forms applies.

The Integral of 1-Forms Over Curves

We shall begin with 1-forms. Let

$$\omega = P(x, y, z)\,dx + Q(x, y, z)\,dy + R(x, y, z)\,dz$$

be a 1-form on K and let C be an oriented simple curve as in Figure 8.6.2. The real number that ω assigns to C is given by the formula

$$\int_C \omega = \int_C P(x, y, z)\,dx + Q(x, y, z)\,dy + R(x, y, z)\,dz. \tag{1}$$

Recall (see Section 7.2) that this integral is evaluated as follows. Suppose that $\mathbf{c}\colon [a, b] \to K$, $\mathbf{c}(t) = (x(t), y(t), z(t))$ is an orientation-preserving parametrization of C. Then

$$\int_C \omega = \int_{\mathbf{c}} \omega = \int_a^b \left[P(x(t), y(t), z(t)) \cdot \frac{dx}{dt} \right.$$
$$\left. + Q(x(t), y(t), z(t)) \cdot \frac{dy}{dt} + R(x(t), y(t), z(t)) \cdot \frac{dz}{dt} \right] dt.$$

Theorem 1 of Section 7.2 guarantees that $\int_C \omega$ does not depend on the choice of the parametrization $\mathbf{c}$.

We can thus interpret a 1-form ω on K as a rule assigning a real number to each oriented curve $C \subset K$; a 2-form η will similarly be seen to be a rule assigning a real number to each oriented surface $S \subset K$; and a 3-form ν will be a rule assigning a real number to each elementary subregion of K. The rules for associating real numbers with curves, surfaces, and regions are completely contained in the formal expressions we have defined.

EXAMPLE 5 Let $\omega = xy\,dx + y^2\,dy + dz$ be a 1-form on $\mathbb{R}^3$ and let C be the oriented simple curve in $\mathbb{R}^3$ described by the parametrization $\mathbf{c}(t) = (t^2, t^3, 1)$, $0 \le t \le 1$. C is oriented by choosing the positive direction of C to be the direction in which $\mathbf{c}(t)$ traverses C as t goes from 0 to 1. Then, by formula (1),

$$\int_C \omega = \int_0^1 [t^5(2t) + t^6(3t^2) + 0]\,dt = \int_0^1 (2t^6 + 3t^8)dt = \frac{13}{21}.$$

Thus, this 1-form ω assigns to each oriented simple curve and each oriented simple closed curve C in $\mathbb{R}^3$ the number $\int_C \omega$. ▲

The Integral of 2-Forms over Surfaces

A 2-form η on an open set $K \subset \mathbb{R}^3$ can similarly be interpreted as a function that associates with each oriented surface $S \subset K$ a real number. This is accomplished by means of the notion of integration of 2-forms over surfaces. Let

$$\eta = F(x, y, z)\,dx\,dy + G(x, y, z)\,dy\,dz + H(x, y, z)\,dz\,dx$$

be a 2-form on K, and let $S \subset K$ be an oriented surface parametrized by a function $\mathbf{\Phi}\colon D \to \mathbb{R}^3$, $D \subset \mathbb{R}^2$, $\mathbf{\Phi}(u, v) = (x(u, v), y(u, v), z(u, v))$ (see Section 7.3).

DEFINITION If S is such a surface and η is a 2-form on K, we define $\iint_S \eta$ by the formula

$$\begin{aligned}\iint_S \eta &= \iint_S F\,dx\,dy + G\,dy\,dz + H\,dz\,dx\\ &= \iint_D \left[F(x(u, v), y(u, v), z(u, v)) \cdot \frac{\partial(x, y)}{\partial(u, v)}\right.\\ &\quad + G(x(u, v), y(u, v), z(u, v)) \cdot \frac{\partial(y, z)}{\partial(u, v)}\\ &\quad \left. + H(x(u, v), y(u, v), z(u, v)) \cdot \frac{\partial(z, x)}{\partial(u, v)}\right] du\,dv,\end{aligned} \tag{2}$$

where

$$\frac{\partial(x,y)}{\partial(u,v)} = \begin{vmatrix} \dfrac{\partial x}{\partial u} & \dfrac{\partial x}{\partial v} \\ \dfrac{\partial y}{\partial u} & \dfrac{\partial y}{\partial v} \end{vmatrix}, \qquad \frac{\partial(y,z)}{\partial(u,v)} = \begin{vmatrix} \dfrac{\partial y}{\partial u} & \dfrac{\partial y}{\partial v} \\ \dfrac{\partial z}{\partial u} & \dfrac{\partial z}{\partial v} \end{vmatrix}, \qquad \frac{\partial(z,x)}{\partial(u,v)} = \begin{vmatrix} \dfrac{\partial z}{\partial u} & \dfrac{\partial z}{\partial v} \\ \dfrac{\partial x}{\partial u} & \dfrac{\partial x}{\partial v} \end{vmatrix}.$$

If S is composed of several pieces $S_i, i = 1, \dots, k$, as in Figure 8.4.4, each with its own parametrization $\mathbf{\Phi}_i$, we define

$$\iint_S \eta = \sum_{i=1}^{k} \iint_{S_i} \eta.$$

One must verify that $\iint_S \eta$ does not depend on the choice of parametrization $\mathbf{\Phi}$. This result is essentially (but not obviously) contained in Theorem 4, Section 7.6.

EXAMPLE 6 Let $\eta = z^2\, dx\, dy$ be a 2-form on $\mathbb{R}^3$, and let S be the upper unit hemisphere in $\mathbb{R}^3$. Find $\iint_S \eta$.

SOLUTION Let us parametrize S by

$$\mathbf{\Phi}(u, v) = (\sin u \cos v, \sin u \sin v, \cos u),$$

where $(u, v) \in D = [0, \pi/2] \times [0, 2\pi]$. By formula (2),

$$\iint_S \eta = \iint_D \cos^2 u \left[\frac{\partial(x,y)}{\partial(u,v)}\right] du\, dv,$$

where

$$\frac{\partial(x,y)}{\partial(u,v)} = \begin{vmatrix} \cos u \cos v & -\sin u \sin v \\ \cos u \sin v & \sin u \cos v \end{vmatrix}$$
$$= \sin u \cos u \cos^2 v + \cos u \sin u \sin^2 v = \sin u \cos u.$$

Therefore,

$$\iint_S \eta = \iint_D \cos^2 u \cos u \sin u\, du\, dv$$
$$= \int_0^{2\pi} \int_0^{\pi/2} \cos^3 u \sin u\, du\, dv = \int_0^{2\pi} \left[-\frac{\cos^4 u}{4}\right]_0^{\pi/2} dv = \frac{\pi}{2}. \quad \blacktriangle$$

EXAMPLE 7 Evaluate $\iint_S x\, dy\, dz + y\, dx\, dy$, where S is the oriented surface described by the parametrization $x = u + v, y = u^2 - v^2, z = uv$, where $(u, v) \in D = [0, 1] \times [0, 1]$.

SOLUTION By definition, we have

$$\frac{\partial(y,z)}{\partial(u,v)} = \begin{vmatrix} 2u & -2v \\ v & u \end{vmatrix} = 2\,(u^2+v^2);$$

$$\frac{\partial(x,y)}{\partial(u,v)} = \begin{vmatrix} 1 & 1 \\ 2u & -2v \end{vmatrix} = -2\,(u+v).$$

Consequently,

$$\begin{aligned}\iint_S x\,dy\,dz + y\,dx\,dy &= \iint_D [(u+v)(2)(u^2+v^2) + (u^2-v^2)(-2)(u+v)]\,du\,dv \\ &= 4\iint_D (v^3+uv^2)\,du\,dv = 4\int_0^1\int_0^1 (v^3+uv^2)\,du\,dv \\ &= 4\int_0^1 \left[uv^3 + \frac{u^2v^2}{2}\right]_0^1 dv = 4\int_0^1 \left(v^3+\frac{v^2}{2}\right) dv \\ &= \left[v^4 + \frac{2v^3}{3}\right]_0^1 = 1+\frac{2}{3} = \frac{5}{3}. \quad \blacktriangle\end{aligned}$$

The Integral of 3-Forms over Regions

Finally, we must interpret 3-forms as functions on the elementary subregions of K. Let $\nu = f(x,y,z)\,dx\,dy\,dz$ be a 3-form and let $R \subset K$ be an elementary subregion of K. Then to each such $R \subset K$ we assign the number

$$\iiint_R \nu = \iiint_R f(x,y,z)\,dx\,dy\,dz, \tag{3}$$

which is just the ordinary triple integral of f over R, as described in Section 5.5.

EXAMPLE 8 Suppose $\nu = (x+z)\,dx\,dy\,dz$ and $R = [0,1]\times[0,1]\times[0,1]$. Evaluate $\iiint_R \nu$.

SOLUTION We compute:

$$\begin{aligned}\iiint_R \nu &= \iiint_R (x+z)\,dx\,dy\,dz = \int_0^1\int_0^1\int_0^1 (x+z)\,dx\,dy\,dz \\ &= \int_0^1\int_0^1 \left[\frac{x^2}{2}+zx\right]_0^1 dy\,dz = \int_0^1\int_0^1 \left(\frac{1}{2}+z\right) dy\,dz = \int_0^1 \left(\frac{1}{2}+z\right) dz \\ &= \left[\frac{z}{2}+\frac{z^2}{2}\right]_0^1 = 1. \quad \blacktriangle\end{aligned}$$

The Algebra of Forms

We now discuss the algebra (or rules of multiplication) of forms that, together with differentiation of forms, will enable us to state Green's, Stokes', and Gauss' theorems in terms of differential forms.

If ω is a k-form and η is an l-form on K, $0 \leq k + l \leq 3$, there is a product called the ***wedge product*** $\omega \wedge \eta$ of ω and η that is a $k + l$ form on K. The wedge product satisfies the following laws:

(i) For each k there is a zero k-form 0 with the property that $0 + \omega = \omega$ for all k-forms ω and $0 \wedge \eta = 0$ for all l-forms η if $0 \leq k + l \leq 3$.

(ii) (*Distributivity*) If f is a 0-form, then

$$(f\omega_1 + \omega_2) \wedge \eta = f(\omega_1 \wedge \eta) + (\omega_2 \wedge \eta).$$

(iii) (*Anticommutativity*) $\omega \wedge \eta = (-1)^{kl}(\eta \wedge \omega)$.

(iv) (*Associativity*) If $\omega_1, \omega_2, \omega_3$ are k_1, k_2, k_3 forms, respectively, with $k_1 + k_2 + k_3 \leq 3$, then

$$\omega_1 \wedge (\omega_2 \wedge \omega_3) = (\omega_1 \wedge \omega_2) \wedge \omega_3.$$

(v) (*Homogeneity with respect to functions*) If f is a 0-form, then

$$\omega \wedge (f\eta) = (f\omega) \wedge \eta = f(\omega \wedge \eta).$$

Notice that rules (ii) and (iii) actually imply rule (v).

(vi) The following multiplication rules for 1-forms hold:

$$\begin{aligned}
dx \wedge dy &= dx\, dy \\
dy \wedge dx &= -\, dx\, dy = (-1)(dx \wedge dy) \\
dy \wedge dz &= dy\, dz = (-1)(dz \wedge dy) \\
dz \wedge dx &= dz\, dx = (-1)(dx \wedge dz) \\
dx \wedge dx &= 0,\ dy \wedge dy = 0,\ dz \wedge dz = 0 \\
dx \wedge (dy \wedge dz) &= (dx \wedge dy) \wedge dz = dx\, dy\, dz.
\end{aligned}$$

(vii) If f is a 0-form and ω is any k-form, then $f \wedge \omega = f\omega$.

Using laws (i) to (vii), we can now find a unique product of any l-form η and any k-form ω, if $0 \leq k + l \leq 3$.

EXAMPLE 9 Show that $dx \wedge dy\,dz = dx\,dy\,dz$.

SOLUTION By rule (vi), $dy\,dz = dy \wedge dz$. Therefore,

$$dx \wedge dy\,dz = dx \wedge (dy \wedge dz) = dx\,dy\,dz. \quad \blacktriangle$$

EXAMPLE 10 If $\omega = x\,dx + y\,dy$ and $\eta = zy\,dx + xz\,dy + xy\,dz$, find $\omega \wedge \eta$.

SOLUTION Computing $\omega \wedge \eta$, to get

$$\begin{aligned}
\omega \wedge \eta &= (x\,dx + y\,dy) \wedge (zy\,dx + xz\,dy + xy\,dz) \\
&= [(x\,dx + y\,dy) \wedge (zy\,dx)] + [(x\,dx + y\,dy) \wedge (xz\,dy)] \\
&\quad + [(x\,dx + y\,dy) \wedge (xy\,dz)] \\
&= xyz(dx \wedge dx) + zy^2(dy \wedge dx) + x^2z(dx \wedge dy) + xyz(dy \wedge dy) \\
&\quad + x^2y(dx \wedge dz) + xy^2(dy \wedge dz) \\
&= -zy^2\,dx\,dy + x^2z\,dx\,dy - x^2y\,dz\,dx + xy^2\,dy\,dz \\
&= (x^2z - y^2z)\,dx\,dy - x^2y\,dz\,dx + xy^2\,dy\,dz. \quad \blacktriangle
\end{aligned}$$

EXAMPLE 11 If $\omega = x\,dx - y\,dy$ and $\eta = x\,dy\,dz + z\,dx\,dy$, find $\omega \wedge \eta$.

SOLUTION

$$\begin{aligned}
\omega \wedge \eta &= (x\,dx - y\,dy) \wedge (x\,dy\,dz + z\,dx\,dy) \\
&= [(x\,dx - y\,dy) \wedge (x\,dy\,dz)] + [(x\,dx - y\,dy) \wedge (z\,dx\,dy)] \\
&= (x^2\,dx \wedge dy\,dz) - (xy\,dy \wedge dy\,dz) + (xz\,dx \wedge dx\,dy) \\
&\quad - (yz\,dy \wedge dx\,dy) \\
&= [x^2\,dx \wedge (dy \wedge dz)] - [xy\,dy \wedge (dy \wedge dz)] + [xz\,dx \wedge (dx \wedge dy)] \\
&\quad - [yz\,dy \wedge (dx \wedge dy)] \\
&= x^2\,dx\,dy\,dz - [xy(dy \wedge dy) \wedge dz] + [xz(dx \wedge dx) \wedge dy] \\
&\quad - [yz(dy \wedge dx) \wedge dy] \\
&= x^2\,dx\,dy\,dz - xy(0 \wedge dz) + xz(0 \wedge dy) + [yz(dy \wedge dy) \wedge dx] \\
&= x^2\,dx\,dy\,dz. \quad \blacktriangle
\end{aligned}$$

The last major step in the development of this theory is to show how to differentiate forms. The derivative of a k-form is a $(k+1)$-form if $k < 3$, and the derivative of a

3-form is always zero. If ω is a k-form, we shall denote the derivative of ω by $d\omega$. The operation d has the following properties:

(1) If $f\colon K \to \mathbb{R}$ is a 0-form, then

$$df = \frac{\partial f}{\partial x}\,dx + \frac{\partial f}{\partial y}\,dy + \frac{\partial f}{\partial z}\,dz.$$

(2) *(Linearity)* If ω_1 and ω_2 are k-forms, then

$$d(\omega_1 + \omega_2) = d\omega_1 + d\omega_2.$$

(3) If ω is a k-form and η is an l-form,

$$d(\omega \wedge \eta) = (d\omega \wedge \eta) + (-1)^k(\omega \wedge d\eta).$$

(4) $d(d\omega) = 0$ and $d(dx) = d(dy) = d(dz) = 0$ or, simply, $d^2 = 0$.

Properties (1) to (4) provide enough information to allow us to uniquely differentiate any form.

EXAMPLE 12 Let $\omega = P(x, y, z)\,dx + Q(x, y, z)\,dy$ be a 1-form on some open set $K \subset \mathbb{R}^3$. Find $d\omega$.

SOLUTION

$$\begin{aligned}
&d[P(x, y, z)\,dx + Q(x, y, z)\,dy]\\
&\quad = d[P(x, y, z) \wedge dx] + d[Q(x, y, z) \wedge dy] && \text{(using 2)}\\
&\quad = (dP \wedge dx) + [P \wedge d(dx)] + (dQ \wedge dy) + [Q \wedge d(dy)] && \text{(using 3)}\\
&\quad = (dP \wedge dx) + (dQ \wedge dy) && \text{(using 4)}\\
&\quad = \left(\frac{\partial P}{\partial x}\,dx + \frac{\partial P}{\partial y}\,dy + \frac{\partial P}{\partial z}\,dz\right) \wedge dx\\
&\qquad + \left(\frac{\partial Q}{\partial x}\,dx + \frac{\partial Q}{\partial y}\,dy + \frac{\partial Q}{\partial z}\,dz\right) \wedge dy && \text{(using 1)}\\
&\quad = \left(\frac{\partial P}{\partial x}\,dx \wedge dx\right) + \left(\frac{\partial P}{\partial y}\,dy \wedge dx\right) + \left(\frac{\partial P}{\partial z}\,dz \wedge dx\right)\\
&\qquad + \left(\frac{\partial Q}{\partial x}\,dx \wedge dy\right) + \left(\frac{\partial Q}{\partial y}\,dy \wedge dy\right) + \left(\frac{\partial Q}{\partial z}\,dz \wedge dy\right)\\
&\quad = -\frac{\partial P}{\partial y}\,dx\,dy + \frac{\partial P}{\partial z}\,dz\,dx + \frac{\partial Q}{\partial x}\,dx\,dy - \frac{\partial Q}{\partial z}\,dy\,dz\\
&\quad = \left(\frac{\partial Q}{\partial x} - \frac{\partial P}{\partial y}\right)dx\,dy + \frac{\partial P}{\partial z}\,dz\,dx - \frac{\partial Q}{\partial z}\,dy\,dz. \quad \blacktriangle
\end{aligned}$$

EXAMPLE 13 Let f be a 0-form. Using only differentiation rules (1) to (3) and the fact that $d(dx) = d(dy) = d(dz) = 0$, show that $d(df) = 0$.

SOLUTION By rule (1),

$$df = \frac{\partial f}{\partial x}\, dx + \frac{\partial f}{\partial y}\, dy + \frac{\partial f}{\partial z}\, dz,$$

and so

$$d(df) = d\left(\frac{\partial f}{\partial x}\, dx\right) + d\left(\frac{\partial f}{\partial y}\, dy\right) + d\left(\frac{\partial f}{\partial z}\, dz\right).$$

Working only with the first term, using rule (3), we get

$$\begin{aligned} d\left(\frac{\partial f}{\partial x}\, dx\right) &= d\left(\frac{\partial f}{\partial x} \wedge dx\right) = d\left(\frac{\partial f}{\partial x}\right) \wedge dx + \frac{\partial f}{\partial x} \wedge d(dx) \\ &= \left(\frac{\partial^2 f}{\partial x^2}\, dx + \frac{\partial^2 f}{\partial y \partial x}\, dy + \frac{\partial^2 f}{\partial z \partial x}\, dz\right) \wedge dx + 0 \\ &= \frac{\partial^2 f}{\partial y \partial x}\, dy \wedge dx + \frac{\partial^2 f}{\partial z \partial x}\, dz \wedge dx \\ &= -\frac{\partial^2 f}{\partial y \partial x}\, dx\, dy + \frac{\partial^2 f}{\partial z \partial x}\, dz\, dx. \end{aligned}$$

Similarly, we find that

$$d\left(\frac{\partial f}{\partial y}\, dy\right) = \frac{\partial^2 f}{\partial x \partial y}\, dx\, dy - \frac{\partial^2 f}{\partial z \partial y}\, dy\, dz$$

and

$$d\left(\frac{\partial f}{\partial z}\, dz\right) = -\frac{\partial^2 f}{\partial x \partial z}\, dz\, dx + \frac{\partial^2 f}{\partial y \partial z}\, dy\, dz.$$

Adding these up, we get $d(df) = 0$ by the equality of mixed partial derivatives. ▲

EXAMPLE 14 Show that $d(dx\, dy)$, $d(dy\, dz)$, and $d(dz\, dx)$ are all zero.

SOLUTION To prove the first case, we use property (3):

$$d(dx\, dy) = d(dx \wedge dy) = [d(dx) \wedge dy - dx \wedge d(dy)] = 0.$$

The other cases are similar. ▲

EXAMPLE 15 If $\eta = F(x, y, z)\,dx\,dy + G(x, y, z)\,dy\,dz + H(x, y, z)\,dz\,dx$, find $d\eta$.

SOLUTION By property (2),

$$d\eta = d(F\,dx\,dy) + d(G\,dy\,dz) + d(H\,dz\,dx).$$

We shall compute $d(F\,dx\,dy)$. Using property (3) again, we get

$$d(F\,dx\,dy) = d(F \wedge dx\,dy) = dF \wedge (dx\,dy) + F \wedge d(dx\,dy).$$

By Example 14, $d(dx\,dy) = 0$, so we are left with

$$\begin{aligned} dF \wedge (dx\,dy) &= \left(\frac{\partial F}{\partial x}\,dx + \frac{\partial F}{\partial y}\,dy + \frac{\partial F}{\partial z}\,dz\right) \wedge (dx \wedge dy) \\ &= \left[\frac{\partial F}{\partial x} dx \wedge (dx \wedge dy)\right] + \left[\frac{\partial F}{\partial y} dy \wedge (dx \wedge dy)\right] \\ &\quad + \left[\frac{\partial F}{\partial z} dz \wedge (dx \wedge dy)\right]. \end{aligned}$$

Now

$$\begin{aligned} dx \wedge (dx \wedge dy) &= (dx \wedge dx) \wedge dy = 0 \wedge dy = 0, \\ dy \wedge (dx \wedge dy) &= -dy \wedge (dy \wedge dx) \\ &= -(dy \wedge dy) \wedge dx = 0 \wedge dx = 0, \end{aligned}$$

and

$$dz \wedge (dx \wedge dy) = (-1)^2 (dx \wedge dy) \wedge dz = dx\,dy\,dz.$$

Consequently,

$$d(F\,dx\,dy) = \frac{\partial F}{\partial z}\,dx\,dy\,dz.$$

Analogously, we find that

$$d(G\,dy\,dz) = \frac{\partial G}{\partial x}\,dx\,dy\,dz \qquad \text{and} \qquad d(H\,dz\,dx) = \frac{\partial H}{\partial y}\,dx\,dy\,dz.$$

Therefore,

$$d\eta = \left(\frac{\partial F}{\partial z} + \frac{\partial G}{\partial x} + \frac{\partial H}{\partial y}\right) dx\, dy\, dz. \quad \blacktriangle$$

We have now developed all the concepts needed to reformulate Green's, Stokes', and Gauss' theorems in the language of forms.

THEOREM 13: Green's Theorem Let D be an elementary region in the xy plane, with ∂D given the counterclockwise orientation. Suppose $\omega = P(x, y)\, dx + Q(x, y)\, dy$ is a 1-form on some open set K in $\mathbb{R}^3$ that contains D. Then

$$\int_{\partial D} \omega = \iint_D d\omega.$$

Here $d\omega$ is a 2-form on K and D is in fact a surface in $\mathbb{R}^3$ parametrized by $\mathbf{\Phi}\colon D \to \mathbb{R}^3$, $\mathbf{\Phi}(x, y) = (x, y, 0)$. Because P and Q are explicitly *not* functions of z, then $\partial P/\partial z$ and $\partial Q/\partial z = 0$, and by Example 12, $d\omega = (\partial Q/\partial x - \partial P/\partial y)\, dx\, dy$. Consequently, Theorem 13 means nothing more than that

$$\int_{\partial D} P\, dx + Q\, dy = \iint_D \left(\frac{\partial Q}{\partial x} - \frac{\partial P}{\partial y}\right) dx\, dy,$$

which is precisely Green's theorem. Hence, Theorem 13 holds. Likewise, we have the following theorems.

THEOREM 14: Stokes' Theorem Let S be an oriented surface in $\mathbb{R}^3$ with a boundary consisting of a simple closed curve ∂S (Figure 8.6.3) oriented as the boundary of S (see Figure 8.2.1). Suppose that ω is a 1-form on some open set K that contains S. Then

$$\int_{\partial S} \omega = \iint_S d\omega.$$

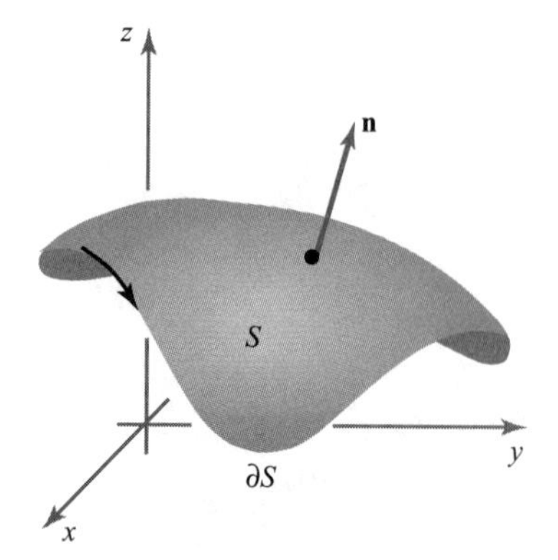

Figure 8.6.3 An oriented surface to which Stokes' theorem applies.

THEOREM 15: Gauss' Theorem Let $W \subset \mathbb{R}^3$ be an elementary region with ∂W given the outward orientation (see Section 8.4). If η is a 2-form on some region K containing W, then

$$\iint_{\partial W} \eta = \iiint_W d\eta.$$

The reader has probably noticed the strong similarity in the statements of these theorems. In the vector-field formulations, we used divergence for regions in $\mathbb{R}^3$ (Gauss' theorem) and the curl for surfaces in $\mathbb{R}^3$ (Stokes' theorem) and regions in $\mathbb{R}^2$ (Green's theorem). Here we just use the unified notion of the derivative of a differential form for all three theorems; and, in fact, we can state all theorems as one by introducing a little more terminology.

By an ***oriented 2-manifold with boundary*** in $\mathbb{R}^3$ we mean a surface in $\mathbb{R}^3$ whose boundary is a simple closed curve with orientation as described in Section 8.2. By an ***oriented 3-manifold*** in $\mathbb{R}^3$ we mean an elementary region in $\mathbb{R}^3$ (we assume its boundary, which is a surface, is given the outward orientation discussed in Section 8.4). We call the following unified theorem "Stokes' theorem," according to the current convention.

THEOREM 16: General Stokes' Theorem Let M be an oriented k-manifold in $\mathbb{R}^3$ ($k = 2$ or 3) contained in some open set K. Suppose ω is $(k-1)$-form on K. Then

$$\int_{\partial M} \omega = \int_M d\omega.$$

Here the integral is interpreted as a single, double, or triple integral, as is appropriate. In fact, it is this form of Stokes' theorem that generalizes to spaces of arbitrary dimension.

EXERCISES

1. Evaluate $\omega \wedge \eta$ if

(a) $\omega = 2x\,dx + y\,dy$
$\eta = x^3\,dx + y^2\,dy$

(b) $\omega = x\,dx - y\,dy$
$\eta = y\,dx + x\,dy$

(c) $\omega = x\,dx + y\,dy + z\,dz$
$\eta = z\,dx\,dy + x\,dy\,dz + y\,dz\,dx$

(d) $\omega = xy\,dy\,dz + x^2\,dx\,dy$
$\eta = dx + dz$

(e) $\omega = e^{xyz}\,dx\,dy$
$\eta = e^{-xyz}\,dz$

2. Prove that

$$(a_1\,dx + a_2\,dy + a_3\,dz) \wedge (b_1\,dy\,dz + b_2\,dz\,dx + b_3\,dx\,dy) = \left(\sum_{i=1}^{3} a_i b_i\right) dx\,dy\,dz.$$

3. Find $d\omega$ in the following examples:

(a) $\omega = x^2 y + y^3$
(b) $\omega = y^2 \cos x\,dy + xy\,dx + dz$
(c) $\omega = xy\,dy + (x+y)^2\,dx$
(d) $\omega = x\,dx\,dy + z\,dy\,dz + y\,dz\,dx$
(e) $\omega = (x^2 + y^2)\,dy\,dz$
(f) $\omega = (x^2 + y^2 + z^2)\,dz$
(g) $\omega = \dfrac{-x}{x^2+y^2}\,dx + \dfrac{y}{x^2+y^2}\,dy$
(h) $\omega = x^2 y\,dy\,dz$

4. Let $\mathbf{V}$: $K \to \mathbb{R}^3$ be a vector field defined by $\mathbf{V}(x, y, z) = G(x, y, z)\mathbf{i} + H(x, y, z)\mathbf{j} + F(x, y, z)\mathbf{k}$, and let η be the 2-form on K given by

$$\eta = F\,dx\,dy + G\,dy\,dz + H\,dz\,dx.$$

Show that $d\eta = (\mathrm{div}\,\mathbf{V})\,dx\,dy\,dz$.

5. If $\mathbf{V} = A(x, y, z)\mathbf{i} + B(x, y, z)\mathbf{j} + C(x, y, z)\mathbf{k}$ is a vector field on $K \subset \mathbb{R}^3$, define the operation Form_2: Vector Fields $\to$ 2-forms by

$$\mathrm{Form}_2(\mathbf{V}) = A\,dy\,dz + B\,dz\,dx + C\,dx\,dy.$$

(a) Show that $\mathrm{Form}_2(\alpha\mathbf{V}_1 + \mathbf{V}_2) = \alpha\,\mathrm{Form}_2(\mathbf{V}_1) + \mathrm{Form}_2(\mathbf{V}_2)$, where α is a real number.
(b) Show that $\mathrm{Form}_2(\mathrm{curl}\,\mathbf{V}) = d\omega$, where $\omega = A\,dx + B\,dy + C\,dz$.

6. Using the differential form version of Stokes' theorem, prove the vector field version in Section 8.2. Repeat for Gauss' theorem.

7. Interpret Theorem 16 in the case $k = 1$.

8. Let $\omega = (x+y)\,dz + (y+z)\,dx + (x+z)\,dy$, and let S be the upper part of the unit sphere; that is, S is the set of (x, y, z) with $x^2 + y^2 + z^2 = 1$ and $z \geq 0$. ∂S is the unit circle in the xy plane. Evaluate $\int_{\partial S} \omega$ both directly and by Stokes' theorem.

9. Let T be the triangular solid bounded by the xy plane, the xz plane, the yz plane, and the plane $2x + 3y + 6z = 12$. Compute

$$\iint_{\partial T} F_1\,dx\,dy + F_2\,dy\,dz + F_3\,dz\,dx$$

directly and by Gauss' theorem, if

(a) $F_1 = 3y,\ F_2 = 18z,\ F_3 = -12$; and
(b) $F_1 = z,\ F_2 = x^2,\ F_3 = y$.

10. Evaluate $\iint_S \omega$ where $\omega = z\,dx\,dy + x\,dy\,dz + y\,dz\,dx$ and S is the unit sphere, directly and by Gauss' theorem.

11. Let R be an elementary region in $\mathbb{R}^3$. Show that the volume of R is given by the formula

$$v(R) = \frac{1}{3}\iint_{\partial R} x\,dy\,dz + y\,dz\,dx + z\,dx\,dy.$$

12. In Section 4.2, we saw that the length $l(\mathbf{c})$ of a curve $\mathbf{c}(t) = (x(t), y(t), z(t)), a \le t \le b$, was given by the formula

$$l(\mathbf{c}) = \int d\mathbf{s} = \int_a^b \left(\frac{ds}{dt}\right) dt$$

where, loosely speaking, $(ds)^2 = (dx)^2 + (dy)^2 + (dz)^2$, that is,

$$\frac{ds}{dt} = \sqrt{\left(\frac{dx}{dt}\right)^2 + \left(\frac{dy}{dt}\right)^2 + \left(\frac{dz}{dt}\right)^2}.$$

Now suppose a surface S is given in parametrized form by $\mathbf{\Phi}(u, v) = (x(u, v), y(u, v), z(u, v))$, where $(u, v) \in D$. Show that the area of S can be expressed as

$$A(S) = \iint_D dS$$

where formally $(dS)^2 = (dx \wedge dy)^2 + (dy \wedge dz)^2 + (dz \wedge dx)^2$, a formula requiring interpretation. (HINT:

$$dx = \frac{\partial x}{\partial u}\,du + \frac{\partial x}{\partial v}\,dv,$$

and similarly for dy and dz. Use the law of forms for the basic 1-forms du and dv. Then dS turns out to be a function times the basic 2-form $du\,dv$, which we can integrate over D.)

REVIEW EXERCISES FOR CHAPTER 8

1. Let $\mathbf{F} = 2yz\mathbf{i} + (-x + 3y + 2)\mathbf{j} + (x^2 + z)\mathbf{k}$. Evaluate $\iint_S (\nabla \times \mathbf{F}) \cdot d\mathbf{S}$, where S is the cylinder $x^2 + y^2 = a^2, 0 \le z \le 1$ (without the top and bottom). What if the top and bottom are included?

2. Let W be a region in $\mathbb{R}^3$ with boundary ∂W. Prove the identity

$$\iint_{\partial W} [\mathbf{F} \times (\nabla \times \mathbf{G})] \cdot dS = \iiint_W (\nabla \times \mathbf{F}) \cdot (\nabla \times \mathbf{G})\,dV - \iiint_W \mathbf{F} \cdot (\nabla \times \nabla \times \mathbf{G})\,dV.$$

3. Let $\mathbf{F} = x^2y\mathbf{i} + z^8\mathbf{j} - 2xyz\mathbf{k}$. Evaluate the integral of $\mathbf{F}$ over the surface of the unit cube.

4. Verify Green's theorem for the line integral

$$\int_C x^2 y\,dx + y\,dy,$$

when C is the boundary of the region between the curves $y = x$ and $y = x^3$, $0 \le x \le 1$.

5. (a) Show that $\mathbf{F} = (x^3 - 2xy^3)\mathbf{i} - 3x^2y^2\mathbf{j}$ is a gradient vector field.
(b) Evaluate the integral of $\mathbf{F}$ along the path $x = \cos^3\theta$, $y = \sin^3\theta$, $0 \le \theta \le \pi/2$.

6. Can you derive Green's theorem in the plane from Gauss' theorem?

7. (a) Show that $\mathbf{F} = 6xy(\cos z)\mathbf{i} + 3x^2(\cos z)\mathbf{j} - 3x^2y(\sin z)\mathbf{k}$ is conservative (see Section 8.3).
(b) Find f such that $\mathbf{F} = \nabla f$.
(c) Evaluate the integral of $\mathbf{F}$ along the curve $x = \cos^3\theta$, $y = \sin^3\theta$, $z = 0$, $0 \le \theta \le \pi/2$.

8. Let $\mathbf{r}(x, y, z) = (x, y, z)$, $r = \|\mathbf{r}\|$. Show that $\nabla^2(\log r) = 1/r^2$ and $\nabla^2(r^n) = n(n+1)r^{n-2}$.

9. Let the velocity of a fluid be described by $\mathbf{F} = 6xz\mathbf{i} + x^2y\mathbf{j} + yz\mathbf{k}$. Compute the rate at which fluid is leaving the unit cube.

10. Let $\mathbf{F} = x^2\mathbf{i} + (x^2y - 2xy)\mathbf{j} - x^2z\mathbf{k}$. Does there exist a $\mathbf{G}$ such that $\mathbf{F} = \nabla \times \mathbf{G}$?

11. Let $\mathbf{a}$ be a constant vector and $\mathbf{F} = \mathbf{a} \times \mathbf{r}$ [as usual, $\mathbf{r}(x, y, z) = (x, y, z)$]. Is $\mathbf{F}$ conservative? If so, find a potential for it.

12. (For students who have read Section 8.5.) Consider the case of incompressible fluid flow with velocity field $\mathbf{F}$ and density ρ.

(a) If ρ is constant for each fixed t, then show that ρ is constant in t as well.
(b) If ρ is constant in t, then show that $\mathbf{F} \cdot \nabla\rho = 0$.

13. (a) Let $f(x, y, z) = 3xye^{z^2}$. Compute ∇f.
(b) Let $\mathbf{c}(t) = (3\cos^3 t, \sin^2 t, e^t)$, $0 \le t \le \pi$. Evaluate

$$\int_{\mathbf{c}} \nabla f \cdot d\mathbf{s}.$$

(c) Verify directly Stokes' theorem for gradient vector fields $\mathbf{F} = \nabla f$.

14. Using Green's theorem, or otherwise, evaluate $\int_C x^3\,dy - y^3\,dx$, where C is the unit circle $(x^2 + y^2 = 1)$.

15. Evaluate the integral $\iint_S \mathbf{F} \cdot d\mathbf{S}$ where $\mathbf{F} = x\mathbf{i} + y\mathbf{j} + 3\mathbf{k}$ and where S is the surface of the unit sphere $x^2 + y^2 + z^2 = 1$.

16. (a) State Stokes' theorem for surfaces in $\mathbb{R}^3$.

(b) Let $\mathbf{F}$ be a vector field on $\mathbb{R}^3$ satisfying $\nabla \times \mathbf{F} = \mathbf{0}$. Use Stokes' theorem to show that $\int_C \mathbf{F} \cdot d\mathbf{s} = 0$ where C is a closed curve.

17. Use Green's theorem to find the area of the loop of the curve $x = a \sin\theta \cos\theta$, $y = a \sin^2\theta$, for $a > 0$ and $0 \le \theta \le \pi$.

18. Evaluate $\int_C yz\, dx + xz\, dy + xy\, dz$ where C is the curve of intersection of the cylinder $x^2 + y^2 = 1$ and the surface $z = y^2$.

19. Evaluate $\int_C (x + y)\, dx + (2x - z)\, dy + (y + z)\, dz$ where C is the perimeter of the triangle connecting $(2, 0, 0)$, $(0, 3, 0)$, and $(0, 0, 6)$ in that order.

20. Which of the following are conservative fields on $\mathbb{R}^3$? For those that are, find a function f such that $\mathbf{F} = \nabla f$.

(a) $\mathbf{F}(x, y, z) = 3x^2 y\mathbf{i} + x^3\mathbf{j} + 5\mathbf{k}$
(b) $\mathbf{F}(x, y, z) = (x + z)\mathbf{i} - (y + z)\mathbf{j} + (x - y)\mathbf{k}$
(c) $\mathbf{F}(x, y, z) = 2xy^3\mathbf{i} + x^2z^3\mathbf{j} + 3x^2yz^2\mathbf{k}$

21. Consider the following two vector fields in $\mathbb{R}^3$:

(i) $\mathbf{F}(x, y, z) = y^2\mathbf{i} - z^2\mathbf{j} + x^2\mathbf{k}$
(ii) $\mathbf{G}(x, y, z) = (x^3 - 3xy^2)\mathbf{i} + (y^3 - 3x^2y)\mathbf{j} + z\mathbf{k}$

(a) Which of these fields (if any) are conservative on $\mathbb{R}^3$? (That is, which are gradient fields?) Give reasons for your answer.

(b) Find potential for the fields that are conservative.

(c) Let α be the path that goes from $(0, 0, 0)$ to $(1, 1, 1)$ by following edges of the cube $0 \le x \le 1, 0 \le y \le 1, 0 \le z \le 1$ from $(0, 0, 0)$ to $(0, 0, 1)$ to $(0, 1, 1)$ to $(1, 1, 1)$. Let β be the path from $(0, 0, 0)$ to $(1, 1, 1)$ directly along the diagonal of the cube. Find the values of the line integrals

$$\int_\alpha \mathbf{F} \cdot d\mathbf{s}, \qquad \int_\alpha \mathbf{G} \cdot d\mathbf{s}, \qquad \int_\beta \mathbf{F} \cdot d\mathbf{s}, \qquad \int_\beta \mathbf{G} \cdot d\mathbf{s}.$$

22. Consider the *constant* vector field $\mathbf{F}(x, y, z) = \mathbf{i} + 2\mathbf{j} - \mathbf{k}$ in $\mathbb{R}^3$.

(a) Find a scalar field $\phi(x, y, z)$ in $\mathbb{R}^3$ such that $\nabla\phi = \mathbf{F}$ in $\mathbb{R}^3$ and $\phi(0, 0, 0) = 0$.
(b) On the sphere Σ of radius 2 about the origin find all the points at which
  (i) ϕ is a maximum, and
  (ii) ϕ is a minimum.
(c) Compute the maximum and minimum values of ϕ on Σ.

23. Let $\mathbf{F}$ be a C^1 vector field and suppose $\nabla \cdot \mathbf{F}(x_0, y_0, z_0) > 0$. Show that for a sufficiently small sphere S centered at (x_0, y_0, z_0), the flux of $\mathbf{F}$ out of S is positive.

24. Let $B \subset \mathbb{R}^3$ be a planar region, and let $\mathrm{O} \in \mathbb{R}^3$ be a point. If we connect all points in B to O we get a cone, say C, with vertex O and base B. Show that

$$\text{volume } (C) = \frac{1}{3} \text{ area } (B)\, h,$$

where h is the distance of O from the plane of B, using the following steps.

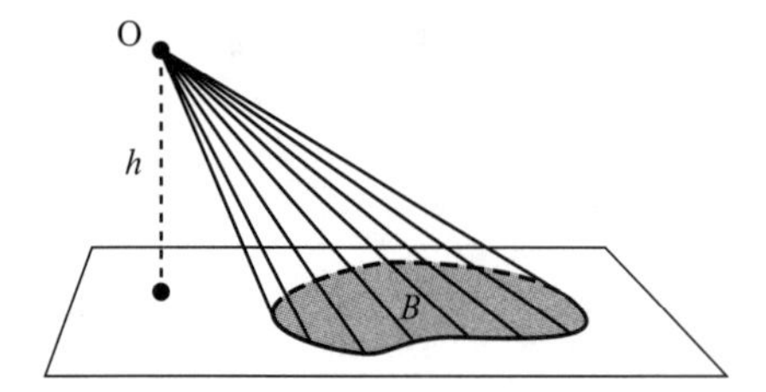

Figure 8.R.1

1. Let O be the origin of the coordinate system. Define $\mathbf{r}(x, y, z) := (x, y, z)$. Evaluate the flux of $\mathbf{r}$ through the boundary of C, that is, $\iint_{\partial C} \mathbf{r} \cdot \mathbf{n}\, dA$, where $\mathbf{n}$ is the outward unit normal to ∂C.
2. Evaluate the divergence of $\mathbf{r}$ in C, that is, $\iiint_C \nabla \cdot \mathbf{r}\, dV$.
3. Use Gauss' theorem, which states that the total divergence of a vector field within a region enclosed by a surface is equal to the flux of that vector field through the surface:

$$\iiint_C \nabla \cdot \mathbf{r}\, dV = \iint_{\partial C} \mathbf{r} \cdot \mathbf{n}\, dA.$$

Answers to Odd-Numbered Exercises

Some solutions requiring proofs may be incomplete or be omitted.

Chapter 1

Section 1.1

1. 4; 17

3. $(-104 + 16a, -24 - 4b, -22 + 26c)$

5.

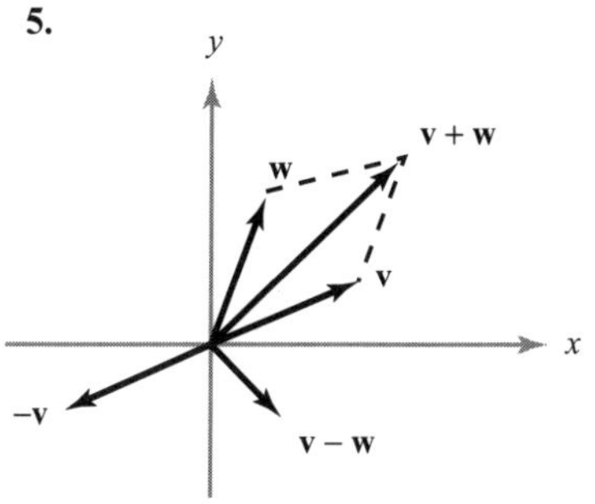

7.

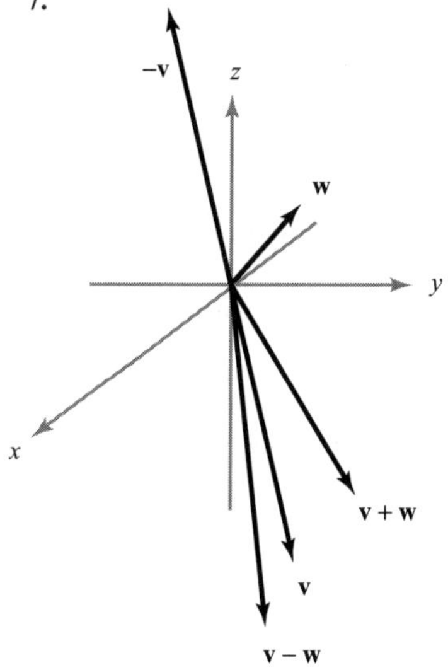

9. $x = 0, z = 0, y \in \mathbb{R}; x = 0, y = 0, z \in \mathbb{R}; y = 0, x, z \in \mathbb{R}; x = 0, y, z \in \mathbb{R}$

11. $\{(2s, 7s + 2t, 7t) \mid s \in \mathbb{R}, t \in \mathbb{R}\}$

13. $\mathbf{l}(t) = -\mathbf{i} + (t - 1)\mathbf{j} - \mathbf{k}$

15. $\mathbf{l}(t) = (2t - 1)\mathbf{i} - \mathbf{j} + (3t - 1)\mathbf{k}$

17. $\{s\mathbf{i} + 3s\mathbf{k} - 2t\mathbf{j} \mid 0 \le s \le 1, 0 \le t \le 1\}$

19. If (x, y, z) lies on the line, then $x = 2 + t$, $y = -2 + t$, and $z = -1 + t$. Therefore, $2x - 3y + z - 2 = 4 + 2t + 6 - 3t - 1 + t - 2 = 7$, which is not zero. Hence, no (x, y, z) satisfies both conditions.

21. Yes.

23. The set of vectors of the form

$$\mathbf{v} = p\mathbf{a} + q\mathbf{b} + r\mathbf{c}$$

where $0 \le p \le 1, 0 \le q \le 1$, and $0 \le r \le 1$.

25. All points of the form

$$(x_0 + t(x_1 - x_0) + s(x_2 - x_0), y_0 + t(y_1 - y_0) + s(y_2 - y_0), z_0 + t(z_1 - z_0) + s(z_2 - z_0))$$

for real numbers t and s.

27. If one vertex is placed at the origin and the two adjacent sides are **u** and **v**, the new triangle has sides $b\mathbf{u}$, $b\mathbf{v}$, and $b(\mathbf{u} - \mathbf{v})$.

29. $(1, 0, 1) + (0, 2, 1) = (0, 2, 0) + (1, 0, 2)$

31. Two such lines (there are many others) are $x = 1, y = t, z = t$ and $x = 1, y = t, z = -t$.

Section 1.2

1. 6

3. 99°

5. No, it is 75.7; it would be zero only if the vectors were parallel.

7. $\|\mathbf{u}\| = \sqrt{5}, \|\mathbf{v}\| = \sqrt{2}, \mathbf{u} \cdot \mathbf{v} = -3$

9. $\|\mathbf{u}\| = \sqrt{11}, \|\mathbf{v}\| = \sqrt{62}, \mathbf{u} \cdot \mathbf{v} = -14$

11. $\|\mathbf{u}\| = \sqrt{14}, \|\mathbf{v}\| = \sqrt{26}, \mathbf{u} \cdot \mathbf{v} = -17$

13. In Exercise 9, $\cos^{-1}(-14/\sqrt{11}\sqrt{62})$; in Exercise 10, $\pi/2$; and in 11, $\cos^{-1}(-17/\sqrt{14}\sqrt{26})$.

15. $-4(-\mathbf{i} + \mathbf{j} + \mathbf{k})/3$

17. Any (x, y, z) with $x + y + z = 0$; for example, $(1, -1, 0)$ and $(0, 1, -1)$.

19. $\mathbf{i} + 4\mathbf{j}$, $\theta \approx 0.24$ radian east of north

21. (a) 12:03 P.M. (b) 4.95 km

23.

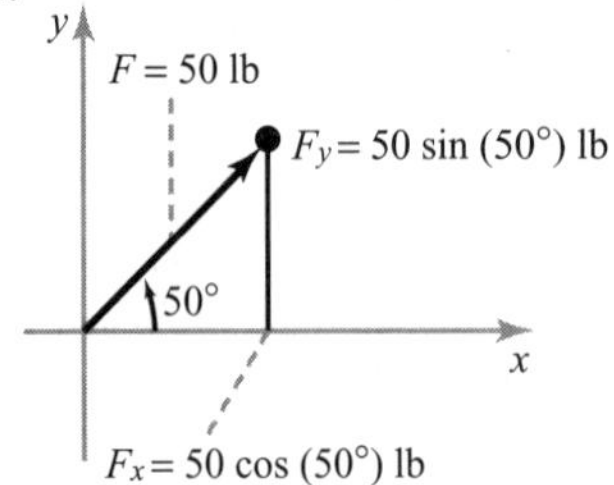

25. (4.9, 4.9, 4.9) and (−4.9, −4.9, 4.9) N.

27. (a) $\mathbf{F} = (3\sqrt{2}\mathbf{i} + 3\sqrt{2}\mathbf{j})$ (b) $\approx$0.322 radian or 18.4° (c) $18\sqrt{2}$

Section 1.3

1. $\begin{vmatrix} 1 & 2 & 1 \\ 3 & 0 & 1 \\ 2 & 0 & 2 \end{vmatrix} = -8, \quad \begin{vmatrix} 3 & 0 & 1 \\ 1 & 2 & 1 \\ 2 & 0 & 2 \end{vmatrix} = 8.$

3. $-3\mathbf{i} + \mathbf{j} + 5\mathbf{k}$

5. $\sqrt{35}$

7. 10

9. $\pm\mathbf{k}$

11. $\pm(113\mathbf{i} + 17\mathbf{j} - 103\mathbf{k})/\sqrt{23,667}$

13. $\mathbf{u} + \mathbf{v} = 3\mathbf{i} - 3\mathbf{j} + 3\mathbf{k}$; $\mathbf{u} \cdot \mathbf{v} = 6$; $\|\mathbf{u}\| = \sqrt{6}$; $\|\mathbf{v}\| = 3$; $\mathbf{u} \times \mathbf{v} = -3\mathbf{i} + 3\mathbf{k}$

15. (a) $x + y + z - 1 = 0$
(b) $x + 2y + 3z - 6 = 0$
(c) $5x + 2z = 25$
(d) $x + 2y - 3z = 13$

17. (a) The parallel planes $Ax + By + Cz + D = 0$ and $\sigma Ax + \sigma By + \sigma Cz + D' = 0$ are identical when $D' = \sigma D$ and otherwise never intersect.
(b) In a line.

19. The line $x = t, y = 2t, z = -5t$.

21. (a) Do the first by working out each side in coordinates, and then use that and $\mathbf{a} \times (\mathbf{b} \times \mathbf{c}) = -(\mathbf{b} \times \mathbf{c}) \times \mathbf{a}$ to get the second.
(b) Use the identities in part (a) to write the quantity in terms of inner products.
(c) Use the identities in part (a) and collect terms.

23. Compute the results of Cramer's rule and check that they satisfy the equation.

25. $x - 2y + 3z + 12 = 0$

27. $4x - 6y - 10z = 14$

29. $10x - 17y + z + 25 = 0$

31. For Exercise 19, note that $(2, -3, 1) \cdot (1, 1, 1) = 0$, and so the line and plane are parallel and $(2, -2, -1)$ does not lie in the plane. For Exercise 20, the line and plane are parallel and $(1, -1, 2)$ *does* lie in the plane.

33. $\sqrt{2}/13$

35. (a) Show that $\mathbf{M}$ satisfies the geometric properties of $\mathbf{R} \times \mathbf{F}$. (b) $2\sqrt{3}$

37. Show that $n_1(\mathbf{N} \times \mathbf{a})$ and $n_2(\mathbf{N} \times \mathbf{b})$ have the same magnitude and direction.

39. One method is to write out all terms in the left-hand side and see that the terms involving λ all cancel. Another method is to first observe that the determinant is linear in each row or column and that if any row or column is repeated, the answer is zero. Then

$$\begin{vmatrix} a_1 & b_1 & c_1 \\ a_2 + \lambda a_1 & b_2 + \lambda b_1 & c_2 + \lambda c_1 \\ a_3 & b_3 & c_3 \end{vmatrix} = \begin{vmatrix} a_1 & b_1 & c_1 \\ a_2 & b_2 & c_2 \\ a_3 & b_3 & c_3 \end{vmatrix} + \lambda \begin{vmatrix} a_1 & b_1 & c_1 \\ a_1 & b_1 & c_1 \\ a_3 & b_3 & c_3 \end{vmatrix} = \begin{vmatrix} a_1 & b_1 & c_1 \\ a_2 & b_2 & c_2 \\ a_3 & b_3 & c_3 \end{vmatrix}.$$

Section 1.4

1. (a)

Cylindrical			Rectangular			Spherical		
r	θ	z	x	y	z	ρ	θ	ϕ
1	45°	1	$\sqrt{2}/2$	$\sqrt{2}/2$	1	$\sqrt{2}$	45°	45°
2	$\pi/2$	-4	0	2	-4	$2\sqrt{5}$	$\pi/2$	$\pi - \arccos(2\sqrt{5}/5)$
0	45°	10	0	0	10	10	45°	0
3	$\pi/6$	4	$3\sqrt{3}/2$	3/2	4	5	$\pi/6$	$\arccos \frac{4}{5}$
1	$\pi/6$	0	$\sqrt{3}/2$	$\frac{1}{2}$	0	1	$\pi/6$	$\pi/2$
2	$3\pi/4$	-2	$-\sqrt{2}$	$\sqrt{2}$	-2	$2\sqrt{2}$	$3\pi/4$	$3\pi/4$

(b)

Rectangular			Spherical			Cylindrical		
x	y	z	ρ	θ	ϕ	r	θ	z
2	1	-2	3	$\arctan \frac{1}{2}$	$\pi - \arccos(2/3)$	$\sqrt{5}$	$\arctan \frac{1}{2}$	-2
0	3	4	5	$\pi/2$	$\arccos(4/5)$	3	$\pi/2$	4
$\sqrt{2}$	1	1	2	$\arctan(\sqrt{2}/2)$	$\pi/3$	$\sqrt{3}$	$\arctan(\sqrt{2}/2)$	1
$-2\sqrt{3}$	-2	3	5	$7\pi/6$	$\arccos \frac{3}{5}$	4	$7\pi/6$	3

3. (a) Rotation by π around the z axis
(b) Reflection across the xy plane
(c) Rotation by $\pi/2$ about the z axis together with a radial expansion by a factor of 2

5. No; (ρ, θ, ϕ) and $(-\rho, \theta + \pi, \pi - \phi)$ represent the same point.

7. (a) $\mathbf{e}_\rho = (x\mathbf{i} + y\mathbf{j} + z\mathbf{k})/\sqrt{x^2 + y^2 + z^2} = (x\mathbf{i} + y\mathbf{j} + z\mathbf{k})/\rho$
$\mathbf{e}_\theta = (-y\mathbf{i} + x\mathbf{j})/\sqrt{x^2 + y^2} = (-y\mathbf{i} + x\mathbf{j})/r$
$\mathbf{e}_\phi = (xz\mathbf{i} + yz\mathbf{j} - (x^2 + y^2)\mathbf{k})/r\rho$
(b) $\mathbf{e}_\theta \times \mathbf{j} = -y\mathbf{k}/\sqrt{x^2 + y^2}$, $\mathbf{e}_\phi \times \mathbf{j} = (xz/r\rho)\mathbf{k} + (r/\rho)\mathbf{i}$

9. (a) The length of $x\mathbf{i} + y\mathbf{j} + z\mathbf{k}$ is $(x^2 + y^2 + z^2)^{1/2} = \rho$
(b) $\cos\phi = z/(x^2 + y^2 + z^2)^{1/2}$ (c) $\cos\theta = x/(x^2 + y^2)^{1/2}$

11. $0 \le r \le a$, $0 \le \theta \le 2\pi$ means that (r, θ, z) is inside the cylinder with radius a centered on the z axis, and $|z| \le b$ means that it is no more than a distance b from the xy plane.

13. $-d/(6\cos\phi) \le \rho \le d/2, 0 \le \theta \le 2\pi,$ and $\pi - \cos^{-1}(\frac{1}{3}) \le \phi \le \pi$

15. This is a surface whose cross section with each surface $z = c$ is four-petaled rose. The petals shrink to zero as $|c|$ changes from 0 to 1.

Section 1.5

1. 7

3. $|\mathbf{x}\cdot\mathbf{y}| = 10 = \sqrt{5}\sqrt{20} = \|\mathbf{x}\|\,\|\mathbf{y}\|$, so $|\mathbf{x}\cdot\mathbf{y}| \le \|\mathbf{x}\|\,\|\mathbf{y}\|$ is true.
$\|\mathbf{x}+\mathbf{y}\| = 3\sqrt{5} = \|\mathbf{x}\| + \|\mathbf{y}\|$, so $\|\mathbf{x}+\mathbf{y}\| \le \|\mathbf{x}\| + \|\mathbf{y}\|$ is true.

5. $|\mathbf{x}\cdot\mathbf{y}| = 5 < \sqrt{65} = \|\mathbf{x}\|\,\|\mathbf{y}\|$, so $|\mathbf{x}\cdot\mathbf{y}| \le \|\mathbf{x}\|\,\|\mathbf{y}\|$ is true.
$\|\mathbf{x}+\mathbf{y}\| = \sqrt{28} < \sqrt{5} + \sqrt{13} = \|\mathbf{x}\| + \|\mathbf{y}\|$, so $\|\mathbf{x}+\mathbf{y}\| \le \|\mathbf{x}\| + \|\mathbf{y}\|$ is true.

7. $AB = \begin{bmatrix} -1 & -1 & 3 \\ -1 & 11 & 3 \\ -6 & 5 & 8 \end{bmatrix}$, $\det A = -5$, $\det B = -24$,
$\det AB = 120(=\det A \det B)$, $\det (A+B) = -61(\neq \det A + \det B)$

9. HINT: For $k = 2$ use the triangle inequality to show that $\|\mathbf{x}_1 + \mathbf{x}_2\| \le \|\mathbf{x}_1\| + \|\mathbf{x}_2\|$; then for $k = i+1$ note that $\|\mathbf{x}_1 + \mathbf{x}_2 + \cdots + \mathbf{x}_{i+1}\| \le \|\mathbf{x}_1 + \mathbf{x}_2 + \cdots + \mathbf{x}_i\| + \|\mathbf{x}_{i+1}\|$.

11. (a) Check $n = 1$ and $n = 2$ directly. Then reduce an $n \times n$ determinant to a sum of $(n-1)\times(n-1)$ determinants and use induction.

(b) The argument is similar to that for part (a). Suppose the first row is multiplied by λ. The first term of the sum will be λa_{11} times an $(n-1)\times(n-1)$ determinant with no factors of λ. The other terms obtained (by expanding across the first row) are similar.

13. Not necessarily. Try $A = \begin{bmatrix} 0 & 1 \\ 0 & 0 \end{bmatrix}$ and $B = \begin{bmatrix} 1 & 0 \\ 0 & 0 \end{bmatrix}$.

15. (a) The sum of two continuous functions and a scalar multiple of a continuous function are continuous.

(b) (i)
$$\begin{aligned}(\alpha f + \beta g)\cdot h &= \int_0^1 (\alpha f + \beta g)(x) h(x)\,dx \\ &= \int_0^1 f(x)h(x)\,dx + \beta \int_0^1 g(x)h(x)\,dx \\ &= \alpha f\cdot h + \beta g\cdot h.\end{aligned}$$

(ii) $f\cdot g = \int_0^1 f(x)g(x)\,dx = \int_0^1 g(x)f(x)\,dx = g\cdot f$.

In conditions (iii) and (iv), the integrand is a perfect square. Therefore, the integral in nonnegative and can be 0 only if the integrand is 0 everywhere. If $f(x) \neq 0$ for some x, then it would be positive in a neighborhood of x by continuity, and the integral would be positive.

17. Compute the matrix product in both orders and check that you get the identity.

19. $(\det A)(\det A^{-1}) = \det (AA^{-1}) = \det (I) = 1$

Review Exercises for Chapter 1

1. $\mathbf{v}+\mathbf{w} = 4\mathbf{i} + 3\mathbf{j} + 6\mathbf{k}$; $3\mathbf{v} = 9\mathbf{i} + 12\mathbf{j} + 15\mathbf{k}$; $6\mathbf{v} + 8\mathbf{w} = 26\mathbf{i} + 16\mathbf{j} + 38\mathbf{k}$; $-2\mathbf{v} = -6\mathbf{i} - 8\mathbf{j} - 10\mathbf{k}$; $\mathbf{v}\cdot\mathbf{w} = 4$; $\mathbf{v}\times\mathbf{w} = 9\mathbf{i} + 2\mathbf{j} - 7\mathbf{k}$. Your sketch should display $\mathbf{v}$, $\mathbf{w}$, $3\mathbf{v}$, $6\mathbf{v}$, $8\mathbf{w}$, $6\mathbf{v}+8\mathbf{w}$, $\mathbf{v}\cdot\mathbf{w}$ as the projection of $\mathbf{v}$ along $\mathbf{w}$ and $\mathbf{v}\times\mathbf{w}$ as a vector perpendicular to both $\mathbf{v}$ and $\mathbf{w}$.

3. (a) $\mathbf{l}(t) = -\mathbf{i} + (2+t)\mathbf{j} - \mathbf{k}$ (c) $-2x + y + 2z = 9$
(b) $\mathbf{l}(t) = (3t-3)\mathbf{i} + (t+1)\mathbf{j} - t\mathbf{k}$

5. (a) 0 (b) 5 (c) -10

7. (a) $\pi/2$ (b) $5/(2\sqrt{15})$ (c) $-10/(\sqrt{6}\sqrt{17})$

9. $\{st\mathbf{a} + s(1-t)\mathbf{b} \mid 0 \le t \le 1 \text{ and } 0 \le s \le 1\}$

11. Let $\mathbf{v} = (a_1, a_2, a_3)$, $\mathbf{w} = (b_1, b_2, b_3)$, and apply the CBS inequality.

13. The area is the absolute value of

$$\begin{vmatrix} a_1 & a_2 \\ b_1 & b_2 \end{vmatrix} = \begin{vmatrix} a_1 & a_2 \\ b_1 + \lambda a_1 & b_2 + \lambda a_2 \end{vmatrix}.$$

(A multiple of one row of a determinant can be added to another row without changing its value.) Your sketch should show two parallelograms with the same base and height.

15. The cosines of the two parts of the angle are equal, because
$\mathbf{a} \cdot \mathbf{v}/\|\mathbf{a}\|\|\mathbf{v}\| = (\mathbf{a} \cdot \mathbf{b} + \|\mathbf{a}\|\|\mathbf{b}\|)/\|\mathbf{v}\| = \mathbf{b} \cdot \mathbf{v}/\|\mathbf{b}\|\|\mathbf{v}\|$.

17. $\mathbf{i} \times \mathbf{j} = \begin{vmatrix} \mathbf{i} & \mathbf{j} & \mathbf{k} \\ 1 & 0 & 0 \\ 0 & 1 & 0 \end{vmatrix} = \mathbf{k}$; etc.

19. (a) HINT: The length of the projection of the vector connecting any pair of points, one on each line, onto $(\mathbf{a}_1 \times \mathbf{a}_2)/\|\mathbf{a}_1 \times \mathbf{a}_2\|$ is d.
(b) $\sqrt{2}$

21. (a) Note that

$$\frac{1}{2}\begin{vmatrix} 1 & 1 & 1 \\ x_1 & x_2 & x_3 \\ y_1 & y_2 & y_3 \end{vmatrix} = \frac{1}{2}\begin{vmatrix} 1 & 0 & 0 \\ x_1 & x_2 - x_1 & x_3 - x_1 \\ y_1 & y_2 - y_1 & y_3 - y_1 \end{vmatrix} = \frac{1}{2}\begin{vmatrix} x_2 - x_1 & x_3 - x_1 \\ y_2 - y_1 & y_3 - y_1 \end{vmatrix}.$$

(b) $\frac{1}{2}$

23.

Rectangular	Spherical	(plot omitted)
(a) $(\sqrt{2}/2, \sqrt{2}/2, 1)$	(a) $(\sqrt{2}, \pi/4, \pi/4)$	
(b) $(3\sqrt{3}/2, 3/2, -4)$	(b) $(5, \pi/6, \arccos(-4/5))$	
(c) $(0, 0, 1)$	(c) $(1, \pi/4, 0)$	
(d) $(0, -2, 1)$	(d) $(\sqrt{5}, 3\pi/2, \arccos(\sqrt{5}/5))$	
(e) $(0, 2, 1)$	(e) $(\sqrt{5}, \pi/2, \arccos(\sqrt{5}/5))$	

25. $z = r^2 \cos 2\theta$; $\cos\phi = \rho \sin^2\phi \cos 2\theta$

27. $|\mathbf{x}\cdot\mathbf{y}| = 6 < \sqrt{98} = \|\mathbf{x}\|\,\|\mathbf{y}\|$; $\|\mathbf{x}+\mathbf{y}\| = \sqrt{33} < \sqrt{14}+\sqrt{7} = \|\mathbf{x}\| + \|\mathbf{y}\|$

29. (a) The associative law for matrix multiplication may be checked as follows:

$$\begin{aligned}[(AB)C]_{ij} &= \sum_{k=1}^{n}(AB)_{ik}C_{kj} = \sum_{k=1}^{n}\sum_{l=1}^{n} A_{il}B_{lk}C_{kj}\\ &= \sum_{l=1}^{n} A_{i}(BC)_{lj} = [A(BC)]_{ij}.\end{aligned}$$

Use this with C taken to be a column vector.

(b) The matrix for the composition is the product matrix.

31. $\mathbb{R}^n$ is spanned by the vectors $\mathbf{e}_1, \mathbf{e}_2, \ldots, \mathbf{e}_n$. If $\mathbf{v} \in \mathbb{R}^n$, then

$$T\mathbf{v} = T\left[\sum_{i=1}^{n}(\mathbf{v}\cdot\mathbf{e}_i)\mathbf{e}_i\right] = \sum_{i=1}^{n}(\mathbf{v}\cdot\mathbf{e}_i)T\mathbf{e}_i.$$

Let $a_{ij} = (T\mathbf{e}_j \cdot \mathbf{e}_i)$, so that

$$T\mathbf{e}_j = \sum_{j=1}^{n} a_{ij}\mathbf{e}_i.$$

Then

$$T\mathbf{v}\cdot\mathbf{e}_k = \sum_{i=1}^{n}(\mathbf{v}\cdot\mathbf{e}_i)a_{ki}.$$

That is, if

$$\mathbf{v} = \begin{bmatrix} v_1\\ \vdots\\ v_n\end{bmatrix}, \qquad \text{then} \qquad T\mathbf{v} = \begin{bmatrix} a_{11} & \cdots & a_{1n}\\ \vdots & & \\ a_{n1} & \cdots & a_{nn}\end{bmatrix}\begin{bmatrix} v_1\\ \vdots\\ v_n\end{bmatrix},$$

as desired.

33. (a) $70\cos\theta + 20\sin\theta$ (b) $(21\sqrt{3}+6)$ ft · lb

35. Each side equals $2xy - 7yz + 5z^2 - 48x + 54y - 5z - 96$. (Or switch the first two columns and then subtract the first row from the second.)

37. Add the last row to the first and subtract it from the second.

39. (a) $\dfrac{1}{6}\begin{vmatrix} a_1 & a_2 & a_3\\ b_1 & b_2 & b_3\\ c_1 & c_2 & c_3\end{vmatrix}$ (b) $1/3$

41. Use the fact that $\|\mathbf{a}\|^2 = \mathbf{a} \cdot \mathbf{a}$, expand both sides, and use the definition of $\mathbf{c}$.

43. $(1/\sqrt{38})\mathbf{i} - (6/\sqrt{38})\mathbf{j} + (1/\sqrt{38})\mathbf{k}$

45. $(2/\sqrt{5})\mathbf{i} - (1/\sqrt{5})\mathbf{j}$

47. $(\sqrt{3}/2)\mathbf{i} + (1/2\sqrt{2})\mathbf{j} + (1/2\sqrt{2})\mathbf{k}$

Chapter 2

Section 2.1

1. The level curves and graphs are sketched below. The graph in part (c) is a hyperbolic paraboloid like that of Example 4, but rotated 45° and vertically compressed by a factor of 1/4. To see this, use the variables $u = x + y$ and $v = x - y$. Then $z = (v^2 - u^2)/4$.

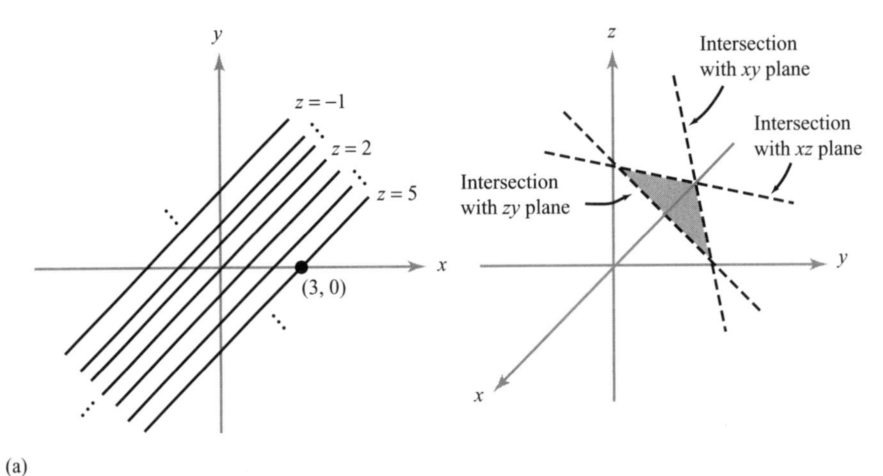

(a)

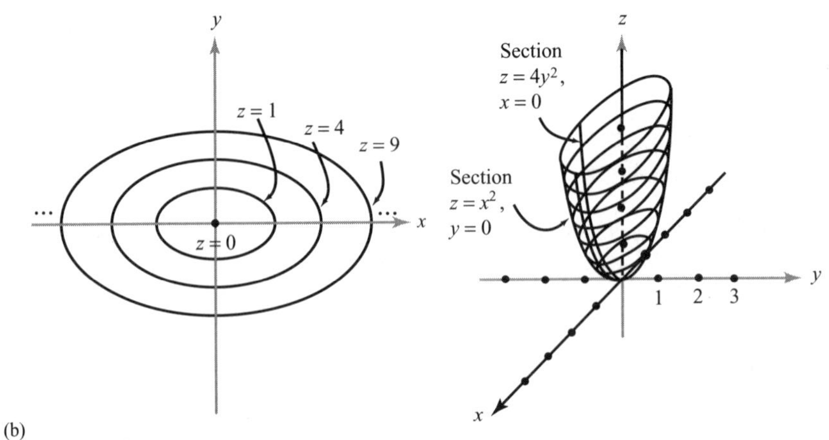

(b)

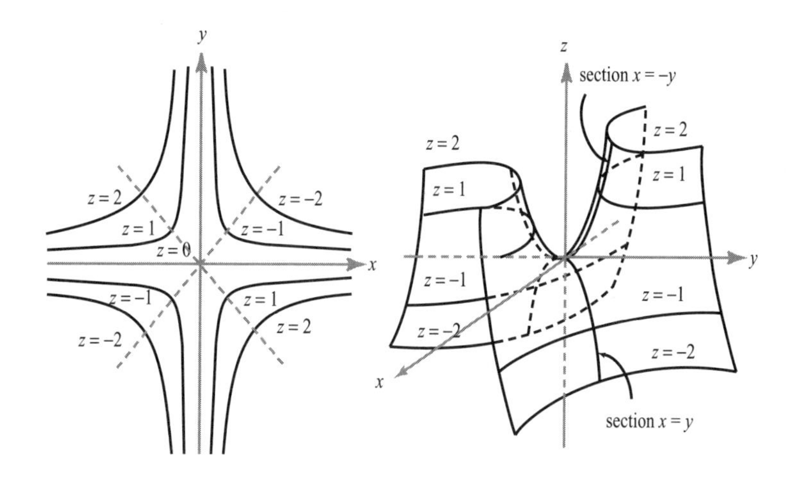

(c) $z = -xy$

3. For Example 2, $z = r(\cos\theta + \sin\theta) + 2$, shape depends on θ; for Example 3, $z = r^2$, shape is independent of θ; for Example 4, $z = r^2(\cos^2\theta - \sin^2\theta)$, shape depends on θ.

5. The level curves are circles $x^2 + y^2 = 100 - c^2$ when $c \leq 10$. The graph is the upper hemisphere of $x^2 + y^2 + z^2 = 100$.

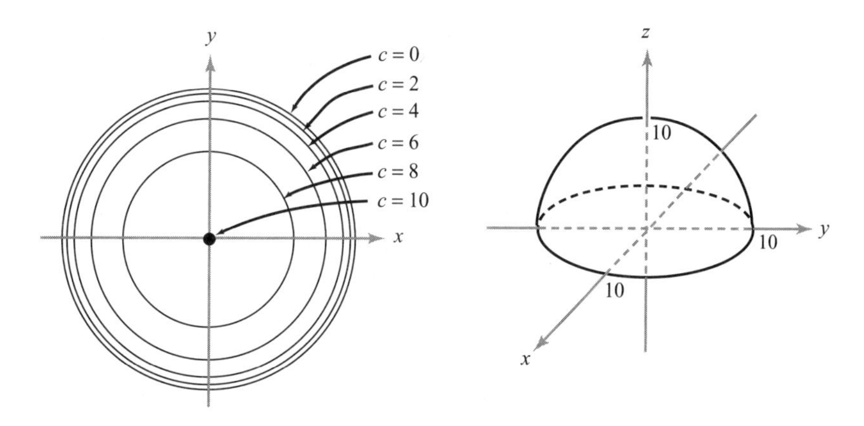

7. The level curves are circles, and the graph is a paraboloid of revolution. See Example 3 of this section.

9. If $c = 0$, the level curve is the straight line $y = -x$ together with the line $x = 0$. If $c \neq 0$, then $y = -x + (c/x)$. The level curve is a hyperbola with the y axis and the line $y = -x$ as asymptotes. The graph is a hyperbolic paraboloid. Sections along the line $y = ax$ are the parabolas $z = (1 + a)x^2$.

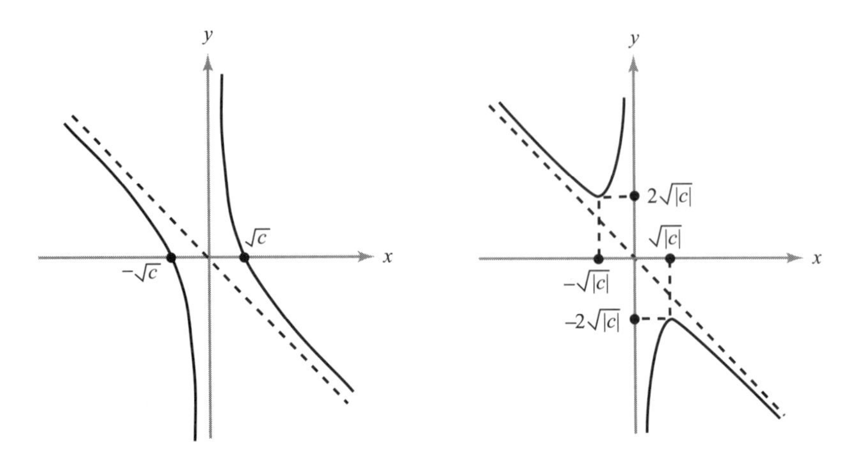

11. If $c > 0$, the level surface $f(x, y, z) = c$ is empty. If $c = 0$, the level surface is the point $(0, 0, 0)$. If $c < 0$, the level surface is the sphere of radius $\sqrt{-c}$ centered at $(0, 0, 0)$. A section of the graph determined by $z = a$ is given by $t = -x^2 - y^2 - a^2$, which is a paraboloid of revolution opening down in xyt space.

13. If $c < 0$, the level surface is empty. If $c = 0$, the level surface is the z axis. If $c > 0$, it is the right-circular cylinder $x^2 + y^2 = c$ of radius $\sqrt{c}$ whose axis is the z axis. A section of the graph determined by $z = a$ is the paraboloid of revolution $t = x^2 + y^2$. A section determined by $x = b$ is a "trough" with parabolic cross section $t(y, z) = y^2 + b^2$.

15. Setting $u = (x - z)/\sqrt{2}$ and $v = (x + z)/\sqrt{2}$ gives the u and v axes rotated 45° around the y axis from the x and z axes. Because $f = vy\sqrt{2}$, the level surfaces $f = c$ are "cylinders" perpendicular to the vy plane ($z = -x$) whose cross sections are the hyperbolas $vy = c/\sqrt{2}$, so the section $S_{x=a} \cap$ graph f is the hyperbolic paraboloid $t = (z + a)y$ in yzt space [see Exercise 1(c)]. The section $S_{y=b} \cap$ graph f is the plane $t = bx + bz$ in xzt space. The section $S_{z=b} \cap$ graph f is the hyperbolic paraboloid $t = y(x + b)$ in xyt space.

17. If $c < 0$, the level curve is empty. If $c = 0$, the level curve is the x axis. If $c > 0$, it is the pair of parallel lines $|y| = c$. The sections of graph with x constant are V-shaped curves $z = |y|$ in yz space. The graph is shown in the accompanying figure.

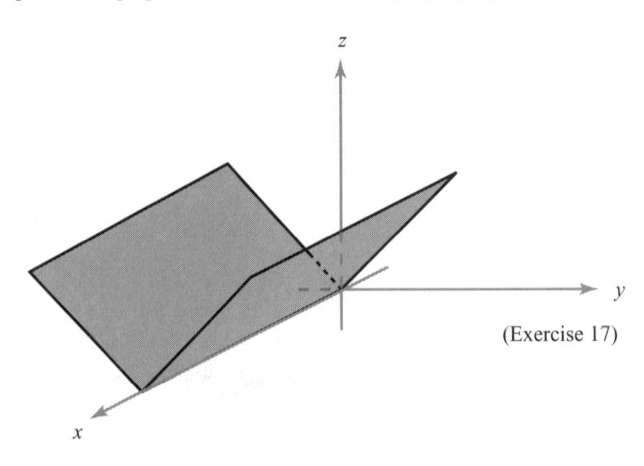

(Exercise 17)

19. The value of z does not matter, so we get a "cylinder" of elliptic cross section parallel to the z axis and intersecting the xy plane in the ellipse $4x^2 + y^2 = 16$.

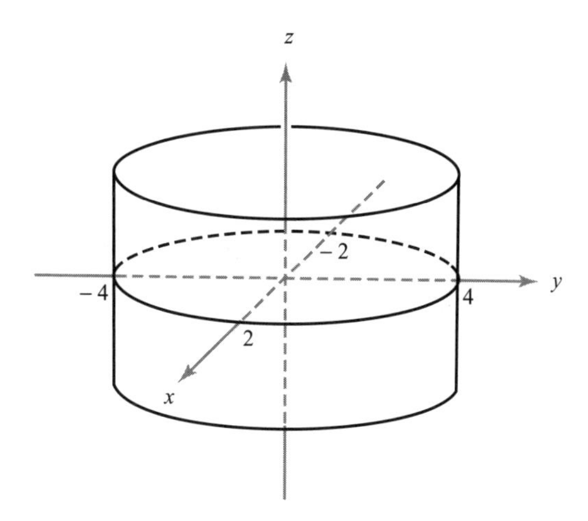

21. The value of x does not matter, so we get a "cylinder" parallel to the x axis of hyberbolic cross section intersecting the yz plane in the hyperbola $z^2 - y^2 = 4$.

23. An elliptic paraboloid with axis along the x axis.

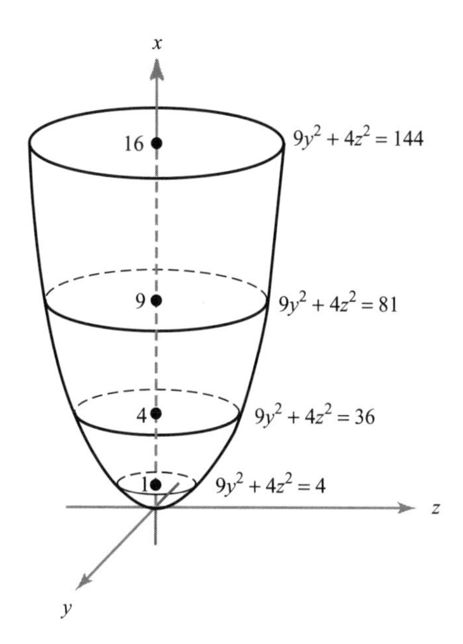

25. The value of y does not matter, so we get a "cylinder" of parabolic cross section.

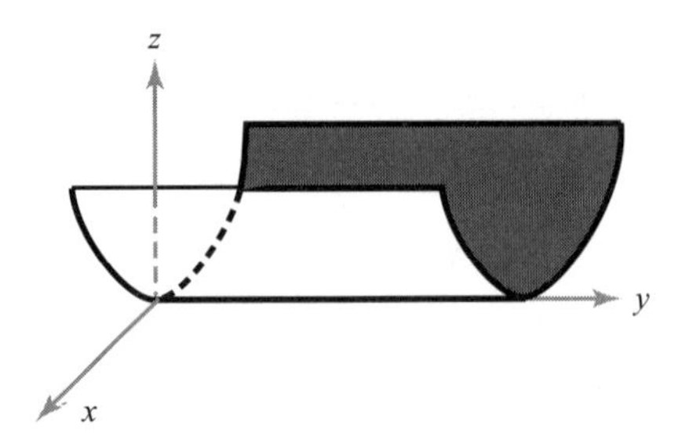

27. This is a saddle surface similar to that of Example 4, but the hyperbolas, which are level curves, no longer have perpendicular asymptotes.

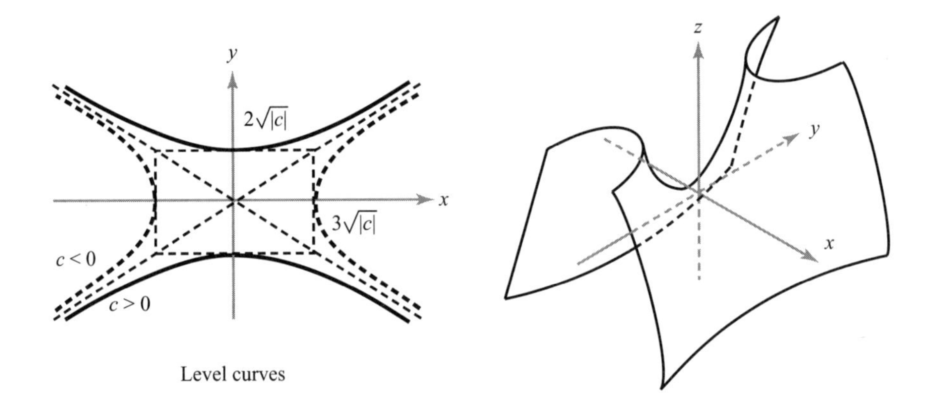

29. A double cone with axis along the y axis and elliptical cross sections

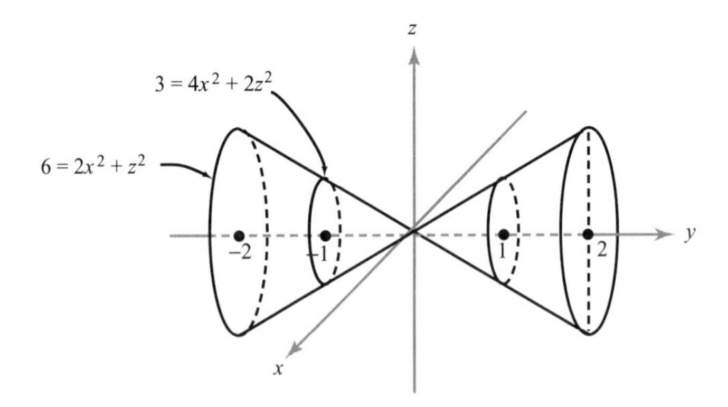

31. Complete the square to get $(x+2)^2 + (y - b/2)^2 + (z + \frac{9}{2})^2 = (b^2 + 4b + 97)/4$. This is an ellipsoid with center at $(-2, b/2, -\frac{9}{2})$ and axes parallel to the coordinate axes.

33. Level curves are described by $\cos 2\theta = cr^2$. If $c > 0$, then $-\pi/4 \le \theta \le \pi/4$ or $3\pi/4 \le \theta \le 5\pi/4$. If $c < 0$, then $\pi/4 \le \theta \le 3\pi/4$ or $5\pi/4 \le \theta \le 7\pi/4$. In either case you get a figure-eight shape, called a *lemniscate*, through the origin. (Such shapes were first studied by Jacques Bernoulli and are sometimes called Bernoulli's lemniscates.)

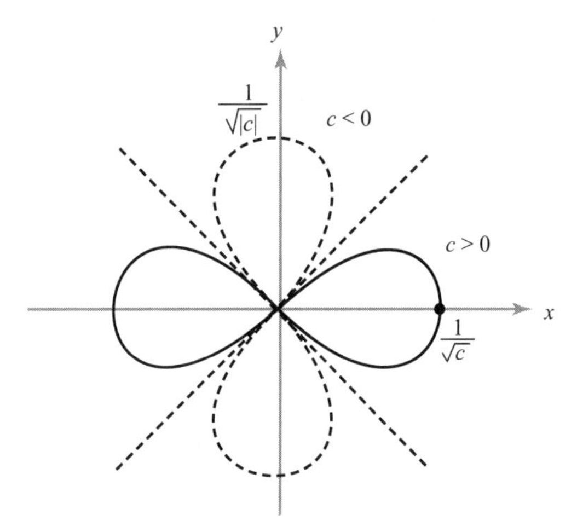

Section 2.2

1. If $(x_0, y_0) \in A$, then $|x_0| < 1$ and $|y_0| < 1$. Let $r = \min(1 - |x_0|, 1 - |y_0|)$. Prove that $D_r(x_0, y_0) \subset A$ either analytically or by drawing a figure.

3. Let $r = \min(2 - \sqrt{x_0^2 + y_0^2}, \sqrt{x_0^2 + y_0^2} - \sqrt{2})$.

5. (a) 0 (b) $-1/2$ (c) 1

7. (a) 5 (b) 0 (c) $2x$

9. (a) 0 (b) $-1/2$ (c) 0

11. (a) Compose $f(x, y) = xy$ with $g(t) = (\sin t)/t$ for $t \ne 0$ and $g(0) = 1$.
(b) 0 (c) 0

13. (a) 1 (b) $\|\mathbf{x}_0\|$ (c) $(1, e)$
(d) Limit doesn't exist (look at the limits for $x = 0$ and $y = 0$ separately).

15. Use parts (ii) and (iii) of Theorem 4.

17. (a) Let the value of the function be 1 at $(0, 0)$. (b) No.

19. For $|x - 2| < \delta = \sqrt{\varepsilon + 4} - 2$, we have $|x^2 - 4| = |x - 2||x + 2| < \delta(\delta + 4) = \varepsilon$. By Theorem 3(iii), $\lim_{x\to 2} x^2 = (\lim_{x\to 2} x)^2 = 2^2 = 4$.

21. Let $r = \|\mathbf{x} - \mathbf{y}\|/2$. If $\|\mathbf{z} - \mathbf{y}\| \le r$, let $f(\mathbf{z}) = \|\mathbf{z} - \mathbf{y}\|/r$. If $\|\mathbf{z} - \mathbf{y}\| > r$, let $f(\mathbf{z}) = 1$.

23. (a) $\lim_{x\to b^+} f(x) = L$ if for every $\varepsilon > 0$ there is a $\delta > 0$ such that $x > b$ and $0 < x - b < \delta$ imply $|f(x) - L| < \varepsilon$.

(b) $\lim_{x\to 0^-}(1/x) = -\infty$, $\lim_{t\to-\infty} e^t = 0$, and so $\lim_{x\to 0^-} e^{1/x} = 0$. Hence $\lim_{x\to 0^-} 1/(1+e^{1/x}) = 1$. The other limit is 0.

(c)

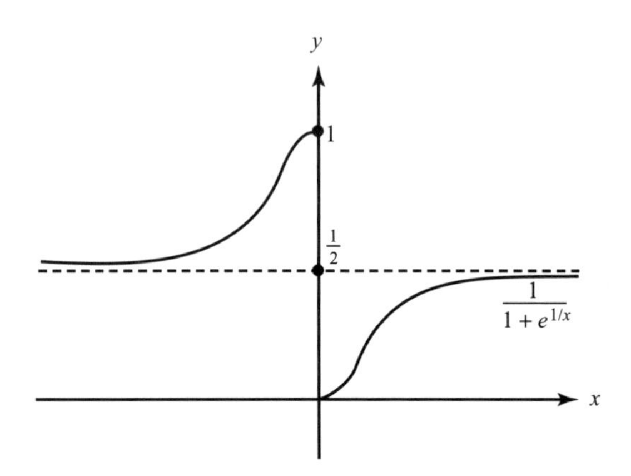

25. If $\varepsilon > 0$ and $\mathbf{x}_0$ are given, let $\delta = (\varepsilon/K)^{1/\alpha}$. Then $\|f(\mathbf{x}) - f(\mathbf{x}_0)\| < K\delta^\alpha = \varepsilon$ whenever $\|\mathbf{x} - \mathbf{x}_0\| < \delta$. Notice that the choice of δ does not depend on $\mathbf{x}_0$. This means that f is *uniformly continuous*.

27. (a) Choose $\delta < 1/500$. (b) Choose $\delta < 0.002$.

Section 2.3

1. (a) $\partial f/\partial x = y$; $\partial f/\partial y = x$
(b) $\partial f/\partial x = ye^{xy}$; $\partial f/\partial y = xe^{xy}$
(c) $\partial f/\partial x = \cos x \cos y - x \sin x \cos y$; $\partial f/\partial y = -x \cos x \sin y$
(d) $\partial f/\partial x = 2x[1 + \log(x^2+y^2)]$; $\partial f/\partial y = 2y[1 + \log(x^2+y^2)]$; $(x, y) \neq (0, 0)$

3. (a) $\partial w/\partial x = (1+2x^2)\exp(x^2+y^2)$; $\partial w/\partial y = 2xy\exp(x^2+y^2)$
(b) $\partial w/\partial x = -4xy^2/(x^2-y^2)^2$; $\partial w/\partial y = 4yx^2/(x^2-y^2)^2$
(c) $\partial w/\partial x = ye^{xy}\log(x^2+y^2) + 2xe^{xy}/(x^2+y^2)$;
$\partial w/\partial y = xe^{xy}\log(x^2+y^2) + 2ye^{xy}/(x^2+y^2)$
(d) $\partial w/\partial x = 1/y$; $\partial w/\partial y = -x/y^2$
(e) $\partial w/\partial x = -y^2e^{xy}\sin ye^{xy}\sin x + \cos ye^{xy}\cos x$;
$\partial w/\partial y = (xye^{xy} + e^{xy})(-\sin ye^{xy}\sin x)$

5. $z = 6x + 3y - 11$

7. (a) $\begin{bmatrix} 1 & 0 \\ 0 & 1 \end{bmatrix}$ (c) $\begin{bmatrix} 1 & 1 & e^z \\ 2xy & x^2 & 0 \end{bmatrix}$

(b) $\begin{bmatrix} e^y & xe^y - \sin y \\ 1 & 0 \\ 1 & e^y \end{bmatrix}$ (d) $\begin{bmatrix} (y+xy^2)e^{xy} & (x+x^2y)e^{xy} \\ \sin y & x\cos y \\ 5y^2 & 10xy \end{bmatrix}$

9. At $z = 1$.

11. Both are xye^{xy}.

13. (a) $\nabla f = (e^{-x^2-y^2-z^2}(-2x^2+1), -2xye^{-x^2-y^2-z^2}, -2xze^{-x^2-y^2-z^2})$
(b) $\nabla f = (x^2+y^2+z^2)^{-2}(yz(y^2+z^2-x^2), xz(x^2+z^2-y^2), xy(x^2+y^2-z^2))$
(c) $\nabla f = (z^2e^x \cos y, -z^2e^x \sin y, 2ze^x \cos y)$

15. $2x + 6y - z = 5$

17. $-2\mathbf{k}$

19. They are constant. Show that the derivative is the zero matrix.

Section 2.4

1. This curve is the ellipse $(y^2/16) + x^2 = 1$:

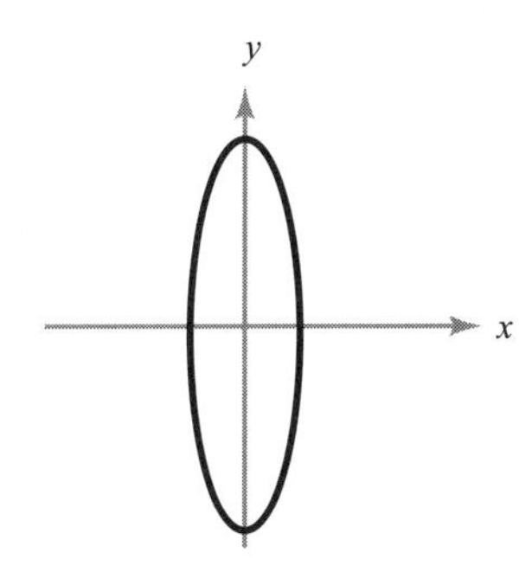

3. This curve is the straight line through $(-1, 2, 0)$ with direction $(2, 1, 1)$:

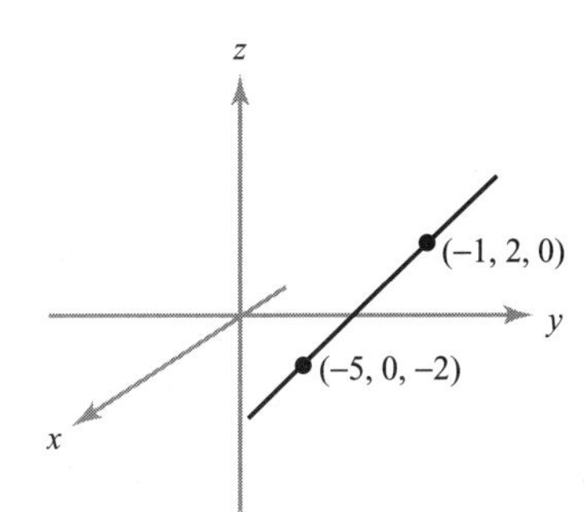

5. $6\mathbf{i} + 6t\mathbf{j} + 3t^2\mathbf{k}$

7. $(-2\cos t \sin t, 3 - 3t^2, 1)$

9. $\mathbf{c}'(t) = (e^t, -\sin t)$

11. $\mathbf{c}'(t) = (t\cos t + \sin t, 4)$

13. Horizontal when $t = (R/v)n\pi$, n an integer; speed is zero if n is even; speed is $2v$ if n is odd.

15. $(\sin 3, \cos 3, 2) + (3\cos 3, -3\sin 3, 5)(t - 1)$

17. $(8, 8, 0)$

19. $(8, 0, 1)$

Section 2.5

1. Use parts (i), (ii), and (iii) of Theorem 10. The derivative at $\mathbf{x}$ is $2(f(\mathbf{x})+1)\mathbf{D}f(\mathbf{x})$.

3. (a) $h(x,y)=f(x,u(x,y))=f(p(x),u(x,y))$. We use p here solely as notation: $p(x)=x$.

Written out: $\dfrac{\partial h}{\partial x}=\dfrac{\partial f}{\partial p}\dfrac{dp}{dx}+\dfrac{\partial f}{\partial u}\dfrac{\partial u}{\partial x}=\dfrac{\partial f}{\partial p}+\dfrac{\partial f}{\partial u}\dfrac{\partial u}{\partial x}$ because $\dfrac{dp}{dx}=\dfrac{dx}{dx}=1$

JUSTIFICATION: Call (p,u) the variables of f. To use the chain rule we must express h as a composition of functions; that is, first find g such that $h(x,y)=f(g(x,y))$. Let $g(x,y)=(p(x),u(x,y))$. Therefore, $\mathbf{D}h=(\mathbf{D}f)(\mathbf{D}g)$. Then

$$\begin{bmatrix}\dfrac{\partial h}{\partial x} & \dfrac{\partial h}{\partial y}\end{bmatrix}=\begin{bmatrix}\dfrac{\partial f}{\partial p} & \dfrac{\partial f}{\partial u}\end{bmatrix}\begin{bmatrix}\dfrac{\partial g_1}{\partial x} & \dfrac{\partial g_1}{\partial y}\\ \dfrac{\partial g_2}{\partial x} & \dfrac{\partial g_2}{\partial y}\end{bmatrix}=\begin{bmatrix}\dfrac{\partial f}{\partial p} & \dfrac{\partial f}{\partial u}\end{bmatrix}\begin{bmatrix}1 & 0\\ \dfrac{\partial u}{\partial x} & \dfrac{\partial u}{\partial y}\end{bmatrix}$$

$$=\begin{bmatrix}\dfrac{\partial f}{\partial p}+\dfrac{\partial f}{\partial u}\dfrac{\partial u}{\partial x} & \dfrac{\partial f}{\partial u}\dfrac{\partial u}{\partial y}\end{bmatrix},$$

and so $\dfrac{\partial h}{\partial x}=\dfrac{\partial f}{\partial p}+\dfrac{\partial f}{\partial u}\dfrac{\partial u}{\partial x}$. You may see $\dfrac{\partial h}{\partial x}=\dfrac{\partial f}{\partial x}+\dfrac{\partial f}{\partial u}\dfrac{\partial u}{\partial x}$ as an answer. This requires careful interpretation because of possible ambiguity about the meaning of $\partial f/\partial x$, which is why the name p was used.

(b) $\dfrac{\partial h}{\partial x}=\dfrac{\partial f}{\partial x}+\dfrac{\partial f}{\partial u}\dfrac{du}{dx}+\dfrac{\partial f}{\partial v}\dfrac{dv}{dx}$ (c) $\dfrac{\partial h}{\partial x}=\dfrac{\partial f}{\partial u}\dfrac{\partial u}{\partial x}+\dfrac{\partial f}{\partial v}\dfrac{\partial v}{\partial x}+\dfrac{\partial f}{\partial w}\dfrac{dw}{dx}$

5. Compute each in two ways; the answers are
(a) $(f\circ\mathbf{c})'(t)=e^t(\cos t-\sin t)$
(b) $(f\circ\mathbf{c})'(t)=15t^4\exp(3t^5)$
(c) $(f\circ\mathbf{c})'(t)=(e^{2t}-e^{-2t})[1+\log(e^{2t}+e^{-2t})]$
(d) $(f\circ\mathbf{c})'(t)=(1+4t^2)\exp(2t^2)$

7. Use Theorem 10(iii) and replace matrices by vectors.

9. $(f\circ g)(x,y)=(\tan(e^{x-y}-1)-e^{x-y},e^{2(x-y)}-(x-y)^2)$ and
$\mathbf{D}(f\circ g)(1,1)=\begin{bmatrix}0 & 0\\ 2 & -2\end{bmatrix}$.

11. $\frac{1}{2}\cos(1)\cos(\log\sqrt{2})$

13. $-2\cos t\sin te^{\sin t}+\sin^4 t+\cos^3 te^{\sin t}-3\cos^2 t\sin^2 t$ for both (a) and (b).

15. $(2,0)$

17. (a) $G(x, y(x)) = 0$ and so $\dfrac{\partial G}{\partial x} + \dfrac{\partial G}{\partial y}\dfrac{dy}{dx} = 0.$

(b) $$\begin{bmatrix} \dfrac{dy_1}{dx} \\ \dfrac{dy_2}{dx} \end{bmatrix} = -\begin{bmatrix} \dfrac{\partial G_1}{\partial y_1} & \dfrac{\partial G_1}{\partial y_2} \\ \dfrac{\partial G_2}{\partial y_1} & \dfrac{\partial G_2}{\partial y_2} \end{bmatrix}^{-1} \begin{bmatrix} \dfrac{\partial G_1}{\partial x} \\ \dfrac{\partial G_2}{\partial x} \end{bmatrix}$$ where $^{-1}$ means the inverse matrix.

The first component of this equation reads

$$\frac{dy_1}{dx} = \frac{-\dfrac{\partial G_1}{\partial x}\dfrac{\partial G_2}{\partial y_2} + \dfrac{\partial G_2}{\partial x}\dfrac{\partial G_1}{\partial y_2}}{\dfrac{\partial G_1}{\partial y_1}\dfrac{\partial G_2}{\partial y_2} - \dfrac{\partial G_2}{\partial y_1}\dfrac{\partial G_1}{\partial y_2}}.$$

(c) $\dfrac{dy}{dx} = \dfrac{-2x}{3y^2 + e^y}$

19. Apply the chain rule to $\partial G/\partial T$ where $G(t(T, P), p(T, P), V(T, P)) = P(V - b)e^{a/RVT} - RT$ is identically 0; $t(T, P) = T$; and $p(T, P) = P$.

21. Define $R_1(\mathbf{h}) = f(\mathbf{x}_0 + \mathbf{h}) - f(\mathbf{x}_0) - [\mathbf{D}f(\mathbf{x}_0)]\mathbf{h}$.

23. Let g_1 and g_2 be C^1 functions from $\mathbb{R}^3$ to $\mathbb{R}$ such that $g_1(\mathbf{x}) = 1$ for $\|\mathbf{x}\| < \sqrt{2}/3$; $g_1(\mathbf{x}) = 0$ for $\|\mathbf{x}\| > 2\sqrt{2}/3$; $g_2(\mathbf{x}) = 1$ for $\|\mathbf{x} - (1, 1, 0)\| < \sqrt{2}/3$; and $g_2(\mathbf{x}) = 0$ for $\|\mathbf{x} - (1, 1, 0)\| > 2\sqrt{2}/3$. (See Exercise 22.) Let

$$h_1(\mathbf{x}) = \begin{bmatrix} 1 & 0 & 0 \\ 0 & -1 & 0 \\ 0 & 0 & 0 \end{bmatrix}\begin{bmatrix} x_1 \\ x_2 \\ x_3 \end{bmatrix} + \begin{bmatrix} 1 \\ 1 \\ 0 \end{bmatrix} \quad \text{and} \quad h_2(\mathbf{x}) = \begin{bmatrix} 0 & 0 & -1 \\ 0 & 0 & 0 \\ 0 & 0 & 1 \end{bmatrix}\begin{bmatrix} x_1 \\ x_2 \\ x_3 \end{bmatrix},$$

and put $f(\mathbf{x}) = g_1(\mathbf{x})h_1(\mathbf{x}) + g_2(\mathbf{x})h_2(\mathbf{x})$.

25. Proof of rule (iii) follows:

$$\begin{aligned} &\frac{|h(\mathbf{x}) - h(\mathbf{x}_0) - [f(\mathbf{x}_0)\mathbf{D}g(\mathbf{x}_0) + g(\mathbf{x}_0)\mathbf{D}f(\mathbf{x}_0)](\mathbf{x} - \mathbf{x}_0)|}{\|\mathbf{x} - \mathbf{x}_0\|} \\ &\qquad \le |f(\mathbf{x}_0)|\frac{|g(\mathbf{x}) - g(\mathbf{x}_0) - \mathbf{D}g(\mathbf{x}_0)(\mathbf{x} - \mathbf{x}_0)|}{\|\mathbf{x} - \mathbf{x}_0\|} \\ &\qquad\quad + |g(\mathbf{x}_0)|\frac{|f(\mathbf{x}) - f(\mathbf{x}_0) - \mathbf{D}f(\mathbf{x}_0)(\mathbf{x} - \mathbf{x}_0)|}{\|\mathbf{x} - \mathbf{x}_0\|} \\ &\qquad\quad + \frac{|f(\mathbf{x}) - f(\mathbf{x}_0)|}{\|\mathbf{x} - \mathbf{x}_0\|}\frac{|g(\mathbf{x}) - g(\mathbf{x}_0)|}{\|\mathbf{x} - \mathbf{x}_0\|}\|\mathbf{x} - \mathbf{x}_0)\|. \end{aligned}$$

As $\mathbf{x} \to \mathbf{x}_0$, the first two terms go to 0 by the differentiability of f and g. The third does so because $|f(\mathbf{x}) - f(\mathbf{x}_0)|/\|\mathbf{x} - \mathbf{x}_0\|$ and $|g(\mathbf{x}) - g(\mathbf{x}_0)|/\|\mathbf{x} - \mathbf{x}_0\|$ are bounded by a constant, say M, on some ball $D_r(\mathbf{x}_0)$. To see this, choose r small enough that $[f(\mathbf{x}) - f(\mathbf{x}_0)]/\|\mathbf{x} - \mathbf{x}_0\|$ is within 1 of $\mathbf{D}f(\mathbf{x}_0)(\mathbf{x} - \mathbf{x}_0)/\|\mathbf{x} - \mathbf{x}_0\|$ if $\|\mathbf{x} - \mathbf{x}_0\| < r$. Then we have

$|f(\mathbf{x}) - f(\mathbf{x}_0)|/\|\mathbf{x}-\mathbf{x}_0\| \le 1 + |\mathbf{D}f(\mathbf{x}_0)(\mathbf{x}-\mathbf{x}_0)|/\|\mathbf{x}-\mathbf{x}_0\| = 1 + |\nabla f(\mathbf{x}_0)\cdot(\mathbf{x}-\mathbf{x}_0)|/\|\mathbf{x}-\mathbf{x}_0\| \le 1 + \|\nabla f(\mathbf{x}_0)\|$ by the Cauchy–Schwarz inequality.

The proof of rule (iv) follows from rule (iii) and the special case of the quotient rule, with f identically 1; that is, $\mathbf{D}(1/g)(\mathbf{x}_0) = [-1/g(\mathbf{x}_0)^2]\mathbf{D}g(\mathbf{x}_0)$. To obtain this answer, note that on some small ball $D_r(\mathbf{x}_0)$, $g(\mathbf{x}) > m > 0$. Use the triangle inequality and the Schwarz inequality to show that

$$\frac{\left|\dfrac{1}{g(\mathbf{x})} - \dfrac{1}{g(\mathbf{x})} + \dfrac{1}{g(\mathbf{x}_0)^2}\mathbf{D}g(\mathbf{x}_0)(\mathbf{x}-\mathbf{x}_0)\right|}{\|\mathbf{x}-\mathbf{x}_0\|}$$

$$\le \frac{1}{|g(\mathbf{x})|}\frac{1}{|g(\mathbf{x}_0)|}\frac{|g(\mathbf{x})-g(\mathbf{x}_0)-\mathbf{D}g(\mathbf{x}_0)(\mathbf{x}-\mathbf{x}_0)|}{\|\mathbf{x}-\mathbf{x}_0\|}$$

$$+\frac{|g(\mathbf{x})-g(\mathbf{x}_0)|}{|g(\mathbf{x})|g(\mathbf{x}_0)^2}\frac{|\mathbf{D}g(\mathbf{x}_0)(\mathbf{x}-\mathbf{x}_0)|}{\|\mathbf{x}-\mathbf{x}_0\|}$$

$$\le \frac{1}{m^2}\frac{|g(\mathbf{x})-g(\mathbf{x}_0)-\mathbf{D}g(\mathbf{x}_0)(\mathbf{x}-\mathbf{x}_0)|}{\|\mathbf{x}-\mathbf{x}_0\|} + \frac{\|\nabla g(\mathbf{x}_0)\|}{m^3}|g(\mathbf{x})-g(\mathbf{x}_0)|.$$

These last two terms both go to 0, because g is differentiable and continuous.

27. First find formula for $(\partial/\partial x)(F(x,x))$, using the chain rule. Let $F(x,z) = \int_0^x f(z,y)\,dy$ and use the fundamental theorem of calculus.

29. By Exercise 26 and Theorem 10(iii) (Exercise 25), each component of k is differentiable and $\mathbf{D}k_i(\mathbf{x}_0) = f(\mathbf{x}_0)\mathbf{D}g_i(\mathbf{x}_0) + g_i(\mathbf{x}_0)\mathbf{D}f(\mathbf{x}_0)$. Because $[\mathbf{D}g_i(\mathbf{x}_0)]\mathbf{y}$ is the ith component of $[\mathbf{D}g(\mathbf{x}_0)]y$ and $[\mathbf{D}f(\mathbf{x}_0)]\mathbf{y}$ is the number $\nabla f(\mathbf{x}_0)\cdot\mathbf{y}$, we get $[\mathbf{D}k(\mathbf{x}_0)]\mathbf{y} = f(\mathbf{x}_0)[\mathbf{D}g(\mathbf{x}_0)]\mathbf{y} + [\mathbf{D}f(\mathbf{x}_0)]\mathbf{y}[g(\mathbf{x}_0)] = f(x_0)[\mathbf{D}g\,(x_0)]\mathbf{y} + [\nabla f(\mathbf{x}_0)\cdot\mathbf{y}]g(\mathbf{x}_0)$.

Section 2.6

1. $\nabla f(1,1,2)\cdot\mathbf{v} = (4,3,4)\cdot(1/\sqrt{5}, 2/\sqrt{5}, 0) = 2\sqrt{5}$

3. (a) $17e^e/13$ (b) $e/\sqrt{3}$ (c) 0

5. (a) $z + 9x = 6y - 6$ (b) $z + y = \pi/2$ (c) $z = 1$

7. (a) $-\dfrac{1}{3\sqrt{3}}(\mathbf{i}+\mathbf{j}+\mathbf{k})$ (b) $2\mathbf{i}+2\mathbf{j}+2\mathbf{k}$ (c) $-\frac{2}{9}(\mathbf{i}+\mathbf{j}+\mathbf{k})$

9. $\mathbf{k}$

11. The graph of f is the level surface $0 = F(x,y,z) = f(x,y) - z$. Therefore, the tangent plane is given by

$$\begin{aligned} 0 &= \nabla F(x_0,y_0,z_0)\cdot(x-x_0, y-y_0, z-z_0) \\ &= \left(\frac{\partial f}{\partial x}(x_0,y_0), \frac{\partial f}{\partial y}(x_0,y_0), -1\right)\cdot(x-x_0, y-y_0, z-z_0). \end{aligned}$$

Because $z_0 = f(x_0, y_0)$, this is $z = f(x_0, y_0) + (\partial f/\partial x)(x_0, y_0)(x - x_0) + (\partial f/\partial y)(x_0, y_0)(y - y_0)$.

13. (a) $\nabla f = (z + y, z + x, x + y)$, $\mathbf{g}'(t) = (e^t, -\sin t, \cos t)$, $(f \circ \mathbf{g})'(1) = 2e\cos 1 + \cos^2 1 - \sin^2 1$

(b) $\nabla f = (yze^{xyz}, xze^{xyz}, xye^{xyz})$, $\mathbf{g}'(t) = (6, 6t, 3t^2)$, $(f \circ \mathbf{g})'(1) = 108e^{18}$

(c) $\nabla f = [1 + \log(x^2 + y^2 + z^2)](x\mathbf{i} + y\mathbf{j} + z\mathbf{k})$, $\mathbf{g}' = (e^t, -e^{-t}, 1)$, $(f \circ \mathbf{g})'(1) = [1 + \log(e^2 + e^{-2} + 1)](e^2 - e^{-2} + 1)$

15. Let $f(x, y, z) = 1/r = (x^2 + y^2 + z^2)^{-1/2}$; $\mathbf{r} = (x, y, z)$. Then we calculate $\nabla f = -(x^2 + y^2 + z^2)^{-3/2}(x, y, z) = -(1/r^3)\mathbf{r}$.

17. $\nabla f = (g'(x), 0)$.

19. $\mathbf{D}f(0, 0, \ldots, 0) = [0, \ldots, 0]$

21. $\mathbf{d}_1 = [-(0.03 + 2by_1)/2a]\mathbf{i} + y_1\mathbf{j}$, $\mathbf{d}_2 = [-(0.03 + 2by_2)/2a]\mathbf{i} + y_2\mathbf{j}$, where y_1 and y_2 are the solutions of $(a^2 + b^2)y^2 + 0.03by + \left(\frac{0.03^2}{4} - a^2\right) = 0$.

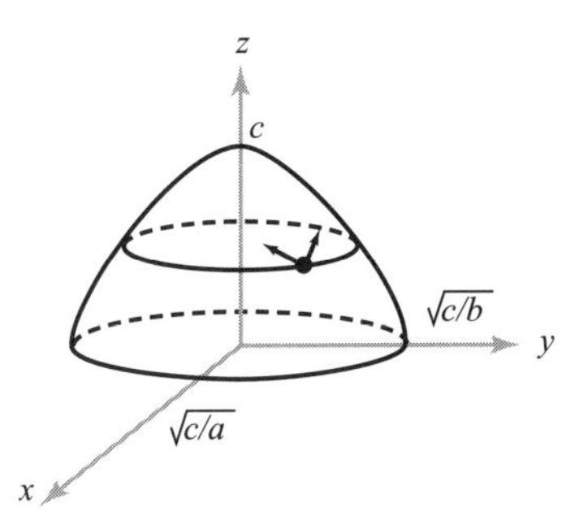

23. $$\nabla V = \frac{\lambda}{2\pi\varepsilon_0}\left[\left(\frac{x + x_0}{r_2^2} - \frac{x - x_0}{r_1^2}\right)\mathbf{i} + 2y\left(\frac{1}{r_2^2} - \frac{1}{r_1^2}\right)\mathbf{j}\right]$$

25. Crosses at $(2, 2, 0)$, $\sqrt{5}/10$ seconds later.

Review Exercises for Chapter 2

1. (a) Elliptic paraboloid.

(b) Let $y' = y + 3$ and write $z = xy'$. This is a (shifted) hyperbolic paraboloid.

3. (a) $\mathbf{D}f(x, y) = \begin{bmatrix} 2xy & x^2 \\ -ye^{-xy} & -xe^{-xy} \end{bmatrix}$

(b) $\mathbf{D}f(x) = \begin{bmatrix} 1 \\ 1 \end{bmatrix}$

(c) $\mathbf{D}f(x, y, z)] = [e^x \quad e^y \quad e^z]$

(d) $\mathbf{D}f(x, y, z) = \begin{bmatrix} 1 & 0 & 0 \\ 0 & 1 & 0 \\ 0 & 0 & 1 \end{bmatrix}$

5. The plane tangent to a sphere at (x_0, y_0, z_0) is normal to the line from the center to (x_0, y_0, z_0).

7. (a) $z = x - y + 2$ (d) $10x + 6y - 4z = 6 - \pi$
(b) $z = 4x - 8y - 8$ (e) $2z = \sqrt{2}x + \sqrt{2}y$
(c) $x + y + z = -1$ (f) $x + 2y - z = 2$

9. (a) The level curves are hyperbolas $xy = 1/c$:

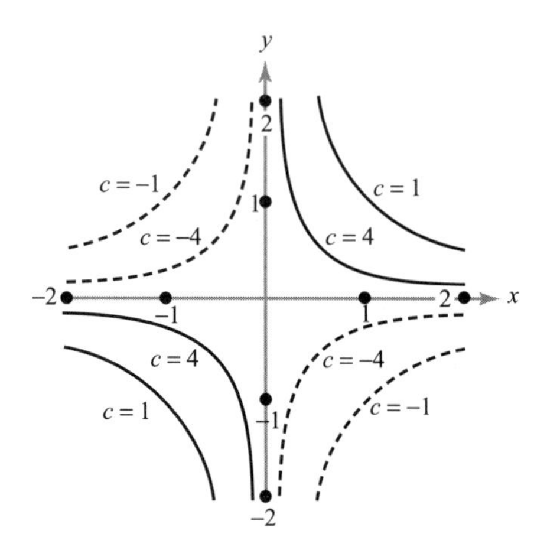

(b) $c = x^2 - xy - y^2 = \left(x - \frac{1+\sqrt{5}}{2}y\right)\left(x - \frac{1-\sqrt{5}}{2}y\right)$

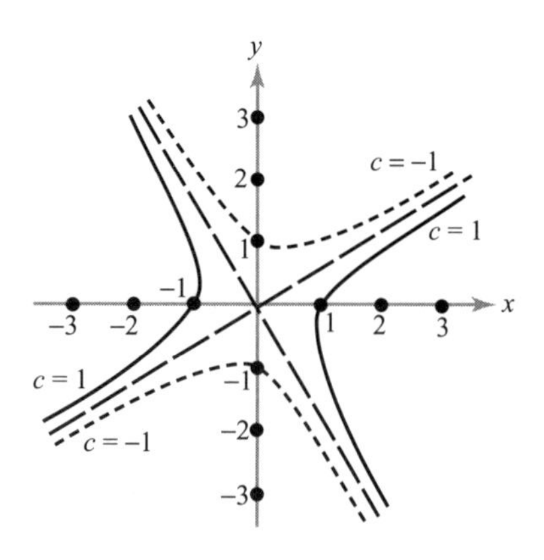

11. (a) 0 (b) Limit does not exist.

13. $(1 + 2x^2)\exp(1 + x^2 + y^2)$

15. (a) The line $\mathbf{L}(t) = (x_0, y_0, f(x_0, y_0)) + t(a, b, c)$ lies in the plane $z = f(x_0, y_0)$ if $c = 0$ and is perpendicular to $\nabla f(x_0, y_0)$ if $a(\partial f/\partial x)(x_0, y_0) + b(\partial b/\partial y)(x_0, y_0) = 0$. On **L**, we have

$$f(x_0, y_0) + \left[\frac{\partial f}{\partial x}(x_0, y_0)\right](x - x_0) + \left[\frac{\partial f}{\partial y}(x_0, y_0)\right](y - y_0)$$

$$= f(x_0, y_0) + at\left[\frac{\partial f}{\partial x}(x_0, y_0)\right] + bt\left[\frac{\partial f}{\partial y}(x_0, y_0)\right]$$

$$= f(x_0, y_0) = z$$

Therefore, **L** lies in the tangent plane. An upward unit normal to the tangent plane is $\mathbf{p} = (1 + \|\nabla f\|^{-1/2}(-(\partial f/\partial x)(x_0, y_0), -(\partial f/\partial y)(x_0, y_0), 1)$. Therefore, $\cos\theta = \mathbf{p}\cdot\mathbf{k} = (1 + \|\nabla f\|^2)^{-1/2}$, and $\tan\theta = \sin\theta/\cos\theta = \{\|\nabla f\|^2/(1 + \|\nabla f\|^2)\}^{1/2}/(1 + \|\nabla f\|^2)^{-1/2} = \|\nabla f\|$ as claimed.

(b) The tangent plane contains the horizontal line through $(1, 0, 2)$ perpendicular to $\nabla f(1, 0) = (5, 0)$, that is, parallel to the y axis. It makes an angle of $\arctan(\|\nabla f(1, 0)\|) = \arctan 5 \approx 78.7^\circ$ with respect to the xy plane.

17. $(1/\sqrt{2}, 1/\sqrt{2})$ or $(-1/\sqrt{2}, -1/\sqrt{2})$

19. A unit normal is $(\sqrt{2}/10)(3, 5, 4)$. The tangent plane is $3x + 5y + 4z = 18$.

21. $4\mathbf{i} + 16\mathbf{j}$

23. (a) Because g is the composition $\lambda \mapsto \lambda x \mapsto f(\lambda x)$, the chain rule gives

$$g'(\lambda) = \mathbf{D}f(\lambda\mathbf{x})\begin{bmatrix} x_1 \\ \vdots \\ x_n \end{bmatrix}.$$

Thus,

$$g'(1) = \mathbf{D}f(\mathbf{x})\begin{bmatrix} x_1 \\ \vdots \\ x_n \end{bmatrix} = \nabla f(\mathbf{x})\cdot\mathbf{x}$$

But also $g(\lambda) = \lambda^p f(\mathbf{x})$, so $g'(\lambda) = p\lambda^{p-1} f(\mathbf{x})$ and $g'(1) = pf(\mathbf{x})$.

(b) $p = 1$.

25. Differentiate directly using the chain rule, or use Exercise 23(a) with $p = 0$.

27. (a) If $(x, y) \neq (0, 0)$, then one calculates for (i) that $\partial f/\partial x = (y^3 - yx^2)/(x^2 + y^2)^2$ and $\partial f/\partial y = (x^3 - xy^2)/(x^2 + y^2)^2$. If $x = y = 0$, use the definition directly to find that both partial derivatives are 0. For (ii), if $(x, y) \neq (0, 0)$, then $\partial f/\partial x = 2xy^6/(x^2 + y^4)^2$ and $\partial f/\partial y = (2x^4y - 2x^2y^5)/(x^2 + y^4)^2$. The partials at the origin are zero.

(b) The function (i) is not continuous at (0, 0); the function (ii) is differentiable, but the derivative is not continuous.

29. (a) $\sqrt{2}\pi/8$ (b) $-\sin\sqrt{2}$ (c) $-2\sqrt{2}e^{-2}$

31. $(-4e^{-1}, 0)$

33. (a) See Theorem 11.

(b) $g(u) = (\sin 3u)^2 + \cos 8u$
$g'(u) = 6\sin 3u\cos 3u - 8\sin 8u$
$g'(0) = 0$

$\nabla f = (2x, 1)$
$\nabla f(\mathbf{h}(0)) = \nabla f(0, 1) = (0, 1)$
$\mathbf{h}'(u) = (3\cos 3u, -8\sin 8u)$
$g'(0) = \nabla f(\mathbf{h}(0))\cdot\mathbf{h}'(0) = (0, 1)\cdot(3, 0) = 0$

35. $t = \sqrt{14}(-3+\sqrt{359})/70 = (-3+\sqrt{359})/5\sqrt{14}$

37. $\partial z/\partial x = 4(e^{-2x-2y+2xy})(1+y)/(e^{-2x-2y} - e^{2xy})^2$
$\partial z/\partial y = 4(e^{-2y-2x+2xy})(1+x)/(e^{-2x-2y} - e^{2xy})^2$

39. Notice that $y = x^2$, so that if y is constant, x cannot be a variable.

41. $[f'(t)g(t) + f(t)g'(t)]\exp[f(t)g(t)]$

43. $d[f(\mathbf{c}(t))]/dt = 2t/[(1+t^2+2\cos^2 t)(2-2t^2+t^4)]$
$-4t(t^2-1)\ln(1+t^2+2\cos^2 t)/(2-2t^2+t^4)^2$
$-4\cos t\sin t/[(1+t^2+2\cos^2 t)(2-2t^2+t^4)]$

45. Let $x = f(t)$, $y = t$, and use the chain rule to differentiate $u(x, y)$ with respect to t.

47. (a) $n = PV/RT; P = nRT/V; T = PV/nR; V = nRT/P$.

(b) $\partial V/\partial T = nR/P; \partial T/\partial P = V/nR; \partial P/\partial V = -nRT/V^2$. Multiply, remembering that $PV = nRT$.

49. (a) One can solve for any of the variables in terms of the other two.

(b) $\partial T/\partial P = (V-\beta)/R;$
$\partial P/\partial V = -RT/(V-\beta)^2 + 2\alpha/V^3;$
$\partial V/\partial T = R/[(V-\beta)(RT/(V-\beta)^2 - 2\alpha/V^3)]$

(c) Multiply and cancel factors.

51. (a) $(1/\sqrt{2}, 1/\sqrt{2})$

(b) The directional derivative is 0 in the direction

$$(x_0\mathbf{i} + y_0\mathbf{j})/\sqrt{x_0^2 + y_0^2}.$$

(c) The level curve through (x_0, y_0) must be tangent to the line through (0, 0) and (x_0, y_0). The level curves are lines or half-lines emanating from the origin.

53. $G(x, y) = x - y$

Chapter 3

Section 3.1

1. $\dfrac{\partial^2 f}{\partial x^2} = 24\dfrac{x^3y - xy^3}{(x^2+y^2)^4}, \dfrac{\partial^2 f}{\partial y^2} = 24\dfrac{-x^3y + xy^3}{(x^2+y^2)^4},$

$$\frac{\partial^2 f}{\partial x \partial y} = \frac{\partial^2 f}{\partial y \partial x} = \frac{-6x^4 + 36x^2y^2 - 6y^4}{(x^2+y^2)^4}$$

3. $\dfrac{\partial^2 f}{\partial x^2} = -y^4\cos(xy^2), \dfrac{\partial^2 f}{\partial y^2} = -2x\sin(xy^2) - 4x^2y^2\cos(xy^2),$

$$\frac{\partial^2 f}{\partial x\, \partial y} = \frac{\partial^2 f}{\partial y \partial x} = -2y\sin(xy^2) - 2xy^3\cos(xy^2)$$

5. $\dfrac{\partial^2 f}{\partial x^2} = \dfrac{2(\cos^2 x + e^{-y})\cos 2x + 2\sin^2 2x}{(\cos^2 x + e^{-y})^3},$

$$\frac{\partial^2 f}{\partial y^2} = \frac{e^{-y} - \cos^2 x}{e^y(\cos^2 x + e^{-y})^3}$$

$$\frac{\partial^2 f}{\partial x\, \partial y} = \frac{\partial^2 f}{\partial y\, \partial x} = \frac{2\sin 2x}{e^y(\cos^2 x + e^{-y})^3}$$

7. (a) $\partial^2 z/\partial x^2 = 6, \partial^2 z/\partial y^2 = 4,$ $\partial^2 z/\partial x\, \partial y = \partial^2 z/\partial y\, \partial x = 0$ (b) $\partial^2 z/\partial x^2 = 0, \partial^2 z/\partial y^2 = 4x/3y^3,$ $\partial^2 z/\partial x\, \partial y = \partial^2 z/\partial y\, \partial x = -2/3y^2$

9. $f_{xy} = 2x + 2y,\ f_{yz} = 2z,\ f_{zx} = 0,\ f_{xyz} = 0$

11. Because f and $\partial f/\partial z$ are both of class C^2, we have

$$\frac{\partial^3 f}{\partial x\, \partial y\, \partial z} = \frac{\partial^2}{\partial x\, \partial y}\frac{\partial f}{\partial z} = \frac{\partial^2}{\partial y\, \partial x}\frac{\partial f}{\partial z} = \frac{\partial}{\partial y}\left(\frac{\partial^2 f}{\partial x\, \partial z}\right) = \frac{\partial}{\partial y}\left(\frac{\partial^2 f}{\partial z\, \partial x}\right) = \frac{\partial^3 f}{\partial y\, \partial z\, \partial x}.$$

13. $f_{xzw} = f_{zwx} = e^{xyz}[2xy\cos(xw) + x^2y^2z\cos(xw) - x^2yw\sin(xw)]$

15. (a) $\dfrac{\partial f}{\partial x} = \arctan\dfrac{x}{y} + \dfrac{xy}{x^2+y^2}$

$$\frac{\partial f}{\partial y} = \frac{-x^2}{x^2+y^2}$$

$$\frac{\partial^2 f}{\partial x^2} = \frac{2y^3}{(x^2+y^2)^2}, \frac{\partial^2 f}{\partial y^2} = \frac{2x^2y}{(x^2+y^2)^2}$$

$$\frac{\partial^2 f}{\partial x\, \partial y} = \frac{\partial^2 f}{\partial y\, \partial x} = \frac{-2xy^2}{(x^2+y^2)^2}$$

(b) $\dfrac{\partial f}{\partial x} = \dfrac{-x \sin\sqrt{x^2+y^2}}{\sqrt{x^2+y^2}}, \dfrac{\partial f}{\partial y} = \dfrac{-y \sin\sqrt{x^2+y^2}}{\sqrt{x^2+y^2}}$

$$\frac{\partial^2 f}{\partial x^2} = \frac{x^2 \sin\sqrt{x^2+y^2}}{(x^2+y^2)^{3/2}} - \frac{x^2 \cos\sqrt{x^2+y^2}}{x^2+y^2} - \frac{\sin\sqrt{x^2+y^2}}{(x^2+y^2)^{1/2}}$$

$$\frac{\partial^2 f}{\partial y^2} = \frac{y^2 \sin\sqrt{x^2+y^2}}{(x^2+y^2)^{3/2}} - \frac{y^2 \cos\sqrt{x^2+y^2}}{x^2+y^2} - \frac{\sin\sqrt{x^2+y^2}}{(x^2+y^2)^{1/2}}$$

$$\frac{\partial^2 f}{\partial x\,\partial y} = \frac{\partial^2 f}{\partial y\,\partial x} = xy\left[\frac{\sin\sqrt{x^2+y^2}}{(x^2+y^2)^{3/2}} - \frac{\cos\sqrt{x^2+y^2}}{x^2+y^2}\right]$$

(c) $\dfrac{\partial f}{\partial x} = -2x \exp(-x^2-y^2), \dfrac{\partial f}{\partial y} = -2y \exp(-x^2-y^2),$

$$\frac{\partial^2 f}{\partial x^2} = (4x^2-2)\exp(-x^2-y^2), \frac{\partial^2 f}{\partial y^2} = (4y^2-2)\exp(-x^2-y^2),$$

$$\frac{\partial^2 f}{\partial x\,\partial y} = \frac{\partial^2 f}{\partial y\,\partial x} = 4xy \exp(-x^2-y^2)$$

17. $\dfrac{\partial^2 f}{\partial x^2}\left(\dfrac{dx}{dt}\right)^2 + 2\dfrac{\partial^2 f}{\partial x\,\partial y}\dfrac{dx}{dt}\dfrac{dy}{dt} + \dfrac{\partial^2 f}{\partial y^2}\left(\dfrac{dy}{dt}\right)^2 + \dfrac{\partial f}{\partial x}\dfrac{d^2x}{dt^2} + \dfrac{\partial f}{\partial y}\dfrac{d^2y}{dt^2},$
where $\mathbf{c}(t) = (x(t), y(t))$

19. Evaluate the derivatives $\partial^2 u/\partial x^2$ and $\partial^2 u/\partial y^2$ and add.

21. (a) Evaluate the derivatives and compare.
(b) 

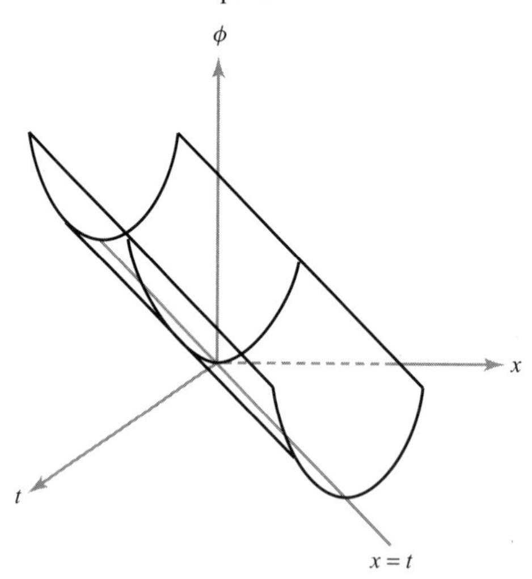

23. $V = -GmM/r = -GmM(x^2+y^2+z^2)^{-1/2}$. Check that
$$\frac{\partial^2 V}{\partial x^2} + \frac{\partial^2 V}{\partial y^2} + \frac{\partial^2 V}{\partial z^2} = GmM(x^2+y^2+z^2)^{-3/2}[3 - 3(x^2+y^2+z^2)(x^2+y^2+z^2)^{-1}] = 0$$

Section 3.2

1. $f(h_1, h_2) = h_1^2 + 2h_1h_2 + h_2^2$ [$R_2(\mathbf{0}, \mathbf{h}) = 0$ in this case]

3. $f(h_1, h_2) = 1 + h_1 + h_2 + \frac{h_1^2}{2} + h_1h_2 + \frac{h_2^2}{2} + R_2(\mathbf{0}, \mathbf{h})$

5. $f(h_1, h_2) = 1 + h_1h_2 + R_2(\mathbf{0}, \mathbf{h})$

7. (a) Show that $|R_k(x, a)| \le AB^{k+1}/(k+1)!$ for constants A, B, and x in a fixed interval $[a, b]$. Prove that $R_k \to 0$ as $k \to \infty$. (Use convergence of the series $\sum c^k/k! = e^c$ and use Taylor's theorem.)

(b) The only possible trouble is at $x = 0$. Use L'Hôpital's rule to show that

$$\lim_{t\to\infty} p(t)e^t = \infty$$

for every polynomial $p(t)$. Using this, establish that $\lim_{x\to 0^+} p(x)e^{-1/x} = 0$ for every rational function $p(x)$, and conclude that $f^{(k)}(0) = 0$ for every k.

(c) $f: \mathbb{R}^n \to \mathbb{R}$ is analytic at $\mathbf{x}_0$ if the series

$$f(\mathbf{x}_0) + \sum_{i=1}^{n} h_i \frac{\partial f}{\partial x_i}(\mathbf{x}_0) + \frac{1}{2}\sum_{i,j=1}^{n} h_ih_j \frac{\partial^2 f}{\partial x_i\,\partial x_j}(\mathbf{x}_0) + \cdots$$
$$+ \frac{1}{k!}\sum_{i_1,\ldots,i_k=1}^{n} h_{i_1}h_{i_2}\cdots h_{i_k} \frac{\partial^k f}{\partial x_{i_1}\cdots\partial x_{i_k}}(\mathbf{x}_0) + \cdots$$

converges to $f(\mathbf{x}_0 + \mathbf{h})$ for all $\mathbf{h} = (h_1, \ldots, h_n)$ in some sufficiently small disk $\|\mathbf{h}\| < \varepsilon$. The function f is analytic if for every $R > 0$ there is a constant M such that $|(\partial^k f/\partial x_{i_1}\cdots\partial x_{i_k})(\mathbf{x})| < M^k$ for each kth-order derivative at every x satisfying $\|\mathbf{x}\| \le R$.

(d) $f(x, y) = 1 + x + y + \frac{1}{2}(x^2 + 2xy + y^2) + \cdots + \frac{1}{k!}\sum_{j=0}^{k}\binom{k}{j}x^jy^{k-j} + \cdots$

Section 3.3

1. (0, 0); saddle point.

3. The critical points are on the line $y = -x$; they are local minima, because $f(x, y) = (x + y)^2 \ge 0$, equaling zero only when $x = -y$.

5. (0, 0); saddle point.

7. $(-\frac{1}{4}, -\frac{1}{4})$; local minimum.

9. (0, 0); local maximum. (The tests fail, but use the fact that $\cos z \le 1$.)
$(\sqrt{\pi/2}, \sqrt{\pi/2})$, local minimum
$(0, \sqrt{\pi})$, local minimum.

11. No critical points.

13. $(1, 1)$ is a local minimum.

15. $(0, n\pi)$; critical points, no local maxima or minima.

17. Minimum at $(0, 0)$ and maxima at $(0, \pm 1)$ [and saddles at $(\pm 1, 0)$].

19. (a) $\partial f/\partial x$ and $\partial f/\partial y$ vanish at $(0, 0)$.
(b) Show that $f(g(t)) = 0$ at $t = 0$ and that $f(g(t)) \geq 0$ if $|t| < |b|/3a^2$.
(c) f is negative on the parabola $y = 2x^2$.

21. The critical points are on the line $y = x$ and they are local minima (see Exercise 3).

23. Minimize $S = 2xy + 2yz + 2xz$ with $z = V/xy$, V the constant volume.

25. 40, 40, 40

27. The only critical point is $(0, 0, 0)$. It is a minimum, because

$$f(x, y, z) \geq \frac{x^2 + y^2}{2} + z^2 + xy = \frac{1}{2}(x + y)^2 + z^2 \geq 0.$$

29. $(1, \frac{3}{2})$ is a saddle point; $(5, \frac{27}{2})$ is a local minimum.

31. $\frac{3}{2}$ is the absolute maximum and 0 is the absolute minimum.

33. -2 is the absolute minimum; 2 is the absolute maximum.

35. $(\frac{1}{2}, 4)$ is a local minimum.

37. If $u_n(x, y) = u(x, y) + (1/n)e^x$, then $\nabla^2 u_n = (1/n)e^x > 0$. Thus, u_n is strictly subharmonic and can have its maximum only on ∂D, say, at $\mathbf{p}_n = (x_n, y_n)$. If $(x_0, y_0) \in D$, check that this implies $u(x_n, y_n) > u(x_0, y_0) - e/n$. Thus, there must be a point $\mathbf{q} = (x_\infty, y_\infty)$ on ∂D such that arbitrarily close to $\mathbf{q}$ we can find an (x_n, y_n) for n as large as we like. Conclude from the continuity of u that $u(x_\infty, y_\infty) \geq u(x_0, y_0)$.

39. Follow the methods of Exercise 37.

41. (a) If there were an x_1 with $f(x_1) < f(x_0)$, then the maximum of f on the interval between x_0 and x would be another critical point.

(b) Verify (i) by the second derivative test; for (ii), f goes to $-\infty$ as $y \to \infty$ and $x = -y$.

Section 3.4

1. Maximum at $\sqrt{\frac{2}{3}}(1, -1, 1)$, minimum at $\sqrt{\frac{2}{3}}(-1, 1, -1)$

3. Maximum at $(\sqrt{3}, 0)$, minimum at $(-\sqrt{3}, 0)$

5. Maximum at $(\frac{9}{\sqrt{70}}, \frac{4}{\sqrt{70}})$, minimum at $(-\frac{9}{\sqrt{70}}, -\frac{4}{\sqrt{70}})$

7. The minimum value 4 is attained at (0, 2). Use a geometric picture rather than Lagrange multipliers.

9. $(0, 0, 2)$ is a minimum of f.

11. $\frac{3}{2}$ is the absolute maximum and 0 is the absolute minimum.

13. The diameter should equal the height, $20/\sqrt[3]{2\pi}$ cm.

15. Maximum value $\sqrt{3}$ at $(\frac{1}{\sqrt{3}}, \frac{1}{\sqrt{3}}, -\frac{1}{\sqrt{3}})$ and minimum value $-\sqrt{3}$ at $(-\frac{1}{\sqrt{3}}, -\frac{1}{\sqrt{3}}, \frac{1}{\sqrt{3}})$.

17. Horizontal length is $\sqrt{qA/p}$, vertical length is $\sqrt{pA/q}$.

19. For Exercise 1, the bordered Hessians required are

$$|\bar{H}_2| = \begin{vmatrix} 0 & 2x & 2y \\ 2x & -2\lambda & 0 \\ 2y & 0 & -2\lambda \end{vmatrix} = 8\lambda(x^2 + y^2),$$

$$|\bar{H}_3| = \begin{vmatrix} 0 & 2x & 2y & 2z \\ 2x & -2\lambda & 0 & 0 \\ 2y & 0 & -2\lambda & 0 \\ 2z & 0 & 0 & -2\lambda \end{vmatrix} = -16\lambda(x^2 + y^2 + z^2).$$

At $\sqrt{\frac{2}{3}}(1, -1, 1)$ the Lagrange multiplier is $\lambda = \sqrt{6}/4 > 0$, indicating a maximum at $\sqrt{\frac{2}{3}}(1, -1, 1)$, and $\lambda = -\sqrt{6}/4 < 0$ indicates a minimum at $\sqrt{\frac{2}{3}}(-1, 1, -1)$. In Exercise 5, $|\bar{H}| = 24\lambda(4x^2 + 6y^2)$, and so $\lambda = \sqrt{70}/12 > 0$ indicates a maximum at $(9/\sqrt{70}, 4/\sqrt{70})$ and $\lambda = -\sqrt{70}/12 < 0$ indicates a minimum at $(-9/\sqrt{70}, -4/\sqrt{70})$.

21. 11,664 in^3

23. (a) $\nabla f(\mathbf{x}) = A\mathbf{x}$.
(b) S is defined by the constraint function $g(\mathbf{x}) = x_1^2 + x_2^2 + x_3^2 - 1$. Because $\nabla g(\mathbf{x}) = 2\mathbf{x}$ is not $\mathbf{0}$, Theorem 9 applies. At an $\mathbf{x}$ where f is extreme, there is a $\lambda/2$ such that $\nabla f(\mathbf{x}) = (\lambda/2)/\nabla g(\mathbf{x})$. That is, $A\mathbf{x} = \lambda\mathbf{x}$.

25. Minimum is $(-1/\sqrt{2}, 0)$ maximum is $(\frac{1}{4}, \pm\sqrt{7/8})$, local minimum at $(1/\sqrt{2}, 0)$.

27. No critical points; no maximum or minimum

29. $(-1, 0, 1)$

31. The point $(K, L) = (\alpha B/q, (1-\alpha)B/p)$ optimizes the profit.

Section 3.5

1. Use the special implicit function theorem with $n = 1$. (See Example 1.) Line (i) is given by $0 = (x - x_0, y - y_0) \cdot \nabla F(x_0, y_0) = (x - x_0)(\partial F/\partial x)(x_0, y_0) + (y - y_0)(\partial F/\partial y)(x_0, y_0)$. For line (ii), Theorem 11 gives $dy/dx = -(\partial F/\partial x)/(\partial F/\partial y)$, and so the lines agree and are given by

$$y = y_0 - \frac{(\partial F/\partial x)(x_0, y_0)}{(\partial F/\partial y)(x_0, y_0)}(x - x_0).$$

3. (a) If $x < -\frac{1}{4}$, we can solve for y in terms of x using the quadratic formula.
(b) $\partial F/\partial y = 2y + 1$ is nonzero for $\{y \mid y < -\frac{1}{2}\}$ and $\{y \mid > -\frac{1}{2}\}$. These regions correspond to the upper and lower halves of a horizontal parabola with vertex at $(-\frac{1}{4}, -\frac{1}{2})$ and to the choice of sign in the quadratic formula. The derivative $dy/dx = -3/(2y + 1)$ is negative on the top half of the parabola, positive on the bottom.

5. Let $F(x, y, z) = x^3z^2 - z^3yx$; $\partial F/\partial z = 2x^3z - 3z^2yx \neq 0$ at $(1, 1, 1)$. Near the origin, with $x = y \neq 0$, we get solutions $z = 0$ and $z = x$, and so there is no unique solution. At $(1, 1)$, $\partial z/\partial x = 2$ and $\partial z/\partial y = -1$.

7. With $F_1 = y + x + uv$ and $F_2 = uxy + v$, the determinant in the general implicit function theorem is

$$\begin{vmatrix} \partial F_1/\partial u & \partial F_1/\partial v \\ \partial F_2/\partial u & \partial F_2/\partial v \end{vmatrix} = v - uxy,$$

which is 0 at $(0, 0, 0, 0)$. Thus, the implicit function theorem does not apply. If we try directly, we find that $v = -uxy$, so $x + y = u^2xy$. For a particular choice of (x, y) near $(0, 0)$, either there are no solutions for (u, v) or else there are two.

9. No. $f(x, y) = (-1, 0)$ has infinitely many solutions, namely, $(x, y) = (0, y)$ for any y.

11. (a) $x_0^2 + y_0^2 \neq 0$.
(b) $f'(z) = -z(x + 2y)/(x^2 + y^2)$; $g'(z) = z(y - 2x)/(x^2 + y^2)$.

13. Multiply and equate coefficients to get a_0, a_1, and a_2 as functions of r_1, r_2, and r_3. Then compute the Jacobian determinant $\partial(a_0, a_1, a_2)/\partial(r_1, r_2, r_3) = (r_3 - r_2)(r_1 - r_2)(r_1 - r_3)$. This is not zero if the roots are distinct. Thus, the inverse function theorem shows that the roots may be found as functions of the coefficients in some neighborhood of any point at which the roots are distinct. That is, if the roots r_1, r_2, r_3 of $x^3 + a_2x^2 + a_1x + a_0$ are all different, then there are neighborhoods V of (r_1, r_2, r_3) and W of (a_0, a_1, a_2) such that the roots in V are smooth functions of the coefficients in W.

Review Exercises for Chapter 3

1. (a) Saddle point.
(b) Saddle point for any C.

3. (a) 1 (b) $\sqrt{83}/6$

5. Use the second derivative test; $(0, 0)$ is a local maximum; $(-1, 0)$ is a saddle point; $(2, 0)$ is a local minumum.

7. Saddle points at $(n, 0)$, $n =$ integer.

9. Maximum ≈ 2.618, minimum ≈ 0.382.

11. Maximum 1, minimum cos 1

13. $z = 1/4$

15. $(0, 0, \pm 1)$

17. If $b \geq 2$, the minimum distance is $2\sqrt{b-1}$; if $b \leq 2$, the minimum distance is $|b|$.

19. Not stable.

21. $f(-\frac{3}{2}, -\sqrt{3}/2) = 3\sqrt{3}/4$

23. $x = (20/3)\sqrt[3]{3}; y = 10\sqrt[3]{3}; z = 5\sqrt[3]{3}$

25. The determinant required in the general implicit function theorem is not zero, and so we can solve for u and v; $(\partial u/\partial x)(2, -1) = 13/32$.

27. A new orthonormal basis may be found with respect to which the quadratic form given by the matrix

$$A = \begin{bmatrix} a & b \\ b & c \end{bmatrix}$$

takes diagonal form. This change of basis defines new variables ξ and η, which are linear functions of x and y. Manipulations of linear algebra and the chain rule show that $Lv = \lambda(\partial^2 v/\partial \xi^2) + \mu(\partial^2 v/\partial \eta^2)$. The numbers λ and μ are the eigenvalues of A and are positive, because the quadratic form is positive-definite. At a maximum, $\partial v/\partial \xi = \partial v/\partial \eta = 0$. Moreover, $\partial^2 v/\partial \xi^2 \leq 0$ and $\partial^2 v/\partial \eta^2 \leq 0$, because if either were greater than 0, the cross section of the graph in that direction would have a minimum. Then $Lv \leq 0$, thus contradicting strict subharmonicity.

29. Reverse the inequalities in Exercises 27 and 28.

31. The equations for a critical point, $\partial s/\partial m = \partial s/\partial b = 0$, when solved for m and b give $m = (y_1 - y_2)/(x_1 - x_2)$ and $b = (y_2 x_1 - y_1 x_2)/(x_1 - x_2)$. The line $y = mx + b$ then goes through (x_1, y_1) and (x_2, y_2).

33. At a minimum of s, we have $0 = \partial s/\partial b = -2\sum_{i=1}^{n}(y_i - mx_i - b)$.

35. $y = \frac{9}{10}x + \frac{6}{5}$

Chapter 4

Section 4.1

1. $\mathbf{r}'(t) = -(\sin t)\mathbf{i} + 2(\cos 2t)\mathbf{j}$, $r'(0) = 2\mathbf{j}$, $\mathbf{a}(t) = -(\cos t)\mathbf{i} - 4(\sin 2t)\mathbf{j}$, $\mathbf{a}(0) = -\mathbf{i}$, $\mathbf{l}(t) = \mathbf{i} + 2t\mathbf{j}$

3. $\mathbf{r}'(t) = \sqrt{2}\mathbf{i} + e^t\mathbf{j} - e^{-t}\mathbf{k}$, $\mathbf{r}'(0) = \sqrt{2}\mathbf{i} + \mathbf{j} - \mathbf{k}$, $\mathbf{a}(t) = e^t\mathbf{j} + e^{-t}\mathbf{k}$, $\mathbf{a}(0) = \mathbf{j} + \mathbf{k}$, $\mathbf{l}(t) = \sqrt{2}t\mathbf{i} + (1+t)\mathbf{j} + (1-t)\mathbf{k}$

5. $(e^t - e^{-t}, \cos t - \sin t, -3t^2)$

7. $[-3t^2(2\sin t + \cos t) - t^3(2\cos t - \sin t)]\mathbf{i} + [3t^2(2e^t + e^{-t}) + t^3(2e^t - e^{-t})]\mathbf{j} + [e^t(\cos t - \sin t) - e^{-t}(-\sin t + \cos t)]\mathbf{k}$

9. $m(0, 6, 0)$

11. $-24\pi^2(\cos(2\pi t/5), \sin(2\pi t/5))/25$

13. $\dfrac{d}{dt}(\|\mathbf{v}\|^2) = \dfrac{d}{dt}(\mathbf{v}\cdot\mathbf{v}) = 2\mathbf{v}\cdot\dfrac{d\mathbf{v}}{dt} = 2\mathbf{v}\cdot\mathbf{a} = 0$

15. 6129 seconds

17. $\mathbf{c}(t) = \left(\dfrac{t^2}{2}, e^t - 6, \dfrac{t^3}{3} + 1\right)$

19. (a) $\mathbf{c}(t) = (t, e^t)$, $-\infty < t < \infty$. The image of this path is the graph $y = e^x$.
(b) $\mathbf{c}(t) = (\frac{1}{2}\cos t, \sin t)$, $0 \le t \le 2\pi$, an ellipse.
(c) $\mathbf{c}(t) = (at, bt, ct)$
(d) $\mathbf{c}(t) = (\frac{2}{3}\cos t, \frac{1}{2}\sin t)$, $0 \le t \le 2\pi$, an ellipse.

21. $\mathbf{c}(t) \times \mathbf{c}'(t)$ is normal to the plane of the orbit at time t. As in Exercise 20, its derivative is 0, and so the orbital plane is constant.

Section 4.2

1. $2\sqrt{5}\pi$

3. $2(2\sqrt{2} - 1)$

5. $\dfrac{6-\sqrt{3}}{\sqrt{2}} + \dfrac{1}{2}\log\left[\dfrac{2\sqrt{2}+3}{\sqrt{2}+\sqrt{3}}\right]$

7. $2\pi(\sqrt{5} + \sqrt{2})$

9. $3 + \log 2$

11. (a) Because α is strictly increasing, it maps $[a, b]$ one-to-one onto $[\alpha(a), \alpha(b)]$. By definition, $\mathbf{v}$ is the image of $\mathbf{c}$ if and only if there is a t in $[a, b]$ with $\mathbf{c}(t) = \mathbf{v}$. There is one point s in $[\alpha(a), \alpha(b)]$ with $s = \alpha(t)$, so $\mathbf{d}(s) = \mathbf{c}(t) = \mathbf{v}$. Therefore, the image of $\mathbf{c}$ is contained in that of $\mathbf{d}$. Use α^{-1} similarly for the opposite inclusion.

(b)

$$l_{\mathbf{d}} = \int_{\alpha(a)}^{\alpha(b)} \|\mathbf{d}'(s)\|\, ds = \int_{s=\alpha(a)}^{s=\alpha(b)} \|\mathbf{d}'(\alpha(t))\|\alpha'(t)\, dt$$

$$= \int_{t=a}^{t=b} \|\mathbf{d}'(\alpha(t))\alpha'(t)\| dt = \int_a^b \|\mathbf{c}'(t)\|\, dt = l_{\mathbf{c}}.$$

(c) Differentiate $\mathbf{d}$ using the chain rule.

13. (a) $l_{\mathbf{c}} = \int_a^b \|\mathbf{c}'(s)\|\, ds = \int_a^b ds = b - a$

(b) $\mathbf{T}(s) = \mathbf{c}'(s)/\|\mathbf{c}'(s)\| = \mathbf{c}'(s)$, so $\mathbf{T}'(s) = \mathbf{c}''(s)$. Then $k = \|\mathbf{T}'\| = \|\mathbf{c}''(s)\|$.

(c) Show that if $\mathbf{v}$ and $\mathbf{w}$ are in $\mathbb{R}^3$, $\|\mathbf{v} \times \mathbf{w}\| = \|\mathbf{w} - (\mathbf{v} \cdot \mathbf{w}/\|\mathbf{v}\|^2)\mathbf{v}\| \cdot \|\mathbf{v}\|$. Use this to show that if $\rho(t) = (x(t), y(t), z(t))$ is never $(0, 0, 0)$ and $\mathbf{f}(t) = \rho(t)/\|\rho(t)\|$, then

$$\frac{d\mathbf{f}}{dt} = \frac{1}{\|\rho(t)\|}\left[\rho'(t) - \frac{\rho(t) \cdot \rho'(t)}{\|\rho(t)\|^2}\rho(t)\right] \quad \text{and} \quad \frac{d\mathbf{f}}{dt} = \frac{\|\rho(t) \times \rho'(t)\|}{\|\rho(t)\|^2}.$$

With $\rho(t) = \mathbf{c}'(t)$, this gives

$$\mathbf{T}'(t) = \frac{\mathbf{c}''(t)}{\|\mathbf{c}'(t)\|} - \frac{\mathbf{c}'(t) \cdot \mathbf{c}''(t)}{\|\mathbf{c}'(t)\|^3}\mathbf{c}'(t) \quad \text{and} \quad \|\mathbf{T}'(t)\| = \frac{\|\mathbf{c}'(t) \times \mathbf{c}''(t)\|}{\|\mathbf{c}'(t)\|^2}.$$

If s is the arc length of $\mathbf{c}$, $ds/dt = \|\mathbf{c}'(t)\|$, and therefore

$$\left\|\frac{d\mathbf{T}}{dt}\right\| = \left\|\frac{d\mathbf{T}}{ds}\frac{ds}{dt}\right\| = k\|\mathbf{c}'(t)\|.$$

Thus,

$$k = \frac{1}{\|\mathbf{c}'(t)\|}\frac{d\mathbf{T}}{dt} = \frac{\|\mathbf{c}'(t) \times \mathbf{c}''(t)\|}{\|\mathbf{c}'(t)\|^3}.$$

(This result is useful in Exercise 15.)

(d) $1/\sqrt{2}$

15. (a) Because $\mathbf{c}$ is parametrized by arc length, $\mathbf{T}(s) = \mathbf{c}'(s)$, and $\mathbf{N}(s) = \mathbf{c}''(s)/\|\mathbf{c}''(s)\|$. Use Exercise 13 to show that

$$\frac{d\mathbf{B}}{ds} = \left(\mathbf{c}'' \times \frac{\mathbf{c}''}{\|\mathbf{c}''\|}\right) + \mathbf{c}' \times \left(\frac{\mathbf{c}'''}{\|\mathbf{c}''\|} - \frac{\mathbf{c}'' \cdot \mathbf{c}'''}{\|\mathbf{c}''\|^3}\mathbf{c}''\right)$$

and

$$\tau = -\frac{d\mathbf{B}}{ds}\cdot\mathbf{N} = -\frac{(\mathbf{c}'\times\mathbf{c}''')\cdot\mathbf{c}''}{\|\mathbf{c}''\|^2} = \frac{(\mathbf{c}'\times\mathbf{c}'')\cdot\mathbf{c}'''}{\|\mathbf{c}''\|^2}.$$

(b) Obtain $\mathbf{T}'(t)$ and $\|\mathbf{T}'(t)\|$ as in Exercise 13. $\mathbf{B}$ is a unit vector in the direction of $\mathbf{c}'\times\mathbf{T}' = (\mathbf{c}'\times\mathbf{c}'')/\|\mathbf{c}'\|$, so $\mathbf{B} = (\mathbf{c}'\times\mathbf{c}'')/\|\mathbf{c}'\times\mathbf{c}''\|$. Use the solution of Exercise 13 with $\rho = \mathbf{c}'\times\mathbf{c}''$ to obtain

$$d\mathbf{B}/dt = (\mathbf{c}'\times\mathbf{c}''')/\|\mathbf{c}'\times\mathbf{c}''\| - \{[(\mathbf{c}'\times\mathbf{c}'')\cdot(\mathbf{c}'\times\mathbf{c}''')]/\|\mathbf{c}'\times\mathbf{c}''\|^3\}(\mathbf{c}'\times\mathbf{c}''),$$

and the values of $\mathbf{T}'$ and $\|\mathbf{T}'\|$ to get

$$\mathbf{N} = (\|\mathbf{c}'\|/\|\mathbf{c}'\times\mathbf{c}''\|)(\mathbf{c}'' - (\mathbf{c}'\times\mathbf{c}'')/\|\mathbf{c}'\|^2).$$

Finally, use the chain rule and the inner product of these to obtain

$$\tau = -\left[\frac{d\mathbf{B}}{ds}(s(t))\right]\cdot\mathbf{N}(s(t)) = -\frac{1}{|ds/dt|}\frac{d\mathbf{B}}{dt}\cdot\mathbf{N} = \frac{(\mathbf{c}'\times\mathbf{c}'')\cdot\mathbf{c}'''}{\|\mathbf{c}'\times\mathbf{c}''\|^2}.$$

(c) $\sqrt{2}/2$

17. (a) $\mathbf{N}$ is defined as $\mathbf{T}'/\|\mathbf{T}'\|$, so $\mathbf{T}' = \|\mathbf{T}'\|\mathbf{N} = k\mathbf{N}$. Because $\mathbf{T}\cdot\mathbf{T}' = 0$, $\mathbf{T}$, $\mathbf{N}$, and $\mathbf{B}$ are an orthonormal basis for $\mathbb{R}^3$. Differentiating $\mathbf{B}(s)\cdot\mathbf{B}(s) = 1$ and $\mathbf{B}(s)\cdot\mathbf{T}(s) = 0$ shows that $\mathbf{B}'\cdot\mathbf{B} = 0$ and $\mathbf{B}'\cdot\mathbf{T} + \mathbf{B}\cdot\mathbf{T}' = 0$. But $\mathbf{T}'\cdot\mathbf{B} = \|\mathbf{T}'\|\mathbf{N}\cdot\mathbf{B} = 0$, so $\mathbf{B}'\cdot\mathbf{T} = 0$ also. Thus, $\mathbf{B}' = (\mathbf{B}'\cdot\mathbf{T})\mathbf{T} + (\mathbf{B}'\cdot\mathbf{N})\mathbf{N} + (\mathbf{B}'\cdot\mathbf{B})\mathbf{B} = (\mathbf{B}'\cdot\mathbf{N})\mathbf{N} = -\tau\mathbf{N}$. Also, $\mathbf{N}'\cdot\mathbf{N} = 0$, because $\mathbf{N}\cdot\mathbf{N} = 1$. Thus, $\mathbf{N}' = (\mathbf{N}'\cdot\mathbf{T})\mathbf{T} + (\mathbf{N}'\cdot\mathbf{B})\mathbf{B}$. But differentiating $\mathbf{N}\cdot\mathbf{T} = 0$ and $\mathbf{N}\cdot\mathbf{B} = 0$ gives $\mathbf{N}'\cdot\mathbf{T} = -\mathbf{N}\cdot\mathbf{T}' = -k$ and $\mathbf{N}'\cdot\mathbf{B} = -\mathbf{N}\cdot\mathbf{B}' = \tau$, and so the middle equation follows.

(b) $\omega = \tau\mathbf{T} + k\mathbf{B}$

19. Follow the hint in the text.

Section 4.3

1. **3.**

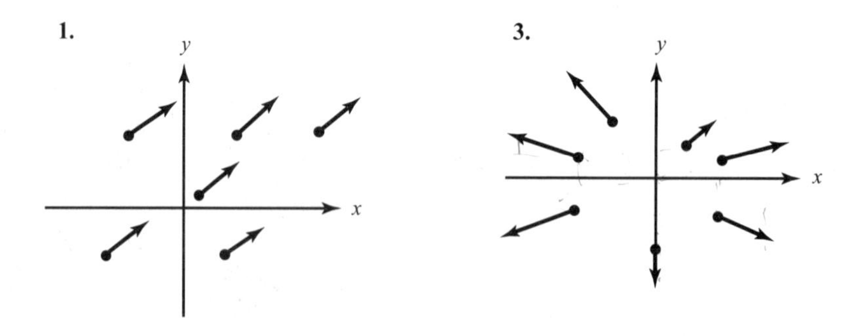

5. $\mathbf{F} = (2y, x)$:

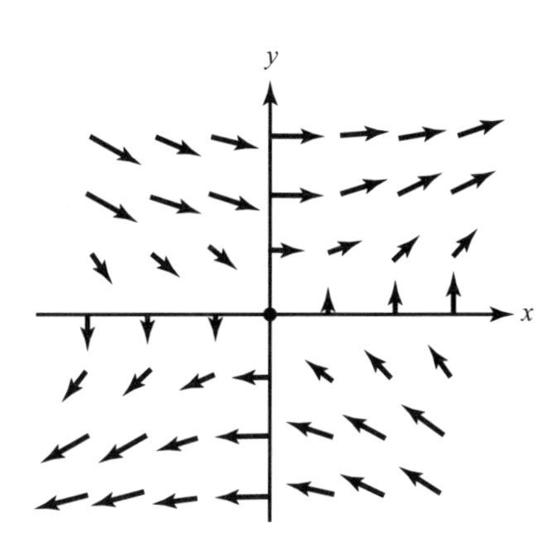

7.

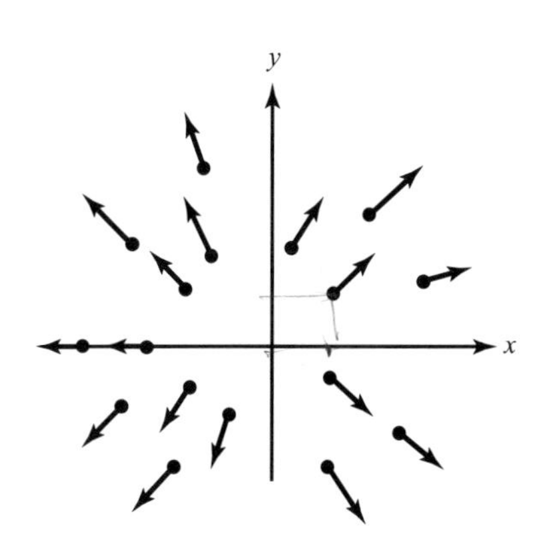

9. The flow lines are concentric circles:

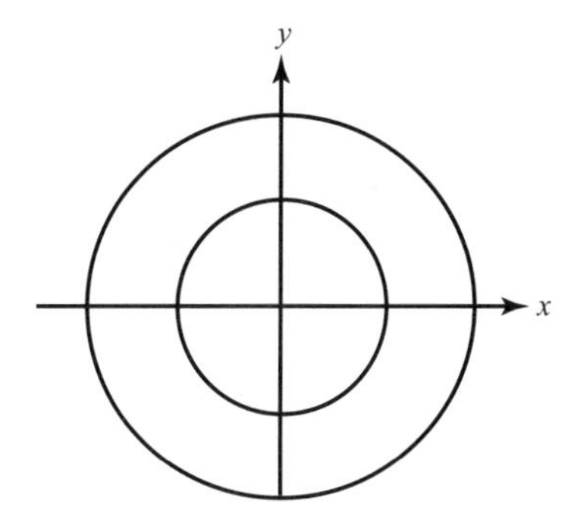

11. The flow lines for $t > 0$:

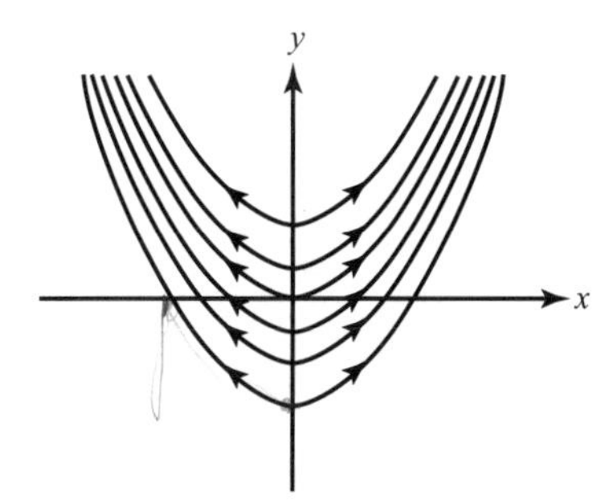

13. $\mathbf{c}'(t) = (2e^{2t}, 1/t, -1/t^2) = \mathbf{F}(\mathbf{c}(t))$

15. $\mathbf{c}'(t) = (\cos t, -\sin t, e^t) = \mathbf{F}(\mathbf{c}(t))$

17. Compare $\frac{1}{2}mv^2$ for the escape velocity $v_e = \sqrt{2gR_0}$ and the velocity in an orbit of radius R_0 given in Section 4.1. (Ignore the rotation of the earth.)

19. Use the fact that $-\nabla T$ is perpendicular to the surface T = constant.

Section 4.4

1. $ye^{xy} - xe^{xy} + ye^{yz}$

3. 3

5. div $\mathbf{V} > 0$ in the first and third quadrants,
div $\mathbf{V} < 0$ in the second and fourth quadrants.

7. $\nabla \cdot \mathbf{F} = 0$; if $\mathbf{F}$ represents a fluid, there is neither expansion nor compression; the area of a small rectangle remains the same.

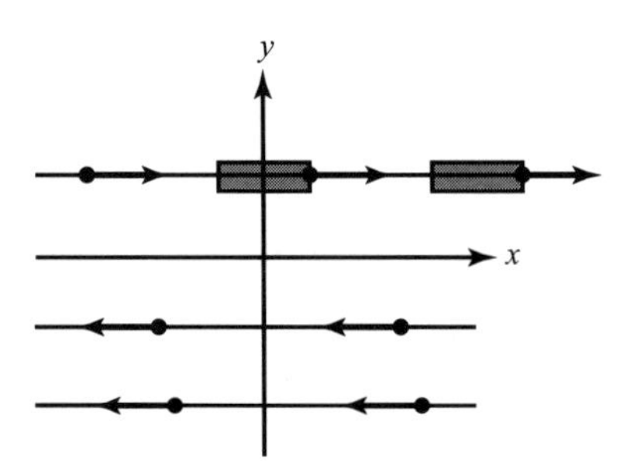

9. $3x^2 - x^2 \cos(xy)$

11. $y \cos(xy) + x^2 \sin(x^2 y)$

13. $\mathbf{0}$

15. $(10y - 8z)\mathbf{i} + (6z - 10x)\mathbf{j} + (8x - 6y)\mathbf{k}$

17. $-\sin x$

19. x

21. $\nabla \times \nabla f = \mathbf{0}$

23. $\nabla \times \nabla f = \mathbf{0}$

25. $\nabla \times \mathbf{F} \neq \mathbf{0}$

27. Let $\mathbf{F} = F_1\mathbf{i} + F_2\mathbf{j} + F_3\mathbf{k}$ and compute both sides of the identity.

29. (a) $2xy\mathbf{i} + x^2\mathbf{j}$
(b) $(3y^2xz, 4xz - y^3z, 0)$
(c) $(-y^3zx^3, 2x^2y^4z, 2x^3z^2 - 2xy)$
(d) $4x^2yz^2 + x^2$

31. No.

33. Separate each expression into its real and imaginary parts and then treat the resulting quantity as a vector field on $\mathbb{R}^2$. Directly calculate its curl and divergence.

In (a), $\mathbf{F} = (x^2 - y^2)\mathbf{i} - 2xy\mathbf{j}$; in (b), $\mathbf{F} = (x^3 - 3xy^2)\mathbf{i} + (y^3 - 3x^2y)\mathbf{j}$; and in (c), $\mathbf{F} = (e^x \cos y)\mathbf{i} - (e^x \sin y)\mathbf{j}$. Show that $\nabla \cdot \mathbf{F} = 0$ and $\nabla \times \mathbf{F} = 0$ in each case.

Review Exercises for Chapter 4

1. $\mathbf{v}(1) = (3, -e^{-1}, -\pi/2)$; $\mathbf{a}(1) = (6, e^{-1}, 0)$;
$s(1) = \sqrt{9 + e^{-2} + \frac{\pi^2}{4}}$; $\mathbf{l}(t) = (2, e^{-1}, 0) + (t-1)(3, -e^{-1}, -\pi/2)$

3. $\mathbf{v}(0) = (1, 1, 0)$; $\mathbf{a}(0) = (1, 0, -1)$; $s = \sqrt{2}$; $l(t) = (1, 0, 1) + t(1, 1, 0)$

5. Tangent vector: $\mathbf{v} = -(1/\sqrt{2})\mathbf{i} + (1/\sqrt{2})\mathbf{j} + \mathbf{k}$
Acceleration vector: $\mathbf{a} = -(1/\sqrt{2})(\mathbf{i} + \mathbf{j})$

7. $m(2, 0, -1)$

9. $\displaystyle\int_1^4 \sqrt{1 + \tfrac{4}{9}t^{-2/3} + \tfrac{4}{25}t^{-6/5}}\, dt$

11. (a) $\mathbf{v} = (-2t \sin(t^2), 2t \cos(t^2), 0)$; $s = 2t$
(b) $\left(\frac{1}{2}, -\frac{\sqrt{3}}{2}, 0\right)$
(c) $\sqrt{5\pi/3}$
(d) $\mathbf{v} = 2\sqrt{5\pi/3}(\sqrt{3}/2, 1/2, 0)$; $s = 2\sqrt{5\pi/3}$
(e) $\left(\frac{3}{2} + \frac{5\pi}{\sqrt{3}}\right)\Big/\sqrt{5\pi}$

13. $x = 1 + t,\ y = -\frac{1}{2} + \frac{t}{2},\ z = -\frac{2}{3} + \frac{t}{3}$

15. Compute $\mathbf{c}'(t)$ and check that it equals $\mathbf{F}(\mathbf{c}(t))$.

17. 9; $\mathbf{0}$

19. 3; $-\mathbf{i} - \mathbf{j} - \mathbf{k}$

21. 0; $-\mathbf{i} - \mathbf{j} - \mathbf{k}$

23. $\nabla f = (ye^{xy} - y \sin xy, xe^{xy} - x \sin xy, 0)$; verify that $\nabla \times \nabla f = 0$ in this case.

25. $\nabla f = (2xe^{x^2} + y^2 \sin xy^2, 2xy \sin xy^2, 0)$; check that $\nabla \times \nabla f = 0$ from this.

27. (a) $(yz^2, xz^2, 2xyz)$; (b) $(z - y, 0, -x)$
(c) $(2xyz^3 - 3xy^2z^2, 2x^2y^2z - y^2z^3, y^2z^3 - 2x^2yz^2)$

29. $\operatorname{div} \mathbf{F} = 0$; $\operatorname{curl} \mathbf{F} = (0, 0, 2(x^2 + y^2)f'(x^2 + y^2) + 2f(x^2 + y^2))$

31. (a) A cone about $\mathbf{i}'$ making an angle of $\pi/3$ with $\mathbf{i}'$.
(b) $\nabla g = (3x^2, 5z, 5y + 2z)$

33. (a) $[\partial P/\partial x)^2 + (\partial P/\partial y)^2]^{1/2}$
(b) A small packet of air would obey $\mathbf{F} = m\mathbf{a}$.

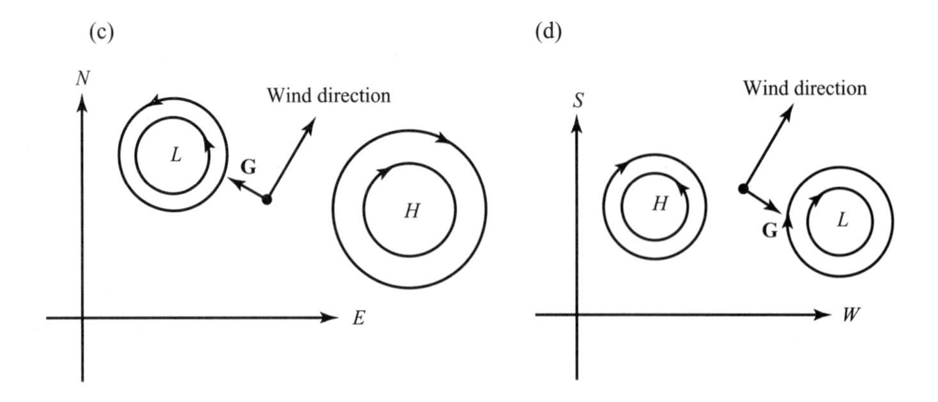

35. (a) $\dfrac{\sqrt{R^2+\rho^2}}{\rho}(z_0 - z_1)$ (b) $\sqrt{\dfrac{2(R^2+\rho^2)z_0}{g\rho^2}}$

37. 680 miles per hour

Chapter 5

Section 5.1

1. (a) $\frac{13}{15}$ (b) $\pi + \frac{1}{2}$ (c) 1 (d) $\log 2 - \frac{1}{2}$

3. To show that the volumes of the two cylinders are equal, show that their area functions are equal.

5. $2r^3(\tan\theta)/3$

7. $\frac{26}{9}$

9. $(2/\pi)(e^2+1)$

11. $\frac{196}{15}$

Section 5.2

1. (a) $\frac{7}{12}$ (b) $e-2$ (c) $\frac{1}{9}\sin 1$ (d) $2\ln 4 - 2$

3. $1/4$

5. Use Fubini's theorem to write

$$\int\!\!\int_R [f(x)g(y)]\,dx\,dy = \int_c^d g(y)\left[\int_a^b f(x)\,dx\right]dy,$$

and notice that $\int_a^b f(x)\,dx$ is a constant and so may be pulled out.

7. 11/6

9. Because $\int_0^1 dy = \int_0^1 2y\,dy = 1$, we have $\int_0^1[\int_0^1 f(x,y)\,dy]\,dx = 1$. In any partition of $R = [0,1] \times [0,1]$, each rectangle R_{jk} contains points $\mathbf{c}_{jk}^{(1)}$ with x rational and $\mathbf{c}_{jk}^{(2)}$ with x irrational. If in the regular partition of order n we choose $\mathbf{c}_{jk} = \mathbf{c}_{jk}^{(1)}$ in those rectangles with $0 \le y \le \frac{1}{2}$ and $\mathbf{c}_{jk} = \mathbf{c}_{jk}^{(2)}$ when $y > \frac{1}{2}$, the approximating sums are the same as those for

$$g(x,y) = \begin{cases} 1 & 0 \le y \le \frac{1}{2} \\ 2y & \frac{1}{2} < y < 1. \end{cases}$$

Because g is integrable, the approximating sums must converge to $\int_R g\,dA = 7/8$. However, if we had picked all $\mathbf{c}_{ij} = \mathbf{c}_{jk}^{(1)}$, all approximating sums would have the value 1.

11. Fubini's theorem does not apply because the integrand is not continuous nor bounded at $(0,0)$.

Section 5.3

1. (a) 1/3, both (b) 5/2, both (c) $(e^2 - 1)/4$, both (d) 1/35, both

3. $A = \int_{-r}^{r} \int_{-\sqrt{r^2-x^2}}^{\sqrt{r^2-x^2}} dy\,dx = 2\int_{-r}^{r} \sqrt{r^2 - x^2}\,dx = r^2[\arcsin 1 - \arcsin(-1)] = \pi r^2.$

5. 28,000 ft^3

7. 0

9. y-simple; $\pi/2$

11. $50\,\pi$

13. $\pi/24$

15. Compute the integral with respect to y first. Split that into integrals over $[-\phi(x), 0]$ and $[0, \phi(x)]$ and change variables in the first integral, or use symmetry.

17. Let $\{R_{ij}\}$ be a partition of a rectangle R containing D and let f be 1 on D. Thus, f^* is 1 on D and 0 on $R\backslash D$. Let $\mathbf{c}_{jk} \in R\backslash D$ if R_{ij} is not wholly contained in D. The approximating Riemann sum is the sum of the areas of those rectangles of the partition that are contained in D.

Section 5.4

1. (a) 1/8 (b) $\pi/4$ (c) 17/12
(d) $G(b) - G(a)$, where $dG/dy = F(y,y) - F(a,y)$ and $\partial F/\partial x = f(x,y)$

3. Note that the maximum value of f on D is e and the minimum value of f on D is $1/e$. Use the ideas in the proof of Theorem 4 to show that

$$\frac{1}{e} \le \frac{1}{4\pi^2} \iint f(x,y)\,dA \le e.$$

5. The smallest value of $f(x, y) = 1/(x^2 + y^2 + 1)$ on D is $\frac{1}{6}$, at (1, 2), and so

$$\iint_D f(x, y)\, dx\, dy \geq \frac{1}{6} \cdot \text{area } D = 1.$$

The largest value is 1, at (0, 0), and so

$$\iint_D f(x, y)\, dx\, dy \leq 1 \cdot \text{area } D = 6.$$

7. $\frac{4}{3}\pi abc$

9. $\pi(20\sqrt{10} - 52)/3$

11. $\sqrt{3}/4$

13. D looks like a slice of pie.

$$\int_0^1 \left[\int_0^x f(x, y)\, dy\right] dx + \int_1^{\sqrt{2}} \left[\int_0^{\sqrt{2-x^2}} f(x, y)\, dy\right] dx.$$

15. Use the chain rule and the fundamental theorem of calculus.

Section 5.5

1. $1/3$

3. 10

5. $x^2 + y^2 \leq z \leq \sqrt{x^2 + y^2}$, $-\sqrt{1 - y^2} \leq x \leq \sqrt{1 - y^2}$, $-1 \leq y \leq 1$

7. $0 \leq z \leq \sqrt{1 - x^2 - y^2}$, $-\sqrt{1 - y^2} \leq x \leq \sqrt{1 - y^2}$, $-1 \leq y \leq 1$

9. $50\pi/\sqrt{6}$

11. $1/2$

13. 0

15. $a^5/20$

17. 0

19. $3/10$

21. $1/6$

23. $\int_{-1}^{1} \int_{-\sqrt{1-x^2}}^{\sqrt{1-x^2}} \int_{\sqrt{x^2+y^2}}^{1} f(x, y, z)\, dz\, dy\, dx$

25. $\int_{-1}^{1} \int_{-\sqrt{1-x^2}}^{\sqrt{1+x^2}} \int_0^{\sqrt{4-x^2-y^2}} f(x, y, z)\, dz\, dy\, dx$

27. $\iint_D \int_0^{f(x,y)} dz\, dx\, dy = \iint_D f(x, y)\, dx\, dy$

29. Let M_ϵ and m_ϵ be the maximum and minimum of f on B_ϵ. Then we have the inequality $m_\epsilon \operatorname{vol}(B_\epsilon) \leq \iiint_{B_\epsilon} f\, dV \leq M_\epsilon \operatorname{vol}(B_\epsilon)$. Divide by $\operatorname{vol}(B_\epsilon)$, let $\epsilon \to 0$ and use continuity of f.

Review Exercises for Chapter 5

1. 81/2

3. $\frac{1}{4}e^2 - e + \frac{9}{4}$

5. 81/2

7. $\frac{1}{4}e^2 - e + \frac{9}{4}$

9. 7/60

11. 1/2

13. In the notation of Figure 5.3.1,

$$\iint_D dx\,dy = \int_a^b [\phi_2(x) - \phi_1(x)]\,dx.$$

15. (a) 0 (b) $\pi/24$ (c) 0

17. y-simple; $2\pi + \pi^2$

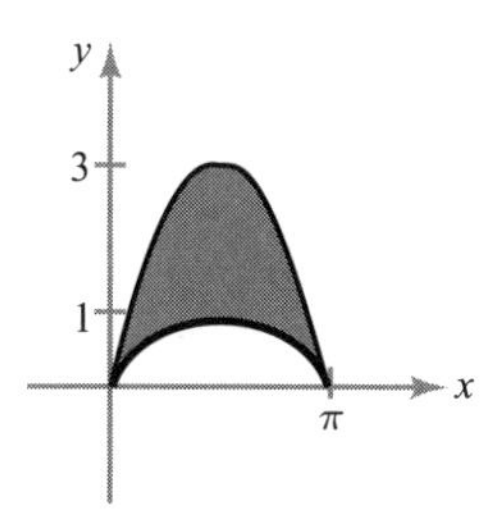

19. x-simple; 73/3

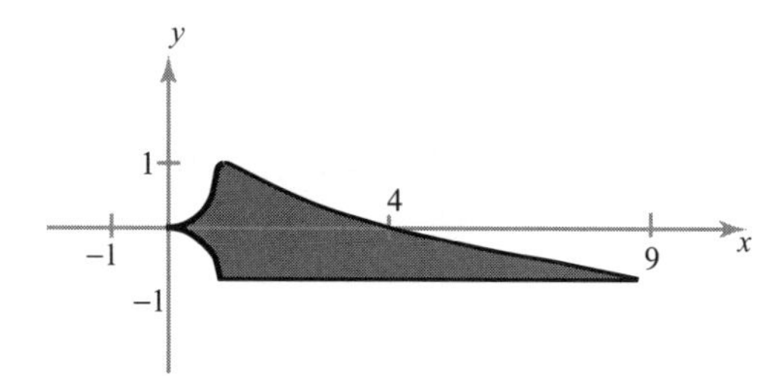

21. y-simple; 33/140

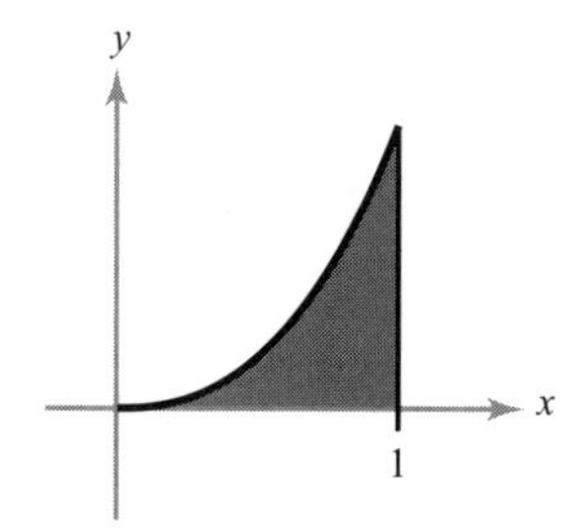

23. y-simple; $71/420$.

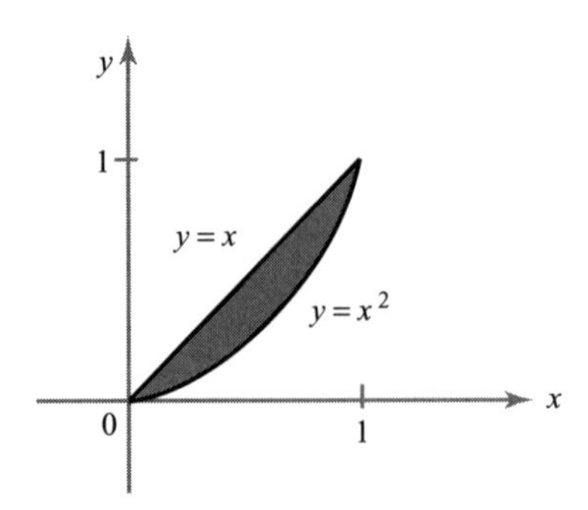

25. $1/3$

27. $19/3$

29. $7/12$

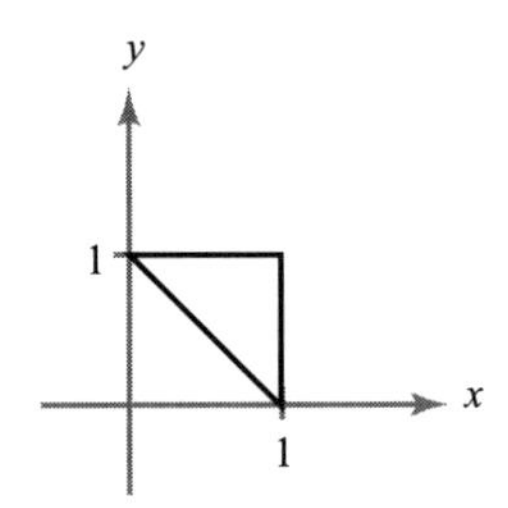

31. The function $f(x, y) = x^2 + y^2 + 1$ lies between 1 and $2^2 + 1 = 5$ on D, and so the integral lies between these values times 4π, the area of D.

33. Interchange the order of integration (the reader should draw a sketch in the (u, t) plane);

$$\int_0^x \int_0^t F(u)\, du\, dt = \int_0^x \int_u^x F(u)\, dt\, du = \int_0^x (x - u)F(u)\, du.$$

35. $\pi/12$

37. The region is the shaded region W in the figure.

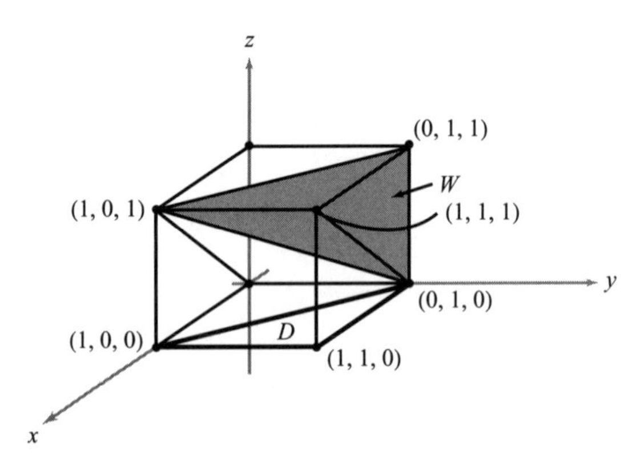

The integral in the order $dy\,dx\,dz$, for example, is

$$\int_0^1 \int_z^1 \int_{1-x}^1 f(x, y, z)\,dy\,dx\,dz.$$

Chapter 6

Section 6.1

1. $S =$ the unit disk minus its center.

3. $D = [0, 3] \times [0, 1]$; yes

5. The image is the triangle with vertices (0, 0), (0, 1), and (1, 1). T is not one-to-one, but becomes so if we eliminate the portion $x^* = 0$.

7. D is the set of (x, y, z) with $x^2 + y^2 + z^2 \leq 1$ (the unit ball). T is not one-to-one, but is one-to-one on $(0, 1] \times (0, \pi) \times (0, 2\pi]$.

9. Showing that T is onto is equivalent in the 2×2 case to showing that the system $ax + by = e$, $cx + dy = f$ can always be solved for x and y, where

$$A = \begin{bmatrix} a & b \\ c & d \end{bmatrix}.$$

When you do this by elimination or by Cramer's rule, the quantity by which you must divide is det A. Thus, if det $A \neq 0$, the equations can always be solved.

11. Because det $A \neq 0$, T maps $\mathbb{R}^2$ one-to-one onto $\mathbb{R}^2$. Let T^{-1} be the inverse transformation. Show that T^{-1} has matrix A^{-1} and det $(A^{-1}) = 1/\det A$, where det $A \neq 0$. By Exercise 10, $P^* = T^{-1}(P)$ is a parallelogram.

Section 6.2

1. $\pi(e - 1)$

3. D is the region $0 \leq x \leq 4$, $\frac{1}{2}x + 3 \leq y \leq \frac{1}{2}x + 6$. (a) 140 (b) -42

5. D^* is the region $0 \leq u \leq 1, 0 \leq v \leq 2$; $\frac{2}{3}(9 - 2\sqrt{2} - 3\sqrt{3})$.

7. π

9. $\dfrac{64\pi}{5}$

11. $3\pi/2$

13. $\dfrac{5\pi}{2}(e^4 - 1)$

15. $2a^2$

17. $\dfrac{21}{2}\left(e - \dfrac{1}{e}\right)$

19. $100\pi/3$

21. $2\pi[\sqrt{3} - 2\log(1+\sqrt{3}) + \log\sqrt{2}]$

23. $4\pi\log(a/b)$

25. $2\pi[(b^2+1)e^{-b^2} - (a^2+1)e^{-a^2}]$

27. 24 (use the change of variables $x = 3u - v + 1$, $y = 3u + v$).

29. (a) $\frac{4}{3}\pi abc$ (b) $\frac{4}{5}\pi abc$

31. (a) Check that if $T(u_1, v_1) = T(u_2, v_2)$, then $u_1 = u_2$ and $v_1 = v_2$.
(b) $160/3$

33. $\frac{4}{9}a^{2/3}\iint_{D^*}[f((au^2)^{1/3}, (av^2)^{1/3})u^{-1/3}v^{-1/3}]\,du\,dv$

Section 6.3

1. $[\pi^2 - \sin(\pi^2)]/\pi^3$

3. $\left(\frac{11}{18}, \frac{65}{126}\right)$

5. \$503.64

7. (a) δ, where δ is the (constant) mass density. (b) $37/12$

9. $\left(\frac{1}{2}, \frac{1}{2}, \frac{1}{2}\right)$

11. $1/4$

13. Letting δ be density, the moment of inertia is $\delta\int_0^k\int_0^{2\pi}\int_0^{a\sec\phi}(\rho^4\sin^3\phi)\,d\rho\,d\theta\,d\phi$.

15. $(1.00\times10^8)m$

17. (a) The only plane of symmetry for the body of an automobile is the one dividing the left and right sides of the car.

(b) $\bar{z}\cdot\iiint_W \delta(x, y, z)\,dx\,dy\,dz$ is the z coordinate of the center of mass times the mass of W. Rearrangement of the formula for $\bar{z}$ gives the first line of the equation. The next step is justified by the additivity property of integrals. By symmetry, we can replace z by $-z$ and integrate in the region above the xy plane. Finally, we can factor the minus sign outside the second integral, and because $\delta(x, y, z) = \delta(u, v, -w)$, we are subtracting the second integral from itself. Thus, the answer is 0.

(c) In part (b), we showed that $\bar{z}$ times the mass of W is 0. Because the mass must be positive, $\bar{z}$ must be 0.

(d) By part (c), the center of mass must lie in both planes.

19. $V = -(4.71\times10^{19})Gm/R \approx -(3.04\times10^9)m/R$, where m is the mass of a test particle at distance R from the planet's center.

Section 6.4

1. 4

3. $3/16$

5. (a) 3π (b) $\lambda < 1$

7. Integration of $\iint e^{-xy}\,dx\,dy$ with respect to x first and then y gives log 2. Reversing the order gives the integral on the left side of the equality stated in the exercise.

9. Integrate over $[\varepsilon, 1] \times [\varepsilon, 1]$ and let $\varepsilon \to 0$ to show the improper integral exists and equals 2 log 2.

11. $\frac{2\pi}{9}[(1 + a^3)^{3/2} - a^{9/2} - 1]$

13. Use the fact that

$$\frac{\sin^2(x - y)}{\sqrt{1 - x^2 - y^2}} \le \frac{1}{\sqrt{1 - x^2 - y^2}}.$$

15. Use the fact that $e^{x^2+y^2}/(x - y) \ge 1/(x - y)$ on the given region.

17. Each integral equals 1/4, and Theorem 3 (Fubini's theorem) does apply.

Review Exercises for Chapter 6

1. (a) $T\binom{u}{v} = \begin{pmatrix} 2 & 1 \\ 0 & 2 \end{pmatrix}\binom{u}{v} = \binom{2u+v}{2v} = \binom{x}{y}$
(b) $\iint_P f(x, y)\,dx\,dy = 4\iint_S f(2u + v, 2v)\,du\,dv$

3. 3 (Use the change of variables $u = x^2 - y^2$, $v = xy$.)

5. $\frac{1}{3}\pi(4\sqrt{2} - \frac{7}{2})$

7. $(5\pi/2)\sqrt{15}$

9. $abc/6$

11. Cut with the planes $x + y + z = \sqrt[3]{k/n}$, $1 \le k \le n - 1$, k an integer.

13. $(25 + 10\sqrt{5})\pi/3$

15. $(e - e^{-1})/4$ (Use the change of variables $u = y - x$, $v = y + x$.)

17. $(9.92 \times 10^6)\pi$ grams

19. (a) 32
(b) This occurs at the point of the unit sphere $x^2 + y^2 + z^2 = 1$ inscribed in the cube.

21. $(0, 0, 3a^{4/8})$

23. $4\pi \ln(a/b)$

25. $\pi/2$

27. (a) $9/2$ (b) 64π

29. Work the integral with respect to y first on the region $D_{\varepsilon,L} = \{(x, y) | \varepsilon \le x \le L,$ $0 \le y \le x\}$ to obtain $I_{\varepsilon,L} = \iint_{D_{\varepsilon,L}} f\,dx\,dy = \int_\varepsilon^L x^{-3/2}(1 - e^{-x})\,dx$. The integrand is positive, and so $I_{\varepsilon,L}$ increases as $\varepsilon \to 0$ and $L \to \infty$. Bound $1 - e^{-x}$ above by x for $0 < x < 1$ and by 1 for $1 < x < \infty$ to see that $I_{\varepsilon,L}$ remains bounded and so must converge. The improper integral does exist.

31. (a) $1/6$ (b) $16\pi/3$

33. 2π

Chapter 7

Section 7.1

1. $\int_{\mathbf{c}} f(x, y, z)\,ds = \int_I f(x(t), y(t), z(t)) \|\mathbf{c}'(t)\|\,dt = \int_0^1 0 \cdot 1\,dt = 0.$

3. (a) 2 (b) $52\sqrt{14}$

5. $-\frac{1}{3}(1 + 1/e^2)^{3/2} + \frac{1}{3}(2^{3/2})$

7. (a) The path follows the straight line from (0, 0) to (1, 1) and back to (0, 0) in the xy plane. Over the path, the graph of f is a straight line from (0, 0, 0) to (1, 1, 1). The integral is the area of the resulting triangle covered twice and equals $\sqrt{2}$.

(b)
$$s(t) = \begin{cases} \sqrt{2}(1 - t^4) & \text{when } -1 \le t \le 0 \\ \sqrt{2}(1 + t^4) & \text{when } 0 < t \le 1. \end{cases}$$

The path is

$$\mathbf{c}(s) = \begin{cases} (1 - s/\sqrt{2})(1, 1) & \text{when } 0 \le s \le \sqrt{2} \\ (s/(\sqrt{2} - 1))(1, 1) & \text{when } \sqrt{2} \le s \le 2\sqrt{2} \end{cases}$$

and $\int_{\mathbf{c}} f\,ds = \sqrt{2}$.

9. $2a/\pi$

11. (a) $[2\sqrt{5} + \log(2 + \sqrt{5})]/4$ (b) $(5\sqrt{5} - 1)/[6\sqrt{5} + 3\log(2 + \sqrt{5})]$

13. The path is a unit circle centered at (0, 0, 0) in the plane $x + y + z = 0$ and so may be parametrized by $\mathbf{c}(\theta) = (\cos\theta)\mathbf{v} + (\sin\theta)\mathbf{w}$, where $\mathbf{v}$ and $\mathbf{w}$ are orthogonal unit vectors laying in that plane. For example, $\mathbf{v} = (1/\sqrt{2})(-1, 0, 1)$ and $\mathbf{w} = (1/\sqrt{6})(1, -2, 1)$ will do. The total mass is $2\pi/3$ grams.

15. Choosing either $\mathbf{c}(t) = (t^2, 1, 0)$ or $\mathbf{c}(t) = (1, t^2, 0)$, $0 \le t \le 1$ will do.

Section 7.2

1. (a) 3/2 (b) 0 (c) 0 (d) 147

3. 9

5. By the Cauchy–Schwarz inequality, $|\mathbf{F}(\mathbf{c}(t)) \cdot \mathbf{c}'(t)| \le \|\mathbf{F}(\mathbf{c}(t))\| \, \|\mathbf{c}'(t)\|$ for every t. Thus,

$$\left| \int_{\mathbf{c}} \mathbf{F} \cdot d\mathbf{s} \right| = \left| \int_a^b \mathbf{F}(\mathbf{c}(t)) \cdot \mathbf{c}'(t)\, dt \right| \le \int_a^b |\mathbf{F}(\mathbf{c}(t)) \cdot \mathbf{c}'(t)|\, dt$$
$$\le \int_a^b \|\mathbf{F}(\mathbf{c}(t))\| \, \|\mathbf{c}'(t)\|\, dt \le M \int_a^b \|\mathbf{c}'(t)\|\, dt = Ml.$$

7. $\frac{3}{4} - (n-1)/(n+1)$

9. 0

11. The length of $\mathbf{c}$.

13. If $\mathbf{c}'(t)$ is never 0, then the unit vector $\mathbf{T}(t) = \mathbf{c}'(t)/\|\mathbf{c}'(t)\|$ is a continuous function of t and so is a smoothly turning tangent to the curve. The answer is no.

15. 7

17. Use the fact that $\mathbf{F}$ is a gradient to show that the work done is $\frac{1}{R_2} - \frac{1}{R_1}$, independent of the path.

19. (a) $\|\mathbf{c}'(x)\|$

(b) f has a positive derivative; it is one-to-one and onto $[0, L]$ by the mean-value and intermediate-value theorems. It has a differentiable inverse by the inverse function theorem.

(c) $g'(s) = 1/\|\mathbf{c}'(x)\|$ where $s = f(x)$.

(d) By the chain rule, $\mathbf{b}'(s) = \mathbf{c}'(x) \cdot g'(s)$, which has unit length by part (c).

Section 7.3

1. $z = 2(y-1) + 1$

3. $18(z-1) - 4(y+2) - (x-13) = 0$ or $18z - 4y - x - 13 = 0$.

5. The vector $\mathbf{n} = (\cos v \sin u, \sin v \sin u, \cos u) = (x, y, z)$. The surface is the unit sphere centered at the origin.

7. $\mathbf{n} = -(\sin v)\mathbf{i} - (\cos v)\mathbf{k}$; the surface is a cylinder.

9. (a) $x = x_0 + (y - y_0)(\partial h/\partial y)(y_0, z_0) + (z - z_0)(\partial h/\partial z)(y_0, z_0)$ describes the plane tangent to $x = h(y, z)$ at (x_0, y_0, z_0), $x_0 = h(y_0, z_0)$.

(b) $y = y_0 + (x - x_0)(\partial k/\partial x)(x_0, z_0) + (z - z_0)(\partial k/\partial z)(x_0, z_0)$.

11. $z - 6x - 8y + 3 = 0$

13. (a) The surfaces is a helicoid. It looks like a spiral ramp winding around the z axis. (See Figure 7.4.2.) It winds twice around, since θ goes up to 4π.

(b) $\mathbf{n} = \pm(1/\sqrt{1+r^2})(\sin\theta, -\cos\theta, r)$.

(c) $y_0 x - x_0 y + (x_0^2 + y_0^2)z = (x_0^2 + y_0^2)z_0$.

(d) If $(x_0, y_0, z_0) = (r_0, \cos\theta_0, r_0 \sin\theta_0, \theta_0)$, then representing the line segment in the form $\{(r\cos\theta_0, r\sin\theta_0, \theta_0) | 0 \le r \le 1\}$ shows that the line lies in the surface. Representing the line as $\{(x_0, ty_0, z_0) | 0 \le t \le 1/(x_0^2 + y_0^2)\}$ and substituting into the results of part (c) shows that it lies in the tangent plane at (x_0, y_0, z_0).

15. (a) Using cylindrical coordinates leads to the parametrization

$$\Phi(z,\theta) = (\sqrt{25+z^2}\cos\theta, \sqrt{25+z^2}\sin\theta, z), \qquad -\infty < z < \infty, 0 \le \theta \le 2\pi$$

as one possible solution.

(b) $\mathbf{n} = (\sqrt{25+z^2}\cos\theta, \sqrt{25+z^2}\sin\theta, -z)/\sqrt{25+2z^2}$.

(c) $x_0 x + y_0 y = 25$.

(d) Substitute the coordinates along these lines into the defining equation of the surface and the result of part (c).

17. (a) $u \mapsto u$, $v \mapsto v$, $u \mapsto u^3$, and $v \mapsto v^3$ all map $\mathbb{R}$ onto $\mathbb{R}$.

(b) $\mathbf{T}_u \times \mathbf{T}_v = (0,0,1)$ for $\mathbf{\Phi}_1$, and this is never $\mathbf{0}$. For the surface $\mathbf{\Phi}_2$, $\mathbf{T}_u \times \mathbf{T}_v = 9u^2v^2(0,0,1)$, and this is $\mathbf{0}$ along the u and v axes.

(c) We want to show that any two parametrizations of a surface that are smooth near a point will give the same tangent plane there. Thus, suppose $\mathbf{\Phi}: D \subset \mathbb{R}^2 \to \mathbb{R}^3$ and $\mathbf{\Psi}: B \subset \mathbb{R}^2 \to \mathbb{R}^3$ are parametrized surfaces such that

$$\mathbf{\Phi}(u_0, v_0) = (x_0, y_0, z_0) = \mathbf{\Psi}(s_0, t_0) \tag{i}$$

and

$$\left(\mathbf{T}_u^{\Phi} \times \mathbf{T}_v^{\Phi}\right)\Big|_{(u_0,v_0)} \neq \mathbf{0} \quad \text{and} \quad \left(\mathbf{T}_s^{\Psi} \times \mathbf{T}_t^{\Psi}\right)\Big|_{(s_0,t_0)} \neq \mathbf{0}, \tag{ii}$$

so that $\mathbf{\Phi}$ and $\mathbf{\Psi}$ are smooth and one-to-one in neighborhoods of (u_0, v_0) and (s_0, t_0), which we may as well assume are D and B. Suppose further that they "describe the same surface," that is, $\mathbf{\Phi}(D) = \mathbf{\Phi}(B)$. To see that they give the same tangent plane at (x_0, y_0, z_0), show that they have parallel normal vectors. To do this, show that there is an open set C with $(u_0, v_0) \in C \subset D$ and a differentiable map $f: C \to B$ such that $\mathbf{\Phi}(u,v) = \mathbf{\Psi}(f(u,v))$ for $(u,v) \in C$. Once you have done this, computation shows that the normal vectors are related by $\mathbf{T}_u^{\Phi} \times \mathbf{T}_v^{\Phi} = [\partial(s,t)/\partial(u,v)]\mathbf{T}_s^{\Psi} \times \mathbf{T}_t^{\Psi}$.

To see that there is such an f, notice that since $\mathbf{T}_s^{\Psi} \times \mathbf{T}_t^{\Psi} \neq \mathbf{0}$, at least one of the 2×2 determinants in the cross product is not zero. Assume, for example, that

$$\begin{vmatrix} \dfrac{\partial x}{\partial s} & \dfrac{\partial y}{\partial s} \\ \dfrac{\partial x}{\partial t} & \dfrac{\partial y}{\partial t} \end{vmatrix} \neq 0.$$

Now use the inverse function theorem to write (s,t) as a differentiable function of (x,y) in some neighborhood of (x_0, y_0).

(d) No.

Section 7.4

1. 4π

3. $\frac{3}{2}\pi[\sqrt{2} + \log(1 + \sqrt{2})]$

5. $\frac{1}{3}\pi(6\sqrt{6} - 8)$

7. The integral for the volume converges, whereas that for the area diverges.

9. $A(E) = \int_0^{2\pi} \int_0^{\pi} \sqrt{a^2b^2 \sin^2\phi \cos^2\phi + b^2c^2 \sin^4\phi \cos^2\theta + a^2c^2 \sin^4\phi \sin^2\theta}\, d\phi\, d\theta$

11. $(\pi/6)(5\sqrt{5} - 1)$

13. $(\pi/2)\sqrt{6}$

15. $4\sqrt{5}$; for fixed θ, (x, y, z) moves along the horizontal line segment $y = 2x$, $z = \theta$ from the z axis out to a radius of $\sqrt{5}|\cos\theta|$ into quadrant 1 if $\cos\theta > 0$ and into quadrant 3 if $\cos\theta < 0$.

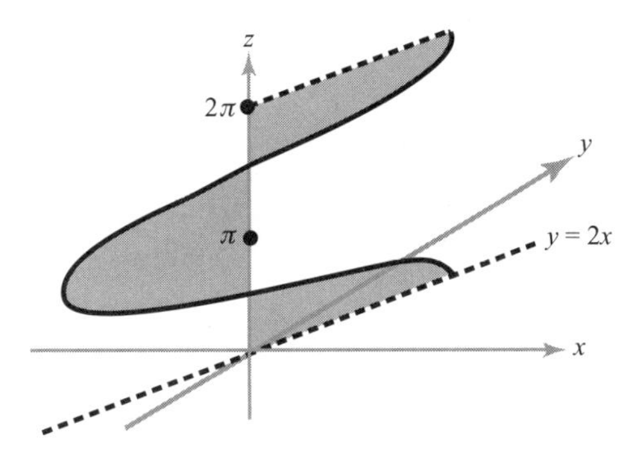

17. $(\pi + 2)/(\pi - 2)$

19. $\pi(a + b)\sqrt{1 + m^2}(b - a)$

21. $\frac{4}{15}(9\sqrt{3} - 8\sqrt{2} + 1)$

23. With $f(x, y) = \sqrt{R^2 - x^2 - y^2}$, (4) becomes

$$A(S') = \iint_D \sqrt{\frac{x^2 + y^2}{R^2 - x^2 - y^2} + 1}\, dx\, dy = \iint_D \frac{R}{\sqrt{R^2 - x^2 - y^2}}\, dx\, dy$$

where D is the disk of radius R. Evaluate using polar coordinates, noting it is improper at the boundary, to get $2\pi R^2$.

Section 7.5

1. $\frac{3\sqrt{2}+5}{24}$

3. πa^3

5. (a) $\sqrt{2}\pi R^2$ (b) $2\pi R^2$

7. $\frac{\pi}{4}\left(\frac{5\sqrt{5}}{3}+\frac{1}{15}\right)$

9. $16\pi R^3/3$

11. (a) The sphere looks the same from all three axes, so these three integrals should be the same quantity with different labels on the axes.
(b) $4\pi R^4/3$
(c) $4\pi R^4/3$

13. $(R/2, R/2, R/2)$

15. (a) Directly compute the vector cross product $\mathbf{T}_u \times \mathbf{T}_v$ and then calculate its length and compare your answer to the left-hand side.
(b) In this case, $F = 0$, so $A(s) = \iint_D \sqrt{EG}\, du\, dv$.
(c) $4\pi a^2$.

17. Let $a = \partial x/\partial u$, $b = \partial y/\partial u$, $c = \partial x/\partial v$, and $d = \partial y/\partial v$. The conditions (a) and (b) in Exercise 16 are then $a^2 + b^2 = c^2 + d^2$ and $ac + bd = 0$. Show that $a \neq 0$ and, by a normalization argument, show that you can assume $a = 1$. Now calculate further.

19. $2a^2$

Section 7.6

1. $\pm 48\pi$ (the sign depends on orientation)

3. 4π

5. 2π (or -2π, if you choose a different orientation)

7. 2π

9. $12\pi/5$

11. With the usual spherical coordinate parametrization, $\mathbf{T}_\theta \times \mathbf{T}_\phi = -\sin\phi\, \mathbf{r}$ (see Example 1). Thus,

$$\iint_S \mathbf{F}\cdot d\mathbf{S} = \iint \mathbf{F}\cdot(\mathbf{T}_\phi \times \mathbf{T}_\theta)\, d\phi\, d\theta = \iint (\mathbf{F}\cdot\mathbf{r}) \sin\phi\, d\phi\, d\theta$$

$$= \int_0^{2\pi}\int_0^{\pi} F_r \sin\phi\, d\phi\, d\theta$$

and

$$\iint_S f\,dS = \int_0^{2\pi}\int_0^{\pi} f \sin\phi\, d\phi\, d\theta.$$

13. For a cylinder of radius $R = 1$ and normal component F_r,

$$\iint_S \mathbf{F}\cdot d\mathbf{S} = \int_a^b \int_0^{2\pi} F_r\, d\theta\, dz.$$

15. $2\pi/3$

17. $\frac{2}{5}a^3bc\pi$

Section 7.7

1. Apply formula (3) of this section and simplify; $H = 0$ and $K = -b^2/(u^2+b^2)^2$.

3. Apply formula (3) of this section and simplify.

5. $K = \dfrac{-4a^6b^6}{(a^4b^4+4b^4u^2+4a^4v^2)^2}$

7. Apply formula (3) of this section and simplify.

9. Apply formula (2) of this section and simplify; $K = -h''/[(1+(h')^2)^2h]$.

Review Exercises for Chapter 7

1. (a) $3\sqrt{2}(1-e^{6\pi})/13$
(b) $-\pi\sqrt{2}/2$
(c) $(236, 158\sqrt{26}-8)/35\cdot(25)^3$
(d) $8\sqrt{2}/189$

3. (a) $2/\pi + 1$ (b) $-1/2$

5. $2a^3$

7. (a) A sphere of radius 5 centered at $(2, 3, 0)$; $\mathbf{\Phi}(\theta,\phi) = (2+5\cos\theta\sin\phi,$ $3+5\sin\theta\sin\phi,\ 5\cos\phi)$; $0\le\theta\le 2\pi; 0\le\phi\le\pi$
(b) An ellipsoid with center at $(2, 0, 0)$; $\mathbf{\Phi}(\theta,\phi) = (2+(1/\sqrt{2})3\cos\theta\sin\phi,$ $3\sin\theta\sin\phi,\ 3\cos\phi)$; $0\le\theta\le 2\pi,\ 0\le\phi\le\pi$
(c) An elliptic hyperboloid of one sheet; $\mathbf{\Phi}(\theta, z) = \left(\frac{1}{2}\sqrt{8+2z^2}\cos\theta,\ \frac{1}{3}\sqrt{8+2z^2}\sin\theta,\ z\right)$; $0\le\theta\le 2\pi,\ -\infty < z < \infty$

9. $A(\mathbf{\Phi}) = \frac{1}{2}\int_0^{2\pi}\sqrt{3\cos^2\theta+5}\,d\theta$; $\mathbf{\Phi}$ describes the upper nappe of a cone with elliptical horizontal cross sections.

11. $11\sqrt{3}/6$

13. $\sqrt{2}/3$

15. $5\sqrt{5}/6$

17. (a) $(e^y \cos \pi z, xe^y \cos \pi z, -\pi x e^y \sin \pi z)$ (b) 0

19. $\frac{1}{2}(e^2 + 1)$

21. $\mathbf{n} = (1/\sqrt{5})(-1, 0, 2)$, $2z - x = 1$

23. 0

25. If $\mathbf{F} = \nabla\phi$, then $\nabla \times \mathbf{F} = \mathbf{0}$ (at least if ϕ is of class C^2; see Theorem 1, Section 3.4). Theorem 3 of Section 7.2 shows that $\int_{\mathbf{c}} \nabla\phi \cdot d\mathbf{s} = 0$ because $\mathbf{c}$ is a closed curve.

27. (a) 24π (b) 24π (c) 60π

29. (a) $[\sqrt{R^2 + p^2}(z_0 - z_1)]/p$ (b) $\sqrt{2z_0(R^2 + p^2)/p^2 g}$

Chapter 8

Section 8.1

1. -8

3. (a) 0 (b) $-\pi R^2$ (c) 0 (d) $-\pi R^2$

5. $3\pi a^2$

7. $3\pi/2$

9. $3\pi(b^2 - a^2)/2$

11. (a) Both sides are 2π. (b) 0

13. 0

15. πab

17. A horizontal line segment divides the region into three regions of which Green's theorem applies; now use Exercise 8 or the technique in Figure 8.1.5.

19. $9\pi/8$

21. If $\varepsilon > 0$, there is a $\delta > 0$ such that $|u(\mathbf{q}) - u(\mathbf{p})| < \varepsilon$ whenever $\|\mathbf{p} - \mathbf{q}\| = \rho < \delta$. Parametrize $\partial B_\rho(\mathbf{p})$ by $\mathbf{q}(\theta) = \mathbf{p} + \rho(\cos\theta, \sin\theta)$. Then

$$|I(\rho) - 2\pi u(\mathbf{p})| \leq \int_0^{2\pi} |u(\mathbf{q}(\theta)) - u(\mathbf{p})|\, d\theta \leq 2\pi\varepsilon.$$

23. Parametrize $\partial B_\rho(\mathbf{p})$ as in Exercise 21. If $\mathbf{p} = (p_1, p_2)$, then $I(\rho) = \int_0^{2\pi} u(p_1 + \rho\cos\theta, p_2 + \rho\sin\theta)\, d\theta$. Differentiation under the integral sign gives

$$\frac{dI}{d\rho} = \int_0^{2\pi} \nabla u \cdot (\cos\theta, \sin\theta)\, d\theta = \int_0^{2\pi} \nabla u \cdot \mathbf{n}\, d\theta = \frac{1}{\rho}\int_{\partial B_\rho} \frac{\partial u}{\partial \mathbf{n}}\, ds = \frac{1}{\rho}\iint_{B_\rho} \nabla^2 u\, dA$$

(the last equality uses Exercise 22).

25. Using Exercise 24,

$$\begin{aligned}\iint_{B_R} u\, dA &= \int_0^R \int_0^{2\pi} u[\mathbf{p} + \rho(\cos\theta, \sin\theta)]\rho\, d\theta\, d\rho \\ &= \int_0^R \left(\int_{\partial B_\rho} u\, ds\right) d\rho = \int_0^R 2\pi\rho u(\mathbf{p})\, d\rho = \pi R^2 u(\mathbf{p}).\end{aligned}$$

27. Suppose u is subharmonic. We establish the assertions corresponding to Exercises 26(a) and (b). The argument for superharmonic functions is similar, with inequalities reversed.

Suppose $\nabla^2 u \geq 0$ and $u(\mathbf{p}) \geq u(\mathbf{q})$ for all $\mathbf{q}$ in $B_R(\mathbf{p})$. By Exercise 23, $I'(\rho) \geq 0$ for $0 < \rho \leq R$, and so Exercise 24 shows that $2\pi u(\mathbf{p}) \leq I(\rho) \leq I(R)$ for $0 < \rho \leq R$. If $u(\mathbf{q}) < u(\mathbf{p})$ for some $\mathbf{q} = \mathbf{p} + \rho(\cos\theta_0, \sin\theta_0) \in B_R(\mathbf{p})$, then, by continuity, there is an arc $[\theta_0 - \delta, \theta_0 + \delta]$ on $\partial B_\rho(\mathbf{p})$ where $u < u(\mathbf{p}) - d$ for some $d > 0$. This would mean that

$$\begin{aligned}2\pi u(\mathbf{p}) \leq I(\rho) &= \frac{1}{\rho}\int_0^{2\pi} u[\mathbf{p} + \rho(\cos\theta, \sin\theta)]\rho\, d\theta \\ &\leq (2\pi - 2\delta)u(\mathbf{p}) + 2\delta[u(\mathbf{p}) - d] \leq 2\pi u(\mathbf{p}) - 2\delta d.\end{aligned}$$

This contradiction shows that we must have $u(\mathbf{q}) = u(\mathbf{p})$ for every $\mathbf{q}$ in $B_B(\mathbf{p})$.

If the maximum at $\mathbf{p}$ is absolute for D, the last paragraph shows that $u(\mathbf{x}) = u(\mathbf{p})$ for all $\mathbf{x}$ in some disk around $\mathbf{p}$. If $\mathbf{c}\colon [0, 1) \to D$ is a path from $\mathbf{p}$ to $\mathbf{q}$, then $u(\mathbf{c}(t)) = u(\mathbf{p})$ for all t in some interval $[0, b)$. Let b_0 be the largest $b \in [0, 1]$ such that $u(\mathbf{c}(t)) = u(\mathbf{p})$ for all $t \in [0, b)$. (Strictly speaking, this requires the notion of the least upper bound from a good calculus text.) Because u is continuous, $u(\mathbf{c}(b_0)) = u(\mathbf{p})$. If $b_0 \neq 1$, then the last paragraph would apply at $\mathbf{c}(b_0)$ and u is constantly equal to $u(\mathbf{p})$ on a disk around $\mathbf{c}(b_0)$. In particular, there is a $\delta > 0$ such that $u(\mathbf{c}(t)) = u(\mathbf{c}(b_0)) = u(\mathbf{p})$ on $[0, b_0 + \delta)$. This contradicts the maximality of b_0, so we must have $b_0 = 1$. That is, $\mathbf{c}(\mathbf{q}) = \mathbf{c}(\mathbf{p})$. Because $\mathbf{q}$ was an arbitrary point in D, u is constant on D.

29. Assume $\nabla^2 u_1 = 0$ and $\nabla^2 u_2 = 0$ are two solutions. Let $\phi = u_1 - u_2$. Then $\nabla^2\phi = 0$ and $\phi(x) = 0$ for all $x \in \partial D$. Consider the integral $\iint_D \phi\nabla^2\phi\, dA = -\iint_D \nabla\phi \cdot \nabla\phi\, dA$. Thus, $\iint_D \nabla\phi \cdot \nabla\phi\, dA = 0$, which implies that $\nabla\phi = \mathbf{0}$, and so ϕ is a constant function and hence must be identically zero.

Section 8.2

1. -2π

3. Each integral in Stokes' theorem is zero.

5. 0

7. $-4\pi/\sqrt{3}$

9. 0

11. $\pm 2\pi$

13. Using Faraday's law, $\iint_S [\nabla \times \mathbf{E} + \partial \mathbf{H}/\partial t] \cdot d\mathbf{S} = 0$ for any surface S. If the integrand were a nonzero vector at some point, then by continuity the integral over some small disk centered at that point and lying perpendicular to that vector would be nonzero.

15. The orientations of $\partial S_1 = \partial S_2$ must agree.

17. Suppose C is a closed loop on the surface drawn so that it divides the surface into two pieces S_1 and S_2. For the surface of a doughnut (torus) you must use two closed loops; can you see why? Then C bounds both S_1 and S_2, but with positive orientation with respect to one and negative with respect to the other. Therefore,

$$\iint_S \nabla \times \mathbf{F} \cdot d\mathbf{S} = \iint_{S_1} \nabla \times \mathbf{F} \cdot d\mathbf{S} + \iint_{S_2} \nabla \times \mathbf{F} \cdot d\mathbf{S} = \int_C \mathbf{F} \cdot d\mathbf{s} - \int_C \mathbf{F} \cdot d\mathbf{s} = 0.$$

19. (a) If $C = \partial S$, $\int_C \mathbf{v} \cdot d\mathbf{s} = \iint_S (\nabla \times \mathbf{v}) \cdot d\mathbf{S} = \iint_S \mathbf{0} \cdot d\mathbf{s} = 0$.
(b) $\int_C \mathbf{v} \cdot d\mathbf{s} = \int_a^b \mathbf{v} \cdot \mathbf{c}'(t) dt = \mathbf{v} \cdot \int_a^b \mathbf{c}'(t) dt = \mathbf{v} \cdot (\mathbf{c}(b) - \mathbf{c}(a))$, where $\mathbf{c}\colon [a, b] \to \mathbb{R}^3$ is a parametrization of C. (The vector integral is the vector whose components are the integrals of the component functions.) If C is closed, the last expression is 0.

21. Both integrals give $\pi/4$.

23. (a) 0 (b) π (c) π

25. -20π (or 20π if the opposite orientation is used)

27. One possible answer: The Möbius curve C is also the boundary of an *oriented* surface $\tilde{S}$; the equation in Faraday's law *is* valid for this new surface.

Section 8.3

1. If $\mathbf{F} = \nabla f = \nabla g$ and C is a curve from $\mathbf{v}$ to $\mathbf{w}$, then $(f - g)(\mathbf{w}) - (f - g)(\mathbf{v}) = \int_C \nabla (f - g) \cdot d\mathbf{s} = 0$ and so $f - g$ is constant.

3. $x^2yz - \cos x + C$

5. Yes, it is the gradient of $g(x, y) = F(x) + F(y)$, where $F'(x) = f(x)$.

7. No; $\nabla \times \mathbf{F} = (0, 0, -x) \neq \mathbf{0}$.

9. $e \sin 1 + \frac{1}{3}e^3 - \frac{1}{3}$

11. 3.5×10^{29} ergs

13. (a) $f = x^2/2 + y^2/2 + C$ (c) $f = \frac{1}{3}x^3 + xy^2 + C$
(b) **F** is not a gradient field.

15. Use Theorem 7 in each case.

(a) $-3/2$ (b) -1 (c) $\cos(e^2) - \cos(1/e)/e$

17. (a) No. (b) $\left(\frac{1}{2}z^2, xy - z, x^2y\right)$ or $\left(\frac{1}{2}z^2 - 2xyz - \frac{1}{2}y^2, -x^2z - z, 0\right)$

19. $\frac{1}{3}(z^3\mathbf{i} + x^3\mathbf{j} + y^3\mathbf{k})$

21. $(-z\sin y + y\sin x, xz\cos y, 0)$ (Other answers are possible.)

23. (a) $\nabla \times \mathbf{F} = (0, 0, 2) \neq \mathbf{0}$
(b) Let $\mathbf{c}(t)$ be the path of an object in the fluid. Then $\mathbf{F}(\mathbf{c}(t)) = \mathbf{c}'(t)$. Let $\mathbf{c}(t) = (x(t), y(t), z(t))$. Then $x' = -y$, $y' = x$, and $z' = 0$, and so z is constant and the motion is parallel to the xy plane. Also, $x'' + x = 0$, $y'' + y = 0$. Thus, $x = A\cos t + B\sin t$ and $y = C\cos t + D\sin t$. Substituting these values in $x' = -y$, $y' = x$, we get $C = -B$, $D = A$, so that $x^2 + y^2 = A^2 + B^2$ and we have a circle.
(c) Counterclockwise

25. (a) $\mathbf{F} = -\dfrac{GmM}{(x^2 + y^2 + z^2)^{3/2}}(x, y, z)$;

$$\nabla \cdot \mathbf{F} = -GmM\left[\frac{x^2 + y^2 + z^2 - 3x^2}{(x^2 + y^2 + z^2)^{5/2}} + \frac{x^2 + y^2 + z^2 - 3y^2}{(x^2 + y^2 + z^2)^{5/2}} + \frac{x^2 + y^2 + z^2 - 3z^2}{(x^2 + y^2 + z^2)^{5/2}}\right]$$
$$= 0$$

(b) Let S be the unit sphere, S_1 the upper hemisphere, S_2 the lower hemisphere, and C the unit circle. If $\mathbf{F} = \nabla \times \mathbf{G}$, then

$$\iint_S \mathbf{F} \cdot d\mathbf{S} = \iint_{S_1} \mathbf{F} \cdot d\mathbf{S} + \iint_{S_2} \mathbf{F} \cdot d\mathbf{S} = \int_C \mathbf{G} \cdot d\mathbf{s} - \int_C \mathbf{G} \cdot d\mathbf{s} = 0.$$

But $\iint_S \mathbf{F} \cdot d\mathbf{S} = -GmM \iint_S (\mathbf{r}/\|\mathbf{r}\|^3) \cdot \mathbf{n}\, dS = -4\pi GmM$, because $\|\mathbf{r}\| = 1$ and $\mathbf{r} = \mathbf{n}$ on S. Thus, $\mathbf{F} = \nabla \times \mathbf{G}$ is impossible. This does not contradict Theorem 8 because **F** is not smooth at the origin.

Section 8.4

1. 4π

3. 3

5. (a) 0 (b) $4/15$ (c) $-4/15$

7. 6

9. 1

11. Apply the divergence theorem to $f\mathbf{F}$ using $\nabla\cdot(f\mathbf{F}) = \nabla f\cdot\mathbf{F} + f\nabla\cdot\mathbf{F}$.

13. If $\mathbf{F} = \mathbf{r}/r^2$, then $\nabla\cdot\mathbf{F} = 1/r^2$. If $(0,0,0)\notin\Omega$, the result follows from Gauss' theorem. If $(0,0,0)\in\Omega$, we compute the integral by deleting a small ball $B_\varepsilon = \{(x,y,z)|(x^2+y^2+z^2)^{1/2} < \varepsilon\}$ around the origin and then letting $\varepsilon\to 0$:

$$\begin{aligned}\iiint_\Omega \frac{1}{r^2}\,dV &= \lim_{\varepsilon\to 0}\iiint_{\Omega\setminus B_\varepsilon}\frac{1}{r^2}\,dV = \lim_{\varepsilon\to 0}\iint_{\partial(\Omega\setminus B_\varepsilon)}\frac{\mathbf{r}\cdot\mathbf{n}}{r^2}\,dS\\ &= \lim_{\varepsilon\to 0}\left(\iint_{\partial\Omega}\frac{\mathbf{r}\cdot\mathbf{n}}{r^2}\,dS - \iint_{\partial B_\varepsilon}\frac{\mathbf{r}\cdot\mathbf{n}}{r^2}\,dS\right) = \lim_{\varepsilon\to 0}\left(\iint_{\partial\Omega}\frac{\mathbf{r}\cdot\mathbf{n}}{r^2}\,dS - 4\pi\varepsilon\right)\\ &= \iint_{\partial\Omega}\frac{\mathbf{r}\cdot\mathbf{n}}{r^2}\,dS.\end{aligned}$$

The integral over ∂B_ε is obtained from Theorem 10 (Gauss' law), because $r=\varepsilon$ everywhere on B_ε.

15. Use the vector identity for $\operatorname{div}(f\mathbf{F})$ and the divergence theorem for part (a). Use the vector identity $\nabla\cdot(f\nabla g - g\nabla f) = f\nabla^2 g - g\nabla^2 f$ for part (b).

17. (a) If $\phi(\mathbf{p}) = \iiint_W \rho(\mathbf{q})/(4\pi\|\mathbf{p}-\mathbf{q}\|)\,dV(\mathbf{q})$, then

$$\begin{aligned}\nabla\phi(\mathbf{p}) &= \iiint_W [\rho(\mathbf{q})/4\pi]\nabla_{\mathbf{p}}(1/\|\mathbf{p}-\mathbf{q}\|)\,dV(\mathbf{q})\\ &= -\iiint_W [\rho(\mathbf{q})/4\pi][(\mathbf{p}-\mathbf{q})/\|\mathbf{p}-\mathbf{q}\|^3]\,dV(\mathbf{q}),\end{aligned}$$

where $\nabla_{\mathbf{p}}$ means the gradient with respect to the coordinates of $\mathbf{p}$ and the integral is the vector whose components are the three component integrals. If $\mathbf{p}$ varies in $V\cup\partial V$ and $\mathbf{n}$ is the outward unit normal to ∂V, we can take the inner product using these components and collect the pieces as

$$\nabla\phi(\mathbf{p})\cdot\mathbf{n} = -\iiint_W \frac{\rho(\mathbf{q})}{4\pi}\frac{1}{\|\mathbf{p}-\mathbf{q}\|^3}(\mathbf{p}-\mathbf{q})\cdot\mathbf{n}\,dV(\mathbf{q}).$$

Thus,

$$\iint_{\partial V}\nabla\phi(\mathbf{p})\cdot\mathbf{n}\,dV(\mathbf{p}) = -\iint_{\partial V}\left(\iiint_W \frac{\rho(\mathbf{q})}{4\pi}\frac{1}{\|\mathbf{p}-\mathbf{q}\|^3}(\mathbf{p}-\mathbf{q})\cdot\mathbf{n}\,d\mathbf{q}\right)dV(\mathbf{p}).$$

There are essentially five variables of integration here, three placing $\mathbf{q}$ in W and two placing $\mathbf{p}$ on ∂V. Use Fubini's theorem to obtain

$$\iint_{\partial V}\nabla\phi\cdot\mathbf{n}\cdot d\mathbf{S} = -\iiint_W \frac{\rho(\mathbf{q})}{4\pi}\left[\iint_{\partial V}\frac{(\mathbf{p}-\mathbf{q})\cdot\mathbf{n}}{\|\mathbf{p}-\mathbf{q}\|^3}\,dS(\mathbf{p})\right]dV(\mathbf{q}).$$

If V is a symmetric elementary region, Theorem 10 says that the inner integral is 4π if $\mathbf{q} \in V$ and 0 if $\mathbf{q} \notin V$. Thus,

$$\iint_{\partial V} \nabla\phi \cdot \mathbf{n}\, dS = -\iiint_{W\cap V} \rho(\mathbf{q})\, dV(\mathbf{q}).$$

Because $\rho = 0$ outside W,

$$\iint_{\partial V} \nabla\phi \cdot \mathbf{n}\, dS = -\iiint_{V} \rho(\mathbf{q})\, dV(\mathbf{q}).$$

If V is not a symmetric elementary region, subdivide it into a sum of such regions. The equation holds on each piece, and, upon adding them together, the boundary integrals along appropriately oriented interior boundaries cancel, leaving the desired result.

(b) By Theorem 9, $\iint_{\partial V} \nabla\phi \cdot d\mathbf{S} = \iiint_V \nabla^2\phi\, dV$, and so $\iiint_V \nabla^2\phi\, dV = -\iiint_V \rho\, dV$. Because both ρ and $\nabla^2\phi$ are continuous and this holds for arbitrarily small regions, we must have $\nabla^2\phi = -\rho$.

19. If the charge Q is spread evenly over the sphere S of radius R centered at the origin, the density of charge per unit area must be $Q/4\pi R^2$. If $\mathbf{p}$ is a point not on S and $\mathbf{q} \in S$, then the contribution to the electric field at $\mathbf{p}$ due to charge near $\mathbf{q}$ is directed along the vector $\mathbf{p} - \mathbf{q}$. Because the charge is evenly distributed, the tangential component of this contribution will be canceled by that from a symmetric point on the other side of the sphere at the same distance from $\mathbf{p}$. (Draw the picture.) The total resulting field must be radial. Because S looks the same from any point at a distance $\|\mathbf{p}\|$ from the origin, the field must depend only on radius and be of the form $\mathbf{E} = f(r)\mathbf{r}$.

If we look at the sphere Σ of radius $\|\mathbf{p}\|$, we have

$$\begin{aligned}(\text{charge inside } \Sigma) &= \iint_\Sigma \mathbf{E} \cdot d\mathbf{S} = \iint_\Sigma f(\|\mathbf{p}\|)\mathbf{r} \cdot \mathbf{n}\, dS \\ &= f(\|\mathbf{p}\|)\|\mathbf{p}\| \text{ area } \Sigma = 4\pi \|\mathbf{p}\|^3 f(\|\mathbf{p}\|).\end{aligned}$$

If $\|\mathbf{p}\| < R$, there is no charge inside Σ; if $\|\mathbf{p}\| > R$, the charge inside Σ is Q, and so

$$\mathbf{E}(\mathbf{p}) = \begin{cases} \dfrac{1}{4\pi} \dfrac{Q}{\|\mathbf{p}\|^3}\mathbf{p} & \text{if } \|\mathbf{p}\| > R \\ 0 & \text{if } \|\mathbf{p}\| < R. \end{cases}$$

21. By Theorem 10, $\iint_{\partial M} \mathbf{F} \cdot d\mathbf{S} = 4\pi$ for any surface enclosing the origin. But if $\mathbf{F}$ were the curl of some field, then the integral over such a closed surface would have to be 0.

23. If $S = \partial W$, then $\iint_S \mathbf{r} \cdot \mathbf{n}\, dS = \iiint_W \nabla \cdot \mathbf{r}\, dV = \iiint_W 3dV = 3$ volume (W). For the geometric explanation, assume $(0, 0, 0) \in W$ and consider the skew cone with its vertex at $(0, 0, 0)$ with base ΔS and altitude $\|\mathbf{r}\|$. Its volume is $\frac{1}{3}(\Delta S)(\mathbf{r} \cdot \mathbf{n})$.

Section 8.5

1. Write the components of φ as $\xi(\mathbf{x}, t)$, $\eta(\mathbf{x}, t)$, and $\zeta(\mathbf{x}, t)$. First, observe that by definition of φ,

$$\frac{\partial}{\partial t}\varphi(\mathbf{x}, t) = \mathbf{F}(\varphi(\mathbf{x}, t), t).$$

The determinant J can be differentiated by recalling that the determinant of a matrix is multilinear in the columns (or rows). Thus, holding $\mathbf{x}$ fixed,

$$\frac{\partial}{\partial t}J = \begin{bmatrix} \frac{\partial}{\partial t}\frac{\partial \xi}{\partial x} & \frac{\partial \eta}{\partial x} & \frac{\partial \zeta}{\partial x} \\ \frac{\partial}{\partial t}\frac{\partial \xi}{\partial y} & \frac{\partial \eta}{\partial y} & \frac{\partial \zeta}{\partial y} \\ \frac{\partial}{\partial t}\frac{\partial \xi}{\partial z} & \frac{\partial \eta}{\partial z} & \frac{\partial \zeta}{\partial z} \end{bmatrix} + \begin{bmatrix} \frac{\partial \xi}{\partial x} & \frac{\partial}{\partial t}\frac{\partial \eta}{\partial x} & \frac{\partial \zeta}{\partial x} \\ \frac{\partial \xi}{\partial y} & \frac{\partial}{\partial t}\frac{\partial \eta}{\partial y} & \frac{\partial \zeta}{\partial y} \\ \frac{\partial \xi}{\partial z} & \frac{\partial}{\partial t}\frac{\partial \eta}{\partial z} & \frac{\partial \zeta}{\partial z} \end{bmatrix} + \begin{bmatrix} \frac{\partial \xi}{\partial x} & \frac{\partial \eta}{\partial x} & \frac{\partial}{\partial t}\frac{\partial \zeta}{\partial x} \\ \frac{\partial \xi}{\partial y} & \frac{\partial \eta}{\partial y} & \frac{\partial}{\partial t}\frac{\partial \zeta}{\partial y} \\ \frac{\partial \xi}{\partial z} & \frac{\partial \eta}{\partial z} & \frac{\partial}{\partial t}\frac{\partial \zeta}{\partial z} \end{bmatrix}.$$

Now write

$$\frac{\partial}{\partial t}\frac{\partial \xi}{\partial x} = \frac{\partial}{\partial x}\frac{\partial \xi}{\partial t} = \frac{\partial}{\partial x}F_1(\varphi(\mathbf{x}, t), t),$$

$$\frac{\partial}{\partial t}\frac{\partial \xi}{\partial y} = \frac{\partial}{\partial y}\frac{\partial \xi}{\partial t} = \frac{\partial}{\partial y}F_2(\varphi(\mathbf{x}, t), t),$$

$$\frac{\partial}{\partial t}\frac{\partial \zeta}{\partial z} = \frac{\partial}{\partial z}\frac{\partial \zeta}{\partial t} = \frac{\partial}{\partial z}F_3(\varphi(\mathbf{x}, t), t).$$

The components F_1, F_2, and F_3 of $\mathbf{F}$ in this expression are functions of x, y, and z through $\varphi(\mathbf{x}, t)$; therefore,

$$\frac{\partial}{\partial x}F_1(\varphi(\mathbf{x}, t), t) = \frac{\partial F_1}{\partial \xi}\frac{\partial \xi}{\partial x} + \frac{\partial F_1}{\partial \eta}\frac{\partial \eta}{\partial x} + \frac{\partial F_1}{\partial \zeta}\frac{\partial \zeta}{\partial x},$$

$$\vdots$$

$$\frac{\partial}{\partial z}F_3(\varphi(\mathbf{x}, t), t) = \frac{\partial F_3}{\partial \xi}\frac{\partial \xi}{\partial z} + \frac{\partial F_3}{\partial \eta}\frac{\partial \eta}{\partial z} + \frac{\partial F_3}{\partial \zeta}\frac{\partial \zeta}{\partial z}.$$

When these are substituted into the previous expression for $\partial J/\partial t$, one gets for the respective terms

$$\frac{\partial F_1}{\partial x}J + \frac{\partial F_2}{\partial y}J + \frac{\partial F_3}{\partial z}J = (\operatorname{div} \mathbf{F})J.$$

3. HINTS: By the transport equation from Theorem 12, with $\mathbf{V}$ in place of $\mathbf{F}$,

$$\frac{d}{dt}\iiint_{W_t} \rho\, dx\, dy\, dz = \iiint_{W_t} \left(\frac{D\rho}{Dt} + \rho \operatorname{div} \mathbf{V}\right) dx\, dy\, dz.$$

Now use the fact that

$$\frac{D\rho}{Dt} + \rho \operatorname{div} \mathbf{V} = \operatorname{div} \mathbf{J} + \frac{\partial \rho}{\partial t},$$

where $\mathbf{J} = \rho\mathbf{V}$, as in the text.

5. If v_i is the ith component of a vector $\mathbf{v}$, then by the transport equation (Exercise 2),

$$\left[\frac{d}{dt}\iiint_{W_t} f\mathbf{F}\,dx\,dy\,dz\right]_i = \frac{d}{dt}\iiint_{W_t}(f\mathbf{F})_i\,dx\,dy\,dz = \frac{d}{dt}\iiint_{W_t} fF_i\,dx\,dy\,dz$$

$$= \iiint_{W_t}\left[\frac{D(fF_i)}{Dt} + (fF_i)\operatorname{div}\mathbf{F}\right]dx\,dy\,dz$$

$$= \iiint_{W_t}\left[\frac{\partial}{\partial t}(fF_i) + \mathbf{D}_x(fF_i)\cdot\mathbf{F} + (f\mathbf{F}_i)\operatorname{div}\mathbf{F}\right]dx\,dy\,dz$$

$$= \iiint_{W_t}\left[\frac{\partial}{\partial t}(f\mathbf{F}_i) + \nabla(fF_i)\cdot\mathbf{F} + (fF_i)\operatorname{div}\mathbf{F}\right]dx\,dy\,dz$$

$$= \iiint_{W_t}\left\{\frac{\partial}{\partial t}(f\mathbf{F}_i) + [\mathbf{D}(f\mathbf{F})\mathbf{F}]_i + [(f\mathbf{F})\operatorname{div}\mathbf{F}]_i\right\}dx\,dy\,dz$$

$$= \iiint_{W_t}\left[\frac{\partial}{\partial t}(f\mathbf{F}) + \mathbf{D}(f\mathbf{F})\mathbf{F} + (f\mathbf{F})\operatorname{div}\mathbf{F}\right]_i dx\,dy\,dz$$

$$= \left[\iiint_{W_t}\frac{\partial}{\partial t}(f\mathbf{F}) + \mathbf{D}(f\mathbf{F})\mathbf{F} + (f\mathbf{F})\operatorname{div}\mathbf{F}\,dx\,dy\,dz\right]_i$$

$$= \left[\iiint_{W_t}\left(\frac{\partial}{\partial t}(f\mathbf{F}) + (\mathbf{F}\cdot\nabla)(f\mathbf{F}) + (f\mathbf{F})\operatorname{div}\mathbf{F}\right)dx\,dy\,dz\right]_i.$$

7. (a) Because $\mathbf{V} = \nabla\phi$, $\nabla\times\mathbf{V} = \mathbf{0}$, and therefore $(\mathbf{V}\cdot\nabla)\mathbf{V} = \dfrac{1}{2}\nabla(\|\mathbf{V}\|^2)$, Euler's equation becomes

$$-\frac{\nabla p}{\rho} = \frac{d\mathbf{V}}{dt} + \frac{1}{2}\nabla(\|\mathbf{V}\|^2) = \nabla\left(\frac{d\phi}{dt} + \frac{1}{2}\|\mathbf{V}\|^2\right).$$

If $\mathbf{c}$ is a path from P_1 to P_2, then

$$-\int_{\mathbf{c}}\frac{1}{\rho}\,dp = -\int\frac{1}{\rho}\nabla p\cdot\mathbf{c}'(t)\,dt = \int_{\mathbf{c}}\nabla\left(\frac{d\phi}{dt} + \frac{1}{2}\|\mathbf{V}\|^2\right)\cdot\mathbf{c}'(t)\,dt$$

$$= \left(\frac{d\phi}{dt} + \frac{1}{2}\|\mathbf{V}\|^2\right)\Bigg|_{P_1}^{P_2}.$$

(b) If $d\mathbf{V}/dt = \mathbf{0}$ and ρ is constant, then $\frac{1}{2}\nabla(\|\mathbf{V}\|^2) = -(\nabla p)/\rho = -\nabla(p/\rho)$, and therefore $\nabla\left(\dfrac{1}{2}\|\mathbf{V}\|^2 + p/\rho\right) = \mathbf{0}$.

9. By Ampère's law, $\nabla\cdot\mathbf{J} = \nabla\cdot(\nabla\times\mathbf{H}) - \nabla\cdot(\partial\mathbf{E}/\partial t) = -\nabla\cdot(\partial\mathbf{E}/\partial t) = -(\partial/\partial t)(\nabla\cdot\mathbf{E})$. By Gauss' law this is $-\partial\rho/\partial t$. Thus, $\nabla\cdot\mathbf{J} + \partial\rho/\partial t = 0$.

Section 8.6

1. (a) $(2xy^2 - yx^3)\,dx\,dy$
(b) $(x^2 + y^2)\,dx\,dy$
(c) $(x^2 + y^2 + z^2)\,dx\,dy\,dz$
(d) $(xy + x^2)\,dx\,dy\,dz$
(e) $dx\,dy\,dz$

3. (a) $2xy\,dx + (x^2 + 3y^2)\,dy$
(b) $-(x + y^2 \sin x)\,dx\,dy$
(c) $-(2x + y)\,dx\,dy$
(d) $dx\,dy\,dz$
(e) $2x\,dx\,dy\,dz$
(f) $2y\,dy\,dz - 2x\,dz\,dx$
(g) $-\dfrac{4xy}{(x^2 + y^2)^2}\,dx\,dy$
(h) $2xy\,dx\,dy\,dz$

5. (a)
$$\begin{aligned}
\mathrm{Form}_2\,(\alpha \mathbf{V}_1 + \mathbf{V}_2) &= \mathrm{Form}_2\,(\alpha A_1 + A_2, \alpha B_1 + B_2, \alpha C_1 + C_2)\\
&= (\alpha A_1 + A_2)\,dy\,dz + (\alpha B_1 + B_2)\,dz\,dx\\
&\quad + (\alpha C_1 + C_2)\,dx\,dy\\
&= \alpha(A_1\,dy\,dz + B_1\,dz\,dx + C_1\,dx\,dy)\\
&\quad + (A_2\,dy\,dz + B_2\,dz\,dx + C_2\,dx\,dy)\\
&= \alpha\,\mathrm{Form}_2\,(\mathbf{V}_1) + \mathrm{Form}_2(\mathbf{V}_2)
\end{aligned}$$

(b)
$$\begin{aligned}
d\omega = &\left(\frac{\partial A}{\partial x}\,dx + \frac{\partial A}{\partial y}\,dy + \frac{\partial A}{\partial z}\,dz\right) \wedge dx + A(dx)^2\\
&+ \left(\frac{\partial B}{\partial x}\,dx + \frac{\partial B}{\partial y}\,dy + \frac{\partial B}{\partial z}\,dz\right) \wedge dy + B(dy)^2\\
&+ \left(\frac{\partial C}{\partial x}\,dx + \frac{\partial C}{\partial y}\,dy + \frac{\partial C}{\partial z}\,dz\right) \wedge dz + C(dz)^2.
\end{aligned}$$

But $(dx)^2 = (dy)^2 = (dz)^2 = dx \wedge dx = dy \wedge dy = dz \wedge dz = 0$, $dy \wedge dx = -dx \wedge dy$, $dz \wedge dy = -dy \wedge dz$, and $dx \wedge dz = -dz \wedge dx$. Hence,

$$\begin{aligned}
d\omega &= \left(\frac{\partial C}{\partial y} - \frac{\partial B}{\partial z}\right) dy\,dz + \left(\frac{\partial A}{\partial z} - \frac{\partial C}{\partial y}\right) dz\,dx + \left(\frac{\partial B}{\partial x} - \frac{\partial A}{\partial y}\right) dx\,dy\\
&= \mathrm{Form}_2\,(\mathrm{curl}\,\mathbf{V}).
\end{aligned}$$

7. An oriented 1-manifold is a curve. Its boundary is a pair of points that may be considered a 0-manifold. Therefore, ω is a 0-form or function, and $\int_{\partial M} d\omega = \omega(b) - \omega(a)$ if the curve M runs from a to b. Furthermore, $d\omega$ is the 1-form $(\partial\omega/\partial x)\,dx + (\partial\omega/\partial y)\,dy$. Therefore, $\int_M d\omega$ is the line integral $\int_M (\partial\omega/\partial x)\,d\omega + (\partial\omega/\,dy)\,dy = \int_M \nabla\omega \cdot d\mathbf{s}$. Thus, we obtain Theorem 3 of Section 7.2, $\int_M \nabla\omega \cdot d\mathbf{s} = \omega(b) - \omega(a)$.

9. Put $\omega = F_1\,dx\,dy + F_2\,dy\,dz + F_3\,dz\,dx$. The integral becomes

$$\iint_{\partial T} \omega = \iiint_T d\omega = \iiint_T \left(\frac{\partial F_1}{\partial z} + \frac{\partial F_2}{\partial x} + \frac{\partial F_3}{\partial y}\right) dx\,dy\,dz.$$

(a) 0 (b) 40

11. Consider $\omega = x\,dy\,dz + y\,dz\,dx + z\,dx\,dy$. Compute that $d\omega = 3\,dx\,dy\,dz$, so that $\frac{1}{3}\iint_{\partial R} \omega = \frac{1}{3}\iiint_R d\omega = \iiint_R dx\,dy\,dz = v(R)$.

Review Exercises for Chapter 8

1. (a) $2\pi a^2$ (b) 0

3. 0

5. (a) $f = x^4/4 - x^2y^3$ (b) $-1/4$

7. (a) Check that $\nabla \times \mathbf{F} = \mathbf{0}$ (b) $f = 3x^2 y \cos z + C$ (c) 0

9. 23/6

11. No: $\nabla \times (\mathbf{a} \times \mathbf{r}) = 2\mathbf{a}$

13. (a) $\nabla f = 3ye^{z^2}\mathbf{i} + 3xe^{z^2}\mathbf{j} + 6xyze^{z^2}\mathbf{k}$ (b) 0 (c) Both sides are 0.

15. $8\pi/3$

17. $\pi a^2/4$

19. 21

21. (a) $\mathbf{G}$ is conservative; $\mathbf{F}$ is not.
(b) $\mathbf{G} = \nabla\phi$ if $\phi = (x^4/4) + (y^4/4) - \frac{3}{2}x^2y^2 + \frac{1}{2}z^2 + C$, where C is any constant.
(c) $\int_\alpha \mathbf{F} \cdot d\mathbf{s} = 0$; $\int_\alpha \mathbf{G} \cdot d\mathbf{s} = -\frac{1}{2}$; $\int_\beta \mathbf{F} \cdot d\mathbf{s} = \frac{1}{3}$; $\int_\beta \mathbf{G} \cdot d\mathbf{s} = -\frac{1}{2}$

23. Use $(\nabla \cdot \mathbf{F})(x_0, y_0, z_0) = \text{limit}_{\rho \to 0} \frac{1}{V(\Omega_\rho)} \iint_{\partial\Omega_\rho} \mathbf{F} \cdot \mathbf{n}\, dS$ from Section 8.4.

Index

Illustration Credits

FM Opener: David Gifford/Photo Researchers
FM Fig. 3: Erich Lessing/Art Resource, NY
FM Fig. 4: From *Theoricae novae planetarum*
FM Fig. 5: Science Photo Library/Photo Researchers, Inc.
FM Fig. 6: J. L. Charmet/Photo Researchers, Inc.
FM Fig. 7: From the *Codex Vigilanus*, Escurial Library, Madrid
FM Fig. 8: Michael Pasdzior/Getty
FM Fig. 9: Alinari/Art Resource
FM Fig. 10: Corbis
FM Fig. 11: Corbis
FM Fig. 12: From Kepler's *Astronomia Nova* (1609), p. 4
FM Fig. 14: Corbis
FM Fig. 15: From *Principia*
Fig. 1.3.8: Corbis
Fig. 1.3.9: Getty Images
Fig. 1.5.1: Corbis
Fig. 1.5.2: The Granger Collection, NY
CO3: Science Photo Library/Photo Researchers, Inc.
Fig. 3.3.1: From *Parsimonious Universe*/Courtesy of Anthony Tromba
Fig. 3.3.2: American Institute of Physics/Science Photo Library/Photo Researchers, Inc.
Fig. 4.1.5: Corbis
Fig. 4.1.6: Cern/Science Photo Library/Photo Researchers, Inc.
Fig. 4.1.11: Roby Wilson, JPL
Fig. 4.4.9: Corbis
Fig. 6.3.3: From Leonardo's treatise on painting, *Tractat vond der Mahlerey*, 2nd ed., Nuremburg, 1747
Fig. 7.4.4: Städtische Galerie; Liebieghaus Frankfurt am Main
Fig. 7.4.5: Fritz Goro
Fig. 7.6.4: Moebius Strip II, 1963, by M. C. Escher. Escher Foundation, Haags Gemeentemuseum, The Hague.
Fig. 7.7.1: Sinclair Stammers/Science Photo Library/Photo Researchers, Inc.
Fig. 7.7.2: Courtesy of Anthony Tromba
Fig. 7.7.3: © 1997 Michael Dalton/Fundamental Photographs, NYC
Fig. 7.7.5: UCO/Lick Observatory (Crossley reflector)
Fig. 7.7.6: Science Photo Library/Photo Researchers, Inc.
Fig. 7.7.12: Corbis

48. $$\int \frac{x}{\sqrt{a+bx}}\,dx = \frac{2(bx-2a)\sqrt{a+bx}}{3b^2}$$

49. $$\int \frac{1}{x\sqrt{a+bx}}\,dx = \begin{cases} \dfrac{1}{\sqrt{a}} \log\left|\dfrac{\sqrt{a+bx}-\sqrt{a}}{\sqrt{a+bx}+\sqrt{a}}\right| & (a>0) \\ \dfrac{2}{\sqrt{-a}} \arctan\left|\sqrt{\dfrac{a+bx}{-a}}\right| & (a<0) \end{cases}$$

50. $$\int \frac{\sqrt{a^2-x^2}}{x}\,dx = \sqrt{a^2-x^2} - a\log\left|\frac{a+\sqrt{a^2-x^2}}{x}\right|$$

51. $$\int x\sqrt{a^2-x^2}\,dx = -\tfrac{1}{3}(a^2-x^2)^{3/2}$$

52. $$\int x^2\sqrt{a^2-x^2}\,dx = \frac{x}{8}(2x^2-a^2)\sqrt{a^2-x^2} + \frac{a^4}{8}\arcsin\frac{x}{a} \quad (a>0)$$

53. $$\int \frac{1}{x\sqrt{a^2-x^2}}\,dx = -\frac{1}{a}\log\left|\frac{a+\sqrt{a^2-x^2}}{x}\right|$$

54. $$\int \frac{x}{\sqrt{a^2-x^2}}\,dx = -\sqrt{a^2-x^2}$$

55. $$\int \frac{x^2}{\sqrt{a^2-x^2}}\,dx = -\frac{x}{2}\sqrt{a^2-x^2} + \frac{a^2}{2}\arcsin\frac{x}{a} \quad (a>0)$$

56. $$\int \frac{\sqrt{x^2+a^2}}{x}\,dx = \sqrt{x^2+a^2} - a\log\left|\frac{a+\sqrt{x^2+a^2}}{x}\right|$$

57. $$\int \frac{\sqrt{x^2-a^2}}{x}\,dx = \sqrt{x^2-a^2} - a\arccos\frac{a}{|x|} \quad (a>0)$$

58. $$\int \frac{x^2}{\sqrt{x^2+a^2}}\,dx = \frac{x\sqrt{x^2+a^2}}{2} - \frac{a^2}{2}\log\left(x+\sqrt{x^2+a^2}\right)$$

59. $$\int \frac{1}{x\sqrt{x^2+a^2}}\,dx = \frac{1}{a}\log\left|\frac{x}{a+\sqrt{x^2+a^2}}\right|$$

60. $$\int \frac{1}{x\sqrt{x^2-a^2}}\,dx = \frac{1}{a}\arccos\frac{a}{|x|} \quad (a>0)$$

61. $$\int \frac{1}{x^2\sqrt{x^2\pm a^2}}\,dx = \mp\frac{\sqrt{x^2\pm a^2}}{a^2x}$$

62. $$\int \frac{1}{\sqrt{x^2\pm a^2}}\,dx = \sqrt{x^2\pm a^2}$$

63. $$\int \frac{1}{ax^2+bx+c}\,dx = \begin{cases} \dfrac{1}{\sqrt{b^2-4ac}}\log\left|\dfrac{2ax+b-\sqrt{b^2-4ac}}{2ax+b+\sqrt{b^2-4ac}}\right| & (b^2>4ac) \\ \dfrac{2}{\sqrt{4ac-b^2}}\arctan\dfrac{2ax+b}{\sqrt{4ac-b^2}} & (b^2<4ac) \end{cases}$$

64. $$\int \frac{x}{ax^2+bx+c}\,dx = \frac{1}{2a}\log|ax^2+bx+c| - \frac{b}{2a}\int \frac{1}{ax^2+bx+c}\,dx$$

65. $$\int \frac{1}{\sqrt{ax^2+bx+c}}\,dx = \begin{cases} \dfrac{1}{\sqrt{a}}\log|2ax+b+2\sqrt{a}\sqrt{ax^2+bx+c}| & (a>0) \\ \dfrac{1}{\sqrt{-a}}\arcsin\dfrac{-2ax-b}{\sqrt{b^2-4ac}} & (a<0) \end{cases}$$

66. $$\int \sqrt{ax^2+bx+c}\,dx = \frac{2ax+b}{4a}\sqrt{ax^2+bx+c} + \frac{4ac-b^2}{8a}\int \frac{1}{\sqrt{ax^2+b+c}}\,dx$$

67. $$\int \frac{x}{\sqrt{ax^2+bx+c}}\,dx = \frac{\sqrt{ax^2+bx+c}}{a} - \frac{b}{2a}\int \frac{1}{\sqrt{ax^2+bx+c}}\,dx$$

68. $$\int \frac{1}{x\sqrt{ax^2+bx+c}}\,dx = \begin{cases} \dfrac{-1}{\sqrt{c}}\log\left|\dfrac{2\sqrt{c}\sqrt{ax^2+bx+c}+bx+2c}{x}\right| & (c>0) \\ \dfrac{1}{\sqrt{-c}}\arcsin\dfrac{bx+2c}{|x|\sqrt{b^2-4ac}} & (c<0) \end{cases}$$

69. $$\int x^3\sqrt{x^2+a^2}\,dx = \left(\tfrac{1}{5}x^2 - \tfrac{2}{15}a^2\right)\sqrt{(a^2+x^2)^3}$$

70. $$\int \frac{\sqrt{x^2 \pm a^2}}{x^4}\,dx = \frac{\mp\sqrt{(x^2 \pm a^2)^3}}{3a^2x^3}$$

71. $$\int \sin ax \sin bx\,dx = \frac{\sin(a-b)x}{2(a-b)} - \frac{\sin(a+b)x}{2(a+b)} \quad (a^2 \neq b^2)$$

72. $$\int \sin ax \cos bx\,dx = -\frac{\cos(a-b)x}{2(a-b)} - \frac{\cos(a+b)x}{2(a+b)} \quad (a^2 \neq b^2)$$

73. $$\int \cos ax \cos bx\,dx = \frac{\sin(a-b)x}{2(a-b)} + \frac{\sin(a+b)x}{2(a+b)} \quad (a^2 \neq b^2)$$

74. $$\int \sec x \tan x\,dx = \sec x$$

75. $$\int \csc x \cot x\,dx = -\csc x$$

76. $$\int \cos^m x \sin^n x\,dx = \frac{\cos^{m-1} x \sin^{n+1} x}{m+n} + \frac{m-1}{m+n}\int \cos^{m-2} x \sin^n x\,dx$$
$$= -\frac{\sin^{n-1} x \cos^{m+1} x}{m+n} + \frac{n-1}{m+n}\int \cos^m x \sin^{n-2} x\,dx$$

77. $$\int x^n \sin ax\,dx = -\frac{1}{a}x^n \cos ax + \frac{n}{a}\int x^{n-1}\cos ax\,dx$$

78. $$\int x^n \cos ax\,dx = \frac{1}{a}x^n \sin ax - \frac{n}{a}\int x^{n-1}\sin ax\,dx$$

79. $$\int x^n e^{ax}\,dx = \frac{x^n e^{ax}}{a} - \frac{n}{a}\int x^{n-1}e^{ax}\,dx$$

80. $$\int x^n \log ax\,dx = x^{n+1}\left[\frac{\log ax}{n+1} - \frac{1}{(n+1)^2}\right]$$

81. $$\int x^n(\log ax)^m\,dx = \frac{x^{n+1}}{n+1}(\log ax)^m - \frac{m}{n+1}\int x^n(\log ax)^{m-1}\,dx$$

82. $$\int e^{ax}\sin bx\,dx = \frac{e^{ax}(a\sin bx - b\cos bx)}{a^2+b^2}$$

83. $$\int e^{ax}\cos bx\,dx = \frac{e^{ax}(b\sin bx + a\cos bx)}{a^2+b^2}$$

84. $$\int \operatorname{sech} x \tanh x\,dx = -\operatorname{sech} x$$

85. $$\int \operatorname{csch} x \coth x\,dx = -\operatorname{csch} x$$